AF323838

Global Attractors of
Non-Autonomous Dissipative
Dynamical Systems

INTERDISCIPLINARY MATHEMATICAL SCIENCES

Series Editor: Jinqiao Duan *(Illinois Inst. of Tech., USA)*

Published

Vol. 1:   Global Attractors of Nonautonomous Dissipative Dynamical Systems
        David N. Cheban

Forthcoming

Mathematica in Finance
    Michael Kelly

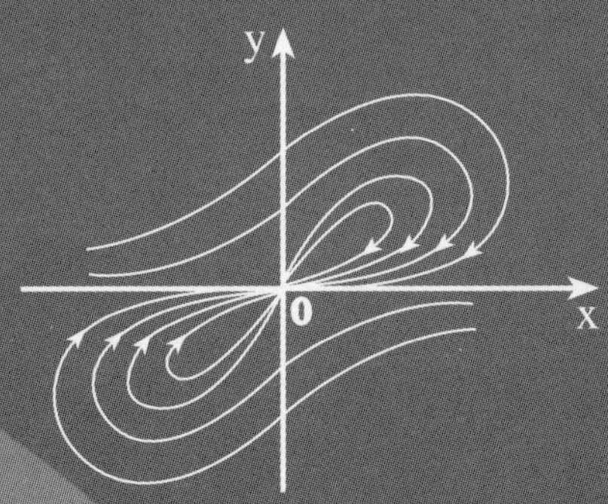

Global Attractors of Non-Autonomous Dissipative Dynamical Systems

David N Cheban

State University of Moldova, Moldova

World Scientific

NEW JERSEY · LONDON · SINGAPORE · BEIJING · SHANGHAI · HONG KONG · TAIPEI · CHENNAI

Published by

World Scientific Publishing Co. Pte. Ltd.

5 Toh Tuck Link, Singapore 596224

USA office: 27 Warren Street, Suite 401-402, Hackensack, NJ 07601

UK office: 57 Shelton Street, Covent Garden, London WC2H 9HE

British Library Cataloguing-in-Publication Data
A catalogue record for this book is available from the British Library.

GLOBAL ATTRACTORS OF NON-AUTONOMOUS DISSIPATIVE DYNAMICAL SYSTEMS

ISBN 981-256-028-9

Printed in Singapore by World Scientific Printers (S) Pte Ltd

Dedicated to my wife *Ivanna*
and my children *Olga* and *Anatoli*

Preface

In the qualitative theory of differential equations non-local problems play an important role, especially in regard to questions of boundedness, periodicity, almost periodicity, Poisson stability, asymptotic behaviour, dissipativity, etc.

The present work takes a similar approach and is dedicated to the study of abstract non-autonomous dissipative dynamical systems and their application to differential equations.

In applications there often occur systems

$$u' = f(t, u), \tag{0.1}$$

which have every one of their solutions driven into fixed bounded domain and kept there under further increase of time, becouse of natural dissipation. Such systems are called dissipative ones [135]-[137],[270]-[272],[325], [326]. Solutions of dissipative systems are called limit (finally) bounded [325, 326].

Dynamical systems occur in hydrodynamics studying turbulent phenomena, meteorology, oceanography, theory of oscillations, biology, radio engineering and other domains of sciences and engineering technics related to the study of asymptotic behaviour. Lately interest in dissipative systems increased even more because of intensive elaboration of strange attractors (see, e.g.,[143, 239, 296, 306]).

The study of the dissipative systems there are dedicated plenty of works, beginning from the classical works of N. Levinson. Among works on dissipative systems of ordinary differential equations two directions can be made out.

To the first belong works which contain some conditions assuring the dissipativity of system (0.1), some class or concrete system, representing theoretical or applied interest. Examples are works of P. V. Atrashenok [9], B. P. Demidovich [135]-[137], V. I. Zubov [336]-[338], V. M. Matrosov [251], V. V. Nemytski [259]-[260], V. A. Pliss [270], V. N. Schennikov [298, 299], C. Corduneanu [125], N. Levinson [237], N. Pavel [264]-[266], R. Reissig [272], T. Talpalaru [309] and a lot of other authors.

To the second direction belong works in which are studied inner conditions of dissipative systems, that is, conditions relating to the character of behaviour of solution of the system when assuming its dissipativity, for different classes of differential equations. Among them are the works of V. M. Gershtein [159]-[163], V. V. Zhikov [331]-[332], I. L. Zinchenko [334], M. A. Krasnoselsky [159], S. Yu. Pilyugin [269], V. A. Pliss [270]-[271], M. L. Cartwright and J. E. Littlewood [44]-[46], N. Levinson [237], G. Fusco and M. Oliva [157], J. Skowronski and S. Ziemba [307] and other authors.

For a compact map on a Banach space, V. M. Gerstein [162], V. M. Gerstein and M. A. Krasnoselskii [159] investigated the existence of maximal compact invariant set and studied some of its properties. In works of G. Billoti, G. Cooperman, J. LaSalle, O. Lopes, P. Massat, M. Slemrod, J. Hale [175, 26, 124], [171]-[176],[235],[246]-[248] and a lot of other authors many important results obtained for ordinary differential equations [270]-[271] are generalized to functional-differential equations.

Of late in the theory of partial differential equations there appeared works of A. V. Babin and M. I. Vishik [12]-[16], Ju. S. Ilyashenko [194]-[197], O. A. Ladyzhenskaya [230], A. N. Sharkovsky [296], R. Temam [314] and other authors, in which, are studied evolutionary equations with maximal attractors (dissipative evolutionary equations).

We note that all works mentioned above (with rare exceptions) studied periodical or autonomous systems. These results are presented in monographs [175], [270]-[272],[296].

If the right hand side f of the equation (0.1) is non-periodic, e.g. quasi-periodic (almost periodic by Bohr, recurrent in sense the of Birkhoff, almost periodic by Levitan, stable by Poisson) or depending on time in more complicated way, then the situation essentially complicates already in the class of almost periodic systems. It is caused at least by two reasons.

First, the definition of dissipativity in the non-autonomous case needs to be made more precise because Levinson's definition in the class of non-periodical systems divides on some non equivalent notions and we need to choose one which allows us develop a general theory which would contain as particular case most essential results obtained for periodical dissipative systems.

Second, in the study of periodic dissipative systems an important role is played by discrete dynamical system (cascade) generated by degrees of Poincaré's transformation (mapping). For non-periodic systems there is no Poincaré's transformation and, consequently, the approach created for research on periodical dissipative systems is not useful in the more general case. That is why to study non-periodical dissipative systems we need new ideas; that is, making a theory of non-autonomous dissipative dynamical systems demands making corresponding methods of research.

Our approach to the study of dissipative systems of differential equations consists of drawing to the study of non-autonomous dissipative systems ideas and methods developed in the theory of abstract dynamical systems. We select one class of dynamical systems (called in this work, dissipative) modelling the properties of dissipative differential equations. The selected class is systematically researched and then the general results obtained are applied to the study of dissipative systems of differential and some other classes of equations.

The idea of applying methods of the theory of dynamical systems to the study of non-autonomous differential equations is not new. It has been successfully applied to the resolution of different problems in the theory of linear and non-linear non-autonomous differential equations for more than thirty years. First this approach to non-autonomous differential equations was introduced in works of V. M. Millionshchikov [253]-[255], B. A. Shcherbakov [300, 302], L. G. Deyseach and G. R. Sell [139], R. K. Miller [252], G. Seifert [289], G. R. Sell [290, 291], later in works of V. V. Zhikov [331], I. U. Bronshtein [32], R. A. Johnson [203]-[204] and many other authors. This approach consists of naturally linking with equation (0.1) a pair of dynamical systems and a homomorphism of the first onto the second. In one dynamical system is put the information about right hand side of equation (0.1) and in the other about the solutions of equation (0.1).

We note that there exists the another approach offered in the works of V. I. Zubov [336] and then developed in works of C. M. Dafermos [128]-[131], J. K. Hale [175], I. Hitoshi [189] and many other authors (for details, see survey [303]). It consists of linking with every non-autonomous differential equation a two-parametric family of mappings (by terms of some authors – process).

In the beginning a these two approaches developed independently but later there was found a relation between them (see, e.g. [130],[187]).

This author adheres to the first approach because in his opinion it is better adapted for resolving those problems which are studied in this work.

The proposed work consists of fourteen chapters.

In the first chapter for autonomous dynamical systems there are introduced and studied different kinds of dissipativity: point, compact, local, bounded and weak one. Criteria of point, compact and local dissipativity are given. It is shown that for dynamical systems in locally compact spaces all three types of dissipativity are equivalent. Examples are given showing that in the general case the notions of point, compact and local dissipativity are different. The notion of Levinson's center, which is an important characteristic of compact dissipative system, is introduced. The solution of one J. K. Hale's problem for locally bounded dynamical systems is given.

The second chapter is dedicated to non-autonomous dissipative dynamical systems. It is noted that in the general case Levinson's center of non-autonomous dissipative dynamical system is not orbitally stable. The question of stability of Levinson's center of non-autonomous system is studied. A simple geometric description ensuring its stability is given, as is a description of Levinson's center of non-autonomous systems satisfying the condition of uniform positive stability. There are pointed conditions of keeping the property of dissipativity under homomorphisms. It is selected a class of dynamical systems which allow full description of Levinson's center's structure, called in this work systems with convergence. Some criteria of convergence in terms of Lyapunov's functions depending on two space variables are given. It is shown that for non-autonomous dissipative dynamical systems with finite-dimensional phase space all three types of dissipativity are equivalent. There are given series of conditions that are equivalent to dissipativity in finite-dimensional space. At last, it is proved that for linear systems dissipativity reduces to convergence. Also there are given series of conditions equivalent to dissipativity of linear systems.

The third chapter deals mostly with one special class of non-autonomous dissipative dynamical systems called in the work $\mathbb{C}$-analytic. It is proved that $\mathbb{C}$-analytic dissipative dynamical system has the property of uniform positive stability on compact subsets. Full description of Levinson's center of these systems is given. One general construction allowing to connect with given non-autonomous dynamical system an autonomous dynamical system in space of continuous sections is given. With the help of these constructions are studied quasi-periodic solutions of analytic systems with quasi-periodic coefficients. In conclusion conditions are given which guarantee the dissipativity of weakly nonlinear systems of differential equations, as is a condition which assure the existence of almost periodic solution of weakly nonlinear system with almost periodic coefficients in Levinson's center.

The fourth chapter is dedicated to a study of Levinson's center's structure with condition of hyperbolicity on closure of recurrent motion's set. There we establish some topological properties of Levinson's center of compact dissipative dynamical system. In particular, it is shown that Levinson's center is indecomposable if the phase space of dynamical systems is also decomposable. It is proved that in connected and locally connected space Levinson's center of compact dissipative dynamical system both with continuous and discrete time is a connected set. There we establish some properties of a set of chain recurrent motions of dissipative system. A theorem is proved about the spectral decomposition of Levinson's center which is analogous to known Smale's theorem. For one-dimensional dissipative dynamical systems it was proved theorem precising theorem about spectral decomposition of Levinson's center and, particularly, it was shown that Levinson's center of such systems contains a local maximal hyperbolic Markov set. In the end of the chapter an application of obtained results to periodic systems is given.

In the fifth chapter we develop the method of Lyapunov's functions for research of non-autonomous dissipative dynamical systems in finite-dimensional space. Criteria of dissipativity in terms of Lyapunov's functions, with the help of which we can get sufficient tests for dissipativity suitable in applications, are discassed. With the help of Lyapunov's functions there were proved series of tests for dissipativity of multi-dimensional non-autonomous differential equations. On the basis developed for research of non-autonomous dissipative systems methods a criterion of asymptotic stability of zero section of non-autonomous systems has been obtained. In particular, it is proved the analog of known Barbashin–Krasovsky's theorem for non-autonomous dynamical systems. There are established some tests for convergence of systems of differential equations with the help of Lyapunov's functions depending on two space variable. There were proved tests for dissipativity and convergence of some systems of differential equations of the 2nd and 3rd order appearing in applications.

The sixth chapter is dedicated to some applications of general results obtained in previous chapters to difference equations, equations with impulses, functional-differential equations and evolutionary equations $x' + Ax = f$ with uniformly monotone operator A. In particular, there are given tests for dissipativity and convergence of weakly nonlinear systems of difference equations and equations with impulses. It is proved the criterion of asymptotic stability of linear functional-differential equations. It is established test for convergence of evolutionary equation $x' + Ax = f$ with uniformly monotone operator A.

In the seventh chapter we systematically study of the problem of upper semi-continuity of compact global attractors and compact pullback attractors of abstract non-autonomous dynamical systems for small perturbations. Several applications of our results are given for different classes of evolutionary equations.

The eighth chapter is devoted to the study of the relationship between the global attractor of the skew–product system and the pullback and forward attractors of the cocycle system. We also note that forward attractors are stronger than global attractors if we suppose a compact set of non-autonomous perturbations. An example is presented in which the cartesian product of the component subsets of a pullback attractor is not a global attractor of the skew–product flow. This set is, however, a maximal compact invariant subset of the skew–product flow. By a generalization of some stability results of V. I. Zubov [336] it is asymptotically stable. Thus a pullback attractor always generates a local attractor of the skew–product system, but this need not be a global attractor. If, however, the pullback attractor generates a global attractor in the skew–product flow and if, in addition, its component subsets depend lower continuously on the parameter, then the pullback attractor is also a forward attractor. Several examples illustrating these results are presented in the final section.

In the ninth chapter we systematically study the global pullback attractors of $\mathbb{C}$−analytic cocycles. For the large class of $\mathbb{C}$−analytic cocycles we give the description of the structure of their pullback attractors. Particularly we prove that it is trivial, i.e. the fibers of these attractors contain only one point. Several applications of these results are given (ODEs, Caratheodory's equations with almost periodic coefficients, almost periodic ODEs with impulse).

The tenth chapter is dedicated to the investigation of the effect of time discretization on the pullback attractor of a non-autonomous ordinary differential equation for which the vector fields depend on a parameter that varies in time rather than depending directly on time itself. The parameter space is assumed to be compact so the skew product flow formalism as well as cocycle formalism also applies and the vector fields have a strong dissipative structure that implies the existence of a compact set that absorbs all compact sets under the resulting non-autonomous dynamics. The numerical scheme considered is a general 1–step scheme such as the Euler scheme with variable time-steps. Our main result is to show that the numerical scheme interpreted as a discrete time non-autonomous dynamical system, hence discrete time cocycle mapping and skew product flow on an extended parameter space, also possesses a cocycle attractor and that its component subsets converge upper semi–continuously to those of the cocycle attractor of the original system governed by the differential equation. We will also see that the corresponding skew product flow systems have global attractors with the cocycle attractor component sets as their cross-sectional sets in the original state space. Finally, we investigate the periodicity and almost periodicity of the discretized pullback attractor when the parameter dynamics in the ordinary differential equation is periodic or almost periodic and the pullback attractor consists of singleton valued component sets, i.e. the pullback attractor is a single trajectory.

In the eleventh chapter we study the non-autonomous Navier-Stokes equations. It is proved that such systems admit compact global attractors. This problem is formulated and solved in the terms of general non-autonomous dynamical systems. We give conditions of convergence of non-autonomous Navier-Stokes equations . A test of existence of almost periodic (quasi periodic, recurrent, pseudo recurrent) solutions of non-autonomous Navier-Stokes equations is given. We prove the global averaging principle for non-autonomous Navier-Stokes equations.

The twelfth chapter is devoted to the investigation of the global attractors of general V- monotone non-autonomous dynamical systems and their applications to different class of differential equations (ODEs, ODEs with impulse, some class of evolution partial differential equations).

In the thirteenth chapter we study the linear almost periodic dynamical systems. The bounded solutions, relation between different types of stability and uniform exponential stability for those systems are studied. We give several applications the obtained results for ODEs, PDEs and functional-differential equations.

Chapter 14 is devoted to the study of quasi-linear triangular maps: chaos, almost periodic and recurrent solutions, integral manifolds, chaotic sets etc. This problem is formulated and solved in the framework of non-autonomous dynamical systems with discrete time. We prove that such systems admit an invariant continuous section (an invariant manifold). Then, we obtain the conditions of the existence of a compact global attractor and characterize its structure. We give a criterion for the existence of almost periodic and recurrent solutions of the quasi-linear triangular maps. Finally, we prove that quasi-linear maps with chaotic base admits a chaotic compact invariant set.

The results given in this work belong mostly to the author. The results of chapters eight, ten and twelve are jointly obtained by the author, Peter Kloeden and Bjoern Schmalfuss [96, 97, 100, 106]. The results of chapter eleven and paragraph six of chapter six are obtained jointly with Jinqiao Duan [104, 107]. The results of chapter fourteen are obtained jointly with Cristiana Mammana [108].

The reader needs no deep knowledge of special branches of mathematics, although it should easier for readers who know the fundamentals of the qualitative theory of differential equations.

The quality of English of this exposition is adversely affected by its not being the native tongue of the author, for which fact the author asks for the reader's kind indulgence and understanding.

Acknowledgment: The research described in this publication, in part, was possible due to Award No. MM1-3016 of the Moldovan Research and Development Association (MRDA) and the U.S. Civilian Research & Development Foundation for the Independent States of the Former Soviet Union (CRDF).

June 2003

D. N. Cheban

cheban@usm.md

http://www.usm.md/davidcheban

Contents

Notations

$\forall$	for all;
$\exists$	there exists;
$:=$	is equal (coincide) by the definition;
0	the number 0 and also the null element of all additive group (semigroup);
$\mathbb{N}$	the set of all natural numbers;
$\mathbb{Z}$	the set of all integers;
$\mathbb{Q}$	the set of all rational numbers;
$\mathbb{R}$	the set of all real numbers;
$\mathbb{C}$	the set of all complex numbers;
$\mathbb{S}$	one of the sets $\mathbb{R}$ or $\mathbb{Z}$;
$\mathbb{S}_+(\mathbb{S}_-)$	the set of all nonnegative (non-positive) numbers from the set $\mathbb{S}$;
$X \times Y$	the product space of two sets;
M^n	the product space of n copies of the set M;
E^n	the real or complex n-dimensional Euclidean space;
$\{x_n\}$	a sequence;
$x \in X$	x is an element of the set X;
∂X	the boundary of the set X;
$X \subseteq Y$	the set X is a part of or coincides with the set Y;
$X \bigcup Y$	the union of the two sets X and Y;
$X \setminus Y$	the complement of the set Y in the set X;
$X \bigcap Y$	the intersection of the sets X and Y;
$\emptyset$	the empty-set;
(X, ρ)	the complete metric space with the distance ρ;

2^X	the space of all closed bounded subsets of the metric space X;
$\overline{M}$	the closure of the set M;
f^{-1}	the inverse mapping of the mapping f;
$f(M)$	the image of the set $M \subseteq X$ by the mapping $f : X \to Y$, i.e., $\{y \in Y : y = f(x),\ x \in M\}$;
$f \circ g$	the composition of the mappings f and g, i.e., $(f \circ g)(x) = f(g(x))$;
$f\vert_M$	the restriction of the mapping f on the set M;
$f(\cdot, x)$	the partial mapping, defined by the function f if the second variable takes a value x;
Id_X	the identity mapping from the X into itself;
$Im(f)$	range of values of the function f;
$D(f)$	the domain of the definition of the function f;
$\|x\|$ or $\|\|x\|\|$	the norm of the element x;
(x, y)	the ordered pair;
$C(X, Y)$	the set of all continuous mappings from X into Y with compact-open topology;
$C^k(U, M)$	the set of all k-time continuously differentiable functions $f : U \to M$;
$f : X \to Y$	the mapping from X into Y;
$B(M, \varepsilon)$	the open ε-neighborhood of the set M in the metric space X;
$B[M, \varepsilon]$	the closed ε-neighborhood of the set M in the metric space X;
$\{x, y, ..., z\}$	the set consisting from the elements $x, y, ..., z$;
$\overline{1, n}$	the set consisting from the elements $1, 2, ..., n$;
$\{x \in X \vert \Re(x)\}$	the set of all elements of the set X, possessing the property $\Re$;
$f^{-1}(M)$	the pro-image of the set $M \subseteq Y$ by the mapping $f : X \to Y$, i.e. $\{x \in X : f(x) \in M\}$;
$F(t, \cdot) := f^t$	the partial mapping, defined by the function f if the first variable takes the value t;
$\rho(\xi, \eta)$	the distance in the metric space X;
$\lim\limits_{n \to +\infty} x_n$	the limit of the sequence;

$\varepsilon_k \downarrow 0$	the monotone decreasing to the 0 sequence;
$\lim\limits_{x \to a} f(x)$	the limit of the mapping f as $x \to a$;
$\bigcup\{M_\lambda : \lambda \in \Lambda\}$	the union of the family of sets $\{M_\lambda\}_\lambda \in \Lambda$;
$\bigcap\{M_\lambda : \lambda \in \Lambda\}$	the intersection of the family of sets $\{M_\lambda\}_\lambda \in \Lambda$;
$(H, \langle \cdot, \cdot \rangle)$	Hilbert space with the scalar product $\langle \cdot, \cdot \rangle$;
$\dot\varphi := \frac{d\varphi}{dt}$	the derivative of the function φ;
$cardM$	the cardinality of the set M;
$C(X)$	the family of all compacts from X;
$B(X)$	the family of all bounded subsets from X;
$\lambda(A)$	the measure of non-compactness of the set $A \in B(X)$;
$(X, \mathbb{T}, \pi)$	a dynamical system;
(X, P)	cascade, generated by the mapping P;
$\omega_x(\alpha_x)$	$\omega(\alpha)$-limit set of the point x;
φ_x	the whole trajectories of $(X, \mathbb{T}, \pi)$ with condition $\varphi_x(0) = x$.
Φ_x	the set of all whole trajectories φ_x of $(X, \mathbb{T}, \pi)$.
α_{φ_x}	α-limit set of the whole trajectory $\varphi_x \in \Phi_x$;
$W^s(M)$	the stable manifold of the set M;
M st.L^+	the set M is stable in sense of Lagrange in the positive direction;
$D_x^+ (J_x^+)$	positive extension (positive limit extension) of the point x;
$xt = \pi^t x = \pi(t, x)$	the state of point x at the moment t;
$pr_i : X_1 \times X_2 \to X_i$	the projection of $X_1 \times X_2$ on the i-th $(i = 1, 2)$ component X_i;
$D^+(M)$	the positive extension of the set M;
$J^+(M)$	the positive limit extension of the set M;
Σ_x^+	the positive semi-trajectory of the point x;
$\Sigma^+(M)$	the positive semi-trajectory of the set M;
$H^+(x)$	the closure of positive semi-trajectory of the point x;
Σ_x	the trajectory of the point x;
$H(x)$	the closure of trajectory of the point x;
Ω	the closure of the union of all ω-limit points of $(X, \mathbb{T}, \pi)$;
$\beta(A, B)$	the semi-deviation of the set A from the set B $(A, B \in 2^X)$;

Chapter 1

Autonomous dynamical systems

1.1 Some notions, notations and facts from theory of dynamical systems

1. Below we give some notions, denotations and facts from theory of dynamical systems $[32, 33, 258, 261, 300, 302, 304]$ which we will use in this book.

Let X be a topological space, $\mathbb{R}$ $(\mathbb{Z})$ be a group of real (integer) numbers, $\mathbb{R}_+$ $(\mathbb{Z}_+)$ be a semi-group of the nonnegative real (integer) numbers, $\mathbb{S}$ be one of the two sets $\mathbb{R}$ or $\mathbb{Z}$ and $\mathbb{T} \subseteq \mathbb{S}$ $(\mathbb{S}_+ \subseteq \mathbb{T})$ be a sub-semigroup of additive group $\mathbb{S}$.

Definition 1.1 Triplet $(X, \mathbb{T}, \pi)$, where $\pi : \mathbb{T} \times X \to X$ is a continuous mapping satisfying the following conditions:

$$\pi(0, x) = x; \tag{1.1}$$

$$\pi(s, \pi(t, x)) = \pi(s + t, x); \tag{1.2}$$

is called a dynamical system. If $\mathbb{T} = \mathbb{R}$ $(\mathbb{R}_+)$ or $\mathbb{Z}$ $(\mathbb{Z}_+)$, then the dynamical system $(X, \mathbb{T}, \pi)$ is called a group (semi-group). In the case, when $\mathbb{T} = \mathbb{R}_+$ or $\mathbb{R}$ the dynamical system $(X, \mathbb{T}, \pi)$ is called a flow, but if $\mathbb{T} \subseteq \mathbb{Z}$, then $(X, \mathbb{T}, \pi)$ is called a cascade (discrete flow).

Sometimes, briefly, we will write xt instead of $\pi(t, x)$.

Below X will be a complete metric space with metric ρ.

Definition 1.2 The function $\pi(\cdot, x) : \mathbb{T} \to X$ is called a motion passing through the point x at the moment $t = 0$ and the set $\Sigma_x := \pi(\mathbb{T}, x)$ is called a trajectory of this motion.

Definition 1.3 A nonempty set $M \subseteq X$ is called positively invariant (negatively invariant, invariant) with respect to dynamical system $(X, \mathbb{T}, \pi)$ or, simple, positively invariant (negatively invariant, invariant), if $\pi(t, M) \subseteq M$ $(M \supseteq \pi(t, M), \pi(t, M) = M)$ for every $t \in \mathbb{T}$.

1

Definition 1.4   A closed positively invariant set, which does not contain own closed positively invariant subset, is called minimal.

It easy to see that every positively invariant minimal set is invariant.

Definition 1.5   A closed positively invariant (invariant) set is called indecomposable, if it can not be represented in the form of union of two nonempty disjoint positively invariant (invariant) subsets.

Definition 1.6   Let $M \subseteq X$. The set

$$\omega(M) := \bigcap_{t \geq 0} \overline{\bigcup_{\tau \geq t} \pi(\tau, M)}$$

is called ω-limit for M. If $\mathbb{T} = \mathbb{S}$, then the set

$$\alpha(M) := \bigcap_{t \leq 0} \overline{\bigcup_{\tau \leq t} \pi(\tau, M)}$$

is called α-limit for M.

Denote by $\Sigma(M) := \pi(\mathbb{S}, M)$ $(\Sigma^+(M) := \pi(\mathbb{S}_+, M))$ and $H(M) := \overline{\pi(\mathbb{S}, M)}$ $(H^+(M) := \overline{\pi(\mathbb{S}_+, M)})$. If $M = \{x\}$, then we put $\omega_x := \omega(\{x\})$, $\alpha_x := \alpha(\{x\})$, $\Sigma_x^+ := \Sigma^+(\{x\})$, $H^+(x) := H^+(\{x\})$ and $H(x) := H(\{x\})$.

Let $(X, \mathbb{T}, \pi)$ be a dynamical system.

Definition 1.7   The point $x \in X$ is called a τ-periodic $(\tau > 0, \tau \in \mathbb{T})$ point, if $xt = x$ $(x\tau = x$ respectively$)$ for all $t \in \mathbb{T}$, where $xt := \pi(t, x)$.

Definition 1.8   The number $\tau \in \mathbb{T}$ is called $\varepsilon > 0$ shift (almost period) if $\rho(x\tau, x) < \varepsilon$ (respectively $\rho(x(\tau + t), xt) < \varepsilon$ for all $t \in \mathbb{T}$).

Definition 1.9   The point $x \in X$ is called almost recurrent (almost periodic) if for any $\varepsilon > 0$ there exists a positive number l such that on any segment of length l, will be found an ε shift (almost period) of point $x \in X$.

Definition 1.10   If a point $x \in X$ is almost recurrent and the set $H(x) = \overline{\{xt \mid t \in \mathbb{T}\}}$ is compact, then x is called recurrent.

Definition 1.11   The point $x \in X$ is called stable in the sense of Poisson in the positive direction if $x \in \omega_x$. If the dynamical system $(X, \mathbb{T}, \pi)$ is a two-sided one and $x \in \alpha_x$, then the point x is called stable in the sense of Poisson in the negative direction. If the point x is stable in the sense of Poisson in both directions, it is called Poisson stable.

Definition 1.12   The continuous mapping $\varphi : \mathbb{S} \to X$ with the properties:

(1) $\varphi(0) = x$;

(2) $\pi^t \varphi(s) = \varphi(t+s)$ for all $t \in \mathbb{T}$ and $s \in \mathbb{S}$

is called a whole trajectory of one-sided dynamical system $(X, \mathbb{T}_1, \pi)$.

Denote by Φ_x the family of all whole trajectory of dynamical system $(X, \mathbb{T}_1, \pi)$ passing through the point $x \in X$ at the moment $t = 0$.

Definition 1.13 The motion $\pi(\cdot, x) : \mathbb{T}_1 \to X$ of one-sided dynamical system $(X, \mathbb{T}_1, \pi)$ is called extendable on $\mathbb{S}$, if there exists a whole trajectory φ passing through point x.

We denote by $\alpha_{\varphi_x} := \{y : \exists t_n \to -\infty, \ t_n \in \mathbb{S} \text{ and } \varphi(t_n) \to y\}$, where φ is an extension on $\mathbb{S}$ the motion $\pi(\cdot, x)$.

Definition 1.14 The set $W^s(\Lambda)$ $(W^u(\Lambda))$, defined by equality

$$W^s(\Lambda) := \{x \in X \mid \lim_{t \to +\infty} \rho(xt, \Lambda) = 0\}$$

$$(W^u(\Lambda) := \{x \in X \mid \lim_{t \to -\infty} \rho(xt, \Lambda) = 0\}),$$

is called a stable (unstable) manifold of set $\Lambda \subseteq X$.

Definition 1.15 The set M is called:

- orbital stable, if for every $\varepsilon > 0$ there exists $\delta = \delta(\varepsilon) > 0$ such that $\rho(x, M) < \delta$ implies $\rho(xt, M) < \varepsilon$ for all $t \geq 0$;
- attracting, if there exists $\gamma > 0$ such that $B(M, \gamma) \subset W^s(M)$;
- asymptotic stable, if it is orbital stable and attracting;
- global asymptotic stable, if it is asymptotic stable and $W^s(M) = X$;
- uniform attracting, if there exists $\gamma > 0$ such that

$$\lim_{t \to +\infty} \sup_{x \in B(M, \gamma)} \rho(xt, M) = 0.$$

Theorem 1.1 [305] *Let $M \subseteq X$ be a nonempty compact invariant set, then every motion in M is extendable on $\mathbb{S}$.*

Theorem 1.2 [305] *The compact set $M \subseteq X$ is negative invariant, if and only if for every point $x \in M$ there exists a whole trajectory φ passing through point $x \in M$ such that $\varphi(t) \in M$ for all $t \leq 0$.*

Theorem 1.3 [305] *Union and closure of positively invariant (negatively invariant, invariant) sets is positively invariant (negatively invariant, invariant).*

Theorem 1.4 [305] *The following statements are true:*

(1) $\forall\ x \in X$ *the set* ω_x *is positively invariant;*

(2) $\forall\ x \in X$ *and* $\varphi \in \Phi_x$ *the set* α_{φ_x} *is positively invariant.*

2. Below we indicate one general method of construction of dynamical systems on the space of continuous functions. In this way we will get many well known dynamical systems on the functional spaces (see, for example, [32, 300]).

Let $(X, \mathbb{T}, \pi)$ be a dynamical system on X, Y be a complete pseudo metric space and $\mathcal{P}$ be a family of pseudo metric on Y. Denote by $C(X, Y)$ the family of all continuous functions $f : X \to Y$ equipped with the compact-open topology. This topology is given by the following family of pseudo metric $\{d_K^p\}$ $(p \in \mathcal{P},\ K \in C(X))$, where

$$d_K^p(f, g) := \sup_{x \in K} p(f(x), g(x))$$

and $C(X)$ a family of all compact subsets of X. We define for all $\tau \in \mathbb{T}$ the mapping $\sigma_\tau : C(X, Y) \to C(X, Y)$ by the following equality: $(\sigma_\tau f)(x) := f(\pi(\tau, x))$ $(x \in X)$. We note that the family of mappings$\{\sigma_\tau : \tau \in \mathbb{T}\}$ possesses the following properties:

a. $\sigma_0 = id_{C(X,Y)}$;

b. $\forall \tau_1, \tau_2 \in \mathbb{T}\ \sigma_{\tau_1} \circ \sigma_{\tau_2} = \sigma_{\tau_1 + \tau_2}$;

c. $\forall \tau \in \mathbb{T}\ \sigma_\tau$ is continuous.

Lemma 1.1 *The mapping* $\sigma : \mathbb{T} \times C(X, Y) \to C(X, Y)$, *defined by the equality* $\sigma(\tau, f) := \sigma_\tau f$ $(f \in C(X, Y),\ \tau \in \mathbb{T})$, *is continuous.*

Proof. Let $f \in C(X, Y)$, $\tau \in \mathbb{T}$ and $\{f_\nu\}, \{\tau_\nu\}$ are arbitrary directions which converge to f and τ respectively. Then

$$d_K^p(\sigma(\tau_\nu, f_\nu), \sigma(\tau, f)) = \sup_{x \in K} p(\sigma(\tau_\nu, f_\nu)(x),\ \sigma(\tau, f)(x)) =$$

$$\sup_{x \in K} p((f_\nu(\pi(\tau_\nu, x)),\ f(\pi(\tau, x)))) : x \in K\} \leq$$

$$\sup_{x \in K} p(f_\nu(\pi(\tau_\nu, x)),\ f(\pi(\tau_\nu, x))) + \sup_{x \in K} p(f(\pi(\tau_\nu, x)),\ f(\pi(\tau, x))) \leq$$

$$\sup_{x \in K,\ s \in Q} p(f_\nu(\pi(s, x)),\ f(\pi(s, x))) + \sup_{x \in K} p(f(\pi(\tau_\nu, x)),\ f(\pi(\tau, x)))$$

$$\leq \sup_{m \in \pi(Q, K)} p(f_\nu(m), f(m)) + \sup_{x \in K} p(f(\pi(\tau_\nu, x)),\ f(\pi(\tau, x))) =$$

$$d_{\pi(Q,K)}^p(f_\nu, f) + \sup_{x \in K} p(f(\pi(\tau_\nu, x)),\ f(\pi(\tau, x))), \tag{1.3}$$

where $Q = \overline{\{\tau_\nu\}}$. Passing to limit in the inequality (1.3) we obtain the necessary statement. $\square$

Corollary 1.1 *The triple $(C(X,Y), \mathbb{T}, \sigma)$ is a dynamical system on $C(X,Y)$.*

Remark 1.1 *Lemma 1.1 is true, if X is an arbitrary Hausdorff topological space.*

Consider now same examples of dynamical systems of form $(C(X,Y), \mathbb{T}, \sigma)$ useful in the applications.

Example 1.1 Let $X = \mathbb{T}$ and we denote by $(X, \mathbb{T}, \pi)$ a dynamical system on $\mathbb{T}$, where $\pi(t, x) := x + t$. Dynamical system $(C(\mathbb{T}, Y), \mathbb{T}, \sigma)$ is called *Bebutov's dynamical system* [300] (dynamical system of translations or shifts dynamical system).

Definition 1.16 We will say that the function $\varphi \in C(\mathbb{T}, Y)$ possesses the property A, if with this property possesses the motion $\sigma(\cdot, \varphi) : \mathbb{T} \to C(\mathbb{T}, Y)$ in the dynamical system of Bebutov $(C(\mathbb{T}, Y), \mathbb{T}, \sigma)$, generated by the function φ. In quality of property A may be periodicity, almost periodicity, recurrence, asymptotic almost periodicity etc.

Example 1.2 Let $X := \mathbb{T} \times W$, where W some metric space and by $(X, \mathbb{T}, \pi)$ we denote a dynamical system on X defined in the following way: $\pi(t, (s, w)) := (s + t, w)$. Using the general method proposed above we can define on $C(\mathbb{T} \times W, Y)$ a dynamical system of translations $(C(\mathbb{T} \times W, Y), \mathbb{T}, \sigma)$.

The function $f \in C(\mathbb{T} \times W, Y)$ is called almost periodic (recurrent, asymptotic almost periodic etc) with respect to $t \in \mathbb{T}$ uniform on w on every compacts from W, if the motion $\sigma(\cdot, f)$ is almost periodic (recurrent, asymptotic almost periodic, etc.) in the dynamical system $(C(\mathbb{T} \times W, Y), \mathbb{T}, \sigma)$.

Example 1.3 Let $W := \mathbb{C}^n$, $Y := \mathbb{C}^m$ and $A(\mathbb{T} \times \mathbb{C}^n, \mathbb{C}^m)$ be a family of all functions $f \in C(\mathbb{T} \times \mathbb{C}^n, \mathbb{C}^m)$ which are holomorphic with respect to second variable. It easy to verify that the set $A(\mathbb{T} \times \mathbb{C}^n, \mathbb{C}^m)$ is closed and invariant subset of dynamical system $(C(\mathbb{T} \times \mathbb{C}^n, \mathbb{C}^m), \mathbb{T}, \sigma)$ and, consequently, on $A(\mathbb{T} \times \mathbb{C}^n, \mathbb{C}^m)$ it is induced a dynamical system of translations $(A(\mathbb{T} \times \mathbb{C}^n, \mathbb{C}^m), \mathbb{T}, \sigma)$.

Example 1.4 Let W and X be complete pseudo metric spaces [213]. Denote by $\mathcal{P}(\mathbb{T} \times W, X)$ a set of all continuous functions $f : \mathbb{T} \times W \to X$ which are bounded on every bounded subset from $\mathbb{T} \times W$ and continuous w.r.t. $t \in \mathbb{T}$ uniformly on w on bounded subsets W equipped with the topology of uniform convergence on bounded subsets from $\mathbb{T} \times W$. This topology can by generate by the following family of pseudo metrics $\{d_B^p\}$ $(p \in \mathcal{P}, B \in B(\mathbb{T} \times W))$, where

$$d_B^p(f, g) = \sup_{(t,w) \in B} p(f(t, w), g(t, w)), \tag{1.4}$$

$B(\mathbb{T} \times W)$ is a family of all bounded subsets of $\mathbb{T} \times W$ and $\mathcal{P}$ is a family of pseudo metric on X. We note that the set $\mathcal{P}(\mathbb{T} \times W, X)$ endowed with the family of pseudo metric $\{d_B^p | p \in \mathcal{P}, \ B \in B(\mathbb{T} \times W)\}$ become a complete pseudo metric space, invariant with respect to translations on $t \in \mathbb{T}$. For each $\tau \in \mathbb{T}$ we denote by f_τ the translation of function $f \in \mathcal{P}(\mathbb{T} \times W, X)$ on τ w.r.t. variable $t \in \mathbb{T}$, i.e. $f_\tau(t, w) = f(t + \tau, w)$ ($(t, w) \in \mathbb{T} \times W$). Now we define a mapping σ : $\mathbb{T} \times \mathcal{P}(\mathbb{T} \times W, X) \to \mathcal{P}(\mathbb{T} \times W, X)$ as following: $\sigma(\tau, f) := f_\tau$ for all $f \in \mathcal{P}(\mathbb{T} \times W, X)$ and $\tau \in \mathbb{T}$. It is easy to see that $\sigma(0, f) = f$ and $\sigma(\tau_2, \sigma(\tau_1, f)) = \sigma(\tau_1 + \tau_2, f)$ for all $f \in \mathbb{P}(\mathbb{T} \times W, X)$ and $\tau_1, \tau_2 \in \mathbb{T}$. Using the same reasoning as in the lemma 1.1 it is possible to verify the mapping σ is continuous and, consequently, the triplet $(\mathcal{P}(\mathbb{T} \times W, X), \mathbb{T}, \sigma)$ is a dynamical system on $\mathcal{P}(\mathbb{T} \times W, X)$.

Remark 1.2 *If the function $f \in C(\mathbb{T} \times W, X)$ is uniform continuous on every bounded subset from $\mathbb{T} \times W$, then $f \in \mathcal{P}(\mathbb{T} \times W, X)$. Denote by $U(\mathbb{T} \times W, X)$ the set of all functions $f \in C(\mathbb{T} \times W, X)$ which are bounded and uniform continuous on every bounded subset from $\mathbb{T} \times W$. Then the set $U(\mathbb{T} \times W, X)$ is invariant w.r.t. translations on $t \in \mathbb{T}$ and it is closed in $\mathcal{P}(\mathbb{T} \times W, X)$ and, consequently, on $U(\mathbb{T} \times W, X)$ it is induced a dynamical system of translations $(U(\mathbb{T} \times W, X), \mathbb{T}, \sigma)$ (see, for example, [32, 300] and [302]).*

Definition 1.17 Let $(X, \mathbb{T}_1, \pi)$ and $(Y, \mathbb{T}_2, \sigma)$ ($\mathbb{S}_+ \subseteq \mathbb{T}_1 \subseteq \mathbb{T}_2 \subseteq \mathbb{S}$) be two dynamical systems. The mapping $h : X \to Y$ is called a homomorphism (respectively isomorphism) of dynamical system $(X, \mathbb{T}_1, \pi)$ on $(Y, \mathbb{T}_2, \sigma)$, if the mapping h is continuous (respectively homeomorphic) and $h(\pi(x, t)) = \sigma(h(x), t)$ ($t \in \mathbb{T}_1$, $x \in X$). In this case a dynamical system $(X, \mathbb{T}_1, \pi)$ is an extension of dynamical system $(Y, \mathbb{T}_2, \sigma)$ by homomorphism h, but a dynamical system $(Y, \mathbb{T}_2, \sigma)$ is called a factor of dynamical system $(X, \mathbb{T}_1, \pi)$ by homomorphism h. Dynamical system $(Y, \mathbb{T}_2, \sigma)$ is called also a base of extension $(X, \mathbb{T}_1, \pi)$.

Definition 1.18 The triplet $\langle (X, \mathbb{T}_1, \pi), (Y, \mathbb{T}_2, \sigma), h \rangle$, where h is a homomorphism from $(X, \mathbb{T}_1, \pi)$ on $(Y, \mathbb{T}_2, \sigma)$ and (X, h, Y) is a local-trivial fibering [190], is called a non-autonomous dynamical system.

Remark 1.3 *In the latter years in the works of I.U.Bronsteyn and his collaborators (see, for example, [32, 33, 34, 158]) an extension is called a triplet $\langle (X, \mathbb{T}, \pi), (Y, \mathbb{T}, h), h \rangle$, i.e. the object which we name here a non-autonomous dynamical system.*

Definition 1.19 The triplet $\langle W, \varphi, (Y, \mathbb{T}_2, \sigma) \rangle$ (or shortly φ), where $(Y, \mathbb{T}_2, \sigma)$ is a dynamical system on Y, W is a complete metric space and φ is a continuous mapping from $\mathbb{T}_1 \times W \times Y$ in W, possessing the following conditions:

a. $\varphi(0, u, y) = u$ ($u \in W, y \in Y$);

b. $\varphi(t+\tau,u,y) = \varphi(\tau,\varphi(t,u,y),\sigma(t,y))$ $(t,\tau \in \mathbb{T}_1,\ u \in W, y \in Y)$,

is called [6, 290, 291] a cocycle on $(Y,\mathbb{T}_2,\sigma)$ with fiber W.

Definition 1.20 Let $X := W \times Y$ and we define a mapping $\pi : X \times \mathbb{T}_1 \to X$ as following: $\pi((u,y),t) := (\varphi(t,u,y),\sigma(t,y))$ (i.e. $\pi = (\varphi,\sigma)$). Then it easy to see that $(X,\mathbb{T}_1,\pi)$ is a dynamical system on X which is called a skew-product dynamical system [2, 292] and $h = pr_2 : X \to Y$ is a homomorphism from $(X,\mathbb{T}_1,\pi)$ on $(Y,\mathbb{T}_2,\sigma)$ and, consequently, $\langle (X,\mathbb{T}_1,\pi),(Y,\mathbb{T}_2,\sigma),h\rangle$ is a non-autonomous dynamical system.

Thus, if we have a cocycle $\langle W,\varphi,(Y,\mathbb{T}_2,\sigma)\rangle$ on dynamical system $(Y,\mathbb{T}_2,\sigma)$ with the fiber W, then it generates a non-autonomous dynamical system $\langle (X,\mathbb{T}_1,\pi),$ $(Y,\mathbb{T}_2,\sigma),h\rangle$ $(X := W \times Y)$, which is called a non-autonomous dynamical system, generated by cocycle $\langle W,\varphi,(Y,\mathbb{T}_2,\sigma)\rangle$ on $(Y,\mathbb{T}_2,\sigma)$.

Non-autonomous dynamical systems (cocycles) play a very important role in the study of non-autonomous evolutionary differential equations. Under appropriate assumptions every non-autonomous differential equation generates some cocycle (non-autonomous dynamical system). Below we give an example of this type.

Example 1.5 Let E^n be an n-dimensional real or complex Euclidean space. Let us consider a differential equation

$$u' = f(t,u), \tag{1.5}$$

where $f \in C(\mathbb{R} \times E^n, E^n)$. Along with equation (1.5) we consider its H-class [32, 137, 238, 300, 302], i.e., the family of equations

$$v' = g(t,v), \tag{1.6}$$

where $g \in H(f) = \overline{\{f_\tau : \tau \in \mathbb{R}\}}$, $f_\tau(t,u) = f(t+\tau,u)$ for all $(t,u) \in \mathbb{R} \times E^n$ and by bar we denote the closure in $C(\mathbb{R} \times E^n, E^n)$. We will suppose also that the function f is regular, i.e. for every equation (1.6) the conditions of existence, uniqueness and extendability on $\mathbb{R}_+$ are fulfilled. Denote by $\varphi(\cdot,v,g)$ the solution of equation (1.6), passing through the point $v \in E^n$ at the initial moment $t = 0$. Then it is correctly defined a mapping $\varphi : \mathbb{R}_+ \times E^n \times H(f) \to E^n$, verifying the following conditions (see, for example, [32, 290, 291]):

1) $\varphi(0,v,g) = v$ for all $v \in E^n$ and $g \in H(f)$;
2) $\varphi(t,\varphi(\tau,v,g),g_\tau) = \varphi(t+\tau,v,g)$ for every $v \in E^n$, $g \in H(f)$ and $t,\tau \in \mathbb{R}_+$;
3) the mapping $\varphi : \mathbb{R}_+ \times E^n \times H(f) \to E^n$ is continuous.

Denote by $Y := H(f)$ and $(Y,\mathbb{R}_+,\sigma)$ a dynamical system of translations (semigroup system) on Y, induced by dynamical system of translations $(C(\mathbb{R} \times E^n, E^n),\mathbb{R},\sigma)$. The triplet $\langle E^n,\varphi,(Y,\mathbb{R}_+,\sigma)\rangle$ is a cocycle on $(Y,\mathbb{R}_+,\sigma)$ with the fiber E^n. Thus the equation (1.5) generates a cocycle $\langle E^n, \varphi, (Y, \mathbb{R}_+, \sigma) \rangle$ and a non-autonomous dynamical system $\langle (X, \mathbb{R}_+, \pi), (Y, \mathbb{R}_+, \sigma), h \rangle$, where $X := E^n \times Y$, $\pi := (\varphi, \sigma)$ and $h := pr_2 : X \to Y$.

1.2   Limit properties of dynamical systems

Let $M \subseteq X$ and $\omega(M)$ be ω–limit set of M. Directly from the definition of $\omega(M)$ follows

Lemma 1.2   *The following statements take place:*

(1) the point $y \in \omega(M)$ if and only if there exist sequences $\{x_n\} \subseteq M$ and $\{t_n\} \subseteq \mathbb{T}$ such that $t_n \to +\infty$ and $y = \lim\limits_{n \to +\infty} x_n t_n$;

(2) the set $\omega(M)$ is closed and positively invariant;

(3) if $A \subseteq B$ then $\omega(A) \subseteq \omega(B)$;

(4) $\omega(A \cup B) \subseteq \omega(A) \cup \omega(B)$ for any pair of subsets A and B from X;

(5) if set A is positively invariant (negatively invariant, invariant) then $\omega(A) \subseteq \overline{A}$ ($\overline{A} \subseteq \omega(A)$, $\omega(A) = \overline{A}$);

(6) $\overline{\cup\{\omega_x \mid x \in M\}} \subseteq \omega(M)$;

(7) if $\Sigma^+(M)$ is relatively compact, then $\omega(M)$ is invariant.

Lemma 1.3   *Let $B \subseteq X$ then the following conditions are equivalent:*

(1) for every $\{x_k\} \subseteq B$ and $t_k \to +\infty$ sequence $\{x_k t_k\}$ is relatively compact;

(2)(a) $\omega(B)$ is not empty and is compact;

(b) $\omega(B)$ is invariant and the equality below takes place

$$\lim\limits_{t \to +\infty} \sup\limits_{x \in B} \rho(xt, \omega(B)) = 0; \tag{1.7}$$

(3) there exists a non empty compact $K \subseteq X$ such that

$$\lim\limits_{t \to +\infty} \sup\limits_{x \in B} \rho(xt, K) = 0.$$

Proof.   Let us show that from 1. follows 2. Let $\{x_k\} \subseteq B$ and $t_k \to +\infty$, then according to 1. the sequence $\{x_k t_k\}$ can be considered convergent. Assume $\overline{x} = \lim\limits_{k \to +\infty} x_k t_k$, then $\overline{x} \in \omega(B)$ and consequently $\omega(B) \neq \emptyset$. Let us show $\omega(B)$ is compact. Let $\varepsilon_k \downarrow 0$ and $\{y_k\} \subseteq \omega(B)$, then there exist $x_k \in B$ and $t_k \geq k$ such that

$$\rho(x_k t_k, y_k) < \varepsilon_k.$$

According to the condition (1), the sequence $\{x_k t_k\}$ is relatively compact and since $\varepsilon_k \downarrow 0$, $\{y_k\}$ also is relatively compact. From the definition of $\omega(B)$ follows its positive invariance and consequently for $\omega(B)$ to be invariant it is sufficient to show that it is negatively invariant. Let $y \in \omega(B)$ and $t \in \mathbb{T}$, then there exist $\{x_k\} \subseteq B$ and $t_k \to +\infty$ such that $y = \lim\limits_{k \to +\infty} x_k t_k = \lim\limits_{k \to +\infty} x_k(t_k - t + t) = \lim\limits_{k \to +\infty} \pi^t(x_k(t_k - t))$. As $t_k - t \to +\infty$ then according to condition (1), the sequence $\{x_k(t_k - t)\}$ can be considered convergent. Assume $y_t = \lim\limits_{k \to +\infty} x_k(t_k - t)$, then $y = \pi^t y_t$ and $y_t \in \omega(B)$, i.e. $y \in \pi^t \omega(B)$. Like so the invariance of $\omega(B)$ is proved. Now let us show that the equality (1.7) takes place. If we suppose that (1.7) does not take place, then there will exist $\varepsilon_0 > 0$, $t_k \to +\infty$ and $x_k \in B$ such that

$$\rho(x_k t_k, \omega(B)) \geq \varepsilon_0. \tag{1.8}$$

According to the condition (2), the sequence $\{x_k t_k\}$ can be considered convergent. Let $y = \lim\limits_{k \to +\infty} x_k t_k$, then $y \in \omega(B)$. On the other hand, passing to the limit in (1.8) when $k \to +\infty$ we will obtain $y \notin \omega(B)$. This contradiction completes the proof of the implication condition (1) $\Rightarrow$ condition (2).

It is evident that (2) $\Rightarrow$ (3) and (3) $\Rightarrow$ (1). The lemma is proved. $\qquad\square$

Corollary 1.2 *Let $M \subseteq X$ be nonempty and $\Sigma^+(M)$ be relatively compact, then $\omega(M) \neq \emptyset$, is compact, invariant and*

$$\lim_{t \to +\infty} \sup_{x \in M} \rho(xt, \omega(M)) = 0. \tag{1.9}$$

The statement inverse to the corollary 1.2 takes place, but before formulating it we will review the notion of the measure of non-compactness.

Definition 1.21 The mapping $\lambda : B(X) \to \mathbb{R}_+$, satisfying the following conditions:

(1) $\lambda(A) = 0$ if and only if $A \in B(X)$ is relatively compact;
(2) $\lambda(A \cup B) = \max(\lambda(A), \lambda(B))$ for every $A, B \in B(X)$.

is called $[175, 179, 278]$ a measure of non-compactness on X.

Definition 1.22 The measure of non-compactness of Kuratowsky $\lambda : B(X) \to \mathbb{R}_+$ is defined by the equality $\lambda(A) := \inf\{\varepsilon > 0 \mid A \text{ admits finite } \varepsilon\text{-covering }\}$.

Lemma 1.4 *Let $M \subseteq X$ be nonempty and relatively compact. If set $\omega(M)$ is nonempty, compact and the equality (1.9) tales place, then $\Sigma^+(M)$ is relatively compact.*

Proof. Let $\varepsilon > 0$, then from the equality (1.9) follows the existence of a positive number $L(\varepsilon)$ such that

$$M_\varepsilon := \bigcup_{t \geq L(\varepsilon)} \pi^t M \subseteq B(\omega(M), \varepsilon). \tag{1.10}$$

Denote by $\lambda(K)$ the measure of non-compactness of Kuratowsky of the set K. From the inclusion (1.10) follows that

$$\lambda(\Sigma^+(M)) = \lambda(\pi(M, [0, L(\varepsilon)]) \cup M_\varepsilon) =$$

$$\max(\lambda(\pi(M, [0, L_\varepsilon]), \lambda(M_\varepsilon)) = \lambda(M_\varepsilon) \leq 2\varepsilon.$$

From this we obtain the equality $\lambda(\Sigma^+(M)) = 0$. The lemma is proved. $\qquad\square$

Theorem 1.5 *Let $M \subseteq X$ be nonempty and relatively compact. For the set $\Sigma^+(M)$ to be compact it is necessary and sufficient the fulfillment of the following 3 conditions:*

(1) $\omega(M) \neq \emptyset$;
(2) $\omega(M)$ is compact;
(3) the equality (1.9) takes place.

The formulated statement follows from the Corollary 1.2 and Lemma 1.4.

1.3   Center of Levinson

Let $\mathfrak{M}$ be some family of subsets from X.

Definition 1.23 Dynamical system $(X, \mathbb{T}, \pi)$ will be called $\mathfrak{M}$-dissipative if for every $\varepsilon > 0$ and $M \in \mathfrak{M}$ there exists $L(\varepsilon, M) > 0$ such that $\pi^t M \subseteq B(K, \varepsilon)$ for any $t \geq L(\varepsilon, M)$, where K is a certain fixed subset from X depending only on $\mathfrak{M}$. In this case K we will call the attractor for $\mathfrak{M}$.

For the applications the most important ones are the cases when K is bounded or compact and $\mathfrak{M} = \{\{x\} \mid x \in X\}$ or $\mathfrak{M} = C(X)$, or $\mathfrak{M} = \{B(x, \delta_x) \mid x \in X, \ \delta_x > 0\}$, or $\mathfrak{M} = B(X)$.

Definition 1.24 The system $(X, \mathbb{T}, \pi)$ is called:

- point dissipative if there exist $K \subseteq X$ such that for every $x \in X$

$$\lim_{t \to +\infty} \rho(xt, K) = 0; \tag{1.11}$$

- compact dissipative if the equality (1.11) takes place uniformly w.r.t. x on the compacts from X;
- locally dissipative if for any point $p \in X$ there exist $\delta_p > 0$ such that the equality (1.11) takes place uniformly w.r.t. $x \in B(p, \delta_p)$;
- bounded dissipative if the equality (1.11) takes place uniformly w.r.t. x on every bounded subset from X.

As we have already mentioned above, for applications our main interest rests on cases in which K is compact and bounded. According to this the system $(X, \mathbb{T}, \pi)$ is called point k (b)-dissipative if $(X, \mathbb{T}, \pi)$ is point dissipative and the set K from (1.11) is compact (bounded). Also like that is defined the notion of the compact k (b)-dissipative system and of other types of k (b)-dissipativity.

From the definitions above follows that bounded k (b)-dissipativity implies local k (b)-dissipativity (see Lemma 1.5), local k (b)-dissipativity implies compact k (b)-dissipativity, and compact k (b)-dissipativity implies point $k(b)$-dissipativity.

As it will be shown below in general case introduced by us classes are different. We will take interest in the following tasks:

(1) What connection does exist between different types of dissipativity?
(2) Conditions under which a dissipative system admits maximal compact invariant set.
(3) Structure of the center of Levinson (maximal compact invariant attractor) for different classes of dynamical systems.
(4) Conditions of asymptotic and uniform asymptotic stability of the center of Levinson.

Lemma 1.5 *Let* $(X, \mathbb{T}, \pi)$ *be local k (b)-dissipative, then it is compact k (b)-dissipative.*

Proof. Let $(X, \mathbb{T}, \pi)$ be local k (b)-dissipative. Then there exist a nonempty compact (bounded) set $K \subseteq X$, such that for any $\varepsilon > 0$ and $x \in X$ there exist $\delta(x) > 0$ and $l = l(\varepsilon, x) > 0$ for which

$$\rho(yt, K) < \varepsilon \qquad (1.12)$$

for every $t \geq l$ and $y \in B(x, \delta(x))$.

Let M is a nonempty compact from X. Then for every $x \in M$ there exist $\delta = \delta(x) > 0$ and $l = l(\varepsilon, x) > 0$ such that the inequality (1.12) takes place. Consider the open covering $\{B(x, \delta(x)) \mid x \in M\}$ of the set M. By virtue of the compactness of M and completeness of the space X, from the constructed covering of the set M we can extract $\{B(x_i, \delta(x_i)) \mid i \in \overline{1, m}\}$. Assume $L(\varepsilon, M) = \max\{l(\varepsilon, x_i) \mid i \in \overline{1, m}\}$, then from (1.12) follows that $\rho(xt, K) < \varepsilon$ for all $x \in M$ and $t \geq L(\varepsilon, M)$. The lemma is proved. $\qquad \square$

Lemma 1.6 *Let $M \subseteq X$ and the set $\Sigma^+(M)$ be relatively compact. If $\omega(M) \subseteq M$, then*

$$\omega(M) = \cap\{\pi^t M \mid t \in \mathbb{T}\}. \qquad (1.13)$$

Proof. Denote by $I(M) := \cap\{\pi^t M \mid t \in \mathbb{T}\}$. It is clear that $I(M) \subseteq \omega(M)$. Let us show that the inverse inclusion also takes place if $\omega(M) \subseteq M$. In fact, according to

Lemma 1.3 and Corollary 1.2 the set $\omega(M)$ is invariant and, consequently, $\omega(M) = \pi^t\omega(M) \subseteq \pi^t M$ for every $t \in \mathbb{T}$. From what follows the inclusion $\omega(M) \subseteq I(M)$. The lemma is proved. $\qquad\square$

Remark 1.4 *Below we will mainly consider k-dissipative dynamical systems. That's why the sign k- we will drop out everywhere where it won't lead to misunderstanding.*

Let $(X, \mathbb{T}, \pi)$ be compact dissipative and K is a nonempty compact set that is an attractor for compact subsets X. Then for every compact $M \subseteq X$ the equality

$$\lim_{t \to +\infty} \sup_{x \in M} \rho(xt, K) = 0$$

holds. Hence $\omega(K) \subseteq K$ and, consequently,

$$J := \omega(K) = \cap\{\pi^t K \mid t \in \mathbb{T}\}. \tag{1.14}$$

Let us show that the set J does not depend on the choice of the set K attracting all compact subsets of the space X. In fact, if we denote by $J(K) := \omega(K)$ and K_1 every other compact set attracting all compacts from X, then there will be $L = L(K, K_1, \varepsilon) > 0$ such that $\pi^t J(K) \subseteq K_1$ and $\pi^t J(K_1) \subseteq K$ for all $t \geq L$. As $J(K) = \omega(K) \subseteq K$ and $J(K_1) = \omega(K_1) \subseteq K_1$ then from the invariance of $J(K_1)$ and $J(K)$ follows that $J(K) \subseteq K_1$, $J(K_1) \subseteq K$, $J(K) \subseteq \pi^t K_1$ and $J(K_1) \subseteq \pi^t K$ for all $t \in \mathbb{T}$ and consequently $J(K) = J(K_1)$.

So, the set J defined by the equality (1.14) does not depend on the choice of the attractor K but is defined only by inner properties of the dynamical system $(X, \mathbb{T}, \pi)$.

Definition 1.25 The set J defined by the equality (1.14), according to [332] we will call the center of Levinson of the compact dissipative dynamical system $(X, \mathbb{T}, \pi)$.

Theorem 1.6 [102, 175, 232] *Let $(X, \mathbb{T}, \pi)$ be compact dissipative system and let J be its center of Levinson. Then:*

(1) J is compact invariant set;
(2) J is orbitally stable;
(3) J is the attractor of the family of all compacts of X;
(4) J is the maximal compact invariant set in $(X, \mathbb{T}, \pi)$.

Proof. The first statement of the Theorem directly follows from the definition of J and from Lemma 1.3. Let us show that the set J is orbitally stable. If we suppose the contrary we will obtain the existence of $\varepsilon_0 > 0$, $\delta_n \to 0$ $(\delta_n > 0)$, $x_n \in B(J, \delta_n)$

and $t_n \to +\infty$ such that

$$\rho(x_n t_n, J) \geq \varepsilon_0. \tag{1.15}$$

As $x_n \in B(J, \delta_n)$, $\delta_n \to 0$ and J is compact, without loss of generality the sequence $\{x_n\}$ can be considered convergent. By virtue of the compact dissipativity of the system $(X, \mathbb{T}, \pi)$ the set $\Sigma^+(\{x_n\})$ is relatively compact. Note that together with the set K the set $K' = K \cup H^+(\{x_n\})$ is the attractor for the family of all compacts from X and consequently $\omega(K') = \omega(K) = J$. In particular, $\omega(H^+(\{x_n\})) \subseteq \omega(K) = J$. By compactness of $H^+(\{x_n\})$ the sequence $\{x_n t_n\}$ can be considered convergent. Let $p = \lim\limits_{n \to +\infty} x_n t_n$, then $p \in \omega(H^+(\{x_n\}))$. On the other hand from (1.15) follows that $p \notin J$. The obtained contradiction proves the second statement of the theorem.

Now let $M \in C(X)$. Then by virtue of the compact dissipativity of $(X, \mathbb{T}, \pi)$ the set $\Sigma^+(M)$ is relatively compact and according to Theorem 1.5 the conditions (1)-(2) are fulfilled. In particular, for every $\varepsilon > 0$ there exists $L(\varepsilon) > 0$ such that $\pi^t M \subseteq B(\omega(M), \varepsilon)$ for all $t \geq L(\varepsilon)$. Together with the set K the set $K' = K \cup \omega(M)$ also is the attractor of the compact subsets from X and consequently $\omega(K') = \omega(K) = J$. So, $\omega(M) \subseteq \omega(K') = J$, and hence

$$\beta(\pi^t M, J) \leq \beta(\pi^t M, \omega(M)) < \varepsilon,$$

i.e. $\lim\limits_{t \to +\infty} \beta(\pi^t M, J) = 0$ for all $M \subseteq C(X)$ where $\beta(A, B)$ is semi-deviation of the set A from the set B.

At last let us prove the 4th statement of the theorem. Let J_1 be compact invariant subset from X. Then according to the 3rd statement of the theorem we have

$$\lim\limits_{t \to +\infty} \beta(\pi^t J_1, J) = 0. \tag{1.16}$$

By virtue of the invariance of J_1 the equality $\pi^t J_1 = J_1$ holds for all $t \in \mathbb{T}$. From this and from the equality (1.16) we obtain $J_1 \subseteq J$. The theorem is completely proved. $\qquad\square$

Denote by $\{K_\lambda | \lambda \in \Lambda\}$ the family of all nonempty compact positive invariant sets that attract all compacts from X.

The following theorem holds.

Theorem 1.7 *Let $(X, \mathbb{T}, \pi)$ be compact dissipative dynamical system and let J be its Levinson center.*

$$J = J' := \cap\{K_\lambda | \lambda \in \Lambda\},$$

i.e. J is the least compact positive invariant set attracting all compacts from X.

Proof. Assume

$$K := \cap\{K_\lambda | \lambda \in \Lambda\}.$$

First of all note that $J \subseteq K$ and consequently $K \neq \emptyset$. In fact, for every $\lambda \in \Lambda$ we have $J = \omega(K_\lambda) \subseteq K_\lambda$, i.e. $J \subseteq K$.

Let us show now that the reverse inclusion also holds. As J attracts all compacts from X and is nonempty and positive invariant, then $J \in \{K_\lambda | \lambda \in \Lambda\}$ and consequently $K \subseteq J$. The theorem is proved. $\square$

Lemma 1.7 *Let $(X, \mathbb{T}, \pi)$ be compact dissipative dynamical system, J be its Levinson's center and K be a nonempty compact attracting all compacts from X, then*

$$J = \bigcap_{t \in T} \pi^t K.$$

Proof. As K is the attractor of all compacts from X, then $\omega(K) \subseteq K$ and according to Lemma 1.6

$$J := \omega(K) = \bigcap \pi^t K.$$

The lemma is proved. $\square$

Lemma 1.8 *Let M be compact positive invariant asymptotically stable set, then the following statements hold:*

(1) the domain of attraction $W^s(M)$ of the set M is open;
(2) the equality

$$\lim_{t \to +\infty} \beta(\pi^t K, M) = 0 \tag{1.17}$$

takes place for every compact K from $W^s(M)$.

Proof. Show that $W^s(M)$ is open. Really, as M is attracting set, then there exists $\delta > 0$ such that $B(M, \delta) \subset W^s(M)$. Let us show that for every point $p \in W^s(M) \backslash B(M, \delta)$ there exists $\eta > 0$ such that $B(p, \eta) \subset W^s(M)$. As $p \in W^s(M)$, there will be $t_p > 0$ such that $pt_p \in B(M, \delta)$. By openness of $B(M, \delta)$ there is $\gamma > 0$ for which $B(pt_p, \gamma) \subset B(M, \delta)$. According to continuity of the mapping $\pi(t_p, \cdot) : X \longrightarrow X$ there exists $\eta > 0$ such that the inclusion $\pi^{t_p} B(p, \eta) \subset B(M, \delta) \subset W^s(M)$ holds and consequently the set $W^s(M)$ is open.

Let us prove the second statement of the lemma. Let $\varepsilon > 0$ and K be a compact from $W^s(M)$. For the number $\varepsilon > 0$ we choose $\delta(\varepsilon) > 0$ taking into account the condition of the stability of M. As M attracts points from $W^s(M)$, for every point $x \in K$ there are $\gamma(x, \varepsilon) > 0$ and $l(x, \varepsilon) > 0$ such that

$$\pi^t B(x, \gamma(x, \varepsilon)) \subseteq B(M, \varepsilon) \tag{1.18}$$

for all $t \geq l(x, \varepsilon)$. By the compactness of K from the open covering $\{B(x, \gamma(x, \varepsilon)) \mid x \in K\} \supseteq K$ we can extract sub-covering $\cup\{B(x_i, \gamma(x, \varepsilon)) \mid i = \overline{1, n}\} \supseteq K$. Assume $L(M, \varepsilon) := \max\{l(x_i, \varepsilon) \mid i = \overline{1, n}\}$, then $\pi^t M \subseteq B(K, \varepsilon)$ for all $t \geq L(M, \varepsilon)$, i.e. M attracts K. The lemma is proved. $\square$

Denote by $\{M_\lambda \mid \lambda \in \Lambda\}$ the family of all nonempty compact positively invariant and globally asymptotically stable sets from X and $J'' := \bigcap\{M_\lambda \mid \lambda \in \Lambda\}$.

Theorem 1.8 *Let $(X, \mathbb{T}, \pi)$ be compact dissipative and J be its center of Levinson, then $J = J''$, i.e. the center of Levinson J of the compact dissipative system $(X, \mathbb{T}, \pi)$ is the least compact positively invariant globally asymptotically stable set in X.*

Proof. According to Lemma 1.8 the inclusion $J' \subseteq J''$ takes place and according to Theorem 1.7 $J = J' \subseteq J''$. The reverse inclusion also holds. For that it is sufficient to note that for certain $\lambda_0 \in \Lambda$ the equality $M_{\lambda_0} = J$ holds and consequently $J'' = \cap\{M_\lambda \mid \lambda \in \Lambda\} \subseteq J$. The theorem is proved. $\square$

Theorem 1.9 *Let $(X, \mathbb{T}, \pi)$ be compact dissipative dynamical system and K be a nonempty compact invariant set from X, then the following statements are equivalent:*

1. *K is the center of Levinson of $(X, \mathbb{T}, \pi)$;*
2. *K is globally asymptotically stable;*
3. *K is maximal compact invariant set in X.*

Proof. According to Theorem 1.6 from 1. follows 2. Now we will show that the inverse statement holds. Denote by J the center of Levinson of $(X, \mathbb{T}, \pi)$, then according to Theorem 1.8 $J \subseteq K$. On the other hand, the center of Levinson according to Theorem 1.6 is the attractor of all compact subsets from X and by invariance of K we have $K \subseteq J$. So, $J = K$.

From Theorem 1.6 follows that 1. implies 3. To complete the proof of the theorem it is sufficient to show that the inverse implication also holds. Let J be the center of Levinson of $(X, \mathbb{T}, \pi)$. Then from one hand in virtue of Theorem 1.6 the set J is compact and invariant and by the maximality of K we have $J \subseteq K$. On the other hand, according to Theorem 1.6 the center of Levinson is the attractor of the family $C(X)$ and by the invariance of K we have $K \subseteq J$ and consequently $K = J$. The theorem is proved. $\square$

1.4   Dissipative systems on the local compact spaces

As it was mentioned above from the local dissipativity of the system $(X, \mathbb{T}, \pi)$ follows its compact dissipativity. On the other hand if the space X is locally compact then is is easy to see that the converse also is true. So, in locally compact spaces compact and local dissipativity are equivalent. In fact, as it will be shown bellow a stronger statement holds, namely: in locally compact spaces point dissipativity implies local dissipativity.

Definition 1.26   The dynamical system $(X, \mathbb{T}, \pi)$ we will call:

- locally completely continuous if for every point $p \in X$ there exist $\delta = \delta(p) > 0$ and $l = l(p) > 0$ such that $\pi^l B(p, \delta)$ is relatively compact;
- weakly dissipative if there exist a nonempty compact $K \subseteq X$ such that for every $\varepsilon > 0$ and $x \in X$ there is $\tau = \tau(\varepsilon, x) > 0$ for which $x\tau \in B(K, \varepsilon)$. In this case we will call K weak attractor.

Note that every dynamical system $(X, \mathbb{T}, \pi)$ defined on the locally compact metric space X is locally completely continuous.

Lemma 1.9   *Let $K \subseteq X$ be a nonempty compact, $p_i \in X$ and $\delta_i > 0$ $(i = \overline{1, m})$. If $K \subseteq \cup\{B(p_i, \delta_i) \mid i = \overline{1, m}\}$, then there exists $\gamma > 0$ such that*

$$B(K, \gamma) \subseteq \cup\{B(p_i, \delta_i) \mid i = \overline{1, m}\}. \tag{1.19}$$

Proof.   Assume that the inclusion (1.19) does not hold for any $\gamma > 0$. Then there are $\gamma_n \downarrow 0$ and $r_n \in B(K, \gamma_n)$ such that $r_n \notin \cup\{B(p_i, \delta_i) \mid i = \overline{1, m}\}$. As $\gamma_n \downarrow 0$ then for every point r_n there exists the point $q_n \in K$ such that

$$\rho(r_n, q_n) < \gamma_n. \tag{1.20}$$

By the compactness of K the sequence $\{q_n\}$ can be considered converging to some point $q \in K$. Then according to (1.20) $r_n \to q$, which contradicts the choice of the sequence $\{r_n\}$. $\square$

Lemma 1.10   *Let $(X, \mathbb{T}, \pi)$ be weakly dissipative and locally completely continuous and $K \subset X$ be the weak attractor of $(X, \mathbb{T}, \pi)$, then:*

1) *there exist $a_0 > 0$ and $l_0 > 0$ such that $\pi^t B(K, a_0)$ is relatively compact for every $t \geq l_0$;*
2) *there exist $L_0 \geq l_0$ such that for all $t \geq L_0$*

$$\pi^t B(K, a_0) \subseteq \pi^{l_0} B(K, a_0).$$

Proof. Let $x \in K$. By virtue of the local complete continuity of $(X, \mathbb{T}, \pi)$ for the point $x \in K$ there exist $l(x) > 0$ and $\delta_x > 0$ such that $\pi^t B(x, \delta_x)$ is relatively compact for all $t \geq l(x)$. It is clear that $\{B(x, \delta_x) \mid x \in K\}$ is an open covering of K and by its compactness we can extract finite sub-covering $\{B(x_i, \delta_{x_i}) \mid i \in \overline{1, n}\}$ from the constructed covering. Let $l_0 = \max\{l(x_i) \mid i \in \overline{1, n}\}$. According to Lemma 1.9 there exists $a_0 > 0$ such that

$$K \subseteq B(K, a_0) \subseteq \cup\{B(x_i, \delta_{x_i}) \mid i \in \overline{1, n}\}.$$

Consequently

$$\pi^t B(K, a_0) \subseteq \cup\{\pi^t B(x_i, \delta_{x_i}) \mid i \in \overline{1, n}\},$$

that is why the set $\pi^t B(K, a_0)$ is relatively compact for all $t \geq l_0$.

Let us now prove the 2nd statement of the lemma. Let a_0 and l_0 be positive numbers from the previous point. Supposing that the 2nd statement of the lemma is not true, we will find that there exist $\{x_k\} \subset B(K, a_0)$ and $t_k \to +\infty$ such that

(1) $x_k \in \overline{B(K, a_0)}$.
(2) $x_k t \in X \setminus B(K, a_0)$ for all $0 < t < t_k$.
(3) $x_k t_k \in \overline{B(K, a_0)}$.

From 2. follows that

4. $x'_k t \in X \setminus B(K, a_0)$ for all $0 < t < t_k - l_0$, where $x'_k = x_k l_0$.

By virtue of the relative compactness of $\pi^{l_0} B(K, a_0)$ the sequence $x'_k = x_k l_0$ can be considered convergent. Assume $x_0 = \lim_{k \to +\infty} x'_k$, then from 4. follows that $x_0 t \in X \setminus B(K, a_0)$ for all $0 < t < +\infty$ and consequently $\emptyset \neq \omega_{x_0} \subset X \setminus B(K, a_0)$. So $\omega_{x_0} \cap K = \emptyset$ that contradicts the weak dissipativity of $(X, \mathbb{T}, \pi)$ and to the fact that K is a weak attractor. The contradiction obtained completes the proof of the lemma. $\square$

Theorem 1.10 *For the locally completely continuous dynamical systems the weak, point, compact and local dissipativity are equivalent.*

Proof. It is clear that for the proof of the formulated theorem it is sufficient to show that from the weak dissipativity of $(X, \mathbb{T}, \pi)$ follows its local dissipativity if $(X, \mathbb{T}, \pi)$ is locally completely continuous. Let $K \neq \emptyset$ be compact and be the weak attractor of $(X, \mathbb{T}, \pi)$. Denote by a_0 the number from Lemma 1.10. If $x \in X$ then there exists $\tau > 0$ such that $x\tau \in B(K, a_0)$. Let $\gamma > 0$ be such that $B(x\tau, \gamma) \subset B(K, a_0)$. By the continuity of the mapping $\pi^\tau : X \to X$ in the point x there exists $\alpha > 0$ such that

$$\pi^\tau B(x, \alpha) \subset B(x\tau, \gamma) \subset B(K, a_0).$$

Assume $M = \overline{B(K, a_0)}$. According to Lemmas 1.3 and 1.10 the set $M \neq \emptyset$ is compact and

$$\lim_{k \to +\infty} \beta(\pi^t B(x, \alpha), M) = 0.$$

So, we constructed the nonempty compact $M \subset X$ attracting every point with some α-neighborhood that is $(X, \mathbb{T}, \pi)$ is locally dissipative. The theorem is proved. $\quad\square$

1.5   Criterions of compact dissipativity

Assume $\Omega := \overline{\cup\{\omega_x | x \in X\}}$. Let $(X, \mathbb{T}, \pi)$ be compact dissipative dynamical system and J be its Levinson center. It is clear that $\Omega \subseteq J$. At once plain examples show that in general case $\Omega \neq J$. At the same time the set Ω is an important characteristic of dissipative dynamical system. For example, from Theorem 1.10 follows that in the locally compact space X the dynamical system $(X, \mathbb{T}, \pi)$ is point dissipative if and only if Ω is not empty, compact and $\omega_x \neq \emptyset$ for all $x \in X$. In view of that has been said above, we take on interest in such questions:

a. Which are the conditions necessary for the coincidence of the sets Ω and J?
b. What is the connection between Ω and J in general case?

From the results given bellow follow the answers to the formulated questions.
   Denote by

$$D^+(M) := \bigcap_{\varepsilon > 0} \overline{\bigcup \{\pi^t B(M, \varepsilon) | t \geq 0\}},$$

$$J^+(M) := \bigcap_{\varepsilon > 0} \bigcap_{t \geq 0} \overline{\bigcup \{\pi^\tau B(M, \varepsilon) | \tau \geq t\}},$$

$D_x^+ := D^+(\{x\})$ and $J_x^+ := J^+(\{x\})$.
   Directly from the definition $D^+(M)$ and $J^+(M)$ follows the following lemma.

Lemma 1.11   *The following statements hold:*

(1) $p \in D^+(M)$ $(p \in J^+(M))$ if and only if there exist $\{x_n\}$ and $\{t_n\}$ $(t_n \to +\infty)$ such that $\rho(x_n, M) \to 0$ and $x_n t_n \to p$;
(2) the set $D^+(M)$ $(J^+(M))$ is closed and positive invariant.

Lemma 1.12   *Let $y \in \omega_x$, then $J_x^+ \subseteq J_y^+$.*

Proof. Let $y \in \omega_x$ and $p \in J_x^+$, then there exist sequences $\{\tau_n'\} \to +\infty$, $x\tau_n' \to y$, $\{t_n'\}$ and $\{x_n\}$ such that $x_n \to x$, $t_n' \to +\infty$ and $x_n t_n' \to p$. According Lemma 1.11 we can consider that $t_n' - \tau_n' > n$ for all $n \in \mathbb{N}$. For every $k \in \mathbb{N}$ consider the sequence $\{x_n \tau_k'\}$. By the axiom of continuity $x_n \tau_k' \to x\tau_k'$ for $n \to +\infty$ (for every $k \in \mathbb{N}$) and consequently for every $k \in \mathbb{N}$ there is $n_k \geq k$ such that $\rho(x_n \tau_k', x\tau_k') \leq k^{-1}$ for all $n \geq n_k$. As $x\tau_n' \to y$, we have $\rho(y, x_{n_k} \tau_k') \leq \rho(y, x\tau_k') + \rho(x\tau_k', x_{n_k} \tau_k') \leq \rho(y, x\tau_k') + k^{-1}$. Note that $x_{n_k} t_{n_k}' = x_{n_k} \tau_{n_k}' (t_{n_k}' - \tau_{n_k}')$, $x_{n_k} \tau_{n_k}' \to y$ and $t_{n_k}' - \tau_{n_k}' > n_k \geq k$. From this follows that $p \in J_y^+$, i.e. $J_x^+ \subseteq J_y^+$. The lemma is proved. $\qquad\square$

Corollary 1.3 *If $y \in \omega_x$ then $J_y^+ = D_y^+$.*

Proof. In fact, as $J_y^+ \subseteq D_y^+$, it is sufficient to show that $D_y^+ \subseteq J_y^+$. Note that $D_y^+ = \Sigma_y^+ \cup J_y^+$. Since $y \in \omega_x$, $\Sigma_y^+ \subseteq \omega_x \subseteq J_x^+ \subseteq J_y^+$. Hence, $D_y^+ = \Sigma_y^+ \cup J_y^+ \subseteq J_y^+ \cup J_y^+ = J_y^+$. $\qquad\square$

Lemma 1.13 *If $x_n \to x$, $y_n \to y$ for $n \to +\infty$ and $x_n \in D_{y_n}^+$ $(x_n \in J_{y_n}^+)$, then $x \in D_y^+$ $(x \in J_y^+)$.*

Proof. Let $\varepsilon > 0$ and $\delta > 0$. From the inclusion $x_n \in D_{y_n}^+$ $(x_n \in J_{y_n}^+)$ for every n follows the existence of the point $z_n \in B(y_n, \frac{\delta}{2})$ and of the number $t_n \geq 0$ $(t_n \geq n)$ such that

$$\rho(x_n, z_n t_n) < \frac{\varepsilon}{2}. \tag{1.21}$$

From $x_n \to x$ and $y_n \to y$ follows the existence of such integer n_0 that for all $n > n_0$ simultaneously the inequalities

$$\rho(y, y_n) < \frac{\delta}{2} \quad \text{and} \quad \rho(x_n, x) < \frac{\varepsilon}{2} \tag{1.22}$$

hold. From $z_n \in B(y_n, \frac{\delta}{2})$ and (1.22) we obtain $\rho(y, z_n) < \delta$, and from (1.21) and (1.22) follows the inequality $\rho(x, z_n t_n) < \varepsilon$, i.e. $x \in D_y^+$ $(x \in J_y^+)$. The lemma is proved. $\qquad\square$

Lemma 1.14 *If the set $M \subseteq X$ is compact, then $D^+(M) = \cup\{D_x^+ | x \in M\}$ and $J^+(M) = \cup\{J_x^+ | x \in M\}$.*

Proof. Since $\cup\{D_x^+ | x \in M\} \subseteq D^+(M)$, then to prove the lemma it is sufficient to show that $D^+(M) \subseteq \cup\{D_x^+ | x \in M\}$. Let $y \in D^+(M)$, then there exist $\{x_n\}$ and $t_n \geq 0$ such that $\rho(x_n, M) \to 0$ and $y = \lim_{n \to +\infty} x_n t_n$. As M is compact the sequence $\{x_n\}$ can be considered convergent. Assume $x := \lim_{n \to +\infty} x_n$, then $x \in M$. So, $x_n \to x$, $y \in D_x^+$ and $y_n \to y$ $(y_n := x_n t_n)$. According to Lemma 1.13 $y \in D_x^+ \subseteq \cup\{D_x^+ | x \in M\}$. The second statement is established in the same way. The lemma is proved. $\qquad\square$

Lemma 1.15   *If M is nonempty compact negative invariant set, then $M \subseteq J^+(\Omega)$.*

Proof. Let M be nonempty compact negatively invariant set and $x \in M$, then according to Theorem 1.2 there exists an entire trajectory $\varphi : S \longrightarrow X$ such that $\varphi(0) = x$ and $\varphi(t) \in M$ for all $t \geq 0$. As $\alpha_{\varphi_x} \neq \emptyset$, is closed and $\alpha_{\varphi_x} \subseteq M$, then it is compact. Let $y \in \alpha_{\varphi_x}$, then $\omega_y \subseteq \alpha_{\varphi_x}$. If $p \in \omega_y \subseteq \Omega$, then there exists $t_n \to +\infty$ such that $x_k = \varphi(-t_k) \to p$, $x = \pi^{t_k}\varphi(-t_k)$ and consequently $x \in J_p^+ \subseteq J^+(\Omega)$. The lemma is proved. $\qquad\square$

Corollary 1.4   *If $(X, \mathbb{T}, \pi)$ is point dissipative, $\Omega \neq \emptyset$ and it is compact, then $\Omega \subseteq J^+(\Omega)$.*

Proof. In fact, since ω_x is invariant for all $x \in X$ and $\Omega = \overline{\cup\{\omega_x \mid x \in X\}}$, then Ω is also invariant and according to Lemma 1.15 $\Omega \subseteq J^+(\Omega)$. $\qquad\square$

Lemma 1.16   *If the dynamical system $(X, \mathbb{T}, \pi)$ is point dissipative and $D^+(\Omega)$ $(J^+(\Omega))$ is compact, then $D^+(\Omega) = D^+(D^+(\Omega))$ $(J^+(\Omega) = J^+(J^+(\Omega)))$.*

Proof. As $\Omega \subseteq D^+(\Omega)$, then $D^+(D^+(\Omega)) \subseteq D^+(\Omega)$ and, consequently, to prove Lemma 1.16 it is sufficient to establish the inverse inclusion $D^+(D^+(\Omega)) \subseteq D^+(\Omega)$. Let $x \in D^+(\Omega)$, then $\omega_x \subseteq \Omega$, and if $y \in \omega_x \subseteq \Omega$ then according to Lemma 1.12 and Corollary 1.3 $J_x^+ \subseteq J_y^+ = D_y^+ \subseteq D^+(\Omega)$. Since $D_x^+ = \Sigma_x^+ \cup J_x^+ \subset D^+(\Omega) \cup D^+(\Omega) = D^+(\Omega)$ for all $x \in D^+(\Omega)$ and $D^+(\Omega)$ is compact, then according to Lemma 1.14 $D^+(D^+(\Omega)) = \cup\{D_x^+ | x \in D^+(\Omega)\} \subseteq D^+(\Omega)$. By analogy with this the proof of the second equation is executed but we should keep in mind that $\Omega \subseteq J^+(\Omega)$. The lemma is proved. $\qquad\square$

Lemma 1.17   *If $(X, \mathbb{T}, \pi)$ is compact dissipative, then $D^+(D^+(\Omega)) = D^+(\Omega)$ $(J^+(J^+(\Omega)) = J^+(\Omega))$.*

Proof. Let $(X, \mathbb{T}, \pi)$ be compact dissipative and let J be its Levinson's center. Then $\Omega \subseteq J$. By virtue of orbital stability of J we have $D^+(J) = J$ $(J^+(J) \subseteq J)$ and, consequently, $D^+(\Omega) \subseteq J$ $(J^+(\Omega) \subseteq J)$. As $D^+(\Omega)$ $(J^+(\Omega))$ is closed and J is compact, $D^+(\Omega)$ $(J^+(\Omega))$ is also compact and to finish the proof of the lemma it is sufficient to refer to Lemma 1.16. $\qquad\square$

It is known [25] that for compact positive invariant set M the equality $D^+(M) = M$ holds if and only if M is orbital stable (X is locally compact and $\mathbb{T} = \mathbb{R}$). Below we will show that this statement is valid also for compact dissipative systems on arbitrary metric space both with continuous and discrete time. Namely, the following lemma holds.

Lemma 1.18 *Let $(X, \mathbb{T}, \pi)$ be compact dissipative. For the compact positive invariant set $M \subseteq X$ to be orbital stable, it is necessary and sufficient that the equality $D^+(M) = M$ holds.*

Proof. If M is orbitally stable, then by standard reasoning (see, for example, [25]) is shown that $D^+(M) = M$ for every systems (including dissipative ones).

Let now $D^+(M) = M$ and show that M is orbital stable. If we suppose the contrary then there exist $\varepsilon_0 > 0$, $x_n \to x \in M$, and $t_n \geq 0$ such that

$$\rho(x_n t_n, M) \geq \varepsilon_0. \tag{1.23}$$

Since the system $(X, \mathbb{T}, \pi)$ is compact dissipative, the set $\Sigma^+(K)$ is relatively compact where $K := \{x_n\}$ and, consequently, the sequence $\{x_n t_n\}$ can be considered convergent. Assume $y := \lim\limits_{n \to +\infty} x_n t_n$. Then on the one hand $y \in D^+(M) = M$. On the other hand, from (1.23) follows that $\rho(y, M) \geq \varepsilon_0 > 0$. The obtained contradiction proves the lemma. $\square$

Theorem 1.11 *If the dynamical system $(X, \mathbb{T}, \pi)$ is compact dissipative, then $J = J^+(\Omega)$.*

Proof. As $\Omega \subseteq J$ and J is asymptotically stable, then $J^+(\Omega) \subseteq J^+(\Omega)$. From Lemmas 1.17 and 1.18 follows that the set $J^+(\Omega)$ is orbital stable because $D^+(J^+(\Omega)) = J^+(\Omega)$. Let $x \in J \setminus J^+(\Omega)$ and $d_x := \rho(x, J^+(\Omega)) \geq 0$. If $d_x = 0$ for all $x \in J \setminus J^+(\Omega)$, then the theorem is proved. Suppose that for some $x_0 \in J \setminus J^+(\Omega)$ we have $d_{x_0} > 0$. For the number $0 < \varepsilon < 2^{-1} d_{x_0}$, choose $\delta(\varepsilon) > 0$ out of the condition of orbital stability of $J^+(\Omega)$. Since $x_0 \in J$, according to Theorem 1.2 there exists a continuous mapping $\varphi : \mathbb{S} \longrightarrow J$ such that $\pi^t \varphi(s) = \varphi(t+s)$ for all $t \in \mathbb{T}$, $s \in \mathbb{S}$ and $\varphi(0) = x_0$. Since J is compact, the set α of limit points $\alpha_{\varphi_{x_0}}$ of the motion φ is not empty, compact and $\alpha_{\varphi_{x_0}} \cap \Omega \neq \emptyset$ and, consequently, there exists $t_n \to -\infty$ such that $\rho(x_0 t_n, \Omega) \to 0$. Choose n_0 such that $\rho(x_0 t_n, \Omega) < \delta$ $(n \geq n_0)$. Then we have $\rho(x_0 t_n, A, AA?, J^+(\Omega)) < \delta$ and, consequently, $\rho(x_0(t+t_n), J^+(\Omega)) < \varepsilon$ for all $t_0 \geq 0$ and $n \geq n_0$. In particular, when $t = -t_n$, we have $d_{x_0} = \rho(x_0, J^+(\Omega)) < \varepsilon < 2^{-1} d_{x_0}$. The obtained contradiction completes the proof of the theorem. $\square$

Corollary 1.5 *Under the conditions of Theorem 1.11 the equality $J = D^+(\Omega)$ holds.*

Proof. The formulated statement follows from Theorem 1.11 taking into account that $J \subseteq J^+(\Omega) \subseteq D^+(\Omega) \subseteq J$. $\square$

Corollary 1.6 *If $(X, \mathbb{T}, \pi)$ is compact dissipative, then $J = \Omega$ if and only if Ω is orbital stable.*

Proof. This statement follows from Theorem 1.11 and Lemma 1.18. $\qquad\square$

Theorem 1.12 *For the dynamical system $(X, \mathbb{T}, \pi)$ to be compact dissipative, it is necessary and sufficient that there exists a nonempty compact set $K \subseteq X$ satisfying the condition: for every $\varepsilon > 0$ and $x \in X$, there exist $\delta(\varepsilon, x) > 0$ and $l(\varepsilon, x) > 0$ such that*

$$\pi^t B(x, \delta(\varepsilon, x)) \subseteq B(K, \varepsilon) \tag{1.24}$$

for all $t \geq l(\varepsilon, x)$.

Proof. Let $(X, \mathbb{T}, \pi)$ be compact dissipative, J be its center of Levinson, $\varepsilon > 0$ and $x \in X$. According to Theorem 1.6 J is orbital stable. Denote by $\gamma(\varepsilon) > 0$ a number defined by ε out of the condition of orbital stability of J. The set J is globally asymptotically stable, hence for $x \in X$ and $\gamma(\varepsilon) > 0$ there exists $l(\varepsilon, x)$ such that $xt \in B(J, \gamma)$ for all $t \geq l(\varepsilon, x)$. Since $B(J, \gamma)$ is open, there exists $\alpha = \alpha(\varepsilon, x) > 0$ for which $B(\pi(x, l(\varepsilon, x)), \alpha) \subseteq B(J, \gamma)$. By virtue of continuity of the mapping $\pi(l(\varepsilon, x), \cdot) : X \longrightarrow X$ for all $x \in X$ and $\alpha > 0$, there exists $\delta = \delta(\varepsilon, x) > 0$ such that

$$\pi(l(\varepsilon, x), B(x, \delta)) \subseteq B(\pi(l(\varepsilon, x), x), \alpha). \tag{1.25}$$

From the inclusion (1.25) and by the choice of γ we have $\pi^t B(x, \delta) \subseteq B(J, \varepsilon)$ for all $t \geq l(\varepsilon, x)$.

Let now $K \subseteq X$ be a nonempty compact satisfying the condition (1.24). If $\varepsilon > 0$ and M is a nonempty compact from X, then for every $x \in M$ there exist $\delta(\varepsilon, x) > 0$ and $l(\varepsilon, x) > 0$ such that (1.24) holds. Consider open covering $\{B(x, \delta(\varepsilon, x)) \mid x \in M\}$ of the set M. By virtue of compactness of M and of the completeness of the space X we can extract from this covering a finite sub-covering $\{B(x_i, \delta(\varepsilon, x_i)) \mid i \in \overline{1, n}\}$. Assume $L(\varepsilon, M) := \max\{l(\varepsilon, x_i) \mid i = \overline{1, n}\}$. From (1.24) follows that $\pi^t M \subseteq B(K, \varepsilon)$ for all $t \geq L(\varepsilon, M)$. The theorem is proved. $\qquad\square$

Theorem 1.13 *Let $(X, \mathbb{T}, \pi)$ be point dissipative. For $(X, \mathbb{T}, \pi)$ to be compact dissippative it is necessary and sufficient that there exists a nonempty compact set M possessing the following properties:*

(1) $\Omega \subseteq M$;
(2) M is orbital stable.

In this case $J \subseteq M$ where J is the center of Levinson of $(X, \mathbb{T}, \pi)$.

Proof. Necessity of the conditions of the theorem is obvious. To prove the formulated statement it is enough to show the sufficiency of the conditions (1) and (2). Reasoning in the same way as in the first part of Theorem 1.12 we establish that for every $\varepsilon > 0$ and $x \in X$ there exist $\delta(\varepsilon, x) > 0$ and $l(\varepsilon, x) > 0$ satisfying the condition (1.24). In this case as K we should take the set M from Theorem 1.13. Applying Theorem 1.12 we obtain compact dissipativity of $(X, \mathbb{T}, \pi)$.

Now let us show that $J \subseteq M$. Since $\Omega \subseteq M$ and M is orbital stable, $D^+(\Omega) \subseteq D^+(M) = M$. To finish the proof of the theorem it is sufficient to refer to Corollary 1.5. $\qquad\square$

Theorem 1.14 *Let $(X, \mathbb{T}, \pi)$ be point dissipative. For $(X, \mathbb{T}, \pi)$ to be compact dissipative it is necessary and sufficient that the set $D^+(\Omega)$ $(J^+\Omega))$ be compact and orbital stable. In this case $J = D^+(\Omega)$ $(J = J^+(\Omega))$ where J is the center of Levinson of the dynamical system $(X, \mathbb{T}, \pi)$.*

Proof. The formulated statement follows from Theorems 1.11, 1.12, 1.13 and Corollary 1.5. $\qquad\square$

Theorem 1.15 *Let $(X, \mathbb{T}, \pi)$ be point dissipative. For $(X, \mathbb{T}, \pi)$ to be compact dissipative it is necessary and sufficient that $\Sigma^+(K)$ be relatively compact for any compact $K \subseteq X$.*

Proof. The necessity of the conditions of the theorem follows from Theorem 1.5. Let us prove the sufficiency. First of all, note that under the conditions of the theorem the set $J^+(\Omega) \neq \emptyset$ and is compact. In fact, according to Corollary 1.4 $\Omega \subseteq J^+(\Omega)$ and, consequently, $J^+(\Omega) \neq \emptyset$. Show now that $J^+(\Omega)$ is compact. Let $\{y_k\} \subseteq J^+(\Omega)$ and $\varepsilon_k \downarrow 0$, then there exist $p_k \in \Omega$, $y_k \in J^+_{p_k}$, $\overline{p}_k \in B(p_k, \varepsilon_k)$ and $t_k > 0$ such that

$$\rho(y_k, \overline{p}_k t_k) < \varepsilon_k. \tag{1.26}$$

Since $\{p_k\} \subseteq \Omega$, Ω is compact and $\varepsilon_k \downarrow 0$, the sequence $\{p_k\}$ is relatively compact and, consequently, under the conditions of the theorem the sequence $\{\overline{p}_k t_k\}$ is also compact. Then from the inequality (1.26) follows that the sequence $\{y_k\}$ is relatively compact, i.e. $J^+(\Omega)$ is compact.

Let us show that $J^+(\Omega)$ attracts all compacts from X. Let K be an empty compact from X. Then under the conditions of the theorem the set $\Sigma^+(K)$ is relatively compact and according to Theorem 1.5 the set $\Omega(K)$ is not empty and the equality

$$\lim_{t \to +\infty} \beta(\pi^t K, \Omega(K)) = 0 \tag{1.27}$$

holds. According to Lemma 1.3 the set $\Omega(K)$ is invariant and by Lemma 1.15 $\Omega(K) \subseteq J^+(\Omega)$ and, consequently, from (1.27) follows that $J^+(\Omega)$ attracts K. The theorem is proved. $\qquad\square$

At the end of this chapter, we give an example of point dissipative dynamical system that is not compact dissipative.

Example 1.6   Let $\varphi : \mathbb{R} \longrightarrow \mathbb{R}$ be a continuous function equal to zero on $\mathbb{R}_-$ and defined on $\mathbb{R}_+$ by the equality

$$\varphi(t) = \begin{cases} \exp\left\{ \left[(t-1)^2 - 1\right]^{-1} + 1 \right\} & 0 \le t < 2 \\ 0, & 2 \le t < +\infty. \end{cases} \tag{1.28}$$

Assume $X = \{\varphi(at+b)|\ a,b; t \in \mathbb{R}\}$. Note that $X \subset C(\mathbb{R},\mathbb{R})$ is closed and invariant w.r.t. the shifts subset of $C(\mathbb{R},\mathbb{R})$ and, consequently, the dynamical system of Bebutov induced on X a group dynamical system of shifts $(X,\mathbb{R},\sigma)$. We remark on some properties of the constructed dynamical system:

(1) For any function $\psi \in X$, the set $\{\sigma(t,\psi)\ : t \in \mathbb{R}\}$ is relatively compact and there exists $c = c(\psi) \in [0,1]$ such that $\omega_\psi = \alpha_\psi = \{\varphi_c\}$ where $\varphi_c(t) = c$ for all $t \in \mathbb{R}$;

(2) $(X,\mathbb{R},\sigma)$ is point dissipative and $\Omega = \{\varphi_c\ |\ 0 \le c \le 1\}$;

(3) $D^+(\Omega) = X$.

Note that X is not compact. In fact, if that were not so, then according to Theorem of Arzela-Ascoli the functions from X would be equicontinuous on the segment $[0,2]$. Let us construct a subset from X that does not possess this property. Let $\varphi_n(t) = \varphi(nt)$. Assume $t_n^1 = n^{-1}$ and $t_n^2 = 2n^{-1}$. It is clear that $t_n^1, t_n^2 \in [0,2]$, $t_n^1 - t_n^2 \to 0$ for $n \to +\infty$ and $|\varphi_n(t_n^1) - \varphi_n(t_n^2)| = |\varphi(1) - \varphi(2)| = 1$, i.e. the sequence $\{\varphi_n\}$ is not equicontinuous on the segment $[0,2]$. So, $D^+(\Omega) = X$ is not compact and according to Theorem 1.14 the dynamical system $(X,\mathbb{R},\sigma)$ is not compact dissipative.

In relation to Example 1.6, note that according to Theorem 1.14 examples of point dissipative but not compact dissipative dynamical systems can be either of two types: $D^+(\Omega)$ is not compact (as in Example 1.6), or $D^+(\Omega)$ is compact but is not orbital stable. Let us show the latter case.

Example 1.7   Let φ be the function from the previous example. Assume $X = \{\varphi(at+b)|\ a,b \in \mathbb{R}\ \text{ and }\ t \in \mathbb{R}\}$. By $(X,\mathbb{R},\sigma)$ denote a semigroup dynamical system of shifts on X. We note the following properties of the constructed dynamical system:

(1) For any function $\psi \in X$, the set $\{\sigma(t,\psi)\ |\ t \in \mathbb{R}_+\}$ is relatively compact and there exists $c \in [0,1]$ such that $\omega_\psi = \{\varphi_c\}$;

(2) $D^+(\Omega) = \Omega$ and, consequently, $D^+(\Omega)$ is compact, as $\Omega = \{\varphi_c\ |\ 0 \le c \le 1\}$.

Show that $D^+(\Omega)$ is not orbital stable. It is clear that to prove this statement it is sufficient to construct the sequence $\{\psi_n\} \subseteq X$ and $t_n \ge 0$ such that $\psi_n \to \psi_0 \in \Omega$ and $\inf\{\rho(\sigma(t_n,\psi_n),\Omega)|\ n \ge 0\} > 0$. Assume $\psi_n(t) = \varphi(nt + 1 - n^2)$ $(t \in \mathbb{R})$ and $t_n = n$. Then $\sigma(t_n,\psi_n)(s) = \varphi(ns + 1)$ is not convergent in X, hence

$\inf\{\rho(\sigma(t_n, \psi_n), \Omega) \mid n \geq 0\} > 0$ (in fact, the sequence, $\varphi(ns + 1)$ does not contain sub-sequences convergent in X). So, for semigroup point dissipative dynamical system $(X, \mathbb{R}_+, \sigma)$ the set $D^+(\Omega)$ is compact but not orbital stable. According to Theorem 1.14 $(X, \mathbb{R}_+, \sigma)$ is not compact dissipative.

Remark 1.5 *As it was proved in Lemma 1.18, for the compact positively invariant set $M \subseteq X$ the equality $D^+(M) = M$ assures the orbital stability of M if the dynamical system $(X, \mathbb{T}, \pi)$ is compact dissipative. Example 1.7 shows that for point dissipative systems the corresponding fact does not hold, i.e., in this case the well known Theorem of Ura [25] does not hold.*

1.6 Local dissipative systems

Theorem 1.16 *For the compact dissipative dynamical system $(X, \mathbb{T}, \pi)$ to be locally dissipative is necessary and sufficient that for every point $p \in X$ there exists $\delta_p > 0$ such that*

$$\lim_{t \to +\infty} \beta(\pi^t B(p, \delta_p), J) = 0, \tag{1.29}$$

where J is the center of Levinson of $(X, \mathbb{T}, \pi)$.

Proof. The sufficiency of the theorem is obvious. Let $(X, \mathbb{T}, \pi)$ be locally dissipative dynamical system. Then there exists a nonempty compact $K \subseteq X$ such that for every point $p \in X$ there exists $\delta_p > 0$ for which the following equality holds:

$$\lim_{t \to +\infty} \beta(\pi^t B(p, \delta_p), K) = 0. \tag{1.30}$$

From Lemma 1.3 follows that the set $K_p := \omega(B(p, \delta_p))$ is nonempty, compact, invariant and

$$\lim_{t \to +\infty} \beta(\pi^t B(p, \delta_p), K_p) = 0. \tag{1.31}$$

From the equality (1.30) follows the inclusion $K_p \subseteq K$ and consequently $\omega(K_p) \subseteq \omega(K) \subseteq J$. By virtue of the invariance of the set K_p we have $K_p = \omega(K_p)$; hence $K_p \subseteq J$. From the last inclusion and the equality (1.31) follows the equality (1.29). The theorem is proved. $\square$

Theorem 1.17 *For the compact dissipative dynamical system $(X, \mathbb{T}, \pi)$ to be local dissipative, it is necessary and sufficient that its Levinson center J would be uniformly attracting set.*

Proof. Let $(X, \mathbb{T}, \pi)$ be local dissipative, J be its center of Levinson and $p \in J$. Then according to Theorem 1.16 there exists $\delta_p > 0$ such that the equality (1.29) holds. By compactness of J from its open covering $\{B(p, \delta_p)|\ p \in J\}$, it is possible to extract finite sub-covering $\{B(p_i, \delta_{p_i})|\ i \in \overline{1, m}\}$. According to Lemma 1.9 there exists $\gamma > 0$ such that $B(J, \gamma) \subset U\{B(p_i, \delta_{p_i})|\ i \in \overline{1, m}\}$. It is clear that for the $\gamma > 0$ the equality

$$\lim_{t \to +\infty} \beta(\pi^t B(J, \gamma), J) = 0 \tag{1.32}$$

holds, i.e., J is uniformly attracting set.

Let now $(X, \mathbb{T}, \pi)$ be compact dissipative and its Levinson center be uniformly attracting set, i.e., there exists $\gamma > 0$ such that the equality (1.32) holds. For $x \in X$ there exists $l = l(x) > 0$ for which

$$\rho(xt, J) < \gamma \tag{1.33}$$

for all $t \geq l$. According to (1.32) for every positive number ε we can choose $L(\varepsilon) > 0$ such that

$$\rho(yt, J) < \varepsilon \tag{1.34}$$

for all $t \geq L(\varepsilon)$ and $y \in B(J, \gamma)$. Since $B(J, \gamma)$ is open by virtue of (1.33) we can select $\eta = \eta(x) > 0$ such that the inclusion $B(xl, \eta) \subset B(J, \gamma)$ holds. By the continuity of the mapping $\pi(l, \cdot) : X \to X$ for η we can find $\delta = \delta_x > 0$ such that $yl \in B(xl, \eta)$ and $yl \in B(J, \gamma)$ for all $y \in B(x, \delta_x)$. By virtue of (1.34) $y(t + l) \in B(J, \varepsilon)$ for all $t \geq L(\varepsilon)$ and $y \in B(x, \delta_x)$. Let $L(\varepsilon, x) := l(x) + L(\varepsilon)$, then $yt \in B(J, \varepsilon)$ for all $t \geq L(\varepsilon, x)$ and $y \in B(x, \delta_x)$, i.e. (X, T, π) is local dissipative. The theorem is completely proved. $\qquad\square$

Lemma 1.19 *Let $M \subseteq X$ be a nonempty compact positive invariant set in $(X, \mathbb{T}, \pi)$. If M is uniformly attracting it is orbital stable.*

Proof. Suppose that the conditions of the lemma are fulfilled but M is not orbital stable. then there exist $\varepsilon_0 > 0$, $\delta_n \downarrow 0$, $x_n \in B(M, \delta_n)$ and $t_n \to +\infty$ such that

$$\rho(x_n t_n, M) \geq \varepsilon_0. \tag{1.35}$$

Since M is uniformly attracting, for the number ε_0 there exists positive number $L(\varepsilon_0)$ such that

$$\rho(xt, M) < \frac{\varepsilon_0}{2} \tag{1.36}$$

for all $x \in B(M, \gamma)$ and $t \geq L(\varepsilon_0)$ where $\gamma > 0$ is such that

$$\lim_{t \to +\infty} \beta(\pi^t B(M, \gamma), M) = 0.$$

As $x_n \in B(M, \delta_n)$ and $\delta_n \downarrow 0$, the sequence $\{x_n\}$ can be considered convergent. Assume $x_0 = \lim\limits_{n \to +\infty} x_n$. Then $x_0 \in M$ and $t_n \geq L(\varepsilon_0)$ for sufficiently large n. According to (1.36) holds the equality

$$\rho(x_n t_n, M) < \frac{\varepsilon_0}{2}. \tag{1.37}$$

The inequalities (1.35) and (1.37) are contradictory. The obtained contradiction completes the proof of the lemma. $\qquad\square$

Theorem 1.18 *For a point dissipative dynamical system $(X, \mathbb{T}, \pi)$ to be local dissipative, it is necessary and sufficient that the following two conditions hold:*

1. *$D^+(\Omega)$ $(J^+(\Omega))$ is compact;*
2. *$D^+(\Omega)$ $(J^+(\Omega))$ is uniformly attracting set.*

Proof. Let $(X, \mathbb{T}, \pi)$ be point dissipative and the conditions 1. and 2. of the theorem are fulfilled. Since according to Lemma 1.19 the set $D^+(\Omega)$ $(J^+(\Omega))$ is orbital stable, then according to Theorem 1.14 the dynamical system $(X, \mathbb{T}, \pi)$ is compact dissipative and $D^+(\Omega)$ $(J^+(\Omega))$ coincides with its center of Levinson J. Now to finish the proof of the theorem it is sufficient to refer to Theorem 1.17.

Let $(X, \mathbb{T}, \pi)$ be local dissipative. As according to Lemma 1.5 the dynamical system $(X, \mathbb{T}, \pi)$ is compact dissipative. Then according to Theorem 1.14 $D^+(\Omega)$ $(J^+(\Omega))$ is compact and orbital stable. By Theorem 1.11 and Corollary 1.5 the equality $J = D^+(\Omega)$ $(J = J^+(\Omega))$ holds, where J is the center of Levinson of $(X, \mathbb{T}, \pi)$. From Theorem 1.17 follows that $D^+(\Omega)$ $(J^+(\omega))$ is a uniformly attracting set. The theorem is completely proved. $\qquad\square$

Definition 1.27 We will call the dynamical system $(X, \mathbb{T}, \pi)$ local asymptotically condensing if for every point $p \in X$ there exist $\delta_p > 0$ and a nonempty compact $K_p \subseteq X$ such that

$$\lim\limits_{t \to +\infty} \beta(\pi^t B(p, \delta_p), K_p) = 0. \tag{1.38}$$

The following holds:

Theorem 1.19 *Let $(X, \mathbb{T}, \pi)$ be point dissipative. For $(X, \mathbb{T}, \pi)$ to be local dissipative it is necessary and sufficient that $(X, \mathbb{T}, \pi)$ be locally asymptotically condensing.*

Proof. If $(X, \mathbb{T}, \pi)$ is locally dissipative, then it is clear that $(X, \mathbb{T}, \pi)$ is asymptotically condensing too. Suppose that $(X, \mathbb{T}, \pi)$ is point dissipative and locally asymptotically condensing and let us show that $(X, \mathbb{T}, \pi)$ is local dissipative. First of all, let us show that for every compact $K \in C(X)$ the set $\Sigma^+(K)$ is relatively compact. Let $p \in K$, $\delta_p > 0$ and $K_p \in C(X)$ be such that the equality (1.38) is fulfilled. By compactness of K from its open covering $\{B(p, \delta_p)|\ p \in K\}$ we can extract finite subcovering $\{B(p_i, \delta_{p_i})|\ 1 \le i \le n\}$. Assume $W = K_{p_1} \cup K_{p_2} \cup \ldots \cup K_{p_n}$. Then W is compact and

$$\lim_{t \to +\infty} \beta(\pi^t K, W) = 0. \tag{1.39}$$

From the equality (1.39) follows relative compactness of $\Sigma^+(K)$. According to Theorem 1.15 the dynamical system $(X, \mathbb{T}, \pi)$ is compact dissipative. Let now J be the center of Levinson of $(X, \mathbb{T}, \pi)$, and that $p \in J$, $\delta_p > 0$, and $K_p \in C(X)$ are such that the equality (1.38) holds. By Lemma 1.3 the set $\omega(B(p, \delta_p)) \ne \emptyset$, is compact, invariant and the equality

$$\lim_{t \to +\infty} \beta(\pi^t B(p, \delta_p), \omega(B(p, \delta_p))) = 0 \tag{1.40}$$

holds. Since J is maximal compact invariant set of $(X, \mathbb{T}, \pi)$, then $\omega(B(p, \delta_p)) \subseteq J$ and from the equality (1.40) follows the equality (1.29). According to Theorem 1.16 the dynamical system $(X, \mathbb{T}, \pi)$ is local dissipative. The theorem is proved. $\qquad\square$

At the end of this paragraph, we give an example of compact dissipative dynamical system that is not local dissipative.

Example 1.8   Consider a linear differential equation

$$\dot{x} = Ax \tag{1.41}$$

in the Hilbert space $H := L_2[0, 1]$ with the continuous operator $A : L_2[0, 1] \to L_2[0, 1]$ defined by the equality $(A\varphi)(\tau) = -\tau\varphi(\tau)$ for all $\tau \in [0, 1]$ and $\varphi \in L_2[0, 1]$. Note that the spectrum of the operator A coincides with the segment $[-1, 0]$. Denote by $U(t)$ the Cauchy operator of the equation (1.41). It is clear that $(U(t)\varphi)(\tau) = e^{-\tau t}\varphi(\tau)$ for all $t \in \mathbb{R}$, $\tau \in [0, 1]$ and $\varphi \in L_2[0, 1]$ (see [250, p.404]). Denote by $(H, \mathbb{R}, \pi)$ the dynamical system generated by the equation (1.41) i.e. $\pi(\varphi, t) := U(t)\varphi$ for every $t \in \mathbb{R}$ and $\varphi \in L_2[0, 1]$. According to Theorem of Lebesgue on the limit passage under the sign of integral we have

$$||\pi(t, \varphi)||^2 = \int_0^1 e^{-2\tau t}|\varphi(\tau)|^2 d\tau \to 0$$

as $t \to +\infty$. Hence, the dynamical system $(H, \mathbb{R}, \pi)$ is point dissipative and $\omega_\varphi = \{0\}$ for every $\varphi \in H$ and consequently $\Omega = \overline{\cup\{\omega_\varphi | \varphi \in H\}} = \{0\}$. Further note that

$$||\pi(t, \varphi)||^2 = \int_0^1 e^{-2\tau t}|\varphi(\tau)|^2 d\tau \le \int_0^1 |\varphi(\tau)|^2 d\tau = ||\varphi||^2$$

for all $t \ge 0$. According to Theorem 1.14 the dynamical system $(H, \mathbb{R}, \pi)$ is compact dissipative and its Levinson center $J = \{0\}$.

Let us show that the constructed dynamical system $(H, \mathbb{R}, \pi)$ is not local dissipative. In fact, if $(H, \mathbb{R}, \pi)$ were local dissipative then according to Theorem 1.17 there would be $\gamma > 0$ such that

$$\lim_{t \to +\infty} \sup_{\|\varphi\| \le \gamma} \|\pi(t, \varphi)\| = 0. \tag{1.42}$$

By virtue of the linearity of the system $(H, \mathbb{R}, \pi)$ the equality (1.42) is equivalent to the condition

$$\lim_{t \to +\infty} \sup_{\|\varphi\| \le 1} \|\pi(t, \varphi)\| = 0. \tag{1.43}$$

Define the function $\varphi_n \in H$ $(n = 1, 2, \ldots)$ by the following rule: $\varphi_n(\tau) := \sqrt{n} \chi_n(\tau)$ for all $\tau \in [0, 1]$ where χ_n is the characteristic function of the set $[0, n^{-1}] \subseteq [0, 1]$. Note that $\|\varphi_n\| = 1$ and

$$\|\pi(t_n, \varphi_n)\|^2 = \int_0^{\frac{1}{n}} n e^{-2\tau t} d\tau$$
$$= \frac{n}{2t_n} \left[1 - e^{-\frac{2t_n}{n}} \right] = 1 - e^{-1} \not\to 0, \tag{1.44}$$

where $t_n = \frac{n}{2}$. However, (1.43) and (1.44) cannot take place simultaneously. The obtained contradiction shows that our assumption about the local dissipativity of $(H, \mathbb{R}, \pi)$ is not true. The necessary example is constructed.

1.7 Global attractors

Definition 1.28 A nonempty compact set $I \subset X$ we will call the global attractor of the dynamical system $(X, \mathbb{T}, \pi)$ if the following conditions are fulfilled:

a. I is invariant w.r.t. $(X, \mathbb{T}, \pi)$;
b. I attracts all the bounded subsets from X.

Theorem 1.20 *The following conditions are equivalent:*

1. *the dynamical system $(X, \mathbb{T}, \pi)$ admits a compact global attractor;*
2. *the dynamical system $(X, \mathbb{T}, \pi)$ is bounded k-dissipative;*
3. *the dynamical system $(X, \mathbb{T}, \pi)$ is compact k-dissipative and its center of Levinson attracts all bounded subsets of X.*

Proof. The implication 1. $\Rightarrow$ 2. is obvious. Let us show that from 2. follows 3. Let $(X, \mathbb{T}, \pi)$ be bounded k-dissipative, then $(X, \mathbb{T}, \pi)$ is compact k-dissipative. Denote by J the center of Levinson of $(X, \mathbb{T}, \pi)$ and show that J attracts all bounded subsets from X. Let $B \in B(X)$ and let K be a nonempty compact from X attracting all bounded subsets from X, then

$$\lim_{t \to +\infty} \beta(\pi^t B, K) = 0. \tag{1.45}$$

From the equality (1.45) and Lemma 1.3 follows that $\omega(B) \neq \emptyset$, is compact, invariant and

$$\lim_{t \to +\infty} \beta(\pi^t B, \omega(B)) = 0. \tag{1.46}$$

According to Theorem 1.6 J is maximal compact invariant set of $(X, \mathbb{T}, \pi)$ and consequently $\omega(B) \subseteq J$. From the last inclusion and the equality (1.46) follows that

$$\lim_{t \to +\infty} \beta(\pi^t B, J) = 0. \tag{1.47}$$

At last, it is clear that from 3. follows 1. The theorem is proved. $\qquad\square$

Definition 1.29   Following to [206] we will say that the dynamical system $(X, \mathbb{T}, \pi)$ satisfies the condition of Ladyzhenskaya if for every bounded set $B \subseteq X$ there exists a nonempty compact $K \subseteq X$ such that the equality (1.45) holds.

Theorem 1.21   *The following conditions are equivalent:*

1. $(X, \mathbb{T}, \pi)$ is bounded k-dissipative;

2. $(X, \mathbb{T}, \pi)$ is point b-dissipative and satisfies the condition of Ladyzhenskaya.

Proof. According to Theorem 1.20 from 1. follows 2. Let us now prove that from 2. follows 1. First of all note that from the condition 2. follows compact k-dissipativity of $(X, \mathbb{T}, \pi)$. Let $K \in C(X)$, then from the condition of Ladyzhenskaya, Lemma 1.3 and Theorem 1.5 follows relative compactness of $\Sigma^+(K)$. According to Theorem 1.15 the dynamical system $(X, \mathbb{T}, \pi)$ is compact k-dissipative. Denote by J the center of Levinson of $(X, \mathbb{T}, \pi)$ and let $B \in B(X)$. Then by virtue of the condition of Ladyzhenskaya and Lemma 1.3 $\omega(B) \neq \emptyset$, is compact, invariant and the equality (1.46) holds. As $\omega(B) \subseteq J$, from the equality (1.46) follows that J attracts B. The theorem is proved. $\qquad\square$

Definition 1.30   We will call the dynamical system $(X, \mathbb{T}, \pi)$:

- completely continuous if for every $B \in B(X)$ there exists $l = l(B) > 0$ such that $\pi^l B$ is relatively compact;

- weakly b-dissipative if there exists a nonempty bounded set $B_0 \subseteq X$ such that $\Sigma_x^+ \cap B_0 \neq \emptyset$ for every point x from X (i.e. for every $x \in X$ there is $\tau = \tau(x) \geq 0$ such that $x\tau \in B_0$). In this case the set B_0 we will call a weak b-attractor of the dynamical system $(X, \mathbb{T}, \pi)$.

Theorem 1.22 *Let $(X, \mathbb{T}, \pi)$ be weakly b-dissipative and completely continuous. Then $(X, \mathbb{T}, \pi)$ admits a compact global attractor.*

Proof. Let $B_0 \in B(X)$ be a weak attractor of the system $(X, \mathbb{T}, \pi)$, $l_0 = l(B_0) > 0$ be such that $\pi^{l_0} B_0$ is relatively compact and let $K := \overline{\pi^{l_0} B_0}$. Under the conditions of Theorem 1.22 for every point $x \in X$ there exists $\tau = \tau(x) \geq 0$ such that $x\tau \in B_0$ and consequently $x(\tau + l_0) \in K$. So, the dynamical system $(X, \mathbb{T}, \pi)$ is weakly k-dissipative and the compact K is its weak k-attractor. According to Theorem 1.10 the dynamical system $(X, \mathbb{T}, \pi)$ is compact k-dissipative. Let J be the center of Levinson of $(X, \mathbb{T}, \pi)$, $B \in B(X)$ and $l = l(B) > 0$ be such that $\pi^l B$ is relatively compact. According to Theorem 1.6 the center of Levinson J attracts the compact $\overline{\pi^l B}$ and consequently holds the equality (1.47). The theorem is proved. $\square$

Definition 1.31 The dynamical system $(X, \mathbb{T}, \pi)$ is called (see [179, 234]) asymptotically compact if for every bounded positive invariant set $B \subseteq X$ (i.e. $\pi^t B \subseteq B$ for all $t \geq 0$) there exists a nonempty compact $K \subseteq X$ such that the equality (1.45) holds.

Theorem 1.23 *Let $(X, \mathbb{T}, \pi)$ be compact k-dissipative and asymptotically compact. Then $(X, \mathbb{T}, \pi)$ is local k-dissipative.*

Proof. Let J be the center of Levinson of $(X, \mathbb{T}, \pi)$. According to Theorem 1.17 for $(X, \mathbb{T}, \pi)$ to be local k-dissipative it is necessary and sufficient to show that the set J is uniformly attracting. Let $\varepsilon_0 = 1$ and $\delta_0 = \delta(1) > 0$ be such numbers that from $x \in B(J, \delta_0)$ follows $xt \in B(J, 1)$ for all $t \geq 0$ (according to Theorem 1.6 such δ_0 exists). It is clear that the set $B := \cup\{\pi^t B(J, \delta_0) \mid t \geq 0\}$ is bounded and positively invariant. By virtue of the asymptotic compactness of $(X, \mathbb{T}, \pi)$ there exists a nonempty compact $K \subseteq X$ such that the equality (1.45) holds . According to Lemma 1.3 the set $\omega(B) \neq \emptyset$, is compact, invariant and the equality (1.46) holds . Since J is the maximal compact invariant set of $(X, \mathbb{T}, \pi)$, then $\omega(B) \subseteq J$ and, in particular, from (1.46) follows (1.47). From the equality (1.47) and from the inclusion $B(J, \delta_0) \subseteq B$ follows the equality below

$$\lim_{t \to +\infty} \beta(\pi^t B(J, \delta_0), J) = 0.$$

The theorem is proved. $\square$

Lemma 1.20 *Let $(X, \mathbb{T}, \pi)$ be b-dissipative w.r.t. the family $\mathfrak{M}$ and asymptotically compact. Then $(X, \mathbb{T}, \pi)$ is also k-dissipative w.r.t. the family $\mathfrak{M}$.*

Proof. There exists a bounded closed set $B_0 \subseteq X$ such that for every $M \in \mathfrak{M}$ there is $L > 0$ for which $\pi^t M \subseteq B_0$ for all $t \geq L$. Assume $B := \{x \in B_0 \mid xt \in B_0$ for all $t \geq 0\}$. It is clear that $B \neq \emptyset$, is bounded and positively invariant. By the asymptotic compactness of $(X, \mathbb{T}, \pi)$ there exists a compact set $K \subseteq B_0$ such that the equality (1.45) holds. According to Lemma 1.3 the set $\omega(B) \neq \emptyset$, is compact and the equality (1.46) holds. Note that $M_L := \cup\{\pi^t M \,|\, t \geq L\} \subseteq B$ and M_L is positively invariant. By virtue of the asymptotic compactness of $(X, \mathbb{T}, \pi)$ and according to Lemma 1.3 the set $\omega(M_L) \neq \emptyset$, is compact, $\omega(M_L) \subseteq \omega(B)$ and $\omega(M_L)$ attracts M_L. Now to finish the proof of the lemma it is sufficient to note that the set M is attracted by the compact $\omega(M_L)$, hence by the compact $\omega(B)$ too. In such way we constructed the nonempty compact $\omega(B)$ attracting all set from the family $\mathfrak{M}$. The lemma is proved. $\qquad\qquad\square$

Corollary 1.7 *Let $(X, \mathbb{T}, \pi)$ be point (compact, local, bounded) b-dissipative and asymptotically compact. Then $(X, \mathbb{T}, \pi)$ is point (compact, local, bounded) k-dissipative.*

Corollary 1.8 *Let $(X, \mathbb{T}, \pi)$ be asymptotically compact. Then local b-dissipativity and compact b-dissipativity are equivalent.*

Proof. The formulated statement follows from Corollary 1.7 and Theorem 1.23. $\square$

Theorem 1.24 *Let $(X, \mathbb{T}, \pi)$ satisfy the condition of Ladyzhenskaya. Then the following conditions are equivalent:*

1. *the dynamical system $(X, \mathbb{T}, \pi)$ is weakly b-dissipative;*
2. *there exists a bounded set $B_1 \subseteq X$ absorbing all points from X, i.e. for every $x \in X$ there exists $\tau = \tau(x) \geq 0$ such that $xt \in B_1$ for all $t \geq \tau$;*
3. *the dynamical system $(X, \mathbb{T}, \pi)$ is point k-dissipative;*
4. *the dynamical system $(X, \mathbb{T}, \pi)$ is weakly k-dissipative;*
5. *the dynamical system $(X, \mathbb{T}, \pi)$ is bounded k-dissipative;*
6. *there exists a bounded set $B_2 \subseteq X$ absorbing all bounded subsets of X, i.e. for every $B \in B(X)$ there exists $L = L(B) > 0$ such that $\pi^t B \subseteq B_2$ for all $t \geq L(B)$.*

Proof. It is clear that from 2. follows 1.. Let us show from 1. follows 3.. Let B_0 be a weak attractor of the system $(X, \mathbb{T}, \pi)$. As $(X, \mathbb{T}, \pi)$ satisfies the condition of Ladyzhenskaya, the set $K_1 := \omega(B_0) \neq \emptyset$, is compact and attracts the set B_0. Since for every $x \in X$ there exists $\tau = \tau(x) > 0$ such that $x\tau \in B_0$ and consequently $\lim\limits_{t \to +\infty} \rho(xt, K_1) = 0$.

Obviously, from 3. follows 4. Let the condition 4. be fulfilled and $K_2 \subseteq X$ be a nonempty compact such that $\omega_x \cap K_2 \neq \emptyset$ for every $x \in X$, $\gamma > 0$ and $K_3 := \omega(B(K_2, \gamma))$. Under the conditions of the theorem the set $K_3 \neq \emptyset$, is compact and attracts $B(K_2, \gamma)$. According to the condition 4. for every $x \in X$ there exists $\tau(x) > 0$ such that $x\tau \in B(K_2, \gamma)$ and consequently x is attracted by the set K_3.

So, $(X, \mathbb{T}, \pi)$ is point k-dissipative and according to Theorem 1.21 it is bounded k-dissipative.

To finish the proof of the theorem it is sufficient to note that from 5. follows 6. and from 6. follows 2. The theorem is proved. $\qquad\square$

1.8 On a Problem of J. Hale

Let $P : X \to X$ be a continuous mapping and (X, P) be a discrete dynamical system, generated by the positive powers of mapping P.

Definition 1.32 A continuous mapping $P : X \to X$ is called a λ-contraction of order $k \in]0, 1[$, if $\lambda(P(A)) \leq k\lambda(A)$ for all $A \in B(X)$, where λ is some measure of non-compactness on X.

In the work [174] it was formulated the following problem

Problem (Problem of J. Hale). Let a discrete dynamical system (X, P) be pointwise b-dissipative and the mapping P be a λ-contraction. In the dynamical system (X, P), does a maximal compact invariant set exist?

In the relation to the problem of J. Hale it is interesting to note the following problem.

Problem. Let a discrete dynamical system (X, P) be pointwise b-dissipative. To find the conditions which guarantee existence in the dynamical system (X, P) a maximal compact invariant set.

Below we give some results in relation to this problem.

The example 1.6 shows, that there exist pointwise k-dissipative dynamical systems which do not admit a maximal compact invariant set. On the other hand, by Theorem 1.6 for the compact k-dissipative dynamical system, its Levinson center is a maximal compact invariant set.

The following assertion holds.

Lemma 1.21 *Let $(X, \mathbb{T}, \pi)$ be a pointwise k-dissipative dynamical system and $J^{+}(\Omega)$ be a compact set, then in $(X, \mathbb{T}, \pi)$ there is a maximal negatively invariant compact set $I \subseteq J^{+}(\Omega)$.*

Proof. Denote by I^{*} the family of all nonempty negatively invariant compact subsets $A \subseteq J^{+}(\Omega)$. We note that $I^{*} \neq \emptyset$, because according to the corollary 1.4 $\Omega \in I^{*}$.

Denote by I a closure of union of all negatively invariant compact subsets of $J^{+}(\Omega)$. By Theorem 1.3 the set I is negative invariant and, consequently, it is unknown. The lemma is proved. $\qquad\square$

Corollary 1.9 *Under the conditions of Lemma 1.21 the set I is invariant.*

Proof. In fact, since the set I is negatively invariant, then for every $t \in \mathbb{T}$ the set $\pi^t I$ also is negatively invariant and compact and by Lemma 1.15 $\pi^t I \subseteq J^+(\Omega)$. According to Lemma 1.21 I is a maximal compact negatively invariant set and, consequently, $\pi^t I \subseteq I$ i.e. $\pi^t I = I$ for all $t \in \mathbb{T}$. $\qquad\square$

Thus, if $(X, \mathbb{T}, \pi)$ is pointwise k-dissipative and $J^+(\Omega)$ is compact, then in $(X, \mathbb{T}, \pi)$ there exists a maximal compact invariant set $I \subseteq J^+(\Omega)$. On the other hand according to the example 1.7, there are pointwise k-dissipative dynamical systems, but not compact k-dissipative, for which the set $J^+(\Omega)$ is compact.

Lemma 1.22   *Let $\mathbb{T} = \mathbb{R}_+$. Dynamical system $(X, \mathbb{T}, \pi)$ is asymptotic compact, if and only if for some $t_0 > 0$ the discrete dynamical system (X, P) is asymptotic compact, where $P := \pi^{t_0} : X \longrightarrow X$.*

Proof. It easy to see that if the dynamical system $(X, \mathbb{T}, \pi)$ is asymptotic compact, then the discrete dynamical system (X, P) is asymptotic compact too, where $P := \pi^{t_0} : X \longrightarrow X$ and t_0 is a arbitrary positive number. Let $M \subseteq X$ be a bounded positively invariant set of $(X, \mathbb{T}, \pi)$. Then $P(M) \subseteq M$ and according to asymptotic compactness of discrete dynamical system (X, P) there exists a compact $K_0 \in C(X)$ such that

$$\lim_{n \to +\infty} \beta(P^n(M), K_0) = 0. \tag{1.48}$$

Let $K := \pi([0, t_0], K_0)$. Then by the continuity of mapping $\pi : \mathbb{T} \times X \longrightarrow X$ and by compactness of the set $[0, t_0] \times K_0$ the set K is also compact. We will show that the following equality $\lim\limits_{t \to +\infty} \beta(\pi^t M, K) = 0$ holds. In fact, if we suppose that it is false, then there exist $\varepsilon_0 > 0$, $t_n \to +\infty$ and $\{x_n\} \subseteq M$ such that

$$\rho(x_n t_n, K) \geq \varepsilon_0. \tag{1.49}$$

Let $t_n = k_n t_0 + \tau_n$, where k_n is a integer part of the number t_n/t_0 and $\tau_n \in [0, t_0]$, then $\pi(t_n, x_n) = \pi(\tau_n, \pi(k_n t_0, x_n))$. Taking into account the equality (1.48) we may suppose that the sequence $\{\pi(k_n t_0, x_n)\}$ is convergent. Let $\tau_0 := \lim\limits_{n \to +\infty} \tau_n$ and $y_0 := \lim\limits_{n \to +\infty} \pi(k_n t_0, x_n)$, then $\lim\limits_{n \to +\infty} \pi(t_n, x_n) = \pi(\tau_0, y_0)$. Since $y_0 \in K_0$, then $\pi(\tau_0, y_0) \in K$. On the other hand, passing to limit in the inequality (1.49) as $n \to +\infty$, we obtain $\rho(\pi(\tau_0, y_0), K) \geq \varepsilon_0$, i.e. $y_0 \tau_0 \notin K$. The obtained contradiction completes he proof of Lemma. $\qquad\square$

Definition 1.33   Recall (see [175, 179] and [278]) that the mapping $P : X \longrightarrow X$ is called condensing with respect to measure of the non-compactness λ, if $\lambda(P(A)) < \lambda(A)$ for all $A \in B(X)$ with condition $\lambda(A) > 0$.

Lemma 1.23   [175] *Let $P : X \to X$ be a continuous λ-condensing mapping. Then the discrete dynamical system (X, P) is asymptotic compact.*

Proof. At first we will show, that for every bounded set $B \subset X$ with condition $P(B) \subseteq B$ and every sequence $\{k_j\}$ with integer $k_j \to +\infty$ the set $\{P^{k_j}x_j\}$ is relatively compact. Let $C = \{\{P^{k_j}x_j\} : \{x_j\} \subseteq B, \{k_j\}$ are integers, $0 \le k_j \to +\infty\}$. We put $\eta := \sup\{\lambda(h) : h \in C\}$. Since $h \subseteq B$, if $h \in C$, then $\eta < \lambda(B)$. We will prove the existence $h^* \in C$ such that $\lambda(h^*) = \eta$. Let $\{h_l\}$ be a sequence of elements of the set C for which $\lambda(h_l) \to \eta$. We will define $\widetilde{h}_l := \{P^{k_j}x_j : P^{k_j}x_j \in h_l, k_j > l\}$. Then $\lambda(\widetilde{h}_l) = \lambda(h_l)$. If $h^* = \cup\{\widetilde{h}_l | \ l = 1, 2, \dots\}$ is ordered in arbitrary way, then $h^* \in C$ and $\lambda(h^*) \ge \lambda(h_l)$ for every l. Thus $\lambda(h^*) = \eta$. We put $\widetilde{h}^* := \{P^{k_j-1}x_j : P^{k_j}x_j \in h^*\}$. Then $\widetilde{h}^* \in C$ and $\lambda(\widetilde{h}^*) \le \eta$. But $P(\widetilde{h}^*) = h^*$, so that $\lambda(P(\widetilde{h}^*)) \ge \lambda(\widetilde{h}^*)$ and, consequently, $\lambda(\widetilde{h}^*) = 0$ because the mapping P is λ-condensing. Therefore $\widetilde{h}^*$ and h^* are relatively compact and $\eta = 0$. Thus, each sequence in C is relatively compact.

By Lemma 1.3 the set $\omega(B) \ne \emptyset$, compact, invariant and the equality

$$\lim_{n \to +\infty} \beta(P^n(B), \omega(B)) = 0 \tag{1.50}$$

holds. Lemma is proved. $\qquad\square$

Definition 1.34 A dynamical system $(X, \mathbb{T}, \pi)$ is called locally bounded, if for every $x \in X$ there are $\delta_x > 0$ and $l_x > 0$ such that the set $\cup\{\pi^t B(x, \delta_x) : t \ge l_x\}$ is bounded.

The example 1.6 shows, that there exist pointwise dissipative and locally bounded dynamical systems that are not compact dissipative.

Definition 1.35 A dynamical system $(X, \mathbb{T}, \pi)$ is called λ-condensing, if $\lambda(\pi^t B) < \lambda(B)$ for all $B \in B(X), t \in \mathbb{T}, \lambda(B) > 0$ and $\pi^t B \in B(X)$.

Theorem 1.25 *Let* $(X, \mathbb{T}, \pi)$ *be a pointwise b-dissipative, locally bounded and λ-condensing dynamical system, then* $(X, \mathbb{T}, \pi)$ *is locally k-dissipative.*

Proof. Let $(X, \mathbb{T}, \pi)$ be a λ-condensing dynamical system. Then by Lemmas 1.22 and 1.23 the dynamical system $(X, \mathbb{T}, \pi)$ is asymptotic compact. From the corollary 1.7 it follows that the dynamical system $(X, \mathbb{T}, \pi)$ is pointwise k-dissipative. Let $p \in X$, $\delta_p > 0$ and $l_p > 0$ be numbers from the definition of local boundedness of dynamical system $(X, \mathbb{T}, \pi)$. Consider a set $M_p := \cup\{\pi^t B(p, \delta_p) : t \ge l_p\}$. By asymptotic compactness of dynamical system $(X, \mathbb{T}, \pi)$ there exists a compact set $K_p \in C(X)$ such that

$$\lim_{t \to +\infty} \beta(\pi^t B(p, \delta_p), K_p) = 0. \tag{1.51}$$

According to Theorem 1.19 the dynamical system $(X, \mathbb{T}, \pi)$ is local dissipative. The theorem is proved. $\qquad\square$

Theorem 1.26 *Let $(X, \mathbb{T}, \pi)$ be a pointwise b-dissipative and λ-condensing dynamical system. If for any $p \in \Omega$ there exist $\delta_p > 0$ and $l_p > 0$ such that $\cup\{\pi^t B(p, \delta_p)| \ t \geq l_p\} \in B(X)$, then $(X, \mathbb{T}, \pi)$ is local k-dissipative.*

Proof. Let $(X, \mathbb{T}, \pi)$ be pointwise dissipative and λ-condensing. Then by Lemmas 1.22 and 1.23 the dynamical system $(X, \mathbb{T}, \pi)$ is asymptotic compact. According to Corollary 1.7 the dynamical system $(X, \mathbb{T}, \pi)$ is pointwise k-dissipative and, consequently, the set Ω is nonempty and compact. From the open covering $\{B(p, \delta_p)| \ p \in \Omega\}$ of the compact set Ω we choice a finite sub-covering $\{B(p_i, \delta_{p_i}) \ | \ i \in \overline{1, m}\}$. We put $l_0 := \max\{l_{p_i} \ | \ i \in \overline{1, m}\}$. By Lemma 1.9, there exists $\gamma > 0$ such that $B(\Omega, \gamma) \subset \cup\{B(p_i, \delta_{p_i})| \ i = \overline{1, m}\}$. If $x \in X$, then there exists $l(x) > 0$ such that $xl \in B(\Omega, \gamma)$. Since the set $B(\Omega, \gamma)$ is open, then there exists a number $\varepsilon > 0$ such that $B(xl, \varepsilon) \subset B(\Omega, \gamma)$. According to the continuity of mapping $\pi^l : X \longrightarrow X$, there exists a number $\delta_x > 0$ such that $yl \in B(xl, \varepsilon)$ for all $y \in B(x, \delta)$ and, consequently, the set $\cup\{\pi^t B(x, \delta)| \ t \geq l_0 + l(x)\}$ is bounded. To finish the proof it is sufficient to use Theorem 1.25. $\square$

Theorems 1.25 and 1.26 give a solution of the problem of J. Hale for the local bounded dynamical systems.

Definition 1.36 A dynamical system $(X, \mathbb{T}, \pi)$ is called compact bounded, if for every compact $K \in C(X)$ the set $\Sigma^+(K)$ is bounded, i.e. $\Sigma^+(K) \in B(X)$.

The following theorem holds:

Theorem 1.27 *Let $(X, \mathbb{T}, \pi)$ be a pointwise b-dissipative, compact bounded and λ-condensing dynamical system. Then $(X, \mathbb{T}, \pi)$ is compact b-dissipative.*

Proof. By Lemmas 1.22 and 1.23 the dynamical system $(X, \mathbb{T}, \pi)$ is asymptotic compact and, consequently, $(X, \mathbb{T}, \pi)$ is pointwise k-dissipative. Since the dynamical system $(X, \mathbb{T}, \pi)$ is compact bounded, asymptotic compact and the set $\Sigma^+(K)$ is positively invariant, then the set $\Sigma^+(K)$ is relatively compact for every $K \in C(X)$ and according to Theorem 1.15 the dynamical system $(X, \mathbb{T}, \pi)$ is compact k-dissipative. $\square$

Definition 1.37 A dynamical system $(X, \mathbb{T}, \pi)$ is called bounded, if for every $B \in B(X)$ the set $\Sigma^+(B)$ is bounded, i.e. $\Sigma^+(B) \in B(X)$.

Theorem 1.28 *Let $(X, \mathbb{T}, \pi)$ be a bounded and λ-condensing dynamical system. Then $(X, \mathbb{T}, \pi)$ is bounded k-dissipative.*

Proof. This assertion can be proved using the same arguments as in the proof of Theorem 1.25. $\square$

At the conclusion of this paragraph we give an example of pointwise dissipative dynamical system that is local bounded and does not admit a maximal compact invariant set.

Example 1.9 Let $\varphi \in C(\mathbb{R}, \mathbb{R})$ be a function possessing the following properties:

1. $\varphi(0) = 0$;
2. $\operatorname{supp}(\varphi) = [0, 2]$;
3. $\varphi \in C^{\infty}(\mathbb{R}, \mathbb{R})$;
4. $\varphi(1) = 1$;
5. the function φ is monotone creasing from 0 to 1 and it is decreasing from 1 to 2;
6. $x\varphi(x^{-1}) \to 0$ as $x \to +\infty$.

A function φ with the properties $1. - 6.$ can be constructed as following. Let

$$\varphi_0(t) := \begin{cases} \exp\left([t^2 - 1]^{-1} + 1\right), & |t| < 1 \\ 0, & |t| \geq 1. \end{cases}$$

Then the function $\varphi(t) := \varphi_0(t + 1)$ is unknown. We put $X := \{a\varphi\left(\frac{t}{a} + h\right) \mid h \in \mathbb{R}, \ a > 0\} \cup \{\theta\}$, where θ is function from $C(\mathbb{R}, \mathbb{R})$ identical equal to 0. It is possible to show that the set X is closed in $C(\mathbb{R}, \mathbb{R})$ and it is invariant with respect to shifts. Thus on the set X is induced a dynamical system (on $C(\mathbb{R}, \mathbb{R})$ is defined a dynamical system of translations or Bebutov's dynamical system), which we denote by $(X, \mathbb{R}, \sigma)$. We will indicate some properties of this system:

(1) for every function $\psi \in X$ the set $\{\sigma(t, \psi) : t \in \mathbb{R}\}$ is relatively compact and $\omega_\psi = \alpha_\psi = \{\theta\}$;
(2) the dynamical system $(X, \mathbb{R}, \sigma)$ is pointwise dissipative and $\Omega := \overline{\cup\{\omega_x \mid x \in X\}} = \{\theta\}$;
(3) $D^+(\Omega) = X$ and, consequently, the dynamical system $(X, \mathbb{R}, \sigma)$ is not compact dissipative because the set X, evidently, is not compact;
(4) the dynamical system $(X, \mathbb{R}, \sigma)$ is not local bounded;

In fact, for every $\delta > 0$ we have $\Sigma^+(B(\theta, \delta)) := \cup\{\pi^t B(\theta, \delta) : t \geq 0\} = X$. From this equality it follows the set $\Sigma^+(B(\theta, \delta))$ is not bounded for all $\delta > 0$.

5. the dynamical system $(X, \mathbb{R}, \sigma)$ does not admit a maximal compact invariant set.

The necessary example is constructed.

1.9 Connectedness of the Levinson's center

Below we give the conditions of of the Levinson center for a compact dissipative dynamical system.

Lemma 1.24 *Let $M_1, M_2 \in C(X)$, $W^s(M_1) \neq \emptyset$ and $W^s(M_2) \neq \emptyset$, then:*

(1) if $W^s(M_1) \cap W^s(M_2) \neq \emptyset$, then $M_1 \cap M_2 \neq \emptyset$.
(2) if $M_1 \cap M_2 = \emptyset$, then $W^s(M_1) \cap W^s(M_2) = \emptyset$.
(3) if the sets M_1 and M_2 are attracting and $W^s(M_1) \cap W^s(M_2) = \emptyset$, then $M_1 \cap M_2 = \emptyset$.
(4) if the sets M_1 and M_2 are attracting and $M_1 \cap M_2 \neq \emptyset$, then $W^s(M_1) \cap W^s(M_2) \neq \emptyset$.

Proof. The formulated affirmation follows from the corresponding definitions. $\square$

Definition 1.38 A closed (open) positively invariant set $M \subseteq X$ of a dynamical system $(X, \mathbb{T}, \pi)$ is called indecomposable if it cannot be represented in the form of a union of two own closed (open) positively invariant subsets.

Lemma 1.25 *Let M be a nonempty compact positively invariant and indecomposable subset of X and $\mathbb{T} = \mathbb{R}_+$. Then M is connected.*

Proof. Suppose that the set M is not connected. Then there are nonempty closed subsets M_1 and $M_2 \subseteq X$ such that $M = M_1 \sqcup M_2$. Let $x \in M_1$ and $\mathbb{T}_i = \{t \in \mathbb{T} \mid xt \in M_i\}$ $(i = 1, 2)$. Then according to the continuity of the mapping $\pi(\cdot, x) : \mathbb{T} \to X$ the sets $\mathbb{T}_1$ and $\mathbb{T}_2$ are closed, $\mathbb{T}_1 \cap \mathbb{T}_2 = \emptyset$ and $\mathbb{T} = \mathbb{T}_1 \cup \mathbb{T}_2$. From the connectedness of the set $\mathbb{R}_+$, it follows that $\mathbb{T}_2 = \emptyset$ and, consequently, the set M_1 is positively invariant. Using the same arguments we can prove that the set M_2 is also connected. But this fact contradicts to the indecomposability of the set M. The lemma is proved. $\square$

Lemma 1.26 *Let $M \in C(X)$ be orbitally stable, $M = M_1 \sqcup M_2$ and each of the sets M_1 and M_2 be nonempty and positively invariant. Then*

1) the sets M_1 and M_2 are orbitally stable;
2) $W^s(M) = W^s(M_1) \cup W^s(M_2)$.
3) if the set M is asymptotically stable, then the sets M_1 and M_2 are also asymptotically stable.

Proof. Let $d = \rho(M_1, M_2) > 0$. For the number $d/2$ we will choice a number δ_2 such that the inequality $\rho(x, M_i) < \delta_2$ $(i = 1, 2)$ implies $\rho(xt, M_i) < d/2$ for all $t \in [0, 1]$. We take positive numbers $\varepsilon > 0$ $(\varepsilon < d/2)$ and $\delta_1 = \delta_1(\varepsilon, M)$ out of the condition of orbital stability of the set M. Let $\delta(\varepsilon) := \min(\delta_1, \delta_2)$ and show that from the inequality $\rho(x, M_1) < \delta(\varepsilon)$ follows $\rho(xt, M_1) < \varepsilon$ for all $t \geq 0$. In fact, if $\rho(x, M_1) < \delta(\varepsilon)$, then $\rho(x, M) < \delta(\varepsilon)$ and, consequently, $\rho(xt, M) < \varepsilon$ for all $t \geq 0$. Denote by

$$\mathbb{T}_i = \{t \in \mathbb{T} \mid \rho(xt, M_i) < \varepsilon\} \ (i = 1, 2).$$

Note, that if the set $\mathbb{T}$ is connected, then $\mathbb{T}_2 = \emptyset$ and, consequently, $\mathbb{T}_1 = \mathbb{T}$.

Let us consider the case when the set $\mathbb{T}$ is discrete. We will show that then the set $\mathbb{T}_2$ is also empty. Suppose that $\mathbb{T}_2 \neq \emptyset$ and denote by $t_0 = \inf \mathbb{T}_2$. It is clear that $t_0 > 1$, $t_0 \in \mathbb{T}_2$ and $t_0 - 1 \in \mathbb{T}_1$. Note, that for the point $y := x(t_0 - 1)$ we have $\rho(y, M_1) < \varepsilon$ and, consequently,

$$\rho(y, M_1) < d/2. \tag{1.52}$$

On the other hand,

$$\rho(y, M_1) \geq \rho(M_1, M_2) - \rho(y, M_2) > d/2. \tag{1.53}$$

The inequalities (1.52) and (1.53) are contradictory. Thus, our assumption that $\mathbb{T}_2 \neq \emptyset$ is not true and, consequently, $\mathbb{T} = \mathbb{T}_1$, i.e. the set M_1 is orbitally stable. On the analogy with this we can prove the orbital stability of the set M_2. Hence, the first statement of the theorem is proved.

Now let us prove that $W^s(M) = W^s(M_1) \cup W^s(M_2)$. Obviously, it is sufficient to show that $W^s(M) \subset W^s(M_1) \cup W^s(M_2)$. Let $0 < \varepsilon < d/2$ and $\delta(\varepsilon) = \min(\delta(\varepsilon), \delta_1(\varepsilon), \delta_2(\varepsilon))$, where the number $\delta(\varepsilon)$ $(\delta_1(\varepsilon), \delta_2(\varepsilon))$ is chosen out of the orbital stability of the sets M (M_1, M_2). Then for every point $x \in W^s(M)$ there exists $t_0 \geq 0$ such that $\rho(xt_0, M) < \delta(\varepsilon)$. One of the following two conditions is true:

$$\rho(xt_0, M_1) < \delta(\varepsilon) \quad \text{or} \quad \rho(xt_0, M_2) < \delta(\varepsilon).$$

If the first condition is true, then $\rho(xt, M_1) < \varepsilon$ for all $t \geq t_0$, $\omega_x \subseteq B(M_1, \varepsilon)$ and $\omega_x \cap M_2 = \emptyset$, i.e. $\omega_x \subseteq M_1$ and $x \in W^s(M_1)$. If the second condition holds, then $\rho(xt, M_2) < \varepsilon$ for all $t \geq t_0$, $\omega_x \subseteq B(M_2, \varepsilon)$ and $\omega_x \cap M_1 = \emptyset$, i.e. $\omega_x \subseteq M_2$ and $x \in W^s(M_2)$.

Now it remains to prove the third statement. For this it is sufficient to prove that each of the sets M_1 and M_2 is attracting. Since the set M is asymptotic stable, then there exists a number $\gamma > 0$ such that $B(M, \gamma) \subset W^s(M)$, $B(M, \gamma) = B(M_1, \gamma) \sqcup B(M_2, \gamma)$ and $\gamma < d/2$, where $d = \rho(M_1, M_2) > 0$. For the number γ we will choose $\delta = \delta(M_1, \gamma) > 0$ $(\delta = \delta(M_2, \gamma) > 0)$ out of the condition of orbital stability of the set M_1 (M_2). Then if $x \in B(M_1, \delta)$ $(x \in B(M_2, \delta))$, we have $xt \in B(M_1, \gamma)$ $(xt \in B(M_2, \gamma))$ for all $t \geq 0$ and, consequently, $\omega_x \subseteq B(M_1, \gamma)$

$(\omega_x \subseteq B(M_2, \gamma)$, i.e. $x \in W^s(M_1)$ $(x \in W^s(M_2))$. Thus $B(M_1, \delta) \subseteq W^s(M_1)$ $(B(M_2, \delta) \subseteq W^s(M_2))$. The lemma is completely proved. $\square$

Theorem 1.29 *Let $(X, \mathbb{T}, \pi)$ be a compact dissipative dynamical system and J be its Levinson center. If the space X is indecomposable, then the set J is also indecomposable.*

Proof. We will show that if the space X is indecomposable, then the set J possesses the same property. Suppose that it is not true. Then there exist nonempty compact positively invariant subsets J_1, $J_2 \subseteq J$ such that $J = J_1 \sqcup J_2$. Since the set J is asymptotically stable, then by Lemma 1.26 each of the sets J_1 and J_2 possesses the same property, $X = W^s(J) = W^s(J_1) \cup W^s(J_2)$ and, in addition, by Lemmas 1.8 and 1.26 each of the sets $W^s(J_1)$ and $W^s(J_2)$ is open. From Lemma 1.24 it follows that $W^s(J_1) \cap W^s(J_2) = \emptyset$. The last equality contradicts the indecomposability of the set J. $\square$

Lemma 1.27 *Let M be a nonempty compact positively invariant attracting subset of X. If the set M is indecomposable and the set $W^s(M)$ is closed, then the set $W^s(M)$ is indecomposable too.*

Proof. Suppose that the conditions of the lemma are fulfilled, but the set $W^s(M)$ is decomposable. Then the set $W^s(M)$ can be presented in the form of a disjoint union of two own closed positively invariant subsets A_1 and A_2. It is clear that $M \subseteq A_1 \cup A_2$. Note, that for every point $x \in A_1$ $\emptyset \neq \omega_x \subseteq M \cap A_1$ and for every point $y \in A_2$ $\emptyset \neq \omega_x \subset M \cap A_2$ and, consequently, $M_i = M \cap A_i \neq \emptyset$ $(i = 1, 2)$ and $M = M_1 \cup M_2$. The last equality contradicts the indecomposability of the set M. $\square$

Corollary 1.10 *If the set $M \in C(X)$ is attracting and connected, then the set $W^s(M)$ is connected too.*

Proof. The formulated statement follows from Lemmas 1.25 and 1.27. $\square$

Corollary 1.11 *Let $(X, \mathbb{R}_+, \pi)$ be a compact dissipative dynamical system and J be its Levinson center. The set J is connected if and only if the space X is connected.*

Proof. This statement follows dierectly from Theorem 1.29, Lemma 1.25 and Corollary 1.10. $\square$

Remark 1.6 *Different versions of this statement are contained in the works [20, 175, 179, 232, 294] (see also the references within).*

Corollary 1.12 *For a compact dissipative dynamical system $(X, \mathbb{T}, \pi)$ from the indecomposability of its Levinson center follows the indecomposability of the space X.*

Theorem 1.30 *Let $(X, \mathbb{T}, \pi)$ be a compact dissipative dynamical system and J be its Levinson center. The set J is indecomposable if and only if the space X is indecomposable.*

Proof. This affirmation follows from Theorem 1.29 and Corollary 1.10. $\qquad\square$

Lemma 1.28 *Let $M \in C(X)$ be an asymptotically stable set. If the set $W^s(M)$ is indecomposable, then the set M is indecomposable too.*

Proof. Suppose that the statement of the lemma is not true, i.e. the compact set M is asymptotically stable and the set $W^s(M)$ is indecomposable, but the set M is decomposable. Then there exist nonempty subsets M_1 and M_2 of the set M such that $M = M_1 \sqcup M_2$. By Lemma 1.24 $W^s(M_1) \cap W^s(M_2) = \emptyset$. Since the set M is asymptotically stable, then according to Lemma 1.26 the sets M_1, M_2 and $W^s(M) = W^s(M_1) \cup W^s(M_2)$ possess the same property. The sets $W^s(M_1)$ and $W^s(M_2)$ are positively invariant and by Lemmas 1.8 and 1.26 each of the sets $W^s(M_1)$ and $W^s(M_2)$ is open. That contradicts the indecomposability of the set $W^s(M)$. $\qquad\square$

Corollary 1.13 *Let $M \in C(X)$ be an asymptotically stable set. The set M is connected if and only if the set $W^s(M)$ is connected.*

Theorem 1.31 *Let X be locally connected. If $M \in C(X)$ is a nonempty positively invariant uniform attracting set, then the set M has a finite number of components of connectivity.*

Proof. By Lemma 1.8 the set $W^s(M)$ is open. Since the space X is locally connected, then the set $W^s(M)$ is also locally connected. We will cover the set M by a finite number of open connected sets $H_1, H_2, \ldots, H_m$ satisfying the condition $H = H_1 \cup H_2 \cup \cdots \cup H_m \subset W^s(M)$. Let $V_1, V_2, \ldots, V_k$ be the components of connectivity of the open set H. Then the sets $V_1, V_2, \ldots, V_k$ are mutually disjoint and they are contained in $W^s(M)$. Denote by $M_i := M \cap V_i$ $(i = \overline{1, k})$ and note that $M = M_1 \cup M_2 \cup \cdots \cup M_k$, the sets $M_1, M_2, \ldots, M_k$ are mutually disjoint and in virtue of Lemma 1.24 $W^s(M_1), W^s(M_2), \ldots, W^s(M_k)$ also are mutually disjoint.

If $\mathbb{T} = \mathbb{R}_+$, then each of the sets M_i is closed, positively invariant and in view of Lemma 1.26 each of the sets M_i is asymptotically stable, $W^s(M) = W^s(M_1) \cup W^s(M_2) \cup \cdots \cup W^s(M_k)$ and $V_i \subset W^s(M_i)$. Every set M_i $(i = \overline{1.k})$ is positively invariant and by Lemma 1.8 they are open. We will prove the connectedness of the sets M_i $(i = \overline{1.k})$. Assuming the contrary, we obtain that the set M_i can be presented in the form of a union of two closed nonempty disjoint positively invariant sets M_i' and M_i'' in V_i. By virtue of Lemma 1.24 $W^s(M_i') \cap W^s(M_i'') = \emptyset$. Note, that the sets M_i' and M_i'' are asymptotically stable and $W^s(M_i) = W^s(M_i') \cup W^s(M_i'')$. The sets M_i' and M_i'' are open and positively invariant. That contradicts the connectivity of the set V_i.

Let now (X, f) be a discrete dynamical system. Since the set M is uniformly attracting, then for a sufficiently large number m we have $f^m(H) \subset H$ and, consequently, $f^m(V_i) \subset V_1 \cup V_2 \cup \cdots \cup V_k$. According to the connectivity of the sets V_1, V_2, ..., V_k and the continuity of the mapping f, the sets $f^m(V_i)$ $i = \overline{1, k}$ are connected and, consequently, for every number i there exists a unique number j such that $f^m(V_i) \subset V_j$. Let as above $M_i := M \cap V_i$ $(i = \overline{1, k})$, then $f^m(M_i) \subseteq f^m(M) \cap f^m(V_i) \subseteq M \cap V_j = M_j$. Since the set M is positively invariant, then $f^m(M) \subseteq M$ and, consequently, $f^m(M_i) \subseteq M_j$. Thus, the mapping f^m permutes the sets M_1, M_2, ..., M_k and, consequently, there exists a number n such that $f^n(M_i) \subset M_i$ $(i = \overline{1, k})$. Thus, every set M_1, M_2, ..., M_k is positively invariant with respect to mapping f^n. Then in view of Lemma 1.26 each of the sets M_1, M_2, ..., M_k is asymptotically stable with respect to the discrete dynamical system (X, f^n) and $W^s(M) = W^s(M_1) \cup W^s(M_2) \cup \cdots \cup W^s(M_k)$. Note, that $f^n(V_i) \subseteq V_i$ and $V_i \subset W^s(M_i)$ $(i = \overline{1, k})$. Using the similar arguments as ones above, for the case $\mathbb{T} = \mathbb{R}_+$ we obtain the connectivity of the set M_i. Theorem is completely proved. $\qquad\square$

Theorem 1.32   *Let $(X, \mathbb{Z}_+, f)$ be a locally dissipative dynamical system and J be its Levinson center. If the space X is connected and locally connected, then the set J is also connected.*

Proof.   If the space X is connected and locally connected, then by Theorems 1.17 and 1.31 the center of Levinson of the discrete dynamical system $(X, \mathbb{Z}_+, f)$ consists of finite numbers of components of the connectivity J_1, J_2, ..., J_k. In the proof of Theorem 1.31 there has been established the existence of a number n such that $f^n(J_i) \subseteq J_i$ $(i = \overline{1, k})$. Put $P := f^n$ and consider the discrete dynamical system (X, P) generated by the the positive powers of the mapping P. Note, that the system (X, P) is compactly dissipative and satisfies the following condition: $J = J_1 \sqcup J_2 \sqcup \cdots \sqcup J_k$. In addition, we have $P(J_i) \subseteq J_i$ $(i \in \overline{1, k})$. Thus, the center of Levinson $\tilde{J}$ of the discrete dynamical system (X, P) is decomposable, but this fact contradicts Theorem 1.29 because the space X is connected and, consequently, indecomposable. The theorem is proved. $\qquad\square$

Remark 1.7   *We note that only the connectedness of the space X, without local connectivity, does not guarantee the connectivity of the center of Levinson J of the discrete dynamical system (X, f) (see [166]).*

Definition 1.39 A metric space X is said to be possessing the property (S) if for every compact $K \subseteq X$ there exists a connected compact $I \subseteq X$ such that $K \subseteq I$.

Remark 1.8 *a) If the space X possesses the property (S), then it is connected.*

b) It is possible to construct an example of a connected space that does not possess the property (S).

c) Every linear metric space possesses the property (S).

In fact, if $K \subseteq X$ is a compact set, then the set $L(K) = \{\lambda x + (1 - \lambda)y \mid x, y \in K, 0 \le \lambda \le 1\}$ is a connected compact and $K \subseteq L(K)$.

Theorem 1.33 *Let $(X, \mathbb{Z}_+, \pi)$ be a compact dissipative dynamical system, J be its Levinson center and the space X possesses the property (S). Then the set J is connected.*

Proof. Let X be a space with the property (S). Suppose that the set J is not connected. Then there are such subsets U_1 and U_2 that $J \subset U_1 \cup U_2$, $J \cap U_i \ne \emptyset$ $(i \in \overline{1,2})$ and $U_1 \cap U_2 = \emptyset$. Let I be some connected compact containing the set J. Then the set $\pi(t, I)$ is a connected compact for every $t \in \mathbb{Z}_+$. As $J \subset I$ and $\pi(t, J) = J$ $(t \in \mathbb{Z}_+)$, then $J \subset \pi(t, I)$ and, consequently, $\pi(t, I) \cap U_i \ne \emptyset$ $(i = \overline{1,2}, t \in \mathbb{Z}_+$. Let $t_n \to +\infty$. By the connectivity of the set $\pi(t_n, I)$ there exists $x_n \in \pi(t_n, I)$ such that $x_n \notin U_1 \cup U_2$. Since the dynamical system $(X, \mathbb{Z}_+, \pi)$ is compactly dissipative, then we can suppose that the sequence $\{x_n\}$ is convergent. Let $x_n \to x$ as $t_n \to +\infty$. Thus, we have $x \notin U_1 \cup U_2$ and, on the other hand, $x \in J \subseteq U_1 \cup U_2$. The obtained contradiction proves the theorem. $\square$

Corollary 1.14 *Let X be a complete linear metric space. The Levinson center of the compact dissipative dynamical system $(X, \mathbb{Z}_+, f)$ is a connected set.*

Remark 1.9 *a) In the case when X is a Banach space, the statement containing in Corollary 1.14 was proved in the work [248].*

b) There exist examples of compact dissipative dynamical systems $(X, \mathbb{Z}_+, \pi)$ with a disconnected phase space X and with a connected Levinson center.

1.10 Weak attractors and center of Levinson

Let $(X, \mathbb{T}, \pi)$ be a compact dissipative dynamical system. Recall that a compact set $M \subseteq X$ is called a weak attractor of the dynamical system $(X, \mathbb{T}, \pi)$ if $\omega_x \cap M \ne \emptyset$ for all $x \in X$. In this paragraph we establish the relationship between weak attractors of the dynamical system $(X, \mathbb{T}, \pi)$ and its Levinson center.

Definition 1.40   The point $p \in X$ is uniformly stable in the sense of Lagrange in positive direction if the sequence $\{x_k t_k\}$ is relatively compact for each $\{x_k\} \to p$ and $\{t_k\} \to +\infty$.

Theorem 1.34   *The point $p \in X$ is uniformly stable in the sense of Lagrange in positive direction if and only if the following conditions are fulfilled:*

1. *the set $J_p^+ := \bigcap\limits_{\varepsilon > 0} \bigcap\limits_{t \geq 0} \overline{\bigcup\limits_{\tau \geq t} \pi^\tau B(p, \varepsilon)} \neq \emptyset$ and is compact;*
2. *the set J_p^+ is invariant;*
3.

$$\lim_{x \to p, t \to +\infty} \rho(xt, J_p^+) = 0. \tag{1.54}$$

Proof.   Let $p \in X$, $x_n \to p$ and $t_n \to +\infty$. Then under the conditions of the theorem the sequence $\{x_n t_n\}$ is relatively compact. If y is a limit point of the sequence $\{x_k t_k\}$, then $y \in J_p^+ \neq \emptyset$. We will show that the set J_p^+ is compact. Let $\{y_m\} \subseteq J_p^+$, $\varepsilon_m \downarrow 0, k_m \in \mathbb{N}$ and $x_k^m \to p$ and $t_k^m \to +\infty$ as $k \to +\infty$ (for every $m \in \mathbb{N}$) such that:

a. $x_k^m t_k^m \to y_m$ as $k \to +\infty$ ($m \in \mathbb{N}$);
b. $\rho(x_{k_m}^m, p) < \varepsilon_m$;
c. $t_{k_m}^m > m$ and $\rho(x_{k_m}^m t_{k_m}^m, y_m) < \varepsilon_m$.

Let $\overline{x}_m := x_{k_m}^m$ and $\overline{t}_m := t_{k_m}^m > m$, then $\overline{x}_m \to p$ and $\overline{t}_m \to +\infty$ and, consequently, under the conditions of the theorem we may suppose that the sequence $\{\overline{x}_m \overline{t}_m\}$ is convergent and $\overline{y} := \lim\limits_{m \to +\infty} \overline{x}_m \overline{t}_m$. Then

$$\rho(y_m, \overline{y}) \leq \rho(y_m, \overline{x}_m \overline{t}_m) + \rho(\overline{x}_m \overline{t}_m, \overline{y}) < \varepsilon_m + \rho(\overline{x}_m \overline{t}_m, \overline{y}). \tag{1.55}$$

Passing to limit in the inequality (1.55) when $m \to +\infty$, we obtain $\overline{y} = \lim\limits_{m \to +\infty} y_m$, i.e., the set J_p^+ is compact.

We will show now that the set J_p^+ is invariant, i.e. $\pi^t J_p^+ = J_p^+$ for all $t \in T$. First, we note that from the definition of J_p^+ follows its positive invariance, i.e., $\pi^t J_p^+ \subseteq J_p^+$ for all $t \in T$. Thus, to prove that the set J_p^+ is invariant is sufficient to prove that $J_p^+ \subseteq \pi^t J_p^+$ for all $t \in T$. Let $q \in J_p^+$ and $t \in T$, then there exist $x_k \to p$ and $t_k \to +\infty$ such that $q = \lim\limits_{k \to +\infty} x_k t_k = \lim\limits_{k \to +\infty} \pi^t(\pi^{t_k - t} x_k)$. Since $t_k - t \to +\infty$ and $x_k \to p$, the sequence $\{x_k(t_k - t)\}$ is relatively compact. Without loss of generality we can suppose that the sequence $\{x_k(t_k - t)\}$ is convergent. We put $z = \lim\limits_{k \to +\infty} x_k(t_k - t)$, then $q = \pi^t z$, i.e. $J_q^+ \subseteq \pi^t J_q^+$.

Finally, let us show that the equality (1.54) is true. If we suppose that it is not so, then there exist $\varepsilon_0 > 0$, $x_k \to p$ and $t_k \to +\infty$ such that

$$\rho(x_k t_k, J_p^+) \geq \varepsilon_0. \tag{1.56}$$

According to the stability in the sense of Lagrange in positive direction of the point p, we may suppose that the sequence $\{x_k t_k\}$ is convergent. Let $q := \lim_{k \to +\infty} x_k t_k$, then $q \in J_p^+$. Passing to limit in the inequality (1.56) as $k \to +\infty$, we obtain $\rho(q, J_q^+) \geq \varepsilon_0$, i.e. $q \notin J_p^+$. The obtained contradiction proves our statement.

To finish the proof of the theorem it is sufficient to note that from the conditions 3. and 1. follows the uniform stability in the sense of Lagrange in positive direction of the point p. The theorem is proved. $\qquad\square$

Lemma 1.29 *Let $M \neq \emptyset$ and be compact. Then the following equality*

$$J^+(M) = \cup\{J_m^+ | m \in M\} \qquad (1.57)$$

holds.

Proof. It is easy to see that $J_m^+ \subseteq J^+(M)$ for all $m \in M$ and, consequently, $\cup\{J_m^+ \mid m \in M\} \subseteq J^+(M)$. We will show that the inverse inclusion also is true. Let $x \in J^+(M)$, then there exist $\{x_n\}$ and $t_n \to \infty$ such that $x = \lim_{n \to +\infty} x_n t_n$ and $\rho(x_n, M) \to 0$ as $n \to +\infty$. By virtue of compactness of the set M we can suppose that the sequence $\{x_n\}$ is convergent. Let $\lim_{n \to +\infty} x_n = m \in M$, then $x \in J_m^+$ and, consequently, $J^+(M) \subseteq \cup\{J_p^+ | p \in M\}$. The lemma is proved. $\qquad\square$

Theorem 1.35 *Let $(X, \mathbb{T}, \pi)$ be compactly dissipative, J be its Levinson center and M be a compact weak attractor of the dynamical system $(X, \mathbb{T}, \pi)$. Then $J = J^+(M)$.*

Proof. Since the set M is compact, then according to Lemma 1.29 the equality (1.57) holds. Thus, the dynamical system $(X, \mathbb{T}, \pi)$ is compactly dissipative, then every point $p \in X$ is uniformly stable in the sense of Lagrange in positive direction. By virtue of Theorem 1.34 for each $m \in M$ the set $J_m^+ \neq \emptyset$, is compact and invariant and, consequently, $J_m^+ \subseteq J$. Hence, $J^+(M) \subseteq J$. Show that the reverse inclusion is also true. Let $x \in J$, then by the invariance of the set J and Theorem 1.1 there exists a whole trajectory φ passing through point x at the initial moment $t = 0$ and $\alpha_{\varphi_x} \neq \emptyset$, is compact and positively invariant. If $p \in \alpha_{\varphi_x}$, then $\omega_p \subseteq \alpha_{\varphi_x}$. Since $\omega_p \cap M \neq \emptyset$, then there are $q \in \omega_p \cap M$ and $t_n \to +\infty$ such that $\varphi(-t_n) \to q$ and, consequently, $x = \pi^{t_n}\varphi(-t_n) \in J_q^+ \subseteq J^+(M)$, i.e. $J \subseteq J^+(M)$. The theorem is proved. $\qquad\square$

Lemma 1.30 *Let M be a weak attractor of the compact dissipative dynamical system $(X, \mathbb{T}, \pi)$. Then $\omega(M)$ is also a weak attractor of this system.*

Proof. As $(X, \mathbb{T}, \pi)$ is compactly dissipative, then according to Lemma 1.3 $\omega(M) \neq \emptyset$, is compact and invariant. We will show that the set $\omega(M)$ is a weak attractor of the system $(X, \mathbb{T}, \pi)$. If we suppose that it is not true, then there exists $x_0 \in X$ such that

$$\omega_{x_0} \cap \omega(M) = \emptyset. \tag{1.58}$$

Since $\omega_{x_0} \cap M \neq \emptyset$, then there is $m \in \omega_{x_0} \cap M$. According to the compactness and invariance of the set ω_{x_0}, we may suppose that the point m is recurrent and, consequently, $m \in \omega_m \subseteq \omega(M)$. The last inclusion contradicts to the condition (1.58). The obtained contradiction finishes the proof of the lemma. $\qquad\square$

Lemma 1.31 *Let M be a weak attractor of the system $(X, \mathbb{T}, \pi)$ and let a point $p \in X$ be recurrent, then $p \in \omega(M)$.*

Proof. If the point $p \in X$ is recurrent, then $\omega_p \cap M \neq \emptyset$. Let $q \in \omega_p \cap M$. Since the point p is recurrent, there exists a sequence $t_n \to +\infty$ such that $qt_n \to p$ and, consequently, $p \in \omega(M)$. The lemma is proved. $\qquad\square$

Lemma 1.32 *Let M be a weak attractor of a dynamical system $(X, \mathbb{T}, \pi)$. Then the closure of the set $\mathfrak{M}$ consisting of all recurrent points of the compact dissipative dynamical system $(X, \mathbb{T}, \pi)$ is also a weak attractor and $\mathfrak{M} \subset M$.*

Proof. First, we note that $\mathfrak{M}$ is a a weak attractor of the system $(X, \mathbb{T}, \pi)$. Really, let $x \in X$. Then ω_x contains at least one recurrent point and, consequently, $\omega_x \cap \mathfrak{M} \neq \emptyset$. Let now M be a closed, invariant and weakly attracting set of the system $(X, \mathbb{T}, \pi))$. Then $\omega(M) = M$ and, consequently, according to Lemma 1.31 $\mathfrak{M} \subseteq \omega(M) = M$. The lemma is proved. $\qquad\square$

Remark 1.10 *For a compact dissipative dynamical system $(X, \mathbb{T}, \pi)$ the maximal compact invariant weak attractor is its Levinson center J.*

Lemma 1.33 *Let $(X, \mathbb{T}, \pi)$ be a compact dissipative dynamical system, J be its center of Levinson and $\Omega(J) = \overline{\bigcup\{\omega_x | x \in J\}}$. Then $\Omega(J)$ is a weak attractor of the system $(X, \mathbb{T}, \pi)$.*

Proof. Let $p \in X$ be a recurrent point, then ω_p is a nonempty compact minimal invariant set. Since J is a maximal compact invariant set of the dynamical system $(X, \mathbb{T}, \pi)$, then $\omega_p \subseteq J$. Thus, $p \in \omega_p \subseteq J$ and, consequently, $p \in \Omega(J)$. This means that $\mathfrak{M} \subseteq \Omega(J)$ ($\mathfrak{M}$ is a closure of all recurrent points) and, consequently, $\mathfrak{M} \subseteq \Omega(J)$ also is a weak attractor. The lemma is proved. $\qquad\square$

Corollary 1.15 $J = J^+(\Omega)$.

Proof. The formulated statement follows from Lemma 1.33 and Theorem 1.35. $\qquad\square$

1.11 Asymptotic stability

In this paragraph we study the problem of asymptotic stability of compact positively invariant sets of the dynamical systems with infinite-dimensional (non locally compact) phase space. The different conditions that are equivalent to asymptotic stability are given. The obtained results are the local version of the results proved before for dissipative dynamical systems.

Lemma 1.34 *Let $M \subseteq X$ be a compact and asymptotically stable set of a dynamical system $(X, \mathbb{T}, \pi)$. Then there exists $\delta_0 > 0$ such that*

$$\lim_{t \to +\infty} \beta(\pi^t K, M) = 0 \qquad (1.59)$$

for every compact $K \in C(B(M, \delta_0))$.

Proof. Since M is asymptotically stable, then there exists δ_0 such that $B(M, \delta_0) \subset W^s(M)$. We will show that for every compact $K \in C(B(M, \delta_0))$ the equality (1.59) holds. For this we note, that from the asymptotic stability of the set M follows the existence for all $\varepsilon > 0$ and $x \in B(M, \delta_0)$ of such positive numbers $\delta(\varepsilon, x)$ and $l(\varepsilon, x)$ that

$$\rho(yt, M) < \varepsilon \qquad (1.60)$$

for all $t \geq l(\varepsilon, x)$ and $y \in B(x, \delta)$. We will show that from (1.60) follows (1.59). If we suppose that it is not true, then there are $K_0 \in C(B(M, \delta_0))$, $t_n \to +\infty$ and $\varepsilon_0 > 0$ such that

$$\beta(\pi_n^t K_0, M) \geq \varepsilon_0. \qquad (1.61)$$

From the inequality (1.61) follows that there are $x_n \in K_0$ such that

$$\rho(x_n t_n, M) \geq \varepsilon_0. \qquad (1.62)$$

Since the set K_0 is compact, then we can suppose that the sequence $\{x_n\}$ is convergent. We put $x_0 := \lim_{n \to +\infty} x_n$, then from the inequality (1.60) follows that for n large enough the inequality

$$\rho(x_n t_n, M) < \varepsilon_0, \qquad (1.63)$$

holds. But the inequality (1.63) contradicts (1.62). The obtained contradiction proves our statement. $\square$

Theorem 1.36 *Let $(X, \mathbb{T}, \pi)$ be a dynamical system, $M \subseteq X$ be a compact and positively invariant subset. Then the following conditions are equivalent:*

1. the set M is asymptotically stable;

2. *the set $W^s(M)$ is open and for each $\varepsilon > 0$ and $x \in W^s(M)$ there are $\delta = \delta(\varepsilon, x) > 0$ and $\tau = \tau(\varepsilon, x) > 0$ such that*

$$\beta(\pi^t B(x, \delta), M) < \varepsilon \qquad (1.64)$$

for all $t \geq \tau$;

3. *the set $W^s(M)$ is open and M attracts all compact subsets from $W^s(M)$, i.e. for every compact $K \in C(W^s(M))$ the equality (1.59) holds.*

4. *the set $W^s(M)$ is open, every compact $K \subseteq W^s(M)$ is stable in the sense of Lagrange in positive direction and $\emptyset \neq J_x^+ \subseteq M$ for all $x \in W^s(M)$.*

Proof. We will show that 1. implies 2. First of all, let us prove that the set $W^s(M)$ is open. In fact, since M is a attractive set, then there exists $\delta > 0$ such that $B(M, \delta) \subset W^s(M)$. Let now $q \in W^s(M) \setminus B(M, \delta)$, then there is $\tau_q > 0$ such that $q\tau_q \in B(M, \delta)$. Since the set $B(M, \delta)$ is open, then there exists $\gamma > 0$ such that

$$B(q\tau_q, \gamma) \subset B(M, \delta). \qquad (1.65)$$

According to the continuity of the mapping $\pi^{\tau_q} : X \to X$ there exists $\eta > 0$ such that the inclusion

$$\pi^{\tau_q} B(q, \eta) \subset B(q\tau_q, \gamma)$$

holds. By Lemma 1.34 $B(q, \eta) \subset W^s(M)$ and, consequently, the set $W^s(M)$ is open.

Denote by $\xi(\varepsilon) > 0$ a number, chosen for $\varepsilon > 0$ out of the condition of orbital stability of the set M. Since $W^s(M) = \{x \in X : \lim_{t \to +\infty} \rho(xt, M) = 0 \text{ for } t \to +\infty\}$, then for all ξ and $x \in W^s(M)$ there exists $\tau = \tau(\varepsilon, x) > 0$ such that $\rho(xt, M) < \varepsilon$ for all $t \geq \tau$. From the continuity of the mapping $\pi^{\tau(\varepsilon, x)} : X \to X$ follows that for $x \in W^s(M)$ and $\xi > 0$ there is $\delta = \delta(\varepsilon, x) > 0$ such that

$$\pi^\tau B(x, \delta) \subseteq B(x\tau, \xi). \qquad (1.66)$$

By the choice of the number ξ and from (1.66) we have $\beta(\pi^t B(x, \delta), M) < \varepsilon$ for all $t \geq \tau$.

Now we will show that from 2. follows 3. Let $\varepsilon > 0$ and $\emptyset \neq K \subset W^s(M)$ be compact. Then for every point $x \in K$ there exist $\delta = \delta(\varepsilon, x) > 0$ and $\tau = \tau(\varepsilon, x) > 0$ such that the inequality (1.64) holds for all $t \geq \tau(\varepsilon, x)$. Consider an open covering $\{B(x, \delta(\varepsilon, x))\}_{x \in K}$ of the set K. Since the set K is compact and the space X is complete, we can extract from this covering a finite sub-covering $\{B(x_i, \delta(\varepsilon, x_i)\}_{i=1}^n$. Let $l(\varepsilon, K) := \max\{\tau(\varepsilon, x_i) \,|\, i \in \overline{1, n}\}$, then from the inequality (1.64) follows that $\beta(\pi^t K, M) < \varepsilon$ for all $t \geq l(\varepsilon, K)$.

Let us prove now that 3. implies 4. Really, since the set M attracts every compact $K \subset W^s(M)$, then by Theorem 1.5 the set $\Sigma^+(K) = \cup\{\pi^t K \,|\, t \geq 0\}$ is relatively compact and, consequently, $J_x^+ \supseteq \omega_x \neq \emptyset$ for every point $x \in W^s(M)$. We note that $J_x^+ \subseteq M$ for all $x \in W^s(M)$. In fact, if $y \in J_x^+$, then there exist $x_n \to x$ and $t_n \to +\infty$ such that $x_n t_n \to y$. Since the set $K_1 := \overline{\{x_n\}}$ is compact and the set M attracts K_1, then $y \in M$, i.e. $J_x^+ \subset M$.

Finally, we will prove that from 4. follows 1. For this aim we note that from the openness of $W^s(M)$ and compactness of M follows the existence of $\delta > 0$ such that $B(M, \delta) \subset W^s(M)$, i.e. the set M is attracting. Now show that the set M is orbitally stable. Since the set M is positively invariant, then

$$M \subseteq D^+(M) = \Sigma^+(M) \cup J^+(M) \subseteq M \cup M = M$$

and, consequently, $D^+(M) = M$. To finish the proof of the theorem is sufficient to note that under the conditions of the theorem from the equality $D^+(M) = M$ follows the orbital stability of the set M. If we suppose that it is not true, then there are $\varepsilon_0 > 0$, $\varepsilon_0 > \delta_n \to 0$, $x_n \in B(M, \delta_n)$ and $t_n \to +\infty$ such that

$$\rho(x_n t_n, M) \geq \varepsilon_0. \tag{1.67}$$

As $x_n \in B(M, \delta_n)$, $\delta_n \to 0$ and the set M is compact, then the set $K := \overline{\{x_n\}} \subseteq B(M, \varepsilon_0)$ is compact and, consequently, the set $\Sigma^+(K)$ is relatively compact. Thus, we may suppose that the sequence $\{x_n t_n\}$ is convergent. Let $y = \lim\limits_{n \to +\infty} x_n t_n$, then $y \in D^+(M) = M$. From the equality (1.67), it follows that $y \notin M$. The obtained contradiction finishes the proof of the theorem. $\qquad\square$

Theorem 1.37 *Let $M \subseteq X$ be a nonempty compact positively invariant and asymptotically stable subset of a dynamical system $(X, \mathbb{T}, \pi)$, then the following affirmations hold:*

(1) $\omega(M) \subseteq M$;

(2) the set $\omega(M)$ is invariant;

(3) $\omega(M) = \bigcap\limits_{t \geq 0} \pi^t M$;

(4) $\omega(M)$ is a maximal compact invariant set in $W^s(M)$.

Proof. The first statement of the theorem is true because the set M is positively invariant and closed.

Let us prove the second statement. Since the set M is stable in the sense of Lagrange in positive direction and $\omega(M) \subseteq M$, then by Lemma 1.3 the set $\omega(M)$ is invariant.

To prove the third statement of the theorem we note that $\bigcap\limits_{t\geq 0} \pi^t M \subseteq \omega(M)$. Since from the inclusion $\omega(M) \subseteq M$ and the invariance of the set $\omega(M)$ follows that $\omega(M) \subseteq \bigcap\limits_{t\geq 0} \pi^t M$, then $\omega(M) = \bigcap\limits_{t\geq 0} \pi^t M$.

Finally we will show that the fourth statement is true. Let $K \subset W^s(M)$ be an arbitrary compact invariant set. By Theorem 1.36 the set M attracts the compact K and, consequently, $K \subseteq \omega(K) \subseteq \omega(M)$. The theorem is proved. $\qquad\square$

Theorem 1.38 *Let $M \subseteq X$ be a nonempty compact asymptotically stable set. Then the following statements hold:*

(1) the set $\omega(M)$ is orbitally stable;
(2) the set $\omega(M)$ attracts every compact from $W^s(M)$;
(3) $J^+(A) = \omega(M)$, where $A = \cup\{\alpha_{\varphi_x} \mid \varphi \in \Phi_x,\ x \in \omega(M)\}$.

Proof. Show that the set $\omega(M)$ is orbitally stable. If we suppose that it is not true, then there exist $\varepsilon_0 > 0$ and $\delta_n \to 0$ such that for certain $x_n \in B(\omega(M), \delta_n)$ and $t_n \to +\infty$

$$\rho(x_n t_n, \omega(M)) \geq \varepsilon_0. \tag{1.68}$$

Since $x_n \in B(\omega(M), \delta_n)$ and the set $\omega(M)$ is compact, then we can suppose that the sequence $\{x_n\}$ is convergent. By Theorem 1.36 the set $H^+(K) := \overline{\cup\{\pi^t K \mid t \geq 0\}}$ is compact. We note that along with the set M the set $M' = M \cup H^+(K)$ also is attracting for the family of compact subsets from $W^s(M)$ and, consequently, $\omega(M') = \omega(M)$. In particular, $\omega(H^+(K)) \subseteq \omega(M') = \omega(M)$. By the compactness of the set $H^+(K)$ we my suppose that the sequence $\{x_n t_n\}$ is convergent. Let $y := \lim\limits_{n\to+\infty} x_n t_n$, then $y \in \omega(H^+(K)) \subseteq \omega(M)$. But from the inequality (1.68) follows that $y \notin \omega(M)$. The obtained contradiction proves the first statement of the theorem.

Let us prove now the second assertion. Let K_1 be an arbitrary compact subset from $W^s(M)$. By Theorem 1.5 the set K_1 is stable in the sense of Lagrange in positive direction and the set $\omega(K_1) \neq \emptyset$, is compact and $\beta(\pi^t K_1, \omega(K_1)) \to 0$ as $t \to +\infty$. According to Lemma 1.3 the set $\omega(K_1)$ is invariant, and by Theorem 1.37 we have $\omega(K_1) \subseteq \omega(M)$ because $\omega(M)$ is a maximal compact invariant set in $W^s(M)$.

We will show that $J^+(A) = \omega(M)$. Since $A \subseteq \omega(M)$, then $J^+(A) \subseteq J^+(\omega(M))$. By the asymptotic stability of the set $\omega(M)$ we have $J^+(\omega(M)) \subseteq \omega(M)$ and, consequently, $J^+(A) \subseteq \omega(M)$. Now we will prove the reverse inclusion. Let $x \in \omega(M)$ and $\varphi \in \Phi_x$, then $\emptyset \neq \alpha_{\varphi_x} \subset A$ and, consequently, there are $y \in \alpha_{\varphi_x}$ and $t_n \to +\infty$ such that $\varphi(-t_n) \to y \in A$ and $x = \pi^{t_n} \varphi(-t_n)$. Thus, $x \in J_y^+ \subseteq J^+(A)$, i.e. $\omega(M) \subseteq J^+(A)$. The theorem is completely proved. $\qquad\square$

Denote by $\{M_\lambda \mid \lambda \in \Lambda\}$ a family of nonempty compact positively invariant sets attracting all the compacts from $W^s(M)$ and $I := \cap\{M_\lambda \mid \lambda \in \Lambda\}$.

Theorem 1.39 *Let M be a nonempty compact positively invariant and asymptotically stable set. Then $I = \omega(M)$ (i.e. the set $\omega(M)$ is the least compact positively invariant set attracting all the compact sets from $W^s(M)$).*

Proof. First of all we will prove that $\omega(M) \subseteq I$ and, consequently, $I \neq \emptyset$. In fact, for all $\lambda \in \Lambda$ we have $\omega(M) = \omega(M_\lambda) \subseteq M_\lambda$, i.e. $\omega(M) \subseteq I$. To finish the proof of the theorem it is sufficient to show that $I \subseteq \omega(M)$. Since the set $\omega(M)$ is nonempty, compact and attracts all the compacts from $W^s(M)$, then $\omega(M) \in \{M_\lambda \mid \lambda \in \Lambda\}$ and, consequently, $I \subseteq \omega(M)$. The theorem is proved. $\qquad\square$

Chapter 2

Non-autonomous dissipative dynamical systems

2.1 On the stability of Levinson's center

Let us consider a non-autonomous dynamical system

$$\langle (X, \mathbb{T}_1, \pi), (Y, \mathbb{T}_2, \sigma), h \rangle. \tag{2.1}$$

Denote by 2^X a family of all bounded closed subsets from X, equipped with the Hausdorff distance. Let $K \subseteq X$ be a compact subset of X.

Definition 2.1 We will say that the set K satisfies the condition (C) if the mapping $F : Y \to 2^X$ defined by the equality

$$F(y) := K_y := \{x \in K \mid h(x) = y\}, \tag{2.2}$$

is continuous.

Note, that the condition (C) means that the mapping $h : X \to Y$ is open on K. This condition plays an important role in our reasoning; therefore we give a simple test, which guarantees that on a compact invariant set we have this property.

Lemma 2.1 *A nonempty compact invariant set $K \subseteq X$ possesses the property* (C), *if at least one of the following two conditions is fulfilled:*

(1) there exist $y_0 \in Y$ and $\tau > 0$ ($\tau \in \mathbb{T}_2$) such that $y_0\tau = y_0$ and $Y = \{\sigma(t, y_0) \mid 0 \le t < \tau\}$, i.e. τ is the least positive period for the point y_0;

(2) the set Y is compact minimal one and the non-autonomous dynamical system (2.1) is distal on the set K, i.e., for all different points $x_1, x_2 \in K$ such that $h(x_1) = h(x_2)$, we have

$$\inf_{t \in \mathbb{T}_1} \rho(x_1 t, x_2 t) > 0. \tag{2.3}$$

Proof. The first statement of the lemma is evident. The second statement follows from the proposal 4 from [238, p.107]. $\qquad\square$

53

Definition 2.2 The non-autonomous dynamical system (2.1) is said to be point-wise (compactly, locally, boundedly) dissipative if the autonomous dynamical system $(X, \mathbb{T}_1, \pi)$ possesses this property.

Definition 2.3 Let the non-autonomous dynamical system (2.1) be compactly di-issipative and the set J be the center of Levinson of the dynamical system $(X, \mathbb{T}_1, \pi)$. The set J is said to be the Levinson's center of the non-autonomous dynamical system (2.1).

Above we have established that the Levinson's center J is orbitally stable with respect to the autonomous dynamical system $(X, \mathbb{T}_1, \pi)$. Considering the set J as the Levinson's center of the non-autonomous dynamical system (2.1), the orbital stability with respect to the non-autonomous system (2.1) is more naturally defined as follows:

Definition 2.4 The set $K \subseteq X$ is said to be orbitally stable with respect to the non-autonomous system (2.1) if for all $\varepsilon > 0$ there exists $\delta(\varepsilon) > 0$ such that $\rho(x, K_y) < \delta$ $(x \in X, y = h(x))$ implies $\rho(xt, K_{yt}) < \varepsilon$ for all $t \geq 0$. If in addition

 a. there is $\gamma > 0$ such that $\rho(xt, K_{yt}) \to 0$ as $t \to +\infty$ for all $x \in K_y$ with the condition that $\rho(x, K_y) \leq \gamma$, then the set K is said to be asymptotically orbitally stable;

 b. if $\rho(xt, K_{yt}) \to 0$ as $t \to +\infty$ for all $x \in X_y$, then the set $K \subseteq X$ is said to be globally asymptotically stable with respect to the non-autonomous system (2.1).

Definition 2.5 The set $K \subseteq X$ is called [332] globally asymptotically stable in the sense of Lyapunov-Barbashin if

$$\lim_{t \to +\infty} \rho(xt, K_{h(x)t}) = 0 \qquad (2.4)$$

for all $x \in X$, moreover the equality (2.4) holds uniformly with respect to the variable x on compact subsets of X.

In the work [332] there it was shown that the Levinson's center of a non-autonomous dynamical system, generally speaking, is not globally asymptotically stable in the sense of Lyapunov-Barbashin. We will establish below that the Levinson's center J of the non-autonomous system (2.1) is globally asymptotically stable if the set J possesses the property (C).

The following assertion may be made.

Lemma 2.2 *Let $K \subseteq X$. If the set $\Sigma^+(K)$ is relatively compact and the set M $(\omega(K) \subseteq M, h(M) = Y)$ possesses the property (C). Then for each $x \in K$ the* *equality*

$$\lim_{t \to +\infty} \sup_{x \in K_y} \rho(xt, M_{yt}) = 0 \qquad (2.5)$$

holds uniformly with respect to $y \in h(H^+(K))$, *where* $M_y := M \cap h^{-1}(y)$.

Proof. If we suppose the contrary, then there are $\varepsilon_0 > 0$, $\{y_n\} \subseteq h(H^+(K))$, $x_n \in M_{y_n}$ and $t_n \to +\infty$ such that

$$\rho(x_n t_n, M_{y_n t_n}) \geq \varepsilon_0. \qquad (2.6)$$

Since the set $\Sigma^+(K)$ is relatively compact, then we may suppose that the sequences $\{x_n t_n\}$ and $\{y_n t_n\}$ are convergent. Let $\bar{x} := \lim_{n \to +\infty} x_n t_n$ and $\bar{y} := \lim_{n \to +\infty} y_n t_n$. We note that

$$h(\bar{x}) = \lim_{n \to +\infty} h(x_n t_n) = \lim_{n \to +\infty} h(x_n) t_n = \bar{y} \qquad (2.7)$$

and, consequently, $\bar{x} \in X_{\bar{y}}$. On the other hand, according to Lemma 1.2 $\bar{x} \in \omega(K)$. Thus, $\bar{x} \in \omega_{\bar{y}}(K) \subseteq M_{\bar{y}}$.

Since the set M satisfies the condition (C), then passing to limit in the inequality (2.6) as $n \to +\infty$, we obtain

$$\rho(\bar{x}, M_{\bar{y}}) \geq \varepsilon_0. \qquad (2.8)$$

The inequality (2.8) contradicts the inclusion $\bar{x} \in M_{\bar{y}}$. The lemma is proved. $\square$

Theorem 2.1 *Let* $\langle (X, \mathbb{T}_1, \pi), (Y, \mathbb{T}_2, \sigma), h \rangle$ *be a compact dissipative non-autonomous dynamical system and* J *be its Levinson center satisfying the condition* (C). *If* $Y = h(J)$, *then:*

(1) the set J *is orbitally stable in the positive direction with respect to the non-autonomous dynamical system (2.1).*

(2) the set J *is a compact attractor of this system, i.e., the following equality*

$$\lim_{t \to +\infty} \rho(xt, J_{h(x)t}) = 0, \qquad (2.9)$$

holds uniformly with respect to x *on every compact subset from* X.

Proof. We will show that the Levinson center J is orbitally stable in positive direction with respect to the non-autonomous dynamical system (2.1). If it is not true, then there exist $\varepsilon_0 > 0$, $\delta_n \downarrow 0$, $x_n \in B(J_{h(x_n)}, \delta_n)$ and $t_n \geq 0$ such that

$$\rho(x_n t_n, J_{h(x_n)t_n}) \geq \varepsilon_0. \qquad (2.10)$$

Under the conditions of the theorem we may suppose that the sequences $\{x_n\}$ and $\{h(x_n)\}$ are convergent. Let $x_0 := \lim_{n \to +\infty} x_n$. Since the non-autonomous dynamical system (2.1) is compactly dissipative, then the set $H^+(\{x_n\})$ is compact and, consequently, we may suppose that the sequences $\{x_n t_n\}$ and $\{h(x_n)t_n\}$ are convergent. Put $\bar{x} := \lim\limits_{n \to +\infty} x_n t_n$ and $\bar{y} := \lim\limits_{n \to +\infty} h(x_n)t_n$.

We will show now that the sequence $\{t_n\}$ figuring in the equality (2.10) tends to $+\infty$. If we suppose that it is not true, then we may suppose that it is convergent. Denote by $t_0 := \lim\limits_{n \to +\infty} t_n$ and, passing to limit in the inequality (2.10) and taking into consideration that the set J satisfies the condition (C), we obtain

$$\rho(x_0 t_0, J_{h(x_0)t_0}) \geq \varepsilon_0. \qquad (2.11)$$

On the other hand, $\rho(x_n, J_{h(x_n)}) < \delta_n$ and, consequently, $x_0 \in J_{h(x_0)}$. Since the set J is invariant, then from the last inclusion follows that $x_0 t_0 \in J_{h(x_0)t_0}$. But it contradicts the inequality (2.11). Thus, $t_n \to +\infty$ and, consequently, $\bar{x} \in \omega(K_0) \subseteq J$. From the equality (2.5) follows that $\bar{x} \in X_{\bar{y}}$, i.e. $\bar{x} \in \omega_{\bar{y}}(K) \subseteq J_{\bar{y}}$. From the inequality (2.6) it follows that $\bar{x} \in X_{\bar{y}}$, i.e., $\bar{x} \in \omega_{\bar{y}}(K) \subseteq J_{\bar{y}}$.

Passing to limit in the inequality (2.10) as $n \to +\infty$ and taking into account that $\bar{x} \in J_{\bar{y}}$ and the fact that the set J possesses the property (C), we obtain $\varepsilon_0 \leq 0$. But this contradicts the choice of the number ε_0. Thus, the orbital stability of the set J in the positive direction is proved.

Now we will prove the second statement of the theorem. Let $K \subseteq X$ be an arbitrary compact, then under the conditions of the theorem the set $H^+(K)$ is compact and $\omega(K) \subseteq J$. To finish the proof of the theorem it is sufficient to refer to Lemma 2.2. $\qquad\qquad\square$

Remark 2.1   *We note, that the problem of the stability in the sense of Lyapunov-Barbashin of the Levinson center of a non-autonomous dynamical system has been studied (in one special case) in the work [332]. We also note that in the work [332] stability means the equality (2.9), but not orbital stability. Using our terminology we can formulate the results of the work [332] in the following way. Let X be a locally compact space, the non-autonomous dynamical system (2.1) be pointwise dissipative and Y be a compact minimal set. If the Levinson center J of the dynamical system $(X, \mathbb{T}, \pi)$ satisfies the condition (C), then the non-autonomous dynamical system $\langle(X, \mathbb{T}_1, \pi), (Y, \mathbb{T}_2, \sigma), h\rangle$ is compactly dissipative and its Levinson center attracts every compact subset from X with respect to the non-autonomous dynamical system (2.1).*

This statement follows from Theorems 1.10 and 2.1. However, from the same theorems follows that the Levinson center J is orbitally stable in the positive direction.

Theorem 2.2   *Let $\langle(X, \mathbb{T}_1, \pi), (Y, \mathbb{T}_2, \sigma), h\rangle$ be a compact dissipative non-autonomous dynamical system, its Levinson center J satisfy the condition (C) and $h(J) = Y$, then the following conditions are equivalent:*

1. *the set J is globally asymptotically stable, i.e., the set J is orbitally stable and the equality*

$$\lim_{t \to +\infty} \rho(xt, J_{h(x)t}) = 0 \tag{2.12}$$

holds for all $x \in X$;

2. *the set J is globally asymptotically stable in the sense of Lyapunov-Barbashin.*

Proof. We will show that under the conditions of the theorem 1. implies 2. In fact, if it is not true, then there are $\varepsilon_0 > 0$, $K_0 \in C(X)$, $x_n \in K_0$ and $t_n \to +\infty$ such that

$$\rho(x_n t_n, J_{h(x_n)t_n}) \geq \varepsilon_0.$$

Since the dynamical system $\langle (X, \mathbb{T}_1, \pi), (Y, \mathbb{T}_2, \sigma), h \rangle$ is compactly dissipative we may suppose that the sequences $\{x_n t_n\}$ and $\{h(x_n)t_n\}$ are convergent. Let $\bar{x} = \lim_{n \to +\infty} x_n t_n$ and $\bar{y} = \lim_{n \to +\infty} h(x_n)t_n$, then $\bar{x} \in J_{\bar{y}} = J \cap X_{\bar{y}}$. On the other hand, passing to limit in the inequality (2.12) as $n \to +\infty$ and taking into account that $J_{h(x_n)t_n} \to J_{\bar{y}}$ in the Hausdorff metric, we obtain $\rho(\bar{x}, J_{\bar{y}}) \geq \varepsilon_0$, i.e. $\bar{x} \notin J_{\bar{y}}$. The obtained contradiction completes the proof of our statement.

Now we will show that 2. implies 1. For this aim it is sufficient to prove that from 2. follows the orbital stability of the set J. If we suppose that it is not so, then there exist $\varepsilon_0 > 0$, $\delta_n \downarrow 0$, $x_n \in B(J, \delta_n)$ and $t_n \to +\infty$ such that

$$\rho(x_n t_n, J_{h(x_n)t_n}) \geq \varepsilon_0. \tag{2.13}$$

Since the set J is compact, then the sequence $\{x_n\}$ is relatively compact and by the condition 2. for the number $\varepsilon_0 > 0$ and the compact $K_0 = \overline{\{x_n\}}$ there exists a number $L = L(\varepsilon_0, K_0) > 0$ such that

$$\rho(x_n t, J_{h(x_n)t}) < \frac{\varepsilon_0}{2} \tag{2.14}$$

for all $t \geq L$. The inequalities (2.13) and (2.14) are contradictory. The obtained contradiction completes the proof of the theorem. $\qquad\square$

Remark 2.2 *Note, that in the proof of the second part of the theorem 2.2 we did not use the fact that the set J satisfies the condition (C). For us it is not clear how important is the condition (C) in the first part of the theorem.*

Below we give an example of the non-autonomous dynamical system with the unstable (orbital) Levinson center.

Example 2.1 Consider Opial's example [263] of the scale almost periodic differential equation

$$\dot{p} = f(t, p), \tag{2.15}$$

which has all the solutions bounded, but does not admit an almost periodic solution. We fix some minimal set $M \subset X = \mathbb{R} \times H(f)$ $(H(f) = \overline{\{f_\tau \mid \tau \in \mathbb{R}\}})$, where by bar is denoted a closure in the space $C(\mathbb{R} \times \mathbb{R}, \mathbb{R})$ and let φ^* and $\varphi_* : H(f) \to M$ be the mappings defined by the equalities: $\varphi^*(g) := \sup\{p \mid (p, g) \in M\}$ and $\varphi_*(g) := \inf\{p \mid (p, g) \in M\}$ $(g \in H(f))$.

Recall that the equation (2.15) may be obtained in the following way. Let $\dot{y} = g(t, y)$ be the equation from the example of Danjua [263],[332] and the function g be periodic with respect to each variable and realizing on the torus a non-ergodic case (the function g can be chosen from the class C^1 on the torus), P be a perfect nowhere dense limit set on $\mathbb{R}$: $t = 0$ and α be the number of rotations. Denote by $p(t) = y(t) - \alpha t$, then we obtain

$$\dot{p}(t) = g(t, p(t) + \alpha t) - \alpha = f(t, p(t)).$$

Using the theory of the equations on the torus (see, for example, [270, Ch.2]), it is possible to show that the solution $p(t)$ of the equation (2.15) is recurrent if and only if when $p(0) \in P$.

We need to change now the function $f(x) = f(p, h)$ outside of the set M so that the new system would be dissipative and the function $p^*(g)$ $(p^*(g) := \sup\{p \mid (p, g) \in J\})$ would be discontinuous.

We put $F(x) = f(x) - \rho(x, M)$; then (conserving the continuity) we will change this function for the negative numbers p large enough by the module such that $F(p, h) \geq 1$ $(h \in H)$. It is clear that the non-autonomous dynamical system generated by the equation $\dot{p} = F(p, \sigma_t h)$ is dissipative ($\sigma_t h$ is an irrational hull of the torus).

Suppose that the function $p^*(h)$ is continuous. Since $p^*(h) \geq \varphi^*(h)$ and the solutions $p^*(\sigma_t h)$ are almost periodic, then $\inf\{p^*(h) - \varphi^*(h) \mid h \in H(f)\} > 0$. From the definition of the function $F(x)$ follows the inequality

$$F(p^*(h), h) \leq f(p^*(h), h) - c_1 \quad (c_1 > 0, \ h \in H). \tag{2.16}$$

We put $p(t) = p^*(\sigma_t h)$ and let $q(t)$ be the solution of the equation (2.15) with the initial condition $q(0) = p(0)$. From the relation (2.16) we obtain the inequalities

$$\inf_{t \geq 1}(q(t) - p(t)) > 0, \qquad \inf_{t \leq -1}(p(t) - q(t)) > 0. \tag{2.17}$$

From the inequalities (2.17), it follows that the function $q(t)$ is not recurrent, i.e. $q(0) \notin P$. Let q_1, q_2 $(q_1 > q_2)$ be the ends of the corresponding adjoining interval and $q_1(t), q_2(t)$ be the solution of the equation (2.15) with initial conditions $q_1(0) = q_1$, $q_2(0) = q_2$. Since the functions $q_1(t), p(t)$ are simultaneously recurrent, then there exists a sequence $t_m \to +\infty$ such that $q_1(t + t_m) \to q_1(t)$, $p(t + t_m) \to p(t)$ for all $t \in R$. Then from the inequality (2.17) we obtain the inequality $\inf\{q_1(t) -$

$p(t) \,|\, t \in R\} > 0$. In the same way we can establish the inequality

$$\inf\{p(t) - q_2(t) \,|\, t \in R\} > 0.$$

But this inequality contradicts the relation $\lim\limits_{t \to +\infty} (q_1(t) - q_2(t)) = 0$ (see [263]). The example needed is constructed.

Note that the example given above belongs to V. V. Zhikov [332]. An analogous example was published also in the work [152].

Lemma 2.3 *Let $M \subseteq X$ be a nonempty compact set and the mapping $F : Y \to 2^M$ $(F(y) = M_y)$ be continuous in the metric of Hausdorff. Then for all $\delta > 0$ there is $\gamma = \gamma(\delta) > 0$ such that the following inclusion*

$$B(M,\gamma) \subset \widetilde{B}(M,\delta) = \bigcup_{y \in Y} B_y(M,\delta) \tag{2.18}$$

holds, where $B_y(M,\delta) = \{x \,|\, x \in X_y,\ \rho(x,M) < \delta\}$.

Proof. Suppose that the lemma is not true, then there are $\delta_0 > 0$, $\gamma_n \downarrow 0$ and $x_n \in B(M,\gamma_n)$ such that $x_n \notin \widetilde{B}(M,\delta)$, i.e.

$$\rho(x_n, M_{y_n}) \geq \delta_0 \tag{2.19}$$

for all n, where $y_n = h(x_n)$. Since the set M is compact and the mapping $F : Y \to 2^M$ is continuous, then we may suppose that $x_n \to x_0$ and $M_{y_n} \to M_{y_0}$ where $y_0 = h(x_0)$. Passing to limit in the inequality (2.19) as $n \to +\infty$, we obtain

$$x_0 \notin B(M_{y_0}, \delta_0). \tag{2.20}$$

On the other hand, $x_n \in B(M,\gamma_n)$ and since $\gamma_n \to 0$, then $x_0 \in M$ and, consequently, $x_0 \in M_{y_0}$. This inclusion contradicts the relation (2.20). The obtained contradiction completes the proof of the lemma. $\square$

Lemma 2.4 *Let $M \subseteq X$ be a nonempty compact orbitally asymptotically stable with respect to the non-autonomous dynamical system (2.1) invariant set, then it is orbitally asymptotically stable also with respect to the autonomous dynamical system $(X, \mathbb{T}_1, \pi)$.*

Proof. This statement follows directly from Lemma 2.3 and the corresponding definitions. $\square$

Theorem 2.3 *Under the conditions of Lemma 2.3, if the set M is orbitally asymptotically stable with respect to the dynamical system $(X, \mathbb{T}_1, \pi)$, then the set M is orbitally asymptotically stable with respect to dynamical system $\langle (X, \mathbb{T}_1, \pi),\ (Y, \mathbb{T}_2, \sigma), h \rangle$ too.*

Proof.   We will show that the set M is orbitally stable with respect to $\langle (X, \mathbb{T}_1, \pi), (Y, \mathbb{T}_2, \sigma), h \rangle$. If we suppose that it is false, then there exist $\varepsilon_0 > 0$, $\delta_n \downarrow 0$, $x_n \in B(M_{y_n}, \delta_n)$ $(y_n = h(x_n))$ and $t_n \to +\infty$ such that

$$\rho(x_n t_n, M_{y_n t_n}) \geq \varepsilon_0. \tag{2.21}$$

According to Theorem 1.36 we may suppose that the sequence $\{x_n t_n\}$ is convergent. Let $\bar{x} := \lim_{n \to +\infty} x_n t_n$, $x_0 := \lim_{n \to +\infty} x_n$, $y_0 := h(x_0)$ and $\bar{y} := \lim_{n \to +\infty} y_n t_n$. Note, that $\bar{x} \in J_{\bar{y}}^+$. According to Theorem 1.36 $\bar{x} \in J_{\bar{y}}^+ \subseteq M$ and, consequently, $\bar{x} \in M_{\bar{y}}$. On the other hand, passing to limit in the inequality (2.21) as $n \to +\infty$, we obtain $\bar{x} \notin M_{\bar{y}}$. The obtained contradiction shows that the set M is orbitally stable with respect to the non-autonomous dynamical system $\langle (X, \mathbb{T}_1, \pi), (Y, \mathbb{T}_2, \sigma), h \rangle$.

Let now $x \in W^s(M)$, then $\lim_{t \to +\infty} \rho(xt, M) = 0$. By Lemma 2.2 the following equality

$$\lim_{t \to +\infty} \rho(xt, M_{h(x)t}) = 0 \tag{2.22}$$

holds and, consequently, $W^s(M) = \bigcup_{y \in Y} W_y^s(M)$. The theorem is proved. $\square$

2.2   The positively stable systems

Let $\langle (X, \mathbb{T}_1, \pi), (Y, \mathbb{T}_2, \sigma), h \rangle$ be a non-autonomous dynamical system. Denote by $C(X, Y; h)$ the set of all functions $\xi : X \to X$, satisfying the following conditions:

(1) $\xi\left(X_{h(x)}\right) \subseteq X_{h(\xi(x))}$ for all $x \in X$,
(2) the mapping $\xi : X_{h(x)} \longrightarrow X_{h(g(x))}$ is continuous on $X_{h(x)}$ for each $x \in X$.

We define the topology on $C(X, Y; h)$ by a family of pseudo-metrics

$$\rho_{y,K}(f, g) := \max_{x \in K_y} \rho(f(x), g(x)) \tag{2.23}$$

where K is an arbitrary nonempty compact from X and $y \in Y$. Note that the family of pseudo-metrics (2.23) defines on $C(X, Y; h)$ a topology of the convergence uniform (on every fiber) on the compact subsets from X.

Lemma 2.5   *The space $C(X, Y; h)$ is a topological semigroup with respect to composition (on the fibers).*

Proof. To prove this statement it is sufficient to verify the continuity of the mapping

$$F : C(X, Y; h) \times C(X, Y; h) \to C(X, Y; h),$$

defined by the equality $F(f, g) := f \circ g$. If we suppose that this statement is not true, then there is a point (f_0, g_0) such that the mapping F in this point is not continuous, i.e., there is a direction $(f_\lambda, g_\lambda) \to (f_0, g_0)$ such that $f_\lambda \circ g_\lambda \nrightarrow f_0 \circ g_0$. Then there are $\varepsilon_0 > 0$, a compact $K^0 \subseteq X$ and $y_0 \in Y$ such that

$$\max_{x \in K^0_{y_0}} \rho(f_\lambda(g_\lambda(x)), f_0(g_0(x))) \geq \varepsilon_0. \tag{2.24}$$

From the inequality (2.24) follows the existence $\{x_\lambda\} \subseteq K^0_{y_0}$ such that

$$\rho(f_\lambda(g_\lambda(x_\lambda)), f_0(g_0(x_\lambda))) \geq \varepsilon_0. \tag{2.25}$$

Since the set K^0 is compact, then we may suppose that the direction $\{x_\lambda\}$ is convergent. Let $x_\lambda \to x_0$, then $g_0(x_\lambda) \to g_0(x_0)$. We will show that $g_\lambda(x_\lambda) \to g_0(x_0)$. In fact,

$$\rho(g_\lambda(x_\lambda), g_0(x_0)) \leq \max_{x \in K^0_{y_0}} \rho(g_\lambda(x), g_0(x)) + \rho(g_0(x_\lambda), g_0(x_0)) \tag{2.26}$$

and passing to the limit in (2.26), we obtain the unknown affirmation. Analogously can be proved that $f_\lambda(g_\lambda(x_\lambda)) \to f_0(g_0(x_0))$. Passing to the limit in (2.25) we obtain $\varepsilon_0 \leq 0$, that contradicts the choice of the number ε_0. The lemma is proved. $\square$

Let $\alpha > 0$ and $\mathbb{T}_\alpha := \{t \in \mathbb{T} : t \geq \alpha\}$. We put $\mathcal{P}^\alpha := \mathcal{P}^\alpha(X, Y; h) := \overline{\{\pi^t : t \in \mathbb{T}_\alpha\}}$, where by the bar we note the closure in $C(X, Y; h)$.

Lemma 2.6 *The set $\mathcal{P}^\alpha$ is a closed subsemigroup of the semigroup $C(X, Y; h)$.*

Proof. The formulated statement immediately follows from the lemma 2.5 and from the definition of the set $\mathcal{P}^\alpha$. $\square$

Definition 2.6 The non-autonomous dynamical system (2.1) is said to be uniform stable in the positive direction on the compact subsets from X, if for all $\varepsilon > 0$ and compact $K \subseteq X$ there is $\delta = \delta(\varepsilon, K) > 0$ such that for all $x_1, x_2 \in K$ with condition $h(x_1) = h(x_2)$, from the inequality $\rho(x_1, x_2) < \delta$ follows $\rho(x_1 t, x_2 t) < \varepsilon$ for all $t \geq 0$.

Let $\mathcal{P}_y := \{\xi \in \mathcal{P}^\alpha : \xi X_y \subseteq X_y\}$. It easy to check that $\mathcal{P}_y$ is a closed subsemigroup of the semigroup $\mathcal{P}^\alpha$.

The following statement holds.

Lemma 2.7 *If the dynamical system (2.1) is compact dissipative and uniform stable in the positive direction on the compact subsets from X, then $\mathcal{P}^\alpha$ is a compact semigroup and $\mathcal{P}_y$ is its nonempty subsemigroup for all Poisson stable, in the positive direction point $y \in Y$.*

Proof. At first we will prove that under the conditions of the lemma the semigroup $\mathcal{P}^\alpha$ is compact. Let $K \subseteq X$ be an arbitrary compact. By the compact dissipativity of the non-autonomous dynamical system (2.1) the set $\Sigma^+(K)$ is compact. In view of the uniform stability in the positive direction of the system (2.1) the family of the mappings $\{\pi^t : t \in \mathbb{T}_\alpha\} \subseteq C(X, Y; h)$ are equicontinuous on K_y $(y \in Y)$. Then according to the theorem of Arzela-Ascoli, taking into account the topology on the set $C(X, Y; h)$, we conclude that $\{\pi^t : t \in \mathbb{T}_\alpha\}$ is relatively compact in $C(X, Y; h)$ and, consequently, its closure is a compact.

We will show that that the set $\mathcal{P}_y$ is nonempty. Let $t_n \to +\infty$ such that $yt_n \to y$. Consider the sequence $\{\pi^{t_n}\} \subseteq \mathcal{P}^\alpha$. By virtue of the compactness of the set $\mathcal{P}^\alpha$ we can suppose that the sequence $\{\pi^{t_n}\}$ is convergent in the space $C(X, Y; h)$. We put $\xi := \lim\limits_{n\to+\infty} \pi^{t_n}$ and we will show that $\xi \in \mathcal{P}_y$. To this end, it is sufficient to prove that $\xi X_y \subseteq X_y$. Let $x \in X_y$, then $\xi(x) = \lim\limits_{n\to+\infty} xt_n$ and, consequently,

$$h(\xi(x)) = h(\lim_{n\to+\infty} xt_n) = \lim_{n\to+\infty} h(xt_n) = \lim_{n\to+\infty} h(x)t_n = \lim_{n\to+\infty} yt_n = y.$$

The lemma is proved. $\qquad\qquad\qquad\qquad\qquad\qquad\qquad\qquad\qquad\qquad\qquad\square$

Lemma 2.8   *Let Y be a compact minimal set and the non-autonomous dynamical system (2.1) satisfies the conditions of the lemma 2.7. Then the Levinson's center J of the system (2.1) is bilateral distal, i.e. for all x_1 and x_2 from J $(x_1 \neq x_2$, $h(x_1) = h(x_2))$ the following inequality $\inf\{\rho(\varphi_1(s), \varphi_2(s)) : s \in \mathbb{S}\} > 0$ holds, where φ_i is an extension on the $\mathbb{S}$ of motion $\pi(\cdot, x_i)$ $(i = 1, 2)$.*

Proof. According to the theorem 1.6 all the motions on the set J are extendable in the negative direction. From the uniform stability in the positive direction of the set J it follows its distality in the negative direction. Since the set Y is minimal, then by the lemma 22.4 [32] (see also [31] and the lemma 1 from [238, p.104]) the set J is bilateral distal. $\qquad\qquad\qquad\qquad\qquad\square$

Theorem 2.4   *Suppose that a non-autonomous dynamical system $\langle(X, \mathbb{T}_1, \pi),$ $(Y, \mathbb{T}_2, \sigma), h\rangle$ is compact dissipative, uniform stable in the positive direction on the compact subsets from X and the set Y is minimal, then:*

(1) all the motions on the Levinson's center J of the system (2.1) can be extended in the negative direction and the set J is bilateral distal;

(2) the set J consists of recurrent trajectories and every pair of points $x_1, x_2 \in J_y := X_y \cap J$ $(y \in Y)$ is mutually recurrent;

(3) if the fibers X_y are connected, then for each $y \in Y$ the set J_y is connected and for the distinct points y_1 and y_2 the sets J_{y_1} and J_{y_2} are homeomorphic;

(4) the set J is orbital stable with respect to (2.1), i.e. for all $\varepsilon > 0$ there is $\delta(\varepsilon) > 0$ such that the inequality $\rho(x, J_{h(x)}) < \delta$ implies $\rho(xt, J_{h(x)}t) < \varepsilon$ for all $t \geq 0$;

(5) the equality $\lim\limits_{t\to+\infty} \rho(xt, J_{h(x)t}) = 0$ holds for each $x \in X$.

Proof. The first statement of theorem coincides with the lemma 2.8.

The second statement of theorem follows from the first one and from the lemma 1 from [238, p.104].

We will prove the third statement of the theorem. Let $y \in Y$, then under the conditions of the theorem the point y is stable in the sense of Poisson and by the lemma 2.7 $\mathcal{P}_y$ is a nonempty compact topological subsemigroup of the semigroup $\mathcal{P}^\alpha$. According to the lemma 4.11 [32] in the semigroup $\mathcal{P}_y$ there exists at least one idempotent, i.e., an element $u \in \mathcal{P}_y$, such that $u \circ u = u$. We wills show that for idempotent u there is a sequence $t_n \to +\infty$ such that $u = \lim\limits_{n \to +\infty} \pi^{t_n}$. In fact since $u \in \mathcal{P}_y \subseteq \mathcal{P}^\alpha$, then there is a sequence $\bar{t}_n \geq \alpha$ such that $u = \lim\limits_{n \to +\infty} \pi^{t_n}$. If the sequence $\{t_n\}$ is unbounded, then from this sequence it is possible to extract a subsequence which converges to $+\infty$ and our statement is proved. Suppose now that the sequence $\{t_n\}$ is bounded, then we can suppose that it is convergent. Let $t_0 := \lim\limits_{n \to +\infty} t_n$ $(t_0 \geq \alpha)$ and $t'_n := nt_0$. It easy to see that as $t'_n \to +\infty$, $\pi^{t'_n} \to u$.

Thus in the semigroup $\mathcal{P}_y$ there is an idempotent u and $t_n \to +\infty$ such that $u = \lim\limits_{n \to +\infty} \pi^{t_n}$. From the first two assertions of the theorem 2.4 it follows that $u(X_y) \subseteq J_y$. We will show that in fact we have the equality $J_y = u(X_y)$. To this end it is sufficient to show that $J_y = u(J_y)$. But this equality follows from the bilateral distality on the set J and, consequently, every idempotent from $\mathcal{P}_y$ on the set J_y acts as a identity mapping. The connectedness of the set J_y follows from the continuity of the mapping u, the connectedness of the fiber X_y and the equality $J_y = u(X_y)$.

We will show that the second part of the third statement of the theorem is true. Let y_1 and $y_2 \in Y$. Since the set Y is minimal, then there is $\{t_n\} \subseteq \mathbb{T}$ such that $y_2 = \lim\limits_{n \to +\infty} y_1 t_n$. Consider the sequence $\{\pi^{t_n}\}$ on J_{y_1}. Since $\pi^{t_n} : J_{y_1} \to J$ and the set J is bilateral uniform stable [238, p.107], then the family of mappings $\{\pi^{t_n}\}$ is equicontinuous and, consequently, we may suppose that it is uniform convergent. Let $\xi := \lim\limits_{n \to +\infty} \pi^{t_n}$. From the third statement of the theorem and from the proposal 4 from [238, p.107] it follows that $\xi(J_{y_1}) = J_{y_2}$. Thus, we will show that the continuous mapping ξ acts from J_{y_1} on J_{y_2}. From the bilateral distality of the set J it follows that $\xi x_1 \neq \xi x_2$, if $x_1 \neq x_2$ $(x_1, x_2 \in J_{y_1})$. This means that the mapping ξ is a homeomorphism from J_{y_1} onto J_{y_2}.

The fourth assertion of the theorem it follows from the uniform stability in the positive direction of the system (2.1) on the compact subsets from X. If we suppose that it is false, then there are $\varepsilon_0 > 0$, $\delta_n \downarrow 0$, $x_n \in X$ and $t_n \to +\infty$ such that

$$\rho(x_n, J_{h(x_n)}) < \delta_n \quad \text{and} \quad \rho(x_n t_n, J_{h(x_n)t_n}) \geq \varepsilon_0. \tag{2.27}$$

From the relation (2.27) follows that the set $K_0 = \overline{\{x_n\}}$ is compact. For the number $\frac{\varepsilon_0}{2}$ and the compact $K = K_0 \cup J$ we will choose $\delta = \delta\left(\frac{\varepsilon_0}{2}, K\right) > 0$ from the condition of the uniform stability in the positive direction of the system (2.1) on the compact subsets from X. Let $\bar{x}_n \in J_{h(x_n)}$ be a sequence such that $\rho(x_n, J_{h(x_n)}) = \rho(x_n, \bar{x}_n)$. Then for sufficiently large n we have $\rho(x_n, \bar{x}_n) < \delta$ and, consequently,

$$\rho(x_n t_n, J_{h(x_n)t_n}) \le \rho(x_n t_n, \bar{x}_n t_n) < \frac{\varepsilon_0}{2}. \tag{2.28}$$

But the inequalities (2.28) and (2.27) are contradictory. The obtained contradiction proves our assertion.

Now we will prove the fifth statement of the theorem, i.e., if $x \in X$, then $\lim\limits_{t \to +\infty} \rho(xt, J_{h(x)t}) = 0$. Suppose that it is not true, then there are $x_0 \in X$, $\varepsilon_0 > 0$ and $t_n \to +\infty$ such that

$$\rho(x_0 t_n, J_{h(x_0)t_n}) \ge \varepsilon_0. \tag{2.29}$$

Under the conditions of the theorem we may suppose that the sequence $\{\pi^{t_n}\}$ is convergent in the space $C(X, Y; h)$. Let $\xi = \lim\limits_{n \to +\infty} \pi^{t_n}$, $\bar{x} = \xi(x_0)$ and $\bar{y} = \xi(y_0)$. Passing to the limit in the inequality (2.29) and taking in the consideration that $\xi(J_{h(x_0)}) = J_{h(\xi(x_0))}$, we obtain

$$\rho(\bar{x}, J_{\bar{y}}) \ge \varepsilon_0. \tag{2.30}$$

On the other hand $\bar{x} \in \omega_x$ and $\bar{x} \in X_{\bar{y}}$ and, consequently, $\bar{x} \in J_{\bar{y}}$. This inclusion contradicts the inequality (2.30). The obtained contradiction completes the proof of the theorem. $\square$

2.3   Behaviour of dissipative dynamical systems under homomorphisms

In this paragraph we determine the conditions under which the properties of the point, compact and local dissipativity are preserved under homomorphisms.

Let X and Y be complete metric spaces, $(X, \mathbb{T}_1, \pi)$ and $(Y, \mathbb{T}_2, \sigma)$ be dynamical systems on X and Y correspondingly, Ω_X (Ω_Y) is closure of the set of all ω-limit points of dynamical system $\langle (X, \mathbb{T}_1, \pi), (Y, \mathbb{T}_2, \sigma), h \rangle$ and J_X (J_Y) is its Levinson's center.

Lemma 2.9   *Let $h : X \to Y$ be a homomorphism of dynamical system $(X, \mathbb{T}_1, \pi)$ onto $(Y, \mathbb{T}_2, \sigma)$, then:*

(1) $h(\Omega_X) \subseteq \Omega_Y$;
(2) if the set M is compact, then $h(D^+(M)) \subseteq D^+(h(M))$ and $h(J^+(M)) \subseteq J^+(h(M))$;
(3) if $(X, \mathbb{T}_1, \pi)$ is point dissipative, then $h(D^+(\Omega_X)) \subseteq D^+(\Omega_Y)$ and $h(J^+(\Omega_X)) \subseteq J^+(\Omega_Y)$.

Proof. Let $x \in \Omega_X$, then there exists $x_n \in \omega_{\tilde{x}_n}$ ($\tilde{x}_n \in X$) such that $x = \lim\limits_{n \to +\infty} x_n$. Since $h(\omega_{\tilde{x}_n}) \subseteq \omega_{h(\tilde{x}_n)}$, $h(x_n) \in \omega_{h(\tilde{x}_n)}$ and, consequently, $h(x) = \lim\limits_{n \to +\infty} h(x_n) \in \Omega_{h(X)} \subseteq \Omega_Y$.

Let us prove the second statement of Lemma. Let $x \in \Omega_X$, then according to Lemma 1.14 there exists $\tilde{x} \in M$ such that $x \in D_{\tilde{x}}^+$ and, consequently, there exist $x_n \to \tilde{x}$ and $t_n \geq 0$ such that $x = \lim\limits_{n \to +\infty} x_n t_n$. Note that $h(x_n) \to h(\tilde{x}) \in h(M)$ and $h(x) = \lim\limits_{n \to +\infty} h(x_n) t_n \in D^+(h(M))$. And analogously we establish the second inclusion $h(J^+(M)) \subseteq J^+(h(M))$.

Finally, note that the first and the second statements of Lemma imply the third.

Theorem 2.5 *The following statements hold:*

(1) if h is a homomorphism $(X, \mathbb{T}_1, \pi)$ onto $(Y, \mathbb{T}_2, \sigma)$ and dynamical system $(X, \mathbb{T}_1, \pi)$ is point dissipative, then $(Y, \mathbb{T}_2, \sigma)$ also is point dissipative and $h(\Omega_X) = \Omega_Y$;

(2) if h is a homomorphism $(X, \mathbb{T}_1, \pi)$ onto $(Y, \mathbb{T}_2, \sigma)$ and dynamical system $(X, \mathbb{T}_1, \pi)$ is compact dissipative, then $(Y, \mathbb{T}_2, \sigma)$ also is compact dissipative and $h(J_X) = J_Y$;

(3) if dynamical system $(X, \mathbb{T}_1, \pi)$ is locally dissipative and h is an open homomorphism $(X, \mathbb{T}_1, \pi)$ onto $(Y, \mathbb{T}_2, \sigma)$, then $(Y, \mathbb{T}_2, \sigma)$ also is locally dissipative.

Proof. As $Y = h(X)$ and all positive semi-trajectories of system $(X, \mathbb{T}_2, \sigma)$ are relatively compact, positive semi-trajectories of system $(Y, \mathbb{T}_2, \pi)$ also are relatively compact. That is why for the point dissipativity it is sufficient to show that $\Omega_Y = h(\Omega_X)$. Let $y \in \Omega_Y$, then there exist $\{y_n\}$ and $\{\tilde{y}_n\}$ such that $y_n \in \omega_{\tilde{y}_n}$ and $\lim\limits_{n \to +\infty} y_n = y$. Since $Y = h(X)$, there exists $\tilde{x}_n \in X$ such that $\tilde{y}_n = h(\tilde{x}_n)$ and, consequently, $h(\omega_{\tilde{x}_n}) = \omega_{\tilde{y}_n}$. So, there is $x_n \in \omega_{\tilde{x}_n} \subseteq \Omega_X$ for which $h(x_n) = y_n$. As $(X, \mathbb{T}_1, \pi)$ is point dissipative, the set Ω_X is compact and, consequently, the sequence $\{x_n\}$ can be considered convergent. Assume that $x := \lim\limits_{n \to +\infty} x_n$, then $x \in \Omega_X$ and $h(x) = y$. Thus, $\Omega_Y \subseteq h(\Omega_X)$. To finish the proof of the first statement of the theorem it is sufficient to refer to Lemma 2.9.

Now let us prove the second statement. If h is a homomorphism of compact dissipative dynamical system $(X, \mathbb{T}_1, \pi)$ onto $(Y, \mathbb{T}_2, \sigma)$, then according to the first statement of the theorem the system $(Y, \mathbb{T}_2, \sigma)$ is point dissipative. According to Theorem 1.15 to prove the compact dissipativity of $(Y, \mathbb{T}_2, \sigma)$ it is sufficient to show, that for every non-empty compact $N \subset Y$ the set $\Sigma_N^+ = \{\sigma(t, y) | t \geq 0, y \in N\}$ is relatively compact. Let $\{\tilde{y}_n\} \subset \Sigma_N^+$ be an arbitrary sequence. Then there exist $\{y_n\} \subseteq N$ and $\{t_n\} \subset \mathbb{T}_2$ ($t_n \geq 0$) such that $\tilde{y}_n = \sigma(t_n, y_n)$. If $\{t_n\}$ is bounded, then the sequence $\{\tilde{y}_n\}$ is relatively compact; so without loss of generality we can consider that $t_n \to +\infty$ and $y_n \to y \in Y$. As (X, h, Y) is locally trivial, then there exists a sequence $\{x_n\} \subseteq X$ such that $x_n \to x$ and $h(x_n) = y_n$. By virtue of compact dissipativity of $(X, \mathbb{T}_1, \pi)$ the sequence $\{x_n t_n\}$ can be considered convergent, and, consequently, $\{h(x_n t_n)\} = \{h(x_n) t_n\} = \{y_n t_n\}$ also is convergent. Thus, $(Y, \mathbb{T}_2, \sigma)$ is compact dissipative. Let us show that $h(J_X) = J_Y$. According to Lemma 2.9 $h(J^+(\Omega_X)) \subseteq J^+(\Omega_Y)$. We will show that the reverse inclusion holds $J^+(\Omega_Y) \subseteq h(J^+(\Omega_X))$. Let $q \in J^+(\Omega_Y)$, then it will exist $y \in \Omega_Y$ such that $q \in J_y^+$. According to the first statement of Theorem 2.5 we have $h(\Omega_X) = \Omega_Y$ and, consequently, there exists $x \in \Omega_X$ such that $h(x) = y$. Let $y_n \to y$ and $t_n \to +\infty$ are such that $y_n t_n \to q$. Since $(X, \mathbb{T}_1, \pi)$ is locally trivial, then there exists $\{x_n\} \to x$ such that $h(x_n) = y_n$. As $(X, \mathbb{T}_1, \pi)$ is compact dissipative, the sequence $\{x_n t_n\}$ can be considered convergent. Assume that $p = \lim\limits_{n \to +\infty} x_n t_n$, then $p \in J_x^+ \subseteq J^+(\Omega_X)$ and, consequently, $h(p) = \lim\limits_{n \to +\infty} h(x_n t_n) = \lim\limits_{n \to +\infty} y_n t_n = q$, i.e $J^+(\Omega_Y) \subseteq h(J^+(\Omega_X))$. Thus, $h(J^+(\Omega_X)) = J^+(\Omega_Y)$, and to finish the proof of the second statement of the theorem it is sufficient to note that according to Theorem 1.11 $J_X = J^+(\Omega_X)$ and $J_Y = J^+(\Omega_Y)$.

To prove the third statement of the theorem let us note that according to the second statement the dynamical system $(Y, \mathbb{T}_2, \sigma)$ is compact dissipative and $h(J_X) = J_Y$. By Theorem 1.18 to demonstrate the local dissipativity of $(Y, \mathbb{T}_2, \sigma)$ it is sufficient to show that J_Y is a uniformly attracting set. First of all note that by virtue of the compactness of J_X and the equality $h(J_X) = J_Y$ for any $\varepsilon > 0$ there is $\delta(\varepsilon) > 0$ such that

$$\rho(h(x), J_Y) < \varepsilon \tag{2.31}$$

for all $x \in B(J_X, \delta)$. In fact, if we suppose that is not true, then there exist $\varepsilon_0 > 0, \delta_n \downarrow 0$ and $x_n \in B(J_X, \delta_n)$ such that

$$\rho(h(x_n), J_Y) < \varepsilon. \tag{2.32}$$

The sequence $\{x_n\}$ can be considered convergent. Assuming $x_0 = \lim\limits_{n \to +\infty} x_n$ and going over to the limit in the inequality (2.32), as $n \to +\infty$, we have

$$\rho(h(x_0), J_Y) \geq \varepsilon_x,$$

i.e., $h(x_0) \bar{\in} J_Y$ and $x_0 \in J_X$. It contradicts the equality $h(J_X) = J_Y$. So, let $\varepsilon > 0$ and $\delta(\varepsilon) > 0$ are such that the equality (2.31) is fulfilled. By virtue of locally dissipativity of $(X, \mathbb{T}_1, \pi)$ it follows from theorem 1.18 that there exists $\gamma > 0$ such that the equality

$$\lim \beta(\pi^t B(J_X, \gamma), J_X) = 0 \tag{2.33}$$

is fulfilled. As the homomorphism h is open, the set $V = h(B(J_X, \gamma)) \supset J_Y$ is open. Let $\alpha > 0$ be such that $B(J_Y, \alpha) \subset V$. From the equality (2.33) it follows that for

$\delta(\varepsilon) > 0$ there exists $L(\varepsilon) = L(\delta(\varepsilon)) > 0$ such that

$$\beta(\pi^t B(J_X, \gamma), J_X) < \delta(\varepsilon) \tag{2.34}$$

for all $t \geq L(\varepsilon)$. The inequalities (2.31) and (2.34) imply

$$\beta(\sigma^t B(J_Y, \alpha), J_Y) < \varepsilon$$

for all $t \geq L(\varepsilon)$. The theorem is completely proved. $\qquad\square$

Corollary 2.1 *Let* $\langle (X, \mathbb{T}_1, \pi), (Y, \mathbb{T}_2, \sigma), h \rangle$ *be a compact dissipative non-autonomous dynamical system and Y be a compact invariant set (i.e., $\sigma^t Y = Y$ for all $t \in \mathbb{T}_2$), then:*

(1) $h(J_X) = Y$ and, consequently, $J_y = J_X \bigcap X_y \neq \emptyset$ for all $y \in Y$, where $X_y = h^{-1}(y)$;

(2) $J_y = \{x \in X_y |$ there exists whole relatively compact trajectory φ_x of the dynamical system $(X, \mathbb{T}_1, \pi)$ such that $\varphi(0) = x\}$.

Remark 2.3 *We don't know whether or not the condition of the openness of h is so essential in Theorem 2.5, but apparently we can't completely reject it.*

Definition 2.7 Recall that $\varphi : Y \to X$ is called a continuous section of the homomorphism $h : X \to Y$, if φ is continuous and $h \circ \varphi = id_Y$.

Definition 2.8 A continuous section is called invariant if $\varphi \circ \sigma^t = \pi^t \circ \varphi$ for all $t \in \mathbb{T}_1$.

Theorem 2.6 *Let the homomorphism $h : X \to Y$ admits a continuous invariant section, then:*

(1) if $(X, \mathbb{T}_1, \pi)$ is point dissipative, then $(Y, \mathbb{T}_2, \sigma)$ also is point dissipative, $h(\Omega_X) = \Omega_Y$ and $h(D^+(\Omega_X)) = D^+(\Omega_Y)$;

(2) if $(X, \mathbb{T}_1, \pi)$ is compact dissipative, then $(Y, \mathbb{T}_2, \sigma)$ also is compact dissipative and $h(J_X) = J_y$;

(3) if $(X, \mathbb{T}_1, \pi)$ is locally dissipative, then $(Y, \mathbb{T}_2, \sigma)$ also is locally dissipative.

Proof. Let φ be a continuous invariant section of h. Since $h \circ \varphi = id_Y$, then $h(X) = Y$ and the first and the second statements follow from Theorem 2.5, except the equality $h(D^+(\Omega_X)) = D^+(\Omega_Y)$. To prove it note that by virtue of compactness Ω_X from Lemma 2.9 the inclusion $h(D^+(\Omega_X)) \subseteq D^+(\omega_Y)$ holds.

Let us show that within the conditions of Theorem 2.6 the reverse inclusion also holds. In fact, according to Theorem 2.5 the dynamical system $(Y, \mathbb{T}_2, \sigma)$ is point dissipative and $\Omega_Y = h(\Omega_X)$ is compact. Since $\varphi : Y \to X$ is a homomorphism

$(Y, \mathbb{T}_2, \sigma)$ onto $(X, \mathbb{T}_1, \pi)$, then according to Lemma 2.9 $\varphi(D^+(\Omega_Y)) \subseteq D^+(\Omega_X)$ and, consequently, $D^+(\Omega_Y) = h \circ \varphi(D^+(\Omega_X)) \subseteq h(D^+(\omega_X))$.

Let $(X, \mathbb{T}_1, \pi)$ be locally dissipative, then from the foregoing proof the dynamical system $(Y, \mathbb{T}_2, \sigma)$ is compact dissipative. According to Theorem 1.18 for the locally dissipativity of $(Y, \mathbb{T}_2, \sigma)$ it is sufficient to show that its Levinson's center J_Y is a uniformly attracting set. As well as in Theorem 2.5, for $\varepsilon > 0$ $(\eta > 0)$ we will select $\delta(\varepsilon) > 0$ $(\xi(\eta) > 0)$ such that

$$\rho(h(x), J_Y) < \varepsilon \quad (\rho(\varphi(y), J_X) < \eta) \tag{2.35}$$

for all $x \in B(J_X, \delta)$ $(y \in B(J_Y, \xi))$. Since $X, \mathbb{T}_1, \pi)$ is locally dissipative then there exists $\gamma > 0$ such that

$$\lim_{t \to +\infty} \beta(\pi^t B(J_X, \gamma), J_X) = 0. \tag{2.36}$$

Assume that $\nu = \xi(\gamma) > 0$ and let us show that

$$\lim_{t \to +\infty} \beta(\sigma^t B(J_Y, \nu), J_Y) = 0.$$

Let $\varepsilon > 0$, then according to (2.36) there exists $L(\varepsilon) > 0$ such that

$$\pi^t B(J_X, \gamma) \subseteq B(J_X, \delta) \tag{2.37}$$

for all $t \geq L(\varepsilon)$. By virtue of the choice of $\nu = \xi(\gamma) > 0$ we have

$$\varphi(B(J_X, \gamma)) \subseteq B(J_X, \gamma). \tag{2.38}$$

From the inclusions (2.37) and (2.38) it follows that

$$\pi^t \varphi(B(J_X, \gamma)) \subseteq B(J_X, \delta) \tag{2.39}$$

for all $t \geq L(\varepsilon)$. From (2.35) and (2.39) we obtain

$$h(\pi^t \varphi(B(J_Y, \nu))) \subset B(J_Y, \varepsilon) \tag{2.40}$$

for all $t \geq L(\varepsilon)$. Since $h \circ \pi^t \circ \varphi = \sigma^t \circ h \circ \varphi = \sigma^t$ $(h \circ \varphi = id_Y)$ for all $t \in \mathbb{T}_1$, then from (2.40) we have

$$\sigma^t B(J_Y, \nu) \subset B(J_Y, \varepsilon)$$

for all $t \geq L(\varepsilon)$. The theorem is completely proved. $\square$

2.4 Non-autonomous dynamical systems with convergence

Let (X, ρ) and (Y, d) be two complete metric spaces, $\mathbb{R}(\mathbb{Z})$ be a group of real (integer) numbers, $\mathbb{S} = \mathbb{R}$ or $\mathbb{Z}, \mathbb{S}_+ = \{t \in \mathbb{S} \mid t \geq 0\}$ and $\mathbb{T}(\mathbb{S}_+ \subseteq \mathbb{T})$ be a subgroup of group $\mathbb{S}$.

By $(X, \mathbb{T}, \pi)$ we denote a dynamical system on X and $xt = \pi(t, x) = \pi^t x$.

Recall, that a dynamical system $(X, \mathbb{T}, \pi)$ is called compactly dissipative if there exists a nonempty compact $K \subseteq X$ such that

$$\lim_{t \to +\infty} \rho(xt, K) = 0 \qquad (2.41)$$

for all $x \in X$; moreover equality (2.41) holds uniformly with respect to $x \in X$ on each compact subset from X. In this case the set K is called an attractor of the family of all compact subsets $C(X)$ from the space X.

We denote by

$$J = \Omega(K) = \bigcap_{t \geq 0} \overline{\bigcup_{\tau \geq t} \pi^\tau K},$$

then the set J does not depend of the choice of the attractor K and is characterized by the properties of the dynamical system $(X, \mathbb{T}, \pi)$. The set J is called a Levinson's center of the dynamical system $(X, \mathbb{T}, \pi)$.

Let us mention some facts that we will use below.

Let Y be a compact metric space, $(X, \mathbb{T}_1, \pi)$ $((Y, \mathbb{T}_2, \sigma))$ be a dynamical system on X (Y), $(\mathbb{T}_1 \subseteq \mathbb{T}_2)$ and $h : X \to Y$ be a homomorphism of $(X, \mathbb{T}_1, \pi)$ onto $(Y, \mathbb{T}_2, \sigma)$. Then the triple $\langle (X, \mathbb{T}_1, \pi), (Y, \mathbb{T}_2, \sigma), h \rangle$ is called a non-autonomous dynamical system.

Let W and Y be complete metric spaces, $(Y, \mathbb{S}, \sigma)$ be a group dynamical system on Y and $\langle W, \varphi, (Y, \mathbb{S}, \sigma) \rangle$ be a cocycle over $(Y, \mathbb{S}, \sigma)$ with the fiber W (or, by short, φ), i.e. φ is a continuous mapping of $W \times Y \times \mathbb{T}$ into W satisfying the following conditions: $\varphi(0, w, y) = w$ and $\varphi(t + \tau, w, y) = \varphi(t, \varphi(\tau, w, y), \sigma(\tau, y))$ for all $t, \tau \in \mathbb{T}, w \in W$ and $y \in Y$.

We denote $X = W \times Y$ and define on X a skew product dynamical system $(X, \mathbb{T}, \pi)$ by the equality $\pi = (\varphi, \sigma)$, i.e. $\pi(t, (w, y)) = (\varphi(t, w, y), \sigma(t, y))$ for all $t \in \mathbb{T}$ and $(w, y) \in W \times Y$. Then the triple $\langle (X, \mathbb{T}, \pi), ((Y, \mathbb{S}, \sigma), h \rangle$, where $h = pr_2$, is a non-autonomous dynamical system.

For any two bounded subsets A and B from X we denote by $\beta(A, B)$ a semi-deviation of A to B, i.e. $\beta(A, B) = \sup\{\rho(a, B) | a \in A\}$ and $\rho(a, B) = \inf\{\rho(a, b) | b \in B\}$.

Definition 2.9 A cocycle φ over $(Y, \mathbb{S}, \sigma)$ with the fiber W is called compactly dissipative if there exists a nonempty compact $K \subseteq W$ such that

$$\lim_{t \to +\infty} \sup\{\beta(U(t, y)M, K) : y \in Y\} = 0$$

for all $M \in C(W)$ where $U(t, y) = \varphi(t, \cdot, y)$.

Definition 2.10 By a whole trajectory of the semigroup dynamical system $(X, \mathbb{T}, \pi)$ (of the cocycle $\langle W, \varphi, (Y, \mathbb{S}, \sigma) \rangle$ over $(Y, \mathbb{T}, \sigma)$ with the fiber W) passing through the point $x \in X$ ($(u,y) \in W \times Y$) we mean a continuous mapping $\gamma : \mathbb{S} \to X(\nu : \mathbb{S} \to W)$ satisfying the conditions : $\gamma(0) = x(\nu(0) = w)$ and $\gamma(t + \tau) = \pi^t \gamma(\tau)$ $(\nu(t + \tau) = \varphi(t, \nu(\tau), y\tau))$ for all $t \in \mathbb{T}$ and $\tau \in \mathbb{S}$.

Definition 2.11 $\langle (X,\mathbb{T}_1,\pi),(Y,\mathbb{T}_2,\sigma),h \rangle$ is said to be convergent if the following conditions are valid:

(1) the dynamical systems $(X, \mathbb{T}_1, \pi)$ and $(Y, \mathbb{T}_2, \sigma)$ are compactly dissipative;
(2) the set $J_X \bigcap X_y$ contains no more than one point for all $y \in J_Y$ where $X_y :=$ $h^{-1}(y) := \{x | x \in X, h(x) = y\}$ and $J_X(J_Y)$ is the Levinson's center of the dynamical system $(X, \mathbb{T}_1, \pi)((Y, \mathbb{T}_2, \sigma))$.

Let $M \subseteq X$ and $M \dot\times M := \{(x_1, x_2) \mid x_1, x_2 \in M, h(x_1) = h(x_2)\}$.

Lemma 2.10   *Let* $\langle (X, \mathbb{T}_1, \pi), (Y, \mathbb{T}_2, \sigma), h \rangle$ *be a non-autonomous dynamical system,* $K \subseteq X$ *be a compact invariant set and* $M := h(K)$. *If the equality*

$$\lim_{t \to +\infty} \sup_{(x_1, x_2) \in K \dot\times K} \rho(x_1 t, x_2 t) = 0 \qquad (2.42)$$

holds, then the set $K_y := K \bigcap X_y$ *contains a single point for all* $y \in M$.

Proof. Suppose that there exists $y_0 \in M$ such that K_{y_0} contains at least two points $\bar{x}_1$ and $\bar{x}_2$ ($\bar{x}_1 \neq \bar{x}_2$). Since the set K is invariant, then there exists a trajectory φ_i passing trough the point $\bar{x}_i (i = 1, 2)$ such that $\varphi_i(\mathbb{S}) \subseteq K$. Let $0 < \varepsilon < \frac{\rho(\bar{x}_1, \bar{x}_2)}{2}$ and $L(\varepsilon) > 0$, so that

$$\rho(x_1 t, x_2 t) < \varepsilon$$

for all $t \geq L(\varepsilon)$ and $(x_1, x_2) \in K \dot\times K$. Thus, we have

$$\rho(\bar{x}_1, \bar{x}_2) = \rho(\pi^t \varphi_1(-t), \pi^t \varphi_2(-t)) < \varepsilon$$

for all $t \geq L(\varepsilon)$. The obtained contradiction shows that K_y contains a single point for all $y \in M$. The lemma is proved. $\square$

Definition 2.12   A dynamical system $(X, \mathbb{T}, \pi)$ is said to be satisfying the condition (A) if the set $\bigcup \{\pi^t K \mid t \geq 0\}$ is relatively compact for every $K \in C(X) = \{K \mid K \subseteq X$ and K is compact $\}$.

We denote by $L_Y := \{x \mid x \in X$ is so that at least one entire trajectory of the dynamical system $(X, \mathbb{T}, \pi)$ passes through $x\}$.

Remark 2.4   *For a compactly dissipative system* $(X, \mathbb{T}, \pi)$ *we have* $L_X = J_X$ *where* J_X *is the Levinson's center of* $(X, \mathbb{T}, \pi)$.

Theorem 2.7 *Let $(X, \mathbb{T}_1, \pi)$ be a dynamical system satisfying the condition (A) and let $(Y, \mathbb{T}_2, \sigma)$ be compactly dissipative, then the following conditions are equivalent:*

1. *the set $L_X \bigcap X_y$ contains no more than one point for all $y \in J_Y$;*
2. *every semi-trajectory $\Sigma_x^+ = \{xt \mid t \geq 0\}$ is asymptotically stable, i.e.*

   2.a. *for all $\varepsilon > 0$ and $p \in X$ there exists $\delta(\varepsilon, p) > 0$ such that $\rho(x, p) < \delta$ $(h(x) = h(p))$ implies $\rho(xt, pt) < \varepsilon$ for any $t \geq 0$;*

   2.b. *there exists $\gamma(p) > 0$ such that $\rho(x, p) < \gamma(p)$ $(h(x) = h(p))$ implies $\lim\limits_{t \to +\infty} \rho(xt, pt) = 0$.*

3. *3.a for all $\varepsilon > 0$ and $K \in C(X)$ there exists $\delta(\varepsilon, K) > 0$ such that $\rho(x_1, x_2) < \delta$ $(h(x_1) = h(x_2); x_1, x_2 \in K)$ implies $\rho(x_1 t, x_2 t) < \varepsilon$ for all $t \geq 0$;*

   3.b $\lim\limits_{t \to +\infty} \rho(x_1 t, x_2 t) = 0$ for all $(x_1, x_2) \in X \dot\times X$.

4. *the equality (2.42) holds for all $K \in C(X)$.*

Proof. We will prove that condition 1. implies condition 2. If we suppose that it is not so, then there are $p_0 \in X, \varepsilon_0 > 0, p_n \to p_0$ $(h(p_n) = h(p_0))$ and $t_n \to +\infty$ such that

$$\rho(p_n t_n, p_0 t_n) \geq \varepsilon_0. \tag{2.43}$$

Since $(X, \mathbb{T}_1, \pi)$ satisfies the condition (A), then we may suppose that the sequences $\{p_n t_n\}$ and $\{p_0 t_n\}$ are convergent. Letting $\bar{p} = \lim\limits_{n \to +\infty} p_n t_n$, $\bar{p}_0 = \lim\limits_{n \to +\infty} p_0 t_n$ and taking into consideration (2.43), we will have $\bar{p} \neq \bar{p}_0$. On the other hand, $h(\bar{p}) = \lim\limits_{n \to +\infty} h(p_n) t_n = \lim\limits_{n \to +\infty} h(p_0) t_n = h(\bar{p}_0) = \bar{y} \in J_Y$ and according to Lemma 1.3 $\bar{p}, \bar{p}_0 \in L_X \bigcap X_{\bar{y}}$, but by virtue of condition 1. we have $\bar{p} = \bar{p}_0$. The obtained contradiction proves the necessary assertion.

Now we will note that condition 1. implies condition 2.b. To prove this implication is sufficient to show that

$$\lim_{t \to +\infty} \rho(x_1 t, x_2 t) = 0$$

for all $(x_1, x_2) \in X \dot\times X$. Assuming the contrary we obtain

$$\rho(x_1^0 t_n, x_2^0 t_n) \geq \varepsilon_0. \tag{2.44}$$

The dynamical system $(X, \mathbb{T}_1, \pi)$ satisfies the condition (A) and, consequently, we may assume that the sequences $\{x_i^0 t_n\}(i = 1, 2)$ and $\{y_0 t_n\}$ $(y_0 = h(x_1^0) = h(x_2^0))$ are convergent. We denote by $\bar{x}_i^0 := \lim\limits_{n \to +\infty} x_i^0 t_n$ and $\bar{y}_0 := \lim\limits_{n \to +\infty} y_0 t_n$, then $\bar{x}_1^0, \bar{x}_2^0 \in L_X \bigcap X_{\bar{y}_0}$ and according to the condition 1. $\bar{x}_1^0 = \bar{x}_2^0$. The last equality and the inequality (2.44) are contradictory. This contradiction proves the necessary assertion.

We will show that condition 2. implies condition 3. Note that

$$\lim_{t\to+\infty} \rho(xt, pt) = 0 \tag{2.45}$$

for all $p \in X$ and $x \in X_q$ $(q = h(p))$. In fact, we denote by $G_q := \{x \mid x \in X$ such that the equality (2.45) holds $\}$ and suppose that $G_q \neq X_q$. By virtue of the Condition 2, G_q is open in X_q. Let $\Gamma_q := \partial G_q (\partial G_q$ is the boundary of $G_q)$ and $\bar{p} \in \Gamma_q$, then $B(\bar{p}, \gamma(\bar{b})) \bigcap (X_q \setminus G_q) \neq \emptyset$ $(B(\bar{p}, \gamma(\bar{b})) := \{x \mid h(x) = h(\bar{p}), \rho(x, \bar{p}) < \gamma(\bar{p})\})$. It is easy to see that the last relations are not satisfied simultaneously and, consequently, $\Gamma_q = \emptyset$ for all $q \in Y$, i.e. $X_q = G_q$. Let $K \in C(X)$ and $\varepsilon > 0$, then there exists $\delta(\varepsilon, K) > 0$ such that $\rho(x_1, x_2) < \delta(h(x_1) = h(x_2); x_1, x_2 \in K)$ implies $\rho(x_1 t, x_2 t) < \varepsilon$ for any $t \geq 0$. Assuming the contrary, we obtain $K_0 \in C(X), \varepsilon_0 > 0, \delta_n \to 0$ $(\delta_n > 0), \{x_n^i\} \subseteq K_0$ $(i = 1, 2)$ and $t_n \to +\infty$ such that $\rho(x_n^1, x_n^2) < \delta_n$ and

$$\rho(x_n^1 t_n, x_n^2 t_n) \geq \varepsilon_0. \tag{2.46}$$

Since K_0 is a compact subset of X we may suppose that the sequences $\{x_n^i\}$ $(i = 1, 2)$ are convergent and we denote by $\bar{x} := \lim_{n\to+\infty} x_n^1 = \lim_{n\to+\infty} x_n^2$ $(\bar{x} \in K_0)$. According to the condition 2., for $\varepsilon_0 > 0$ and $\bar{x} \in K_0$ there exists $\delta(\frac{\varepsilon_0}{3}, \bar{x}) > 0$ such that $\rho(x, \bar{x}) < \delta(\frac{\varepsilon_0}{3}, \bar{x})$ $(h(x) = h(\bar{x}))$ implies $\rho(xt, \bar{x}t) < \frac{\varepsilon_0}{3}$ for all $t \geq 0$. Since $x_n^i \to \bar{x}$ $(i = 1, 2)$, then there exists $\bar{n}$ such that $\rho(x_n^i, \bar{x}) < \delta(\frac{\varepsilon_0}{3}, \bar{x})$ $(n \geq \bar{n})$ and, consequently,

$$\rho(x_n^1 t, x_n^2 t) \leq \frac{2\varepsilon_0}{3} \tag{2.47}$$

for all $t \geq 0$ and $n \geq \bar{n}$. But the inequalities (2.47) and (2.46) are contradictory. Thus, we showed that condition 2. implies condition 3.

We will prove that condition 3. implies condition 4. If we suppose the contrary, then there exist $\varepsilon_0 > 0$, $K_0 \in C(X)$, $t_n \to +\infty$ and $\{x_n^i\} \subseteq K_0$ $(i = 1, 2; h(x_n^1) = h(x_n^2))$ such that the inequality (2.46) holds. We may assume without loss of generality that the sequences $\{x_n^i\}$ $(i = 1, 2)$ are convergent, because K_0 is compact. Let $x^i := \lim_{n\to+\infty} x_n^i$, $0 < \varepsilon < \varepsilon_0$ and $\delta(\frac{\varepsilon}{3}, K_0) > 0$ be chosen according to condition 3.a. Since $h(x^1) = h(x^2)$ and $x^1, x^2 \in K_0$, then for $\frac{\varepsilon}{3}$ there exists $L(\frac{\varepsilon}{3}, x^1, x^2) > 0$ such that $\rho(x^1 t, x^2 t) < \frac{\varepsilon}{3}$ for all $t \geq L(\frac{\varepsilon}{3}, x^1, x^2)$ and, consequently,

$$\rho(x_n^1 t_n, x_n^2 t_n) \leq \rho(x_n^1 t_n, x^1 t_n) + \rho(x^1 t_n, x^2 t_n) + \rho(x^2 t_n, x_n^2 t_n) < \varepsilon \tag{2.48}$$

for sufficiently large n. The inequalities (2.47) and (2.48) are contradictory. Hence, the necessary assertion is proved.

Finally, we note that 4. implies 1. In fact, if we suppose that there exists $y_0 \in J_Y$ such that $L_X \bigcap X_{y_0}$ contains at least two points x_1 and x_2 $(x_1 \neq x_2)$ and denoting by K a compact invariant set such that $x_1, x_2 \in K$, we will have

$x_1, x_2 \in K_{y_0} = K \cap X_{y_0}$. On the other hand, according to Lemma 2.10 K_{y_0} contains no more than one point. The obtained contradiction proves Theorem 2.7. $\square$

Corollary 2.2 *Let $(X, \mathbb{T}_1, \pi)$ and $(Y, \mathbb{T}_2, \sigma)$ be two compactly dissipative dynamical systems, then the following conditions are equivalent:*

(1) the non-autonomous dynamical system $\langle (X, \mathbb{T}_1, \pi), (Y, \mathbb{T}_2, \sigma), h \rangle$ is convergent;
(2) every semi-trajectory $\sum_x^+ (x \in X)$ is asymptotically stable;
(3) 3.a and 3.b from Theorem 2.7 are fulfilled;
(4) the equality (2.42) holds for all $K \in C(X)$.

Definition 2.13 Recall that the dynamical system $(X, \mathbb{T}, \pi)$ is called locally compact if for every $x \in X$ there exist $\delta = \delta_x > 0$ and $l = l_x > 0$ such that $\pi^l B(x, \delta)$ is relatively compact.

Theorem 2.8 *Let $\langle (X, \mathbb{T}_1, \pi), (Y, \mathbb{T}_2, \sigma), h \rangle$ be a non-autonomous dynamical system, $(Y, \mathbb{T}_2, \sigma)$ be compactly dissipative and $(X, \mathbb{T}_1, \pi)$ be locally compact. For $\langle (X, \mathbb{T}_1, \pi), (Y, \mathbb{T}_2, \sigma), h \rangle$ to be convergent, it is necessary and sufficient that every semi-trajectory Σ_x^+ of the system $(X, \mathbb{T}_1, \pi)$ would be relatively compact and that the dynamical system $\langle (h^{-1}(J_Y), \mathbb{T}_1, \pi), (J_Y, \mathbb{T}_2, \sigma), h \rangle$, where J_Y is a Levinson center of $(Y, \mathbb{T}_2, \sigma)$, would be convergent.*

Proof. The necessity of the theorem is evident.

Sufficiency. We will prove that the dynamical system $(X, \mathbb{T}_1, \pi)$ is point dissipative. For this aim under the conditions of the theorem it is sufficient to show that $\Omega_X = \overline{\bigcup\{\omega_x \mid x \in X\}}$ is compact. Note that $h(\omega_x) \subseteq \omega_{h(x)} \subseteq J_Y$ and, consequently, $\omega_x \subseteq h^{-1}(J_Y)$. Since ω_x is compact and invariant, then $\omega_x \subseteq \tilde{J}$ where $\tilde{J}$ is the center of Levinson of the system $(h^{-1}(J_Y), \mathbb{T}_1, \pi)$. Thus, $\Omega_X \subseteq \tilde{J}$, and, consequently, Ω_X is compact. According to Theorem 1.10 for a locally compact dynamical system point dissipativity and compact dissipativity are equivalent. Hence, the dynamical system $(X, \mathbb{T}_1, \pi)$ is compactly dissipative. Let J_X be the center Levinson of $(X, \mathbb{T}_1, \pi)$, then $h(J_X) \subseteq J_Y$ and, consequently, $J_X \subseteq h^{-1}(J_Y)$. Since $\tilde{J}$ is a maximal compact invariant set in $(h^{-1}(J_Y), \mathbb{T}_1, \pi)$, then $J_X \subseteq \tilde{J}$ and $J_X \cap X_y \subseteq \tilde{J} \cap X_y$ for all $y \in J_Y$. From this inclusion follows that the set $J_X \cap X_y$ contains at most one point for any $y \in J_Y$. The theorem is proved. $\square$

Theorem 2.9 *Let $\langle (X, \mathbb{T}_1, \pi), (Y, \mathbb{T}_2, \sigma), h \rangle$ be a non-autonomous dynamical system, $(Y, \mathbb{T}_2, \sigma)$ be a compactly dissipative dynamical system and let there exists $y_0 \in Y$ such that $Y = H^+(y_0)$. For the non-autonomous dynamical system $\langle (X, \mathbb{T}_1, \pi), (Y, \mathbb{T}_2, \sigma), h \rangle$ to be convergent, it is necessary and sufficient that the following conditions would be fulfilled:*

(1) the dynamical system $(X, \mathbb{T}_1, \pi)$ satisfies the condition (A);

(2) $L_X \cap X_y$ contains at most one point for every $y \in J_Y = \omega_{y_0}$.

Proof. The necessity of the conditions 1 and 2 is evident.

Sufficiency. Let $x_0 \in X_{y_0}$, then $h(H^+(x_0)) = H^+(y_0)$ and $h(\omega_{x_0}) = \omega_{y_0}$. Note that $h(\Omega_X) \subseteq \Omega_Y \subseteq J_Y = \omega_{y_0}$ and since $\omega_{x_0} \subseteq \Omega_X$, then $h(\Omega_X) = \omega_{y_0}$. On the other hand, $\Omega_X \subseteq L_X$, $L_X \cap X_y$ contains at most one point for all $y \in J_Y = \omega_{y_0}$. Thus, $\Omega_X = \omega_{x_0}$ is compact and, consequently, $(X, \mathbb{T}_1, \pi)$ is pointwise dissipative. Since $(X, \mathbb{T}_1, \pi)$ is point dissipative and satisfies the condition (A), then by Theorem 1.15 $(X, \mathbb{T}_1, \pi)$ is compactly dissipative. Let J_X be the center of Levinson of the dynamical system $(X, \mathbb{T}_1, \pi)$, then $J_X \subseteq L_X$ and, consequently, $J_X \cap X_y$ contains at most one point for every $y \in J_Y$. The theorem is proved. $\qquad\square$

Definition 2.14   A point $y_0 \in Y$ is called [103],[301] asymptotically stationary (asymptotically τ-periodic, asymptotically almost periodic, asymptotically recurrent) if there exists a stationary ($\tau-$ periodic, almost periodic, recurrent) point $q \in Y$ such that

$$\lim_{t \to +\infty} d(y_0 t, qt) = 0.$$

Remark 2.5   *a. Let $Y = H^+(y_0) = \overline{\{y_0 t \mid t \geq 0\}}$ be compact, then the dynamical system $(Y, \mathbb{T}, \sigma)$ is compactly dissipative and $J_Y = \omega_{y_0}$.*

b. Let y_0 be asymptotically stationary (asymptotically τ-periodic, asymptotically almost periodic, asymptotically recurrent) and $Y = H^+(y_0)$, then $(Y, \mathbb{T}, \sigma)$ is compactly dissipative and the set $J_Y = \omega_{y_0}$ is minimal.

Definition 2.15   A point $x \in X$ is called [49],[103],[301] comparable in limit with regard to the recurrence property with a point $y \in Y$ if the inclusion $\mathcal{L}_y \subseteq \mathcal{L}_x$ holds, where $\mathcal{L}_y := \{\{t_n\} | t_n \to +\infty$ and $\{y t_n\}$ is convergent$\}$.

It is known $[49, 103, 301]$ that if $\mathcal{L}_y \subseteq \mathcal{L}_x$, then the point x possesses the same character of the recurrence property in limit as the point $y \in Y$. In particular, if the point $y \in Y$ is asymptotically stationary (asymptotically τ-periodic, asymptotically almost periodic, asymptotically recurrent) and $\mathcal{L}_y \subseteq \mathcal{L}_x$, then the point x will be asymptotically stationary (asymptotically τ-periodic, asymptotically almost periodic, asymptotically recurrent) .

Theorem 2.10   *Let $y_0 \in Y$ be asymptotically stationary (asymptotically τ-periodic, asymptotically almost periodic, asymptotically recurrent) and $Y = H^+(y_0)$. Then the non-autonomous dynamical system $\langle (X, \mathbb{T}_1, \pi), (Y, \mathbb{T}_2, \sigma), h \rangle$ will be convergent if and only if the following conditions are fulfilled:*

a. the dynamical system $(X, \mathbb{T}_1, \pi)$ satisfies the condition (A);

b. *every point $x \in X$ is comparable in limit with regard to the recurrence property with the point $y = h(x)$ and, in particular, x is asymptotically stationary (asymptotically ω- periodic, asymptotically almost periodic, asymptotically recurrent);*

c. *for any $\varepsilon > 0$ and $K \in C(X)$ there exists $\delta = \delta(\varepsilon, K) > 0$ such that $\rho(x_1, x_2) < \delta$ $(h(x_1) = h(x_2); x_1, x_2 \in K)$ implies $\rho(x_1 t, x_2 t) < \varepsilon$ for all $t \geq 0$;*

d. *the equality $\lim\limits_{t \to +\infty} \rho(x_1 t, x_2 t) = 0$ holds for all $(x_1, x_2) \in X \dot{\times} X$.*

Proof. The necessity of the conditions a,c and d is assured by Remark 2.5. Now, let us show that under the conditions of the theorem the condition b. holds. Let $x \in X$ and $y = h(x)$, then, according to the convergence of the non-autonomous dynamical system $\langle (X, \mathbb{T}_1, \pi), (Y, \mathbb{T}_2, \sigma), h \rangle$, the set $H^+(x) := \overline{\{xt \mid t \geq 0\}}$ is compact. We note that $\omega_x \bigcap X_q \subseteq J_X \bigcap X_q$ for all $q \in \omega_y$ and since $\langle (X, \mathbb{T}_1, \pi), (Y, \mathbb{T}_2, \sigma), h \rangle$ is convergent, then $\omega_x \bigcap X_q$ contains a single point. According to Theorem 1 [49] the point x is comparable in limit with regard to the recurrence property with the point y. If $y \in H^+(y_0)$, then it is evident that the point y will be asymptotically stationary (asymptotically ω- periodic, asymptotically almost periodic, asymptotically recurrent) and, consequently, the point x possesses the same character of recurrence property in limit as the point y does.

We will show that the Conditions a, b, c and d imply the convergence of the non-autonomous dynamical system $\langle (X, \mathbb{T}_1, \pi), (Y, \mathbb{T}_2, \sigma), h \rangle$. First of all, according to the Condition b we have that $\omega_x \neq \emptyset$ is compact, minimal and $h(\omega_x) = \omega_{y_0}$ for all $x \in X$. We note that $\omega_x \bigcap X_q$ contains a single point for every $q \in \omega_{y_0}$. In the opposite case there exist $q_0 \in \omega_{y_0}, p_1, p_2 \in \omega_x \bigcap X_{q_0} (p_1 \neq p_2)$ and $t_n^i \to +\infty$ ($i = 1, 2$) such that $xt_n^i \to p_i$ ($i = 1, 2$) as $n \to +\infty$. We note that $yt_n^i \to q_0$ ($i = 1, 2$) as $n \to +\infty$, where $y = h(x)$. Let $\bar{t}_{2n-1} = t_n^1$ and $\bar{t}_{2n} = t_n^2$ for every $n \in \mathbb{N}$, then $\{\bar{t}_n\} \in \mathcal{L}_y$ and, consequently, $\{\bar{t}_n\} \in \mathcal{L}_x$, i.e. $\{xt_n\}$ is convergent; therefore $p_1 = p_2$. The last equality contradicts the choice of the points p_1 and p_2. The obtained contradiction proves the necessary assertion. Now we will prove that $\omega_{x_1} \bigcap X_q = \omega_{x_2} \bigcap X_q$ for all $x_1, x_2 \in X$ and $q \in \omega_{y_0}$. Let $q \in \omega_{y_0}, \{p_i\} = \omega_{x_i} \bigcap X_q$ ($i = 1, 2$) and $\{t_n\} \in \mathcal{L}_q$ be such that $qt_n \to q$. By virtue of the Condition c and the minimality of $\omega_{x_i} (i = 1, 2)$ we have $\rho(p_1 t_n, p_2 t_n) \to 0$ as $n \to +\infty$ and, consequently, $p_1 = p_2$. Thus, $\omega_{x_1} = \omega_{x_2}$ for all $x_1, x_2 \in X$ and, consequently, $(X, \mathbb{T}_1, \pi)$ is point dissipative and since $(X, \mathbb{T}_1, \pi)$ satisfies the condition (A), then according to Theorem 2.9 $(X, \mathbb{T}_1, \pi)$ is compactly dissipative. To finish the proof of the theorem is sufficient to apply Theorem 2.7 and Remark 2.4. $\qquad \square$

Corollary 2.3 *Under the conditions of Theorem 2.11 if the space X is locally compact, then the Condition a results from the Conditions b, c and d.*

Theorem 2.11　*Let $\langle (X, \mathbb{T}_1, \pi), (Y, \mathbb{T}_2, \sigma), h \rangle$ be a non-autonomous dynamical system, $(Y, \mathbb{T}_2, \sigma)$ be compactly dissipative and let its Levinson's center J_Y be minimal (i.e., every semi-trajectory $\sum_y^+$ ($y \in J_Y$) is dense in J_Y). Then the following conditions are equivalent:*

(1) the non-autonomous dynamical system $\langle (X, \mathbb{T}_1, \pi), (Y, \mathbb{T}_2, \sigma), h \rangle$ is convergent;

(2) the dynamical system $(X, \mathbb{T}_1, \pi)$ satisfies the condition (A) and for every $K \in C(X)$ the equality (2.42) holds.

Proof.　By virtue of Corollary 2.2 Condition 1 implies Condition 2. Let us show that the converse assertion holds. Let $K \in C(X)$, then $\sum_K^+ := \bigcup \{ \sum_x^+ | x \in K \}$ is relatively compact and according to Lemma 1.3 the set

$$\Omega(K) = \bigcap_{t \geq 0} \overline{\bigcup_{\tau \geq t} \pi^\tau K}$$

is non-empty, compact, invariant and, consequently, $h(\Omega(K)) \subseteq \Omega(h(K)) \subseteq J_Y$, because J_Y is the maximal compact invariant set in Y. Thus, J_Y is minimal, then the equality

$$h(\Omega(K)) = J_Y$$

holds. We note that $\Omega(K_1) = \Omega(K_2)$ for all K_1 and K_2 from $C(X)$. In fact, since $M := \Omega(K_1) \bigcup \Omega(K_2)$ is compact and invariant and J_Y is minimal, we have $h(M) = J_Y$. On the other hand, according to Lemma 2.10 the set $M_y = M \bigcap X_y$ contains a single point for every $y \in J_Y$. We have $\Omega(K_i) \bigcap X_y \subseteq M \bigcap X_y (i = 1, 2)$ and $\Omega(K_1) \bigcap X_y = \Omega(K_2) \bigcap X_y = M \bigcap X_y$ for any $y \in J_Y$ and, consequently, $\Omega(K_1) = \Omega(K_2)$ for all K_1 and K_2 from $C(X)$. From this follows that $(X, \mathbb{T}_1, \pi)$ is compactly dissipative and according to Theorem 2.7 $\langle (X, \mathbb{T}_1, \pi), (Y, \mathbb{T}_2, \sigma), h \rangle$ is convergent.　□

Corollary 2.4　*Let $\langle (X, \mathbb{T}_1, \pi), (Y, \mathbb{T}_2, \sigma), h \rangle$ be a non-autonomous dynamical system, $(Y, \mathbb{T}_2, \sigma)$ be compactly dissipative, J_Y be minimal and $(X, \mathbb{T}_1, \pi)$ satisfy the condition (A). Then the Conditions 1-4 from Corollary 2.2 are equivalent.*

Theorem 2.12　*Let $(Y, \mathbb{T}_2, \sigma)$ be compactly dissipative and $h(L_X) = J_Y$. For the non-autonomous dynamical system $\langle (X, \mathbb{T}_1, \pi), (Y, \mathbb{T}_2, \sigma), h \rangle$ to be convergent, it is necessary, and if $J_Y = Y$ it is sufficient, that there would be fulfilled the following conditions:*

1. Σ_x^+ is relatively compact for all $x \in X$ and L_X is relatively compact;

2. $L_X \bigcap X_y$ contains only one point for any $y \in Y$;

3. for every $\varepsilon > 0$ there exists $\delta(\varepsilon) > 0$ such that $\rho(x, x_y) < \delta$ ($\{x_y\} = L_X \bigcap X_y$ and $h(x) = y \in J_Y$) implies $\rho(xt, x_{yt}) < \varepsilon$ for all $t \geq 0$ and $x \in X$.

Proof. Necessity. Let $\langle (X, \mathbb{T}_1, \pi), (Y, \mathbb{T}_2, \sigma), h \rangle$ be convergent, then $(X, \mathbb{T}_1, \pi)$ is compactly dissipative and $J_X = L_X$. It is easy to see that the Conditions 1 and 2 are fulfilled. We will show that Condition 3 is fulfilled too. Suppose that it is not so, then there exist $\varepsilon_0 > 0$, $\delta_n \downarrow 0$, $\{x_n\}$ from X, $\{y_n\}$ from J_Y and $t_n \to +\infty$ such that $\rho(x_n, x_{y_n}) < \delta_n$ $(y_n = h(x_n))$ and

$$\rho(x_n t_n, x_{y_n} t_n) \geq \varepsilon_0. \tag{2.49}$$

Since the sets J_Y and J_X are compact, then we can suppose that the sequences $\{y_n\}$ and $\{x_{y_n}\}$ are convergent. Let $y_0 = \lim_{n \to +\infty} y_n$, then $x_{y_0} = \lim_{n \to +\infty} x_{y_n} = \lim_{n \to +\infty} x_n$. By the compact dissipativity of the dynamical system $(X, \mathbb{T}_1, \pi)$ we can suppose that the sequence $\{x_n t_n\}$ is convergent and let $\bar{x} = \lim_{n \to +\infty} x_n t_n$. Since $y_n t_n \in J_Y$, then the sequence $\{y_n t_n\}$ can be considered convergent too. Let $\bar{y} = \lim_{n \to +\infty} y_n t_n$ and we note that $h(\bar{x}) = \lim_{n \to +\infty} h(x_n) t_n = \lim_{n \to +\infty} y_n t_n = \bar{y}$, $\bar{x} \in J_X$ and hence $\bar{x} \in J_X \cap X_{\bar{y}} = \{x_{\bar{y}}\}$, i.e., $\bar{x} = x_{\bar{y}}$. On the other hand, passing to limit in (2.49) as $n \to +\infty$ we obtain $\rho(\bar{x}, x_{\bar{y}}) \geq \varepsilon_0$. The obtained contradiction proves the statement required.

Sufficiency. Suppose that Conditions 1, 2 and 3 of the theorem are fulfilled. For the convergence of the system $\langle (X, \mathbb{T}_1, \pi), (Y, \mathbb{T}_2, \sigma), h \rangle$ it is sufficient to show that under the conditions of the theorem the dynamical system $(X, \mathbb{T}_1, \pi)$ is compactly dissipative. By the condition of the theorem $(X, \mathbb{T}_1, \pi)$ is point dissipative and $\Omega_X = \overline{\bigcup \{\omega_x \mid x \in X\}} \subseteq L_X$. Note, that the set L_X is closed. We will show that the set L_X is orbitally stable. If we suppose that it is not true, then there exist $\varepsilon_0 > 0$, $x_n \to x_0 \in L_X$ and $t_n \to +\infty$ such that

$$\rho(x_n t_n, L_X) \geq \varepsilon_0.$$

Since $y_n \to y_0 = h(x_0)$, where $y_n = h(x_n)$, then under the conditions of the theorem $x_{y_n} \to x_{y_0} = x_0$ and, consequently, $\rho(x_n, x_{y_n}) \leq \rho(x_n, x_0) + \rho(x_0, x_{y_n}) \to 0$. From this relation and Condition 3 of the theorem follows that $\rho(x_n t_n, x_{y_n} t_n) \to 0$. This relation and the inequality (2.49) are contradictory. Thus, L_X is compact, invariant and orbitally stable. Since $\Omega_X \subseteq L_X$, then $J^+(\Omega_X) \subseteq L_X$. Let us show that $J^+(\Omega_X) = J_X$. In fact, let $\bar{x} \in L_X$ and $\varphi : \mathbb{S} \to L_X$ be a whole trajectory of $(X, \mathbb{T}_1, \pi)$ passing through point $\bar{x}$ for $t = 0$. Denote by $\alpha_{\varphi_{\bar{x}}} := \bigcap_{t \leq 0} \overline{\bigcup_{\tau \leq t} \varphi(\tau)}$, then $\Omega_X \cap \alpha_{\varphi_{\bar{x}}} \neq \emptyset$. Let $p \in \alpha_{\varphi_{\bar{x}}} \cap \Omega_X$, then there exists $t_n \to -\infty$ such that $\varphi(t_n) \to p$ and, consequently, $\pi^{-t_n} \varphi(t_n) = \varphi(0) = \bar{x}$, i.e., $\bar{x} \in J_p^+ \subseteq J^+(\Omega_X)$. Thus, $L_X \subseteq J^+(\Omega_X)$, i.e. $L_X = J^+(\Omega_X)$ is compact and orbitally stable and by Theorem 1.13 the dynamical system $(X, \mathbb{T}_1, \pi)$ is compactly dissipative. The theorem is proved. $\square$

Theorem 2.13 *Let $\langle (X, \mathbb{T}_1, \pi), (Y, \mathbb{T}_2, \sigma), h \rangle$ be a non-autonomous dynamical system and Y be a compact minimal set, then the following conditions are equivalent:*

1. $\langle (X, \mathbb{T}_1, \pi), (Y, \mathbb{T}_2, \sigma), h \rangle$ *is convergent;*
2. *every semi-trajectory $\sum_x^+ (x \in X)$ is relatively compact and asymptotically stable;*
3. (a) *every semi-trajectory $\sum_x^+ (x \in X)$ is relatively compact;*
   (b) $\lim\limits_{t \to +\infty} \rho(x_1 t, x_2 t) = 0$ *for all $(x_1, x_2) \in X \dot\times X$;*
   (c) *for any $\varepsilon > 0$ and $K \in C(X)$ there exists $\delta(\varepsilon, K) > 0$ such that $\rho(x_1, x_2) < \delta(h(x_1) = h(x_2); x_1, x_2 \in K)$ implies $\rho(x_1 t, x_2 t) < \varepsilon$ for all $t \geq 0$.*
4. *every semi-trajectory $\sum_x^+ (x \in X)$ is relatively compact and the equality (2.42) holds for all $K \in C(X)$.*

Proof. The implications 1. $\to$ 2. $\to$ 3. $\to$ 4. are proved by a slight modification of the proof of Theorem 2.7. To finish the proof of our theorem is sufficient to establish that Condition 4 implies Condition 1. Let $x_0 \in X$. Since $\Sigma_{x_0}^+$ is relatively compact, then $\omega_{x_0} \neq \emptyset$, is compact and all the motions in ω_{x_0} are extendable to the left. We set $M := \omega_{x_0}$. Since $h(\Omega_X) \subseteq \Omega_Y \subseteq Y$ and Y is minimal, then $h(M) = Y$ and, consequently, $M_y := M \bigcap X_y \neq \emptyset$ for all $y \in Y$. By Lemma 2.1 the set M_y consists of the one point for any $y \in Y$. We will show now that for every $x \in X$ the equality $\omega_x = M$ is true. In fact, let $K := \omega_x \bigcup M$, then by Lemma 2.10 $K_y := K \bigcap X_y$ contains only one point for every $y \in Y$. Since $h(\omega_x) = h(\omega_{x_0}) = h(K) = Y$, then $\omega_x \bigcap X_y = \omega_{x_0} \bigcap X_y = K \bigcap X_y$ for all $y \in Y$ and, consequently, $\omega_x = \omega_{x_0} = M$ for any $x \in X$. Thus, $(X, \mathbb{T}_1, \pi)$ is point dissipative. Let now $K \in C(X)$, then $\Sigma^+(K)$ is relatively compact. In fact, if $\{x_n\} \subseteq K$ and $t_n \to +\infty$, then by the condition 4. we have

$$\lim_{n \to +\infty} \rho(x_n t_n, M_{h(x_n) t_n}) = 0,$$

and, consequently, the sequence $\{x_n t_n\}$ is relatively compact. According to Theorem 1.15 the dynamical system $(X, \mathbb{T}_1, \pi)$ is compact dissipative. Let J_X be the center of Levinson of the dynamical system $(X, \mathbb{T}_1, \pi)$. By Lemma 2.10 the set $J_X \bigcap X_y$ contains only one point for each $y \in Y$. The theorem is completely proved.

Theorem 2.14   *Let $\langle (X, \mathbb{T}_1, \pi), (Y, \mathbb{T}_2, \sigma), h \rangle$ be a non-autonomous dynamical system, $M \neq \emptyset$ be a compact positively invariant set. Suppose that the following conditions are fulfilled:*

(1) $h(M) = Y$;

(2) $M \bigcap X_y$ contains a single point for all $y \in Y$;

(3) M is globally asymptotically stable, i.e. for any $\varepsilon > 0$ there exists $\delta(\varepsilon) > 0$ such that $\rho(x, p) < \delta \ (x \in X_y, p \in M_y := M \bigcap X_y)$ implies $\rho(xt, pt) < \varepsilon$ for all $t \geq 0$ and $\lim\limits_{t \to +\infty} \rho(xt, M_{h(x)t}) = 0$ for all $x \in X$.

Then the non-autonomous dynamical system $\langle (X, \mathbb{T}_1, \pi), (Y, \mathbb{T}_2, \sigma), h \rangle$ is convergent.

Proof. We note that under the conditions of the theorem the dynamical system $(X, \mathbb{T}_1, \pi)$ is point dissipative and $\Omega_X \subseteq M$. We will show that set M is orbitally stable in $(X, \mathbb{T}_1, \pi)$. Suppose that it is not true, then there exist $\varepsilon_0 > 0, \delta_n \to 0, x_n \in B(M, \delta_n)$ and $t_n \to +\infty$ such that

$$\rho(x_n t_n, M) \geq \varepsilon_0. \tag{2.50}$$

Since M is compact, then we may suppose that the sequence $\{x_n\}$ is convergent. Let $x_0 := \lim_{n \to +\infty} x_n, x_{y_n} \in M_{y_n}, \rho(x_n, M) = \rho(x_n, x_{y_n})$ and $y_0 = h(x_0)$, then $x_0 = \lim_{n \to +\infty} x_{y_n}$ and $x_0 \in M_{y_0}$. Let $q_n = h(x_n)$ and note that

$$\rho(x_n, x_{q_n}) \leq \rho(x_n, x_{y_n}) + \rho(x_{y_n}, x_{q_n}) \to 0 \tag{2.51}$$

as $n \to +\infty$, because $q_n \to y_0$ and $x_{q_n} \to x_0$. Taking into account (2.51) and the asymptotic stability of the set M, we have

$$\rho(x_n t_n, x_{q_n t_n}) = \rho(x_n t_n, x_{q_n} t_n) \to 0. \tag{2.52}$$

But the equalities (2.50) and (2.52) are contradictory. Hence, the set M is orbitally stable in $(X, \mathbb{T}_1, \pi)$ and by virtue of Theorem 1.13 the dynamical system $(X, \mathbb{T}_1, \pi)$ is compactly dissipative and $J_X \subseteq M$. To finish the proof of Theorem it is sufficient to note that $h(J_X) = J_Y$ and for all $y \in J_Y$ we have $J_X \bigcap X_y \subseteq M \bigcap X_y$ and, consequently, $J_X \bigcap X_y$ contains a single point for any $y \in J_Y$. The theorem is proved. $\square$

Remark 2.6 *If there exists $y_0 \in Y$ such that $Y = H^+(y_0)$, then it is evident that Theorem 2.14 is invertible. For this aim we may take set $H^+(x_0)$, where $x_0 \in X_{y_0}$, in the quality of set M appearing in the theorem.*

2.5 Tests for convergence

Note that in Theorems 2.5 and 2.6 there are contained some conditions under which the property of compact dissipativity is kept under the homomorphisms.

Below we make some assertions which ensure dissipativity of the system $(X, \mathbb{T}_1, \pi)$ if the dynamical system $(Y, \mathbb{T}_2, \sigma)$ has this property and there exists a homomorphism h of $(X, \mathbb{T}_1, \pi)$ onto $(Y, \mathbb{T}_2, \sigma)$.

Definition 2.16 Let h be a homomorphism of $(X, \mathbb{T}_1, \pi)$ onto $(Y, \mathbb{T}_2, \sigma)$. A set $M \subseteq X$ is called uniformly stable in the positive direction with respect to the homomorphism h if for all $\varepsilon > 0$ there exists $\delta = \delta(\varepsilon, M) > 0$ such that for any $x_1, x_2 \subseteq M$ for which $h(x_1) = h(x_2)$ the inequality $\rho(x_1, x_2) < \delta$ implies $\rho(x_1 t, x_2 t) < \varepsilon$ for every $t \geq 0$.

Definition 2.17   A dynamical system $(X, \mathbb{T}_1, \pi)$ is called uniformly stable (in the positive direction) with respect to the homomorphism h on compact subsets from X, if every compact $M \in C(X)$ is uniformly stable in the positive direction with respect to the homomorphism h.

Lemma 2.11   *Suppose that a homomorphism h satisfies the following conditions:*

(1) there is a continuous invariant section $\varphi : Y \to X$ of the homomorphism h;
(2) $\lim\limits_{t \to +\infty} \rho(x_1 t, x_2 t) = 0$ for all $x_1, x_2 \in X$ $(h(x_1) = h(x_2))$;
(3) the dynamical system $(X, \mathbb{T}_1, \pi)$ is uniformly stable in the positive direction on compact subsets from X with respect to the homomorphism h.

Then:

(1) if $(Y, \mathbb{T}_2, \sigma)$ is point dissipative then $(X, \mathbb{T}_1, \pi)$ is point dissipative too. Moreover, Ω_X and Ω_Y are homeomorphic;
(2) if $(Y, \mathbb{T}_2, \sigma)$ is compactly dissipative then $(X, \mathbb{T}_1, \pi)$ is also compactly dissipative and, moreover, Ω_X and Ω_Y are homeomorphic.

Proof.   Let $(Y, \mathbb{T}_2, \sigma)$ be point dissipative. Then Ω_Y is a nonempty, compact and invariant set and, consequently, $M := \varphi(\Omega_Y) \subseteq \Omega_X$ is also non-empty, compact and invariant. For $x \in X$ and $y := h(x)$ we have $\lim\limits_{t \to +\infty} \rho(xt, \varphi(y)t) = 0$. Hence, Σ_x^+ is relatively compact compact and $\omega_x \subseteq \omega_{\varphi(y)} \subseteq \varphi(\Omega_Y) = M$. Thus, $\Omega_X \subseteq M$ and $(X, \mathbb{T}_1, \pi)$ is point dissipative and $\varphi(\Omega_Y) = \Omega_X$. Since $\varphi : \Omega_Y \to \Omega_X$ separates points and the set Ω_Y is compact, then Ω_Y and Ω_X are homeomorphic.

Now we will prove the second statement of the lemma. Let $(Y, \mathbb{T}_2, \sigma)$ be compactly dissipative. By the statement above, the dynamical system $(X, \mathbb{T}_1, \pi)$ is point dissipative. Let $M := \varphi(J_Y) = \varphi(D^+(\Omega_Y)) \subseteq D^+(\Omega_X)$. We will show that the set M is orbitally stable. If we suppose that it is not true, then there are $\varepsilon_0, x_n \to x_0 \in M$ and $t_n \to +\infty$ such that

$$\rho(x_n t_n, M) \geq \varepsilon_0. \tag{2.53}$$

Note, that $h(x_n) = y_n \to y_0 = h(x_0) \in h(M) = h(\varphi(J_Y)) = J_Y$ and by compact dissipativity of $(Y, \mathbb{T}_2, \sigma)$ we can suppose that the sequence $\{y_n t_n\}$ is convergent. Let $y := \lim\limits_{n \to +\infty} y_n t_n$, then $y \in J_Y$ and $\varphi(y) = \lim\limits_{n \to +\infty} \varphi(h(x_n))t_n$. Since $\varphi : J_Y \to \varphi(J_Y) = M$ separates points and $h \circ \varphi = id_Y$, then $\varphi : J_Y \to M = \varphi(J_Y)$ is a homomorphism and $\varphi \circ h(x) = x$ for all $x \in M$ and, consequently, $\varphi(h(x_n)) \to \varphi(h(x_0)) = x_0 \in M$. From this relation follows that

$$\lim\limits_{n \to +\infty} \rho(x_n, \varphi \circ h(x_n)) = 0. \tag{2.54}$$

Let $K := M \bigcup \overline{\{x_n\}} \bigcup \overline{\{\varphi \circ h(x_n)\}}, \varepsilon > 0$ and $\delta(\varepsilon, K) > 0$ be the numbers from the condition of uniform stability in the positive direction on compact subsets from

X of the dynamical system $(X, \mathbb{T}_1, \pi)$ with respect to the homomorphism h. The equality (2.54) implies that for sufficiently large n the inequality $\rho(x_n, \varphi \circ h(x_n)) < \delta$ holds and, consequently, $\rho(x_n t, \varphi \circ h(x_n)t) < \varepsilon$ for all $t \geq 0$. In particular,

$$\rho(x_n t_n, \varphi \circ h(x_n)t_n) < \varepsilon \qquad (2.55)$$

for sufficiently large n. Taking into account that ε is arbitrary, we get that from (2.55) follows $\lim\limits_{n \to +\infty} x_n t_n = \lim\limits_{n \to +\infty} \varphi \circ h(x_n)t_n = \varphi(y) \in M$. This equality and (2.53) are contradictory. The obtained contradiction shows that M is orbitally stable. Thus, $(X, \mathbb{T}_1, \pi)$ is point dissipative, $\Omega_X \subseteq M$, the set M is non-empty, compact, invariant and orbitally stable. By Theorem 1.13, the dynamical system $(X, \mathbb{T}_1, \pi)$ is compactly dissipative and $J_X \subseteq M = \varphi(J_Y) \subseteq D^+(\Omega_X)$. According to Theorem 1.11 and Corollary 1.5, we have $J_X = \varphi(J_Y) = D^+(\Omega_X)$. Thus, φ is a homeomorphism of J_Y onto J_X. The lemma is proved. $\qquad\square$

Corollary 2.5 *Under the conditions of Lemma 2.11, if $(Y, \mathbb{T}_2, \sigma)$ is compactly dissipative, then $\langle (X, \mathbb{T}_1, \pi), (Y, \mathbb{T}_2, \sigma), h \rangle$ is convergent.*

Proof. This assertion follows from Lemma 2.11 and Corollary 2.2. $\qquad\square$

Let $(Y, \mathbb{S}, \sigma)$ be a two-sided dynamical system, $(X, \mathbb{S}_+, \pi)$ be a semi-group dynamical system and $h : X \to Y$ be a homomorphism of $(X, \mathbb{S}_+, \pi)$ onto $(Y, \mathbb{S}, \sigma)$. Consider a non-autonomous dynamical system $\langle (X, \mathbb{S}_+, \pi), (Y, \mathbb{S}, \sigma), h \rangle$ and denote by $\Gamma_b(Y, X)$ the family of all continuous and bounded sections of the homomorphism h. By equality

$$d(\varphi_1, \varphi_2) := \sup_{y \in Y} \rho(\varphi_1(y), \varphi_2(y)) \qquad (2.56)$$

there is defined a metric on $\Gamma_b(Y, X)$.

Let $X \dot\times X := \{(x_1, x_2) : x_1, x_2 \in X, h(x_1) = h(x_2)\}$ and let $V : X \dot\times X \to \mathbb{R}_+$ be a mapping satisfying the following conditions:

1. $a(\rho(x_1, x_2)) \leq V(x_1, x_2) \leq b(\rho(x_1, x_2))$ for all $(x_1, x_2) \in X \dot\times X$, where a, b are two functions from $\mathfrak{A}$ ($\mathfrak{A} := \{a \mid a : \mathbb{R}_+ \to \mathbb{R}_+, a$ is continuous, strongly increasing and $a(0) = 0 \}$) and $Im(a) = Im(b)$;
2. $V(x_1, x_2) = V(x_2, x_1)$ for all $(x_1, x_2) \in X \dot\times X$;
3. $V(x_1, x_2) \leq V(x_1, x_3) + V(x_3, x_1)$ for all $x_1, x_2, x_3 \in X$ such that $h(x_1) = h(x_2) = h(x_3)$.

From the conditions 1–3 follows that the function V on each fiber $X_y = h^{-1}(y)$ defines some metric which is topologically equivalent to ρ.

Lemma 2.12   *Suppose that the function $V : X \dot\times X \to \mathbb{R}_+$ satisfies the conditions 1–3. Then by equality*

$$p(\varphi_1, \varphi_2) := \sup\{V(\varphi_1(y), \varphi_2(y))|\ y \in Y\} \qquad (2.57)$$

on $\Gamma_b(Y, X)$ there is defined a complete metric that is topologically equivalent to (2.56).

Proof.   From Condition 1 it follows that the function V vanishes only on the diagonal of $X \dot\times X$, i.e. $V(x_1, x_2) = 0$ if and only if $x_1 = x_2$. Taking into account (2.57), we have $d(\varphi, \psi) = 0$ if and only if $\varphi = \psi$.

From Property 2 of the function V it follows that the function d is symmetric (i.e., $d(\varphi, \psi) = d(\psi, \varphi)$). Finally, from Condition 3. it follows that the function d satisfies the triangle inequality.

From the condition 1. follows that the metrics (2.56) and (2.57) are topologically equivalent.

Thus, to complete the proof of the lemma it is sufficient to establish that the space $(\Gamma_b(Y, X), d)$ is complete. Let $\{\varphi_n\}$ be a Cauchy's sequence in $(\Gamma_b(Y, X), d)$, i.e., $d(\varphi_n, \varphi_m) \to 0$ as $n, m \to +\infty$. For every $\varepsilon > 0$ there exists $\mathcal{N}_1(\varepsilon) > 0$ such that

$$\rho(\varphi_n(y), \varphi_m(y)) < \varepsilon \qquad (2.58)$$

for any $y \in Y$ and $n, m > \mathcal{N}_1(\varepsilon)$. From this relation and from the completeness of (X, ρ) it follows that for each $y \in Y$ there exists $\lim\limits_{n \to +\infty} \varphi_n(y)$ in (X, ρ). We set $\varphi(y) := \lim\limits_{n \to +\infty} \varphi_n(y)$. We will show that this equality holds uniformly with respect to $y \in Y$ and, consequently, $\varphi \in \Gamma_b(Y, X)$. In fact, passing to limit in the equality (2.58) as $m \to +\infty$ we have $\rho(\varphi_n(y), \varphi(y)) \le \varepsilon$ for all $n > \mathcal{N}_1(\varepsilon)$ and $y \in Y$. Thus $\varphi \in (\Gamma_b(Y, X), d)$ and $d(\varphi_n, \varphi) \to 0$ as $n \to +\infty$. The lemma is proved. $\qquad\square$

Denote by $S^t : \Gamma_b(Y, X) \to \Gamma(Y, X)$ the mapping, defined by the equality $(S^t \varphi)(y) = \pi^t \varphi(\sigma^{-1} y)$ for all $t \in \mathbb{S}_+$, $\varphi \in \Gamma_b(Y, X)$ and $y \in Y$. It is easy to check that the family of mappings $\{S^t\}_{t \ge 0}$ is a commutative semigroup.

Lemma 2.13   *If there is a function $V : X \dot\times X \to \mathbb{R}_+$ satisfying Conditions 1–3 and*

4. $V(x_1 t, x_2 t) \le \mathcal{N} e^{-\nu t} V(x_1, x_2)$ *($\forall (x_1, x_2) \in X \dot\times X$, $t \ge 0$), where $\mathcal{N}, \nu > 0$,*

then the semigroup $\{S^t\}_{t \ge 0}$ has a unique fixed point $\varphi \in \Gamma_b(Y, X)$ which is an invariant section of h.

Proof. Note that

$$p(S^t\varphi_1, S^t\varphi_2) = \sup\{V(\pi^t\varphi_1(\sigma^{-t}y), \pi^t\varphi_2(\sigma^{-t}y))|\ y \in Y\}$$

$$\leq \mathcal{N}e^{-\nu t}\sup\{V(\varphi_1(\sigma^{-t}y), \varphi_2(\sigma^{-t}y))\ |\ y \in Y\} \leq \mathcal{N}e^{-\nu t}p(\varphi_1, \varphi_2)$$

and, consequently, S^t is a contraction for sufficiently large t. From this fact and taking into consideration that $\{S^t\}$ is commutative we obtain the existence of a unique fixed point φ of the semigroup $\{S^t\}_{t\geq 0}$, which is an invariant section of h. The lemma is proved. $\qquad\square$

Theorem 2.15 *Let* $\langle(X, \mathbb{S}_+, \pi), (Y, \mathbb{S}, \sigma), h\rangle$ *be a non-autonomous dynamical system and let the following conditions be fulfilled:*

1. *$(Y, \mathbb{S}, \sigma)$ is compactly dissipative;*
2. *$\Gamma_b(Y, X) \neq \emptyset$;*
3. *there exists $V : X\dot\times X \to \mathbb{R}_+$ satisfying Conditions 1–4 of Lemma 2.13.*

Then $(X, \mathbb{S}_+, \pi)$ is compactly dissipative and the sets J_X and J_Y are homeomorphic.

Proof. Under the conditions of the theorem, by Lemma 2.13 the semi-group $\{S^t\}_{t\geq 0}$ has a unique fixed point φ that is an invariant section of h.

In addition,

$$a(\rho(x_1t, x_2t)) \leq V(x_1t, x_2t) \leq \mathcal{N}e^{-\nu t}V(x_1, x_2) \leq \mathcal{N}e^{-\nu t}b(\rho(x_1, x_2)).$$

Therefore, $\lim\limits_{t\to+\infty} a(\rho(x_1t, x_2t)) = 0$ for all $(x_1, x_2) \in X\dot\times X$ and, consequently, $\lim\limits_{t\to+\infty} \rho(x_1t, x_2t) = 0$. Note, that the system $(X, \mathbb{S}_+, \pi)$ is uniformly stable with respect to h. In fact, let $\varepsilon > 0$ and $\delta(\varepsilon) = b^{-1}(\mathcal{N}a(\varepsilon))$. Then $\rho(x_1, x_2) < \delta(\varepsilon)$ $(h(x_1) = h(x_2))$ implies $\rho(x_1t, x_2t) < \varepsilon$ for all $t \geq 0$. To finish the proof of the theorem it is sufficient to refer to Lemma 2.11. $\qquad\square$

Corollary 2.6 *Under the conditions of Theorem 2.15 the non-autonomous dynamical system $\langle(X, \mathbb{S}_+, \pi), (Y, \mathbb{S}, \sigma), h\rangle$ is convergent.*

Proof. This assertion follows from Theorem 2.15 and Corollary 2.5. $\qquad\square$

The function $V : X\dot\times X \to \mathbb{R}_+$ is said to be continuous, if $x_n^i \to x^i$ ($i = 1, 2$ and $h(x_n^1) = h(x_n^2)$) implies $V(x_n^1, x_n^2) \to V(x^1, x^2)$.

Theorem 2.16 *Let $(X, \mathbb{T}_1, \pi)$ and $(Y, \mathbb{T}_2, \sigma)$ be compactly dissipative. The dynamical system $\langle(X, \mathbb{T}_1, \pi), (Y, \mathbb{T}_2, \sigma), h\rangle$ is convergent if and only if there is a continuous function $V : X\dot\times X \to \mathbb{R}_+$ satisfying the following conditions:*

(1) V is positively defined, i.e. $V(x_1, x_2) = 0$ if and only if $x_1 = x_2$;

(2) $V(x_1t, x_2t) \le V(x_1, x_2)$ *for all* $t \ge 0$ *and* $(x_1, x_2) \in X \dot\times X$;

(3) $V(x_1t, x_2t) = V(x_1, x_2)$ *for all* $t \ge 0$ *if and only if* $x_1 = x_2$.

Proof. Suppose that the conditions of the theorem are fulfilled. We will show that for all $\varepsilon > 0$ and compact $K \subseteq X$ there is $\delta(\varepsilon, K) > 0$ such that $\rho(x_1, x_2) < \delta$ $(x_1, x_2 \in K$ and $h(x_1) = h(x_2))$ implies $\rho(x_1t, x_2t) < \varepsilon$ for all $t \ge 0$. Indeed, if we suppose that it is not true, then there exist $\varepsilon_0 > 0$, compact $K_0 \subseteq X$, sequences $\delta_n \downarrow 0$, $\{x_n^i\} \subseteq K_0$ $(i = 1, 2$ and $h(x_n^1) = h(x_n^2))$ and $t_n \to +\infty$ such that

$$\rho(x_n^1, x_n^2) < \delta_n \text{ and } \rho(x_n^1 t_n, x_n^2 t_n) \ge \varepsilon_0. \tag{2.59}$$

By the compact dissipativity of $(X, \mathbb{T}_1, \pi)$ we can assume that the sequences $\{x_n^i t_n\}$ $(i = 1, 2)$ are convergent. Let us put $\bar{x}^i := \lim\limits_{n \to +\infty} x_n^i t_n$ $(i = 1, 2)$. It is clear that $h(\bar{x}_1) = h(\bar{x}_2)$. Since $\{x_n^i\} \subseteq K$, then we may suppose that they are convergent too. By virtue of (2.59), $\lim\limits_{n \to +\infty} x_n^1 = \lim\limits_{n \to +\infty} x_n^2 = \bar{x}$ and

$$0 \le V(\bar{x}_1, \bar{x}_2) = \lim\limits_{n \to +\infty} V(x_n^1 t_n, x_n^2 t_n) \le \lim\limits_{n \to +\infty} V(x_n^1, x_n^2) = V(\bar{x}, \bar{x}) = 0,$$

from which it follows that $\bar{x}_1 = \bar{x}_2$. The last equality contradicts (2.59).

We will show now that for all $(x_1, x_2) \in X \dot\times X$ $(h(x_1) = h(x_2))$ the equality

$$\lim\limits_{t \to +\infty} \rho(x_1t, x_2t) = 0$$

holds. If we suppose that it is not true, then there exist $y_0 \in Y, \bar{x}_1, \bar{x}_2 \in X_{y_0}, \varepsilon_0 > 0$ and $t_n \to +\infty$ such that

$$\rho(\bar{x}_1 t_n, \bar{x}_2 t_n) \ge \varepsilon_0. \tag{2.60}$$

Let $(X \times X, \mathbb{T}_1, \pi \times \pi) := (X, \mathbb{T}_1, \pi) \times (X, \mathbb{T}_1, \pi)$ be a direct product of $(X, \mathbb{T}_1, \pi)$ and $(X, \mathbb{T}_1, \pi)$. In view of the compact dissipativity of $(X, \mathbb{T}_1, \pi)$, the point $(\bar{x}_1, \bar{x}_2) \in X \times X$ is L^+ stable in $(X \times X, \mathbb{T}_1, \pi \times \pi)$, and from Condition 2 of the theorem follows the existence of a finite limit

$$V_0 = \lim\limits_{t \to +\infty} V(\bar{x}_1 t, \bar{x}_2 t). \tag{2.61}$$

Let $(p, q) \in \omega_{(\bar{x}_1, \bar{x}_2)}$, then from (2.61) it follows that $V(p, q) = V_0$. By the invariance of $\omega_{(\bar{x}_1, \bar{x}_2)}$ we have $V(pt, qt) = V(p, q)$ for all $t \in \mathbb{T}$, and, according to Condition 3 of the theorem, $p = q$, i.e.,

$$\omega_{(\bar{x}_1, \bar{x}_2)} \subseteq \Delta_X := \{(x, x) \mid x \in X\}. \tag{2.62}$$

We may suppose that the sequences $\{\bar{x}_i t_n\}$ $(i = 1, 2)$ are convergent. Let us put $\bar{p} = \lim\limits_{n \to +\infty} \bar{x}_1 t_n$ and $\bar{q} = \lim\limits_{n \to +\infty} \bar{x}_2 t_n$, then $(\bar{p}, \bar{q}) \in \omega_{(\bar{x}_1, \bar{x}_2)}$ and from (2.60) it follows that $\bar{p} \ne \bar{q}$. The last equality contradicts the inclusion (2.62). The obtained contradiction proves the required statement. By Theorem 2.7, the non-autonomous dynamical system (2.1) is convergent.

Conversely, let $\langle (X, \mathbb{T}_1, \pi), (Y, \mathbb{T}_2, \sigma), h \rangle$ be convergent. We will define the function $V : X \dot\times X \to \mathbb{R}_+$ by the equality

$$V(x_1, x_2) := \sup\{\rho(x_1 t, x_2 t) \mid t \geq 0\}. \tag{2.63}$$

It is clear that this function satisfies Conditions 1 and 2 of the theorem. Let us show that V satisfies Condition 3 of the theorem. Let $V(x_1 t, x_2 t) = V(x_1, x_2)$ for all $t \geq 0$. If we suppose that $x_1 \neq x_2$, then $\delta := V(x_1, x_2) > 0$. In view of Theorem 2.7, $\rho(x_1 t, x_2 t) \to 0$ as $t \to +\infty$. Therefore, there exists $t_0 = t_0\,(\delta) > 0$ such that $\rho(x_1 t, x_2 t) < \frac{\delta}{2}$ for all $t \geq t_0$. Hence,

$$V(x_1 t_0, x_2 t_0) = \sup\{\rho(x_1 t, x_2 t) \mid t \geq t_0\} \leq \frac{\delta}{2} < V(x_1, x_2).$$

The last inequality contradicts our assumption. We will prove that for every compact $K \subseteq X$

$$\lim_{t \to +\infty} \sup\{\rho(x_1 t, x_2 t) \mid x_1, x_2 \in K, h(x_1) = h(x_2)\} = 0. \tag{2.64}$$

In fact, if suppose that it is not true, then there exit $\varepsilon_0 > 0$, compact $K \subseteq X$, $\delta_n \downarrow 0$ and sequences $\{x_n^i\} \subseteq K$ ($i = 1, 2$ and $h(x_n^1) = h(x_n^2)$) $t_n \to +\infty$ such that Condition (2.46) holds. Without loss of generality we may suppose that the sequences $\{x_n^i\}$ and $\{x_n^i t_n\}$ ($i = 1, 2$) are convergent. Let $\lim\limits_{n \to +\infty} x_n^i = x^i$ ($i = 1, 2$). We put $\bar{x}^i := \lim\limits_{n \to +\infty} x_n^i t_n$ ($i = 1, 2$). Note that $h(\bar{x}_1) = h(\bar{x}_2) = y$, $\bar{x}_1, \bar{x}_2 \in \omega(K) \subseteq J_X$ and, hence, $\bar{x}_1 = \bar{x}_2$. The last equality contradicts (2.46). Thus, the equality (2.64) is proved. From the equalities (2.63) and (2.64), it follows that

$$V(x_1, x_2) = \rho(x_1 \tau, x_2 \tau) \tag{2.65}$$

for some $\tau(x_1, x_2) \in [0, l(K)]$.

Let $(x_n^1, x_n^2) \to (x^1, x^2) \in X \dot\times X$. The sequence $\{V(x_n^1, x_n^2)\} = \{\rho(x_n^1 \tau_n, x_n^2 \tau_n)\}$ is bounded, where $\tau_n = \tau(x_n^1, x_n^2)$. We will show that it has a unique limit point.

Denote by $\tilde{V}$ one of the limit points of the sequence $\{V(x_n^1, x_n^2)\}$. There exists a subsequence $\{V(x_{k_n}^1, x_{k_n}^2)\}$ such that $V(x_{k_n}^1, x_{k_n}^2) \to \tilde{V}$ as $n \to +\infty$. Since $\{\tau_{k_n}\}$ is bounded, then we may suppose that it is convergent. Then $\tilde{V} = \rho(x^1 \tau', x^2 \tau')$, where $\tau' = \lim\limits_{n \to +\infty} \tau_{k_n}$. Let us show that $\rho(x^1 \tau, x^2 \tau) = \rho(x_1 \tau', x^2 \tau')$ ($\tau = \tau(x^1, x^2) \geq 0$ is chosen out of Condition (2.65)). If we suppose that $\rho(x^1 \tau, x^2 \tau) \neq \rho(x_1 \tau', x^2 \tau')$, then $\rho(x^1 \tau, x^2 \tau) > \rho(x_1 \tau', x^2 \tau')$. Let $\varepsilon > 0$ be such that $\rho(x^1 \tau, x^2 \tau) + 2\varepsilon < \rho(x^1 \tau, x^2 \tau)$. Then for a sufficiently large $k \in \mathbb{N}$ we have $|\rho(x_{k_n}^1 \tau, x_{k_n}^2 \tau) - \rho(x^1 \tau, x^2 \tau)| < \varepsilon$ and $|\rho(x_{k_n}^1 \tau_{k_n}, x_{k_n}^2 \tau_{k_n}) - \rho(x^1 \tau', x^2 \tau')| < \varepsilon$, and, consequently,

$$\rho(x_{k_n}^1 \tau, x_{k_n}^2 \tau) > \rho(x_{k_n}^1 \tau_{k_n}, x_{k_n}^2 \tau_{k_n}) = V(x_{k_n}^1, x_{k_n}^2). \tag{2.66}$$

The inequality (2.66) contradicts the choice of the number τ_{k_n}. Thus, $V(x^1, x^2) = \rho(x^1\tau', x^2\tau')$ and, consequently, $\lim\limits_{n \to +\infty} V(x_n^1, x_n^2) = V(x^1, x^2)$, i.e., the function V is continuous. The theorem is proved. $\qquad\square$

Theorem 2.16 is a generalization for the abstract non-autonomous dynamical systems of the theorems 7.2 and 7.3 from [270].

Theorem 2.17 *Let dynamical systems $(X, \mathbb{T}_1, \pi)$ and $(Y, \mathbb{T}_2, \sigma)$ be compactly dissipative. A non-autonomous dynamical system $\langle (X, \mathbb{T}_1, \pi), (Y, \mathbb{T}_2, \sigma), h \rangle$ is convergent if and only if there exists a continuous function $V : X \dot\times X \to \mathbb{R}_+$, satisfying the following conditions:*

(1) V is positively defined;
(2) $V(x_1 t, x_2 t) < V(x_1, x_2)$ for all $t > 0$ and $(x_1, x_2) \in X \dot\times X \setminus \Delta_X$, where $\Delta_X := \{(x, x) \mid x \in X\}$.

Proof. The sufficiency of Conditions of the theorem it follows from the previous theorem. It is sufficient to remark that from the second condition of Theorem 2.17 it follows the second and third conditions of Theorem 2.60.

Necessity. Let $\langle (X, \mathbb{T}_1, \pi), (Y, \mathbb{T}_2, \sigma), h \rangle$ be convergent and the function $\mathcal{V} : X \dot\times X \to \mathbb{R}_+$ be defined by the equality (2.63). Then it is continuous and satisfies Conditions 1–3 of Theorem 2.16. Let us put

$$V(x_1, x_2) := \int_0^{+\infty} \mathcal{V}(x_1 t, x_2 t) e^{-t} dt. \qquad (2.67)$$

From the definition of the function V follow its continuity and positive definiteness. The function V satisfies Condition 2 of Theorem 2.17. In fact, if we suppose the contrary, then there exit $(\bar{x}_1, \bar{x}_2) \in X \dot\times X$ and $t_0 > 0$ such that $V(\bar{x}_1 t_0, \bar{x}_2 t_0) = V(\bar{x}_1, \bar{x}_2)$ and $\bar{x}_1 \neq \bar{x}_2$. Then from (2.67) and from the last equality, it follows that

$$\mathcal{V}(\bar{x}_1(t_0 + t), \bar{x}_2(t_0 + t)) = \mathcal{V}(\bar{x}_1 t, \bar{x}_2 t) \qquad (2.68)$$

for all $t \geq 0$. By virtue of (2.68), the function $\varphi(t) = \mathcal{V}(\bar{x}_1 t, \bar{x}_2 t)$ is t_0 periodic. It is obvious that φ is continuous and non-increasing. Therefore, φ is stationary and, hence, $\mathcal{V}(\bar{x}_1 t, \bar{x}_2 t) = \mathcal{V}(\bar{x}_1, \bar{x}_2)$ for all $t \geq 0$. Since the function $\mathcal{V}$ satisfies Conditions 1–3 of Theorem 2.16, then the equality $\mathcal{V}(\bar{x}_1 t, \bar{x}_2 t) = \mathcal{V}(\bar{x}_1, \bar{x}_2)$ for all $t \geq 0$ implies $\bar{x}_1 = \bar{x}_2$. The last equality contradicts the choice of $(\bar{x}_1, \bar{x}_2)$. The obtained contradiction completes the proof of the theorem. $\qquad\square$

Theorem 2.18 *Let $(X, \mathbb{T}_1, \pi)$ and $(Y, \mathbb{T}_2, \sigma)$ be compactly dissipative and there exist a continuous function $V : X \dot\times X \to \mathbb{R}_+$, satisfying the following conditions:*

(1) V is positively defined;

(2) $V(x_1t, x_2t) \leq \omega(V(x_1, x_2), t)$ *for all* $(x_1, x_2) \in X \dot\times X$ *and* $t \geq 0$, *where* $\omega :$ $\mathbb{R}_+ \times \mathbb{R}_+ \to \mathbb{R}_+$ *is a non-increasing in first variable function and* $\omega(x, t) \to 0$, *as* $t \to +\infty$, *for every* $x \in \mathbb{R}_+$.

Then $\langle (X, \mathbb{T}_1, \pi), (Y, \mathbb{T}_2, \sigma), h \rangle$ *is convergent.*

Proof. According to Corollary 2.2 for the convergence of $\langle (X, \mathbb{T}_1, \pi), (Y, \mathbb{T}_2, \sigma), h \rangle$ it is sufficient to show that the equality (2.42) holds for every compact $K \in C(X)$. First of all, we will show that the equality

$$\lim_{t \to +\infty} \sup_{(x_1, x_2) \in K \dot\times K} V(x_1t, x_2t) = 0 \tag{2.69}$$

holds for all compact $K \in C(X)$. Indeed, since K is compact, then there exists $\alpha > 0$ such that $V(x_1, x_2) \leq \alpha$ for all $(x_1, x_2) \in K \dot\times K$ and $V(\bar{x}_1, \bar{x}_2) = \alpha$ for some $(\bar{x}_1, \bar{x}_2) \in K \dot\times K$, and, consequently,

$$V(x_1t, x_2t) \leq \omega(V(x_1, x_2), t) \leq \omega(\alpha, t). \tag{2.70}$$

Since $\omega(\alpha, t) \to 0$ as $t \to +\infty$, then from (2.70) it follows (2.69). Let us show that (2.69) implies (2.42). If we suppose the contrary, then there exit $K \in C(X)$, $\varepsilon_0 > 0$, $\{x_n^i\} \subseteq K$ $(i = 1, 2)$ and $t_n \to +\infty$ such that

$$\rho(x_n^1 t_n, x_n^2 t_n) \geq \varepsilon_0. \tag{2.71}$$

In view of the compact dissipativity of $(X, \mathbb{T}_1, \pi)$ we may suppose that the sequences $\{x_n^i t_n\}$ $(i = 1, 2)$ are convergent. Let us set $\bar{x}_i = \lim_{n \to +\infty} x_n^i t_n$, then from the inequality (2.71), it follows that $\bar{x}_1 \neq \bar{x}_2$. On the other hand, by the inequality (2.70)

$$0 \leq V(x_n^1 t_n, x_n^2 t_n) \leq V(\alpha, t_n) \to 0$$

as $n \to +\infty$ and, hence,

$$V(\bar{x}_1, \bar{x}_2) = \lim_{n \to +\infty} V(x_n^1 t_n, x_n^2 t_n) = 0.$$

From the last equality, it follows that $\bar{x}_1 = \bar{x}_2$. The obtained contradiction completes the proof of the theorem. $\qquad\square$

Remark 2.7 *a. Condition 2 of Theorem 2.18 holds, if the function* $V : X \dot\times X \to \mathbb{R}_+$ *satisfies one of the following conditions:*

(1) $V(x_1t, x_2t) \leq \mathcal{N}e^{-\nu t}V(x_1, x_2)$ *for all* $t \geq 0$ *and* $(x_1, x_2) \in X \dot\times X$ *($\omega(x, t) = \mathcal{N}xe^{-\nu t}$);*

(2) $V(x_1t, x_2t) \leq \frac{2V(x_1, x_2)}{2 + V(x_1, x_2)t}$ *for all* $t \geq 0$ *and* $(x_1, x_2) \in X \dot\times X$ *($\omega(x, t) = \frac{2x}{2 + xt}$);*

b. All results of this section are true also in the case when the spaces X and Y are not metric but pseudo-metric.

In conclusion, we note that convergent systems are in some sense the simplest dissipative dynamical systems. If $\langle (X, \mathbb{T}_1, \pi), (Y, \mathbb{T}_2, \sigma), h \rangle$ is a convergent non-autonomous dynamical system and J_X (J_Y) is a Levinson center of the dynamical system $(X, \mathbb{T}_1, \pi)$ $((Y, \mathbb{T}_2, \sigma))$, then J_X and J_Y are homeomorphic. Although the center of Levinson of a convergent system can by completely described, it may be sufficiently complicated. We will give an example which illustrates the above cooment.

Example 2.2   Let $Y := \mathbb{R}$ and $(Y, \mathbb{Z}_+, \sigma)$ be a cascade generated by positive powers of the odd function g, defined on $\mathbb{R}_+$ in the following way:

$$g(y) = \begin{cases} -2y & , \quad 0 \le y \le \frac{1}{2} \\ 2(y-1) & , \quad \frac{1}{2} < y \le 1 \\ \frac{1}{2}(y-1) & , \quad 1 < y < +\infty. \end{cases}$$

It is easy to check that $(Y, \mathbb{Z}_+, \sigma)$ is dissipative and $J_Y \subseteq [-1, 1]$. Let us put $X := \mathbb{R}^2$ and denote by $(X, \mathbb{Z}_+, \pi)$ a cascade generated by the positive powers of the mapping $P : \mathbb{R}^2 \to \mathbb{R}^2$

$$P \begin{pmatrix} u \\ y \end{pmatrix} = \begin{pmatrix} f(u, y) \\ g(y) \end{pmatrix}, \tag{2.72}$$

where $f(u, y) := \frac{1}{10}u + \frac{1}{2}y$. Finally, let $h = pr_2 : X \to Y$. From (2.72), it follows that h is a homomorphism of $(X, \mathbb{Z}_+, \pi)$ onto $(Y, \mathbb{Z}_+, \sigma)$ and, consequently, $\langle (X, \mathbb{Z}_+, \pi), (Y, \mathbb{Z}_+, \sigma), h \rangle$ is a non-autonomous dynamical system. Note that

$$|(u_1, y) - (u_2, y)| = |u_1 - u_2| = 10|P(u_1, y) - P(u_2, y)|. \tag{2.73}$$

From (2.73), it follows that

$$|P^n(u_1, y) - P^n(u_2, y)| \le \mathcal{N}e^{-\nu n}|\langle u_1, y \rangle - \langle u_2, y \rangle| \tag{2.74}$$

for all $n \in \mathbb{Z}_+$, where $\mathcal{N} = 1$ and $\nu = \ln 10$. We will show that the dynamical system $\langle (X, \mathbb{Z}_+, \pi), (Y, \mathbb{Z}_+, \sigma), h \rangle$ is convergent. It is possible to check directly that the dynamical system $(X, \mathbb{Z}_+, \pi)$ is dissipative; therefore, for convergence of $\langle (X, \mathbb{Z}_+, \pi), (Y, \mathbb{Z}_+, \sigma), h \rangle$ it is sufficient to show that $J_X \bigcap X_y$ contains only one point for all $y \in J_Y$. If we suppose that it is not true, then there exit $y_0 \in J_Y$ and $x_1, x_2 \in J_X \bigcap X_{y_0}$ with $x_1 \ne x_2$. Let φ_i be a motion of the dynamical system $(X, \mathbb{Z}_+, \pi)$ defined on $\mathbb{Z}$ and passing through the point x_i $(i = 1, 2)$. From (2.74), it follows that

$$|\varphi_1(-n) - \varphi_2(-n)| \geq \mathcal{N}^{-1} e^{\nu n} |u_1 - u_2| \qquad (2.75)$$

for all $n \in \mathbb{Z}_+$, where $(u_i, y_0) = x_i$ $(i = 1, 2)$. On the other hand,

$$\sup_{n \in \mathbb{Z}} |\varphi_1(n) - \varphi_2(n)| < +\infty. \qquad (2.76)$$

The inequalities (2.75) and (2.76) are contradictory. The obtained contradiction proves the required statement. Thus, the dynamical system $\langle (X, \mathbb{Z}_+, \pi),$ $(Y, \mathbb{Z}_+, \sigma), h \rangle$ is convergent and, consequently, J_X and J_Y are homeomorphic.

Note that this example is an just a slight modification of an example from [297, p.39–42]. In addition, they have the same attractors and the corresponding systems on the attractors act equally. Hence, the Levinson centers J_X and J_Y are intermixing (strange) attractors.

Thus, the Levinson center J_X of the convergent dynamical system $\langle (X, \mathbb{T}_1, \pi),$ $(Y, \mathbb{T}_2, \sigma), h \rangle$ is completely defined by the structure of J_Y, but the last one can be organized in a very complicated way.

2.6 Global attractors of non-autonomous dynamical systems

Let Y be a compact metric space, (X, h, Y) be a locally trivial Banach fibering [190] and $|\cdot|$ be a norm on (X, h, Y) coordinated with the metric ρ on X (that is $\rho(x_1, x_2) = |x_1 - x_2|$ for any $x_1, x_2 \in X$ such that $h(x_1) = h(x_2)$).

Theorem 2.19 *Let* $\langle (X, \mathbb{T}_1, \pi), (Y, \mathbb{T}_2, \sigma), h \rangle$ *be a non-autonomous dynamical system and for any bounded set* $M \in \mathbb{B}(X)$ *there exists* $l = l(M) > 0$ *such that* $\pi^l(M)$ *would be relatively compact (that is the dynamical system* $(X, \mathbb{T}_1, \pi)$ *is completely continuous). Then the following conditions are equivalent:*

1. *there exits a positive number* r *such that for any* $x \in X$ *there will be* $\tau = \tau(x) \geq 0$ *for which* $|x\tau| < r$;
2. *the dynamical system* $\langle (X, \mathbb{T}_1, \pi), (Y, \mathbb{T}_2, \sigma), h \rangle$ *is compactly dissipative and*

$$\lim_{t \to +\infty} \sup_{|x| \leq R} \rho(xt, J) = 0 \qquad (2.77)$$

for any $R > 0$, *where* J *is the Levinson center of* $(X, \mathbb{T}_1, \pi)$, *that is the non-autonomous system* $\langle (X, \mathbb{T}_1, \pi), (Y, \mathbb{T}_2, \sigma), h \rangle$ *admits a compact global attractor.*

Proof. That Condition 2 implies Condition 1 is evident. Let us show that under conditions of Theorem 2.19, the converse also holds. Suppose that $A(r) := \{x \in E \mid |x| \leq r\}$ where $r > 0$ is the number appearing in Condition 1. As Y is compact and the Banach fibering (X, h, Y) is locally trivial, then its null section $\Theta = \{\theta_y \mid y \in Y$, where θ_y is the null element of the fiber $E_y := h^{-1}(y)\}$ is compact and, hence, the set $A(r)$ is bounded, as $A(r) \subseteq S(\Theta, r) = \{x \in E \mid \rho(x, \theta) \leq r\}$. According to the condition of the theorem for a bounded set M there exits a positive number l such that $\pi^l M$ is relatively compact. Let $x \in M$ and $\tau = \tau(x) \geq 0$ be such that $x\tau \in M$, then $x(\tau + l) \in K := \overline{\pi^l M}$. Thus, the non-empty compact K is a weak attractor of the system $(X, \mathbb{T}_1, \pi)$ and, according to Theorem 1.10, the dynamical system $(X, \mathbb{T}_1, \pi)$ is compactly dissipative. Let J be the Levinson center of $(X, \mathbb{T}_1, \pi)$ and $R > 0$, then the set $A(R) := \{x \in E \mid |x| \leq R\}$, as it was noticed above, is bounded, and for it there exists a number $l > 0$ such that $\pi^l A(R)$ is relatively compact and since $(X, \mathbb{T}_1, \pi)$ is compactly dissipative, then its Levinson center J, according to Theorem 1.6, attracts the set $\pi^l A(R)$ and, hence, the equality (2.77) holds. The theorem is proved. $\qquad\square$

Corollary 2.7   *Let $\langle (X, \mathbb{T}_1, \pi), (Y, \mathbb{T}_2, \sigma), h \rangle$ be a non-autonomous dynamical system and suppose that the vector fibering of $(X, \mathbb{T}_1, \pi)$ is finite-dimensional, then Conditions 1. and 2. of Theorem 2.19 are equivalent .*

Proof.   This assertion it follows from Theorem 2.19 as for any $r > 0$ the set $\{x \in E \mid |x| \leq r\}$ is compact, if the vector fibering (X, h, Y) is finite-dimensional and Y is compact. $\qquad\square$

Recall that the dynamical system $(X, \mathbb{T}_1, \pi)$ is called asymptotically compact, if for any bounded closed positively invariant set $M \in \mathbb{B}(X)$ there exits a non-empty compact K such that the equality

$$\lim_{t \to +\infty} \beta(\pi^t M, K) = 0$$

holds.

Theorem 2.20   *Let $\langle (X, \mathbb{T}_1, \pi), (Y, \mathbb{T}_2, \sigma), h \rangle$ be a non-autonomous dynamical system and for every bounded set $A \subseteq X$ there exists a nonempty compact $K_A \subseteq X$ such that*

$$\lim_{t \to +\infty} \beta(\pi^t A, K_A) = 0. \tag{2.78}$$

Then the conditions 1 and 2 of Theorem 2.19 are equivalent.

Proof. Since Y is compact and (X, h, Y) is locally trivial, then for any $R > 0$ the set $\{x \in E \mid |x| \leq R\}$ is bounded. According to Condition 1 of Theorem 2.19, for any $x \in E$ there exits $\tau = \tau(x) \geq 0$ such that $x\tau \in A(r) := \{x \in E \mid |x| \leq r\}$. According to Theorem 1.24 the dynamical system $(X, \mathbb{T}_1, \pi)$ is compactly dissipative. Let J be the Levinson center of $(X, \mathbb{T}_1, \pi)$ and $R > 0$. As the set $M := A(R) := \{x \in$

$E \mid |x| \le R\}$ is bounded, then according to Condition of the theorem and of Lemma 1.3 the set $\Omega(M) \ne \emptyset$, is compact, invariant and the equality

$$\lim_{t \to +\infty} \beta(\pi^t A, \omega(A)) = 0$$

holds. As J is the maximal compact invariant set in $(X, \mathbb{T}_1, \pi)$, then $\Omega(M) \subseteq J$ and, hence, the equality

$$\lim_{t \to +\infty} \beta(\pi^t A, J) = 0$$

holds. The theorem is proved. $\qquad\square$

Corollary 2.8 *Let $\langle(X, \mathbb{T}_1, \pi), (Y, \mathbb{T}_2, \sigma), h\rangle$ be a non-autonomous dynamical system and suppose that at least one of the following conditions is fulfilled:*

(1) for every bounded set $A \subseteq X$ there exits a positive number l such that $\pi^l A$ is relatively compact;

(2) the dynamical system $\langle(X, \mathbb{T}_1, \pi)$ is asymptotically compact.

Then Conditions 1 and 2 of Theorem 2.19 are equivalent.

Proof. This assertion it follows directly from Theorems 2.19 and 2.20 . $\qquad\square$

Theorem 2.21 *Let $\langle(X, \mathbb{T}_1, \pi), (Y, \mathbb{T}_2, \sigma), h\rangle$ be a non-autonomous dynamical system and $(X, \mathbb{T}_1, \pi)$ be asymptotically compact, then the following conditions are equivalent:*

1. there exits a positive number R_0 and for any $R > 0$ there exists $l(R) > 0$ such that

$$|\pi^t x| \le R_0 \tag{2.79}$$

for all $t \ge l(R)$ and $|x| \le R$;

2. the dynamical system $\langle(X, \mathbb{T}_1, \pi), (Y, \mathbb{T}_2, \sigma), h\rangle$ admits a compact global attractor, i.e it is compactly dissipative and for its Levinson center J the equality (2.77) holds for any $R > 0$.

Proof. Evidently Condition 2 implies Condition 1, that is why for proving the theorem it is sufficient to show that from Condition 1 follows Condition 2. Let $M_0 \in B(X)$, then there exits $R > 0$ such that $M_0 \subseteq A(R) := \{x \in E \mid |x| \le R\}$. According to Condition 1, for a given R there exists $l = l(R) > 0$ such that (2.79) holds and, in particular, the set $M := \bigcup\{\pi^t M_0 \mid t \ge l(R)\}$ is bounded and positively invariant. As $(X, \mathbb{T}_1, \pi)$ is asymptotically compact, for the set M there exists a nonempty compact K for which the equality (2.78) holds. To finish the proof of the theorem it is sufficient to cite Theorem 2.20. The theorem is proved. $\qquad\square$

Theorem 2.22 *Let $\langle (X, \mathbb{T}_1, \pi), (Y, \mathbb{T}_2, \sigma), h \rangle$ be a non-autonomous dynamical system and let mappings $\pi^t := \pi(t, \cdot) : X \to X (t \in \mathbb{T}_1)$ be represented like a sum $\pi(t, x) = \varphi(t, x) + \psi(t, x)$ for all $t \in \mathbb{T}_1$ and $x \in X$ Let also the following conditions be fulfilled:*

(1) $|\varphi(t, x)| \le m(t, r)$ for all $t \in \mathbb{T}_1, r > 0$ and $|x| \le r$, where $m : \mathbb{T}_1 \times \mathbb{R}_+ \to \mathbb{R}_+$ and $m(t, r) \to 0$ as $t \to +\infty$;

(2) the mappings $\psi(t, \cdot) : X \to X (t > 0)$ are conditionally completely continuous, that is $\psi(t, A)$ is relatively compact for any $t > 0$ and a bounded positively invariant set $A \subseteq X$.

Then the dynamical system $(X, \mathbb{T}_1, \pi)$ is asymptotically compact.

Proof. Let $A \subseteq X$ be a bounded set such that $\Sigma^+(A) := \bigcup \{ \pi^t A \mid t \ge 0 \}$ is also bounded, $r > 0$ and $A \subseteq \{ x \in X \mid |x| \le r \}$. Let us show that for any $\{ x_k \} \subseteq A$ and $t_k \to +\infty$, the sequence $\{ x_k t_k \}$ is relatively compact. We will convince ourselves that the set $M := \{ x_k t_k \}$ may be covered by a compact ε net for any $\varepsilon > 0$. Let $\varepsilon > 0$ and $l > 0$ be such that $m(l, r) < \varepsilon/2$ and let us represent M in the form of the union $M_1 \cup M_2$ where $M_1 := \{ x_k t_k \}_{k=1}^{k_1}$, $M_2 := \{ x_k t_k \}_{k=k_1+1}^{+\infty}$ and $k_1 := \max \{ k \mid t_k < l \}$. The set M_2 is a subset of the set $\pi^l(\Sigma^+(A))$, the elements of which we can represent in the form of $\varphi(l, x) + \psi(l, x)(x \in \Sigma^+(A))$. As the set $\psi(\Sigma^+(A), l)$ is relatively compact, then it may be covered by a finite $\varepsilon/2$ net. Let us notice that for any $y \in \varphi(l, \Sigma^+(A))$ there exits $x \in \Sigma^+(A)$ such that $y = \varphi(l, x)$ and $|y| = |\varphi(l, x)| \le m(l, r) < \varepsilon/2$. That is why the null section Θ of the fibering of (X, h, Y) is an $\varepsilon/2$ net of the set $\varphi(l, \Sigma^+(A))$. Thus, M_2, and, subsequently, M is covered by a compact ε net and as the space X is complete, then the set $M = \{ x_k t_k \}$ is relatively compact. Now to finish the proof of the theorem it is sufficient to cite Lemma 1.3. The theorem is proved. $\square$

At the end of this section we will study dissipative dynamical systems in finite-dimensional space. We will give some conditions which are equivalent to dissipativity.

Let $\langle (X, \mathbb{T}_1, \pi), (Y, \mathbb{T}_2, \sigma), h \rangle$ be a non-autonomous dynamical system, Y be compact, (X, h, Y) be a finite-dimensional vectorial fibering [190] and $|\cdot|$ be a Riemannian metric on (X, h, Y).

Lemma 2.14 *Under the conditions mentioned above all the types of dissipativity of the non-autonomous dynamical system $\langle (X, \mathbb{T}_1, \pi), (Y, \mathbb{T}_2, \sigma), h \rangle$ are equivalent.*

Proof. Since the space Y is compact and (X, h, Y) is finite-dimensional and locally trivial [190], then the space X is locally compact and possesses the Heine-Borel property. To finish the proof of the lemma it is sufficient to cite Theorem 1.10. $\square$

Theorem 2.23 *Let* $\langle(X,\mathbb{T}_1,\pi),(Y,\mathbb{T}_2,\sigma),h\rangle$ *be a finite-dimensional non-autonomous dynamical system, then the following conditions are equivalent:*

1. *there exits a positive number R such that for all $x \in X$ the following inequality*

$$\limsup_{t\to+\infty}|xt| < R \qquad (2.80)$$

holds;

2. *there exits a positive number r such that for all $x \in X$ there exists $\tau \geq 0$ such that $|x\tau| < r$;*

3. *there exists a nonempty compact $K_1 \subset X$ such that $\omega_x \bigcap K_1 \neq \emptyset$ for any $x \in X$;*

4. *there exists a nonempty compact $K_2 \subset X$ such that $\emptyset \neq \omega_x \subset K_2$ for every $x \in X$;*

5. *there exits a positive number R_0 such that for all $R_1 > 0$ there exits $l(R_1) > 0$ such that*

$$|xt| < R_0 \qquad (2.81)$$

for all $t \geq l(R_1)$ and $|x| \leq R_1$.

Proof. It is clear that $5. \Rightarrow 1. \Rightarrow 4. \Rightarrow 3. \Rightarrow 2.$ According to Theorem 2.19 from 2. it follows 5. The theorem is completely proved. $\qquad\square$

2.7 Global attractor of cocycles

Definition 2.18 The family $\{I_y \mid y \in Y\}$ $(I_y \subset W)$ of nonempty compact subsets W is called (see, for example, [6] and [153]) a compact pullback attractor (uniform pullback attractor) of a cocycle φ, if the following conditions hold:

(1) the set $I := \bigcup\{I_y \mid y \in Y\}$ is relatively compact;
(2) the family $\{I_y \mid y \in Y\}$ is invariant with respect to the cocycle φ, i.e. $\varphi(t, I_y, y) = I_{\sigma(t,y)}$ for all $t \in \mathbb{T}_+$ and $y \in Y$;
(3) for all $y \in Y$ (uniformly in $y \in Y$) and $K \in C(W)$

$$\lim_{t\to+\infty}\beta(\varphi(t,K,y_{-t}),I_y) = 0,$$

where $\beta(A,B) := \sup\{\rho(a,B) : a \in A\}$ is a semi-distance of Hausdorff.

Below in this section we suppose that $\mathbb{T}_2 = \mathbb{S}$.

Definition 2.19 The family $\{I_y \mid y \in Y\}(I_y \subset W)$ of nonempty compact subsets is called a compact global attractor of the cocycle φ, if the following conditions are fulfilled:

(1) the set $I := \bigcup\{I_y \mid y \in Y\}$ is relatively compact;

(2) the family $\{I_y \mid y \in Y\}$ is invariant with respect to the cocycle φ;

(3) the equality

$$\lim_{t \to +\infty} \sup_{y \in Y} \beta(\varphi(t, K, y), I) = 0$$

holds for every $K \in C(W)$.

Let $M \subseteq W$ and

$$\omega_y(M) := \bigcap_{t \geq 0} \overline{\bigcup_{\tau \geq t} \varphi(\tau, M, \sigma^{-\tau}y)} \tag{2.82}$$

for all $y \in Y$, where $y^{-t} := \sigma(-t, y)$.

Lemma 2.15 *The following assertions hold:*

(1) The point $p \in \omega_y(M)$ if and only if there exit $t_n \to +\infty$ and $\{x_n\} \subseteq M$ such that $p = \lim_{n \to +\infty} \varphi(t_n, x_n, y^{-t_n})$;

(2) $U(t, y)\omega_y(M) \subseteq \omega_{yt}(M)$ for all $y \in Y$ and $t \in \mathbb{T}_+$, where $U(t, y) := \varphi(t, \cdot, y)$;

(3) for any point $w \in \omega_y(M)$ the motion $\varphi(t, w, y)$ is defined on $\mathbb{S}$;

(4) if there exits a nonempty compact $K \subset W$ such that

$$\lim_{t \to +\infty} \beta(\varphi(t, M, y^{-t}), K) = 0,$$

then $\omega_y(M) \neq \emptyset$, is compact,

$$\lim_{t \to +\infty} \beta(\varphi(t, M, y^{-t}), \omega_y(M)) = 0 \tag{2.83}$$

and

$$U(t, y)\omega_y(M) = \omega_{yt}(M) \tag{2.84}$$

for all $y \in Y$ and $t \in \mathbb{S}_+$.

Proof. The first assertion of the lemma directly it follows from the equality (2.82).

Let $w \in \omega_y(M)$, then there exit $t_n \to +\infty$ and $x_n \subseteq M$ such that $w = \lim_{n \to +\infty} \varphi(t_n, x_n, y^{-t_n})$ and, hence,

$$\varphi(t, w, y) = \lim_{n \to +\infty} \varphi(t, \varphi(t_n, x_n, y^{-t_n}), y) = \lim_{n \to +\infty} \varphi(t + t_n, x_n, y^{-t_n}). \tag{2.85}$$

Thus, $\varphi(t, w, y) \in \omega_{yt}(M)$, that is $U(t, y)\omega_y(M) \subseteq \omega_{yt}(M)$ for all $y \in Y$ and $t \in \mathbb{T}_+$.

From the equality (2.85), it follows that the motion $\varphi(t, w, y)$ is defined on $\mathbb{S}$ like $\varphi(t + t_n, x_n, y^{-t_n})$ is defined on $[-t_n, +\infty)$ and $t_n \to +\infty$.

The fourth assertion of the lemma is proved as in Theorem 2.4 and Lemma 1.1.3 from [217]. $\square$

Definition 2.20 A cocycle φ over $(Y, \mathbb{S}, \sigma)$ with the fiber W is said to be compactly dissipative, if there exits a nonempty compact $K \subseteq W$ such that

$$\lim_{t \to +\infty} \sup\{\beta(U(t, y)M, K) \mid y \in Y\} = 0 \tag{2.86}$$

for any $M \in C(W)$.

Lemma 2.16 *Let Y be compact and $\langle W, \varphi, (Y, \mathbb{S}, \sigma)\rangle$ be a cocycle over $(Y, \mathbb{S}, \sigma)$ with the fiber W. For $\langle W, \varphi, (Y, \mathbb{S}, \sigma)\rangle$ to be compactly dissipative, it is necessary and sufficient that the semigroup autonomous system $(X, \mathbb{S}_+, \pi)$ should be compactly dissipative.*

Proof. This assertion follows from the corresponding definitions. $\square$

Theorem 2.24 *Let Y be compact, $\langle W, \varphi, (Y, \mathbb{S}, \sigma)\rangle$ be compactly dissipative and K be the nonempty compact subset of W appearing in the equality (2.86), then:*

1. $I_y = \omega_y(K) \neq \emptyset$, *is compact,* $I_y \subseteq K$ *and*

$$\lim_{t \to +\infty} \beta(U(t, y^{-t})K, I_y) = 0$$

   for every $y \in Y$;
2. $U(t, y)I_y = I_{yt}$ *for all $y \in Y$ and $t \in \mathbb{S}_+$;*
3.

$$\lim_{t \to +\infty} \beta(U(t, y^{-t})M, I_y) = 0 \tag{2.87}$$

   for all $M \in C(W)$ and $y \in Y$;
4.

$$\lim_{t \to +\infty} \sup\{\beta(U(t, y^{-t})M, I) \mid y \in Y\} = 0 \tag{2.88}$$

   for any $M \in C(W)$, where $I := \cup\{I_y \mid y \in Y\}$;
5. $I_y = pr_1 J_y$ *for all $y \in Y$, where J is the Levinson center of $(X, \mathbb{T}_+, \pi)$, and hence $I = pr_1 J$;*
6. *the set I is compact;*
7. *the set I is connected if one of the next two conditions is fulfilled:*

   (a) $\mathbb{S}_+ = \mathbb{R}_+$ and the spaces W and Y are connected;
   (b) $\mathbb{S}_+ = \mathbb{Z}_+$ and the space $W \times Y$ possesses the (S)-property or it is connected and locally connected.

Proof. The first two assertions of the theorem follow from Lemma 2.15. If we suppose that the equality (2.87) does not hold, then there exist $\varepsilon_0 > 0$, $y_0 \in Y$, $M_0 \in C(W)$, $\{x_n\} \subseteq M_0$ and $t_n \to +\infty$ such that

$$\rho(U(t_n, y_0^{-t_n})x_n, I_{y_0}) \geq \varepsilon_0. \tag{2.89}$$

According to the equality (2.87), for ε_0 and $y_0 \in Y$ there exists $t_0 = t_0(\varepsilon_0, y_0) > 0$ such that

$$\beta(U(t, y_0^{-t})K, I_{y_0}) < \frac{\varepsilon_0}{2} \tag{2.90}$$

for all $t \geq t_0$. Let us notice that

$$U(t_n, y_0^{-t_n})x_n = U(t_0, y_0^{-t_0})U(t_n - t_0, y_0^{-t_n})x_n.$$

As $\langle W, \varphi, (Y, S, \sigma)\rangle$ is compactly dissipative, then the sequence $\{U(t_n - t_0, y_0^{-t_n})x_n\}$ may be considered convergent. Suppose $\overline{x} = \lim\limits_{n \to +\infty} \varphi(t_n - t_0, x_n, y_0^{-t_n})$, then, according to Lemma 2.15, $\overline{x} \in \omega_{y_0^{-t_0}}(M_0)$ and $U(t_0, y_0^{-t_0})\overline{x} \in \omega_{y_0}(M_0)$. From the equality (2.86), it follows that $\overline{x} \in K$. Passing to limit in (2.89) as $n \to +\infty$, we get

$$U(t_0, y_0^{-t_0})\overline{x} \notin B(I_{y_0}, \varepsilon_0). \tag{2.91}$$

On the other hand, as $\overline{x} \in K$, from (2.90) we have

$$U(t_0, y_0^{-t_0})\overline{x} \in B(I_{y_0}, \frac{\varepsilon}{2}).$$

This contradicts (2.91). The obtained contradiction proves the assertion we need.

Let us prove now the equality (2.88). If we suppose that it does not hold, then there exist $\varepsilon_0 > 0, M_0 \in C(W), y_n \in Y, \{x_n\} \subseteq M_0$ and $t_n \to +\infty$ such that

$$\rho(U(t_n, y_n^{-t_n})x_n, I) \geq \varepsilon_0. \tag{2.92}$$

As Y is compact, then we may consider the sequences $\{y_n\}$ and $\{y_n t_n\}$ convergent. Suppose $y_0 := \lim\limits_{n \to +\infty} y_n$ and $\overline{y} := \lim\limits_{n \to +\infty} y_n t_n$. According to (2.87), for given $\varepsilon_0 > 0$ and $y_0 \in Y$, there exists $t_0 = t_0(\varepsilon_0, y_0)$ such that

$$\beta(U(t_0, y_0^{-t})M_0, I_{y_0}) < \frac{\varepsilon_0}{2} \tag{2.93}$$

for all $t \geq t_0(\varepsilon_0, y_0)$. Let us note that

$$U(t_n, y_n^{-t_n})x_n = U(t_0, y_n^{-t_0})U(t_n - t_0, y_n^{-t_n})x_n. \tag{2.94}$$

As $\langle W, \varphi, (Y, \mathbb{T}, \sigma)\rangle$ is compactly dissipative, then the sequence $\{U(t_n - t_0, y_n^{-t_n})x_n\}$ may be considered a convergent one. Suppose $x' = \lim\limits_{n \to +\infty} \varphi(t_n - t_0, x_n, y_n^{t_n})$ and let us notice that according to (2.93) $x' \in K$. From the equality (2.94), it follows that $U(t_n, y_n^{-t_n})x_n \to U(t_0, y^{-t_0})x'$ and, hence, from (2.92) we have

$$U(t_0, y_0^{-t_0})x' \notin B(I_{y_0}, \varepsilon_0).$$

On the other hand, from (2.93) and from $x' \in K$, it follows that

$$U(t_0, y_0^{-t_0})x' \in B(I_{y_0}, \frac{\varepsilon_0}{2}).$$

This last inclusion contradicts (2.93), and this finishes the proof of the fourth assertion of the theorem.

Let us prove the fifth assertion of the theorem. In order to do it let us notice that $w \in I_y$, if $\varphi(t, w, y)$ is defined on $\mathbb{S}$ and $\varphi(\mathbb{S}, w, y)$ is relatively compact. In fact, as $w = \varphi(t, \varphi(-t, w, y), y^{-t})$ for all $t \in S$, then from the equality (2.87) it follows the inclusion we need. Thus, we get the following description of the set I_y : $I_y = \{w \in W|$ there exists at least one whole trajectory of $\langle W, \varphi, (Y, \mathbb{S}, \sigma)\rangle$, passing through the point $(x, y)\}$. Now it remains to notice that the Levinson's center J is compact and consists of the whole trajectories of $(X, \mathbb{S}_+, \pi)$, and, hence, $pr_1 J_y \subseteq I_y$ for all $y \in Y$.

The compactness of the set I it follows from the equality $I = pr_1 J$, from compactness of J and from continuity of $pr_1 : X \to W$.

The last assertion it follows from the next observation: under the conditions of the theorem Levinson's center J of the dynamical system $(X, \mathbb{S}_+, \pi)$, according to Corollary 1.11 and Theorem 1.33, is connected. Hence, I, as a continuous image of a connected set, also is connected. The theorem is completely proved. $\qquad\square$

Remark 2.8 *Theorem 2.24 refines and generalises the main results of* [110] *and* [217].

Theorem 2.25 *Under the conditions of Theorem 2.24 the following statements hold true:*

(1) $w \in I_y$ $(y \in Y)$ if and only if there exits a whole trajectory $\nu : \mathbb{S} \to W$ of the cocycle φ, satisfying the following conditions: $\nu(0) = w$ and $\nu(\mathbb{S})$ is relatively compact;

(2) I_y $(y \in Y)$ is connected, if the space W possesses the property (S).

Proof. To prove the first assertion we note that the continuous function $\nu : \mathbb{S} \to W$ is a whole trajectory of the cocycle $\langle W, \varphi, (Y, \mathbb{S}, \sigma)\rangle$ if and only if $\gamma = (\nu, Id_Y)$ is a whole trajectory of the semi-group dynamical system $(X, \mathbb{S}, \pi)$ ($X = W \times Y, \pi = (\varphi, \sigma)$). By Lemma 2.16, the dynamical system $(X, \mathbb{S}, \pi)$ is compactly dissipative and according to Theorem 1.6 the set J is compact and invariant and, consequently, the point $(w, y) = x \in J$ if and only if through the point $(w, y) = x$ passes the whole trajectory $\gamma = (\nu, Id_Y)$ of the dynamical system $(X, \mathbb{S}, \pi)$, which belongs to J, i.e. $\gamma(0) = (\nu(0), y) = (w, y)$ and $\gamma(s) \in J$ for all $s \in \mathbb{S}$. To finish the proof of the first statement of the theorem it is sufficient to cite Theorem 2.24 (item 5.).

To prove the second statement of the theorem we note that under the conditions of Theorem 2.24 the set $I_y \neq \emptyset$ and is compact. Since the space W possesses the property (S), then there exists a connected compact $V \supseteq I$. By (2.86), the following equality

$$\lim_{t \to +\infty} \beta(U(t, \sigma^{-t}y)(V), I_y) = 0$$

holds. Note, that $I_y \subseteq U(t, \sigma^{-t}y)(V)$ for all $y \in Y$ and $t \in \mathbb{S}_+$; the mapping $U(t, \sigma^{-t}y) : W \to W$ is continuous and, consequently, the set $U(t, \sigma^{-t}y)(V)$ is compact and connected. To finish the proof of the theorem it is sufficient to cite Lemma 3.12 from [126]. $\qquad\square$

2.8 Global attractors of non-autonomous dynamical system with minimal base

Everywhere in this paragraph we suppose that $\langle (X, \mathbb{T}_1, \pi), (Y, \mathbb{T}_2, \sigma), h \rangle$ is a non-autonomous dynamical system, Y is a compact minimal set and (X, h, Y) is a locally trivial Banach fibering.

Theorem 2.26 *Let the following conditions be satisfied:*

(1) $(X, \mathbb{T}_1, \pi)$ is completely continuous, that is, for any bounded set $A \subseteq X$ there exits $l = l(A) > 0$ such that $\pi^l(A)$ is relatively compact;

(2) all motions $(X, \mathbb{T}_1, \pi)$ are bounded on $\mathbb{T}_+$, that is, $\sup\{|xt| \,|t \in \mathbb{T}_+\} < +\infty$ for any $x \in E$;

(3) there exit y_0 and $R_0 > 0$ such that for any $x \in X_{y_0}$ there exist $\tau = \tau(x) \geq 0$ for which

$$|x\tau| < R_0. \tag{2.95}$$

Then the non-autonomous dynamical system $\langle (X, \mathbb{T}_1, \pi), (Y, \mathbb{T}_2, \sigma), h \rangle$ admits a compact global attractor.

Proof.   Let $R > R_0$, then for any $x \in X$ there exits $\tau = \tau(x) \geq 0$ such that $|x\tau| < R$. If it were not so, then there would be $R' > R_0$ any $x_0^1 \in E$ such that

$$|x_0^1 \tau| > R' \tag{2.96}$$

for all $\tau \geq 0$. As the dynamical system $(X, \mathbb{T}_1, \pi)$ is completely continuous and the motion $\pi(t, x_0^1)$ is bounded on $\mathbb{T}_+$, the point x_0^1 is stable L^+ and, as Y is minimal, the set $\omega_{x_0^1} \cap X_{y_0}$ is nonempty. Thus, according to Condition (2.96) we have

$$|xt| \geq R' \tag{2.97}$$

for all $x \in \omega_{x^1} \cap X_{y_0}$ and $t \geq 0$. The inequality (2.97) contradicts (2.95). This contradiction proves the assertion we need. Now to finish the proof of the theorem it is sufficient to cite Theorem 2.19. $\qquad\square$

Remark 2.9 *1. For finite-dimensional systems (that is the vector fibering (X, h, Y) is finite-dimensional) Theorem 2.26 refines Theorem 2.6.1 from [209]. To be exact, the condition of uniform boundedness is changed for ordinary boundedness of trajectories of $(X, \mathbb{T}_1, \pi)$.*

2. If the condition of minimality of Y in Theorem 2.26 is taken away, then Theorem 2.26 is not true even in the class of linear non-autonomous systems. This is proved by the example below.

Example 2.3 Let us consider a linear differential equation

$$x' = a(t)x, \qquad (2.98)$$

where $a \in C(R, R)$ is defined by the equality $a(t) := -1 + \sin t^{1/3}$. Let us remark the next properties of the function a and of the equation (2.98):

1. $a'(t) \to 0$ for $t \to +\infty$;
2. $a(t) \in [-2, 0]$ for all $t \in R$;
3. $\{a_\tau | \tau \geq 0\}$ is relatively compact in $C(R, R)$, where $a_\tau(t) = a(t + \tau)(t \in R)$;
4. $\omega_a \neq \emptyset$ and is compact ;
5. all functions from ω_a are constant and $b(t) = c \in [-2, 0](t \in R)$ for any $b \in \omega_a$;
6. $a(t_n) = 0$ if and only if $t_n = -1 + (\frac{\pi}{2} + 2\pi n)^2$ $(n \in \mathbb{Z})$;
7. there exits $\{t_{n_k}\} \subseteq \{t_n\}$ such that $a(t + t_{n_k}) \to b(t)$ and $b(t) = 0$ for all $t \in R$;
8. for any $b \in H^+(a) = \overline{\{a_\tau | \tau \in R_+\}}$ the inequality

$$|\varphi(t, x, b)| \leq |x| \qquad (2.99)$$

holds for all $x \in R$ and $t \in R_+$, where $\varphi(t, x, b)$ is the solution of the equation

$$y'(t) = b(t)y; \qquad (2.100)$$

passing through the point $x \in R$ for $t = 0$.
9. if $b \in \omega_a \setminus \{0\}$, then $b(t) = c < 0(t \in R)$ and, hence,

$$\lim_{t \to +\infty} |\varphi(t, x, b)| = 0 \qquad (2.101)$$

for all $x \in R$;
10. if $b = 0(b \in \omega_a)$, then $\varphi(t, x, b) = x$ for all $t \in R$.

Suppose $Y := H^+(a)$ and denote by $(Y, \mathbb{R}_+, \sigma)$ the dynamical system of translations on Y. Let $X := \mathbb{R} \times Y$ and $(X, \mathbb{R}_+, \pi)$ be a semigroup dynamical system on X, where $\pi = (\varphi, \sigma)$ (that is $\pi(t, (x, b)) = (\varphi(t, x, b), b_t)$ for all $(x, b) \in X$ and $t \in \mathbb{R}_+$). Then $\langle (X, \mathbb{R}_+, \pi), (Y, \mathbb{R}_+, \sigma) \rangle$ is a non-autonomous dynamical system generated by the equation (2.98), where $h = pr_2 : X \to Y$. From Properties 1-10 it follows that for the non-autonomous dynamical system $\langle (X, \mathbb{R}_+, \pi), (Y, \mathbb{R}_+, \sigma), h \rangle$ generated by the equation (2.98) all the conditions of Theorem 2.26 are carried out excepting the minimality of Y and that it has no compact global attractor.

Corollary 2.9   *Let $(X, \mathbb{T}_1, \pi)$ be completely continuous and for any $y \in Y$ let there exist $R(y) \geq 0$ such that*

$$\limsup_{t \to +\infty} |xt| \leq R(y) \qquad (2.102)$$

for any $x \in X_y$. Then the non-autonomous dynamical system $\langle (X, \mathbb{T}_1, \pi), (Y, \mathbb{T}_2, \sigma), h \rangle$ admits a compact global attractor.

Proof.   This assertion it follows from Theorem 2.26, if we will notice that from condition (2.102) it follows the boundedness of every motion from $(X, \mathbb{T}_1, \pi)$ on $\mathbb{T}$. $\square$

Theorem 2.27   *Let the next conditions are carrying out:*

(1) $(X, \mathbb{T}_1, \pi)$ is asymptotic compact, that is for any bounded positively invariant set $A \subset X$ there exits a nonempty compact K_A such that

$$\lim_{t \to +\infty} \beta(\pi^t A, K_A) = 0; \qquad (2.103)$$

(2) $(X, \mathbb{T}_1, \pi)$ is asymptotic bounded, that is for any bounded set $A \subset E$ there exits $l = l(A) \geq 0$ such that $\cup \{\pi^t A \mid t \geq l\}$ is bounded;

(3) there exit $y_0 \in Y$ and $R_0 > 0$ such that (2.95) is fulfilled.

Then the non-autonomous dynamical system $\langle (X, \mathbb{T}_1, \pi), (Y, \mathbb{T}_2, \sigma), h \rangle$ admits the maximal compact attractor.

Proof.   First, let us notice, that in conditions of Theorem the dynamical system $(X, \mathbb{T}_1, \pi)$ satisfies the condition of Ladyzhenskaya. Let $R > R_0$, then for any $x \in E$ there will be $\tau = \tau(x) \geq 0$ such that $|x\tau| < R$. If we suppose that it is not so, then there exist $x^1 \in E$ and $R' > R_0$ such that

$$|x^1 \tau| \geq R' > R_0 \qquad (2.104)$$

for all $\tau \geq 0$ and, hence, $\omega_{x^1} \cap X_{y_0} \neq \emptyset$. That is why for any $x \in \omega_{x^1} \cap X_{y_0}$ the inequality (2.97) holds, but this contradicts (2.95). Thus the assertion we need is proved. Now for ending the proof of the Theorem it is sufficient to cite theorem 1.24. $\square$

Remark 2.10   *Let us notice, that Theorem 2.27 without demanding the minimality of Y does not take place even in class of linear systems. The last assertion is proved by the example 2.3.*

Theorem 2.28   *Let (X, h, Y) be a finite-dimensional vectorial fibering, all the motions $(X, \mathbb{T}_1, \pi)$ are bounded on $\mathbb{T}_+$, Y be a compact minimal set and $y_0 \in Y$, then the following conditions are equivalent:*

1. the non-autonomous dynamical system $\langle (X, \mathbb{T}_1, \pi), (Y, \mathbb{T}_2, \sigma), h \rangle$ is dissipative;

2. *there exits $R > 0$ such that*

$$\limsup_{t \to +\infty} |xt| < R \qquad (2.105)$$

for all $x \in X_{y_0}$ and all motions $(X, \mathbb{T}_1, \pi)$ are bounded on $\mathbb{T}_+$;
3. *there exits a positive number r such that for any $x \in X_{y_0}$ and $l > 0$ there will be $\tau = \tau(x) \geq l$ for which*

$$|x\tau| < r \qquad (2.106)$$

and all the motions $(X, \mathbb{T}_1, \pi)$ are bounded on $\mathbb{T}_+$;
4. *there exits a nonempty compact $K_1 \subset X$ such that $\omega_x \cap K_1 \neq \emptyset$ for all $x \in X_{y_0}$ and all the motions $(X, \mathbb{T}_1, \pi)$ are bounded on $\mathbb{T}_+$;*
5. *there exits a nonempty compact $K_2 \subseteq X$ such that $\omega_x \neq \emptyset$ and $\omega_x \subseteq K_2$ for all $x \in X_{y_0}$;*
6. *there exits a positive number R_0 such that for any $R_1 > 0$ there will be $l(R_1) > 0$, that*

$$|xt| < R_0 \qquad (2.107)$$

for all $t \geq L(R_1), |x| \leq R_1 (x \in X_{y_0})$.

Proof. Implications $1. \Longrightarrow 6. \Longrightarrow 2. \Longrightarrow 5. \Longrightarrow 4. \Longrightarrow 3.$ are evident. According to Theorem 2.26 from 3. it follows 1. The theorem is proved. $\qquad \square$

2.9 Homogeneous dynamical systems

The goal of the present section is to study the connection between uniform asymptotic stability and exponential asymptotic of the solutions of infinite-dimensional systems. This problem is studied and solved within the framework of general non-autonomous dynamical system with infinite-dimensional phase space.

Let (X, h, Y) be a locally trivial fiber bundle [190] with the fiber E, (X, ρ) be a complete metric space, $\mathbb{T} = \mathbb{S}_+ := \{s \in \mathbb{S} \mid = \ s \geq 0\}$, where $\mathbb{S} := \mathbb{R}$ or $\mathbb{Z}$.

Definition 2.21 $\langle (X, \mathbb{T}_1, \pi), (Y, \mathbb{T}_2, \sigma), h \rangle$ $(\mathbb{T}_1 \subseteq \mathbb{T}_2 \subseteq \mathbb{S})$ is said to be homogeneous of order $k = 1$, if $\pi(t, \lambda x) = \lambda \pi(t, x)$ for all $\lambda > 0$, $x \in X$ and $t \in \mathbb{T}_1$.

Definition 2.22 An autonomous dynamical system $(X, \mathbb{R}_+, \pi)$ is said to be homogeneous of order k $(k \geq 1)$, if $\pi(t, \lambda x) = \lambda^k \pi(\lambda^{k-1} t, x)$ for all $\lambda \geq 0$, $x \in X$ and $t \in \mathbb{T}_1$.

If $x \in X$, then we put $|x| := \rho(x, \theta_{h(x)})$, where θ_y $(y \in Y)$ is null (trivial) element of the linear space X_y and $\Theta := \{\theta_y \mid y \in Y\}$ is a null (trivial) section of the vectorial bundle (X, h, Y). Denote by X^s a stable manifold of $\langle (X, \mathbb{T}_1, \pi), (Y, \mathbb{T}_2, \sigma) \rangle$, i.e. $X^s = \{x \mid x \in X, \ \lim_{t \to +\infty} |\pi(t, x)| = 0\}$.

Lemma 2.17 *Let $\langle (X, \mathbb{T}_1, \pi), (X, \mathbb{T}_2, \sigma), h \rangle$ be a homogeneous dynamical system of order $k = 1$. Then*

(1) if Ω_X is compact, then $\Omega_X \subseteq \Theta$;
(2) if $J^+(\Omega_X)$ is compact, then $J^+(\Omega_X) \subseteq \Theta$.

Proof. To prove the inclusion $\Omega_X \subseteq \Theta$ it is sufficient to show that $\omega_x \subseteq \Theta$ for all $x \in X$. Let $x \in X$ and $p \in \omega_x$, then by the homogeneity of $\langle (X, \mathbb{T}_1, \pi), (Y, \mathbb{T}_2, \sigma), h \rangle$ of order $k = 1$ we have $\lambda p \in \omega_{\lambda x}$ for all $\lambda \geq 0$. If we suppose that for some x_0 the inclusion $\omega_{x_0} \subseteq \Theta$ it is not true, then there exits $p \in \omega_{x_0} \setminus \Theta$ and, hence, $\lambda p \in \Omega$ for all $\lambda \geq 0$. The last fact contradicts the compactness of Ω.

Let now $J^+(\Omega_X)$ be compact. We will show that $J^+(\Omega_X) \subseteq \Theta$. If $p \in J^+(\Omega_X)$, then there exit $q \in \Omega_X$, $q_n \to q$ and $t_n \to \infty$ such that $p = \lim_{n \to +\infty} \pi(t_n, q_n)$ and $\lambda p = \lim_{n \to +\infty} \lambda \pi(t_n, q_n) = \lim_{n \to +\infty} \pi(t_n, \lambda q_n)$. Since $\lambda q \in \Omega_X$ along with the point q, then $\lambda p \in J^+_{\lambda q} \subseteq J^+(\Omega_X)$ for all $\lambda \geq 0$ and $p \in J^+(\Omega_X)$. If $J^+(\Omega_X) \not\subseteq \Theta$, then should be the point $p_0 \in J^+(\Omega_X) \setminus \Theta$ and, consequently, $\lambda p_0 \in J^+(\Omega_X) \ \forall \lambda \geq 0$. The last inclusion contradicts the compactness of $J^+(\Omega_X)$. The lemma is proved$\square$

Corollary 2.10 *Let $\langle (X, \mathbb{T}_1, \pi), (Y, \mathbb{T}_2, \sigma), h \rangle$ be a homogeneous non-autonomous dynamical system of order $k = 1$. Then the following conditions are equivalent:*

(1) $\langle (X, \mathbb{T}_1, \pi), (Y, \mathbb{T}_2, \sigma), h \rangle$ is compactly dissipative;
(2) $\langle (X, \mathbb{T}_1, \pi), (Y, \mathbb{T}_2, \sigma), h \rangle$ is convergent.

Theorem 2.29 *Let $\langle (X, \mathbb{T}_1, \pi), (Y, \mathbb{T}_2, \sigma), h \rangle$ be homogeneous of order $k = 1$ and $(Y, \mathbb{T}_2, \sigma)$ be pointwise k-dissipative. Then the following conditions are equivalent:*

(1) $\langle (X, \mathbb{T}_1, \pi), (Y, \mathbb{T}_2, \sigma), h \rangle$ is pointwise dissipative (i.e. $(X, \mathbb{T}, \pi)$ is pointwise dissipative);
(2) $X^s = X$.

Proof. Let $\langle (X, \mathbb{T}_1, \pi), (Y, \mathbb{T}_2, \sigma), h \rangle$ be pointwise dissipative. To prove the equality $X^s = X$ it is sufficient to show that $\Omega_X \subseteq \Theta$. Since $\langle (X, \mathbb{T}_1, \pi), (Y, \mathbb{T}_2, \sigma), h \rangle$ is pointwsice dissipative, then Ω_X is compact and by Lemma 2.17 we have $\Omega_X \subseteq \Theta$.

Let now $X^s = X$; we will show that $(X, \mathbb{T}_1, \pi)$ is pointwise dissipative. First of all, we note that $\{\pi(t, x) \mid t \geq 0\}$ is relatively compact for all $x \in X$. To convince that it is so it is sufficient to show that from arbitrary sequence $\{\pi(t_n, x)\}$ ($t_n \to +\infty$) there can be chosen a convergent subsequence. Let $y = h(x)$, then according to the pointwise dissipativity of $(Y, \mathbb{T}_2, h)$ we may suppose that the sequence $\{\sigma(t_n, y)\}$

is convergent. Let $q = \lim\limits_{n \to +\infty} \sigma(t_n, y)$, then $q \in \omega_y$. According to the local triviality of the fiber bundle (X, h, Y), the sequence $\{\theta_{\sigma(y,t_n)}\} \to \theta_q$ and, hence,

$$\rho(xt_n, \theta_q) \le \rho\left(xt_n, \theta_{\sigma(y,t_n)}\right) + \rho\left(\theta_{\sigma(y,t_n)}, \theta_q\right) \longrightarrow 0$$

for $n \to +\infty$. In addition, from the reasoning mentioned above it follows that $\omega_x \subseteq \Theta$ for all $x \in X$ and, consequently, $\Omega_X \subseteq \Theta$. Note, that by Lemma 2.9 $h(\Omega_X) \subseteq \Omega_Y$ and, hence, $\Omega_X \subseteq \Theta \cap h^{-1}(\Omega_Y)$. From the compactness of Ω_Y and the local triviality of (X, h, Y) it follows the compactness of $\Theta \cap h^{-1}(\Omega_Y)$ and, consequently, the compactness of Ω_X too. The theorem is proved. $\qquad\square$

Theorem 2.30 *Let $\langle (X, \mathbb{T}_1, \pi), (Y, \mathbb{T}_2, \sigma), h \rangle$ be homogeneous of order $k = 1$ and Y be compact. Then the following conditions are equivalent:*

1. *$\langle (X, \mathbb{T}_1, \pi), (Y, \mathbb{T}_2, \sigma), h \rangle$ is compactly dissipative, i.e. $(X, \mathbb{T}_1, \pi)$ is compactly dissipative;*
2. *$X^s = X$ and the set $\Theta \cap h^{-1}(J_Y)$ is uniformly stable, i.e. for an arbitrary $\varepsilon > 0$ there exists $\delta(\varepsilon) > 0$ such that $|x| < \delta$ implies $|xt| < \varepsilon$ for all $t \ge 0$;*
3. *if the fiber bundle (X, h, Y) is normed (i.e. the metric ρ on the fibers (X, h, Y) is compatible with the norm), then there exists a positive number N such that*

$$|xt| \le N|x| \tag{2.108}$$

for all $x \in X$, $t \ge 0$ and $X^s = X$.

Proof. Let $\langle (X, \mathbb{T}_1, \pi), (Y, \mathbb{T}_2, \sigma), h \rangle$ be compactly dissipative and J be the Levinson center of the dynamical system $(X, \mathbb{T}_1, \pi)$. By Theorem 2.29 $X^s = X$. The inclusion can be proved $J_X \subseteq \Theta$ in the same way that the inclusion $\Omega_X \subseteq \Theta$ (see the proof of Theorem 2.29). We will show that the trivial section Θ of the fiber bundle (X, h, Y) is uniformly stable. If we suppose that it is not true, then there exit $\varepsilon_0 > 0$, $\delta_n \downarrow 0$, $|x_n| < \delta_n$ and $t_n \to +\infty$ such that

$$|x_n t_n| \ge \varepsilon_0. \tag{2.109}$$

By virtue of the compactness of Y and the local triviality of (X, h, Y) the set Θ is compact and, consequently, we may suppose that the sequence $\{x_n\}$ converges. Let $\bar{x} = \lim\limits_{n \to +\infty} x_n$. Since $(X, \mathbb{T}_1, \pi)$ is compactly dissipative, then we may suppose that the sequence $\{x_n t_n\}$ is convergent too and we put $\bar{\bar{x}} = \lim\limits_{n \to +\infty} x_n t_n$. Then $\bar{\bar{x}} \in J_X \subseteq \Theta$ and, consequently, $|\bar{\bar{x}}| = 0$. The last equality contradicts the inequality (2.109). The obtained contradiction proves the required statement.

To prove the implication 2. $\Rightarrow$ 1. it is sufficient to refer to Theorem 1.13.

Let now the fiber bundle (X, h, Y) be normed. We will show that in this case every one of Conditions 1. and 2. is equivalent to Condition 3. We will prove, for example, that 2. implies 3. Since the trivial section Θ is stable, then for $\varepsilon = 1$ there exits $\gamma = \delta(1) > 0$ such that $|\pi(t,x)| \leq 1$ for all $t \geq 0$ and $|x| \leq q$. We put $N = \gamma^{-1}$, then

$$|\pi(t,x)| = ||x|\gamma^{-1}\pi(t,x|x|^{-1}\gamma)| \leq |x|\gamma^{-1} = N|x|.$$

for all $t \geq 0$ and $x \in X \setminus \Theta$.

Inversely. Let $X^s = X$ and let exist $N > 0$ such that the inequality (2.108) holds for all $x \in X$ and $t \geq 0$. Let $\varepsilon > 0$ and $\delta(\varepsilon) < \frac{\varepsilon}{N}$ such that $|x| < \delta$ implies $|xt| < \varepsilon$ for all $t \geq 0$. The theorem is proved. $\qquad\square$

Theorem 2.31 *Let $\langle (X, \mathbb{T}_1, \pi), (Y, \mathbb{T}_2, \sigma), h \rangle$ be a homogeneous non-autonomous dynamical system of order $k = 1$ and Y be compact. Then the following conditions are equivalent:*

1. *$\langle (X, \mathbb{T}_1, \pi), (Y, \mathbb{T}_2, \sigma), h \rangle$ is locally dissipative, i.e. $(X, \mathbb{T}_1, \pi)$ is locally dissipative;*

2. *$X^s = X$ and the trivial section Θ is uniformly attracting, i.e. there exits $\gamma > 0$ such that*

$$\lim_{t \to +\infty} \sup_{|x| \leq \gamma} |\pi(t,x)| = 0; \tag{2.110}$$

3. *if the fiber bundle (X, h, Y) is normed, then there exit positive numbers $\mathcal{N}$ and ν such that*

$$|\pi(t,x)| \leq \mathcal{N}e^{-\nu t}|x| \tag{2.111}$$

for all $t \geq 0$ and $x \in X$.

Proof. We will prove that 1. $\Rightarrow$ 2. $\Rightarrow$ 3. If the non-autonomous dynamical system $\langle (X, \mathbb{T}_1, \pi), (Y, \mathbb{T}_2, \sigma), h \rangle$ is locally dissipative and J is its Levinson's center, then by Theorem 2.30 $X^s = X$ and $J_X \subseteq \Theta$. In addition, according to Theorem 1.17 there exists a positive number γ such that

$$\lim_{t \to +\infty} \beta(\pi^t B(J, \gamma), J) = 0. \tag{2.112}$$

We will show that for the obtained $\gamma > 0$ the equality (2.110) holds. If we suppose that it is not true, then there exit $\varepsilon_0 > 0$, $\{x_n\}$ ($|x_n| \leq \gamma$) and $t_n \to +\infty$ such that the inequality (2.109) holds. From the equality (2.112) it follows that the sequence $\{x_n t_n\}$ can be considered convergent. Let $\bar{x} = \lim_{n \to +\infty} x_n t_n$. By (2.112) $\bar{x} \in J$, and since $J_X \subseteq \Theta$, then $|\bar{x}| = 0$. On the other hand, passing to limit in (2.109) as $n \to +\infty$, we obtain $|\bar{x}| \geq \varepsilon_0 > 0$. The obtained contradiction proves the equality (2.110).

From the homogeneity of the system $\langle (X, \mathbb{T}_1, \pi), (Y, \mathbb{T}_2, \sigma), h \rangle$ and the equality (2.110) it follows that

$$\lim_{t \to +\infty} \sup_{|x| \le 1} |\pi(t, x)| = 0. \tag{2.113}$$

We put $\psi(t) = \sup\{|\pi(t, x)| \ : \ |x| \le 1\}$. Note, that $\psi(t + \tau) \le \psi(t)\psi(\tau)$ for all $t, \tau \in T_1$ and $\psi(t) \to 0$ as $t \to +\infty$. By the equality (2.110), the trivial section Θ of the fiber bundle (X, h, Y) is uniformly stable. Indeed, if we suppose that it is not true, then there exit ε_0, $\delta_n \downarrow 0$, $|x_n| < \delta_n$ and $t_n \to +\infty$ such that the relation (2.109) holds. We can suppose that $\delta_n \le \gamma$, and according (2.110) for $\frac{\varepsilon_0}{2}$ there exits $L_0 > 0$ such that

$$|xt| < \frac{\varepsilon_0}{2} \tag{2.114}$$

for all $t \ge L_0$. Since $t_n \to +\infty$, then for the sufficiently large n the inequalities (2.114) and (2.109) are contradictory. The obtained contradiction proves the required statement. By Theorem 2.30 the function ψ is bounded for $t \ge 0$ and according to Lemma 5.4 [21] there exist positive numbers $\mathcal{N}$ and ν such that $\psi(t) \le \mathcal{N}e^{-\nu t}$ for all $t \ge 0$ and, consequently, the inequality (2.111) holds too.

Now we will show, that the inverse implications 3. $\Rightarrow$ 2. $\Rightarrow$ 1. hold too. It is evident that 3. implies 2., therefore to finish the proof of the theorem it is sufficient to show that from 2. it follows 1. First of all we note that from the equality (2.110), according to the reasoning above, it follows the uniform stability of the trivial section Θ. By Theorem 2.30 the dynamical system $\langle (X, \mathbb{T}_1, \pi), (Y, \mathbb{T}_2, \sigma), h \rangle$ is compactly dissipative and $J_X \subseteq \Theta$. We will show that

$$\lim_{t \to +\infty} \beta(\pi^t B(J, \gamma), J) = 0.$$

If we suppose that it is not true, then there exit $\varepsilon_0 > 0$, $x_n \in B(J, \gamma)$ and $t_n \to +\infty$ such that

$$\rho(x_n t_n, \Theta_y) \ge \varepsilon_0 \tag{2.115}$$

for all $y \in Y$. From (2.110) it follows, that the sequence $\{x_n t_n\}$ is convergent and $\lim_{n \to +\infty} x_n t_n = \overline{x} \in \Theta$. On the other hand, passing to limit in (2.115) as $n \to +\infty$, we obtain $\rho(\overline{x}, \Theta_y) \ge \varepsilon_0$ for all $y \in Y$, i.e. $\overline{x} \notin \Theta$. The obtained contradiction proves the equality (2.111) and, hence, the local dissipativity of (X, T, π). The theorem is proved. $\square$

Corollary 2.11 *Let $\langle (X, \mathbb{T}_1, \pi), (Y, \mathbb{T}_2, \sigma), h \rangle$ be a homogeneous non-autonomous dynamical system and (X, T, π) be asymptotically compact, then the following conditions are equivalent:*

(1) $\langle (X, \mathbb{T}_1, \pi), (Y, \mathbb{T}_2, \sigma), h \rangle$ is compactly dissipative;

(2) $\langle (X, \mathbb{T}_1, \pi), (Y, \mathbb{T}_2, \sigma), h \rangle$ *is locally dissipative;*

(3) $X^s = X$ *and the trivial section* Θ *of the fiber bundle* (X, h, Y) *is uniformly stable;*

(4) $X^s = X$ *and the trivial section* Θ *of the fiber bundle* (X, h, Y) *is uniformly attracting;*

(5) *if the fiber bundle* (X, h, Y) *is normed, then there exits a positive number* $\mathcal{N} > 0$ *such that*

$$|xt| \le \mathcal{N}|x|$$

for all $x \in X$, $t \ge 0$, *and* $X^s = X$.

(6) *if the fiber bundle* (X, h, Y) *is normed, then there exit positive numbers* $\mathcal{N}$ *and* γ *such that*

$$|xt| \le \mathcal{N}e^{-\gamma t}|x|$$

for all $t \ge 0$ *and* $x \in X$.

Proof. This statement it follows from Theorems 1.20, 2.30 and 2.31. $\qquad\square$

Corollary 2.12   *Let* $(X, \mathbb{T}_1, \pi)$ *be completely continuous. Then the following conditions are equivalent:*

(1) $\langle (X, \mathbb{T}_1, \pi), (Y, \mathbb{T}_2, \sigma), h \rangle$ *is pointwise dissipative;*

(2) $\langle (X, \mathbb{T}_1, \pi), (Y, \mathbb{T}_2, \sigma), h \rangle$ *is compactly dissipative;*

(3) $\langle (X, \mathbb{T}_1, \pi), (Y, \mathbb{T}_2, \sigma), h \rangle$ *is locally dissipative;*

(4) $X^s = X$;

(5) $X^s = X$ *and* Θ *is uniformly stable;*

(6) $X^s = X$ *and* Θ *is uniformly attracting;*

(7) *if* (X, h, Y) *is normed, then there exits* $\mathcal{N} > 0$ *such that*

$$|xt| \le \mathcal{N}|x|$$

for all $x \in X$, $t \ge 0$ *and* $X^s = X$;

(8) *if* (X, h, Y) *is normed, then there exit* $\mathcal{N} > 0$ *and* $\gamma > 0$ *such that*

$$|xt| \le \mathcal{N}e^{-\gamma t}|x|$$

for all $t \ge 0$ *and* $x \in X$.

Proof. Corollary 2.12 it follows from Theorems 1.10, 2.29 – 2.31. $\qquad\square$

Remark 2.11   *a) All the results of this section hold also for autonomous homogeneous (of order* $k \ge 1$*) system with the continuous time* $(\mathbb{T} = \mathbb{R}_+$ *or* $\mathbb{R})$.

     b) Note, that the results of this section take place, when the bundle (X, h, Y) *is not necessary linear (vectorial). It is sufficient that there would exist a fiber bundle*

(W, h, Y) such that $X \subseteq W$ and $\lambda x \in X$ for all $\lambda \geq 0$ and $x \in X$ (in this case we will call the bundle (X, h, Y) homogeneous).

2.10 Power-law asymptotic of homogeneous systems

Let (X, h, Y) be a local, trivial normed fiber bundle with the fiber E. We will say that a non-autonomous dynamical system $\langle (X, \mathbb{T}_1, \pi), (Y, \mathbb{T}_2, \sigma), h \rangle$ admits a power-law asymptotic of order $m > 1$, if there exist positive numbers a_i and b_i $(i = 1, 2)$ such that

$$(a_1 t + b_1 |x|^{1-m})^{-\frac{1}{m-1}} \leq |\pi^t x| \leq (a_2 t + b_2 |x|^{1-m})^{-\frac{1}{m-1}} \tag{2.116}$$

for all $t \geq 0$ and $x \in X$.

Theorem 2.32 *In order that a non-autonomous dynamical system $\langle (X, \mathbb{T}_1, \pi),$ $(Y, \mathbb{T}_2, \sigma), h \rangle$ would admit a power-law asymptotic of order $m > 1$, it is necessary and sufficient that there would exist a continuous function $V : X \to \mathbb{R}_+$ satisfying the following conditions:*

a) there exit positive numbers α and β such that

$$\alpha |x| \leq V(x) \leq \beta |x| \tag{2.117}$$

for all $x \in X$;
b) V is differentiable along the trajectories of the system $(X, \mathbb{T}_1, \pi)$ and

$$\left. \frac{dV(\pi^t x)}{dt} \right|_{t=0} = -|x|^m \tag{2.118}$$

for all $x \in X$.

Proof. Necessity. Suppose that the equality (2.116) holds and

$$V(x) := \int_0^{+\infty} |\pi^\tau x|^m d\tau. \tag{2.119}$$

It is easy to see that under Condition (2.116) the equality (2.119) defines correctly the mapping $V : X \to \mathbb{R}_+$. In addition, under Condition (2.116) the integral (2.119) is uniformly (in $|x| \leq r$, for each $r > 0$) convergent and, hence, the function V is continuous. Further, from (2.116) and (2.119) it follows that $\alpha |x| \leq V(x) \leq \beta |x|$ for all $x \in X$, where $\alpha = \frac{m-1}{a_1} b_1^{\frac{1}{1-m}}$ and $\beta = \frac{m-1}{a_2} b_2^{\frac{1}{1-m}}$.

Sufficiency. Let $V : X \to \mathbb{R}_+$ be a continuous function satisfying Conditions a) and b) of the theorem. We will show that the non-autonomous dynamical system

$\langle (X, \mathbb{T}_1, \pi), (Y, \mathbb{T}_2, \sigma), h \rangle$ admits a power-law asymptotic of order m. Indeed, let us put $\varphi(t) := V(\pi^t x)$ and note that by (2.118)

$$\frac{d\varphi(t)}{dt} = -|\pi^t x|^m$$

for all $t \geq 0$ and $x \in X$. From Condition (2.117) it follows that $|\pi^t x| \geq \frac{1}{\beta} V(\pi^t x)$ for all $t \geq 0$ and $x \in X$ and, consequently,

$$\frac{d\varphi(t)}{dt} \leq -\frac{1}{\beta} \varphi^m(t).$$

Taking into account that $\varphi(t) = V(\pi^t x) > 0$ for all $t \geq 0$, if $x \neq 0$, we have

$$V(\pi^t x) \leq \left(\frac{m-1}{\beta^m} t + \alpha^{-m+1} |x|^{-m+1} \right)^{-\frac{1}{m-1}}.$$

On the other hand, $|\pi^t x| \leq \frac{1}{2} V(\pi^t x)$ and, hence,

$$|\pi^t x| \leq \left(a_2 t + b_2 |x|^{1-m} \right)^{-\frac{1}{m-1}}$$

for all $t \geq 0$ and $x \in X$, where $a_2 = \frac{m-1}{\beta^m \alpha^{1-m}}$ and $b_2 = 1$.

By analogy, there can be established a lower power-law asymptotic for $|\pi^t x|$. The theorem is completely proved. $\qquad\square$

Theorem 2.33   *Let $\langle (X, \mathbb{T}_1, \pi), (Y, \mathbb{T}_2, \sigma), h \rangle$ be a non-autonomous dynamical system. Then the following conditions are equivalent:*

(1) there exit positive numbers $\mathcal{N}_i$ and ν_i $(i = 1, 2)$ such that

$$\mathcal{N}_1 e^{-\nu_1 t} |x| \leq |\pi^t x| \leq \mathcal{N}_2 e^{-\nu_2 t} |x|$$

    for all $t \geq 0$ and $x \in X$;

*(2) (a) there exit positive numbers α and β such that the inequality (2.117) holds
        for all $x \in X$;*

    *(b) the function V is differentiable along the trajectories of the system
        $(X, \mathbb{T}_1, \pi)$ and*

$$\left. \frac{dV(\pi^t x)}{dt} \right|_{t=0} = -|x|$$

    for all $x \in X$.

Proof.   This assertion can be proved using the same reasoning as in the proof of Theorem 2.32. $\qquad\square$

Lemma 2.18 *Let* $\langle (X, \mathbb{T}_1, \pi), (Y, \mathbb{T}_2, \sigma), h \rangle$ *be a non-autonomous dynamical system and let exist positive numbers* a, b *and* $m > 1$ *such that*

$$|\pi^t x| \le (at + b|x|^{1-m})^{-\frac{1}{m-1}} \tag{2.120}$$

for all $x \in X$ *and* $t \ge 0$. *Then*

$$\lim_{t \to +\infty} |\pi^t x| = 0 \tag{2.121}$$

uniformly in $|x| \le r$ *for each* $r > 0$ *and*

$$|\pi^t x| \le \mathcal{N}|x| \tag{2.122}$$

for all $t \ge 0$ *and* $x \in X$, *where* $\mathcal{N} = b^{-\frac{1}{1-m}}$.

Proof. Consider the function $\omega(r, t) = (at + br^{1-m})^{-\frac{1}{m-1}}$ defined for all $t \ge 0$ and $r \ge 0$. Note, that

$$\frac{\partial \omega(r, t)}{\partial t} = br^{1-m}(at + br^{1-m})^{\frac{m}{1-m}} \ge 0$$

for all $t \ge 0$ and, consequently, for $|x| \le r$ we have

$$|\pi^t x| \le (at + b|x|^{1-m})^{-\frac{1}{m}} \le (at + br^{1-m})^{-\frac{1}{m}},$$

from which it follows (2.121).

The inequality (2.122) it follows from (2.120) and the equality

$$(at + b|x|^{1-m})^{-\frac{1}{m-1}} = |x|(at|x|^{m-1} + b)^{-\frac{1}{m-1}}.$$

The lemma is proved. $\square$

Lemma 2.19 *Let* $\mathfrak{D}$ *be a family of functions* $m : \mathbb{R}_+ \to \mathbb{R}_+$ *satisfying the conditions:*

a. *there exists* $M > 0$ *such that* $0 < m(t) \le M$ *for all* $t \ge 0$ *and* $m \in \mathfrak{D}$;

b. *$m(t) \to 0$ as $t \to +\infty$ uniformly in $m \in \mathfrak{D}$, i.e. for any $\varepsilon > 0$ there exists $L(\varepsilon) > 0$ such that $m(t) < \varepsilon$ for all $t \ge L(\varepsilon)$ and $m \in \mathfrak{D}$.*

Then we have the following assertions:

(1) *if $m(t + \tau) \le m(t)m(\tau)$ for all $t, \tau \ge 0$ and $m \in \mathfrak{D}$, then there exit positive numbers $\mathcal{N}$ and ν such that*

$$m(t) \le \mathcal{N}e^{-\nu t} \tag{2.123}$$

for all $t \ge 0$ and $m \in \mathfrak{D}$;

(2) if $m(t + \tau) \le m(t)m(\tau m^{\alpha-1}(t))$ $(\alpha > 1)$ for all $t, \tau \ge 0$ and $m \in \mathfrak{D}$, then there exist positive numbers a and b such that

$$m(t) \le M(a + bt)^{-\frac{1}{\alpha-1}}$$

for all $t \ge 0$ and $m \in \mathfrak{D}$.

Proof. We will prove the first statement of the lemma. Let $\gamma \in (0, 1)$ and $\tau > 0$ be such that $m(\tau) \le \gamma < 1$ for all $m \in \mathfrak{D}$. Let us put $\nu := -\tau^{-1} \ln \gamma$ and $\mathcal{N} := Me^{\nu\tau}$, then

$$m(n\tau) \le \gamma^n$$

for all $n = 1, 2, \ldots$. Let $t \ge 0$ and n be a natural number, such that $t \in [n\tau, n\tau + \tau]$. Since

$$m(t) = m(t - n\tau + n\tau) \le m(t - n\tau)m(n\tau) \le M\gamma^n,$$

then the following inequality

$$m(t) \le e^{\nu(t - n\tau)}m(t - n\tau)\gamma^n e^{-\nu t}$$

holds too. Taking into account that $e^{\nu\tau} = \gamma^{-1}$, by the choice of the number $\mathcal{N}$ we have the required inequality (2.123).

To prove the second statement of the lemma we will choose $\tau > 0$ so that the inequality

$$m(t) \le \frac{1}{2} \tag{2.124}$$

would hold for all $t \ge \tau$ and $m \in \mathfrak{D}$. Since $0 < m(t) \le M$ and $m(t + \tau) \le m(t)m(\tau m^{\alpha-1}(t))$ for all $t, \tau \ge 0$ and $m \in \mathfrak{D}$, then

$$m(t) \le M \qquad (0 \le t \le q\tau,\ q = 2^{\alpha-1}) \tag{2.125}$$

and

$$m(t) \le \frac{1}{2} \qquad (\tau \le t < +\infty)$$

for all $m \in \mathfrak{D}$. Set $t_0 = 0$, $t_{i+1} = t_i + \tau q_i$ $(q_i = q^i)$ and note that

$$m(t_i) \le \frac{1}{2} \qquad (i \ge 1) \tag{2.126}$$

for all $m \in \mathfrak{D}$. Indeed, according to (2.124), we have $m(t_1) \le 2^{-1}$. Moreover,

$$m(t_{i+1}) = m(t_i + \tau q_i) \le m(t_i)m(\tau q_i m^{\alpha-1}(t_i)) \tag{2.127}$$

for all $m \in \mathfrak{D}$. Suppose that (2.126) is valid for all $i \leq n$; then from (2.127) it follows that

$$m(t_{n+1}) \leq \frac{1}{2^{n+1}},$$

since $\tau q_n m^{\alpha-1}(t_n) \geq \tau$ (for all $m \in \mathfrak{D}$) in view of the choice of q_n and the inductive assumption (2.126). Thus, from (2.125) and (2.126) it follows that

$$m(t) \leq \frac{1}{2^n}$$

for all $t \geq t_n$ and $n \geq 1$. Note. that $t_{n+1} = \tau(q^{n+1} - 1)(q - 1)^{-1}$ and, hence,

$$2^{-n} = 2\left(\frac{2^{\alpha-1}-1}{\tau}t_{n+1} + 1\right)^{-\frac{1}{\alpha-1}}.$$

Now let $t \in [t_n, t_{n+1})$, then we have

$$2^{-n} < 2\left(1 + \frac{2^{\alpha-1}-1}{\tau}t\right)^{-\frac{1}{\alpha-1}}. \tag{2.128}$$

From (2.125) and (2.128) it follows that

$$m(t) \leq M\left(2^{1-\alpha} + \frac{1-2^{1-\alpha}}{\tau}t\right)^{-\frac{1}{\alpha-1}}$$

for all $t \geq 0$ and $m \in \mathfrak{D}$. By setting $a = 2^{1-\alpha}$ and $b = \tau^{-1}(1 - 2^{1-\alpha})$, we obtain the required assertion. The lemma is proved. $\qquad\square$

Theorem 2.34 *Let a non-autonomous system $\langle (X, \mathbb{T}_1, \pi), (Y, \mathbb{T}_2, \sigma), h \rangle$ be homogeneous, Y be compact, and the fiber bundle (X, h, Y) be normed. Then the following conditions are equivalent:*

a. the trivial section of the fiber bundle (X, h, Y) is uniformly asymptotically stable;

b. there exit positive numbers $\mathcal{N}$ and ν such that

$$|\pi^t x| \leq \mathcal{N}e^{-\nu t}|x| \tag{2.129}$$

for all $x \in X$ and $t \geq 0$.

Proof. To prove the theorem it is sufficient to establish the implication $a. \Rightarrow b.$, since the converse assertion is obvious.

Note, that from the uniform asymptotic stability of the trivial section Θ of the fiber bundle (X, h, Y) and the homogeneity of the non-autonomous system $\langle (X, \mathbb{T}_1, \pi), (Y, \mathbb{T}_2, \sigma), h \rangle$ it follows the equality $X^s = X$. Now, to complete the proof of the theorem it is sufficient to refer to Theorem 2.31. $\qquad\square$

Theorem 2.35 *Let X be a Banach space. For an autonomous homogeneous (of order $k > 1$) dynamical system $(X, \mathbb{R}_+, \pi)$ the following conditions are equivalent:*

1. *the trivial motion of $(X, \mathbb{R}_+, \pi)$ is uniformly asymptotically stable;*
2. *there exit positive numbers α and β such that*

$$|\pi^t x| \le (\alpha |x|^{1-k} + \beta t)^{-\frac{1}{k-1}}$$

for all $t \ge 0$ and $x \in X$.

Proof. Let us show that under the conditions of the theorem assumption 1. implies 2. Let $x \in X$ ($x \ne 0$). Then, in view of the homogeneity of order $k > 1$ of the system $(X, \mathbb{R}_+, \pi)$, we have

$$|\pi(t, x)| = |x| |\pi(t|x|^{k-1}, \frac{x}{|x|})| \le |x| \sup_{|y| \le 1} |\pi(t|x|^{k-1}, y)| = |x| m(t|x|^{k-1})$$

for all $t \ge 0$ and $x \in X$, where

$$m(t) = \sup_{|y| \le 1} |\pi(t, y)|.$$

From the uniform asymptotic stability of the trivial motion of $(X, \mathbb{R}_+, \pi)$ and by virtue of the homogeneity of order $k > 1$ of the system $(X, \mathbb{R}_+, \pi)$ we have $X^s = X$. From Theorem 2.29 it follows the boundedness of the function m, but from the uniform asymptotic stability of the trivial motion of $(X, \mathbb{R}_+, \pi)$ it follows that $m(t) \to 0$ as $t \to +\infty$. In addition, note that

$$m(t + \tau) = \sup_{|y| \le 1} |\pi(t + \tau, y)| = \sup_{|y| \le 1} |\pi(\tau, \pi(t, y))|$$

$$= m(t) \sup_{|y| \le 1} \left| \pi \left(\tau m^{k-1}(t), \frac{\pi(t, y)}{m(t)} \right) \right| \le m(t) m(\tau m^{k-1}(t))$$

for all $t, \tau \ge 0$. According to Lemma 2.19, there exit positive numbers M, a and b such that

$$m(t) \le M(a + bt)^{-\frac{1}{k-1}}$$

for all $t \ge 0$. Let us put $\alpha = M^{1-k} a$ and $\beta = M^{1-k} b$, then

$$|\pi(t, x)| \le |x| m(t|x|^{k-1}) \le |x| M(a + bt|x|^{k-1})^{-\frac{1}{k-1}} = (\alpha |x|^{1-k} + \beta t)^{-\frac{1}{k-1}}$$

for all $x \in X$ and $t \ge 0$. To complete the proof of the theorem it is sufficient to refer to Lemma 2.18. $\qquad\square$

Remark 2.12 *All results of this section hold for dynamical systems with homogeneous phase spaces (see Remark 2.11 b.).*

2.11 Linear systems

Let (X, h, Y) be a local trivial fiber bundle with the fiber E and $\langle (X, \mathbb{T}_1, \pi), (Y, \mathbb{T}_2, \sigma), h \rangle$ be a linear non-autonomous dynamical system, i.e. the mapping $\pi^t : X_y \to X_{\sigma^t y}$ is linear for all $y \in Y$ and $t \in \mathbb{T}_1$. From the results of Chapter 2 (section 10)follow the following statements.

Theorem 2.36 *The following statements are equivalent:*

(1) the dynamical system $\langle (X, \mathbb{T}_1, \pi), (Y, \mathbb{T}_2, \sigma), h \rangle$ is pointwise dissipative;
(2) $X^s = X$.

Theorem 2.37 *Let Y be compact. Then the following statements are equivalent:*

(1) the dynamical system $\langle (X, \mathbb{T}_1, \pi), (Y, \mathbb{T}_2, \sigma), h \rangle$ is compactly dissipative;
(2) $X^s = X$ and the trivial section θ of the fiber bundle (X, h, Y) is uniformly stable, that is for all $\varepsilon > 0$ there exits $\delta(\varepsilon) > 0$ such that $|x| < \delta$ implies $|xt| < \varepsilon$ for all $t \geq 0$;
(3) if the fiber bundle (X, h, Y) is normed, then there exits a positive number N such that

$$|xt| \leq N|x|$$

for all $x \in X$, $t \geq 0$ and $X^s = X$.

Theorem 2.38 *Let Y be compact. Then the following statements are equivalent:*

(1) the dynamical system $\langle (X, \mathbb{T}_1, \pi), (Y, \mathbb{T}_2, \sigma), h \rangle$ is locally dissipative;
(2) $X^s = X$ and the trivial section Θ of the fiber bundle (X, h, Y) is uniformly attracting, i.e. there exits $\gamma > 0$ such that

$$\lim_{t \to +\infty} \sup_{|x| \leq \gamma} |\pi(t, x)| = 0;$$

(3) if the fiber bundle (X, h, Y) is normed, then there exit positive numbers N and ν such that

$$|\pi(t, x)| \leq N e^{-\nu t} |x| \tag{2.130}$$

for all $x \in X$, $t \geq 0$.

Remark 2.13 *a. Using Banach-Steinhaus theorem, it is easy to check that for linear autonomous or periodic system the notions of pointwise and compact dissipativity are equivalent.*

b. In the example 1.8 the linear autonomous dynamical system defined on the Hilbert space $X = L_2[0, 1]$ by the differential equation $x' = Ax$ with bounded operator A is compactly dissipative but not locally dissipative.

Theorem 2.39   *Let $\langle (X, \mathbb{T}, \pi), (Y, \mathbb{T}, \sigma), h \rangle$ be a group (i.e. $\mathbb{T} = \mathbb{R}$ or $\mathbb{Z}$) linear non-autonomous dynamical system, Y be compact, (X, h, Y) be a normed fiber bundle and let exist positive numbers M and a such that $|\pi^t x| \le M e^{a|t|} |x|$ ($x \in X$, $t \in \mathbb{T}$). Then the following statements are equivalent:*

1. *$\langle (X, \mathbb{T}, \pi), (Y, \mathbb{T}, \sigma), h \rangle$ is locally dissipative, that is there exit N, $\nu > 0$ such that $|xt| \le N e^{-\nu t} |x|$ for all $t \ge 0$ and $x \in X$;*
2. *there exits a function $V : X \to \mathbb{R}_+$ possessing the following properties:*

   2.a. *V is some norm on (X, h, Y);*

   2.b. *there exit positive numbers α and β such that $\alpha |x| \le V(x) \le \beta |x|$ for all $x \in X$;*

   2.c. *$\dot{V}_\pi(x) = -|x|$ ($x \in X$), where $\dot{V}_\pi(x) := \lim\limits_{t \downarrow 0} t^{-1}[V(xt) - V(x)]$ for $\mathbb{T} = \mathbb{R}$ and $\dot{V}_\pi(x) := V(\pi(1, x)) - V(x)$ for $\mathbb{T} = \mathbb{Z}$.*

Proof.   Let $\langle (X, \mathbb{T}, \pi), (Y, \mathbb{T}, \sigma), h \rangle$ be locally dissipative. Define a function $V : X \to \mathbb{R}_+$ by the formula

$$V(x) = \int_0^{+\infty} |xt| dt \quad (x \in X), \tag{2.131}$$

if $\mathbb{T} = \mathbb{R}$. Directly from the equality (2.131) it follows that V is some norm on (X, h, Y). We will establish some properties of the function V.

$$a. \quad \frac{1}{Ma} |x| \le V(x) \le \frac{N}{\nu} |x| \quad (x \in X).$$

In fact,

$$V(x) = \int_0^{+\infty} |xt| dt \le N \int_0^{+\infty} e^{-\nu t} |x| dt = \frac{N}{\nu} |x|.$$

On the other hand, $|x| = |(xt)(-t)| \le M e^{at} |xt|$ ($x \in X$, $t \ge 0$) and, hence,

$$V(x) = \int_0^{+\infty} |xt| dt \ge \frac{1}{M} \int_0^{+\infty} e^{-at} |x| dt = \frac{1}{Ma} |x|.$$

$$b. \quad \dot{V}_\pi(x) = -|x| \quad (x \in X).$$

Indeed,

$$V(xt) - V(x) = \int_0^{+\infty} |(xt)s| ds - \int_0^{+\infty} |xs| ds$$

$$= \int_t^{+\infty} |xs| ds \int_0^{+\infty} |xs| ds = - \int_0^t |xs| ds$$

and, consequently,

$$\lim_{t \downarrow 0} t^{-1}[V(xt) - V(x)] = \lim_{t \downarrow 0} -t^{-1} \int_0^t |xs| ds = -|x| \quad (x \in X).$$

If $\mathbb{T} = \mathbb{Z}$, then we put $V(x) = \sum_{n=0}^{+\infty} |xn|$ $(x \in X)$. It is obvious that V defines some norm on (X, h, Y). It is easy to see that the function defined above possesses the required properties. Thus, we established that 1. implies 2..

Now we will establish that 2. implies 1. Let $x \in X$ and $|x| \neq 0$. Consider the function $\varphi(t) := V(xt)$. Since $\alpha|x| \leq V(x) \leq \beta|x|$ and $\dot{V}_\pi(x) = -|x|$ $(x \in X)$, then we have

$$\dot{\varphi}(t) = \dot{V}_\pi(xt) = -|xt| \leq -\frac{1}{\beta} V(xt) = -\frac{1}{\beta} \varphi(t)$$

and, hence,

$$\varphi(t) \leq \varphi(0)e^{-\frac{1}{\beta}t} \quad (t \geq 0). \tag{2.132}$$

On the other hand,

$$\frac{1}{\beta} V(xt) \leq |xt| \leq \frac{1}{\alpha V(xt)} \quad (x \in X, \, t \geq 0). \tag{2.133}$$

From the inequalities (2.132) and (2.133) it follows that

$$|xt| \leq \frac{\beta}{\alpha} e^{-\frac{1}{\beta}t} |x| \quad (x \in X, \, t \geq 0).$$

The theorem is proved. $\qquad\square$

Definition 2.23 Let $\langle (X, \mathbb{T}, \pi), (Y, \mathbb{T}, \sigma), h \rangle$ be a linear non-autonomous dynamical system. A non-autonomous dynamical system $\langle (W, \mathbb{T}, \mu), (Z, \mathbb{T}, \lambda), \varrho \rangle$ is said to be linear non-homogeneous, generated by linear (homogeneous) dynamical system $\langle (X, \mathbb{T}, \pi), (Y, \mathbb{T}, \sigma), h \rangle$, if the following conditions hold:

1. there exits a homomorphism q of the dynamical system $(Z, \mathbb{T}, \lambda)$ onto $(Y, \mathbb{T}, \sigma)$;
2. the space $W_y := (q \circ \rho)^{-1}(y)$ is affine for all $y \in (q \circ \varrho)(W) \subseteq Y$ and the vectorial space $X_y = h^{-1}(y)$ is an associated space to W_y ([287, p.175]). The mapping $\mu^t : W_y \to W_{\sigma^t y}$ is affine and $\pi^t : X_y \to X_{\sigma^t y}$ is its linear associated function ([287, p.179]), i.e. $X_y = \{w_1 - w_2 \mid w_1, w_2 \in W_y\}$ and $\mu^t w_1 - \mu^t w_2 = \pi^t(w_1 - w_2)$ for all $w_1, w_2 \in W_y$ and $t \in T$.

Remark 2.14 *The definition of linear non-homogeneous system, associated by the given linear system, is given in the work [33], but our definition is more general and sometimes more flexible.*

Lemma 2.20   *Let $\langle (W, \mathbb{T}, \mu), (Z, \mathbb{T}, \lambda), \varrho \rangle$ be a linear non-homogeneous non-autonomous dynamical system and $(Z, \mathbb{T}, \lambda)$ be compactly dissipative. Then the following statements are equivalent:*

1. *$\langle (W, \mathbb{T}, \mu), (Z, \mathbb{T}, \lambda), \varrho \rangle$ is compactly dissipative;*
2. *$\langle (W, \mathbb{T}, \mu), (Z, \mathbb{T}, \lambda), \varrho \rangle$ is convergent;*

Proof. Let $\langle (W, \mathbb{T}, \mu), (Z, \mathbb{T}, \lambda), \varrho \rangle$ be compactly dissipative and J_W be the Levinson center of $(W, \mathbb{T}, \mu)$. We will show that $J_W \cap W_z$ contains at most one point for all $z \in J_Z$. If we suppose that it is not true, then there exit $z_0 \in J_Z$ and $w_1, w_2 \in J_W \cap W_{z_0}$ such that $w_1 \neq w_2$. It is easy to see that $w = w_2 + \lambda(w_1 - w_2) = (1 - \lambda)w_2 + \lambda w_1 \in J_W \cap W_{z_0}$ for all $\lambda \in \mathbb{R}$, but this contradicts the compactness of J_W. Thus, 1. implies 2. The reverse implication is obvious. The lemma is proved☐

Theorem 2.40   *Let $\Gamma(Z, W) \neq \emptyset$ and $\langle (X, \mathbb{S}_+, \pi), (Y, \mathbb{S}, \sigma), h \rangle$ be a linear homogeneous dynamical system, $\langle (W, \mathbb{S}_+, \mu), (Z, \mathbb{S}, \lambda), \varrho \rangle$ be a linear nonhomogeneous dynamical system generated by $\langle (X, \mathbb{S}_+, \pi), (Y, \mathbb{S}, \sigma), h \rangle$ and q be a homomorphism from $(Z, \mathbb{S}, \lambda)$ onto $(Y, \mathbb{S}, \sigma)$. If the spaces Y and Z are compact and (X, h, Y) is a normed fiber bundle, then the dynamical system $\langle (W, \mathbb{S}_+, \mu), (Z, \mathbb{S}, \lambda), \rho \rangle$ is locally dissipative if and only if the dynamical system $\langle (X, \mathbb{S}_+, \pi), (Y, \mathbb{S}, \sigma), h \rangle$ is locally dissipative.*

Proof. Let $\langle (W, \mathbb{S}_+, \mu), (Z, \mathbb{S}, \lambda), \rho \rangle$ be locally dissipative and J_W be Levinson center of $(W, \mathbb{S}_+, \mu)$. According to Lemma 2.20, the set $J_W \cap W_z$ contains a single point for all $z \in J_Z$. Denote this point by w_z. Since $(W, \mathbb{S}_+, \mu)$ is locally dissipative, then there exits $\gamma > 0$ such that

$$\lim_{t \to +\infty} \rho(\mu^t B(J_W, \gamma), J_W) = 0. \tag{2.134}$$

We will show that the equality (2.134) implies

$$\lim_{t \to +\infty} \rho(\mu^t w, \mu^t w_z) = 0 \tag{2.135}$$

for all $z \in J_Z$ and $w \in B(w_z, \gamma)$. Moreover, the equality (2.135) holds uniformly in z and w. If we suppose that it is not true, then there exit $\varepsilon_0 > 0$, $t_n \to +\infty$, $z_n \in J_Z$ and $x_n \in B(w_{z_n}, \gamma)$ $(h(x_n) = z_n)$ such that

$$\rho(\mu^{t_n} w_n, \mu^{t_n} w_{z_n}) \geq \varepsilon_0. \tag{2.136}$$

From the equality (2.135) and the compactness of Z it follows that we can suppose that the sequences $\{\mu^{t_n} w_n\}$, $\{z_n t_n\}$ and $\{\mu^{t_n} w_{z_n}\}$ convergent. We put $\bar{z} = \lim_{t \to +\infty} z_n t_n$ and $\bar{w} = \lim_{t \to +\infty} \mu^{t_n} w_n$, then $\lim_{t \to +\infty} \mu^{t_n} w_{z_n} = w_{\bar{z}}$. Passing to limit in (2.136) as $n \to +\infty$, we obtain $\rho(\bar{w}, w_{\bar{z}}) \geq \varepsilon_0$. This contradicts the relation $\bar{w} \in W_{\bar{z}} \cap J_W$. The obtained contradiction proves the equality (2.136) .

Note, that the equality (2.135) implies

$$\lim_{t \to +\infty} \sup_{|x| \leq \gamma} |xt| = 0,$$

if we remark that $xt = \mu^t w_1 - \mu^t w_2 + \mu^t w_z - \mu^t w_z$, where $z = h(x)$, $|x| \leq \gamma$ and $x = w_1 - w_2$. From the equality (2.136) and Theorem 2.38 it follows the local dissipativity of the dynamical system $\langle (X, \mathbb{S}_+, \pi), (Y, \mathbb{S}, \sigma), h \rangle$.

Conversely. Let $\langle (X, \mathbb{S}_+, \pi), (Y, \mathbb{S}, \sigma), h \rangle$ be locally dissipative, then according to Theorem 2.38 there exit positive numbers N and ν such that the equality (2.130) holds. Denote by $\Gamma(Z, W)$ the space of all continuous sections of the homomorphism $\varrho : Z \to W$ endowed with the distance

$$d(\varphi, \psi) = \max_{z \in Z} \rho(\varphi(z), \psi(z)) = \max_{z \in Z} |\varphi(z) - \psi(z)|,$$

which converts it in the complete metric space. Define the mapping $S^t : \Gamma(Z, W) \to \Gamma(Z, W)$ by the equality $(S^t \varphi)(z) := \mu^t \varphi(\lambda^{-t} z)$ $(z \in Z, \varphi \in \Gamma(Z, W))$ for all $t \in \mathbb{S}$. It is easy to check that the family of mappings $\{S^t : t \in \mathbb{S}_+\}$ forms a commutative group with respect to composition. Note, that

$$\begin{aligned}
d(S^t \varphi_1, S^t \varphi_2) &= \max_{z \in Z} |\mu^t \varphi_1(\lambda^{-t} z) - \mu^t \varphi_2(\lambda^{-t} z)| \\
&= \max_{z \in Z} |\mu^t \varphi_1(z) - \mu^t \varphi_2(z)| = \max_{z \in Z} |\pi^t (\varphi_1(z) - \varphi_2(z))| \leq \\
&N e^{-\nu t} \max_{z \in Z} |\varphi_1(z) - \varphi_2(z)|.
\end{aligned} \tag{2.137}$$

From the equality (2.137) it follows that

$$d(S^t \varphi_1, S^t \varphi_2) \leq N e^{-\nu t} \max_{z \in Z} |\varphi_1(z) - \varphi_2(z)|$$

for all $\varphi_1, \varphi_2 \in \Gamma(Z, W)$ and $t \geq 0$ and, consequently, the mapping S^t is contraction for the sufficient large t. From this it follows that the commutative semi-group $\{S^t : t \in \mathbb{S}_+\}$ of mappings has a unique fixed point $\varphi \in \Gamma(Z, W)$, i.e. $(S^t \varphi)(z) = \varphi(z)$ for all $z \in Z$ and $t \in \mathbb{S}_+$ and, consequently, $\mu^t \varphi(z) = \varphi(\lambda^t z)$ $(z \in Z, t \in \mathbb{S})$. Let $w \in W$ and $z = \rho(w)$, then

$$|\mu^t w - \mu^t \varphi(z)| = |\pi^t (w - \varphi(z))| \leq N e^{-\nu t} |w - \varphi(z)| . \tag{2.138}$$

From (2.138) it follows that the compactly invariant set $J = \varphi(Z) \subset W$ is globally uniformly asymptotically stable, and in addition, $J_Z = J \cap W_Z = \{\varphi(z)\}$ for all $z \in Z$. According to Theorem 2.14, the dynamical system $\langle (W, \mathbb{S}, \mu), (Z, \mathbb{S}, \lambda), \varrho \rangle$ is convergent and, consequently, it is compactly dissipative and $J = \varphi(Z)$ is the Levinson center of $(W, \mathbb{S}, \mu)$. To prove that $(W, \mathbb{S}, \mu)$ is locally dissipative it is sufficient to show that the set $J = \varphi(Z)$ is uniformly contracting. We will show that the equality

$$\lim_{t \to +\infty} \beta(\mu^t B(J, \gamma), J) = 0.$$

holds for all $\gamma > 0$. If we suppose that it is not so, then there exit ε_0, γ_0, $w_0 \in B(J, \gamma_0)$ and $t_n \to +\infty$ such that

$$\rho(\mu^{t_n} w_n, J) \geq \varepsilon_0. \tag{2.139}$$

From the inequality (2.138) and the compactness of Z it follows that the sequence $\{\mu^{t_n} w_n\}$ can be considered convergent. Put $\overline{w} = \lim_{t \to +\infty} \mu^{t_n} w_n$, then $\overline{w} \in J$. Passing to limit in (2.139) as $n \to +\infty$, we have $\overline{w} \notin J$. The obtained contradiction proves the theorem. $\qquad\square$

Chapter 3

Analytic dissipative systems

3.1 Skew-product dynamical systems and cocycles

Let Y be a compact metric space and $W = E^n$. Consider the cocycle $\langle E^n, \varphi, (Y, \mathbb{T}_2, \sigma) \rangle$ over the dynamical system $(Y, \mathbb{T}_2, \sigma)$ with the fiber E^n and the non-autonomous dynamical system $\langle (X, \mathbb{T}_1, \pi), (Y, \mathbb{T}_2, \sigma), h \rangle$ associated by it. Then (X, h, Y), where $X := E^n \times Y$, is a finite-dimensional vectorial fibering (a trivial fibering with the fiber E^n). The function $V : X \times X \to \mathbb{R}_+$ defined by the equality $V((u_1, y), (u_2, y)) := |u_1 - u_2|$, where $|\cdot|$ is a norm on E^n defines on X a Riemannian metric.

From Theorem 2.23 follows the following statement.

Theorem 3.1 *Let Y be a compact and $\langle E^n, \varphi, (Y, \mathbb{T}_2, \sigma) \rangle$ be a cocycle over the dynamical system $(Y, \mathbb{T}_2, \sigma)$ with the fiber E^n. Then the following statements are equivalent:*

1. *There exists a positive number R such that*

$$\limsup_{t \to +\infty} |\varphi(t, u, y)| < R$$

   for all $u \in E^n$ " $y \in Y$.
2. *There is a positive number r_1 such that for all $u \in E^n$ and $y \in Y$ there exists $\tau = \tau(u, y) > 0$ for which $|\varphi(\tau, u, y)| < r_1$.*
3. *There is a positive number r_2 such that*

$$\liminf_{t \to +\infty} |\varphi(t, u, y)| < r_2$$

   for all $u \in E^n$ and $y \in Y$.
4. *There exists a positive number R_0 and for all $R > 0$ there is $l(R) > 0$ such that $|\varphi(t, u, y)| \le R_0$ for all $t \ge l(R)$, $u \in E^n$, $|u| \le R$ and $y \in Y$.*

Taking into consideration Lemma 2.14 we obtain that every condition 1.-4. that figures in Theorem 3.1 is equivalent to the dissipativity of the non-autonomous

119

dynamical system $\langle (X, \mathbb{T}_1, \pi),\ (Y, \mathbb{T}_2, \sigma), h \rangle$ associated by the cocycle $\langle E^n, \varphi, (Y, \mathbb{T}_2, \sigma) \rangle$ over $(Y, \mathbb{T}_2, \sigma)$ with the fiber E^n.

We will give below an example of a skew-product dynamical system which plays an important role in the study of non-autonomous differential equations.

Example 3.1   Let E^n be an n-dimensional real or complex Euclidean space. Consider the differential equation

$$u' = f(t, u), \tag{3.1}$$

where $f \in C(\mathbb{R} \times E^n, E^n)$. Along with the equation (3.1) we consider its H-class [32],[137], [238], [300],[302], i.e. the family of the equations

$$v' = g(t, v), \tag{3.2}$$

where $g \in H(f) = \overline{\{f_\tau : \tau \in \mathbb{R}\}}$ and $f_\tau(t, u) = f(t + \tau, u)$, where the bar indicating closure in the compact-open topology. We will suppose that the function f is regular, that is for every equation (3.2) the conditions of existence, uniqueness and extendability on $\mathbb{R}_+$ are fulfilled. Denote by $\varphi(\cdot, v, g)$ the solution of (3.2) passing through the point $v \in E^n$ for $t = 0$. Then the mapping $\varphi : \mathbb{R}_+ \times E^n \times H(f) \to E^n$ satisfies the following conditions (see, for example, [32],[290],[291]):

1) $\varphi(0, v, g) = v$ for all $v \in E^n$ and $g \in H(f)$;
2) $\varphi(t, \varphi(\tau, v, g), g_\tau) = \varphi(t + \tau, v, g)$ for each $v \in E^n$, $g \in H(f)$ and $t, \tau \in \mathbb{R}_+$;
3) $\varphi : \mathbb{R}_+ \times E^n \times H(f) \to E^n$ is continuous.

Denote by $Y := H(f)$ and $(Y, \mathbb{R}_+, \sigma)$ a dynamical system of translations on Y, induced by the dynamical system of translations $(C(\mathbb{R} \times E^n, E^n), \mathbb{R}, \sigma)$. The triple $\langle E^n, \varphi, (Y, \mathbb{R}_+, \sigma) \rangle$ is a cocycle over $(Y, \mathbb{R}_+, \sigma)$ with the fiber E^n. Hence, the equation (3.1) generates a cocycle $\langle E^n, \varphi, (Y, \mathbb{R}_+, \sigma) \rangle$ and the non-autonomous dynamical system $\langle (X, \mathbb{R}_+, \pi),\ (Y, \mathbb{R}_+, \sigma), h \rangle$, where $X := E^n \times Y$, $\pi := (\varphi, \sigma)$ and $h := pr_2 : X \to Y$.

Definition 3.1   Recall that the equation (3.1) is called dissipative [137], [271],[325],[326], if for all $t_0 \in \mathbb{R}$ and $x_0 \in E^n$ there exists a unique solution $x(t; x_0, t_0)$ of the equation (3.1) passing through the point (x_0, t_0) and if there exists a number $R > 0$ such that $\lim\limits_{t \to +\infty} \sup |x(t; x_0, t_0)| < R$ for all $x_0 \in E^n$ and $t_0 \in \mathbb{R}$. In other words, for every solution $x(t; x_0, t_0)$ there is an instant $t_1 = t_0 + l(t_0, x_0)$, such that $|x(t; x_0, t_0)| < R$ for all $t \geq t_1$. If the number $l(t_0, x_0)$ can be chosen independent on t_0, then the equation (3.1) is called uniformly dissipative [137].

Below we will establish the relation between the dissipativity of the equation (3.1) and the dissipativity of the non-autonomous dynamical system generated by the equation (3.1).

Lemma 3.1 *Let $f \in C(\mathbb{R} \times E^n, E^n)$ be regular and $H(f)$ be compact. Then the following affirmations are equivalent:*

(1) Equation (3.1) is uniformly dissipative; that is, there are numbers $R > 0$ and $l(x_0) > 0$ such that

$$|x(t; x_0, t_0)| < R \quad (t \geq t_0 + l(x_0),\ x_0 \in E^n). \tag{3.3}$$

(2) There is a positive number r such that

$$\limsup_{t \to +\infty} |\varphi(t, x_0, g)| < r \quad (x_0 \in E^n,\ g \in H(f)). \tag{3.4}$$

Proof. Let $x_0 \in E^n$ and $g \in H(f)$. If $g = f_\tau$ for some $\tau \in \mathbb{R}$. Then

$$\varphi(t, x_0, f_\tau) = x(t + \tau, x_0, \tau) \tag{3.5}$$

and the inequality (3.4) follows from (3.3). For this aim it is sufficient to take $r > R$.

Consider the case when $g = \lim_{k \to +\infty} f_{\tau_k}$ and $\tau_k \to +\infty$ (or $-\infty$). By Theorem 3.2 from [186] follows that the sequence $\{\varphi(t, x_0, f_{\tau_k})\}$ converges to a certain solution of (3.2) and this solution passes through the point $(x_0, 0) \in E^n \times \mathbb{R}$. Note, that the solution passing through the point $(x_0, 0)$ is unique because the function f is regular. Thus, $\varphi(t, x_0, g) = \lim_{k \to +\infty} \varphi(t, x_0, f_{\tau_k})$ uniformly with respect to t on every compact from $\mathbb{R}_+$. From (3.3) and (3.5) follows that $|\varphi(t, x_0, g)| = \lim_{k \to +\infty} |\varphi(t, x_0, f_{\tau_k})| \leq R$ $(t \geq l(x_0))$ and, consequently, the inequality (3.4) holds.

Conversely. Suppose that the inequality (3.4) holds. Then the non-autonomous dynamical system generated by the equation (3.1) is dissipative and by Theorem 3.1 (the condition 4.) there is $R_0 > 0$ such that for all $x_0 \in E^n$ there exists $l(x_0) = l(|x_0|) > 0$ for which

$$|\varphi(t, x_0, g) \leq R_0 \tag{3.6}$$

for all $t \geq l(x_0)$ and $g \in H(f)$. Note that

$$x(t; x_0, t_0) = \varphi(t - t_0, x_0, f_{t_o}), \tag{3.7}$$

and taking into account (3.6) we obtain (3.3). For this aim it is sufficient to take $R > R_0$. The lemma is proved. $\qquad\square$

Thus, for the equation (3.1) (f is regular and $H(f)$ is compact) we established that it is uniformly dissipative if and only if the non-autonomous dynamical system generated by this equation is dissipative.

We note that the notions of the dissipativity and uniform dissipativity for the equation (3.1) with periodic right hand side are equivalent, but for the equations with almost periodic right hand side they are different. This assertion can be illustrated by the following example.

Example 3.2   Consider the linear scalar equation

$$x' = a(t)x, \qquad\qquad (3.8)$$

where the function $a \in C(\mathbb{R}, \mathbb{R})$ is almost periodic such that all the solutions of
the equation (3.8) tend to 0 as $t \to +\infty$, but the trivial solution of (3.8) is not
uniformly stable. The equations (3.8) with the indicated above properties exist
(see, for example [123] and also the example 13.2). It is clear that along with the
equation (3.8) all translations of the equation (3.8), that is

$$x' = a_\tau(t)x,$$

where $a_\tau(t) = a(t + \tau)$ $(t, \tau \in \mathbb{R})$, possess the same property. Therefore, from
the relation (3.7) follows that $\lim\limits_{t \to +\infty} |x(t; x_0, t_0)| = 0$ $(x(t; x_0, t_0)$ is the solution of
the equation (3.8) passing through the point (x_0, t_0)), i.e. the equation (3.8) is
dissipative.

We will prove that the equation (3.8) is not uniformly dissipative. In fact, if we
suppose that is not true, then by Lemma 3.1 the non-autonomous dynamical system
generated by the equation (3.8) will be dissipative. In virtue of Theorem 2.38 the
trivial solution of (3.8) is uniformly asymptotically stable. But this contradicts to
the choice of the function a. The desired example is constructed.

Taking into account the fact established above, we will call the equation (3.1)
dissipative if the non-autonomous dynamical system generated by this equation is
dissipative.

An important class of dissipative equations are equations with convergence [137],
[270].

The equation (3.1) possesses the property of convergence if it has a unique
bounded on $\mathbb{R}$ solution which is globally uniformly asymptotically stable. This
unique bounded solution is called a limit regime for the equation (3.1).

From the results of [8],[290],[291] follows that, if the equation (3.1) is convergent,
then every equation from the family (3.2) possesses the same property. Thus, the
non-autonomous dynamical system generated by the equation (3.1) is convergent.
On the other hand, it is easy to construct examples of the equation which is not
convergent, but the non-autonomous dynamical system associated by this equation
is convergent. Note, that it is possible that the equation (3.1) has the limit regime
which is not a solution of this equation, but if the right hand side of the equation
(3.1) is stable in the sense of Poisson in positive direction, then the phenomenon
indicated above is not true. That is in this case the equation (3.1) is convergent if
and only if the corresponding non-autonomous dynamical system generated by this
equation is convergent.

Below we will call the equation (3.1) convergent if the non-autonomous
dynamical system generated by this equation is convergent.

Example 3.3 Consider the system of differential equations

$$\begin{cases} u' = F(y, u) \\ y' = G(y), \end{cases} \tag{3.9}$$

where $Y \subseteq E^m, G \in C(Y, E^n)$ and $F \in C(Y \times E^n, E^n)$. Suppose that for the system (3.9) the conditions of the existence, uniqueness and extendability on $\mathbb{R}_+$ are fulfilled. Denote by $(Y, \mathbb{R}_+, \sigma)$ a dynamical system on Y generated by the second equation of the system (3.9) and by $\varphi(t, u, y)$ – the solution of the equation

$$u' = F(\sigma(t, y), u) \tag{3.10}$$

passing through the point $u \in E^n$ for $t = 0$. Then the mapping $\varphi : \mathbb{R}_+ \times E^n \times Y \to E^n$ satisfies the conditions 1) and 2) from Example 3.1 and, consequently, the system (3.9) generates a non-autonomous dynamical system $\langle (X, \mathbb{R}_+, \pi), (Y, \mathbb{R}_+, \sigma), h \rangle$ (where $X := E^n \times Y$, $\pi := (\varphi, \sigma)$ and $h := pr_2 : X \to Y$).

We will give some generalization of the system (3.9). Namely, let $(Y, \mathbb{R}_+, \sigma)$ be a dynamical system on the metric space Y. Consider the system

$$\begin{cases} u' = F(\sigma(y, t), u) \\ y \in Y, \end{cases} \tag{3.11}$$

where $F \in C(Y \times E^n, E^n)$. Suppose that for the equation (3.10) the conditions of the existence, uniqueness and extendability on $\mathbb{R}_+$ are fulfilled. The system $\langle (X, \mathbb{R}_+, \pi), (Y, \mathbb{R}_+, \sigma), h \rangle$, where $X := E^n \times Y$, $\pi := (\varphi, \sigma)$, $\varphi(\cdot, x, y)$ is the solution of (3.10) and $h := pr_2 : X \to Y$ is a non-autonomous dynamical system generated by the equation (3.11).

3.2 $\mathbb{C}$-analytic systems

Definition 3.2 We will call that the non-autonomous dynamical system $\langle (X, \mathbb{T}_1, \pi), (Y, \mathbb{T}_2, \sigma), h \rangle$ is $\mathbb{C}$-analytic, if the following conditions are fulfilled:

a. $X := \mathbb{C}^n \times Y$ and $\pi((z, y), t) := (\varphi(t, z, y), \sigma(t, y))$ for all $t \in \mathbb{T}$ and $(z, y) \in \mathbb{C}^n \times Y$ i.e. the non-autonomous dynamical system $\langle (X, \mathbb{T}_1, \pi), (Y, \mathbb{T}_2, \sigma), h \rangle$ is generated by the cocycle $\langle \mathbb{C}^n, \varphi, (Y, \mathbb{T}_2, \sigma) \rangle$ over $(Y, \mathbb{T}_2, \sigma)$ with the fiber $\mathbb{C}^n$;
b. for each $y \in Y$ and $t \in \mathbb{T}_1$ the mapping $U(t, y) : \mathbb{C}^n \to \mathbb{C}^n$ $(U(t, y) := \varphi(t, \cdot, y))$ is holomorphic.

It follows from the results above that, when Y is compact, all three types of dissipation properties for $\mathbb{C}$-analytic systems are equivalent. In addition a $\mathbb{C}$-analytic system is dissipative if and only if there is a positive number r (independent of both z and y) such that $\limsup\limits_{t \to +\infty} |\varphi(t, z, y)| \le r$, where $|\cdot|$ denotes the norm in $\mathbb{C}^n$.

An important property of $\mathbb{C}$-analytic dissipative systems is described in

Theorem 3.2   *If Y is compact, then the $\mathbb{C}$-analytic dissipative dynamical system $\langle (X, \mathbb{T}_1, \pi), (Y, \mathbb{T}_2, \sigma), h \rangle$ is uniformly stable in the positive direction on compacts of $\mathbb{C}^n$, i.e., for arbitrary $\varepsilon > 0$ and $r > 0$, there is a $\delta = \delta(\varepsilon, r) > 0$ such that the inequality $|z_1 - z_2| < \delta$ ($|z_1|, |z_2| \le r$) implies that $|U(t, y)(z_1) - U(t, y)(z_2)| \le \varepsilon$ for all $y \in Y$ and $t \in \mathbb{T}_+$.*

Proof.   We first prove that, for arbitrary $a > 0$, there is a compact $K_a \subset \mathbb{C}^n$ such that $U(t, y)B[0, a] \subseteq K_a$ for all $t \ge 0$ and $y \in Y$. In fact, Theorem 3.1 implies that there is $b > 0$ such that, for arbitrary $a > 0$, there is $l_a \ge 0$ for which $U(t, y)B[0, a] \subseteq B[0, b]$ for $t \ge l_a$ and $y \in Y$. Let $K_a := \varphi([0, l_a], B[0, a], Y) \cup B[0, b]$; then K_a is the required set.

Now let $r > 0$; then the foregoing implies that the family of holomorphic mappings $\{U(t, y)\}$ of the space C^n into itself is uniformly bounded (with respect to $t \ge 0$ and $y \in Y$) on the ball $B[0, r]$, and the generalized Vitali theorem [167] implies that this family is equicontinuous in $B[0, r]$, i.e. for arbitrary $\varepsilon > 0$ there is $\delta(\varepsilon, r) > 0$ such that $|U(t, y)z_1 - U(t, y)z_2| < \varepsilon$ for all $(t, y) \in \mathbb{T}_+ \times Y$ if $|z_1 - z_2| < \delta$ ($|z_1|, |z_2| < r$). This proves the theorem. $\square$

Theorem 3.3   *Let $\langle (X, \mathbb{T}_1, \pi), (Y, \mathbb{T}_2, \sigma), h \rangle$ be a $\mathbb{C}$-analytic non-autonomous dissipative dynamical system and let Y be a compact minimal set; then*

(1) for each $y \in Y$ the set $J_y := J \cap X_y$ ($X_y := h^{-1}(y)$) contains only the single point x_y, where J is the centre of Levinson of the system $(X, \mathbb{T}_1, \pi)$;

(2) for all $\varepsilon > 0$ there is $\delta(\varepsilon)$ such that the inequality $\rho(x, x_y) < \delta$ ($x \in X_y$) implies that $\rho(xt, x_{yt}) < \varepsilon$ for $t \ge 0$.

(3) $\lim\limits_{t \to +\infty} \rho(xt, x_{h(x)t}) = 0$ for $x \in X$, uniformly with respect to x on compacts from X.

Proof.   Since Y is compact, it follows from the preceding remark that the non-autonomous system $\langle (X, \mathbb{T}_1, \pi), (Y, \mathbb{T}_2, \sigma), h \rangle$ is compact dissipative, and Theorem 3.2 implies that it is uniformly positively stable on compacts from X. Hence, if Y is a compact minimal set, then all conditions of Theorem 2.4 are satisfied, and applying this theorem in the case under consideration we obtain assertions 2. and 3. To complete the proof of the theorem it remains to establish 1.

Thus let $y \in Y$. According to Theorem 2.4, the set J_y is connected because $X_y := \mathbb{C}^n \times \{y\}$ ($y \in Y$) is connected. In proving Theorem 2.4 we established that the semigroup $\mathcal{P}_y$ contains an idempotent element u such that $J_y = u(X_y)$, and that there are $t_k \to +\infty$ for which $u = \lim\limits_{k \to +\infty} \pi^{t_k}$. Let $x \in X_y$, that is, $x := (z, y)$ ($z \in \mathbb{C}^n$); then $u(x) = u(z, y) = (\xi_y(z), y)$, where $\xi_y(z) := \lim\limits_{k \to +\infty} U(t_k, y)z$ and this relation holds uniformly on compacts in $\mathbb{C}^n$. Since the mappings $U(t_k, y) : \mathbb{C}^n \to \mathbb{C}^n$ $(k = 1, 2, \ldots)$ are holomorphisms, the same is true of their limit ξ_y. In proving Theorem 2.4 we showed that the idempotent element u acts on J_y in the same fashion as the identity mapping, i.e. $u(x) = x$ $(x \in J_y)$; hence $\xi_y(z) = z$ for all z such that $(z, y) \in J_y$. Thus, if $(z, y) \in J_y$, then $z \in \mathbb{C}^n$ is a fixed point of ξ_y. Since ξ_y is a holomorphism, it has at most a finite number of fixed points in the interior of each bounded set. It follows that J_y contains a finite number of points, and its connectivity implies that it contains only one point x_y. This proves the theorem.$\square$

Corollary 3.1 *If Y is compact and minimal, then $\mathbb{C}$-analytic non-autonomous dynamical system $\langle (X, \mathbb{T}_1, \pi), (Y, \mathbb{T}_2, \sigma), h \rangle$ is convergent.*

Corollary 3.2 *Under the conditions of Theorem 3.3 the dynamical system $(X, \mathbb{T}_1, \pi)$ induces on the centre of Levinson J of the system $(X, \mathbb{T}_1, \pi)$ a dynamical system $(J, \mathbb{T}_2, \widehat{\pi})$ which is isomorphic to $(Y, \mathbb{T}_2, \sigma)$. In particular, the point x_y is recurrent (almost periodic, quasi-periodic, periodic) if $y \in Y$ is recurrent (almost periodic, quasi-periodic, periodic).*

Proof. This follows from Theorem 3.3 (see also Corollary 3.1). It is sufficient to note that the mapping $p : Y \to J$, defined by the relation $p(y) := x_y$ $(J_y = \{x_y\})$, is a homomorphism of $(Y, \mathbb{T}_2, \sigma)$ onto $(J, \mathbb{T}_2, \widehat{\pi})$. $\square$

The general results obtained in this section can be used in the study of the dissipative differential equations with the recurrent (with respect to time t) right side if it is holomorph with respect to spatial variable z. Before formulate the respectively results we will give the following example

Example 3.4 Denote by $\mathcal{A}(\mathbb{R} \times \mathbb{C}^n, \mathbb{C}^n)$ the set of all continuous with respect to $t \in \mathbb{R}$ and holomorphic with respect to $z \in \mathbb{C}^n$ functions $f : \mathbb{R} \times \mathbb{C}^n \to \mathbb{C}^n$, equipped with the topology of convergence uniform on the compacts from $\mathbb{R} \times \mathbb{C}^n$. Consider the equation (3.1) with right side f from $\mathcal{A}(\mathbb{R} \times \mathbb{C}^n, \mathbb{C}^n)$. Along with equation (3.1) we consider its H - class, i.e. the family of equations

$$\dot{z} = g(t, z) \qquad (g \in H(f)), \tag{3.12}$$

where $H(f) = \overline{\{f_\tau : \tau \in \mathbb{R}\}}$, $f_\tau(t, z) = f(t + \tau, z)$ and the bar indicating closure in the space $\mathcal{A}(\mathbb{R} \times \mathbb{C}^n, \mathbb{C}^n)$. Let $\varphi(\cdot, z, g)$ be the solution of (3.12) passing through the point z for $t = 0$ and defined on $\mathbb{R}_+$.

Note the following properties of the mapping $\varphi : \mathbb{R}_+ \times \mathbb{C}^n \times H(f) \to \mathbb{C}^n$

a. $\varphi(0, z, g) = z$ $(z \in \mathbb{C}^n, \quad g \in H(f))$;

b. $\varphi(\tau, \varphi(t, z, g), g_t) = \varphi(t + \tau, z, g)$ $(\tau \in \mathbb{R}_+, \ t \in \mathbb{R}, \ z \in \mathbb{C}^n$ and $g \in H(f))$;

c. the mapping $\varphi : \mathbb{R}_+ \times \mathbb{C}^n \times H(f) \to \mathbb{C}^n$ is continuous and, for fixed $t \in \mathbb{R}_+$ and $g \in H(f)$ the mapping $U(t, g) := \varphi(t, \cdot, g) : \mathbb{C}^n \to \mathbb{C}^n$ is holomorphic [122].

We write $(Y, \mathbb{R}, \sigma)$ $(Y := H(f))$ for the dynamical system of translations on $H(f)$ and define on $X := \mathbb{C}^n \times Y$ the semi-group dynamical system $(X, \mathbb{R}_+, \pi)$ as follows: $\pi(\tau, (z, g)) := (\varphi(\tau, z, g), g_\tau)$. Put $h := pr_2 : X \to Y$ (h is a homomorphism of $(X, \mathbb{R}_+, \pi)$ onto $(Y, \mathbb{R}, \sigma)$). Equation (3.12) therefore generates the non-autonomous dynamical system $\langle (X, \mathbb{R}_+, \pi), (Y, \mathbb{R}, \sigma), h \rangle$. Property c. of φ implies that the non-autonomous dynamical system constructed is $\mathbb{C}$-analytic and application to it of Theorem 3.3 yields

Theorem 3.4 *If $f : \mathbb{R} \times \mathbb{C}^n \to \mathbb{C}^n$ is analytic in $z \in \mathbb{C}^n$ and continuous in $t \in \mathbb{R}$ and f recurrent with respect to $t \in \mathbb{R}$ uniformly with respect to z on compacts from $\mathbb{C}^n$ (in particular, it is almost periodic or periodic), then the dissipativity of equation (3.12) implies that it is convergent, i.e. (3.12) has a unique solution $p : \mathbb{R} \to \mathbb{C}^n$, bounded on $\mathbb{R}$ and uniformly asymptotically stable in the large; moreover, p is recurrent (almost periodic or periodic) if f has this property.*

Proof. This is a direct consequence of Theorem 3.3 Corollaries 3.1 and 3.2. It should be kept in mind that the recurrence of f means that the set $H(f)$ is compact and minimal with respect to the dynamical system of translations on $H(f)$. $\square$

Remark 3.1 *The "local" variant of Theorem 3.3 (and so of Theorem 3.4) holds too, i.e., Theorem 3.4 remains in force when $\mathbb{C}^n$ is replaced everywhere by a domain $W \subseteq \mathbb{C}^n$ (W can be both bounded and unbounded and can coincide with $\mathbb{C}^n$). Hence Theorem 3.4 (see also Corollary 3.2) implies that the autonomous equation $\dot{z} = f(z)$ cannot have limit cycles in $W \subseteq \mathbb{C}$ if f is holomorphic on W.*

Definition 3.3 $\langle (X, \mathbb{T}_1, \pi), (Y, \mathbb{T}_2, \sigma), h \rangle$ is said to be $\mathbb{C}$-analytic on the set $M \subseteq X$, if M is invariant and the system $\langle (M, \mathbb{T}_1, \pi), (\widetilde{Y}, \mathbb{T}_2, \sigma), h \rangle$ is $\mathbb{C}$-analytic, where $\widetilde{Y} := h(M)$, but $(M, \mathbb{T}_1, \pi)$ is a restriction $(X, \mathbb{T}_1, \pi)$ on M (analogous is defined $(\widetilde{Y}, \mathbb{T}_2, \pi)$).

Theorem 3.5 *Suppose that the following conditions are fulfilled:*

1. *$(Y, \mathbb{T}_2, \sigma)$ is compact dissipative and its Levinson's centre J_Y is a minimal set;*
2. *$(X, \mathbb{T}_1, \pi)$ is compact dissipative and J_X is its Levinson's centre;*
3. *$\langle (X, \mathbb{T}_1, \pi), (Y, \mathbb{T}_2, \sigma), h \rangle$ is $\mathbb{C}$-analytic on J_X.*

Then the set $J_X \cap X_y$ contains only one point for every $y \in J_Y$ and, consequently, the system $\langle (X, \mathbb{T}_1, \pi), (Y, \mathbb{T}_2, \sigma), h \rangle$ is convergent.

Proof. Since J_X is nonempty, compact and J_Y is minimal, then $h(J_X) = J_Y$. From the $\mathbb{C}$-analyticity of $\langle (X, \mathbb{T}_1, \pi), (Y, \mathbb{T}_2, \sigma), h \rangle$ on J_Y follows, according to Theorem 3.3, that $J_X \cap X_y$ contains only one point for any $y \in J_Y$. Thus a non-autonomous dynamical system $\langle (X, \mathbb{T}_1, \pi), (Y, \mathbb{T}_2, \sigma), h \rangle$ is convergent. $\square$

Corollary 3.3 *Suppose that for certain $y_0 \in Y$ $H^+(y_0) = Y$ is compact and ω_{y_0} is minimal. If the conditions 2. and 3. of Theorem 3.5 are fulfilled, then the system $\langle (X, \mathbb{T}_1, \pi), (Y, \mathbb{T}_2, \sigma), h \rangle$ is convergent.*

Proof. This assertion follows from Theorem 3.4. In fact, under the conditions of the Corollary 3.3 the system $(Y, \mathbb{T}_2, \sigma)$ is compact dissipative and $J_Y = \omega_{y_0}$. $\square$

Corollary 3.4 *Let y_0 be asymptotic recurrent (i.e. there is a recurrent point $q_0 \in Y$ such that $\lim\limits_{t \to +\infty} \rho(y_0 t, q_0 t) = 0$). If $Y := H^+(y_0)$ and the conditions of Theorem 3.5 are fulfilled, then the system $\langle (X, \mathbb{T}_1, \pi), (Y, \mathbb{T}_2, \sigma), h \rangle$ is convergent and each point $x \in X$ is asymptotic recurrent.*

Proof. The first statement of Corollary follows from Theorem 3.5 and Corollary 3.3. Let now $x \in X_y$ and $y \in Y = H^+(y_0)$. Since the point y_0 is asymptotic recurrent, then the point y is so asymptotic recurrent, i.e. there is a recurrent point $q \in \omega_{y_0}$ such that $\rho(yt, qt) \to 0$ as $t \to +\infty$. Taking into account that $\langle (X, \mathbb{T}_1, \pi), (Y, \mathbb{T}_2, \sigma), h \rangle$ is convergent, then $J_X \cap X_q$ consists from the single point p. Since in our conditions $\rho(xt, pt) \to 0$ as $t \to +\infty$, then the point x is asymptotic recurrent. $\square$

Consider the equation

$$\dot{z} = f(t, z) + r(t, z) \quad (z \in \mathbb{C}). \tag{3.13}$$

It follows directly from Theorem 3.5 and Corollary 3.4 the following theorem.

Theorem 3.6 *Let $f, r \in C(\mathbb{R} \times \mathbb{C}^n, \mathbb{C}^n)$ and the following conditions are fulfilled:*

(1) equation (3.13) is dissipative;
(2) $\lim\limits_{t \to +\infty} |r(t, z)| = 0$ uniformly in z on compacts from $\mathbb{C}^n$;
(3) f is holomorphic in $z \in \mathbb{C}^n$ and recurrent (almost periodic, periodic) in $t \in \mathbb{R}$ uniformly in z on the compacts from $\mathbb{C}^n$.

Then the equation

$$\dot{z} = f(t, z)$$

has a unique bounded on $\mathbb{R}$ solution $p \in C(\mathbb{R}, \mathbb{C}^n)$, which is recurrent (almost periodic, periodic) in $t \in \mathbb{R}$ and $\lim\limits_{t \to +\infty} |\varphi(t) - p(t)| = 0$ for all solution φ of (3.13).

In the conclusion of this section we will give the following

Example 3.5 Consider the system of differential equations

$$\begin{cases} \dot{z}_1 = -2z_1 + z_2^2 \\ \dot{z}_2 = -z_2 \end{cases} \quad (z_1, z_2 \in \mathbb{C}). \tag{3.14}$$

Dynamical system defined by this system is $\mathbb{C}$-analytic and dissipative. In addition, from [7, p.167-174] follows that the system (3.14) can not linearized by biholomorphic transformation.

Thus, there are the nonlinear $\mathbb{C}$-analytic dissipative systems, which can not linearized by biholomorphic transformation.

Remark 3.2   *For real analytic system the assertion analogous to example 3.5 does not hold, as the example below shows.*

Example 3.6   Consider the system of the differential equations

$$\begin{cases} \dot{x} = (1 - x^2 - y^2)x \\ \acute{y} = (1 - x^2 - y^2)y \end{cases} \tag{3.15}$$

It is possible to check by directly integration that the system (3.15) is dissipative and its Levinson's centre $J = \{(x,y) : \; x^2 + y^2 \leq 1\}$ contains a continuum points. Thus , the analogous of Theorem 3.3 for real analytic system is not true.

3.3   Converse of Lyapunov's theorem for $\mathbb{C}$-analytic systems

Let $\mathbb{R}$ ($\mathbb{C}$) be the set of all real (complex) numbers and $\mathbb{C}^n$ be n-dimensional complex space with norm $\|z\| := \max\{|z_i| \; : \; i = 1, ..., n\}$. Denote by $\mathcal{A}(\mathbb{R} \times W, \mathbb{C}^n)$ the space of all continuous functions $f : \mathbb{R} \times W \to \mathbb{C}^n$ which are holomorphic in the second variable, where W is a domain in $\mathbb{C}^n$ containing $0 \in \mathbb{C}^n$.

We consider the differential equation

$$\frac{dz}{dt} = f(t, z), \tag{3.16}$$

where $f \in \mathcal{A}(\mathbb{R} \times \mathbb{C}^n, \mathbb{C}^n)$ and $f(t, 0) = 0.$ along with (3.16) we consider the variational equation for the zero solution

$$\frac{du}{dt} = A(t)u, \tag{3.17}$$

in which $A(t) := \dfrac{\partial f}{\partial t}(t, 0)$. We denote by $z(t; t_0, v)$ the solution of (3.16) passing through $v \in \mathbb{C}^n$ for $t = t_0$.

Definition 3.4   We recall (see, for example, [273]) that the solution $z = 0$ of (3.16) is said to be stable if for any $\varepsilon > 0$ and $t_0 \in \mathbb{R}$ there exists a $\delta = \delta(\varepsilon, t_0) > 0$ such that for all $z \in B(0, \delta) = \{w| \; \|w\| < \delta\}$ and $t \geq t_0$ one has $\|z(t; t_0, v)\| \leq \varepsilon$. If one can choose $\delta(\varepsilon, t_0)$ independent of $t_0 \in \mathbb{R}$, then the solution $z = 0$ is called uniformly stable.

Definition 3.5 The solution $z = 0$ of (3.16) is said to be attracting if for any $t_0 \in \mathbb{R}$ one can find an $\eta = \eta(t_0) > 0$ and for any $\varepsilon > 0$ and $\|v\| < \eta$ one can find a $\sigma = \sigma(t_0, \varepsilon, v) > 0$ such that $\|z(t; t_0, v)\| < \varepsilon$ for all $t \geq t_0 + \sigma$. If one can choose η independent of t_0 and σ, but dependent only on ε, then the solution $z = 0$ is called uniformly attracting.

Definition 3.6 The solution $z = 0$ is said to be asymptotically stable (uniformly asymptotically stable) if it is stable (uniformly stable) and is attracting (uniformly attracting).

Theorem 3.7 *The following assertions hold:*

(1) if the zero solution of (3.16) is stable (uniformly stable), then the zero solution of (3.17) is also stable (uniformly stable);

(2) if the zero solution of (3.16) is attracting (uniformly attracting), then so is the zero solution of (3.17).

Proof. Let the solution $z = 0$ of (3.16) be stable (uniformly stable), $\varepsilon > 0$, $t_0 \in \mathbb{R}$ and $\delta = \delta(\varepsilon, t_0) > 0$ ($\delta = \delta(\varepsilon) > 0$) be a positive number from the stability (uniform stability) condition. By the theorem on analytic dependency on the initial data [122] the function $z(t; t_0, v)$ is holomorphic in v for fixed t and t_0 in the domain of the definition of $z(t; t_0, v)$. We compute the linear part of the Taylor expansion at the point $v = 0$ of the function $z(t; t_0, v)$. According to the variation of parameters formula we have

$$z(t; t_0, v) = U(t, t_0)v + \int_{t_0}^{t} U(t, \tau)F(\tau, z(\tau; t_0, v))d\tau,$$

where $U(t, \tau) := U(t, A)U^{-1}(\tau, A)$, $U(t, A)$ being the Cauchy operator of (3.17) and $F(t, z) := f(t, z) - A(t)z$.

We note that $z(t; t_0, v) = v + \int_{t_0}^{t} f(\tau, z(\tau; t_0, v))d\tau$ and hence

$$\left. \frac{\partial z}{\partial v}(t; t_0, v) \right|_{v=0} = E + \int_{t_0}^{t} \left. \frac{\partial f}{\partial z}(\tau, z(\tau; t_0, v)) \right|_{v=0} \left. \frac{\partial z}{\partial v}(\tau; t_0, v) \right|_{v=0} d\tau,$$

where E is the identity matrix. Thus, $V(t, t_0) = U(t, t_0)$ satisfies the matrix differential equation $\frac{\partial V}{\partial t} = A(t)V$ and the initial condition $V(t_0, t_0) = E$ so $V(t, t_0) = U(t, t_0)$. Since

$$\frac{\partial}{\partial v}\left[\int_{t_0}^{t} U(t, \tau)F(\tau, z(\tau; t_0, v))d\tau \right]_{v=0} = \int_{t_0}^{t} U(t, \tau)\frac{\partial F}{\partial z}(\tau, 0) \cdot U(\tau, t_0)d\tau = 0,$$

the linear part of the Taylor expansion in v of $z(t; t_0, v)$ for any $t_0 \in \mathbb{R}$ and $t \geq t_0$ is $U(t; t_0)v$. On the other hand, this linear part can be computed from the Cauchy integral formula [207]. Since in $\mathbb{C}^n$ the l_∞ norm is chosen, the domain $B(0, \delta)$ is an open polydisk and one can write

$$U(t, t_0)w = \frac{1}{(2\pi i)^n} \int\limits_{|v_1| = \frac{\delta}{2}} \cdots \int\limits_{|v_n| = \frac{\delta}{2}} \frac{z(t; t_0, v)}{v_1 \cdot v_2 \cdots v_n} \times \sum_{k=1}^{n} \frac{w_k}{v_k} dv_1 \cdots dv_n. \quad (3.18)$$

The stability (uniform stability) of the zero solution of (3.17) follows directly from (3.18) if the zero solution of (3.16) is stable (uniformly stable).

If the zero solution of (3.16) is attracting (uniformly attracting), then passing to the limit in (3.18) and considering Lebesgue's theorem on passage to the limit under the integral sign as $t \to +\infty$ we get the analogous property for the zero solution of (3.17). The theorem is proved. $\qquad\square$

Corollary 3.5   *If the zero solution of (3.16) is asymptotically stable (uniform asymptotically stable), then so is the zero solution of (3.17).*

Remark 3.3   *1. If (3.16) is autonomous, Theorem 3.7 is a result of Lyubich [241].*

2. For a real analytic system the assertion analogous to Theorem 3.7, does not hold, as the example $\dot{y} = -y^3$ shows.

3. Theorem 3.7 is true also for the difference equations.

Along with equation (3.16) we consider its H-class

$$\frac{dz}{dt} = g(t, z) \qquad (g \in H(f)). \quad (3.19)$$

Let $\Theta := \{(0, g) | g \in H(f)\} \subseteq \mathbb{C}^n \times H(f)$, $W^s(\Theta) := \{(v, g) | (v, g) \in \mathbb{C}^n \times H(f), \lim\limits_{t \to +\infty} \|\varphi(t, v, g)\| = 0\}$ and $\langle (X, \mathbb{R}_+, \pi), (Y, \mathbb{R}, \sigma), h \rangle$ be a non-autonomous dynamical system generated by the equation (3.16) (see the example 3.1). If the zero solution of (3.16) is uniform asymptotically stable and the set $H(f)$ is compact, then the compact invariant set $\Theta \subseteq X = \mathbb{C}^n \times H(f)$ is asymptotically stable and its domain of attraction $W^s(\Theta)$ is open. The following theorem holds.

Theorem 3.8   *Let $f \in \mathcal{A}(\mathbb{R} \times \mathbb{C}^n, \mathbb{C}^n)$, $H(f)$ be compact and the zero solution of (3.16) be uniform asymptotically stable, then the set $W^s(\Theta)$ is unbounded.*

Proof.   Let as assume that $W^s(\Theta)$ is bounded, then the set $M = \overline{W^s(\Theta)} \subseteq X$ is compact invariant subset $(X, \mathbb{R}_+, \pi)$. In virtue of Theorem 3.7 the zero solution of (3.17) is also uniform asymptotically stable and hence one can find positive numbers $\mathcal{N}$ and ν such that

$$\|U(t, t_0)\| \leq \mathcal{N} e^{-\nu(t - t_0)} \quad (3.20)$$

for all $t \geq t_0$ $(t, t_0 \in \mathbb{R})$. It follows from (3.20) that

$$\|U(t, A)\| \geq \mathcal{N}^{-1} e^{-\nu t} \qquad (3.21)$$

for all $t \leq 0$.

On the other hand, it follows from (3.18) and the boundedness of $W^s(\Theta)$ that

$$\sup_{t \in \mathbb{R}} \|U(t, A)\| < +\infty,$$

which contradicts (3.21). The theorem is proved. $\qquad\square$

Corollary 3.6 *Let $f \in \mathcal{A}(\mathbb{R} \times \mathbb{C}^n, \mathbb{C}^n)$, $H(f)$ be compact and the zero solution of (3.16) be uniform asymptotically stable. Then the set $W_f^s(0) = \{v \mid v \in \mathbb{C}^n,\ \lim_{t \to +\infty} \|\varphi(t, v, f)\| = 0\}$ is unbounded.*

Let Ω be a metric space and $(\Omega, \mathbb{R}, \sigma)$ be a dynamical system on Ω. Consider the differential equation

$$\dot{x} = f(\omega t, x) \qquad (3.22)$$

where $\omega t := \sigma(t, \omega)$, $f \in \mathcal{A}(\Omega \times \mathbb{C}^n, \mathbb{C}^n)$, $x \in \mathbb{C}^n$ and $f(\omega, 0) = 0$ for all $\omega \in \Omega$. The results presented above are true also for the equation (3.22). We will formulated some of them.

Theorem 3.9 *If the zero solution of (3.22) is uniform asymptotically stable, then the zero solution of*

$$\dot{y} = A(\omega t)y, \qquad (3.23)$$

where $A(\omega) := \left. \dfrac{\partial f(\omega, x)}{\partial x} \right|_{x=0}$, is also uniform asymptotically stable.

Theorem 3.10 *Let the zero solution of (3.22) be uniform asymptotically stable and Ω be compact. Then the set*

$$W^s(\Theta) = \{(x, \omega) \in \mathbb{C}^n \times \Omega \mid \lim_{t \to +\infty} |\varphi(t, x, \omega)| = 0\}$$

is unbounded, where $\varphi(t, x, \omega)$ is the solution of (3.22) passing through the point $x \in \mathbb{C}^n$ for $t = 0$ and defined on $\mathbb{R}_+$.

Proof. These statements can be proved as Theorem 3.7 and 3.8. $\qquad\square$

3.4   On the structure of compact attracting sets of $\mathbb{C}$-analytic systems

The present section is devoted to the structure of compact invariant sets of the system

$$\frac{dz}{dt} = f(t,z) \tag{3.24}$$

with right-hand side f holomorphic with respect to z and almost periodic with respect to t. We show that an asymptotically stable compact invariant set of system (3.24) consists of almost periodic solutions to this system. This problem is studied and solved in the frame of general non-autonomous dynamical systems. We give also the description of the structure of uniform asymptotically stable compact invariant set of arbitrary $\mathbb{C}$–analytic non-autonomous dynamical system with compact minimal base.

We recall that the set $K \subseteq X$ is called orbitally stable with respect to non-autonomous system (2.1), if for any $\varepsilon > 0$ there exist $\delta(\varepsilon) > 0$ such that the inequality $\rho(x, K_y) < \delta$ ($x \in X, y = h(x)$) implies $\rho(xt, K_{yt}) < \varepsilon$ for all $t \geq 0$. In addition, if there exist $\gamma > 0$ such that $\rho(xt, K_{yt}) \to 0$ for $t \to +\infty$ and every $x \in K_y$ such that $\rho(x, K_y) \leq \gamma$ then it is said that K orbitally asymptotically stable.

Theorem 3.11   *Assume that $\langle (X, \mathbb{S}_+, \pi), (Y, \mathbb{S}, \sigma), h \rangle$ is a non-autonomous $\mathbb{C}$-analytic dynamical system and the following conditions are satisfied:*

(1) Y is compact and locally connected and $(Y, \mathbb{S}, \sigma)$ is minimal, i.e., $Y = H(y) = \overline{\{yt \,|\, t \in \mathbb{S}\}}$ for all $y \in Y$;

(2) $M \subset X$ is nonempty, compact, orbitally asymptotically stable, and invariant with respect to $\langle (X, \mathbb{S}_+, \pi), (Y, \mathbb{S}, \sigma), h \rangle$;

(3) M orbitally asymptotically stable.

Then $\langle (M, \mathbb{S}, \pi), (Y, \mathbb{S}, \sigma), h \rangle$ is a distal covering of finite multiplicity, i.e.,

(a) for any $y \in Y$ the set $M_y = \{ m \,|\, m \in M,\ h(m) = y \}$ consists of finitely many points;

(b) $\inf\limits_{t \in \mathbb{S}} \rho(x_1 t, x_2 t) > 0$ for all $x_1, x_2 \in M$ ($x_1 \neq x_2$, $h(x_1) = h(x_2)$).

Proof.   Since the space Y is locally connected and the bundle (X, h, Y) is locally trivial, the space X is locally connected as well. By Proposition 1.18 [33], the set M has finitely many connected components $M_1, M_2, \ldots, M_p$. We write

$$W_y^s(M) := \{ x \,|\, x \in X_y = h^{-1}(y),\ \lim_{t \to +\infty} \rho(xt, M_{yt}) = 0 \}$$

and note that $W_y^s(M)$ is open in X_y. By theorem 1.36, for any compact set $K \subset W^s(M) := \cup \{ W_y^s(M) \,|\, y \in Y \}$, the set $\Sigma^+(K) := \cup \{ \pi^t K \,|\, t \geq 0 \}$ is relatively compact.

We set $K_i = W^s(M_i) \cap K$ and note that K_i $(i \in \overline{1,p})$ are closed, $K_i \cap K_j = \emptyset$, if $i \neq j$, and $K = K_1 \cup K_2 \cup \ldots K_p$. Let $\gamma > 0$ $(\gamma < \alpha = \min\limits_{i \neq j} \rho(K_i, K_j))$ and $l > 0$ such that we have $B(M, \gamma)$, $B(K, \gamma) \subset W^s(M)$ and $|xt| \leq l$ for all $t \geq 0$ and $x \in \overline{B(K, \gamma)}$. By the Cauchy integral formula [167] there exists a number $L = L(l, \gamma) > 0$ such that

$$\rho(\pi^t x_1, \pi^t x_2) \leq L\rho(x_1, x_2) \tag{3.25}$$

for all $t \geq 0$ and $x_1, x_2 \in K_i$ $(i \in \{1, 2, \ldots, p\}, h(x_1) = h(x_2))$.

Let us now show that for any $\varepsilon > 0$ and any compact set $K \subset W^s(M)$, there exists $\delta(\varepsilon, K) > 0$ such that $\rho(x_1, x_2) < \delta$ $(x_1, x_2 \in K, h(x_1) = h(x_2))$ the relation implies

$$\rho(x_1 t, x_2 t) < \varepsilon \tag{3.26}$$

for all $t \geq 0$. If we suppose the contrary, then there exist ε_0, $\delta_n \downarrow 0$, $K_0 \subset W^s(M)$, $\{x_n^i\} \subset K_0$ $(i = 1, 2; h(x_n^1) = h(x_n^2))$ and $t_n \to +\infty$ such that

$$\rho(x_n^1, x_n^2) < \delta_n \quad \text{and} \quad \rho(x_n^1 t_n, x_n^2 t_n) \geq \varepsilon_0. \tag{3.27}$$

Since K_0 is compact we can assume that the sequences $\{x_n^i\}$ $(i = 1, 2)$ are convergent. Let $x_0 := \lim\limits_{n \to +\infty} x_n^1 = \lim\limits_{n \to +\infty} x_n^2$ and $x_0 \in K_{y_0} := K \cap X_{y_0}$. Hence $W_{y_0}^s(M) = \bigcup\limits_{i=1}^{p} W_{y_0}^s(M_i)$, there exists $i_0 \in \{1, 2, \ldots, p\}$ such that $x_0 \in W_{y_0}^s(M_{i_0})$. Since $W_{y_0}^s(M_{i_0})$ is open set in X_y, there exists $\delta_0 > 0$ such that $B_{y_0}(x_0, \delta_0) := \{x \mid x \in X_{y_0}, \rho(x, x_0) < \delta_0\} \subset W_{y_0}^s(M_{i_0})$. For sufficiently large n we have $\rho(x_n^1, x_n^2) < \frac{\varepsilon_0}{2L}$. Without loss of generality we can assume that $\{x_n^i\} \subset K_{i_0}$ $(i = 1, 2)$; then by (3.25) we have the estimate

$$\rho(x_n^1 t_n, x_n^2 t_n) \leq L\rho(x_n^1, x_n^2) < \frac{\varepsilon_0}{2}.$$

which contradicts (3.27). The obtained contradiction completes the proof of our statement.

Now we will prove the first statement of the theorem. Note that it follows from inequality (3.25) implies that an invariant set M is distal in the negative direction with respect to h. Now, since Y is minimal, it follows from Lemma 2.8 [32] (see also Lemma 1 [238, p.104]) that M is two-sided distal.

Let $\alpha > 0, y \in Y$. We set $E_y^\alpha := \overline{\{\pi^t|_{W_y^s(M)} \mid t \geq \alpha\}}$, where the bar denotes the closure in the compact-open topology, and $P_y := \{\xi \mid \xi \in E_y^\alpha, \xi(W_y^s(M)) \subseteq W_y^s(M)\}$. As in Lemma 2.7, we can establish that P_y is a nonempty compact topological semigroup. By Lemma 4.11 [32], there exists an idempotent $u \in P_y$, i.e., $u^2 = u$. Since $u \in P_y \subseteq E_y^\alpha$ and Y is minimal, there exists $t_n \to +\infty$ such that $u = \lim\limits_{n \to +\infty} \pi^{t_n}|_{W_y^s(M)}$ and hence $u(W_y^s(M)) \subseteq M_y$. Indeed, we have $M_y = u(W_y^s(M))$, because M is two-sided distal and the idempotent u actcs on M_y

as the identity mapping. Now let $x \in W_y^s(M)$ and $p = u(x) \in M_y$. Let us show that

$$\lim_{t \to +\infty} \rho(xt, pt) = 0. \tag{3.28}$$

Indeed, since $u(p) = u(u(x)) = u(x)$, such that $\rho(xt_n, pt_n) \to 0$ as $n \to +\infty$. Let $\varepsilon > 0$, $K = H^+(x) = \overline{\{xt \mid t \geq 0\}}$ and $\delta = \delta(\varepsilon, K) > 0$ be such that (3.26) holds for $\rho(x_1, x_2) < \delta$ ($x_1, , x_2 \in K$ and $h(x_1) = h(x_2)$). Choose n_0 such that $\rho(xt_n, pt_n) < \delta$ for all $n \geq n_0$; then we have

$$\rho(x(t_n + t), p(t_n + t)) < \varepsilon$$

for all $t \geq 0$ and thus relation (3.28) holds.

Consider the open covering $\{B(p, \gamma) \mid p \in M_y\}$ of the set M_y. Since M_y is compact, this covering contains a finite sub-covering $\{B(p_i, \gamma) \mid i = \overline{1, m}\}$. Note that $u(B(p_i, \gamma)) \subseteq M_y$ and $p_i \in u(B(p_i, \gamma))$. Since $M_y \subseteq \cup\{B(p_i, \gamma) \mid i \in \overline{1, m}\}$, we have $m_y = u(M_y) \subseteq \cup\{u(B(p_i, \gamma)) \mid i \in \overline{1, m}\} \subseteq M_y$, and therefore $\cup\{u(B(p_i, \gamma)) \mid i = \overline{1, m}\} = M_y$. Since u is holomorphic and $u(x) = x$ for all $x \in M_y$, the set M_y is analytic. Since M_y is compact, it consists of finitely many points [167]. Let $M_y = \{x_1, x_2, \ldots, x_{m(y)}\}$. Consider the non-autonomous dynamical system $\langle (M, \mathbb{S}, \pi), (Y, \mathbb{S}, \sigma), h \rangle$. As proved above, this system is two-sided distal. By Proposition 4 [238], the mapping $H : Y \to 2^M$ ($H(y) := M_y$) is continuous in the Hausdorff metric, and therefore $card M_y = n(y)$ does not depend on $y \in Y$. Thus, the theorem is completely proved. $\square$

Corollary 3.7 *Assume that the conditions of Theorem 3.11 are satisfied and that Y consists of recurrent (Bohr almost periodic, quasi-periodic, periodic, stationary) motions. Then the set M consists of recurrent (Bohr almost periodic, quasi-periodic, periodic, stationary respectively) motions.*

Proof. This assertion follows from Theorem 3.11, Theorem 3 [238] and Lemma 1[238]. $\square$

Theorem 3.12 *Under the conditions of Theorem 3.11 the following assertions hold:*

(1) the set M is uniform stable in the sense of Lyapunov, i.e. $\forall \varepsilon > 0 \ \exists \ \delta(\varepsilon) > 0$ such that $\rho(x, p_0) < \delta$ ($x \in X_y, p \in M_y$) implies $\rho(xt, pt) < \varepsilon$ for all $t \geq 0$;
(2) for each $x \in W^s(M)$ there is $p \in M_y$ ($y := h(x)$) such that

$$\lim_{t \to +\infty} \rho(xt, pt) = 0.$$

In particular, if Y consists from the recurrent (almost periodic, quasi-periodic, periodic, stationary) points, then the set $W^s(M)$ consists of asymptotically recurrent *(asymptotically almost periodic, asymptotically quasi-periodic, asymptotically periodic, asymptotically stationary) points.*

Proof. The first statement of Theorem follows directly from (3.26). The second assertion follows from Theorem 3.11 and Lemma 4 [48]. □

Corollary 3.8 *Let $(\mathbb{C}^n, \mathbb{S}, \pi)$ be an autonomous $\mathbb{C}$-analytic dynamical system, $M \subset \mathbb{C}$ be a nonempty compact invariant orbitally asymptotically stable set, then M consists of finitely many stationary points $\{x_1, x_2, \ldots, x_p\}$.*

Let $f \in A(\mathbb{R} \times \mathbb{C}^n, \mathbb{C}^n)$. Consider the differential equation

$$\frac{dz}{dt} = f(t, z). \tag{3.29}$$

Along with equation (3.29), we consider its H-class

$$\frac{dz}{dt} = g(t, z) \quad (g \in H(f)). \tag{3.30}$$

Definition 3.7 The set $M \subseteq \mathbb{C}^n$ is said to be invariant for equation (3.29), if for any point $v \in M$ there exists $g \in H(f)$, such that the solution $\varphi(t, v, g)$ to equation (3.30) is defined on the entire $\mathbb{R}$ and $\varphi(t, v, g) \in M$ for all $t \in \mathbb{R}$.

Definition 3.8 A compact set $M \subset \mathbb{C}^n$ invariant with respect to (3.29) is said to be orbitally stable if for any $\varepsilon > 0$, there exists $\delta(\varepsilon) > 0$ such that the relation $\rho(v, M_g) < \delta$ implies $\rho(\varphi(t, v, g), M_{g_t}) < \varepsilon$ for all $t \geq 0$, where $M_g = \{x \mid x \in M, \varphi(t, x, g) \in M \text{ for all } t \in \mathbb{R}\}$ and g_t - is a translation of g in first variable on t.

We write $W_g^s(M) = \{v \mid v \in \mathbb{C}^n, \lim_{t \to +\infty} \rho(\varphi(t, v, g), M_{g_t}) = 0\}$ and $W^s(M) = \cup\{W_g^s(M) \mid g \in H(f)\}$.

Definition 3.9 A compact set M, invariant with respect to equation (3.29), is said to be asymptotically orbitally stable if it is orbitally stable and there exists $\gamma > 0$ such that $B(M, \gamma) = \{x \mid \rho(x, M) < \gamma\} \subseteq W^s(M)$.

Applying Theorem 3.11 and Corollary 3.7, 3.8 to the constructed non-autonomous $\mathbb{C}$-analytic system generated by equation (3.29), we obtain the following statement.

Theorem 3.13 *Let $f \in A(\mathbb{R} \times \mathbb{C}^n, \mathbb{C}^n)$ be regular and recurrent (Bohr almost periodic, quasi-periodic, periodic, stationary) with respect to $t \in \mathbb{R}$ uniformly with respect to z on compact sets from $\mathbb{C}^n, M \subseteq \mathbb{C}^n$ and let the following conditions be satisfied:*

(1) M is a compact set invariant with respect to equation (3.29);

(2) M is orbitally asymptotically stable,

Then for any $g \in H(f)$, the set M_g consists of finitely many recurrent (Bohr almost periodic, quasi-periodic, periodic, stationary respectively) solutions of equation (3.30).

Remark 3.4   *A statement similar to Theorem 3.13, is valid for difference equations as well.*

3.5   Dynamical systems in spaces of sections

Let $\langle (X, \mathbb{S}_+, \pi), (Y, \mathbb{S}, \sigma), h \rangle$ be a non-autonomous dynamical system. We recall that a mapping $\gamma : Y \to X$ is called a section (selector) of the homomorphism h, if $h(\gamma(y)) = y$ for all $y \in Y$. The section γ of the homomorphism h is called invariant if $\gamma(\sigma(t, y)) = \pi(t, \gamma(y))$ for all $y \in Y$ and $t \in \mathbb{S}$.

Denote by $\Gamma = \Gamma(Y, X)$ family of all continuous sections of h, i.e. $\Gamma(Y, X) = \{\gamma \in C(Y, X) : h \circ \gamma = Id_Y\}$. We will suppose that $\Gamma(Y, X) \neq \emptyset$. This condition is fulfilled in the many important case for the applications.

We endow the space $\Gamma = \Gamma(Y, X)$ with the uniform-convergence on compacts in Y.

Let $K \subseteq Y$ be a nonempty compact from Y, then the relation

$$p_K(\gamma_1, \gamma_2) := \max_{y \in K} \rho(\gamma_1(y), \gamma_2(y))$$

defines a pseudo-metric p_K on Γ, and the family of pseudo-metrics $\{p_K : K \in C(Y)\}$, where $C(Y)$ is the family of nonempty compact subsets of Y, defines a uniform structure on Γ (a uniform-convergence topology on compacts from Y).

We consider the special case in which $Y = \bigcup_{k=1}^{\infty} Q_k$, where $\{Q_k\}$ is an expanding sequence of nested compacts. In this situation it is sufficient to use a countable number of pseudo-metrics $\mathcal{P}' = \{p_k : k = 1, 2 \ldots\}$, where

$$p_k(\gamma_1, \gamma_2) := \max_{y \in Q_k} \rho(\gamma_1(y), \gamma_2(y)).$$

It can be easily verified that this family of pseudo-metrics separates the points of Γ, and $p_1(\gamma_1, \gamma_2) \leq p_2(\gamma_1, \gamma_2) \leq \ldots p_k(\gamma_1, \gamma_2) \leq \ldots$. Hence the topology defined by this family of pseudo-metrics is metrizable. For example the topolgy defined by the metric

$$d(\gamma_1, \gamma_2) := \sum_{1}^{\infty} \frac{1}{2^k} \frac{p_k(\gamma_1, \gamma_2)}{1 + p_k(\gamma_1, \gamma_2)},$$

is consistent with the topology of Γ defined by the family of pseudo-metrics $\mathcal{P}'$.

Remark 3.5 *a. It is easily shown that Γ is a complete space if and only if X is a complete space.*

b. If (X, h, Y) is a vector fibering, then the section space $\Gamma = \Gamma(Y, X)$ is naturally endowed with a vector-space structure, and all operations are continuous with respect to the topology defined on Γ, i.e., in this case Γ is covered into a topological vector space.

c. If Y is compact and (X, h, Y) is a Banach fibering, then the relation

$$||\gamma|| := \max_{y \in Y} |\gamma(y)|$$

defines a norm on Γ (here $|\cdot| : X \to \mathbb{R}_+$ is a norm on (X, h, Y), and consistent with the metric) and the topology generated by this norm is consistent with the original topology on Γ.

Let $\mu : \mathbb{S}_+ \times \Gamma \to \Gamma$ denote the mapping defined by the relation: $\mu(t, \gamma)(y) := \pi^t \gamma(\sigma^{-t} y)$ for all $y \in Y$. It is easily verified that $\mu(t, \gamma) \in \Gamma$.

The following assertion holds.

Theorem 3.14 *The triple $(\Gamma, \mathbb{S}_+, \mu)$ is a (semigroup) dynamical system on Γ, i.e.,*

(1) $\mu(0, \gamma) = \gamma$ for all $\gamma \in \Gamma$.
(2) $\mu(\tau, \mu(t, \gamma)) = \mu(t + \tau, \gamma)$ for all $\gamma \in \Gamma$ and $t, \tau \in \mathbb{S}_+$.
(3) $\mu : \mathbb{S}_+ \times \Gamma \to \Gamma$ is continuous.

Proof. The first two assertions can be established directly. The nontrivial part of Theorem 3.14 is the continuity of μ. Let $t_\nu \to t_0$ and $\gamma_\nu \to \gamma_0$ ($\{t_\nu\}$ (respectively $\{\gamma_\nu\}$) is a directedness in $\mathbb{S}_+$ (respectively Γ)). We shall prove that $\mu(t_\nu, \gamma_\nu) \to \mu(t_0, \gamma_0)$ in the space Γ, i.e.,

$$\lim_\nu \max_{y \in K} \rho(\pi^{t_\nu} \gamma_\nu(\sigma^{-t_\nu} y),\, \pi^{t_0} \gamma_0(\sigma^{-t_0} y)) = 0.$$

for each compact $K \subseteq Y$. Assume the contrary: There is $\varepsilon_0 > 0$ and a compact $K_0 \subseteq Y$ such that

$$\max_{y \in K_0} \rho(\pi^{t_\nu} \gamma_\nu(\sigma^{-t_\nu} y),\, \pi^{t_0} \gamma_0(\sigma^{-t_0} y)) \geq \varepsilon_0.$$

Since K_0 is compact, there is $\{y_\nu\} \subseteq K_0$ for which

$$\rho(\pi^{t_\nu} \gamma_\nu(\sigma^{-t_\nu} y_\nu),\, \pi^{t_0} \gamma_0(\sigma^{-t_0} y_\nu)) \geq \varepsilon_0.$$

Again since K_0 is compact, we assume that the sequence $\{y_\nu\}$ is convergent. Let $y_0 := \lim_\nu y_\nu$, then $\sigma^{-t_\nu} y_\nu \to \sigma^{-t_0} y_0$ and hence

$$\lim_\nu \gamma_0(\sigma^{-t_\nu} y_\nu) = \gamma_0(\sigma^{-t_0} y_0). \tag{3.31}$$

From (3.31) it follows that

$$\lim_{\nu} \pi^{t_\nu}\gamma_0(\sigma^{-t_\nu}y_\nu) = \pi_0^t\gamma_0(\sigma_0^{-t}y_0). \qquad (3.32)$$

However

$$\varepsilon_0 \le \rho(\pi^{t_\nu}\gamma_\nu(\sigma^{-t_\nu}y_\nu),\ \pi^{t_0}\gamma_0(\sigma^{-t_0}y_\nu)) \le \qquad (3.33)$$
$$\rho(\pi^{t_\nu}\gamma_\nu(\sigma^{-t_\nu}y_\nu),\ \pi^{t_0}\gamma_0(\sigma^{-t_0}y_0))$$
$$+\rho(\pi^{t_0}\gamma_0(\sigma^{-t_0}y_0),\ \pi^{t_0}\gamma_0(\sigma^{-t_0}y_\nu)).$$

The second term on the right here tend to zero when $y_\nu \to y_0$. We shall prove that the first term also tends to zero $y_\nu \to y_0$ and $t_\nu \to t_0$. At first we shall establish that $\lim_{\nu}\gamma_\nu(\sigma^{-t_\nu}y_\nu) = \gamma_0(\sigma^{-t_0}y_0)$. In fact

$$\rho(\gamma_\nu(\sigma^{-t_\nu}y_\nu),\ \gamma_0(\sigma^{-t_0}y_0)) \le \rho(\gamma_\nu(\sigma^{-t_\nu}y_\nu),\gamma_0(\sigma^{-t_\nu}y_\nu)) +$$
$$\rho(\gamma_0(\sigma^{-t_\nu}y_\nu),\gamma_0(\sigma^{-t_0}y_0)) \le \max_{y\in K'}\rho(\gamma_\nu(y),\gamma_0(y)) + \qquad (3.34)$$
$$\rho(\gamma_0(\sigma^{-t_\nu}y_\nu),\gamma_0(\sigma^{-t_0}y_0)),$$

where $K' = \sigma(K_0,\mathbb{T}_0)$ and $\mathbb{T}_0 = \overline{\{-t_\nu\}}$. Letting $n \to +\infty$ in (3.34) and using (3.32) and the fact that $\gamma_\nu \to \gamma_0$ in topology Γ, we obtain the required result.

Hence $\gamma_\nu(\sigma^{-t_\nu}y_\nu) \to \gamma_0(\sigma^{-t_0}y_0)$ and $t_\nu \to t_0$, and thus

$$\rho(\pi^{t_\nu}\gamma_\nu(\sigma^{-t_\nu}y_\nu,\ \pi^{t_0}\gamma_0(\sigma^{-t_0}y_0)) \to 0. \qquad (3.35)$$

Letting $t_\nu \to t_0$ and $y_\nu \to y_0$ in (3.33) and employing (3.35), we find that $\varepsilon_0 \le 0$, in contradiction with the choice of ε_0. The mapping $\mu : \Gamma \times \mathbb{S}_+ \to \Gamma$ is therefore continuous, and this completes the proof of the theorem. $\qquad\qquad\qquad \square$

Remark 3.6   *a. If $\langle(X,\mathbb{S},\pi),(Y,\mathbb{S},\sigma),h\rangle$ is a group non-autonomous dynamical system, then the group dynamical system $(\Gamma,\mathbb{S},\mu)$ is defined on Γ.*

b. A continuous section $\gamma \in \Gamma$ is invariant if and only if $\gamma \in \Gamma$ is a stationary point of the system $(\Gamma,\mathbb{S}_+,\mu)$.

We consider a special case of the foregoing construction. Let $\langle W,\varphi,(Y,\mathbb{S},\sigma)\rangle$ be a cocycle over $(Y,\mathbb{S},\sigma)$ with fiber W and $\langle(X,\mathbb{S}_+,\pi),(Y,\mathbb{S},\sigma),h\rangle$ be a non-autonomous dynamical system generated by this cocycle. Then $h \circ \gamma = Id_Y$ and because $h = pr_2$ it follows: $\gamma = (\psi, Id_Y)$, where $\gamma \in \Gamma(Y,X)$ and $\psi : Y \to W$. Hence to each section γ there corresponds a mapping $\psi : Y \to W$ and conversely. There being a one-to-one relation between $\Gamma(Y,W \times Y)$ and $C(Y,W)$, where $C(Y,W)$ is the space of continuous functions $\psi : Y \to W$, with the uniform-convergence topology on compacts of Y, we identify these two objects from now on. The dynamical system $(\Gamma,\mathbb{S}_+,\mu)$ induces, in a natural fashion a dynamical system $(C(Y,W),\mathbb{S}_+,Q)$

on $C(Y, W)$. Namely

$$(\mu^t \gamma)(y) = \pi^t \gamma(\sigma^{-t} y) = \pi^t(\psi, Id_Y)(\sigma^{-t} y) =$$
$$\pi^t(\psi(\sigma^{-t} y), \sigma^{-t} y) = (U(t, \sigma^{-t} y)\psi(\sigma^{-t} y), y) = ((Q^t \psi)(y), y).$$

Hence $\mu^t(\psi, Id_Y) = (Q^t \psi, Id_Y)$, and so the mapping $Q : C(Y, W) \times \mathbb{S}_+ \to C(Y, W)$, defined by the relation $Q(\psi, t) = Q^t \psi$, with $(Q^t \psi)(y) = U(t, \sigma^{-t} y)\psi(\sigma^{-t} y)$ $(y \in Y)$, has the following properties:

a. $Q^0 = Id_{C(Y,W)}$;
b. $Q^t Q^\tau = Q^{t+\tau}$ $(t, \tau \in \mathbb{S}_+)$
c. Q is continuous,

i.e., $(C(Y, W), \mathbb{S}_+, Q)$ is a semigroup dynamical system on $C(Y, W)$. By virtue of the foregoing, we identify the systems (Γ, S_+, μ) and $(C(Y, W), \mathbb{S}_+, Q)$ in the sequel.

3.6 Quasi-periodic solutions

In this section we investigate dissipative differential equation (3.1) with quasi-periodic coefficients, and prove that, if its right side is analytic, then there is always at least one quasi-periodic solution with the same frequency basis as the right side.

Consider the system

$$\begin{cases} \dot{x} = F(x, \varphi), \\ \dot{\varphi} = i\omega\varphi, \end{cases} \tag{3.36}$$

where $\omega = (\omega_1, \omega_2, \ldots, \omega_m) \in \mathbb{R}^m$ $(\omega_1, \omega_2, \ldots, \omega_m$ are linearly independent with respect to the integers), $i^2 = -1$, $\varphi = (\varphi_1, \varphi_2, \ldots, \varphi_m) \in \mathbb{C}^m$, $\omega\varphi = \mathrm{col}(w_1\varphi_1, \ldots, w_m\varphi_m)$, $x \in \mathbb{C}^m$ and $F \in C(\mathbb{C}^n \times \mathbb{C}^m, \mathbb{C}^m)$. We denote by T^m m-dimensional torus, i.e. $T^m := \{\varphi \in \mathbb{C}^m : |\varphi_k| = 1, k = \overline{1, m}\}$. We are interested in the system (3.36), when the parameter φ runs over the points of the torus T^m and a small neighborhood

$$D_\delta := \{\varphi \in C^m : |\varphi_k| < 1 + \delta, k \in \overline{1, m}\},$$

where $\delta > 0$ is sufficiently small.

The system (3.36) is clearly equivalent to the equation

$$\dot{x} = F(x, \varphi_0 e^{i\omega\varphi}), \tag{3.37}$$

where $\varphi_0 \in \mathbb{C}^m$, or to a subset (for example D_δ).

Definition 3.10 We will say that the system (3.36) (or equation (3.37)) is analytic on T^m if there is a positive number δ, for which $F : \mathbb{C}^n \times D_\delta \to \mathbb{C}^n$ is analytic.

Let $(\mathbb{C}^m, \mathbb{R}, \sigma)$ denote the dynamical system on $\mathbb{C}^m$ determined by the second equation of the system (3.36). Then

$$\sigma(\varphi, t) := \mathrm{col}\left(\varphi_1 e^{i\omega_1 t}, \varphi_2 e^{i\omega_2 t}, \ldots, \varphi_m e^{i\omega_m t}\right).$$

It is easily seen that, for arbitrary $\delta > 0$ the sets D_δ and $\overline{D}_\delta$, where the bar indicates the closure of D_δ in $\mathbb{C}^m$, are invariant with respect to $(\mathbb{C}^m, \mathbb{R}, \sigma)$.

Definition 3.11　The system (3.36) (and the equation (3.37)) is said to be dissipative on $\mathcal{T}^m$, if there exists $\delta > 0$ (sufficiently small) such that the dynamical system $(\mathbb{C}^m, \mathbb{R}, \sigma)$, generated by (3.36) ($\varphi \in \overline{D}_\delta$) is dissipative.

We sharpen this definition. Suppose that, for the system (3.36), conditions for existence, uniqueness, and continuability to the right are satisfied. Then it is known that (3.36) determines a semigroup system $(\mathbb{C}^n \times \overline{D}_\delta, \mathbb{R}_+, \pi)$ on $\mathbb{C}^n \times \overline{D}_\delta$; then $\pi((x, \varphi_0), t) := (\psi(t, x, \varphi_0), \varphi_0 e^{i\omega t})$, where $\psi(\cdot, x, \varphi_0)$ is the solution of equation (3.37) passing through the point $x \in \mathbb{C}^n$ for $t = 0$. It is easily to verify that the triple $\langle(\mathbb{C}^n \times \overline{D}_\delta, \mathbb{R}_+, \pi), (\overline{D}_\delta, \mathbb{R}, \sigma), h\rangle$ ($h := pr_2 : \mathbb{C}^n \times \overline{D}_\delta \to \overline{D}_\delta$) is a non-autonomous dynamical system. Hence, if the system of differential equations (3.36) is dissipative on $\mathcal{T}^m$, then the non-autonomous dynamical system $\langle(\mathbb{C}^n \times \overline{D}_\delta, \mathbb{R}_+, \pi), (\overline{D}_\delta, \mathbb{R}, \sigma), h\rangle$ generated by it (for sufficiently small δ) is dissipative, i.e. there is a positive number r for which

$$\limsup_{t \to +\infty} |\psi(t, x, \varphi)| < r$$

for $x \in \mathbb{C}^n$ and $\varphi \in \overline{D}_\delta$, and r is independent of both $x \in \mathbb{C}^n$ and $\varphi \in \overline{D}_\delta$.

Theorem 3.15　*If the equation (3.37) is dissipative on $\mathcal{T}^m$ and the right side of (3.37) is analytic on $\mathcal{T}^m$, then the equation (3.37) has at least one quasi-periodic solution with the same frequency basis $\omega_1, \omega_2, \ldots, \omega_m$ as the right side F.*

Proof. If (3.37) is dissipative and F is analytic on $\mathcal{T}^m$, then there is $\delta > 0$ (sufficiently small) such that $F : \mathbb{C}^n \times D_\delta \to \mathbb{C}^n$ is analytic and the non-autonomous dynamical system $\langle(\mathbb{C}^n \times \overline{D}_\delta, \mathbb{R}_+, \pi), (\overline{D}_\delta, \mathbb{R}, \sigma), h\rangle$ is dissipative. Let $\mathcal{A}(D_\delta, \mathbb{C}^n)$ denote the set of all analytic functions $\gamma : D_\delta \to \mathbb{C}^n$ endowed with the uniform-convergence topology on compacts from D_δ. Clearly $\mathcal{A}(D_\delta, \mathbb{C}^n)$ is a closed subset of the space $C(D_\delta, \mathbb{C}^n)$ of all continuous functions $\gamma : D_\delta \to \mathbb{C}^n$ with the uniform-convergence topology on compacts from D_δ. Since D_δ is an invariant subset of $(\overline{D}_\delta, \mathbb{R}, \sigma)$, the non-autonomous system $\langle(\mathbb{C}^m \times D_\delta, \mathbb{R}_+, \pi), (D_\delta, \mathbb{R}, \sigma), h\rangle$ is correctly defined and, by virtue of Remark 3.6, $C(D_\delta, \mathbb{C}^n)$ is the space of all continuous sections for this system. Theorem 3.14 implies that the triple $(C(D_\delta, \mathbb{C}^n), \mathbb{R}_+, Q)$, where $(Q^t\gamma)(\varphi) = \psi(t, \gamma(\sigma^{-t}\varphi), \sigma^{-t}\varphi)$, is a dynamical system on $C(D_\delta, \mathbb{C}^n)$. We shall prove that $\mathcal{A}(D_\delta, \mathbb{C}^n)$ is an invariant subset of the dynamical system

$(C(D_\delta, \mathbb{C}^n), \mathbb{R}_+, Q)$. In fact, if $\gamma \in \mathcal{A}(D_\delta, \mathbb{C}^n)$, then general properties [122] of solutions of equations with analytic right sides imply that $Q^t\gamma \in \mathcal{A}(D_\delta, \mathbb{C}^n)$ for all $t \geq 0$. Hence a semigroup dynamical system $(\mathcal{A}(D_\delta, \mathbb{C}^n), \mathbb{R}_+, Q)$ is defined on $\mathcal{A}(D_\delta, \mathbb{C}^n)$. We note one property of this system following from the dissipativity of equation (3.37) on T^m. Let $\mathcal{A}_r := \{\gamma \in \mathcal{A}(D_\delta, \mathbb{C}^n) : \gamma(D_\delta) \subseteq B(0, r)\}$, where $B(0, r)$ is the open ball in $\mathbb{C}^n$ of radius r centered at the origin. It follows from Vitali's theorem [167] that, for each $r > 0$, the sets $\mathcal{A}_r$ are precompact in $\mathcal{A}(D_\delta, \mathbb{C}^n)$. It is clear that the sets $\mathcal{A}_r$ are convex. Theorem 3.1 implies that there is $b > 0$ such that, for each $r > 0$, there is $l_r \geq 0$ for which

$$Q^t \mathcal{A}_r \subseteq \mathcal{A}_b$$

for all $t \geq l_r$. Let $r > b$ (for example $r = b + 1$), and $U = \mathrm{Int}\mathcal{A}_r$ (the interior of $\mathcal{A}_r$), then $Q^t\overline{U} \subseteq \overline{\mathcal{A}_b} \subset U$. According to Theorem 1 [205] there exists a stationary point $\gamma \in \mathcal{A}(D_\delta, \mathbb{C}^n)$ of the dynamical system $(\mathcal{A}(D_\delta, \mathbb{C}^n), \mathbb{R}_+, Q)$, i.e.

$$\psi(t, \gamma(\sigma^{-t}\varphi), \sigma^{-t}\varphi) = \gamma(\varphi) \qquad (\varphi \in D_\delta, \quad t \geq 0). \tag{3.38}$$

Hence $\psi(t, \gamma(\varphi), \varphi) = \gamma(\sigma^t\varphi)$ $(\varphi \in D_\delta,\ t \geq 0)$, that is, $\psi(t, \gamma(\varphi_0), \varphi_0) = \gamma(\sigma^t\varphi_0) = \gamma(\varphi_0 e^{i\omega t})$ is a quasi-periodic solution of (3.37) with the frequency basis $\omega_1, \omega_2, \ldots, \omega_m$. This proves the theorem. $\qquad\square$

Remark 3.7 *a. We note that the proof of Theorem 3.15 is of a general nature, and it can be directly applied to equations of more general form. For example, instead of (3.36) we could consider the system*

$$\begin{cases} \dot{x} = F(x, \varphi) \\ \dot{\varphi} = G(\varphi) \end{cases} \qquad (x \in \mathbb{C}^n,\ \varphi \in \mathbb{C}^m). \tag{3.39}$$

We must assume that T^m is a minimal quasi-periodic set for the second equation of (3.39), and that there is a neighborhood D_δ of T^m that is an invariant set of the dynamical system generated by second equation of (3.39). The other conditions, in Theorem 3.15, can clearly be modified for this case.

b. Theorem 3.15 (see also Theorem 3.4) established a conjecture of Bronstein (see [32, p.278]). It should be added that the hypothesis was concerning equation with almost periodic right sides, satisfying the uniform positive stability condition. Hence the latter condition is superfluous for analytic systems.

c. In the author's opinion, the importance of the condition that the right side F be analytic is doubtful in Theorem 3.15. We are inclined to believe that Theorem 3.15 still applies when F is sufficiently smooth. In this connection we note two examples in [152, 332]. One is an example of a dissipative equation of the form (3.1) with quasi-periodic coefficients but with no almost-periodic solutions. Both in

[152, 332] *the construction is based on Denjoy's results concerning flows on a two-dimensional torus, and analysis of the examples shows that, if the right hand side of equation (3.1) is sufficiently smooth, then this situation does not prevail.*

Theorem 3.16   *Suppose that the conditions of Theorem 3.15 are fulfilled and $\overline{F(x,\varphi)} = F(\overline{x},\overline{\varphi})$ (where by bar we note the complex conjugation). Then there exists $\gamma \in \mathcal{A}(D_\delta, \mathbb{C}^n)$ such that*

(1) $\overline{\gamma(x)} = \gamma(\overline{x})$ for all $x \in D_\delta$;
(2) $\gamma(\varphi_0 e^{i\omega t}) = \psi(t, \gamma(\varphi_0), \varphi_0)$ for all $\varphi_0 \in D_\delta$ and $t \in \mathbb{R}$.

Proof. The proof of this statement uses the same argument as Theorem 3.15. $\square$

3.7   The analogy of Cameron-Johnson's theorem

The well-known Cameron-Johnson's theorem [39], [204] asserts that the equation

$$x' = A(t)x \tag{3.40}$$

with a recurrent (almost periodic by Bhor) matrix may be reduced by the Lyapunov-Perron transformation to the equation $y' = B(t)y$ with a skew-symmetric matrix $B(t)$, if all solutions of the equation (3.40) and all its limit equations are bounded on the whole real axis. The generalization of this result on the linear $\mathbb{C}$-analytic equations in the Hilbert space is our main scope in this section.

    Let $\mathbb{R} = (-\infty, +\infty), \mathbb{C}^m$ is an m-dimensional complex Euclidean space, G be a domain from $\mathbb{C}^m, H$ be a real or complex Hilbert space with the scalar product $\langle \cdot, \cdot \rangle$ and with the norm $| \cdot |^2 = \langle \cdot, \cdot \rangle$. Denote by H_w a space H equipped with weak topology, and by $[H]$ (respectively $[H_w]$) the family of all linear continuous operators acting into H (respectively H_w) and equipped with operational norm (with weak topology). Assume that $\mathcal{H}(G, [H]))$ (resp. $\mathcal{H}(G, [H_w]), \mathcal{H}(G, \mathbb{C}^m))$ is the family of all holomorphic functions $h : G \to [H]$ (resp. $h : G \to [H_w], h : G \to \mathbb{C}^m$) equipped with compact-open topology.

    Consider the system

$$\begin{cases} x' = \mathcal{A}(z)x \\ z' = \Phi(z) \end{cases} \quad (z \in G), \tag{3.41}$$

where $\Phi \in \mathcal{H}(G, \mathbb{C}^m)$ and $\mathcal{A} \in \mathcal{H}(G, [H])$. We suppose that the second equation of the system (3.41) generates the dynamical system $(G, \mathbb{R}, \sigma)$ on G. Denote by $U(t, z)$ the Cauchy's operator of the equation

$$x' = \mathcal{A}(zt)x \quad (z \in G), \tag{3.42}$$

where $zt = \sigma(t, z)$. From the general property of solutions of differential equations (see, for example, [122] and [132],[137]), it follows that the family of operators $\{U(t, z)|t \in \mathbb{R}, z \in G\}$ satisfies the following conditions:

1. $U(0, z) = I$ $(\forall z \in G)$, where I is a unit operator on H ;
2. $U(t + \tau, z) = U(t, z\tau)U(\tau, z)$ $(\forall t, \tau \in \mathbb{R}$ and $z \in G$);
3. the mapping $U : \mathbb{R} \times G \to [H]$ $(U : (t, z) \mapsto U(t, z))$ is continuous and for every $t \in \mathbb{R}$ the mapping $U(t, \cdot) : G \to [H]$ is holomorphic.

The following assertion holds.

Theorem 3.17 *Suppose that there exists a positive constant C such that*

$$\|U(t, z)\| \leq C \tag{3.43}$$

for all $t \in \mathbb{R}$ and $z \in G$. Then there exists $P \in \mathcal{H}(G, [H])$ such that

a. *P- is bi-holomorphic, i.e. the operator $P(z)$ is invertible for all $z \in G$ and the mapping $P^{-1} : G \to [H]$ $(P^{-1} : z \mapsto P^{-1}(z))$ is holomorphic;*
b. *$P_*(z) = P(z)$ for all $z \in G$ $(P_*(z)$ is an adjoint for $P(z)$ operator);*
c. *the operator $P(z)$ is positively defined for all $z \in G$;*
d. *$C^{-1}|x| \leq |P(z)x| \leq C|x|$ for all $z \in G$ and $x \in H$;*
e. *the change of variables $x = P(zt)y$ alters the equation (3.42) into*

$$y' = \mathcal{B}(zt)y \tag{3.44}$$

with a skew-Hermitian operator $B \in \mathcal{H}(G, [H])$, i.e. $\mathcal{B}_*(z) = -\mathcal{B}(z)$ for all $z \in G$.

Proof. Let $S^t : \mathcal{H}(G, [H]) \to \mathcal{H}(G, [H])$ (resp. $\mathcal{H}(G, [H_w]) \to \mathcal{H}(G, [H_w]))$ be a mapping defined by the equality $(S^t f)(z) = U_*(t, z)f(zt)U(t, z)$ for all $z \in G$ and $f \in \mathcal{H}(G, [H])$ (resp. $f \in \mathcal{H}(G, [H_w])$). It is easy to verify that the family of mappings $\{S^t|t \in \mathbb{R}\}$ satisfies the following conditions:

4. the operator S^t maps $\mathcal{H}(G, [H])$ (or $f \in \mathcal{H}(G, [H_w])$) into itself for each $t \in \mathbb{R}$;
5. $S^0 = I$, where I is a unit mapping on $\mathcal{H}(G, [H])$ (resp. $\mathcal{H}(G, [H_w])$);
6. $S^t S^\tau = S^{t+\tau}$ for all $t, \tau \in \mathbb{R}$;
7. the mapping S^t ($t \in \mathbb{R}$) is linear and continuous on $\mathcal{H}(G, [H])$ (resp. $\mathcal{H}(G, [H_w])$).

From 4.-7. follows that the family of mappings $\{S^t|t \in \mathbb{R}\}$ is a commutative group of linear continuous operators on $\mathcal{H}(G, [H])$ (or $\mathcal{H}(G, [H_w])$).

Denote by $K := \{\mathcal{A} \in [H]| \ \|\mathcal{A}\| \leq C^2\}$ and note that the set K is weakly closed. According to Tihonoff theorem, the set K is weakly compact because every bounded and closed set in H is weakly compact. From Cauchy's integral formula (see, for example, [288, p.339]) and the Arzela-Ascoli theorem [288] follows that the set $\Gamma = \{\gamma \in \mathcal{H}(G, [H]|\ \ \|\gamma(z)\| \le C^2\}$ is compact in the space $\mathcal{H}(G, [H_w])$. We suppose that $V := \overline{conv}\{S^t Q \mid t \in \mathbb{R}\}$, where $Q(z) = I(\forall z \in G), I$ is a unit operator in H and $\overline{conv}$ is a weak closure of the convex hull of $\{S^t Q \mid t \in \mathbb{R}\}$. Note, that the set V is invariant, i.e. $S^t V \subseteq V$ for all $t \in \mathbb{R}$ because the family of operators $\{S^t\}$ is linear and continuous in $\mathcal{H}(G, [H_w])$. It is easy to see that

$$\langle S^t Q(z)x, x\rangle = \langle U_*(t,z)U(t,z)x, x\rangle = |U(t,z)x|^2 \tag{3.45}$$

for all $z \in G, t \in \mathbb{R}$ and $x \in H$. From (3.43) and (3.45) follows that

$$C^{-2}|x|^2 \le \langle S^t Q(z)x, x\rangle \le C^2|x|^2$$

and, consequently,

$$C^{-2}|x|^2 \le \langle \mathcal{R}(z)x, x\rangle \le C^2|x|^2 \tag{3.46}$$

for all $\mathcal{R} \in V, z \in G$ and $x \in H$. In addition, the equality $\mathcal{R}_*(z) = \mathcal{R}(z)$ holds for all $z \in G$ and $\mathcal{R} \in V$. According to the equality (3.46) we have $\|\mathcal{R}(z)\| \le C^2$ for all $z \in G$ and, consequently, $V \subseteq \Gamma$. According to the theorem of Markov-Kakutany [28], the group $\{S^t \mid t \in \mathbb{R}\}$ admits in V at least one fix point, i.e. there exists $\mathcal{R} \in V$ such that $U_*(t,z)\mathcal{R}(zt)U(t,z) = \mathcal{R}(z)$ for all $z \in G$ and $t \in \mathbb{R}$. We note that the operator $\mathcal{R}(z)$ is self-adjoint and, according to the equality (3.46), is positively defined. From Theorem 12.33 [274] (see also [182, p.65]) follows that there exists a unique reversible self-adjoint and positively defined operator $M(z)$ such that $M^2(z) = \mathcal{R}(z)$ for all $z \in G$ and, according to the inequality (3.46), we have

$$C^{-1}|x| \le |M(z)x| \le C|x|$$

for all $z \in G$ and $x \in H$. Finally, according to the equality

$$\mathcal{R}^\alpha(z) = -\frac{1}{2\pi i}\oint_L \lambda^\alpha (\mathcal{R}(z) - \lambda I)^{-1}d\lambda$$

$(\alpha = \pm\frac{1}{2})$, where L is a simple contour, enclosing the spectrum of the operator $\mathcal{R}(z)$, we have that the mapping $M : G \to [H]$ is biholomorhic.

Note, that $M(zt)U(t,z)M^{-1}(z) = M^{-1}(zt)U_*^{-1}(t,z)M(z)$ since

$$U_*(t,z)M^2(zt)U(t,z) = M^2(z)$$

for all $t \in \mathbb{R}$ and $z \in G$. We set $\mathcal{V}(t,z) = M(zt)U(t,z)M^{-1}(z)$ and note that the family of operators $\{\mathcal{V}(t,z) \mid t \in \mathbb{R}, z \in G\}$ satisfies the conditions 1.-3. and in addition:

8. $\mathcal{V}_*(t,z) = \mathcal{V}^{-1}(t,z)$ for all $t \in \mathbb{R}$ and $z \in G$;
9. $U(t,z)P(z) = P(zt)\mathcal{V}(t,z)$ for all $t \in \mathbb{R}$ and $z \in G$, where $P(z) = M^{-1}(z)$.

Let

$$B(z) := \frac{d}{dt}\mathcal{V}(t,z)|_{t=0} = [\dot{M}(z) + M(z)\mathcal{A}(z)]P(z),$$

where $\dot{M}(z) := \frac{d}{dt}M(zt)|_{t=0} = \frac{dM(z)}{dz}\Phi(z)$. Then $\mathcal{B} \in \mathcal{H}(G,[H])$ and

$$\frac{d\mathcal{V}(t,z)}{dt} = \mathcal{B}(zt)\mathcal{V}(t,z) \qquad (3.47)$$

for all $t \in \mathbb{R}$ and $z \in G$. Thus, $\mathcal{V}(t,z)$ is a Cauchy operator of (3.44). On the other hand, by the condition 8., we have

$$\begin{cases} \mathcal{V'}_*^{-1}(t,z) = -\mathcal{B}_*(zt)\mathcal{V}_*^{-1}(t,z) \\ \mathcal{V}_*^{-1}(0,z) = I \end{cases} \qquad (3.48)$$

for all $t \in \mathbb{R}$ and $z \in G$. From the equalities (3.47), (3.48) and condition 8. follows that $\mathcal{B}_*(z) = -\mathcal{B}(z)$ for all $z \in G$. Finally, to finish the proof of the theorem it is sufficient to remark that the change of variables $x = P(zt)y$ reduces equation (3.42) to the equation (3.44). The theorem is proved. $\qquad\square$

Remark 3.8   *a. In the case, when the space H is finite-dimensional, the statement close to Theorem 3.17 was obtained by V. V. Glavan [165].*

b. If the point $z \in G$ is stationary (ω-periodic, quasi-periodic, almost periodic etc.), then according to [300], the operator-function $P(zt)$ will also be stationary (ω-periodic, quasi-periodic, almost periodic etc.).

c. Theorem 3.17 holds for the system of difference equations

$$\begin{cases} x(k+1) = \mathcal{A}(zk)x(k) \\ z(k+1) = \Phi(zk) \end{cases} \qquad (k \in \mathbb{Z}),$$

where $\mathcal{A} \in \mathcal{H}(G,[H])$ and $\Phi \in \mathcal{H}(G,\mathbb{C}^m)$ and also for the system of differential and difference equations with multi-dimensional time.

3.8   Almost periodic solutions of the weak nonlinear dissipative systems

1. In this item we will study bounded and uniformly compatible solutions of the linear and nonlinear system of differential equations.

Denote by $C_b(\mathbb{R}, E^n)$ the set of all bounded on $\mathbb{R}$ functions $\varphi : \mathbb{R} \to E^n$ with the sup-norm. Consider a linear homogeneous equation

$$\dot{x} = A(t)x, \qquad (3.49)$$

where $A \in C_b(\mathbb{R}, [E^n])$. Along with the equation (3.49) we consider the non-homogeneous equation

$$\dot{x} = A(t)x + f(t) \ . \tag{3.50}$$

Definition 3.12 Recall, that the equation (3.49) is weakly regular if for every function $f \in C_b(\mathbb{R}, E^n)$ the equation (3.50) has at least one solution $\varphi \in C_b(\mathbb{R}, E^n)$.

Let $\varphi \in C(\mathbb{R}, E)$ (respectively $f \in C(\times U, E)$), denote by $\mathfrak{M}_\varphi$ (respectively $\mathfrak{M}_f$) the family of all sequences $\{t_n\}$ such that the functional sequence $\{\varphi_{t_n}\}$ (respectively $\{f_{t_n}\}$) converges in the space $C(\mathbb{R}, E)$ (respectively $C(\times U, E)$), where $\varphi_\tau(t) :=$ $\varphi(t + \tau)$ for all $t \in \mathbb{R}$ (respectively $f_\tau(t, x) := f(t + \tau, x)$ for all $(t, x) \in \mathbb{R} \times U$).

Definition 3.13 Recall $[300, 302]$ that the solution $\varphi \in C(\mathbb{R}, E^n)$ of equation

$$x' = f(t, x) \ \ (f \in C(\mathbb{R} \times E^n, E^n)) \tag{3.51}$$

is called uniformly compatible if $\mathfrak{M}_{f_Q} \in \mathfrak{M}_\varphi$, f_Q is a restriction on $\mathbb{R} \times Q$ of the function f and $Q := \overline{\varphi(\mathbb{R})}$.

It is known $[300, 302]$ that, if the function f is τ-periodic (quasi-periodic, almost periodic, recurrent) w.r.t. variable $t \in \mathbb{R}$ uniformly w.r.t. x on the compacts from E^n and φ is a uniformly compatible solution of equation (3.51), then the function φ is is τ-periodic (quasi-periodic, almost periodic, recurrent) too.

Theorem 3.18 *Let $A \in C_b(\mathbb{R}, [E^n])$. If the set $H(A)$ is compact, then the following conditions are equivalent:*

1. *the equation (3.49) is weakly regular;*
2. *the equation (3.50) has at least one bounded on $\mathbb{R}$ uniformly compatible solution for every function $f \in C_b(\mathbb{R}, E^n)$.*

Proof. It is easy to see that from 2. follows 1.. Therefore, to prove Theorem 3.18 it is sufficient to show, that the condition 1. implies 2. We will consider two cases:

a. Let $A \in C(\mathbb{R}, [E^n])$ be a Poisson stable at least in one direction matrix, i.e. there a sequence $|t_k| \to +\infty$ such that $A_{t_k} \to A$ in $C(\mathbb{R}, [E^n])$. In this case from Theorem 3.18 and Lemma 4.2.1 $[51]$ follows that the equation (3.49) satisfies the condition of exponential dichotomy on $\mathbb{R}$ and, according to Theorem 22.23 $[32]$, the equation (3.50) has a unique bounded on $\mathbb{R}$ and uniformly compatible solution.

b. The matrix-function $A \in C(\mathbb{R}, [E^n])$ is not stable in the sense of Poisson neither in positive nor in negative direction. Let φ be a bounded on $\mathbb{R}$ solution of the equation (3.50) and $\{t_k\} \in \mathfrak{M}_{(A,f)}$. There is $(B, g) \in H(A, f)$ such that $A_{t_k} \to B$ and $f_{t_k} \to g$. Note, that in this case for the sequence $\{t_k\}$ it is possible only the following two cases:

(1) the sequence $\{t_k\}$ converges to $t_0 \in \mathbb{R}$ and, consequently, $\{t_k\} \in \mathfrak{M}_\varphi$;

(2) $|t_k| \to +\infty$.

Then the matrix-function $B := \lim\limits_{k \to +\infty} A_{t_k}$ is ω-limit or α-limit for A and in this case we have $\{t_k\} \in \mathfrak{M}_\varphi$. In fact, the sequence $\{\varphi_{t_k}\}$ is compact in $C(\mathbb{R}, E^n)$ because the function φ is bounded. In addition, every limit function for the sequence $\{\varphi_{t_k}\}$ is a bounded on the $\mathbb{R}$ solution of the equation

$$\dot{x} = B(t)x + g(t) . \tag{3.52}$$

By Lemma 4.2.1 from [51] the equation

$$\dot{y} = B(t)y$$

satisfies the condition of the exponential dichotomy on $\mathbb{R}$ and, consequently, the equation (3.52) cannot have more than one bounded on $\mathbb{R}$ solution. From this fact follows that the sequence $\{\varphi_{t_k}\}$ converges in the space $C(\mathbb{R}, E^n)$, i.e. $\{t_k\} \in \mathfrak{M}_\varphi$.

Finally, we will shows that, in case the sequence $\{t_k\}$ is unbounded, it does not contain bounded sub-sequences. If we suppose that it is not true, then there is a subsequence $\{t'_k\} \subseteq \{t_k\}$ such that $t'_k \to t' \in \mathbb{R}$ and in this case we have

$$B := \lim\limits_{k \to +\infty} A_{t_k} = \lim\limits_{k \to +\infty} A_{t'_k} = A_{t'} \quad (t' \in \mathbb{R}).$$

From this equality follows that the matrix-function A is stable in the sense of Poisson at least in one direction, but this contradicts to our condition. The theorem is proved. $\qquad\square$

Denote by $E_0 := \{u \in E^n : \sup\{\|\varphi(t, u, A)\| : t \in \mathbb{R}\} < +\infty\}$ and by $\mathbb{P}_0$ a projection mapping the space E^n onto E_0. The following statement holds.

Lemma 3.2 *Let $A \in C(\mathbb{R}, [E^n])$. If the matrix-function A is bounded on $\mathbb{R}$ and the equation (3.49) is weakly regular, then for every $f \in C_b(\mathbb{R}, E^n)$ there exists a unique solution $\varphi \in C_b(\mathbb{R}, E^n)$ of the equation (3.50) satisfying the following conditions:*

(1) $\mathbb{P}_0\varphi(0) = 0$;

(2) there exists a positive constant K (a constant of the weak regularity of (3.49)), which does not depend on the function f, such that $\|\varphi\| \leqslant K\|f\|$.

Proof. We prove the formulated statement using the same arguments as in Lemma 6.3 [186, p.515]. $\qquad\square$

Corollary 3.9 *Let $A \in C(\mathbb{R}, [E^n])$. If the set $H(A)$ is compact, then the following conditions are equivalent:*

(1) the equation (3.49) is weakly regular;

(2) for every function $f \in C_b(\mathbb{R}, E^n)$ there is a unique bounded on $\mathbb{R}$ uniformly compatible solution φ of the equation (3.50) satisfying the following conditions:

$$a. \ \mathbb{P}_0\varphi(0) = 0 \quad and \quad b. \ \|\varphi\| \leq K\|f\|,$$

where K is a constant of the weak regularity of the equation (3.49).

Theorem 3.19　*Let $A \in C(\mathbb{R}, [E^n])$ be a bounded on $\mathbb{R}$ matrix-function, the equation (3.49) be weakly regular and $F \in C(\mathbb{R} \times E^n, E^n)$ be a function with the following property: $|F(t, u)| \leq c(|u|)$　$(\forall t \in \mathbb{R}$ and $u \in E^n)$, where $c : \mathbb{R}_+ \to \mathbb{R}_+$ is a nondecreasing function. If $\{r > 0 \ : \ Kc(r) \leq r\} \neq \varnothing$, where K is a constant of the weak regularity of the equation (3.49), then the equation*

$$\dot{u} = A(t)u + F(t, u) \tag{3.53}$$

has at least one bounded on $\mathbb{R}$ solution.

Proof.　For every function $f \in C_b(\mathbb{R}, E^n)$ the equation (3.50) has a unique solution $\psi \in C_b(\mathbb{R}, E^n)$ such that

$$\mathbb{P}_0\psi(0) = 0 \quad and \quad \|\psi\| \leq K\|f\| \quad , \tag{3.54}$$

where K is a constant of the weak regularity of the equation (3.49). Let $\varphi \in C_b(\mathbb{R}, E^n)$. Consider a differential equation

$$\dot{u} = A(t)u + F(t, \varphi(t)) \ . \tag{3.55}$$

Since the function $f(t) := F(t, \varphi(t))$ is bounded on the $\mathbb{R}$ ($|f(t)| = |F(t, \varphi(t))| \leq c(|\varphi(t)|) \leq c(\|\varphi\|)$), then the equation (3.55) has a unique bounded on $\mathbb{R}$ solution ψ_φ satisfying the condition (3.54) and, consequently,

$$\|\psi_\varphi\| \leq K\|f\| = K \sup_{t \in \mathbb{R}} |F(t, \varphi(t))| \leq Kc(\|\varphi\|) \ . \tag{3.56}$$

Let us define an operator $\Phi : C_b(\mathbb{R}, E^n) \to C_b(\mathbb{R}, E^n)$ in the following way: $(\Phi\varphi)(t) = \psi_\varphi(t)$ ($\varphi \in C_b(\mathbb{R}, E^n), t \in \mathbb{R}$). We will show that if the number $r_0 > 0$ satisfies the inequality $Kc(r_0) \leq r_0$, then the ball $B[0, r_0] := \{\varphi \in C_b(\mathbb{R}, E^n) \ : \ \|\varphi\| \leq r_0\}$ is invariant with respect to the mapping Φ. In fact, if $\varphi \in B[0, r_0]$, then $\|\Phi\varphi\| \leq Kc(\|\varphi\|) \leq Kc(r_0) \leq r_0$. Consider now the space $C_b(\mathbb{R}, E^n)$ in the quality of the subset embedded in the space $C(\mathbb{R}, E^n)$. First of all, we note that every ball $B[0, r] \subset C_b(\mathbb{R}, E^n)$ is a convex, bounded and closed subset of the space $C(\mathbb{R}, E^n)$.

　　Note, that the mapping $\Phi : B[0, r_0] \to B[0, r]$ is continuous in the topology of the space $C(\mathbb{R}, E^n)$. In fact, let $\{\varphi_k\} \subseteq B[0, r]$ and $\varphi_k \to \varphi$ in $C(\mathbb{R}, E^n)$. Consider the sequence $(\Phi\varphi_k)(t) := \psi_{\varphi_k}$ ($t \in \mathbb{R}$). Note, that $f_k(t) := F(t, \varphi_k(t)) \to f(t) :=$

$F(t, \varphi(t))$ in the topology of the space $C(\mathbb{R}, E^n)$. If we suppose the contrary, then there are $\varepsilon_0 > 0$ and $L_0 > 0$ such that

$$\max_{|t| \leq L_0} |F(t, \varphi_k(t)) - F(t, \varphi(t))| \geq \varepsilon_0 .$$

Hence, there exists a sequence $\{t_k\} \subseteq [-L_0, L_0]$ such that

$$|F(t_k, \varphi_k(t_k)) - F(t_k, \varphi(t_k))| \geq \varepsilon_0 . \tag{3.57}$$

Since the sequence $\{t_k\}$ is bounded, then we may suppose that it is convergent. Note, that $\varphi_k(t_k) \to \varphi(t_0)$, where $t_0 := \lim_{k \to +\infty} t_k$. In fact,

$$|\varphi_k(t_k) - \varphi(t_0)| \leq |\varphi_k(t_k) - \varphi(t_k)| + |\varphi(t_k) - \varphi(t_0)|$$
$$\leq \max_{|t| \leq L_0} |\varphi_k(t) - \varphi(t)| + |\varphi(t_k) - \varphi(t_0)| .$$

Passing to limit in this inequality, as $k \to +\infty$, we obtain the necessary statement. From the inequality (3.57) (taking into account that the sequence $\{\varphi_k(t_k)\}$ converges to $\varphi(t_0)$ as $k \to +\infty$) we obtain $\varepsilon_0 \leq 0$ and this contradicts to the choice of the number ε_0. Thus, $f_k \to f$ in the space $C(\mathbb{R}, E^n)$ and, in addition, $\|f_k\| = \sup_{t \in \mathbb{R}} |F(t, \varphi_k(t))| \leq c(\|\varphi_k\|) \leq c(r_0)$, i.e. $|f_k(t)| \leq c(r_0)$ $(t \in \mathbb{R})$ and, consequently, $|f(t)| \leq c(r_0)$ $(t \in \mathbb{R})$. On the other hand, the function $\Phi\varphi_k$ is a bounded on $\mathbb{R}$ solution of the equation

$$\dot{u} = A(t)u + f_k(t) .$$

In addition, the functions $\{\Phi\varphi_k\}$ and their derivatives are uniformly bounded on $\mathbb{R}$ and, consequently, the sequence $\{\Phi\varphi_k\}$ is relatively compact in the space $C(\mathbb{R}, E^n)$. Since $f_k \to f$ in $C(\mathbb{R}, E^n)$, then every limit function of the sequence $\{\Phi\varphi_k\}$ is a bounded on $\mathbb{R}$ solution of the equation (3.50), satisfying the conditions (3.54). But by Corollary 3.9 the equation (3.50) has at most one bounded on $\mathbb{R}$ solution satisfying the conditions (3.54). From this fact follows that the sequence $\{\Phi\varphi_k\}$ is convergent in the space $C(\mathbb{R}, E^n)$ and the continuity of the mapping $\Phi : B[0, r_0] \to B[0, r_0]$ is established.

Now we will show that the mapping $\Phi : B[0, r_0] \to B[0, r_0]$ is completely continuous in the topology of the space $C(\mathbb{R}, E^n)$. For this aim we note that

$$(\Phi\varphi)'(t) = A(t)(\Phi\varphi)(t) + F(t, \varphi(t))$$

and, consequently,

$$|(\Phi\varphi)'(t)| \leq a|(\Phi\varphi)(t)| + |F(t, \varphi(t))| \leq ar_0 + c(r_0) \quad (t \in \mathbb{R}) ,$$

where $a = \sup\{\|A(t)\| : t \in \mathbb{R}\}$. From this follows that the set $\Phi(B[0, r_0])$ is relatively compact in the topology of the space $C(\mathbb{R}, E^n)$. According to the theorem of Tihonoff-Shauder, the mapping Φ has at least one fixed point $\varphi \in B[0, r_0]$. It is easy to see that the function $\varphi \in B[0, r_0]$ is a bounded on $\mathbb{R}$ solution of the equation (3.53). The theorem is proved. $\square$

Remark 3.9   *The statement close to Theorem 3.19 was proved in the work [257], but under more strongly assumptions on the nonlinear perturbation F.*

Theorem 3.20   *Let $A \in C(\mathbb{R}, [E^n])$, $F \in C(\mathbb{R} \times E^n, E^n)$ and the following conditions be hold:*

    a. *$a = \sup\{\|A(t)\| : t \in \mathbb{R}\} < +\infty$;*

    b. *the equation (3.49) is weakly regular;*

    c. *$|F(t, u)| \leq c(|u|)$ ($t \in \mathbb{R}$, $u \in E^n$) and $\{r > 0 : Kc(r) \leq r\} \neq \varnothing$, where $c : \mathbb{R}_+ \to \mathbb{R}_+$ is a non-decreasing function and K is a constant of the weak regularity of the equation (3.49);*

    d. *the restriction F_0 of the function F on $\mathbb{R} \times B[0, r_0]$ satisfies the condition of Lipschitz and $Lip(F_0) < K^{-1}$, where r_0 is some positive number satisfying the inequality $Kc(r_0) \leq r_0$.*

Then the equation (3.53) has at least one bounded on $\mathbb{R}$ and uniformly compatible solution.

Proof.   Denote by $B_{r_0}(\mathfrak{M}) := \{\varphi \in C_b(\mathbb{R}, E^n) : |\varphi(t)| \leq r_0 \ (t \in \mathbb{R} \ )\}$ and $\mathfrak{M} \subseteq \mathfrak{M}_\varphi$, where $\mathfrak{M} := \mathfrak{M}_{(A,F)}$. $B_{r_0}(\mathfrak{M})$ is a closed subset of $C_b(\mathbb{R}, E^n)$ [300, p.60]. The operator $\Phi : C_b(\mathbb{R}, E^n) \to C_b(\mathbb{R}, E^n)$, defined as well as in the proof of Theorem 3.19, maps $B_{r_0}(\mathfrak{M})$ into itself. In fact, if $\varphi \in B_{r_0}(\mathfrak{M})$, then the function $f(t) := F(t, \varphi(t))$ ($t \in \mathbb{R}$) belongs also to $B_{r_0}(\mathfrak{M})$. According to Corollary 3.9, $\Phi\varphi$ is a unique bounded on $\mathbb{R}$ uniformly compatible solution of the equation (3.50), satisfying the conditions (3.54). In addition, as well as in Theorem 3.14, $\Phi(B[0, r_0]) \subseteq B[0, r_0]$ and, consequently, $\Phi(B_{r_0}(\mathfrak{M})) \subseteq B_{r_0}(\mathfrak{M})$. Let φ_1 and φ_2 be the function from $B_{r_0}(\mathfrak{M})$. Note, that

$$(\Phi\varphi_1 - \Phi\varphi_2)'(t) = A(t)[(\Phi\varphi_1 - \Phi\varphi_2)(t)] + F(t, \varphi_1(t)) - F(t, \varphi_2(t))$$

and, therefore,

$$\|\Phi\varphi_1 - \Phi\varphi_2\| \leq K \sup_{t \in \mathbb{R}} |F(t, \varphi_1(t)) - F(t, \varphi_2(t))| \leq K Lip(F_0)\|\varphi_1 - \varphi_2\| \ .$$

From the latter inequality follows that the mapping $\Phi : B_{r_0}(\mathfrak{M}) \to B_{r_0}(\mathfrak{M})$ is a contraction, therefore it has a unique fixed point, which is a uniformly compatible solution of the equation (3.53). The theorem is proved. $\square$

Corollary 3.10   *Under the conditions of Theorem 3.20 if the functions $A \in C(\mathbb{R}, [E^n])$ and $F_0 = F|_{\mathbb{R} \times B[0, r_0]}$ are mutually τ-periodic (almost periodic, recurrent, stable in the sense of Poisson, asymptotically τ-periodic, asymptotically almost* *periodic, asymptotically recurrent). Then the equation (3.53) admits at least one τ-periodic (almost periodic, recurrent, stable in the sense of Poisson, asymptotically τ-periodic, asymptotically almost periodic, asymptotically recurrent) solution.*

Remark 3.10 *The problem of existence of uniformly compatible solutions of weak nonlinear equations was studied before in the works [32],[300]. The principal difference of Theorem 3.20 from the corresponding results [32],[300] consists of the following: under the conditions of Theorem 3.20 and Corollary 3.10 the equation (3.49) does not obligatory satisfy the condition of the exponential dichotomy on $\mathbb{R}$. In addition, the nonlinear term is not obligatory small with respect to the variable $u \in E^n$.*

2. In this item we will study uniformly compatible solutions of weak nonlinear dissipative systems.

Theorem 3.21 *Let $A \in C(\mathbb{R}, [E^n]), F \in C(\mathbb{R} \times E^n, E^n)$ and the following conditions be fulfilled:*

(1) $a = \sup\{\|A(t)\| : t \in \mathbb{R}\} < +\infty;$
(2) there are positive numbers N and ν such that

$$\|U(t, A)U^{-1}(\tau, A)\| \le Ne^{-\nu(t-\tau)} \quad (t \ge \tau,\ t, \tau \in \mathbb{R});$$

(3) $|F(t, u)| \le M + \varepsilon|u| \quad (u \in E^n,\ t \in \mathbb{R})$ and $0 \le \varepsilon \le \varepsilon_0 < \nu^2(Na)^{-1}.$

Then the equation (3.53) is dissipative, i.e. there is a number $R_0 > 0$ such that

$$\limsup_{t \to +\infty} |\varphi(t, \nu, B, G)| < R_0 \quad (\nu \in E^n,\ (B, G) \in H(A, F)) ,$$

where $\varphi(\cdot, \nu, B, G)$ is a solution of the equation

$$\dot{\nu} = B(t)\nu + G(t, \nu) , \tag{3.58}$$

satisfying the initial condition $\varphi(0, \nu, B, G) = \nu$.

Proof. For all $B \in H(A)$ we will define on E^n a norm $|\cdot|_B$ by the equality

$$|u|_B := \int_0^{+\infty} |U(t, B)u|dt .$$

As well as in the proof of Theorem 2.39, it is possible to check that

$$\frac{1}{a}|u| \le |u|_B \le \frac{N}{\nu}|u| \quad (u \in E^n).$$

We put

$$u(t) := |\varphi(t, u, B, G)|_{B_t} = \int_0^{+\infty} |U(s, B_t)\varphi(t, u, B, G)|ds \ .$$

Since

$$\varphi(t, u, B, G) = U(t, B)\Big(u + \int_0^t U^{-1}(\tau, B)G(\tau, \varphi(\tau, u, B, G))d\tau\Big),$$

then

$$u(t) = \int_0^{+\infty} |U(s, B_t)U(t, B)\big[u + \int_0^t U^{-1}(\tau, B)G(\tau, \varphi(\tau, u, B, G))d\tau\big]|ds$$

$$= \int_0^{+\infty} |U(s + t, B)(u + \int_0^t U^{-1}(\tau, B)G(\tau, \varphi(\tau, u, B, G))d\tau)|ds$$

$$\leq \int_0^{+\infty} |U(s + t, B)u|ds + \int_0^{+\infty} |\int_0^t U(s + t, B)U^{-1}(\tau, B) \times$$

$$G(\tau, \varphi(\tau, u, B, G))d\tau)|ds \leq \frac{N}{\nu}e^{-\nu t}|u| +$$

$$\int_0^{+\infty}\int_0^t Ne^{-\nu(s+t-\tau)}(M + \varepsilon|\varphi(\tau, u, B, G)|)d\tau ds$$

$$= \frac{N}{\nu}e^{-\nu t}|u| + \frac{N}{\nu}e^{-\nu t}\Big(\frac{M}{\nu}(e^{\tau t} - 1) + \varepsilon \int_0^t |\varphi(\tau, u, B, G)|e^{\nu\tau}d\tau\Big)$$

$$\leq \frac{N}{\nu}e^{-\nu t}|u| + \frac{NM}{\nu^2}(1 - e^{-\nu t}) + \frac{N\varepsilon}{\nu}e^{-\nu t}\int_0^t a|\varphi(\tau, u, B, G)|_{B_\tau}e^{\nu\tau}d\tau$$

$$= \frac{N}{\nu}e^{-\nu t}|u| + \frac{NM}{\nu^2}(1 - e^{-\nu t}) + \frac{aN\varepsilon}{\nu}e^{-\nu t}\int_0^t u(\tau)e^{\nu\tau}d\tau. \tag{3.59}$$

Let $\varphi(t) := u(t)e^{\nu t}$. From the inequality (3.59) follows that

$$\varphi(t) \leq \frac{N}{\nu}|u| + \frac{NM}{\nu^2}(e^{\nu t} - 1) + \frac{a\varepsilon N}{\nu}\int_0^t \varphi(\tau)d\tau$$

and by Theorem 9.3 from [132] $\varphi(t) \leq \psi(t)$ $(t \in \mathbb{R})$, where ψ is a solution of the integral equation

$$y(t) = \frac{N}{\nu}|u| + \frac{NM}{\nu^2}(e^{\nu t} - 1) + \frac{a\varepsilon N}{\nu}\int_0^t y(\tau)d\tau \ .$$

Solving the latter equation, we find that

$$\psi(t) = \left(\frac{N}{\nu}|u| + \frac{NM}{\nu^2 - a\varepsilon N}\right)e^{\frac{a\varepsilon N}{\nu}t} + \frac{NM}{\nu^2 - a\varepsilon N}e^{\nu t}$$

and, consequently,

$$u(t)e^{\nu t} \leq \left(\frac{N}{\nu}|u| + \frac{NM}{\nu^2 - a\varepsilon N}\right)e^{\frac{a\varepsilon N}{\nu}t} + \frac{NM}{\nu^2 - a\varepsilon N}e^{\nu t} . \tag{3.60}$$

From the inequalities (3.59) and (3.60) we obtain

$$|\varphi(t, u, B, G)| \leq a|\varphi(t, u, B, G)|_{B_t}$$
$$\leq a\left(\frac{N}{\nu}|u| + \frac{NM}{\nu^2 - a\varepsilon N}\right)e^{-\frac{\nu^2 - a\varepsilon N}{\nu}t} + \frac{aNM}{\nu^2 - a\varepsilon N} .$$

Therefore,

$$\lim_{t \to +\infty} \sup |\varphi(t, u, B, G)| \leq \frac{aMN}{\nu^2 - a\varepsilon N}$$

($u \in E^n$, $(B, G) \in H(A, F)$ and $\nu - \frac{a\varepsilon N}{\nu} > 0$). The theorem is proved. $\square$

Remark 3.11 *Simple examples show that under the conditions of Theorem 3.22 dissipativity does not reduce to convergence. This statement can be illustrated by the following example $\dot{x} = -x + 2x(1 + x^2)^{-1}$.*

Theorem 3.22 *Let $A \in C(\mathbb{R}, [E^n])$, $F \in C(\mathbb{R} \times E^n, E^n)$ and the following conditions be fulfilled:*

(1) $a := \sup\{ \|A(t)\| : t \in \mathbb{R}\} < +\infty$
(2) there exist positive numbers N and ν such that

$$\|U(t, A)U^{-1}(\tau, A)\| \leq Ne^{-\nu(t-\nu)} \quad (t \geq \tau, \ t, \tau \in \mathbb{R}) .$$

(3) $|F(t, u)| \leq M + \varepsilon|u|$ $(u \in E^n)$ and $0 \leq \varepsilon < \varepsilon_0 := \min(\frac{\nu}{N}, \frac{\nu^2}{Na})$, where M is some positive number.
(4) the restriction F_0 of the function F on $\mathbb{R} \times B[0, r_0]$, where r_0 is a positive number appearing in Theorem 3.20 (for example, $r_0 = \frac{NM}{\nu - \varepsilon_0 N}$), satisfies the condition of Lipschitz with the Lipschitz constant $Lip(F_0) < N\nu^{-1}$.

Then the equation (3.52) is dissipative and in the ball $B[0, r_0]$ there exists a unique bounded on $\mathbb{R}$ uniformly compatible solution of this equation.

Proof. The formulated assertion immediately follows from Theorems 3.20 and 3.21

$\square$

Remark 3.12 *The theorems 3.20 – 3.22 are true also for the differential equations (3.53) in arbitrary Banach spaces.*

Corollary 3.11 *If under the conditions of Theorem 3.22 the functions A and F_0 are almost periodic, then the equation (3.53) is dissipative and its Levinson center contains at least one almost periodic solution.*

Remark 3.13 *Under the conditions of Theorem 3.22 and Corollary 3.11, dissipativity does not reduce to convergence.*

We will give an example which confirms this statement.

Example 3.7 Consider the scalar equation

$$\dot{x} = -kx + \alpha F(x) , \tag{3.61}$$

where

$$F(x) := \begin{cases} \frac{x^2}{2} & \text{for } |x| \leq 10 \\ 50 + 10[1 - \exp(10 - |x|)] & |x| > 10 \end{cases}$$

for $0 < k \leq 5\alpha$. In this connection we can take $r_0 = \frac{k}{\alpha}$. It is easy to check that for the equation (3.61) all the conditions of Theorem 3.22 are fulfilled for the chosen α, k, and F. In addition, its Levinson center contains at least two fixed points (in reality there are 3) and, consequently, the equation (3.61) is not convergent.

Chapter 4

The structure of the Levinson center of system with the condition of the hyperbolicity

4.1 The chain recurrent motions

Let $\Sigma \subseteq X$ be a compact invariant set, $\Sigma, \varepsilon > 0$ and $t > 0$.

Definition 4.1 The collection $\{x = x_0, x_1, x_2, \ldots, x_k = y; t_0, t_1, \ldots, t_k\}$ of the points $x_i \in \Sigma$ and the numbers $t_i \in \mathfrak{T}$ such that $t_i \geq t$ and $\rho(x_i t_i, x_{i+1}) < \varepsilon$ $(i = 0, 1, \ldots, k-1)$ is called a (ε, t, π)-chain joining the points x and y. We denote by $P(\Sigma)$ the set $\{(x, y) : x, y \in \Sigma, \forall\, \varepsilon > 0 \ \forall\, t > 0 \ \exists\, (\varepsilon, t, \pi)$-chain joining x and $y\}$.

The relation $P(\Sigma)$ is closed, invariant and transitive [33]. The point $x \in \Sigma$ is called chain recurrent, if $(x, x) \in P(\Sigma)$. Let $\mathcal{R}(\Sigma) = \{x \in \Sigma : (x, x) \in P(\Sigma)\}$. On $\mathcal{R}(\Sigma)$ we will introduce a relation $\sim$ as follows: $x \sim y$ if and only if $(x, y) \in P(\Sigma)$ and $(y, x) \in P(\Sigma)$. It is easy to check that the introduced relation $\sim$ on $\mathcal{R}(\Sigma)$ is a relation of equivalence and, consequently, it is easy to decompose it on the classes of equivalence $\{\mathcal{R}_\lambda : \lambda \in \mathcal{L}\}$ i.e. $\mathcal{R}(\Sigma) = \sqcup\{\mathcal{R}_\lambda \in \mathcal{L}\}$. By Proposal 2.6 from [33] the defined above components of the decomposition of the set $\mathcal{R}(\Sigma)$ are closed and invariant.

Lemma 4.1 *If the compact invariant set Σ from X contains only a finite numbers of minimal set, then the relation $\sim$ decomposes the set $\mathcal{R}(\Sigma)$ on the finite numbers of different classes of equivalence.*

Proof. Let $\mathcal{R}(\Sigma) = \sqcup\{\mathcal{R}_\lambda : \lambda \in \mathcal{L}\}$. Since $\mathcal{R}_\lambda \subseteq \Sigma$ are closed and invariant and Σ is compact, then by the theorem of Birkhoff the set $\mathcal{R}_\lambda$ contains at least one minimal set. Consequently, the number of different classes of equivalence is not more that the number of minimal sets. The lemma is proved. $\quad\square$

We will indicate below the condition when the number of classes of equivalence $\{\mathcal{R}_\lambda : \lambda \in \mathcal{L}\}$ is finite in case, if the set Σ contains infinite number of different minimal sets.

155

Definition 4.2 By analogy with the work [2] in the collection $\{\mathcal{R}_1, \mathcal{R}_2, \ldots, \mathcal{R}_k\}$ we will introduce a relation of partial order as follows: $\mathcal{R}_i < \mathcal{R}_j$, if there exist $i_1, i_2, \ldots, i_r$ such that $i_1 = i, i_r = j$ and $W^s(\mathcal{R}_{i_p}) \bigcap W^u(\mathcal{R}_{i_{p+1}}) \neq \varnothing$ for all $p = 1, 2, \ldots, r - 1$.

Definition 4.3 The ordered collection of r $(r \geq 2)$ different indexes $\{i_1, i_2, \ldots, i_r\}$ satisfying the condition $\mathcal{R}_{i_1} < \mathcal{R}_{i_2} < \cdots < \mathcal{R}_{i_r} < \mathcal{R}_{i_1}$ is called an r-cycle in the collection $\{\mathcal{R}_1, \mathcal{R}_2 \ldots, \mathcal{R}_k\}$. The 1-cycle is called such index i that $W^s(\mathcal{R}_i) \bigcap W^u(\mathcal{R}_i) \setminus \mathcal{R}_i \neq \varnothing$.

Note, that the introduced above notion of partial order is a slight modification of the corresponding notion from [33, p.61].

Definition 4.4 Following the works [2],[261],[271], the collection of points $\{x_1, x_2, \ldots, x_n\} \subseteq X$ (or $K := H(x_1) \cup H(x_2) \cup \cdots \cup H(x_n)$) is said to be a generalized homoclinic contour, if $\omega_{x_i} \cap \alpha_{x_{i+1}} \neq \varnothing$ for all $i = 1, 2, \ldots, n$, where $x_{n+1} = x_1$.

Lemma 4.2 *Let $\Sigma \subseteq X$ be a compact invariant set and $\{x_1, x_2, \ldots, x_n\}$ be a generalized homoclinic contour. Then $(x_i, x_j) \in P(\Sigma)$ for all $i, j = 1, 2, \ldots, n$.*

Proof. Let $i, j \in \{1, 2, \ldots, n\}$. Suppose, for example, that $i \leq j$ and $0 \leq p = j - i \leq n$. We will show that $(x_i, x_j) \in P(\Sigma)$. Let $\varepsilon > 0$ and $t > 0$. Since $\omega_{x_k} \cap \alpha_{x_{k+1}} \neq \varnothing$ for all $k = i + 1, \ldots, j - 1$, then for the numbers $\varepsilon > 0$, $t > 0$ and the point x_k $(k = i, i + 1, \ldots, j - 2)$ there exist $t'_k > t$, $t''_k < -t$ and $p_k \in \omega_{x_k} \cap \alpha_{x_k}$ such that $\rho(x_k t'_k, p_k) < \frac{\varepsilon}{3}$ and $\rho(x_k t''_k, p_{k-1}) < \frac{\varepsilon}{3}$. Let $\overline{x}_0 := x_i, \overline{x}_1 := x_{i+1} t'_{i+1}, \ldots, \overline{x}_{p-1} := x_j t''_j, \overline{x}_p := x_j; \overline{t}_0 := t'_i, \overline{t}_1 := t'_{i+1} - t''_{i+1}, \ldots, \overline{t}_{p-1} := -t''_j$. It is clear that $\{\overline{x}_0, \overline{x}_1, \ldots, \overline{x}_p; \overline{t}_0, \overline{t}_1, \ldots, \overline{t}_{p-1}\}$ is a (ε, t, π)-chain joining the points x_i and x_j. The lemma is proved. $\qquad\square$

In Lemmas 4.3-4.5 we will suppose that $\mathcal{R}(\Sigma)$ consists of finite number of different classes of equivalence $\mathcal{R}_1, \mathcal{R}_2 \ldots, \mathcal{R}_k$, i.e. $\mathcal{R}(\Sigma) = \mathcal{R}_1 \sqcup \mathcal{R}_2 \sqcup \cdots \sqcup \mathcal{R}_k$. Let us establish some properties of the sets $\mathcal{R}_i$ $(i = 1, 2, \ldots, k)$.

Lemma 4.3 *In the collection $\{\mathcal{R}_1, \mathcal{R}_2 \ldots, \mathcal{R}_k\}$ there is no r-cycles $(r \geq 1)$.*

Proof. Assuming the contrary, we obtain that there exist $r \geq 1$ and $i_1, i_2, \ldots, i_r$ such that $\mathcal{R}_{i_1} < \mathcal{R}_{i_2} < \cdots < \mathcal{R}_{i_r} < \mathcal{R}_{i_1}$. As well as in Lemma 4.2, it is possible to prove that the set $M := \mathcal{R}_{i_1} \cup \cdots \cup \mathcal{R}_{i_r}$ belongs to one class of equivalence and, consequently, $r = 1$. We will show that in the collection $\{\mathcal{R}_1, \mathcal{R}_2 \ldots, \mathcal{R}_k\}$ 1-cycles are absent too. In fact, if we suppose that there exist $1 \leq i \leq r$ and a point x from $(W^s(\mathcal{R}_i) \cap W^u(\mathcal{R}_i)) \setminus \mathcal{R}_i$, then $H(x_0) \cup \mathcal{R}_i \subseteq (\Sigma)$. Moreover, $H(x_0) \cup \mathcal{R}_i$ belongs to one class of equivalence and, consequently, $H(x_0) \cup \mathcal{R}_i \subseteq \mathcal{R}_i$. The latter contradicts to the choice of the point x_0. The lemma is proved. $\qquad\square$

Lemma 4.4   *The sets $\mathcal{R}_i$ $(i = 1, 2, \ldots, k)$ are locally maximal in Σ, i.e. for $\mathcal{R}_i$ there exists a neighborhood U_i of the set $\mathcal{R}_i$ in Σ such that $\mathcal{R}_i$ is a maximal closed invariant set in U_i.*

Proof.   Since $\mathcal{R}_i \cap \mathcal{R}_j \neq \varnothing$ for all $i \neq j$, then there exist neighborhoods U_i of the sets $\mathcal{R}_i$ such that $\overline{U}_i \cap \overline{U}_j = \varnothing$ for $i \neq j$. Note, that in U_i the set $\mathcal{R}_i$ is maximal. In fact, if we suppose that there exists a compact invariant set $\Lambda_i \subset U_i$ such that $\Lambda_i \not\subseteq \mathcal{R}_i$, then $\Lambda_i \setminus \mathcal{R}_i \neq \varnothing$. Let $x \in \Lambda_i \setminus \mathcal{R}_i$. Since the sets α_x and ω_x are chain recurrent [33, p.33], then $\alpha_x, \omega_x \subseteq \mathcal{R}_i$, i.e. $x \in (W^s(\mathcal{R}_i) \cap W^u(\mathcal{R}_i)) \setminus \mathcal{R}_i$ and, consequently, $x \notin \mathcal{R}(\Sigma)$. On the other hand, reasoning in the same way that in the proof of Lemma 4.2, it is possible to show that x is chain recurrent, i.e. $x \in \mathcal{R}(\Sigma)$. The obtained contradiction finishes the proof of the lemma. $\square$

Lemma 4.5   *The sets $\mathcal{R}_i$ $(i = 1, 2, \ldots, k)$ are indecomposable.*

Proof.   Suppose that for some i the set $\mathcal{R}_i$ can be represented in the form of a union of two own closed invariant subsets A_1 and A_2, i.e. $\mathcal{R}_i = A_1 \sqcup A_2$. Since $A_1 \cap A_2 = \varnothing$ and $\mathcal{R}_i \cap \mathcal{R}_j = \varnothing$ for all $j \neq i$, then there are neighborhoods U_1 and U_2 of the sets A_1 and A_2 respectively, such that $\overline{U}_1 \cap \overline{U}_2 = \varnothing$ and $(\overline{U}_1 \cup \overline{U}_2) \cap \mathcal{R}_j = \varnothing$ for all $j \neq i$. Let $\varepsilon_0 > 0$ be such that $B(A_1, \varepsilon_0) \subset \overline{U}_1$ and $B(A_2, \varepsilon_0) \subset \overline{U}_2$. We will take arbitrary points $a_1 \in A_1, a_2 \in A_2$ and numbers $0 < \varepsilon < \varepsilon_0$ and $t > 0$. Since $a_1, a_2 \in \mathcal{R}_i$, then for the numbers ε and t there exists (ε, t, π)-chain $\{a_1 = x_0, x_1, \ldots, x_n = a_2; t_1, t_2, \ldots, t_n\}$ joining the points a_1 and a_2. By Theorem 2.24 from [33] we can suppose that $x_k \in \mathcal{R}_i$ $(k = 0, 1, \ldots, n)$. According to the choice of the number ε and the invariance of the set A_1, we have $x_k \in A_1$ for all $k = 0, 1, \ldots, n$, i.e. $x_n = a_2 \in A_1$. The latter contradicts to our assumption. The lemma is proved. $\square$

4.2   The spectral decomposition of the Levinson's center

Let $\langle (X, \mathbb{S}, \pi), (Y, \mathbb{S}, \sigma), h \rangle$ be a two-sided non-autonomous dynamical system with the fibers $X_y := \{x \in X \mid h(x) = y\}$ $(y \in Y)$. And let every fiber X_y be homeomorphic to V, where V is some n-dimensional manifold (for example, $V = E^n$) and $p \in X_y$. Following the work [33, p.213], we denote by $W_\delta^s(p) := \{x \in X_y \mid \rho(xt, pt) \leq \delta, t \geq 0\}$ and $W_\delta^u(p) := \{x \in X_y \mid \rho(xt, pt) \leq \delta, t \leq 0\}$ $(\delta > 0)$.

Definition 4.5   The compact invariant set $\Lambda \subseteq X$ is said to be hyperbolic (or non-autonomous dynamical system $\langle (X, \mathbb{S}, \pi), (Y, \mathbb{S}, \sigma), h \rangle$ has a hyperbolic structure on Λ), if there are positive numbers N, ν, δ and γ such that

H1. $W_\delta^s(p) \cap W_\delta^u(p) = \{p\}$ for all $p \in \Lambda$, $W_\delta^s(p)$ and the sets $W_\delta^u(p)$ are sub-
manifolds from X_y $(y = h(p))$, which are homeomorphic to the closed desk

in $\mathbb{R}^k$ and $\mathbb{R}^{n-k}$ respectively. In addition, if $\rho(p,q) \le \gamma$ $(p,q \in X_y)$, then $W_\delta^s(p) \cap W_\delta^u(p) \ne \emptyset$;

H2. $\pi^t W_\delta^s(p) \subseteq W_\delta^s(\pi^t p)$ for all $t \ge 0$ and $\pi^t W_\delta^u(p) \supseteq W_\delta^u(\pi^t p)$ for all $t \le 0$ and and each $p \in \Lambda$;

H3. the manifolds $W_\delta^s(p)$ and $W_\delta^u(p)$ depend continuously on the point $p \in \Lambda$ in the distance of Hausdorff;

H4. (a) $\rho(p_1 t, p_2 t) \le N exp(-\nu t)\rho(p_1, p_2)$ for all $p_1, p_2 \in W_\delta^s(p)$ and $t \ge 0$;

(b) $\rho(p_1 t, p_2 t) \le N exp(\nu t)\rho(p_1, p_2)$ for all $p_1, p_2 \in W_\delta^u(p)$ and $t \le 0$.

Remark 4.1   *a. If the set $\Lambda \subseteq X$ is hyperbolic in usual sense* [33, 261], *then under some additional conditions of smoothness, the set Λ will also be hyperbolic in the sense of the definition above. The inverse statement, generally speaking, is not true.*

b. For a discrete autonomous dynamical system there was introduced a close notion (axiom $A^\#$) in the work [2].

Denote by $\mathfrak{M}(\Sigma)$ a closure of all the recurrent points of the set Σ. The following statement takes place.

Theorem 4.1   *Let Y be a compact minimal set, Σ be a compact invariant set from X and let one of the following two conditions be fulfilled:*

a. the number of minimal sets in Σ is finite;

b. if the set Σ contains an infinite number of minimal sets, then on the set $\mathfrak{M}(\Sigma)$ (except, maybe, a finite number of isolated minimal sets) the non-autonomous dynamical system $\langle (X,\mathbb{S},\pi),(Y,\mathbb{S},\sigma),h\rangle$ has a hyporbolic structure.

Then the relation $\sim$ decomposes the set $\mathfrak{R}(\Sigma)$ on finite number of different classes of equivalence.

Proof. If the set Σ contains only a finite number of minimal sets, then the statement of the theorem coincide with the lemma 4.1

Let now the condition b. of the theorem be fulfilled. And we suppose that the theorem is not true. Then there is a denumerable family of classes of equivalence $\{\mathfrak{R}_k : k = 1, 2, \dots\}$. Since the sets $\mathfrak{R}_k$ are closed, invariant and the set Σ is compact, then every set $\mathfrak{R}_k$ contains at least one compact minimal subset $M_k \subseteq \mathfrak{R}_k$. Since the set Y is minimal, then $h(M_k) = Y$ for all $k = 1, 2, \dots$. Let $y \in Y$ and $p_k \in M_k \cap X_y$. According to the compactness of the set Σ, we can suppose that the sequence $\{p_k\}$ is convergent. Put $p := \lim_{k \to +\infty} p_k$ and we note that $p \in \Lambda := \overline{\cup\{M_k : k = 1, 2, \dots\}}$. Without loss of generality we can suppose that the set Λ is hyperbolic. Let $\gamma > 0$ be the number figuring in the condition **H1.** for the set Λ. The sequence $\{p_k\} \subseteq \Lambda$ is convergent; therefore, starting from some k_0, the manifolds $W_\delta^s(p_{k_1})$ and $W_\delta^u(p_{k_2})$

are a nonempty intersection for all $k_1, k_2 \geq k_0$. Let now $k_1 \neq k_2$ and $k_1, k_2 \geq k_0$. Chose points $x_1 \in W_\delta^s(p_{k_1}) \cap W_\delta^u(p_{k_2})$, $x_2 \in W_\delta^u(p_{k_1}) \cap W_\delta^s(p_{k_2})$ and consider the homoclinic contour $K := H(x_1) \cup H(x_2)$. By Corollary 2.19 [33] and taking in consideration that on an ω-limit (α-limit) set every two points are equivalent in the sense of the relation $\sim$, we obtain the equivalence of every two points of homoclinic contour K, i.e. the set K belongs to one class of equivalence. Thus, beginning with k_0, the points p_{k_1} and p_{k_2} are equivalent for all $k_1, k_2 \geq k_0$. On the other hand, the point $\{p_k\}$ was chosen from the different classes of equivalence. The obtained contradiction proves our theorem. $\qquad\square$

Theorem 4.2 *Under the conditions of Theorem 4.1 on the center of Levinson J of the dynamical system $\langle (X, \mathbb{S}, \pi), (Y, \mathbb{S}, \sigma), h \rangle$ the following assertions hold:*

(1) the relation $\sim$ decomposes the set $\mathfrak{R}(J)$ on finite number of different classes of equivalence, i.e. $\mathfrak{R}(J) = \mathfrak{R}_1 \sqcup \mathfrak{R}_2 \sqcup \cdots \sqcup \mathfrak{R}_k$;

(2) the sets $\mathfrak{R}_i$ $(i = 1, 2, \ldots, k)$ are closed, invariant, indecomposable, locally maximal and in the collection $\{\mathfrak{R}_1, \mathfrak{R}_2, \ldots, \mathfrak{R}_k\}$ there is no r $(r \geq 1)$ cycles;

(3) $J = \cup\{W^u(\mathfrak{R}_i) : i = \overline{1,k}\}$, where $W^u(\mathfrak{R}_i) := \{x \in X : \lim_{t \to -\infty} \rho(xt, \mathfrak{R}_i) = 0\}$ $(i = \overline{1,k})$.

Proof. The first and the second statements of Theorem 4.2 directly follow from Lemmas 4.3-4.5 and Theorem 4.1. Now we will prove the third statement. Let $x \in W^u(\mathfrak{R}_i)$ $(i = \overline{1,k})$, then by the compact dissipativity of the dynamical system $(X, \mathbb{S}, \pi)$ the set $H(x)$ is compact and, consequently, $x \in J$.

Inverse. Let $x \in J$. In view of the compactness and invariance of the set J, the sets α_x and ω_x are nonempty, compact, indecomposable and chain recurrent [33]. Since under the conditions of Theorem 4.1 $\mathfrak{R}(J) = \mathfrak{R}_1 \sqcup \mathfrak{R}_2 \sqcup \cdots \sqcup \mathfrak{R}_k$ and the sets $\mathfrak{R}_i$ $(i = \overline{1,k})$ are disjoint, closed and invariant, then there are $i, j \in \{1, 2 \ldots, k\}$ such that $\alpha_x \subseteq \mathfrak{R}_i$ and $\omega_x \subseteq \mathfrak{R}_j$. Thus $x \in W^u(\mathfrak{R}_j)$. The theorem is proved. $\qquad\square$

Remark 4.2 *Theorems 4.1 and 4.2 are true also in the case, if we replace the condition of the minimality of the set Y by the following condition: the set Y contains only finite number of minimal sets.*

4.3 One-dimensional systems with hyperbolic center

In this section we study the structure of the Levinson center for the scalar dissipative equation $u' = f(t, u)$ with recurrent (almost periodic, periodic) right-hand side. Although the problem under study concerns the equation $u' = f(t, u)$, its solution is given within the framework of general non-autonomous dynamical system.

Let $\Lambda \subseteq X$ be a hyperbolic set of non-autonomous dynamical system $\langle (X, \mathbb{S}, \pi), (Y, \mathbb{S}, \sigma), h \rangle$, (X, h, Y) be a finite-dimensional vector bundle with the fiber $\mathbb{R}^n$, and $|\cdot|$ be a Riemannian metric on (X, h, Y), that is compatible with the metric ρ on X.

Remark 4.3   *a. If $p \in \Lambda$ is such that $k = n$ $(k = 0)$, then the set $H(p)$ is exponentially stable on $\mathbb{S}_+$ $(\mathbb{S}_-)$.*

b. For $n = 1$ the hyperbolic set $H(p)$ is exponentially stable either on $\mathbb{S}_+$ or on $\mathbb{S}_-$.

Everywhere in this section, we will suppose that the set Y is compact minimal and $n = 1$.

Lemma 4.6   *If the compact minimal set M of X is hyperbolic, then $h|_M : M \to Y$ is a homeomorphism (or, what is the same thing, $M_y := M \cap X_y$ consists of exactly one point for each $y \in Y$).*

Proof.   Let $M \subseteq X$ be a compact minimal subset of X. If M is hyperbolic, then, by Remark 4.3 b., we can assume that M is exponentially stable on $\mathbb{S}_+$, and, consequently, M is uniformly stable in the positive direction, i.e. for each $\varepsilon > 0$ there exists a $\gamma(\varepsilon) > 0$ such that $\rho(m_1, m_2) < \gamma$ $(m_1, m_2 \in M$ and $h(m_1) = h(m_2))$ implies the inequality $\rho(m_1 t, m_2 t) < \varepsilon$ for all $t \geq 0$. To complete the proof of the Lemma it is sufficient to refer to Theorem 3 from [331, p.110].   □

Let Σ be a nonempty compact invariant set in X, $\{M_\alpha \mid \alpha \in A\}$ by the family of all minimal subsets of Σ, and $\mathfrak{M}(\Sigma) := \overline{\cup \{M_\alpha \mid \alpha \in A\}}$.

Lemma 4.7   *If $\mathfrak{M}(\Sigma)$ is hyperbolic, then Σ contains a finite number of minimal sets $M_1, M_2, \ldots, M_k$ and, consequently, $\mathfrak{M}(\Sigma) = M_1 \sqcup M_2 \sqcup \cdots \sqcup M_k$.*

Proof.   Let us assume the contrary, i.e. let there exist a countable family $\{M_k : k = 1, 2, \ldots\}$ of distinct minimal sets in Σ. Let $y \in Y$ and $p_k \in M_k \cap X_y$ $(k = 1, 2, \ldots)$. Since $p_k \in \mathfrak{M}(\Sigma)$, we can assume that $\{p_k\}$ is convergent. Let $p_k \to p$, then $p \in \mathfrak{M}(\Sigma)$. Since $\mathfrak{M}(\Sigma)$ is hyperbolic, we can assume that (see Remark 4.3 b.) $H(p)$ is exponentially stable on $\mathbb{S}_+$, and, consequently, there exists a $k_0 \in \mathbb{N}$ such that $\rho(p_{k_1} t, p_{k_2} t) \leq N exp(-\nu t) \rho(p_{k_1}, p_{k_2})$ for all $k_1, k_2 \geq k_0$ and $t \geq 0$. It follows from the last inequality that $\rho(p_{k_1} t, p_{k_2} t) \to 0$ as $t \to +\infty$ for all $k_1, k_2 \geq k_0$. But this is not possible because by Lemma 4.6 p_{k_1} and p_{k_2} are jointly recurrent. This contradiction proves the Lemma 4.7.   □

Lemma 4.8   *Under the conditions of Lemma 4.7, for each point $x \in \Sigma$ there exist recurrent points p_1 and p_2 $(h(p_1) = h(p_2) = h(x))$ such that the following conditions* *are fulfilled:*

$$a. \lim_{t \to +\infty} \rho(xt, p_1 t) = 0 \quad and \quad b. \lim_{t \to -\infty} \rho(xt, p_2 t) = 0. \tag{4.1}$$

Proof. Let $x \in \Sigma$, ω and α-limit sets ω_x and α_x are nonempty, compact and contain the minimal sets M_1 and M_2, respectively. Since Y is minimal, it follows that $h(M_1) = h(M_2) = Y$, and, since $\mathfrak{M}(\Sigma)$ is hyperbolic, the sets M_{1y} and M_{2y} consist of exactly one point each for an arbitrary $y \in Y$.

Let x be non-recurrent (in the contrary case Lemma is obvious) and $M_{iy} = \{p_i\}$ $(i = 1, 2)$. Then there exist $t_{1n} \to +\infty$ and $t_{2n} \to -\infty$ such that $xt_{1n} \to p_1$ and $xt_{2n} \to p_2$. By Lemma 4.6 $p_i t_{in} \to p_i$ $(i = 1, 2)$ as $n \to +\infty$. Hence

$$\rho(xt_{in}, p_i t_{in}) \to 0 \tag{4.2}$$

a. Let M_1 be exponentially stable on $\mathbb{S}_+$. Then (4.2) implies (4.1a). Now we show that (4.1b) holds. Indeed, if M_1 is exponentially stable on $\mathbb{S}_+$, then M_2 is exponentially stable on $\mathbb{S}_-$. If we assume the contrary, then, by (4.2) $\rho(xt, p_2 t) \to 0$ as $t \to +\infty$. It follows from Lemma 4.6 that $p_1 = p_2 = p$. Let $\varepsilon = \rho(x, p) > 0$ and $\gamma(\varepsilon) > 0$ be chosen for ε from the condition of uniform stability of $H(p)$ on $\mathbb{S}_+$. Then it follows from (4.2) that for sufficiently large n we have $\rho(xt_{2n}, p_2 t_{2n}) < \gamma(\varepsilon)$ and, consequently, $\rho(x(t + t_{2n}), p_2(t + t_{2n})) < \varepsilon$ for all $t \geq 0$. In particular, $\rho(x, p) < \varepsilon$, which contradicts the choice of ε. This contradiction proves the exponential stability of the set M_2 on $\mathbb{S}_-$. From (4.2) it follows (4.1b).

b. We will show that no $M_1 \subseteq \omega_x$ cannot be exponentially stable on $\mathbb{S}_-$. Indeed, if we assume the contrary, then it follows from (4.2) that $\rho(xt, p_1 t) \to 0$ as $t \to -\infty$ and, consequently, $\alpha_x = M_1 \subseteq \omega_x$. Thus $M_2 = M_1 = \alpha_x \subseteq \omega_x$. Let $p := p_1 = p_2 \in M_1 \cap X_y = M_2 \cap X_y$ and $\varepsilon = \rho(x, p) > 0$. Further, reasoning in the same manner as at the end of the preceding item, we obtain a contradiction, which proves the exponential stability on $\mathbb{S}_+$ of the minimal set $M_1 \subseteq \omega_x$. The Lemma is completely proved. $\square$

Under the conditions of Lemma 4.7, the family Σ consists of only a finite number of distinct minimal sets $M_1, M_2, \ldots, M_k$ and $\mathfrak{M}(\Sigma) = M_1 \cup M_2 \cup \cdots \cup M_k$. Let $y \in Y$ and $\{\gamma_i(y)\} = M_i \cap X_y$. We will suppose that $M_1, M_2, \ldots, M_k$ are ordered such that $\gamma_1(y) < \gamma_2(y) < \cdots < \gamma_k(y)$. Let us set $\mu_y := \inf \Sigma_y$ an $\nu_y := \sup \Sigma_y$. It is obvious that $\mu_y, \nu_y \in \Sigma_y$.

Lemma 4.9 *Under the conditions of Lemma 4.7 the equalities $\mu_y = \gamma_1(y)$ and $\nu_y = \gamma_k(y)$ hold for all $y \in Y$.*

Proof. We prove, for example, the first equality (the second equality is proved in the same manner). If we suppose that $\mu_y \neq \gamma_1(y)$ for certain $y \in Y$, then $\mu_y < \gamma_1(y)$. By Lemma 4.8, there exist recurrent points p_1 and p_2 from $\mathfrak{M}(\Sigma)$, such that (4.1a) and (4.1b) hold. We show that $p_1, p_2 \leq \gamma_1(y)$. In fact, from (4.1) it follows the existence of $t_{1n} \to +\infty$ and $t_{2n} \to -\infty$ such that $\mu_y t_{1n} \to p_1$ and $\mu_y t_{2n} \to p_2$. Moreover, by Lemma 4.6, $\gamma_i(y) t_{in} \to \gamma_i(y)$ $(i = 1, 2)$. Since $\mu_y < \gamma_1(y)$, it follows that $\mu_y t_{1n} < \gamma_1(y) t_{1n}$ and, consequently, $p_1 \leq \gamma_1(y)$. Analogously, $p_2 \leq \gamma_1(y)$. Thus, $p_1 = p_2 = \gamma_1(y)$. Since $H(\gamma_1(p)) = M_1$ is hyperbolic, it follows that M_1 is exponentially stable either on $\mathbb{S}_+$ or on $\mathbb{S}_-$. Further, reasoning in the same manner as in Lemma 4.8, we obtain a contradiction. The lemma is proved. $\qquad\square$

Theorem 4.3   *Let $\Sigma \subseteq X$ be a nonempty compact invariant subset of X and $\mathfrak{M}(\Sigma) \subseteq \Sigma$ be hyperbolic. Then the following statements are valid:*

(1) Σ consists of only a finite number of distinct minimal sets $M_1, M_2, \ldots, M_k$ and
   $$\mathfrak{M}(\Sigma) = M_1 \cup M_2 \cup \cdots \cup M_k;$$
(2) each minimal set M_i $(i = 1, 2, \ldots, k)$ is homeomorphic to Y and, in particular, M_i is almost periodic (periodic) if Y has this property;
(3) for each point $x \in \Sigma$ there exist points p_1 and p_2 from $\mathfrak{M}(\Sigma)$ such that relations (4.1) are fulfilled;
(4) if $M_1, M_2, \ldots, M_k$ are ordered such that $\gamma_i(y) < \gamma_j(y)$ for all $y \in Y$ (where $\{\gamma_i(y)\} := M_i \cap X_y$) if and only if $i < j$, then $\partial\Sigma = M_1 \cup M_k$, where $\partial\Sigma$ is the boundary of Σ.

Proof.   This assertion follows from Lemmas 4.6-4.9. $\qquad\square$

Theorem 4.4   *Let $\langle (X, \mathbb{S}, \pi), (Y, \mathbb{S}, \sigma), h \rangle$ be compact dissipative and J be its Levinson center. If $\mathfrak{M}(J)$ is hyperbolic, then the following statements are valid:*

(1) J contains only a finite number of the minimal sets $M_1, M_2, \ldots, M_k$, each of which is homeomorphic to Y, and $\mathfrak{M}(J) = M_1 \cup M_2 \cup \cdots \cup M_k;$
(2) for each point $x \in J$ there exist points p_1 and p_2 in $\mathfrak{M}(J)$ such that relations (4.1) are fulfilled;
(3) if $x \notin J$, then

   (a) $\rho(xt, \gamma_1(y)t) \to 0$ as $t \to +\infty$ if $x < \gamma_1(y)$, where $y = h(x)$, and
   (b) $\rho(xt, \gamma_k(y)t) \to 0$ as $t \to +\infty$ if $x > \gamma_k(y)$.

(4) $\partial J = M_1 \cup M_k$ and ∂J is uniformly asymptotically stable in the positive direction and, in particular, k is odd.

Proof.   This statement follows in essence from Theorem 4.4 and properties of the Levinson center. $\qquad\square$

Remark 4.4   *Although Theorems 4.3 and 4.4 are proved for one-dimensional systems, yet in fact we have always used only a consequence of one-dimensionality (see Remark 4.3 b.). Therefore, Theorems 4.3 and 4.4 are valid with insignificant* *changes for $n \geq 2$ also if besides the hyperbolicity of $\mathfrak{M}$ we require that for each $m \in \mathfrak{M}$ the set $H(m)$ is exponentially stable either on $\mathbb{S}_+$ or on $\mathbb{S}_-$. In addition, the minimal sets $M_1, M_2, \ldots, M_k$ in Theorems 4.3 and 4.4, generally speaking, not homeomorphic to Y but are only distal covering of Y of finite multiplicity, which also ensures their almost periodicity (periodicity) if Y is almost periodic (periodic).*

Using the known connection between non-autonomous dynamical system and differential equations, we can obtain statements, analogous to Theorems 4.3 and 4.4, for differential equations.

Let us consider the differential equation

$$u' = f(t, u), \tag{4.3}$$

where $f \in C^{0,2}(\mathbb{R} \times \mathbb{R}^n, \mathbb{R}^n)$ and $C^{0,2}(\mathbb{R} \times \mathbb{R}^n, \mathbb{R}^n)$ is the space of all continuous functions $f : \mathbb{R} \times \mathbb{R}^n \to \mathbb{R}^n$, continuously differentiable twice in $u \in \mathbb{R}^n$, equipped with the topology of uniform convergence of functions and their derivatives up to second order inclusive on compacts from $\mathbb{R} \times \mathbb{R}^n$. Along this equation (4.3) we consider the family of equations

$$u' = g(t, u), \tag{4.4}$$

where $g \in H(f) := \overline{\{f_\tau \mid \tau \in \mathbb{R}\}}$, and $f_\tau(t, u) := f(t+\tau, u)$, the bar denoting closure in $f \in C^{0,2}(\mathbb{R} \times \mathbb{R}^n, \mathbb{R}^n)$. Let us suppose that the condition of non-local extendability is fulfilled for each equation (4.4). Let $\varphi(t, v, g)$ denote the solution of (4.4) that passes through $v \in \mathbb{R}^n$ for $t = 0$. Let $Y := H(f)$ and $(Y, \mathbb{R}, \sigma)$ be a dynamical system of translations on Y. Further, let $X := \mathbb{R}^n \times Y$. Then we can define the dynamical system $(X, \mathbb{R}, \pi)$ by the rule $\pi(\tau, (v, g)) := (\varphi(\tau, v, g), g_\tau)$. Finally, we set $h := pr_2 : X \to Y$. Then we obtain the non-autonomous dynamical system $\langle (X, \mathbb{R}, \pi), (Y, \mathbb{R}, \sigma), h \rangle$. Applying Theorems 4.3 and 4.4 to the so-constructed dynamical system, we get the corresponding statements for equation (4.3). For example, we formulate the statement that follows from Theorem 4.4. Fist of all, let us recall certain notion related to equation (4.3).

We recall that the equation (4.3) is called dissipative if the non-autonomous dynamical system, generated by it, is dyssipative. Let $J \subset \mathbb{R}^n \times H(f)$ be the Levinson center of (4.3) and $\Sigma \subseteq J$.For each point $(v, g) \in \Sigma$ let us consider the equation in variations for (4.4) along the solution $\varphi(t, v, g)$, i.e., the linear equation

$$w' = g'_v(t, \varphi(t, v, g))w. \tag{4.5}$$

The family of equation (4.5), where $(v, g) \in \Sigma$, is said to satisfy the exponential dichotomy on $\mathbb{R}$ [132] and the corresponding constants can be chosen independent of $(v, g) \in \Sigma$.

Theorem 4.5 *Let $f \in C^{0,2}(\mathbb{R} \times \mathbb{R}^n, \mathbb{R}^n)$ and f, f'_u and f''_{u^2} be jointly recurrent (almost periodic, τ-periodic) in $t \in \mathbb{R}$ uniformly in x on compacts from $\mathbb{R}$. Suppose that the equation (4.3) is dissipative, $J \subset \mathbb{R} \times H(f)$, be its Levinson center, and the family of equation (4.5), where $(v, g) \in \mathfrak{M}(J)$, satisfy the exponential dichotomy condition on $\mathbb{R}$. Then for each $g \in H(f)$, (4.4) has only a finite number of uniformly compatible solutions* [300]*: These are $q_1, q_2, \cdots, q_k$ (k is an odd number and depends only of f); all the remaining solutions, bounded on $\mathbb{R}$, converge to the indicated solutions as $t \to +\infty$ or $-\infty$. But if a certain solution of (4.4) is bounded on $\mathbb{R}_+$ only (since (4.3) is dissipative, all solutions of (4.4) is bounded on $\mathbb{R}_+$), then it converges to one of the solutions q_1 and q_k as $t \to +\infty$.*

Remark 4.5 *A statement, analogous to Theorem 4.5, holds also for the difference equations.*

4.4 The dissipative cascades

Let M be a differentiable manifold, U be an open subset of M, $f : U \to f(U) \subseteq M$ be a diffeomorphism of the class C^r ($r \geq 1$), $\Lambda \subset U$ be a maximal compact invariant set of the diffeomorphism f. Suppose that U is a domain of attraction for Λ, i.e. $W^s(\Lambda) = U$. In this case, the dynamical system generated by the positive powers of the diffeomorphism f is dissipative and its Levinson center $J = \Lambda$.

Lemma 4.10 *Let $x_0 \in U$. If ω_{x_0} is a compact and hyperbolic set of the diffeomorphism $f : U \to M$, then for every $x \in \omega_{x_0}$ and $\delta > 0$ there is $p \in Per(f)$ such that $\rho(x, p) < \delta$ and $\Sigma_p \subseteq B(\omega_{x_0}, \delta)$, where $Per(f)$ is a set of all the periodic points of the diffeomorphism f, Σ_p is a trajectory of the point p.*

Proof. Let $\delta > 0$. We may take the number δ to be so small such that $B(\omega_{x_0}, \delta) \subset \widetilde{U}(\omega_{x_0})$, where $\widetilde{U}(\omega_{x_0})$ is a neighborhood of the hyperbolic set ω_{x_0} figuring in the theorem of Anosov about the family of ε-trajectories [261, p.220]. In the cited above theorem we will chose a number ε ($\varepsilon < \frac{\delta}{2}$) that corresponds to the number $\frac{\delta}{2}$. Then for $\frac{\varepsilon}{2}$ there is $l > 0$ such that $f^n x_0 \in B(\omega_{x_0}, \frac{\varepsilon}{2})$ for all $n \geq l$. If $x \in \omega_{x_0}$, then there is a number $n_0 \geq l$ such that $f^n x_0 \in B(x, \frac{\varepsilon}{2})$. Put $x_1 = f^{n_0} x_0$. There exists a number $N > 0$ such that $f^N x_1 \in B(x, \frac{\varepsilon}{2})$. Consider the space $X_N = \{0, 1, \ldots, N-1\}$ with discrete topology, the homeomorphism of the shift $\tau : X_N \to X_N$, $\tau(k) = k + 1 \, mod(N)$, and the mapping $\varphi : X_N \to M$ defined by the equality $\varphi(k) = f^k x_1$ ($k = 0, 1, \ldots, N-1$). Note, that $\varphi : X_N \to B(\omega_{x_0}, \frac{\varepsilon}{2}) \subset \widetilde{U}(\omega_{x_0})$. Further: $P(\varphi \circ \tau, f \circ \varphi) := \sup\limits_{0 \leq k \leq N-1} \rho(\varphi(\tau(k)), f(\varphi(k))) \leq \rho(x_1, f^N x_1)$. Since $\rho(x_1, f^N x_1) \leq \rho(x_1, x) + \rho(x, f^N x_1) < \varepsilon$, then $P(\varphi \circ \tau, f \circ \varphi) < \varepsilon$. According to Anosov's theorem of the family of ε-trajectories, there exists a continuous mapping

$\psi : X_N \to \widetilde{U}(\omega_{x_0})$ such that $\psi \circ \tau = f \circ \psi$ and $P(\varphi, \psi) < \frac{\delta}{2}$. To finish the proof of the lemma it is sufficient to note that $\psi(0) \in Per(f)$, $\rho(\psi(0), x) < \frac{\delta}{2}$ and $\psi(k) \in B(\omega_{x_0}, \delta)$ for all $k = 0, 1, \ldots, N - 1$. $\qquad \square$

Corollary 4.1 *If the set Σ is locally maximal and $\omega_{x_0} \subseteq \Sigma$, then under the conditions of Lemma 4.10, in the set Σ there exists at least one periodic point f.*

Corollary 4.2 *Let Λ be a locally maximal compact invariant set. If $\Sigma \subseteq \omega_{x_0} \subseteq \Lambda$ is a compact minimal set, which does not contain periodic trajectories, then under the conditions of Lemma 4.10 in the set Λ there exists an infinite number of different hyperbolic points $p_k \to p \in \Sigma \subseteq \omega_{x_0}$, moreover, $\Sigma_{p_k} \subseteq B(\omega_{x_0}, \varepsilon_k)$ and $\varepsilon_k \downarrow 0$.*

Definition 4.6 The set $M \subseteq X$ of the dynamical system $(X, \mathbb{S}, \pi)$ is called quasi-minimal, if there is a stable in the sense of Poisson point $p \in M$ such that $M = H(p)$.

Theorem 4.6 *Let $f : U \to M$ be a diffeomorphism of the class C^k $(k \geq 1)$ and Λ be a compact invariant set of the diffeomorphism f. If the set $\mathfrak{M}_0(\Lambda)$ is hyperbolic, then:*

(1) the relation $\sim$ decomposes the set $\mathfrak{R}(\Lambda)$ on finite number of classes of equivalence, i.e. $\mathfrak{R}(\Lambda) = \mathfrak{R}_1 \sqcup \mathfrak{R}_2 \sqcup \cdots \sqcup \mathfrak{R}_k$;

(2) the sets $\mathfrak{R}_i$ $(i = \overline{1, k})$ are closed, invariant, indecomposable, locally maximal and in the collection $\{\mathfrak{R}_1, \mathfrak{R}_2, \ldots, \mathfrak{R}_k\}$ there is no r-cycles, where $r \geq 1$.

Proof. This statement follows from Theorem 4.1. $\qquad \square$

Corollary 4.3 *Let M be a differentiable manifold and $f : M \to M$ be a diffeomorphism of the class C^k $(k \geq 1)$. If the set of the chain recurrent points $\mathfrak{R}(f)$ of the diffeomorphism f is compact and hyperbolic, then:*

(1) the relation $\sim$ decomposes the set $\mathfrak{R}(f)$ on finite number of classes of equivalence, i.e. $\mathfrak{R}(f) = \mathfrak{R}_1 \sqcup \mathfrak{R}_2 \sqcup \cdots \sqcup \mathfrak{R}_k$;

(2) the sets $\mathfrak{R}_i$ $(i = \overline{1, k})$ are closed, invariant, quasi-minimal, locally maximal and in the collection $\{\mathfrak{R}_1, \mathfrak{R}_2, \ldots, \mathfrak{R}_k\}$ there is no r $(r \geq 1)$-cycles;

(3) if, in addition, in the set M there exists an infinte number of different minimal sets, then in the set M there is a nontrivial (non periodic) minimal set.

Proof. The formulated assertion follows from Theorem 4.6, Theorems 7.5 (and its discrete analog) 7.8 from [2] and Lemma 6.4 from [261]. $\qquad \square$

Theorem 4.7 *Let $f : U \to M$ be a dissipative diffeomorphism and J be its Levinson center. If one of the following two condition is fulfilled:*

a. in the set Λ there is only a finite number of minimal sets;

  b. in the set Λ there is an infinite number of different minimal sets and the set $\mathfrak{M}_0(\Lambda)$ is hyperbolic [261],

then:

(1) the relation $\sim$ decomposes the set of the chain recurrent points $\mathfrak{R}(f)$ of the diffeomorphism f on finite number of classes of equivalence, i.e. $\mathfrak{R}(f) = \mathfrak{R}_1 \sqcup \mathfrak{R}_2 \sqcup \cdots \sqcup \mathfrak{R}_k$;

(2) the sets $\mathfrak{R}_i$ ($i = 1, 2, \ldots, k$) are closed, invariant, indecomposable, locally maximal and in the collection $\{\mathfrak{R}_1, \mathfrak{R}_2, \ldots, \mathfrak{R}_k\}$ there is no r ($r \geq 1$)-cycles;

(3) $J = \cup\{W^u(\mathfrak{R}_i) : i \in \overline{1, k}\}$.

Proof. This statement follows from Theorem 4.2. In fact, if the set $\mathfrak{M}_0(\Lambda) \neq \emptyset$ and it is hyperbolic, then from the results [261] (see Theorems 0.1, 5.4 and the theorem from the first appendix) follows, that on the set $\mathfrak{M}(\Lambda)$, except, maybe, a finite number of isolated minimal sets, a dynamical system generated by the positive powers of the diffeomorphism f has a hyperbolic structure in the sense of our definition. $\qquad\qquad\square$

Remark 4.6 *a. Theorem 4.7 is an analog of the well-known theorem of Smale about spectral decomposition [261] for A-diffeomorphisms.*

  b. The analog of the theorem 4.7 is true also for the continuous dynamical system (flows).

Theorem 4.8 *Let $f : U \to M$ be a dissipative diffeomorphism and let its center of Levinson contain an infinite number of minimal sets. If the set $\mathfrak{M}_0(\Lambda)$ is hyperbolic, then the statements 1.-3. of Theorem 4.6 hold. In addition, in the set Λ there is an infinite number of hyperbolic periodic trajectories.*

Proof. According to Theorem 4.7 it is sufficient to show, that in the set Λ there is an infinite number of different hyperbolic periodic trajectories. Denote by $\widetilde{U} \subset U$ a neighborhood of the set $\mathfrak{M}_0(\Lambda)$ such that every compact, invariant with respect to f set $\Lambda' \subset \widetilde{U}$ is hyperbolic. By Theorem 2 [261, p.224] there is such a neighborhood. According to [226, p.52] a metric space 2^Λ is compact. Since in the set Λ there is an infinite number of the minimal sets $\{M_i\}$ and $M_i \subseteq \mathfrak{R} = \mathfrak{R}_1 \sqcup \mathfrak{R}_2 \sqcup \cdots \sqcup \mathfrak{R}_k$, then we can suppose that $M_i \subseteq \mathfrak{R}_1$ ($i = 1, 2, \ldots$). Taking into consideration the compactness of the space 2^Λ, we may suppose that the sequence $\{M_i\}$ is convergent in the metric of Hausdorff. In particular, beginning from a number i_0, we have $M_i \subseteq \widetilde{U}$ for $i \geq i_0$. In view of the choice of the set $\widetilde{U}$, all the minimal sets M_i ($i \geq i_0$) are hyperbolic. The following two cases are possible:

  a. every minimal set M_i ($i \geq i_0$) is periodic and in this case the theorem is proved.

  b. between the sets M_i ($i \geq i_0$) there is a nontrivial minimal set M.

Then, according to Corollary 4.2, there exists an infinite number of different periodic points $p_k \to p \in M$, moreover, $\Sigma_{p_k} \subseteq B(M, \varepsilon_k)$ and $\varepsilon_k \downarrow 0$. Since $M_i \subseteq \mathfrak{R}_1$ $(i \geq i_0)$ $\mathfrak{R}_1$ is closed and locally maximal, then starting from a number m_0 we have $p_m \in \mathfrak{R}_1$. The theorem is proved. $\qquad\square$

Definition 4.7   Recall [2]-[3], that an invariant set Λ of the diffeomorphism $f : U \to M$ is called factor-markovian, if there is a TMC (topological markovian chain [2]) $T : \Omega_\Pi \to \Omega_\Pi$ and a continuous mapping $\varphi : \Omega_\Pi \to \Lambda$ such that $\varphi \circ T = f \circ \varphi$. If the mapping φ is a homeomorphism, then the set Λ is called markovian.

Corollary 4.4   *Under the conditions of Theorem 4.8 in the set $\mathfrak{R}$ there is a locally maximal hyperbolic markovian set $\Lambda \subseteq \mathfrak{R}$ and, in particular, in the set $\mathfrak{R}$ there is a transversal homoclynic trajectory.*

Proof.   The formulated statement follows from Theorem 4.8 and also from Theorems 7.7 [2] and 2.4 [271]. $\qquad\square$

4.5   The periodic dissipative systems

1. Let

$$\langle (X, \mathbb{T}_1, \pi), (Y, \mathbb{T}_2, \sigma), h \rangle \tag{4.6}$$

be a non-autonomous dynamical system. In this paragraph we will suppose that the dynamical system $(Y, \mathbb{T}_2, \sigma)$ is periodic, i.e. there exist $\tau \in \mathbb{T}_2$ $(\tau > 0)$ and $q \in Y$ such that $\sigma(q, \tau) = q$ and $Y = \{\sigma(q, t) : 0 \leq t < \tau\}$. In this connection the non-autonomous dynamical system (4.6) is said to be τ-periodic. Denote by $P : X_q \to X_q$ a mapping defined by the equality $P(x) = \pi(\tau, x)$ and by (X_q, P) denote a cascade generated by the positive powers of the mapping P.

Lemma 4.11   *A non-autonomous dynamical system (4.6) is pointwise (compact,local) dissipative if and only if the cascade (X_q, P) is pointwise (compactly, locally) dissipative. In this case the following relations hold:*

$$J = \{\pi^t J_q : 0 \leq t < \tau\} \quad and \quad J_q := J \cap X_q,$$

where J is the Levinson center of the dynamical system (4.6), but J_q is the Levinson center of the dynamical system (X_q, P).

Proof.   This assertion follows immediately from the corresponding definitions. $\square$

Corollary 4.5   *If (X, h, Y) is a finite-dimensional vectorial fibering, then the center of Levinson J of the non-autonomous dynamical system (4.6) and J_q are connected.*

Proof. This statement follows from Theorems 1.32, 1.33 and Corollary 1.14.   □

Corollary 4.6   *Under the conditions of Corollary 4.5, there exists $r > 0$ and for all $a > 0$ there is $k(a) \in \mathbb{Z}_+$ such that $P^k B[0,a] \subseteq B[0,r]$ for all $k \geq k(a)$, where $B[0,a] := \{x \in X_q \ : \ |x| \leq a\}$.*

Proof. Corollary 4.6 follows from Theorem 2.23 and Lemma 4.11.   □

Corollary 4.7   *Under the conditions of Corollary 4.5 there exists a fixed point of the mapping $P : X_q \to X_q$.*

Proof. Corollary 4.7 follows from the theorem on the fixed point of Brauder [35] and Corollary 4.6.   □

Theorem 4.9   *Let (X, h, Y) be a finite-dimensional fibering. There exists a nonempty compact invariant set J (Levinson center) of the dynamical system (4.6) possessing the following properties:*

(1) for every $y \in Y$ the set $J_y = J \cap X_y$ is connected;
(2) for every $\epsilon > 0$ there exists $\delta(\epsilon) > 0$ such that the inequality $\rho(x, J_y) < \delta$ $(x \in X_y)$ implies $\rho(xt, J_{yt}) < \epsilon$ for all $t \geq 0$;
(3) for all $y \in Y$ and $x \in X_y$ $\lim\limits_{t \to +\infty} \rho(xt, J_{yt}) = 0$;
(4) in the center of Levinson J of the system (4.6) there is at least one τ-periodic point.

Proof. The formulated statement follows from Lemma 2.1, Theorem 2.1 and Corollary 4.7.   □

Theorem 4.10   *If X_q is a complex n-dimensional space (i.e. $X_q = \mathbb{C}^n$) and the mapping $P : X_q \to X_q$ is holomorphic, then $J_q = \{p\}$ and $\lim\limits_{t \to +\infty} \rho(xt, pt) = 0$ for all $x \in X_q$, i.e. the dynamical system (4.6) is convergent.*

Proof. The proof of this assertion can be obtained with the help of a slight modification of the proof of Theorem 3.3.   □

2. Consider a differential equation

$$\dot{u} = f(t, u), \quad (t \in \mathbb{R}, \ u \in E^n) \tag{4.7}$$

where f is a τ-periodic function with respect to the variable $t \in \mathbb{R}$ and f is continuously differentiable w.r.t. $u \in E^n$. If the function $f \in C(\mathbb{R} \times E^n, E^n)$ is regular, then (see Example 3.1) the equation (4.7) generates a non-autonomous dynamical system (and this system is τ-periodic). Applying to this non-autonomous dynamical system Theorems 4.6, 4.7 and 4.8, we will obtain the corresponding statements for the equation (4.7).

Let $P : E^n \to E^n$ be a Poincare transformation for the equation (4.7), i.e. $P(u) = \varphi(\tau, u, f)$.

Definition 4.8 The set of chain recurrent points for the equation (4.7) is said to be a set of chain recurrent points of the cascade (E^n, P).

It is well known (see, for example, [33, p.56]), that for the equation (4.7) with the differentiable (w.r.t. variable u) function f the set of chain recurrent points $\mathfrak{R}$ coincides with the set Π that generates periodic (with the period $k\tau$, where $k \in \mathbb{N}$) solutions of the equation (4.7) (for the definition of the set Π see [271, p.278]).

Theorem 4.11 *Suppose, that for the equation (4.7) the set of chain recurrent points $\mathfrak{R}$ is compact and the set $\mathfrak{M}_0(\mathfrak{R})$ is hyperbolic, then*

(1) the relation $\sim$ decomposes the set $\mathfrak{R}$ on finite number of classes of equivalence, i.e. $\mathfrak{R} = \mathfrak{R}_1 \sqcup \mathfrak{R}_2 \sqcup \cdots \sqcup \mathfrak{R}_k$;

(2) the sets $\mathfrak{R}_i$ $(i = \overline{1,k})$ are closed, invariant, indecomposable, locally maximal and in the collection $\{\mathfrak{R}_1, \mathfrak{R}_2, \ldots, \mathfrak{R}_k\}$ there is no r $(r \geq 1)$-cycles;

Corollary 4.8 *Suppose, that for the equation (4.6) the set Π, which generates periodic solutions (with the period $k\tau$, where $k \in \mathbb{N}$), is compact and hyperbolic, then:*

(1) the relation $\sim$ decomposes the set Π on finite number of classes of equivalence, i.e. $\Pi = \Pi_1 \sqcup \Pi_2 \sqcup \cdots \sqcup \Pi_k$;

(2) the sets Π_i $(i = \overline{1,k})$ are closed, invariant, quasi-minimal, locally maximal and the collection $\{\Pi_1, \Pi_2, \ldots, \Pi_k\}$ does not contain r-cycles;

(3) in addition, if in the set Π there is an infinite number of minimal sets, then in Π there exists a nontrivial minimal set.

This assertion generalizes Theorem 3.3 from [271, p.280].

Theorem 4.12 *Let the equation (4.7) be dissipative and the set J be its Levinson's center. Suppose that one of the following two conditions is fulfilled:*

a. in the set J there is a finite number of minimal sets;

b. in the set J there is an infinite number of minimal sets and the set $\mathfrak{M}_0(J)$ is hyperbolic.

Then:

(1) the relation $\sim$ decomposes the set of chain recurrent points of the equation (4.7) $\mathfrak{R}$ on finite number of classes of equivalence, i.e. $\mathfrak{R} = \mathfrak{R}_1 \sqcup \mathfrak{R}_2 \sqcup \cdots \sqcup \mathfrak{R}_k$;

(2) the sets $\mathfrak{R}_i$ $(i = \overline{1,k})$ are closed, invariant, indecomposable, locally maximal and in the collection $\{\mathfrak{R}_1, \mathfrak{R}_2, \ldots, \mathfrak{R}_k\}$ there is no r-cycles $(r \geq 1)$;

(3) $J = \cup\{W^u(\mathfrak{R}_i) : i \in \overline{1,k}\}$.

Theorem 4.13   *Suppose that the equation (4.7) is dissipative, J is its Levinson center. If the set J contains an infinite number of minimal sets and the set $\mathfrak{M}_0(J)$ is hyperbolic, then the statements 1.-3. of Theorem 4.12 hold. In addition, in the set J there is an infinite number of hyperbolic periodic solutions.*

Corollary 4.9   *Under the conditions of Theorem 4.13 in the set J there is a locally maximal markovian set $\Lambda \subseteq J$ and, in particular, in the set J there is a transversal homoclinic periodic trajectory.*

This statement follows from Corollary 4.4.

Chapter 5

Method of Lyapunov functions

5.1 Criterions of dissipativity in term of Lyapunov functions

In this section we suppose that (X, h, Y) is a Banach fiber bundle.

Theorem 5.1 *Let Y be compact and $(X, \mathbb{T}_1, \pi)$ be completely continuous. The non-autonomous dynamical system $\langle (X, \mathbb{T}_1, \pi), (Y, \mathbb{T}_2, \sigma), h \rangle$ is boundedly k-dissipative if and only if there is a number $r > 0$ and a continuous function $V : E_r \to \mathbb{R}$ $(E_r := \{x \in E \mid |x| \geq r\})$ with the following properties:*

1. *the set $\{x \mid V(x) \leq c\}$ is bounded for any $c \in \mathbb{R}$;*
2. *if $x\tau \in E_r$ for all $\tau \in [0, t]$, then $V(xt) \leq V(x)$;*
3. *level lines of the function V do not contain ω-limit points of the dynamical system $(X, \mathbb{T}_1, \pi)$.*

Proof. Let $K_1 = \{x \in E \mid |x| \leq r\}$, where r is the number from the statement of the theorem. As Y is compact and the fiber bundle (X, h, Y) is locally trivial, then the set K_1 is bounded. As $(X, \mathbb{T}_1, \pi)$ is completely continuous, then for the bounded set K_1 there will be a number l such that $\pi^l K_1$ is relatively compact. Suppose $K = \overline{\pi^l K_1}$ and let us show that $\omega_x \cap K \neq \emptyset$ for any $x \in E$. Suppose it is not so, then there will be $x_0 \in E$ for which $\omega_{x_0} \cap K \neq \emptyset$ and, consequently, for some $t_0 \geq 0$ we will have $x_0 t \notin K_1$ for all $t \geq t_0$. If this were not so, then we should have a sequence $t_n \to +\infty$ such that $|x_0 t_n| \leq r$ and, hence, $\{x_0(t_n + l)\} \subseteq K$. We consider the sequence $\{x_0(t_n + l)\}$ to be convergent. Let $\overline{x} = \lim\limits_{n \to +\infty} x_0(t_n + l)$, then $\overline{x} \in \omega_{x_0} \cap K$, but this contradicts the choice of x_0. Thus $x_0 t \notin K_1$ for all $t \geq t_0$ and, hence, $V(x_0 t) \leq V(x_0 t_0)$ for all $t \geq t_0$. According to condition 1. of the theorem, the set $\Sigma_{x_0}^+ = \{x_0 t \mid t \geq 0\}$ is bounded and by virtue of complete continuity of $(X, \mathbb{T}_1, \pi)$ it is relatively compact. From here it results that $\omega_{x_0} \neq \emptyset$ is compact, invariant and $\omega_{x_0} \subseteq E_r$. Let $p \in \omega_{x_0}$ and $\varphi : \mathbb{S} \to \omega_{x_0}$ is the whole trajectory of the system $(X, \mathbb{T}_1, \pi)$ passing through the point $p \in \omega_{x_0}$. Consider the function $\mu : \mathbb{S}_+ \to \mathbb{R}$ defined by the equality $\mu(t) = V(\varphi(-t))$. It is continuous, bounded

171

and non-decreasing and, consequently, there is $c_0 = \lim\limits_{t\to+\infty} \mu(t)$. From here it results that on the set $\alpha_p^{\varphi} := \{q \mid \exists t_n \to +\infty, \varphi(-t_n) \to q\}$ the function V is continuous, that is $\alpha_p^{\varphi} \subseteq V^{-1}(c_0) \cap \omega_{x_0}$. This contradicts the condition 3. of the theorem. So $\omega_x \cap K \neq \emptyset$ for all $x \in E$. According to Theorems 1.22 the non-autonomous dynamical system $\langle (X, \mathbb{T}_1, \pi), (Y, \mathbb{T}_2, \sigma), h \rangle$ be boundedly k-dissipative.

Let now $\langle (X, \mathbb{T}_1, \pi), (Y, \mathbb{T}_2, \sigma), h \rangle$ be boundedly k-dissipative, J be its Levinson center and $r > r_0 := \sup\{|x| : x \in J\}$. Let us define a function $V : E_r \to \mathbb{R}$ by the equality

$$V(x) := \sup\{|xt| : t \geq 0\}. \tag{5.1}$$

Directly from the definition of V it results that:

(1) $V(x) \geq |x|$ for all $x \in E_r$ and, therefore, it satisfies the condition 1. of the theorem;
(2) if $x\tau \in E_r$ for all $\tau \in [0, t]$, then $V(xt) \leq V(x)$.

Let us show that the function V, defined by the equality (5.1), is continuous. Let $x_n \to x(x, x_n \in E_r)$ and $R := \sup\{|x_n| \mid n \in \mathbb{N}\}$, then from Theorem 2.19 for the number $R > 0$ there exists $l(R) > 0$ such that

$$|xt| \leq r \ (t \geq l(R) \text{ and } |x| \leq R). \tag{5.2}$$

From (5.1) and (5.2) it results that

$$V(x) = |x\tau|$$

for some $\tau(x) \in [0, l(R)]$ $(|x| \leq R)$. The sequence $\{V(x_n)\} = \{|x_n\tau_n|\}$ is relatively compact, where $\tau_n = \tau(x_n)$. Let us show that it has a unique limit point. Let V' be one of the limit points of the sequence $\{V(x_n)\}$, then there is a subsequence $\{V(x_{k_n})\}$ such that $V(x_{k_n}) \to V'$ for $k \to +\infty$. As $\{\tau_{k_n}\}$ is bounded, then we can consider it to be convergent. Then $V' = |x\tau'|$, where $\tau' = \lim\limits_{n\to+\infty} \tau_{k_n}$. Let us show that $|x\tau'| = |x\tau| = V(x)$. If we suppose that $|x\tau'| \neq |x\tau|$, then $|x\tau| > |x\tau'|$. Let $\epsilon > 0$ be such that $|x\tau'| + 2\epsilon < |x\tau|$. Then for k sufficiently large we have $||x_{k_n}\tau| - |x\tau|| < \epsilon$ and $||x_{k_n}\tau_{k_n}| - |x\tau'|| < \epsilon$, hence,

$$|x_{k_n}\tau| > |\tau| - \epsilon > |x\tau'| + \epsilon > |x_{k_n}\tau_{k_n}| = V(x_{k_n}). \tag{5.3}$$

Inequality (5.3) contradicts the choice of the number τ_{k_n}. Thus $V(x) = |x\tau'|$ and, hence $V(x) = \lim\limits_{n\to+\infty} V(x_n)$, so that V is continuous.

Finally, let us show that level lines of the function V do not contain ω-limit points $(X, \mathbb{T}_1, \pi)$. Really, for any $x \in E$, we have $\omega_x \subseteq J$. According to the definition of the set E_r it has not common points with J and therefore, $\omega_x \cap V^{-1}(c) = \emptyset$, for any $x \in E_r$ and $c \in \mathbb{R}$. This proves the theorem. $\qquad\square$

Lemma 5.1 *Let $(X, \mathbb{T}_1, \pi)$ be completely continuous, $\langle (X, \mathbb{T}_1, \pi), (Y, \mathbb{T}_2, \sigma), h \rangle$ be a non-autonomous dynamical system and there exists $r > 0$ and a continuous function $V : E_r \to \mathbb{R}$ which satisfies conditions:*

1. *the set $\{x \in E \ : \ V(x) \leq c\}$ is bounded for any $c \in \mathbb{R}$;*
2. *if $x\tau \in E_r$ for all $\tau \in [0, t]$, then $V(xt) \leq V(x)$.*

Then the following conditions are equivalent:

3. *level lines of the function V do not contain ω-limit points of the dynamical system $(X, \mathbb{T}_1, \pi)$;*
4. *level lines of the function V do not contain positive semi-trajectories of the dynamical system $(X, \mathbb{T}_1, \pi)$.*

Proof. Let us show that Condition 3. implies Condition 4. If we suppose that it is not so, then there will be c_0 and $x_0 \in V^{-1}(c_0)$ such that $x_0 t \in\in V^{-1}(c_0)$ for all $t \geq 0$. In conditions of the lemma the level lines of the function V are bounded sets, therefore $\Sigma_{x_0}^+$ is bounded. As $(X, \mathbb{T}_1, \pi)$ is completely continuous then the set $\Sigma_{x_0}^+$ is relatively compact and, hence, $\omega_{x_0} \neq \emptyset$, is compact and $\omega_{x_0} \subseteq V^{-1}(c_0)$. The last is a contradiction to the condition 3. This contradiction proves the implication we need.

Let us show that from the condition 4. follows the condition 3. Suppose the opposite, that there is $c_0 \in \mathbb{R}$ and $x_0 \in E_r$ such that $\omega_{x_0 t} \cap V^{-1}(c_0) \neq \emptyset$. Reasoning like in Theorem 5.1 we can show that ω_{x_0} is compact and $\alpha_p^\varphi \subseteq \omega_{x_0} \cap V^{-1}(c_0) \neq \emptyset$ for some complete motion φ $(\varphi(0) = p)$. As α_p^φ is positive invariant, then $V^{-1}(c_0)$ contains positive semi-trajectories of the system $(X, \mathbb{T}_1, \pi)$, but this contradicts the condition 4. Lemma is proved. $\square$

Corollary 5.1 *Let Y be compact and the dynamical system $(X, \mathbb{T}_1, \pi)$ be completely continuous. The non-autonomous dynamical system $\langle (X, \mathbb{T}_1, \pi), (Y, \mathbb{T}_2, \sigma), h \rangle$ is boundedly k-dissipative if and only if there are number $r > 0$ and continuous function $V : X_r \to \mathbb{R}$, satisfying the conditions:*

(1) $V(x) \geq a(|x|)$ for all $x \in X_r$, where $a \in A := \{a \mid a : \mathbb{R}_+ \to \mathbb{R}_+, a$ is a continuous and strongly monotonically increasing function $\}$ and $Im\, V \subseteq Im\,(a)$;

(2) if $x\tau \in X_r$ for all $\tau \in [0, t]$, then $V(xt) \leq V(x)$;

(3) level lines of the function V do not contain positive semi-trajectories of dynamical system $(X, \mathbb{T}_1, \pi)$.

Theorem 5.2 *Let $(X, \mathbb{T}_1, \pi)$ be completely continuous. A non-autonomous dynamical system $\langle (X, \mathbb{T}_1, \pi), (Y, \mathbb{T}_2, \sigma), h \rangle$ is boundedly k-dissipative if there is a number $r > 0$ and a continuous function $V : E_r \to \mathbb{R}$ with the following properties:*

(1) the set $\{x \in E \mid V(x) \leq c\}$ is bounded for any $c \in \mathbb{R}$;

(2) if $x\tau \in E_r$ for all $\tau \in [0, t]$, then $V(xt) < V(x)$.

Proof. Let $K_1 = \{x \in E : |x| \le r\}, l > 0$, be such that $\pi^l K_1$ is relatively compact and $K = \overline{\pi^l K_1}$. Let us show that $\omega_x \cap K \ne \emptyset$ for all $x \in E$. Like in Theorem 5.1 it can be shown that in the opposite case we should have a point $x_0 \in E$ such that $\Sigma_{x_0}^+$ is relatively compact, $\omega_{x_0} \ne \emptyset$ is compact and $\omega_{x_0} \cap K = \emptyset$. Furthermore there will be $c_0 \in \mathbb{R}$ and a complete motion $\varphi : S \to \omega_{x_0}$ ($\varphi(0) = p$) such that $\alpha_p^{\varphi} \subseteq \omega_{x_0} \cap V^{-1}(c_0) \ne \emptyset$. As α_p^{φ} is positively invariant, then together with the point x the set α_p^{φ} contains xt for all $t > 0$, and, hence, $V(xt) < V(x) = c_0$. The contradiction we have got proves the assertion we need. Thus $\omega_x \cap K \ne \emptyset$ for all $x \in E$ and according to Theorem 2.19 the non-autonomous dynamical system $\langle (X, \mathbb{T}_1, \pi), (Y, \mathbb{T}_2, \sigma), h \rangle$ is boundedly k-dissipative. The Theorem is proved. $\square$

Theorem 5.3 *Let $\langle (X, \mathbb{T}_1, \pi), (Y, \mathbb{T}_2, \sigma), h \rangle$ be a non-autonomous dynamical system and $\mathbb{T}_1 = \mathbb{R}_+$. Assume that there is a number $r > 0$ and function $V : E_r \to \mathbb{R}_+$ with the following properties:*

1. *the function V is bounded on bounded sets and for any $c \in \mathbb{R}$ the set $\{x \in E_r \mid V(x) \le c\}$ is bounded;*
2. *$V_{\pi}'(x) \le -c(|x|)$ for all $x \in E_r$, where $c : \mathbb{R}_+ \to \mathbb{R}_+$ is strictly positive on $[r, +\infty)$, and $V_{\pi}'(x) = \limsup_{t \to +0} t^{-1}[V(xt) - V(x)]$.*

Then there exists $R_0 > 0$ such that for any $R > 0$ there will be $l(R) > 0$ for which $|xt| \le R_0$ for all $t \ge l(R)$ and $|x| \le R$.

Proof. First let us show that for any $x \in E_r$ there will be $\tau = \tau(x) > 0$ such that $|x\tau| < r$. If that assertion is not true then there exists $x_0 \in E_r$ such that $|x_0\tau| \ge r$ for all $\tau \ge 0$. Consequently, $V(x_0\tau) \le V(x_0)$ for all $\tau \ge 0$ and the set $\Sigma_{x_0}^+ := \cup\{x_0\tau \mid \tau \ge 0\}$ is bounded. Let $b_0 := \sup\{|x_0\tau| \mid |\tau \ge 0\}$ and $\nu = \inf\{c(\alpha) \mid r \le \alpha \le b_0\}$ and we note that $V_{\pi}'(x_0t) \le -\nu$, therefore

$$V(x_0t) \le V(x_0) - \nu t \tag{5.4}$$

for all $t > 0$. The right hand side of (5.4) is negative for t sufficiently large, but this contradicts positive definiteness of V on E_r. This contradiction proves the assertion we need.

Let $b(R) := \sup\{V(x) : x \in E_r, |x| \le R\}$, $M(R) := \{x \in E_r \mid V(x) \le b(R)\}$ and $R_0 := \sup\{|x| \mid x \in M(R)\}$. According to condition 1. of the theorem the set $M(R)$ is bounded and, hence, $R_0 < +\infty$. Let us show that $|xt| \le R_0$ for all $|x| \le r$ and $t \ge 0$. In fact, if $|x| \le r, |x\beta_x| = |x\gamma_x| = r$ and $|xt| > r$ for all $t \in]\beta_x, \gamma_x[$, then $V(xt) \le V(x\beta_x) \le b(R)$ and, hence, $xt \in M(R)$. From this it follows that $|xt| \le R_0$ for all $t \in]\beta_x, \gamma_x[$ and $|x| \le r$, and therefore $|xt| \le R_0$ for all $t \ge 0$ and $|x| \le r$. According to the proved above for any $x \in E_r$ there will be $\tau(x) > 0$ such that $|x\tau| < r$. Suppose $\nu_x := \sup\{t \mid x\tau \in E_r$ for all $\tau \in [0,t]\}$, then $|x\nu_x| = r$ and $|xt| > r$ for all $t \in]0, \nu_x[$. Let us notice that for any $R > 0$ the set $B(R) := \{xt \mid x \in E_r, |x| \le R$ and $t \in]0, \nu_x[\}$ is bounded. Indeed, $V(xt) \le V(x) \le b(R)$ for all $|x| \le R$ and $t \in]0, \nu_x[$, therefore $xt \in M(R)$. According to condition 1. of the theorem the set $M(R)$ is bounded. Let us show now that $L(R) := \sup\{\nu_x \mid |x| \le R\}$ is finite for any $R > 0$. If the assertion is not true, then there will be $\overline{R} > 0$ and $\{x_n\}$ such that $|x_n| \le \overline{R}$ and $\nu_{x_n} \to +\infty$. Suppose $d(R) := \sup\{|x| \mid x \in B(R)\}$. As the set $B(R)$ is bounded, then we have $d(R) < +\infty$. Suppose $\overline{\gamma}(R) := \inf\{c(\nu) \mid r \le \nu \le d(R)\}$ and let us notice that $V'_\pi(xt) \le -\overline{\gamma}$ for all $|x| \le R$ and $t \in]0, \nu_x[$ and, hence, $V(xt) \le V(x) - \overline{\gamma}(R)t$. In particular,

$$V(x_n \nu_{x_n}) \le b(\overline{R}) - \overline{\gamma}(\overline{R})\nu_{x_n} \tag{5.5}$$

for all n. The right hand side of (5.5) is negative for n sufficiently large and this contradicts the positive definiteness of V on E_r. From this contradiction it follows that for every $R > 0$ the number $L(R) > 0$ and is finite. Now for finishing the proof of the theorem it is sufficient to notice that $|xt| \le R_0$ for all $|x| \le R$ and $t \ge L(R)$. $\square$

Corollary 5.2 *Let $\langle (X, \mathbb{T}_1, \pi), (Y, \mathbb{T}_2, \sigma), h \rangle$ be a non-autonomous dynamical system and $\mathbb{T}_1 = \mathbb{R}_+$. Assume that there is a function $V : E \to \mathbb{R}$ which satisfies the following conditions:*

(1) V is bounded on the bounded sets from E;
(2) $V(x) \ge \gamma_1 |x|^m - D_1$ $(\gamma_1, D_1, m > 0)$ for all $x \in E$;
(3) $V'_\pi(x) \le -\gamma_2 V(x) + D_2$ $(\gamma_2, D_2 > 0)$ for all $x \in E$.

Then there exists $R_0 > 0$ such that for any $R > 0$ there will be $l(R) > 0$ for which $|xt| \le R_0$ for all $|x| \le R$ and $t \ge l(R)$.

Proof. This assertion follows from Theorem 5.3. To prove that it is sufficient to notice that for $|x|$ sufficiently large conditions 1.–3. of Corollary 5.1 guarantee the fulfillment of the conditions of Theorem 5.3. $\square$

Corollary 5.3 *Under the conditions of Theorem 5.3, if Y is compact and $(X, \mathbb{T}_1, \pi)$ is asymptotic compact, then the non-autonomous dynamical system $\langle (X, \mathbb{T}_1, \pi), (Y, \mathbb{T}_2, \sigma), h \rangle$ is boundedly k-dissipativ.*

Proof. This statement follows from Theorems 5.3 and 2.21. $\square$

Remark 5.1 *The statement, analogously to Corollary 5.3, holds also for the dynamical system with discrete time $\mathbb{T}_1 = \mathbb{Z}_+$.*

Remark 5.2 *Unlike Theorems 5.1 and 5.2 in Theorem 5.3 we do not demand the continuity of the function V, but instead of that we impose certain condition on the velocity of decreasing of the function V along trajectories of the system $(X, \mathbb{T}_1, \pi)$. In additional, when $\mathbb{T}_1 = \mathbb{R}_+$ the completeness of the space is not necessary (in the applications there are examples, when Y is a fortiori incomplete).*

Theorem 5.4 *Let $\langle (X, \mathbb{T}_1, \pi), (Y, \mathbb{T}_2, \sigma), h \rangle$ be a non-autonomous dynamical system, $\mathbb{T}_1 = \mathbb{R}_+$, $(X, \mathbb{T}_1, \pi)$ be asymptotically compact and there is a continuous function $V : E \to \mathbb{R}$ satisfying the following conditions:*

1. *for all $c \in \mathbb{R}$ the set $\{x \in E \mid V(x) \le c\}$ is bounded;*
2. *along trajectories of the system $(X, \mathbb{T}_1, \pi)$ the function V is non-increasing, i.e. $V(xt) \le V(x)$ for all $x \in E$ and $t \ge 0$;*
3. *there exists $r > 0$ such that $V(xt) < V(x)$, if $x\tau \in E_r$ for all $\tau \in [0, t]$ and $t > 0$.*

Then the non-autonomous dynamical system $\langle (X, \mathbb{T}_1, \pi), (Y, \mathbb{T}_2, \sigma), h \rangle$ is locally k-dissipative.

Proof. First let us notice that as V is continuous, then in a point $p \in X$ there is $C_p \in \mathbb{R}$ and $\delta_p > 0$ such that $V(x) \le C_p$ for any $x \in B(p, \delta_p)$. According to the condition 2. $V(xt) \le V(x) \le C_p$ for all $x \in B(p, \delta_p)$ and $t \ge 0$. According to the condition 1. the set $M_p := \Sigma^+(B(p, \delta_p)) = \cup\{\pi^t B(p, \delta_p) \mid t \ge 0\}$ is bounded and as $(X, \mathbb{T}_1, \pi)$ is asymptotically compact, then there is a nonempty compact K_p such that the equality

$$\lim_{t \to +\infty} \beta(\pi^t M_p, K_p) = 0 \tag{5.6}$$

takes place. Define $K := \{x \in X \mid |x| \le r\}$, where $r > 0$ is the number from the condition 3. of the theorem. As Y is compact and as the fiber bundle (X, h, Y) is locally trivial, then the set K is bounded. Let us notice that $\omega_p \cap K \ne \emptyset$ for any point $p \in X$. In fact, if this were not so, we should have a point $p_0 \in X$ such that $\omega_{p_0} \cap K = \emptyset$. According to the equality (5.6) $\Sigma_{p_0}^+ := \cup\{\pi^t p_0 \mid t \ge 0\}$ is relatively compact and, hence, $\omega_{p_0} \ne \emptyset$ is compact and invariant. The function $\mu : \mathbb{R}_+ \to \mathbb{R}$ defined by the equality $\mu(t) := V(p_0 t)$ is continuous, bounded and monotone non-increasing and, hence, there is $\lim_{t \to +\infty} V(p_0 t) = c_0$. As V is continuous, we have $\omega_{p_0} \subseteq V^{-1}(c_0)$. Let $x \in \omega_{p_0}$, then $V(xt) = V(x) = c_0$ for all $t > 0$. The last contradicts condition 3. of the theorem. This contradiction shows that $\omega_p \cap K \ne \emptyset$ for all $p \in X$. Let us show that as a matter of fact $\omega_p \subseteq K$ for all $p \in X$. Indeed, if it is not so, then there will be $t_0 > 0$ such that $x\tau \in \omega_{p_0} \setminus K$ for all $\tau \in [0, t_0]$ and, hence, $V(xt_0) < V(x)$. On the other hand, reasoning as we do above, we will have $V(x) = c_0$ for some $c_0 \in \mathbb{R}$ and all $x \in \omega_{p_0}$. This contradicts the last inequality. Thus $\Omega \subseteq K$, where $\Omega := \overline{\cup\{\omega_p \mid p \in X\}}$, and, as Ω is invariant and as $(X, \mathbb{T}_1, \pi)$

is asymptotically compact, we conclude that Ω is compact and, hence, the system $(X, \mathbb{T}_1, \pi)$ is pointwise k-dissipative. Taking into consideration (5.6) and Theorem 1.19 we conclude that the system $(X, \mathbb{T}_1, \pi)$ is locally k-dissipative. Theorem is proved. $\qquad\square$

Remark 5.3   *a) From condition 3. it follows that level lines of the function V do not contain ω-limit points of the dynamical system $(X, \mathbb{T}_1, \pi)$.*

b) Theorem 5.4 takes place for discrete dynamical systems, too, if condition 3. is changed by the following: level lines of the function V do not contain ω-limit points of the dynamical system $(X, \mathbb{T}_1, \pi)$.

Proof.   For proving the last assertion let us notice that in conditions of Theorem 11.2 the set K is a weak attractor and that is why $\Omega \subseteq K$. Indeed, if this were not so, then we should have $x_0 \in E_r$ such that $\omega_{x_0} \setminus K \neq \emptyset$. Let $p \in \omega_{x_0} \setminus K$, then there exits c_0 such that $\omega_{x_0} \subseteq V^{-1}(c_0)$. Let $c'_0 = V(p)$, then $c'_0 = V(p) = c_0$ and, hence, $p \in V^{-1}(c_0) \cap E_r \cap \omega_{x_0}$. The last contradicts the condition of the theorem. $\qquad\square$

Definition 5.1 $\langle(X, \mathbb{T}_1, \pi), (Y, \mathbb{T}_2, \sigma), h\rangle$ is said to be bounded, if for arbitrary $R > 0$ there is $C(R) > 0$ such that $|xt| \leq C(R)$ for all $|x| \leq R$ and $t \geq 0$.

Theorem 5.5   *Let $\langle(X, \mathbb{T}_1, \pi), (Y, \mathbb{T}_2, \sigma), h\rangle$ be a non-autonomous dynamical system, $\mathbb{T}_1 = \mathbb{R}_+$ and $(X, \mathbb{T}_1, \pi)$ is asymptotically compact. If there exist $r > 0$ and a continuous bounded on the bounded sets from X function $V : X_r \to \mathbb{R}$ which satisfies the following conditions:*

1. *for any $c \in \mathbb{R}$ the set $\{x \in E_r \mid V(x) \leq c\}$ is bounded;*
2. *if $x\tau \in E_r$ for all $\tau \in [0, t]$, then $V(xt) \leq V(x)$;*
3. *level lines of V do not contain ω-limit points of the dynamical system $(X, \mathbb{T}_1, \pi)$.*

Then the non-autonomous dynamical system $\langle(X, \mathbb{T}_1, \pi), (Y, \mathbb{T}_2, \sigma), h\rangle$ is boundedly k-dissipative and bounded, i.e. for any $R > 0$ there exists $C(R) \geq 0$ such that $|x| \leq C(R)$ for all $|x| \leq R$ and $t \geq 0$.

Proof.   Like in Theorem 5.3 the existence of $C(R) \geq 0$ such that $|xt| \leq C(R)$ for all $t \geq 0$ and $|x| \leq R$ ($R \geq r$) is proved.

Let us show that under the conditions of Theorem 5.5 the non-autonomous system $(X, \mathbb{T}_1, \pi)$, $(Y, \mathbb{T}_2, \sigma)$, $h\rangle$ is boundedly k-dissipative. As $\langle(X, \mathbb{T}_1, \pi), (Y, \mathbb{T}_2, \sigma), h\rangle$ is bounded and asymptotically compact, then according to Lemma 1.4 $\Omega(K_1) \neq \emptyset$, is compact and attracts $K_1 := \{x \in E \mid |x| \leq r\}$. Suppose $K := \Omega(K_1)$ and let us show that $\omega_x \cap K \neq \emptyset$ for any $x \in X$. If this were not so, we should have $x_0 \in X$ such that $\omega_{x_0} \cap K = \emptyset$. Under the conditions of Theorem 5.5 the set $\Sigma_{x_0}^+ := \{x_0 t \mid t \geq 0\}$ is relatively compact and, hence, $\omega_{x_0} \neq \emptyset$ and is compact.

Suppose $d = \rho(\omega_{x_0}, K) > 0$, then

$$\overline{B(\omega_{x_0}, \frac{d}{3})} \cap \overline{B(K, \frac{d}{3})} = \emptyset \tag{5.7}$$

and there will be $l > 0$ such that

$$\pi^t K_1 \subseteq B(K, \frac{d}{3}) \tag{5.8}$$

for all $t \geq l$. Let us show that $x_0 t \notin K_1$ for all $t \geq t_0$, where t_0 is some nonnegative number. If it is not so then there will be $t_n \to +\infty$, such that $x_0 t_n \in K_1$ and according to (5.8) $x_0(t_n + l) \in B(K, \frac{d}{3})$. We consider the sequence $\{x_0(t_n + l)\}$ to be convergent. Suppose $p = \lim\limits_{n \to +\infty} x_0(t_n + l)$, then $p \in \omega_{x_0} \cap \overline{B(K, \frac{d}{3})}$. The last contradicts (5.7), and it proves the assertion we need. Further, reasoning like in Theorem 5.1 we will have that for some $c_0 \in \mathbb{R}$ the set $\omega_{x_0} \cap V^{-1}(c_0) \neq \emptyset$. The last contradicts condition 3. of the theorem. The contradiction shows that $\omega_x \cap K \neq \emptyset$ for all $x \in X$. As the dynamical system $\langle (X, \mathbb{T}_1, \pi), (Y, \mathbb{T}_2, \sigma), h \rangle$ is bounded and asymptotically compact, then $(X, \mathbb{T}_1, \pi)$ satisfies the condition of Ladyzhenskaya. Now for finishing the proof of Theorem 5.5 it is sufficient to refer to Theorem 2.20. Theorem is proved. $\square$

Remark 5.4 *a) The condition 1. of Theorem 5.5 is equivalent to positive definiteness V on X_r and to its boundedness on bounded subsets from X;*

b) The assertion which is the opposite to Theorem 5.5 takes place and can be proved like Theorem 5.1.

c) Lemma 5.1 remains correct if the condition of complete continuity of $(X, \mathbb{T}_1, \pi)$ is changed by the condition of Ladyzhenskaya. That is why Theorem 5.5 is correct in the case when condition 3. is changed by the following: level lines of the function V do not contain ω- limit points of the dynamical system $(X, \mathbb{T}_1, \pi)$.

d) Theorem 5.5 remains valid if conditions 2. and 3. are changed by the following: if $x\tau \in E_r$ for all $\tau \in [0, t]$ $(t > 0)$, then $V(xt) < V(x)$, and it can be proved like Theorem 5.2.

e) Theorem 5.5 takes place for discrete dynamical systems if the function V is defined on the whole space and condition 2. takes place for all $t \geq 0$ and $x \in E$. For autonomous systems with discrete time this assertion is contained in [174].

5.2   Some criterions of dissipativity of differential equations

1. Let E^n be n-dimensional Euclidean space, $|\cdot|$ is the norm on E^n generated by the scalar production of $\langle \cdot, \cdot \rangle$. Consider the differential equation

$$\dot{u} = f(t, u), \tag{5.9}$$

where $f \in C(\mathbb{R} \times E^n, E^n)$ is a regular function. Along with equation (5.9) let us consider its H-class

$$\dot{v} = f(t, v) \quad (g \in H(f)). \tag{5.10}$$

As it was mentioned above, the equation (5.9) naturally generates a non-autonomous dynamical system (see example 3.1). Applying to it the results concerning general non-autonomous dynamical system we obtain a series of statements about the equation (5.9). From theorems 5.1, 5.2, Lemma 5.1 and Corollary 5.1 follow Theorems below.

Let $r > 0$ and $E_r^n := \{u \in E^n \mid |u| \geq r\}$.

Theorem 5.6 *Let $H(f)$ be compact. For the equation (5.9) to be dissipative it is necessary and sufficient that would exist $r > 0$ and continious function $V :$ $H(f) \times E_r^n \to \mathbb{R}_+$ satisfying the conditions:*

(1) $V(g, v) \geq a(|v|)$ for all $v \in E_r^n$ and $g \in H(f)$, where $a \in \mathfrak{A}$ and $\mathrm{Im}\, V \subseteq \mathrm{Im}\, a$;
(2) if $|\varphi(\tau, v, g)| \geq r$ for all $\tau \in [0, t]$ then $V(g_t, \varphi(t, v, g)) \leq V(g, v)$ ($g \in H(f)$);
(3) the lines of the level of the function V do not contain positive semitrajectories of equation (5.10).

Theorem 5.7 *Let $H(f)$ be compact. For the equality (5.9) to be dissipative it is necessary and sufficient that there would exist $r > 0$ and a continious function $V : H(f) \times E_r^n \to \mathbb{R}_+$, satisfying the following conditions:*

(1) $V(g, v) \geq a(|v|)$ for all $v \in E_r^n$ $g \in H(f)$, where $a \in \mathfrak{A}$ and $\mathrm{Im}\, V \subseteq \mathrm{Im}\, a$;
(2) if $|\varphi(\tau, v, g)| \geq r$ for all $\tau \in [0, t]$ then $V(g_t, \varphi(t, v, g)) < V(g, v)$ ($g \in H(f)$).

Theorem 5.8 *Let $H(f)$ be compact. Then the following conditions are equivalent:*

1. *the equation (5.9) is dissipative;*
2. *there exists a number $r > 0$ possessing the following property: for every $v \in E^n$ and $g \in H(f)$ there exists $\tau = \tau(v, g) > 0$ such that, $|\varphi(\tau, v, g)| < r$;*
3. *there exists such number $R_1 > 0$ that $\lim\limits_{t \to +\infty} \inf |\varphi(t, v, g)| < R_1$ for all $v \in E^n$ and $g \in H(f)$;*
4. *there exists a number $R_0 > 0$ and for every $R > 0$ there is $l(R) > 0$, such that $|\varphi(t, v, g)| \leq R_0$ for all $t \geq l(R)$, $|v| \leq R$ and $g \in H(f)$.*

Proof. The statement formulated above follows from Theorems 2.19 and 3.1. In the case when f is periodic the equivalence of the conditions 1., 2. and 4. is established in [270]. The equivalence of the conditions 1. and 2. for nonperiodic equationsis established in [163]. $\qquad\square$

Definition 5.2 We will say that the function $F \in C(\mathbb{R} \times E^n, E^n)$ satisfies the local condition of Lipschitz $u \in E^n$ uniformly on $t \in \mathbb{R}$ or just satisfies the condition of Lipschitz if for every $r > 0$ there exists $L(r) > 0$ such that $|F(t, u_1) - F(t, u_2)| \leq L|u_1 - u_2|$ for all $t \in \mathbb{R}$ and $u_1, u_2 \in B[0, r]$.

Theorem 5.9   *Let $f \in C(\mathbb{R} \times E^n, E^n)$ satisfy the condition of Lipschitz, $H(f)$ be compact and let exist $R > 0$ and continuous differentiable function $V \in C(\mathbb{R} \times E_r^n, \mathbb{R}_+)$ satisfying the following conditions:*

a. V satisfies the condition of Lipschitz;

b. $a(|u|) \leq V(t, u) \leq b(|u|)$ $(a, b \in \mathfrak{A}, \, Im\, a = Im\, b, \, t \in \mathbb{R}$ and $u \in E_r^n)$;

c. $\mathfrak{M}_f \subseteq \mathfrak{M}_V \cap \mathfrak{M}_{\frac{\partial V}{\partial t}} \cap \mathfrak{M}_{grad_u V}$ (where $\mathfrak{M}_f := \{\{t_n\} : \{f_{t_n}\}$ is convergent $\}$ and by analogy for $\mathfrak{M}_V$, $\mathfrak{M}_{\frac{\partial V}{\partial t}}$ and $\mathfrak{M}_{grad_u V}$);

d. $\dot{V}_f(t, u) := \dfrac{\partial V}{\partial t}(t, u) + \langle grad_u V, f(t, u)\rangle \leq 0 \;\; (t \in \mathbb{R}$ and $u \in E_r^n)$;

e. whatever would be $(\widetilde{V}, g) \in H(V, f) := \overline{\{(V_\tau, f_\tau) : \tau \in R\}}$ the lines of level of the function $\widetilde{V}$ do not contain positive semi-trajectories of the equation (5.10) lying out of the ball $B[0, r]$.

Then the equation (5.9) is dissipative.

Proof.   We will execute the proof in few steps.

1. Let $g \in H(f)$. There exists $\{t_k\} \subset \mathbb{R}$ such that $f_{t_k} \to g$. Since $\mathfrak{M}_f \subseteq \mathfrak{M}_V$, the sequence $\{V_{t_k}\}$ is also convergent. Assume $\widetilde{V} := \lim\limits_{k \to +\infty} V_{t_k}$. From the conditions a.– e. follows that the function $\widetilde{V}$ satisfies the condition of Lipschitz, $\dot{\widetilde{V}}_g(t, v) := \dfrac{\partial \widetilde{V}}{\partial t}(t, v) + \langle grad_v \widetilde{V}, g\rangle \leq 0$ and besides $a(|v|) \leq \widetilde{V}(t, v) \leq b(|v|)$ for all $v \in E_r^n$ and $t \in \mathbb{R}$.

2. Let $v \in E_r^n$ and $g \in H(f)$. Denote by $\varphi(\cdot, v, g)$ the solution of the equation (5.10) passing through the point v for $t = 0$. Let us show that this solution is extendable to the right onto the whole semi-axis $\mathbb{R}_+$. Obviously, for this it is sufficient to show that it is bounded on the whole domain of its existence $[0, t_v]$. Let $R = R(r) > 0$ be such that $a(R) > b(r)$, $\mathbb{T}_1(v) := \{t \in [0, t_v[: |\varphi(t, v, g)| \leq r\}$ and $\mathbb{T}_2(v) := [0, t_v[\setminus \mathbb{T}_1(v)$. It is clear that $\mathbb{T}_2(v)$ is open and, consequently, $\mathbb{T}_2(v) = \bigcup\limits_{\alpha}]t_\alpha, t_\beta[\;\; (\beta = \beta(\alpha))$. For every $t \in \mathbb{T}_2(v)$ there exists α such that $t \in]t_\alpha, t_\beta[, \; |\varphi(t_\alpha, v, g)| = |\varphi(t_\beta, v, g)| = r, \; |\varphi(t, v, g)| > r$ and, consequently,

$$a(|\varphi(t, v, g)|) \leq \widetilde{V}(t, \varphi(t, v, g)) \leq \widetilde{V}(t_\alpha, \varphi(t_\alpha, v, g))$$

$$\leq b(|\varphi(t_\alpha, v, g)|) = b(r) \leq a(R) \tag{5.11}$$

From the inequality (5.11) follows that $|\varphi(t, v, g)| \leq R$ and we have

$$\sup\{|\varphi(t, v, g)| : t \in [0, t_v[\} \leq r_0 := \max(r, R(r)).$$

Since $f \in C(\mathbb{R} \times E^n, E^n)$ satisfies the condition of Lipshitz, it is regular.

3. Let $\langle (X, \mathbb{R}_+, \pi), (Y, \mathbb{R}_+, \sigma), h \rangle$ be a non-autonomous dynamical system generated by the equation (5.9). Define on $X_r := E_r^n \times H(f)$ a function $\mathcal{V} : X_r \to \mathbb{R}_+$ by the equality $\mathcal{V}(v, g) = \widetilde{V}(0, v)$ where $\widetilde{V}$ is the function defined in the point 1. Note that the function $\mathcal{V}(\cdot, g)$ is defined uniquely by $g \in H(f)$ in virtue of the inclusion $\mathfrak{M}_f \subseteq \mathfrak{M}_V$.

4. Show that the introduced in the previous point function $\mathcal{V}$ satisfies all the conditions of Corollary 5.1. Let us establish its continuity on $X_r = E_r^n \times H(f)$. Let $g_k \to g$ and $v_k \to v$ $((v_k, g_k) \in E_r^n \times H(f))$. Then

$$|\mathcal{V}(v_k, g_k) - \mathcal{V}(v, g)| \leq |\mathcal{V}(v_k, g_k) - \mathcal{V}(v, g_k)| \tag{5.12}$$
$$+ |\mathcal{V}(v, g_k) - \mathcal{V}(v, g)| \leq L g_k(R)|v_k - v| + |\mathcal{V}(v, g_k) - \mathcal{V}(v, g)|,$$

where $L_{g_k}(R)$ is the constant of Lipschitz for $V(\cdot, g_k)$ on $B(0, R)$ and $R := \sup_{k \in \mathbb{N}} |v_k|$. Since $L_{g_k}(R) \leq L_f(R)$ for all $g \in H(f)$ (see, for example, [95]) and $\mathfrak{M}_f \subseteq \mathfrak{M}_V$, $\mathcal{V}(v, g_k) \to \mathcal{V}(v, g)$ and passing to limit in (5.12) with $k \to +\infty$ we establish the continuity of V in the point $(v, g) \in X_r$.

Other conditions of Corollary 5.1 under the conditions of the theorem are obviously fulfilled. To finish the proof it is sufficient to refer to Corollary 5.1. $\qquad \square$

Note, that in the case of periodicity of f the statement of Theorem 5.9 reinforces Theorem 2.5 from [270].

2. Consider the linear homogeneous equation

$$\dot{u} = A(t)u, \tag{5.13}$$

where $A \in C(\mathbb{R}, [E^n])$ and $[E^n]$ is the set of all linear operators $A : E^n \to E^n$. By $U(t, A)$ we denote the operator of Cauchy of the equation (5.13). Along with the equation (5.13) we will consider the perturbed equation

$$\dot{u} = A(t)u + F(t, u) \quad (F \in C(\mathbb{R} \times E^n, E^n)). \tag{5.14}$$

The following theorem takes place.

Theorem 5.10 *For the equation (5.14) to be dissipative the fulfillment of the following condition is sufficient:*

(1) the function F satisfies the condition of Lipschitz;
(2) $H(A, F) := \overline{\{(A_\tau, F_\tau) : \tau \in \mathbb{R}\}}$ is compact in $C(\mathbb{R}, [E^n]) \times C(\mathbb{R} \times E^n, E^n)$;
(3) there exist positive numbers N and ν such that

$$\|U(t, A)U^{-1}(\tau, A)\| \leq N e^{-\nu(t-\tau)} \quad (t \geq \tau); \tag{5.15}$$

(4) $|F(t, u)| \leq A + \varepsilon |u| \ (\forall u \in E^n, t \in \mathbb{R})$ where $A > 0$ and $0 \leq \varepsilon \leq \varepsilon_0 < \nu N^{-2}$.

Proof. Let $(B, G) \in H(A, F)$. Define $W_B \in [E^n]$ by the equality

$$W_B := \int_0^{+\infty} U^*(\tau, B) U(\tau, B) d\tau. \tag{5.16}$$

From (5.15) and (5.16) follows that $\|W_B\| \le N^2 (2\nu)^{-1}$. Denote by $a :=$ $\sup\{\|A(t)\| : t \in \mathbb{R}\}$ and note that

$$|u| = |U(-\tau, B_\tau) U(\tau, B) u| \le e^{a\tau} |U(\tau, B) u|.$$

From this

$$\langle W_B u, u \rangle = \int_0^{+\infty} |U(\tau, B) u|^2 d\tau \ge \int_0^{+\infty} e^{-2a\tau} |u|^2 d\tau = \frac{1}{2a} |u|^2.$$

So, we established the following inequality

$$\frac{1}{2a} |u|^2 \le \langle W_B u, u \rangle \le \frac{N^2}{2\nu} |u|^2 \quad (u \in E^n, \ B \in H(A)).$$

Let $Y := H(A, F)$, $(Y, \mathbb{R}, \sigma)$ be the dynamical system of shifts on Y, $X := E^n \times Y$ and $\pi^t(v; B, g) := (\varphi(t, v, B, g); B_t, g_t)$ $(v \in E^n, \ (B, G) \in H(A, F))$, where $\varphi(\cdot, v, B, G)$ is the solution of the equation

$$\dot{v} = B(t) v + G(t, v), \tag{5.18}$$

passing through the point v as $t = 0$. Let us consider a non-autonomous dynamical system $\langle (X, \mathbb{R}, \pi), (Y, \mathbb{R}, \sigma), h \rangle$ where $h := pr_2 : X \to Y$. Define on X a function $V : X \to \mathbb{R}_+$ by the following rule:

$$V(v; B, G) := \langle W_B v, v \rangle \tag{5.19}$$

for all $v \in E^n$ and $(B, G) \in H(A, F)$. Note that this function possesses the following properties:

a. $\dfrac{1}{2a} |v|^2 \le V(v; B, G) \le \dfrac{N^2}{2a} |v|^2$ $(v \in E^n, \ (B, g) \in H(A, F))$.
b. V is continuous.

   In fact, if $v_k \to v$ and $(B_k, G_k) \to (B, G)$ then

$$|V(v_k; B_k, G_k) - V(v; B, G)| = |\langle W_{B_k} v_k, v_k \rangle - \langle W_B v, v \rangle| \tag{5.20}$$

$$= |\langle W_{B_k}(v_k - v), v_k \rangle + \langle W_{B_k} v, v_k - v \rangle + \langle (W_{B_k} - W_B) v, v \rangle|$$

$$\le \frac{N^2}{2\nu} |v_k| |v_k - v| + \frac{N^2}{2\nu} |v| |v - v_k| + \|W_{B_k} - W_B\| |v|^2.$$

From the inequality (5.20) we can easily see that for V to be continuous it is sufficient to show that $\|W_{B_k} - W_B\| \to 0$ if $B_k \to B$. Let us establish the latter. From (5.16) follows that W_B is a self-conjugate operator. That is why

$$||W_{B_k} - W_B|| = \sup_{|v|=1} |\langle (W_{B_k} - W_B)v, v \rangle| \leq \tag{5.21}$$

$$\leq \int_0^{+\infty} \sup_{|v|\leq 1} ||[U(\tau, B_k) - U(\tau, B)]v||^2 d\tau = \int_0^l \sup_{|v|\leq 1} |[U(\tau, B_k) - U(\tau, B)]v|^2 d\tau$$

$$+ \int_l^{+\infty} \sup_{|v|\leq 1} |[U(\tau, B_k) - U(\tau, B)]v|^2 d\tau \leq \sup_{0\leq\tau\leq l} ||U(\tau, B_k) - U(\tau, B)||^2 l$$

$$+2 \int_l^{+\infty} (||U(\tau, B_k)||^2 + ||U(\tau, B)||^2) d\tau.$$

From (5.15) follows the inequality $||U(\tau, B)|| \leq Ne^{-\nu\tau}$ for all $\tau \in \mathbb{R}_+$ and $B \in H(A)$ (see [51, p.70]). Hence, from (5.21) we obtain

$$||W_{B_k} - W_B|| \leq \sup_{0\leq\tau\leq l} ||U(\tau, B_k) - U(\tau, B)||^2 l + \frac{2N^2}{v} e^{-2vl}. \tag{5.22}$$

Passing to limit as $k \to +\infty$ and taking in consideration that the first term in the right hand side of (5.22) vanishes with any fixed $l > 0$ we establish that

$$\limsup_{k\to+\infty} ||W_{B_k} - W_B|| \leq \frac{2N^2}{v} e^{-2vl}. \tag{5.23}$$

In this case (5.23) takes place for all $l > 0$. Making $l \to +\infty$ and taking in consideration that the left hand side of the equation (5.23) does not depend on l we will obtain the result needed. So, the continuity of V is established.

c. Let now $|\varphi(\tau; v, B, G)| \geq r$ $(r > N^2 A(v - \varepsilon_0 N^2)^{-1})$ for all $\tau \in [0, t]$. Then $V(\varphi(t; v, B, G), B_t, G_t) < V(v; B, G)$. To establish this inequality we calculate the derivative of the function

$$\mathcal{V}(\tau) := V(\varphi(\tau; v, B, G), B_\tau, G_\tau) = \langle W_B \varphi(\tau; v, B, G), \varphi(\tau; v, B, G) \rangle$$

in $\tau \in \mathbb{R}$. By standard reasoning, taking into account the identity

$$\frac{d}{d\tau} W_{B_\tau} + B^*(\tau) W_{B_\tau} + W_{B_\tau} B(\tau) = -E \tag{5.24}$$

(E is the unit operator in E^n), we have

$$\frac{d}{d\tau} V(\varphi(\tau; v, B, G), B_\tau, G_\tau) = -|\varphi(\tau; v, B, G)|^2 +$$
$$2Re\langle G(\tau, \varphi(\tau; v, B, G)), W_{B_\tau} \varphi(\tau, v, B, G) \rangle. \tag{5.25}$$

From (5.25) we obtain

$$\frac{d}{d\tau}V(\varphi(\tau;v,B,G),B_\tau,G_\tau) \leq -|\varphi(\tau;v,B,G)|^2$$

$$+\frac{N^2}{v^2}|\varphi(\tau;v,B,G)|(A+\varepsilon_0|\varphi(\tau;v;B,G)|)$$

$$\leq |\varphi(\tau;v,B,G)|^2\left(-1+\varepsilon_0\frac{N^2}{v}+\frac{N^2 A}{v}\frac{1}{r}\right).$$

From this follows that $V(\varphi(t;v,B,G),B_t,G_t) < V(v,B,G)$. In the same way as in Theorem 5.9 we prove the possibility of non-local extension to the right of all the solutions of the equation (5.18); their uniqueness follows from the condition of Lipschitz for F. To finish the proof of the theorem it is sufficient to refer to Theorem 5.9 (see also Theorem 5.2).   □

Remark 5.5   *a. Theorem 5.10 coincides with Theorem 3.21, which is obtained with the help of a set of a priori estimations. However, Theorem 3.21 takes place also for the equations defined in some Banach space.*

b. Theorem 5.11 remains valid also in the case when we replace the condition 4. by: $|F(t,u)| \leq A+B|u|^\alpha$ $(\forall u \in E^n, t \in R)$ where A and B are some positive numbers and $0 \leq \alpha \leq 1$. We execute the proof of this statement by the same scheme that Theorem 5.10 choosing r such that the condition $r > r_0$ would take place, where r_0 is the solution of the equation $Ar^{-1} + Br^{\alpha-1} = v^2 N^{-2}$.

c. Note that the statement of the point b., generally speaking, does not take place if $\alpha > 1$. This is proved by the example $\dot{x} = -x + x^3$.

Theorem 5.11   *Let $f \in C(\mathbb{R} \times E^n, E^n)$ satisfies the condition of Lipschitz, $H(f)$ be compact and let exist $A \in C(\mathbb{R}, [E^n])$ satisfying the following conditions:*

1. *$\mathfrak{M}_f \subseteq \mathfrak{M}_A \cap \mathfrak{M}_{\dot{A}}$ $(\dot{A}(t) := \dfrac{dA(t)}{dt})$;*
2. *$A(t) = A^*(t)$ and $\alpha|u|^2 \leq \langle A(t)u,u\rangle \leq \beta|u|^2$ $u \in E^n$ for all $t \in \mathbb{R}$ and $\alpha, \beta > 0$;*
3. *$\langle \dot{A}(t)u,u\rangle \leq 0$ $(u \in E^n, t \in \mathbb{R})$;*
4. *there exists $r > 0$ such that $Re\langle A(t)u, f(t,u)\rangle \leq -\gamma(|u|)$ for all $u \in E^n_r$ $t \in R_+$ where $\gamma(s) > 0$ as $s \geq r$.*

Then the equation (5.9) is dissipative.

Proof.   According to Theorem 5.9 the function f is regular. Let $\langle (X,\mathbb{R}_+,\pi), (Y,\mathbb{R},\sigma), h\rangle$ be a non-autonomous dynamical system generated by the equation (5.9) (see example 3.1) and $g \in H(f)$. There exists $\{t_n\} \in \mathfrak{M}_f$ such that $g = \lim\limits_{k\to+\infty} f_{t_k}$ and under the conditions of the theorem $\{A_{t_k}\}$ is convergent. Denote its limit by B_g. We define now on $X_r := E^n_r \times Y$ a function V by the following rule:

$V(v,g) = \langle B_g(0)v, v \rangle$ for every $v \in E_r^n$ and $g \in H(f)$. Note that B_g is defined uniquely because $\mathfrak{M}_f \subseteq \mathfrak{M}_A$. The continuity of V is proved in the same way that in Theorems 5.7 and 5.8.

Let now $|\varphi(\tau, v, g)| \geq r$ for all $\tau \in [0, t]$ $(t > 0)$. Let us calculate the derivative of $V(\varphi(\tau, v, g), g_\tau)$ in $\tau \in \mathbb{R}$. For this note that from the inclusion $\mathfrak{M}_f \subseteq \mathfrak{M}_A \cap \mathfrak{M}_{\dot{A}}$ follows the equality

$$\frac{dV}{dt}(\varphi(\tau, v, g), g_\tau) = \frac{d}{d\tau}\langle B(\tau)\varphi(\tau, v, g), \varphi(\tau, v, g)\rangle \tag{5.26}$$
$$= \langle \dot{B}(\tau)\varphi(\tau, v, g), \varphi(\tau, v, g)\rangle + 2Re\langle B(\tau)\varphi(\tau, v, g), g(\tau, \varphi(\tau, v, g))\rangle.$$

From the conditions 1.–4. of the theorem we obtain

$$\langle \dot{B}(t)v, v \rangle \leq 0 \quad \text{and} \quad Re\langle B(t)v, g(t, v)\rangle \leq -\gamma(|v|) \tag{5.27}$$

for all $v \in E_r^n$, $B \in H(A)$, $t \in \mathbb{R}$ and $g \in H(f)$. From (5.26) and (5.27) follows that $\frac{d}{d\tau}V(\varphi(\tau, v, g), g_\tau) \leq -2\gamma(|\varphi(\tau, v, g)|)$, hence, $V(\varphi(\tau, v, g), g_\tau) < V(v, g)$ for all $v \in E_r^n$ and $g \in H(f)$. To finish the proof it is sufficient to refer to Corollary 5.2 and Theorem 5.2. $\qquad\square$

Corollary 5.4 *If the equation (5.9) is τ-periodic and the conditions of Theorem 5.9 are fulfilled then the equation (5.9) has at least one τ-periodic solution.*

The particular case of Corollary 5.4 is announced in the work [192].

Corollary 5.5 *Let $a_i \in C(\mathbb{R}, \mathbb{R})$. If $H(a_i)$ is compact $(i = \overline{0, 2n + 1})$ and $a_{2n+1}(t) \leq -\alpha$ $(\alpha > 0)$ for all $t \in \mathbb{R}_+$ then the equation*

$$\dot{v}(t) = a_0(t) + a_1(t)v + \cdots + a_{2n+1}v^{2n+1} \tag{5.28}$$

is dissipative.

Proof. The formulated statement follows from Theorem 5.9. For this it sufficient to take as $A(t)$ a unit operator and to notice that

$$\langle A(t)v, f(t, v)\rangle = v(a_0(t) + a_1(t)v + \cdots + a_{2n+1}(t)v^{2n+1}) =$$
$$v^{2n+2}\left(a_{2n+1}(t) + \frac{a_{2n}(t)}{v} + \cdots + \frac{a_0(t)}{v^{2n+1}}\right) \leq -\frac{\alpha}{2}v^{2n+2}$$

for all $|v| \geq r$ and $t \in \mathbb{R}$ and when $r > 0$ is big enough. $\qquad\square$

Corollary 5.6 *If functions a_i $(i \in \overline{0, 2n + 1})$ are τ-periodic and $a_{2n+1}(t) > 0$ for all $t \in [0, \tau[$ then the equation (5.28) has at least one τ-periodic solution.*

Note that in the case when $a_{2n+1}(t) < 0$ $(t \in [0, \tau[)$ according to Corollary 5.6 the equation (5.28) is dissipative and, consequently, by Theorem 2.3 from [270] it has at least one τ-periodic solution. The case $a_{2n+1}(t) > 0$ $(t \in [0, \tau[)$ is reduced to the previous one by the replacement of t by $-t$.

In the case when $a_{2n+1} \equiv 1$ Corollary 5.6 coincides with the result established in [270, p.126].

Corollary 5.7 *Let $f = (f_1, f_2, \ldots, f_n) \in C(\mathbb{R} \times E^n, E^n)$ be continuously differentiable. If $\dfrac{\partial f_i}{\partial x_i} \leq -m < 0$ $(i \in \overline{1, n})$, $H(f)$ is compact and functions $f_i(t, x_1, x_2, \ldots, x_{i-1}, 0, x_{i+1}, \ldots, x_n)$ are bounded, then the equation (5.9) is dissipative.*

Proof. From the equality

$$f_i(t, x_1, x_2, \ldots, x_n) = \int\limits_0^{x_i} \frac{\partial f_i}{\partial x_i}(t, x_1, x_2, \ldots, x_n) dx_i$$
$$+ f_i(t, x_1, x_2, \ldots, x_{i-1}, 0, x_{i+1}, \ldots, x_n),$$

in the conditions of Corollary 5.7 follows that for all $x \in \mathbb{R}^n$ the inequality $x_i f_i(t, x_1, x_2, \ldots, x_n) \leq -mx_i^2 + M|x_i|$ holds, hence,

$$\langle x, f(t, x) \rangle = \sum_{i=1}^n x_i f_i(t, x) \leq -m|x|^2 + Mn|x| \leq -\frac{m}{2}|x|^2$$

for all $t \in \mathbb{R}$ and $|x| \geq \dfrac{2Mn}{m}$. According to Theorem 5.6 the equation (5.9) is dissipative. $\qquad \square$

Note that in the case of periodicity of f Corollary 5.7 is well known [270]. When f is not periodic the statement close to Corollary 5.7 is established in [264] with some additional restrictions on f.

Theorem 5.12 *Let the condition of uniqueness be fulfilled for every equation (5.10) with $g \in \Sigma_f := \{f_\tau : \tau \in R\}$. For the disiipaivity of the equation (5.9) there is sufficient existence of a function $V : \mathbb{R} \times E_r^n \to \mathbb{R}_+$ $(r > 0)$ satisfying the conditions:*

(1) V satisfies the condition of Lipschitz;
(2) $a(|u|) \leq V(t, u) \leq b(|u|)$ $(a, b \in \mathfrak{A}, \operatorname{Im} a = \operatorname{Im} b)$ when $u \in E_r^n$;
(3) $\dot{V}_f(t, u) := \limsup\limits_{\tau \downarrow 0} \tau^{-1} \left[V(t + \tau, u + \tau f(t, u)) - V(t, u) \right] \leq -c(|u|)$ for $u \in E_n^n$,
*   where $c : \mathbb{R}_+ \to \mathbb{R}_+$ is continuous and $c(s) > 0$ as $s > r$.*

Proof. In the same way as in Theorem 5.9 we can show that the solution $\varphi(\cdot, u, f_\tau)$ of the equation (5.10) $(g = f_\tau)$ passing through u as $t = 0$ is extendable to the right onto the whole semi-axis $\mathbb{R}_+$. Assume $Y := \Sigma_f$ and by $(Y, \mathbb{R}_+, \sigma)$ denote the dynamical system defined by the following rule: $\pi^t(u, f_\tau) := (\varphi(t, u, f_\tau), f_{\tau+t})$ for all $\tau \in \mathbb{R}$, $u \in E^n$ and $t \in \mathbb{R}_+$. Define on $X_r := E_r^n \times Y$ a function $\mathcal{V}$ by the equality $\mathcal{V}(u, f_\tau) := V(\tau, u)$. Then $\mathcal{V}(\pi^t(u, f_\tau)) = \mathcal{V}(\varphi(t, u, f_\tau), f_{t+\tau}) = V(t + \tau, \varphi(t, u, f_\tau)) = V(s, \varphi(s - \tau, u, f_\tau)) = V(s, x(s; u, \tau))$ $(s = t + \tau)$ and, consequently,

$$
\begin{aligned}
\dot{\mathcal{V}}_\pi(u, f_\tau) &= \limsup_{t\downarrow 0} t^{-1}[\mathcal{V}(\pi^t(u, f_\tau)) - \mathcal{V}(u, f_\tau)] \\
&= \limsup_{s-\tau\downarrow 0} (s - \tau)^{-1}[V(s, x(s; u, \tau)) - V(\tau, x[\tau; u, \tau))] \\
&= \dot{V}_f(\tau, x(\tau; u, \tau)) = \dot{V}_f(\tau, u) \le -c(|u|) \quad (|u| \ge r).
\end{aligned}
$$

According to Theorem 5.3 the non-autonomous dynamical system $\langle(X, \mathbb{R}_+, \pi),$ $(Y, \mathbb{R}_+, \sigma), h\rangle$ $(h = pr_2 : X \to Y)$ is dissipative and, consequently, there exist $R > 0$ and $l(u, \tau) > 0$ such that $|\varphi(t, u, f_\tau)| < R$ for all $u \in E^n$, $\tau \in \mathbb{R}$ and $t \ge l(u, \tau)$. As it is in [137] we can show that the number $l(u, \tau) > 0$ can be chosen not depending on $\tau \in \mathbb{R}$. Reasoning in the same way that in the proof of Lemma 3.1, we obtain that the equation (5.9) is dissipative. $\qquad\square$

Note that Theorem 5.12 makes precise the known theorem of T. Yoshizava [137],[325]–[326].

5.3 Theorem of Barbashin–Krasovskii for non-autonomous dynamical systems

The ideas and methods that we applied studying dissipative system in first two paragraphs of Chapter 5 allow formulating and proving of a series of statements about the stability of non-autonomous dynamical system and, in particular, generalizing of the known theorem of Barbashin–Krasovskii [21] on asymptotic stability.

Let (X, h, Y) be a finite-dimensional vectorial bundle fiber, $\langle(X, \mathbb{T}_1, \pi),$ $(Y, \mathbb{T}_2, \sigma), h\rangle$ be non-autonomous dynamical system, $\theta_y \in X_y := h^{-1}(y)$ be a null element $(|\theta_y| = 0)$ and $\Theta := \{\theta_y \mid y \in Y\}$ be a null section of the vectorial bundle fiber (X, h, Y). Below we will suppose that the null section Θ is invariant, i.e. $\Theta \subseteq X$ is an invariant set of the dynamical system $(X, \mathbb{T}_1, \pi)$.

Definition 5.3 the null section Θ is called uniformly stable if for every $\varepsilon > 0$ there exists $\delta(\varepsilon) > 0$ such that $y \in Y$, $x \in X_y$ and $|x| < \delta$ implies $|xt| < \varepsilon$ for all $t \ge 0$.

Definition 5.4   If Θ is uniformly stable and $\lim\limits_{t \to +\infty} |xt| = 0$ for all $x \in X$, then the null section is called globally uniformly asymptotically stable.

Theorem 5.13   *Let Y be compact. For the null section Θ to be globally uniformly asymptotically stable it is necessary and sufficient that there would exist a continuous function $V : X \to \mathbb{R}_+$ satisfying the following conditions:*

1. *$V(x) \geq a(|x|)$ for all $x \in X$, $V(\theta_y) = 0$ for all $y \in Y$ and $Im\, a = Im\, V$, where $a \in \mathfrak{A}$.*
2. *$V(xt) \leq V(x)$ for all $x \in X$ and $t \geq 0$.*
3. *the lines of level of V do not contain non-null ω-limit points of the dynamical system $(X, \mathbb{T}, \pi)$.*

Proof.   Sufficiency. Let the conditions of the theory be satisfied. Show that the null section Θ is uniformly stable. Suppose that it is not true. Then there exist $\varepsilon_0 > 0$, $|x_n| < \delta$, $\delta_n \downarrow 0$ and $t_n \geq 0$ such that

$$|x_n t_n| \geq \varepsilon_0. \tag{5.29}$$

On the other hand, $0 \leq a(|x_n t_n|) \leq V(x_n t_n) \leq V(x_n) \to 0$ as $n \to +\infty$ and, consequently, $|x_n t_n| \to 0$. The last contradicts to the equality (5.29).

Now let us show that $\lim\limits_{t \to +\infty} |xt| = 0$ for all $x \in X$. Really, if we suppose the contrary then there exists $x_0 \in X$ ($|x_0| \neq 0$) such that $\limsup\limits_{t \to +\infty} |x_0 t| > 0$, i.e. there exist $\varepsilon_0 > 0$ and $t_n \to +\infty$ for which

$$|x_0 t_n| \geq \varepsilon_0. \tag{5.30}$$

Note that $\Sigma_{x_0}^+$ is relatively compact. In fact, $a(|x_0 t|) \leq V(x_0 t) \leq V(x_0)$ and, consequently, $|x_0 t| \leq a^{-1}(V(x_0))$ for all $t \geq 0$. So, the sequence $\{x_0 t_n\}$ can be considered convergent. Assume $\tilde{x} := \lim\limits_{n \to +\infty} x_0 t_n$, then $\tilde{x} \in \omega_{x_0}$. Reasoning in the same way that in Theorem 5.1 we can show that there exists $c \geq 0$ for which $V(x) = c$ for all $x \in \omega_{x_0}$. Since $\tilde{x} \in \omega_{x_0}$ and $\tilde{x} = \lim\limits_{n \to +\infty} x_0 t_n$, from (5.30) follows that $|\tilde{x}| > 0$, hence $c = V(\tilde{x}) \geq a(|\tilde{x}|) > 0$. So, the lines of level of the function V contain non-null ω–limit points. This contradicts to the conditions of the theorem. The global uniform asymptotic stability of the null section is proved.

Necessity. Let the null section Θ be globally uniformly asymptotically stable. Let $V : X \to \mathbb{R}_+$ be the function, defined by the equality (5.1). Directly from the definition of V follows that it satisfies to the conditions 1. and 2. of the theorem. We will show that it is continuous. For that we note that under conditions of Theorem the non-autonomous system $\langle (X, \mathbb{T}_1, \pi), (Y, \mathbb{T}_2, \sigma), h \rangle$ is dissipative. Now we will show that $J \subseteq \Theta$, where J is the center of Levinson of $(X, \mathbb{T}_1, \pi)$. Indeed, if $x \in J$ then according to Theorem 1.2 the semi-trajectory Σ_x^+ can be extended to the left, i.e. there exists a continuous mapping $\varphi : \mathbb{S} \to J$ such that $\pi^t \varphi(s) = \varphi(t+s)$ for all $t \in \mathbb{T}_1$, $s \in \mathbb{S}$ and $\varphi(0) = x$. Since J is compact, the set of α–limit points α_{φ_x} of the motion φ is not empty, compact and invariant. Obviously, $\alpha_{\varphi_x} \cap \Theta \neq \emptyset$ and, consequently, there exists $t_n \to -\infty$ such that $|\varphi(t_n)| \to 0$. Let $\varepsilon > 0$ and $\delta(\varepsilon) > 0$ be chosen out of the condition of the uniform stability of Θ. Then for n big enough we have $|\varphi(t_n)| < \delta$ and, consequently, $|\pi^t \varphi(t_n)| = |\varphi(t + t_n)| < \varepsilon$ for all $t \geq 0$. In particular, $|x| = |\varphi(t_n - t_n)| < \varepsilon$. And since ε is arbitrary, we have $|x| = 0$, i.e. $J \subseteq \Theta$.

We will show now that for every $r > 0$ the following equality takes place

$$\lim_{t \to +\infty} \sup_{|x| \leq r} |xt| = 0. \tag{5.31}$$

Suppose that it is not so; then there exist $\varepsilon_0 > 0$, $r_0 > 0$, $|x_n| < r_0$ and $t_n \to +\infty$ such that

$$|x_n t_n| \geq \varepsilon_0. \tag{5.32}$$

In virtue of dissipativity of the system $\langle (X, \mathbb{T}_1, \pi), (Y, \mathbb{T}_2, \sigma), h \rangle$ the sequence $\{x_n t_n\}$ can be considered convergent. Assume $x_0 := \lim_{n \to +\infty} x_n t_n$ and note that $x_0 \in J$, hence $|x_0| = 0$. The last contradicts to (5.32). So, the equality (5.31) takes place. The continuity of V is established literally repeating the reasoning of Theorem 5.1 when instead of (5.2) we use the condition (5.31).

As for the 3rd condition, we verify it in the same way that in Theorem 5.1. The theorem is proved. $\qquad\qquad\square$

Note that Theorem 5.13 is valid if we replace X by some tubular neighborhood $U_r := \{x \in X : |x| \leq r\}$ $(r > 0)$ of the null section Θ.

Definition 5.5 The null section Θ is called uniformly asymptotically stable if it is uniformly stable and there exists $r > 0$ such that $\lim_{t \to +\infty} |xt| = 0$ for all $x \in U_r$.

If we slightly change the proof of Theorem 5.13, we can prove the following theorem.

Theorem 5.14 *Let Y be compact. For the null section Θ to be uniformly asymptotically stable it is necessary and sufficient that there would exist $r > 0$ and a continuous function $V : U_r \to \mathbb{R}_+$ satisfying the following conditions:*

(1) $V(x) \geq a(|x|)$ for all $x \in U_r$, $V(\theta_y) = 0$ for all $y \in Y$;

(2) $V(xt) \leq V(x)$ if $x\tau \in U_r$ for all $\tau \in [0, t]$;

(3) the lines of level of V do not contain non-null ω-limit points of the dynamical system $(X, \mathbb{T}_1, \pi)$.

Remark 5.6 *As well as in Lemma 5.1 we can show that in Theorem 5.14 the condition 3. can be replaced by the following one: the lines of level of the function V do not positive semi-trajectories of the dynamical system $(X, \mathbb{T}_1, \pi)$.*

At last note that the following theorem takes place.

Theorem 5.15 *For the null section Θ would be uniformly stable it is necessary and sufficient that there would exist $r > 0$ and a function $V : U_r \to \mathbb{R}_+$ satisfying the following conditions:*

1. $V(x) \geq a(|x|)$ *for all* $x \in U_r$ *and* $\lim\limits_{|x| \to 0} V(x) = 0$;
2. $V(xt) \leq V(x)$ *if* $x\tau \in U_r$ *for all* $\tau \in [0, t]$.

Proof. Sufficiency. Let us show that the null section is uniformly stable. Suppose that it is not so. Then there exist $\varepsilon_0 > 0$ $(\varepsilon_0 < r)$, $\delta_n \downarrow 0$, $|x_n| < \delta_n$ and $t_n \geq 0$ such that $x_n \tau \in U_r$ for all $\tau \in [0, t_n]$ and

$$|x_n t_n| \geq \varepsilon_0.$$

Note that

$$a(\varepsilon_0) \leq a(|x_n t_n|) \leq V(x_n t_n) \leq V(x_n). \tag{5.33}$$

Passing to limit in (5.33) as $n \to +\infty$, we obtain $a(\varepsilon_0) \leq 0$ and, consequently, $\varepsilon_0 = 0$. The last contradicts to the choice of ε_0. So, the uniform stability of Θ is proved.

Necessity. Let Θ be uniformly stable. For $\varepsilon_0 = 1$ select $\delta_0 = \delta(\varepsilon_0) > 0$ $(\delta_0 \leq 1)$ out of the condition of the uniform stability of Θ. Assume $r := \delta_0$ and define $V : U_r \to \mathbb{R}_+$ by the equality (5.1). It is easy to see that the function V is the one in question, i.e. it satisfies the conditions 1. and 2. of Theorem 5.15. The theorem is proved. $\qquad\square$

Remark 5.7 *By the conditions of Theorem 5.15 the function V, generally speaking, is not continuous, unlike the case in Theorems 5.13 and 5.14.*

From the theorems given in this section we can get the corresponding statements for the equation (5.9). Let us previously bring the following definitions.

Let $f(t, 0) \equiv 0$ and the function $f \in C(\mathbb{R} \times E^n, E^n)$ be regular.

Definition 5.6 We will say that the null solution of the equation (5.9) is uniformly stable if for every $\varepsilon > 0$ there exists $\delta(\varepsilon) > 0$ such that $g \in H(f)$, $|v| < \delta$ implies $|\varphi(t, v, g)| < \varepsilon$ for all $t \geq 0$.

Definition 5.7 If the null solution of the equation (5.9) is uniformly stable and there exists $\gamma > 0$ such that $\lim\limits_{t \to +\infty} |\varphi(t, v, g)| = 0$ for all $g \in H(f)$ and $v \in B(0, \gamma)$, then the null solution of the equation (5.9) is called uniformly asymptotically stable.

Definition 5.8 The null solution of (5.9) is globally uniformly asymptotically stable if it is uniformly stable and $|\varphi(t, v, g)| \to 0$ as $t \to +\infty$ for any $g \in H(f)$ and $v \in E^n$.

Note that from the results of the works [8],[291] follows the equivalence of the standard definition of the uniform stability (global uniform asymptotic stability) and of the one given above.

From Theorems 5.13, 5.14 and Remark 5.6 follows the theorems below.

Theorem 5.16 *Let $H(f)$ be compact and $f(t, 0) \equiv 0$. For the null solution of the equation (5.9) to be globally uniformly stable it is necessary and sufficient that there would exists a continuous function $V : H(f) \times E^n \to \mathbb{R}_+$ satisfying the following conditions:*

(1) $V(g, v) \geq a(|v|)$ for all $v \in E^n$, $V(g, 0) = 0$, $g \in H(f)$ and $Im\, a = Im\, V$ where $a \in \mathfrak{A}$;

(2) $V(g_\tau, \varphi(\tau, v, g)) \leq V(g, v)$ for all $g \in H(f)$, $v \in E^n$ and $\tau \geq 0$;

(3) whatever would be the function $g \in H(f)$, the lines of level of the function V do not contain positive semi-trajectories of the equation (5.10) excepting the trivial one.

Theorem 5.17 *Let $H(f)$ be compact and $f(t, 0) \equiv 0$. For the null solution of the equation (5.9) to be uniformly asymptotically stable it is necessary and sufficient that there would exist a continuous function $V : H(f) \times B[0, r_0] \to \mathbb{R}_+$ satisfying the following conditions:*

(1) $V(g, v) \geq a(|v|)$ and $V(g, 0) = 0$ for all $g \in H(f)$ and $x \in B[0, r_0]$, where $a \in \mathfrak{A}$;

(2) $V(g_t, \varphi(t, v, g)) \leq V(g, v)$ if $\varphi(t, v, g) \in B[0, r_0]$ for all $\tau \in [0, t]$;

(3) whatever would be the function $g \in H(f)$, the lines of level of the function V do not contain positive semi-trajectories of the equation (5.10) excepting the trivial one.

From Theorems 5.16 and 5.17 we can get the sufficient conditions of asymptotic stability suitable for the applications.

Theorem 5.18 *Let $H(f)$ be compact, $f(t, 0) \equiv 0$ and let there be a continuously differentiable function $V \in C(\mathbb{R} \times E^n, \mathbb{R}_+)$ satisfying the conditions:*

(1) $a(|u|) \leq V(t, u) \leq b(|u|)$ $(a, b \in \mathfrak{A}, Im\, a = Im\, b)$ for all $t \in \mathbb{R}$ and $u \in E^n$;

(2) $\mathfrak{M}_f \subseteq \mathfrak{M}_V \cap \mathfrak{M}_{\frac{\partial V}{\partial t}} \cap \mathfrak{M}_{grad_u V}$;

(3) $\dot{V}_f(t,u) := \dfrac{\partial}{\partial t}V(t,u) + \langle grad_u V, f\rangle \le 0 \quad (t \in \mathbb{R},\ u \in E^n)$;

(4) whatever would be the function $g \in H(f)$, the lines of level of the function V do not contain positive semi-trajectories of the equation (5.10) excepting the trivial one.

Then the null solution of the equation (5.9) is uniformly asymptotically stable in general.

Theorem 5.19   *Let $r > 0$, $H(f)$ be compact, $f(t,0) \equiv 0$ and let exist a continuously differentiable function $V \in C(\mathbb{R} \times B[0,r], \mathbb{R}_+)$ satisfying the following conditions:*

(1) $a(|u|) \le V(t,u) \le b(|u|) \quad (a, b \in \mathfrak{A})$ for all $t \in \mathbb{R}$ and $x \in B[0,r]$;

(2) $\mathfrak{M}_f \subseteq \mathfrak{M}_V \cap \mathfrak{M}_{\frac{\partial V}{\partial t}} \cap \mathfrak{M}_{grad_u V}$;

(3) $\dot{V}_f(t,u) := \dfrac{\partial}{\partial t}V(t,u) + \langle grad_u V(t,u), f(t,u)\rangle \le 0 \quad (t \in \mathbb{R},\ u \in E^n)$;

(4) whatever would be the function $g \in H(f)$, the lines of level of the function V do not contain positive semi-trajectories of the equation (5.10) excepting the trivial one.

Then the null solution of the equation (5.9) is uniformly asymptotically stable.

Proof.   The proof of Theorems 5.18 and 5.19 is constructed after the same pattern as that in Theorem 5.9. $\square$

Statements similar to Theorem 5.19 have been obtained in the works [4],[193].

5.4   Equations with convergence

1. Let us give the criterions of the convergence of the nonlinear equation (5.9) resulting from Theorems 2.7, 2.12 and 2.13, Corollary 2.2 and Theorems 2.16 and 2.17 by applying the last theorems to the non-autonomous dynamical system generated by the equation (5.9).

Theorem 5.20   *Let $f \in C(\mathbb{R} \times E^n, E^n)$ be recurrent w.r.t. $t \in \mathbb{R}$ uniformly w.r.t. u on compacts from E^n [300]. For the equation (5.9) to be convergent it is necessary and sufficient that the following conditions would be satisfied:*

(1) the equation (5.9) has at least one bounded on $\mathbb{R}_+$ solution;

(2) $\displaystyle\lim_{t \to +\infty} |\varphi(t,v_1,g) - \varphi(t,v_2,g)| = 0$ for all $g \in H(f)$ and $v_1, v_2 \in E^n$;

(3) for every $\varepsilon > 0$ and $r > 0$ there exists $\delta = \delta(\varepsilon, r) > 0$ such that $|v_1 - v_2| < \delta$ ($|v_1|, |v_2| < \delta$) implies $|\varphi(t,v_1,g) - \varphi(t,v_2,g)| < \varepsilon$ for all $t \ge 0$ and $g \in H(f)$.

Note that in the case of almost periodicity of the function f Theorem 5.20 generalizes and refines the criterion of almost periodic convergence of V.I.Zubov [336].

Theorem 5.21 *Let $f \in C(\mathbb{R} \times E^n, E^n)$ be recurrent w.r.t. $t \in \mathbb{R}$ uniformly w.r.t. u on compacts from E^n. For the equation (5.9) to be convergent it is necessary and sufficient that the following conditions would be fulfilled:*

1) *the equation (5.9) has at least one bounded on $\mathbb{R}_+$ solution;*
2) *whatever would be $g \in H(f)$ and $v \in E^n$, the solution $\varphi(\cdot, v, g)$ of the equation (5.10) is asymptotically stable.*

In the case of periodicity of the function f Theorem 5.21 coincides with the theorem of N.N.Krasovskii-V.A.Pliss [270].

Theorem 5.22 *Let $f \in C(\mathbb{R} \times E^n, E^n)$ be recurrent w.r.t. $t \in \mathbb{R}$ uniformly w.r.t. u on compacts from E^n and the equation (5.9) be dissipative. For the equation (5.9) to be convergent it is necessary and sufficient that there would exists a continuous function $V : H(f) \times E^n \times E^n \to \mathbb{R}_+$ satisfying the following conditions:*

1) *V is positively defined, i.e. $V(g, v_1, v_2) = 0$ $(g \in H(f)$ and $v_1, v_2 \in E^n)$ if and only if $v_1 = v_2$;*
2) *$V(g_t, \varphi(t, v_1, g), \varphi(t, v_2, g)) \leq V(g, v_1, v_2)$ for all $t \geq 0$, $v_1, v_2 \in E^n$ and $g \in H(f)$;*
3) *$V(g_t, \varphi(t, v_1, g), \varphi(t, v_2, g)) = V(g, v_1, v_2)$ for all $t \geq 0$ if and only if $v_1, v_2 \in E^n$.*

In the case of the periodicity of the function f Theorem 5.22 coincides with Theorems 7.2 and 7.3 from [270].

Theorem 5.23 *Let $f \in C(\mathbb{R} \times E^n, E^n)$ be recurrent w.r.t. $t \in \mathbb{R}$ uniformly w.r.t. u on compacts from E^n and let the equation (5.9) be dissipative. For the equation (5.9) to be convergent it is necessary and sufficient that there would exist a continuous function $V : H(f) \times E^n \times E^n \to \mathbb{R}_+$ satisfying the following conditions:*

(1) *V is positively defined;*
(2) *$V(g_t, \varphi(t, v_1, g), \varphi(t, v_2, g)) < V(g, v_1, v_2)$ for all $t > 0$, $g \in H(f)$ and $v_1, v_2 \in E^n$, $v_1 \neq v_2$;*

2. Let E be a Banach space with the norm $|\cdot|$ and $[E]$ be the Banach space of all linear bounded operators acting from E to E and equipped with the operational norm.

Consider linear non-homogeneous equation

$$\dot{u} = A(t)u + f(t), \tag{5.34}$$

where $A \in C(\mathbb{R}, [E])$ and $f \in C(\mathbb{R}, E)$. Along with the equation (5.34) we will consider the corresponding homogeneous equation

$$\dot{u} = A(t)u \tag{5.35}$$

and denote by $U(t, A)$ the Cauchy operator of equation (5.35).

In this paragraph we will suppose that the sets $H(A)$ and $H(f)$ are compact in $C(\mathbb{R}, [E])$ and $C(\mathbb{R}, E)$ respectively, though some of the statements given below take place even without this condition.

Lemma 5.2   *For the equation (5.34) to be convergent it is necessary and sufficient that there would exist positive numbers $\mathcal{N}$ and ν such that*

$$\|U(t, A)U^{-1}(\tau, A)\| \leq \mathcal{N} \exp^{-\nu(t-\tau)} \tag{5.36}$$

for all $t \geq \tau$ $(t, \tau \in \mathbb{R})$.

Proof. The formulated statement follows from the definition of convergence and from the fact that uniform asymptotic stability of the null solution of the equation (5.35) is equivalent to the condition (5.36). $\square$

Lemma 5.3   *The mapping $U : \mathbb{R} \times C(\mathbb{R}, [E]) \to [E]$ $(U : (t, A) \to U(t, A))$ is continuous w.r.t. A uniformly w.r.t. $t \in \mathbb{R}$ on compacts from $\mathbb{R}$, i.e. for every $\ell > 0$ $\lim\limits_{n \to +\infty} \max\limits_{|t| \leq \ell} \|U(t, A_n) - U(t, A)\| = 0$ if $A_n \to A$ in $C(\mathbb{R}, [E])$.*

Proof. Let $A_n \subset C(\mathbb{R}, [E])$, $A_n \to A$ be uniform on compacts from $\mathbb{R}$ and $\ell > 0$. Then there exists positive number $M(\ell)$ such that

$$\max_{|t| \leq \ell} \|A_n(t)\| \leq M(\ell). \tag{5.37}$$

Since $U(t, A_n)$ is the solution of the system

$$\begin{cases} \dot{U}(t, A_n) = A_n(t)U(t, A_n) \\ U(0, A_n) = Id_E, \end{cases}$$

where Id_E is a unit operator in E, then we have

$$\|U(t, A_n)\| \leq \exp\left(\int_{t_0}^{t} \|A_n(s)\| ds\right) \quad (t > t_0, \ t, t_0 \in [-\ell, \ell]).$$

From the last inequality follows that

$$\max_{|t| \leq \ell} \|U(t, A_n)\| \leq \exp(2\ell M(\ell)) \quad (n = 1, 2, \ldots).$$

For every $n \in \mathbb{N}$ the function $V_n(t) := U(t, A) - U(t, A_n)$ satisfies the system

$$\begin{cases} \dot{V}_n(t) = A(t)V_n(t) + [A(t) - A_n(t)]U(t, A_n) \\ V_n(0) = 0. \end{cases}$$

From this follows that

$$V_n(t) = U(t, A) \int_0^t U^{-1}(\tau, A)[A_n(\tau) - A(\tau)]U(\tau, A_n)d\tau. \tag{5.38}$$

Let $K(\ell) := \max\limits_{|t| \leq \ell}\{\|U(t, A)\|, \|U^{-1}(t, A)\|\}$. From (5.37) and (5.38) follows the inequality

$$\max_{|t| \leq \ell} \|V_n(t)\| \leq K^2(l)2\ell \exp(2\ell M(\ell)) \max_{|t| \leq \ell} \|A_n(t) - A(t)\|. \tag{5.39}$$

Passing to limit in (5.39) as $n \to +\infty$, we obtain

$$\lim_{n \to +\infty} \max_{|t| \leq l} \|U(t, A) - U(t, A_n)\| = 0.$$

The lemma is proved. $\qquad\qquad\square$

Lemma 5.4 *Let $\mathcal{N} > 0$ and $\nu > 0$ be such that there holds the inequality (5.36). Then for every $B \in H(A)$, $t \geq \tau$ and $\tau \in \mathbb{R}$, the inequality*

$$\|U(t, B)U^{-1}(\tau, B)\| \leq \mathcal{N} \exp(-\nu(t - \tau)), \tag{5.40}$$

holds where $U(t, B)$ is the Cauchy operator of the equation

$$\dot{y} = B(t)y \quad (B \in H(A)).$$

Proof. Let $\mathcal{N} > 0$ and $\nu > 0$ be such that there takes place the inequality (5.36) and let $B \in H(A)$. Then there exists $\{\tau_n\} \subset \mathbb{R}$ such that $A_{\tau_n} \to B$ in $C(\mathbb{R}, [E])$. Note that $U(t + \tau, A) = U(t, A_\tau)U(\tau, A)$ for all $t, \tau \in \mathbb{R}$ and $A \in C(\mathbb{R}, [E])$, hence

$$U(t, A_{\tau_n})U^{-1}(\tau, A_{\tau_n}) = U(t + \tau_n, A)U^{-1}(t + \tau_n, A). \tag{5.41}$$

From what follows that

$$\|U(t, A_{\tau_n})U^{-1}(\tau, A_{\tau_n}) \leq \mathcal{N} \exp(-\nu(t - \tau)) \quad (t \geq \tau, \, n = 1, 2, \dots).$$

Passing to limit in the last inequality as $n \to +\infty$ and taking into account Lemma 5.3 we will obtain the inequality (5.40). The lemma is proved. $\qquad\square$

Lemma 5.5 *Let the condition (5.36) be fulfilled. Then for every $B \in H(A)$ and $p \geq 1$ by the equality*

$$\|v\|_{B,p} = \{\int_0^{+\infty} |U(t, B)v|^p dt\}^{\frac{1}{p}} \tag{5.42}$$

on E is defined a norm that is topologically equivalent to the old and moreover

$$\|U(\tau,B)v_1 - U(\tau,B)v_2\|_{B_\tau,p} \leq \mathcal{N}(\frac{a}{\nu})^{\frac{1}{p}} \exp(-\nu\tau)\|v_1 - v_2\|_{B,p} \qquad (5.43)$$

for all $\tau \geq 0$.

Proof. From the inequality of Minkovskii follows that by the equality (5.42) is defined some norm on E. Let us show that there exist positive number m_p and M_p that does not depend on $B \in H(A)$ such that $m_p|v| \leq \|v\|_{B,p} \leq M_p|v|$. The first inequality follows from (5.36) and from the inequality

$$\|v\|_{B,p}^p = \int_0^{+\infty} |U(t,B)v|^p dt \leq \int_0^{+\infty} \|U(t,B)\|^p dt |v|^p \leq$$

$$\mathcal{N}^p \int_0^{+\infty} \exp(-\nu pt)dt|v|^p = \frac{\mathcal{N}^p}{\nu p}|v|^p.$$

On the other hand since $U(-\tau,B_\tau)U(\tau,B)v = v$ for all $\tau \in \mathbb{R}$ and $v \in E$, then

$$|v| = |U(-\tau,B_\tau)U(\tau,B)v| \leq \|U(-\tau,B_\tau)\|\|U(\tau,B)v| \leq$$

$$\exp(a|\tau|)|U(\tau,B)v| \quad (a := \sup\{\|A(t)\| : t \in \mathbb{R}\} < +\infty)$$

and, consequently,

$$\|v\|_{B,p}^p = \int_0^{+\infty} |U(t,B)v|^p dt \geq \int_0^{+\infty} (\exp(-at)|v|)^p dt =$$

$$\int_0^{+\infty} \exp(-apt)dt|v|^p = \frac{1}{ap}|v|^p$$

$(m_p = (ap)^{-\frac{1}{p}})$. So,

$$(ap)^{-\frac{1}{p}}|v| \leq \|v\|_{B,p} \leq \mathcal{N}(\nu p)^{-\frac{1}{p}}|v|.$$

Now let us establish the inequality (5.43). For that we will note that

$$\|U(\tau,B)v_1 - U(\tau,B)v_2\|_{B_\tau,p}^p = \int_0^{+\infty} |U(t,B_\tau)[U(\tau,B)v_1 - U(\tau,B)v_2]|^p dt$$

$$= \int_0^{+\infty} |U(t,B_\tau)U(\tau,B)(v_1 - v_2)|^p dt = \int_0^{+\infty} |U(t+\tau,B)(v_1 - v_2)|^p dt$$

$$= \int_\tau^{+\infty} |U(s,B)(v_1 - v_2)|^p ds \leq \int_\tau^{+\infty} \mathcal{N}^p \exp(-\nu ps)|v_1 - v_2|^p ds$$

$$= \frac{\mathcal{N}^p}{\nu p} \exp(-\nu p\tau)|v_1 - v_2|^p \leq \frac{\mathcal{N}^p ap}{\nu p} \exp(-\nu p\tau)\|v_1 - v_2\|_{B,p}^p.$$

From the last inequality follows (5.43). The lemma is proved. $\qquad \square$

As it turned out, if the equation (5.34) is convergent then this property remains valid under small nonlinear perturbations. The following theorem takes place.

Theorem 5.24 *Let A, f and F be such that $H(A)$, $H(f)$ and $H(F)$ are compact in $C(\mathbb{R}, [E])$, $C(\mathbb{R}, E)$ and $C(\mathbb{R} \times E, E)$ respectively, and the function F satisfies the condition of Lipschitz w.r.t. $u \in E$ uniformly w.r.t. $t \in \mathbb{R}$ with the small enough constant of Lipschitz $(Lip(F) < \nu^2(\mathcal{N}a)^{-1})$. Then if the equation (5.34) is convergent, the perturbed equation*

$$\dot{u} = A(t)u + F(t, u) \tag{5.44}$$

is convergent too.

Proof. Denote by $H(A, f, F) := \overline{\{(A_\tau, f_\tau, F_\tau) : \tau \in \mathbb{R}\}}$. Let $v \in B$, $(B, g, G) \in H(A, f, F)$, $\tau \in \mathbb{R}$ and $\varphi(\cdot, v, B, g, G)$ be the solution of equation

$$\dot{v} = B(t)v + g(t) + G(t, v), \tag{5.45}$$

passing through the point v as $t = 0$.

According to the conditions of the theorem the equation (5.45) has a unique solution defined on $\mathbb{R}$ and passing through the point v as $\tau = 0$, whatever would be $(B, g, G) \in H(A, f, F)$ and $v \in E$. To see that is sufficient to note that having the conditions of Theorem 5.24 fulfilled we can apply Theorem 1.2 [132] (global theorem of existence and uniqueness) to the equation (5.45).

Assume $Y := H(A, f, F)$ and by $(Y, \mathbb{R}, \sigma)$ denote the dynamical system of shifts on Y. Let $X := E \times Y$. Define on X a dynamical system by the following rule: $\pi^\tau(v; B, g, G) := (\varphi(\tau, v, B, g, G); B_\tau, g_\tau, G_\tau)$. By the conditions of Theorem 5.24 Y is compact and the triplet $\langle (X, \mathbb{R}, \pi), (Y, \mathbb{R}, \sigma), h \rangle$, where $h := pr_2 : X \to Y$ is a non-autonomous dynamical system. Let us show that we can apply Theorem 2.15 and Corollary 2.6 to the constructed dynamical system. Really, the set Y is compact under the conditions of the theorem. It is clear that the set of the continuous sections $\Gamma(H(A, f, F), B \times H(A, f, F))$ is isomorphic to the set of all the continuous mapping $x : H(A, f, F) \to E$ (see paragraph 3.3) and, consequently, is not empty. Define on $X \dot\times X$ a scalar non-negative function V by the following rule:

$$V((v_1; B, g, G), (v_2; B, g, G)) := \|v_1 - v_2\|_{B,1} = \int_0^{+\infty} |U(t, B)(v_1 - v_2)| dt$$

for every $(B, g, G) \in H(A, f, F)$ and $v_1, v_2 \in E$. From Lemma 5.5 follows that the constructed function satisfies the conditions 1.- 3. of Theorem 2.15. Let $\varphi(\cdot, v_i, B, g, G)$ be the solution of the equation (5.45) passing through the point v_i $(i = 1, 2)$ as $\tau = 0$. Then the identity

$$\dot{\varphi}(\tau, v_i, B, g, G) \equiv B(\tau)\varphi(\tau, v_i, B, g, G) + g(\tau) + G(\tau, \varphi(\tau, v_i, B, g, G)).$$

holds. From this follows that

$$\varphi(\tau, v_i, B, g, G) = U(\tau, B)(v_i + \int_0^\tau U^{-1}(s, B) \qquad (5.46)$$

$$[g(s) + G(s, \varphi(s, v_i, B, g, G))]ds).$$

From (5.46) and Lemmas 5.4 and 5.5 follows

$$\|\varphi(\tau, v_1, B, g, G) - \varphi(\tau, v_2, B, g, G)\|_{B_\tau} =$$

$$\int_0^{+\infty} |U(t, B_\tau)(\varphi(\tau, v_1, B, g, G) - \varphi(\tau, v_2, B, g, G))|dt$$

$$= \int_0^{+\infty} |U(t, B_\tau)U(\tau, B)(v_1 - v_2 + \int_0^\tau U^{-1}(s, B) \times$$

$$[G(s, \varphi(s, v_1, B, g, G)) - G(s, \varphi(s, v_2, B, g, G))]ds|dt \le$$

$$\int_0^{+\infty} |U(t + \tau, B)(v_1 - v_2)|dt +$$

$$\int_0^{+\infty} |\int_0^\tau U(t + \tau, B)U^{-1}(s, B) \times$$

$$[G(s, \varphi(s, v_1, B, g, G)) - G(s, \varphi(s, v_2, B, g, G))]ds|dt \le$$

$$\frac{\mathcal{N}}{\nu} \exp(-\nu\tau)|v_1 - v_2| + \int_0^{+\infty} \int_0^\tau \mathcal{N} \exp(-\nu(t + \tau - s))Lip(G) \times$$

$$|\varphi(s, v_1, B, g, G) - \varphi(s, v_2, B, g, G)|ds\, dt =$$

$$\frac{\mathcal{N}}{\nu} \exp(-\nu\tau)|v_1 - v_2| + \frac{\mathcal{N}}{\nu} \exp(-\nu\tau)Lip(G) \times$$

$$\int_0^\tau \exp(\nu s|\varphi(s, v_1, B, g, G) - \varphi(s, v_2, B, g, G)|ds \le$$

$$\frac{\mathcal{N}}{\nu} \exp(-\nu\tau)a\|v_1 - v_2\|_{B,1} + \frac{\mathcal{N}a}{\nu} \exp(-\nu\tau)Lip(G) \times$$

$$\int_0^\tau \exp(\nu s)\|\varphi(s, v_1, B, g, G) - \varphi(s, v_2, B, g, G)\|_{B,1}ds.$$

Let us multiply the both parts of the last inequality by $\exp(\nu\tau)$ and assuming $\varphi(\tau) := \exp(\nu s)\|\varphi(s, v_1, B, g, G) - \varphi(s, v_2, B, g, G)\|_{B,1}$, we will obtain

$$\varphi(\tau) \le \frac{\mathcal{N}a}{\nu}\|v_1 - v_2\|_{B,1} + \frac{\mathcal{N}a}{\nu}Lip(G) \int_0^\tau \varphi(\tau)ds. \qquad (5.47)$$

Applying to (5.47) the Gronwall-Bellman inequality we will obtain

$$\varphi(\tau) \le \frac{\mathcal{N}a}{\nu}\|v_1 - v_2\|_{B,1} \exp(\frac{\mathcal{N}a}{\nu}Lip(G)\tau). \qquad (5.48)$$

Consequently,

$$\|\varphi(\tau, v_1, B, g, G) - \varphi(\tau, v_2, B, g, G)\|_{B,1} \leq \qquad (5.49)$$
$$\frac{\mathcal{N}a}{\nu}|v_1 - v_2|\exp(-(\nu - \frac{\mathcal{N}a}{\nu}Lip(G))\tau).$$

From the inequality (5.49) we have

$$V((\varphi(\tau, v_1, B, g, G); B_\tau, g_\tau, G_\tau)), (\varphi(\tau, v_2, B, g, G); B_\tau, g_\tau, G_\tau)) \leq$$
$$M\exp(-\lambda\tau)V((v_1; B, g, G), (v_2; B, g, G))$$

(here $M = \frac{\mathcal{N}a}{\nu}$, $\lambda = -\frac{\mathcal{N}a}{\nu}Lip(G) + \nu > 0$) for all $v_1, v_2 \in E$, $\tau \geq 0$ and $(B, g, G) \in H(A, f, F)$. So, all the conditions of Theorem 2.15 and of Corollary 2.6 are fulfilled. Applying them to the constructed non-autonomous dynamical system we complete the proof of the theorem. $\square$

Corollary 5.8 *Under the conditions of Theorem 5.24 if A, f and F are jointly τ-periodic (almost periodic, recurrent) then the equation (5.44) is convergent and every equation (5.45) has a unique τ-periodic (almost periodic, recurrent) globally uniformly asymptotically stable solution.*

Note, that the statements analogous to Lemmas 5.4 and 5.5 in the case when the operator-function $A(t)$ is stationary are proved in [132]. Corollary 5.8 in the case of periodicity of A, f and F ($E = \mathbb{R}^n$) are established in [270, p.99].

3. Let H be a real or complex Hilbert space. Consider a differential equation

$$\dot{u} = f(t, u), \qquad (5.50)$$

where $f \in C(\mathbb{R} \times H, H)$. Along with the equation (5.50) we will consider its H-class

$$\dot{v} = g(t, v) \quad (g \in H(f)). \qquad (5.51)$$

There takes place

Theorem 5.25 *Let $f \in C(\mathbb{R} \times H, H)$ and let the set $H(f)$ be compact in $C(\mathbb{R} \times H, H)$. If there exists a self-adjoint operator-function $A \in C(\mathbb{R}, [H])$ satisfying the following conditions:*

1. $M_f \subset M_A \cap M_{\dot{A}}$;
2. $Re\langle A(t)(u - v), f(t, u) - f(t, v)\rangle \leq -\alpha|u - v|$ for all $t \in \mathbb{R}$ and $u, v \in H$ ($\langle\cdot, \cdot\rangle$ is a scalar product in H, $|\cdot|^2 = \langle\cdot, \cdot\rangle$ and $\alpha > 0$);
3. $\|\dot{A}(t)\| \leq \beta$ for all $t \in \mathbb{R}$ ($\beta \geq 0$);
4. $\langle A(t)u, u\rangle \geq \gamma|u|^2$ for all $t \in \mathbb{R}$ $u \in H$ and $-2\alpha + \beta := -\lambda < 0$.

Then the equation (5.50) is convergent, i.e.

   a. whatever would be $g \in H(f)$ and $v \in H$, there exists a unique solution $\varphi(\cdot, v, g)$ of the equation (5.51) passing through the point v as $t = 0$ and defined on $\mathbb{R}$;

   b. for every $g \in H(f)$ the equation (5.51) has a unique compact solution $\psi_g(t) = \varphi(t, v_g, g)$, defined on $\mathbb{R}$;

   c. there exist positive numbers $\mathcal{N}$ and ν (not depending on $g \in H(f)$ and $v \in H$) such that

$$|\varphi(t, v, g) - \varphi(t, v_g, g)| \leq \mathcal{N} \exp(-\nu t)|v - v_g| \tag{5.52}$$

for all $g \in H(f)$, $v \in H$ and $t \in \mathbb{R}_+$.

Proof. Let us prove the formulated statement by several parts.

1. Since $M_f \subset M_A \cap M_{\dot{A}}$, then from [300] – [302] follows the existence of the continuous mapping $q : H(f) \rightarrow H(A) \times H(\dot{A})$ satisfying the condition

$$q(f) = (A, \dot{A}) \quad \text{and} \quad q(g_\tau) = (q(g)_\tau, q(\dot{g})_\tau)$$

for all $g \in H(f)$ and $\tau \in \mathbb{R}$.

2. Let us show that under the conditions of the theorem every function $g \in H(f)$ satisfies the conditions analogous to 1.–4. Really, let $g \in H(f)$ and $q(g) = (B_g, \dot{B}_g) \in H(A) \times H(\dot{A})$. Then there exists $\{\tau_n\} \subset \mathbb{R}$ such that $g = \lim\limits_{n \to +\infty} f_{\tau_n}$ and $(B_g, \dot{B}_g) = (\lim\limits_{n \to +\infty} A_{\tau_n}, \lim\limits_{n \to +\infty} \dot{A}_{\tau_n})$. From the condition 2. of the theorem follows that

$$Re\langle A(t + \tau_n)(u - v), f(t + \tau_n, u)f(t + \tau_n, v)\rangle \leq -\alpha|u - v|^2 \tag{5.53}$$

for all $t \in \mathbb{R}$, u $v \in H$ (n=1,2,...). Passing to limit in (5.53) as $n \to +\infty$ we will obtain

$$Re\langle B_g(t)(u - v), g(t, u) - g(t, v)\rangle \leq -\alpha|u - v|^2.$$

In the same way we can show that the operator-function $B_g \in H(A)$ $(g \in H(f))$ satisfies the conditions 3. and 4. with the same constants β and γ that has the operator-function $A \in C(\mathbb{R}, [H])$.

3. A slight modification of the reasoning in the work [267] (see also [329]) allows to prove that if the conditions of the theorem are fulfilled then, whatever would be $u \in H$, the equation (5.50) has a unique solution $\varphi(\cdot, u, f)$ defined on $\mathbb{R}$ and satisfying the condition $\varphi(0, u, f) = u$. Taking into account that along with the function f every other function $g \in H(f)$ satisfies the conditions of the theorem, we make the conclusion that for every $v \in H$ and $g \in H(f)$ the equation (5.51) has a unique solution $\varphi(\cdot, v, g)$ defined on $\mathbb{R}$ and satisfying the condition $\varphi(0, v, g) = v$.

4. Assume $Y := H(f)$ and by $(Y, \mathbb{R}, \sigma)$ denote a dynamical system of shifts on Y. Let $X := H \times H(f)$. Define on X a dynamical system by the following rule:

$\pi((v, g), \tau) := (\varphi(\tau, v, g), g_\tau)$. Note that under the conditions of the theorem the set Y is compact. The triplet $\langle (X, \mathbb{R}, \pi), (Y, \mathbb{R}, \sigma), h \rangle$, where $h := pr_2 : X \to Y$, is a non-autonomous dynamical system. Let us show that we can apply to the constructed dynamical system Theorem 2.15 and Corollary 2.6. Really, the set Y is compact under the conditions of the theorem. The set of continuous sections $\Gamma(Y, X)$ is not empty. Define on $X \dot{\times} X$ a non-negative scalar function V by the following rule:

$$V((v_1, g), (v_2, g)) := \langle B_g(0)(v_1 - v_2), v_1 - v_2 \rangle$$

for every $g \in H(f)$ and $v_1, v_2 \in H$ ($B_g = q(g)$).

It is clear that the function V satisfies the conditions 2. and 3. of Theorem 2.15. Really, according to the condition 4.

$$\gamma |v_1 - v_2|^2 \le \langle A(t)(v_1 - v_2), v_1 - v_2 \rangle \le \delta |v_1 - v_2|^2 \qquad (5.54)$$

for all $v_1, v_2 \in H$ and $t \in \mathbb{R}$, where $\delta := \sup\{\|A(t)\| \mid t \in \mathbb{R}\}$ ($\delta < +\infty$, as $M_f \subset M_A$ and $H(f)$ is compact). For $g \in H(f)$ there exists $\{\tau_n\} \subset \mathbb{R}$ such that $A_{\tau_n} \to B_g$. From (5.54) and the last correspondence follows that

$$\gamma |v_1 - v_2|^2 \le \langle B_g(t)(v_1 - v_2), v_1 - v_2 \rangle \le \delta |v_1 - v_2|^2$$

for all $g \in H(f)$, $t \in \mathbb{R}$ and $v_1, v_2 \in H$. In particular, we have

$$\gamma |v_1 - v_2|^2 \le V((v_1, g), (v_2, g)) \le \delta |v_1 - v_2|^2 \qquad (5.55)$$

for all $g \in H(f)$ and $v_1, v_2 \in H$.

At last, let us show that the condition 4. of Lemma 2.13 is fulfilled too. Really,

$$V((\varphi(\tau, v_1, g), g_\tau), (\varphi(\tau, v_2, g), g_\tau)) =$$
$$\langle B_{g_\tau}(0)(\varphi(\tau, v_1, g) - \varphi(\tau, v_2, g)), \varphi(\tau, v_1, g) - \varphi(\tau, v_2, g) \rangle =$$
$$\langle B_g(\tau)(\varphi(\tau, v_1, g) - \varphi(\tau, v_2, g)), \varphi(\tau, v_1, g) - \varphi(\tau, v_2, g) \rangle.$$

By direct calculation we can verify that

$$\dot{V}((\varphi(\tau, v_1, g), g_\tau), (\varphi(\tau, v_2, g), g_\tau)) =$$
$$\langle \dot{B}_g(\tau)(\varphi(\tau, v_1, g) - \varphi(\tau, v_2, g)), \varphi(\tau, v_1, g) - \varphi(\tau, v_2, g) \rangle +$$
$$2Re\langle B_g(\tau)(\varphi(\tau, v_1, g) - \varphi(\tau, v_2, g)), g(\tau, \varphi(\tau, v_1, g)) - g(\tau, \varphi(\tau, v_2, g)) \rangle.$$

From the last equality and (5.53) and also from the condition 3. of Theorem 5.25 follows the inequality

$$\dot{V}((\varphi(\tau, v_1, g), g_\tau), (\varphi(\tau, v_2, g), g_\tau)) \le \qquad (5.56)$$
$$\beta |\varphi(\tau, v_1, g) - \varphi(\tau, v_2, g)|^2 - -2\alpha |\varphi(\tau, v_1, g) - \varphi(\tau, v_2, g)|^2 =$$
$$-\lambda |\varphi(\tau, v_1, g) - \varphi(\tau, v_2, g)|^2.$$

On the other hand, from the inequality (5.55) we obtain

$$V((\varphi(\tau, v_1, g), g_\tau), (\varphi(\tau, v_2, g), g_\tau)) \le \delta |\varphi(\tau, v_1, g) - \varphi(\tau, v_2, g)|^2. \tag{5.57}$$

From the inequality (5.56) and (5.57) we have

$$\dot{V}((\varphi(\tau, v_1, g), g_\tau), (\varphi(\tau, v_2, g), g_\tau)) \le -\lambda \delta V((\varphi(\tau, v_1, g), g_\tau), (\varphi(\tau, v_2, g), g_\tau)).$$

Integrating the last inequality, we get the following inequality

$$V((\varphi(\tau, v_1, g), g_\tau), (\varphi(\tau, v_2, g), g_\tau)) \le \exp(-\lambda \delta \tau) V((v_1, g), (v_2, g))$$

for all $g \in H(f)$, $v_1, v_2 \in H$ and $\tau \in \mathbb{R}_+$. So, all the conditions of Theorem 2.15 are fulfilled and applying it and Corollary 2.6 to the constructed dynamical system we complete the proof of Theorem 5.25. $\square$

Corollary 5.9   *Under the conditions of Theorem 5.25 if the right hand side f of the equation (5.50) is τ-periodic (almost periodic, recurrent), then the equation (5.50) is convergent and every equation (5.51) has a unique τ-periodic (almost periodic, recurrent) solution complying with the estimation (5.52).*

Remark 5.8   *1. In the case when f is τ-periodic, Corollary 5.9 reinforces the main result of the work [192] (in [192] under the same conditions on basis of the principle of Lere-Showder there was proved only the existence of a unique τ-periodic solution in the case when $H = E^n$).*

2. If the right hand side f is almost periodic and the operator-function $A(t)$ does not depend on time $(\dot{A}(t) \equiv 0)$, then the statement of Corollary 5.9 coincides with the main result of the work [267]. Also note, that the results of the work [267] are achieved on basis of the series of delicate estimations. Our method is of purely topological character.

3. Note, at last, that in the work [187] there are studied so-called α-contracting non-autonomous dynamical systems (i.e. such systems that satisfy the condition:

$$\rho(x_1 t, x_2 t) \le \alpha \rho(x_1, x_2)$$

for every $t \in \mathbb{R}_+$, $x_1, x_2 \in X$ and $h(x_1) = h(x_2)$, where $0 \le \alpha \le 1$) in the relation with the problem of stability and asymptotic stability in the sense of Poisson of the solutions of differential equations with monotone right hand side.

In the case of finite-dimensional Hilbert space $(H = E^n)$, when the function f is recurrent and $A(t)$ $(\dot{A}(t) \equiv 0)$ is stationary, Theorem 5.25 admits the following amplification.

Theorem 5.26   *Let $f \in C(\mathbb{R} \times H, H)$ satisfy the condition of Lipschitz and be recurrent w.r.t. $t \in \mathbb{R}$ uniformly w.r.t. u on compacts from H. If there exists a* self-adjoint operator $A \in [H]$ and a positively defined function $c : \mathbb{R}_+ \to \mathbb{R}_+$ such that

$$Re\langle A(u - v), f(t, u) - f(t, v)\rangle \le -c(|u - v|) \quad (\forall \ u, v \in H) \tag{5.58}$$

and $c(r)r^{-1} \to +\infty$ as $r \to +\infty$ (for example, $c(r) = r^{1+\alpha}$, $\alpha > 0$), then Equation (5.50) is convergent.

Proof. In the same way that in Theorem 5.25 we establish that whatever would be $v \in H$ and $g \in H(f)$, the equation (5.51) has a unique solution $\varphi(\cdot, v, g)$ defined on $\mathbb{R}$ and satisfying the condition $\varphi(0, v, g) = v$. Let us define a function $\mathcal{V} : H(f) \times H \to \mathbb{R}_+$ by the following rule

$$\mathcal{V}(g, v) := \langle Av, v\rangle. \tag{5.59}$$

From the equality (5.59) follows that

$$\dot{\mathcal{V}}(g_\tau, \varphi(\tau, v, g)) = 2Re\langle A\varphi(\tau, v, g), \varphi(\tau, v, g)\rangle. \tag{5.60}$$

Since $c(r)r^{-1} \to +\infty$ as $r \to +\infty$, there exists $r_0 > 0$ such that

$$c(r)r^{-1} > |f_0|\|A\| \tag{5.61}$$

for all $r > r_0$, where $|f_0| := \sup\{|f(t, 0)| : t \in \mathbb{R}\}$. Let $t > 0$, $v \in H$ and $g \in H(f)$ be such that $|\varphi(\tau, v, g)| \ge r_0$ for all $\tau \in [0, t]$. From (5.58), (5.60) and (5.61) follows that $\mathcal{V}(g_\tau, \varphi(\tau, v, g)) < \mathcal{V}(g, v)$, and according to Theorem 5.7 the equation (5.50) is dissipative.

Let us denote by $V : H(f) \times H \times H \to \mathbb{R}_+$ a function defined by the equation

$$V(g, u, v)) := \langle A(u - v), (u - v)\rangle. \tag{5.62}$$

By direct calculation we obtain

$$\dot{V}(g_\tau, \varphi(\tau, u, g), \varphi(\tau, v, g)) = 2Re\langle A(\varphi(\tau, u, g) - \varphi(\tau, v, g)),$$

$$g(\tau, \varphi(\tau, u, g)) - g(\tau, \varphi(\tau, v, g))\rangle. \tag{5.63}$$

from (5.58) and (5.63) follows that $V(g_t, \varphi(t, u, g), \varphi(t, v, g)) < V(g, u, v)$ for all $t > 0$, $g \in H(f)$ and $u, v \in H$ $(u \ne v)$. According to Theorem 2.17 the equation (5.50) is convergent. $\square$

5.5 Dissipativity and convergence of some equations of 2nd and 3rd order

1. Let us consider the system of two differential equation of the following kind

$$\begin{cases} \dot{x} = y - F(x) + Q(x, y, t) \\ \dot{y} = -g(x), \end{cases} \tag{5.64}$$

where F and g are continuous functions defined on $\mathbb{R}$ and $Q(x, y, t)$ is continuous w.r.t. the set of variables, is bounded and uniformly continuous w.r.t. $t \in \mathbb{R}$ uniformly w.r.t. (x, y) on compacts from $\mathbb{R}^2$. As we know, we can reduce the equation

$$\ddot{x} + h(x)\dot{x} + g(x) = p(t) \tag{5.65}$$

to the system (5.64) if we introduce a new variable $y = \dot{x} + F(x) - g(t)$ where $F(x) = \int_0^x h(\tau)d\tau$, $g(t) = \int_0^t p(s)ds$. Along with the system (5.64) consider its H-class

$$\begin{cases} \dot{x} = y - F(x) + G(x, y, t) \\ \dot{y} = -g(x), \end{cases} \tag{5.66}$$

$(G \in H(Q) = \overline{\{Q_\tau : \tau \in \mathbb{R}\}})$.

Theorem 5.27 *Let the following conditions be satisfied:*

(1) $xg(x) > 0$ for $|x| \geq 1$;
(2) $\int_0^{+\infty} g(x)dx = - \int_{-\infty}^0 g(x)dx = +\infty$;
(3) there exists constant $N > 0$ such that $|Q(x, y, t)| \leq N$ for $|x| \leq 1$.
(4) $[F(x) - Q(x, y, t)]\operatorname{sgn} x \geq 0$ for $|x| \geq 1$; besides for every $G \in H(Q)$ there exists $\theta_G > 0$ such that $F(x) - G(x, y, \theta_G) \neq 0$ for $|x| \geq 1$ and $y \in \mathbb{R}$.
(5) when $|x| \geq 1$ the inequality $|F(x) - Q(x, y, t)| \geq \alpha(y) \geq 0$ is valid, where $\alpha(y)$ on every interval is a measurable in the sense of Lebesgue function and $\int_{-\infty}^{+\infty} \alpha(y)dy > 0$.

Then the system (5.64) is dissipative.

Proof. In the same way that in the work [160] we show that there exists $\eta > 0$ such that under certain τ every solution of the equation (5.66) gets in the rectangle $\{(x, y) : |x| \leq 1, |y| \leq \eta\}$ (τ depends on the system (5.66) and the according solution). Now to complete the proof of the theorem it is sufficient to refer to Theorem 5.8. $\square$

Let us consider a more general equation

$$\ddot{x} + h(x)\dot{x} + g(x) = \mathcal{R}(x, \dot{x}, t).$$

Assume $y = \dot{x} + F(x)$ where $F(x) = \int_0^x h(\tau)d\tau$. Then we obtain the following system

$$\begin{cases} \dot{x} = y - F(x) \\ \dot{y} = -g(x) + \mathcal{R}(x, y, t) \end{cases} \tag{5.67}$$

Along with the system (5.67) consider its H-class

$$\begin{cases} \dot{x} = y - F(x) \\ \dot{y} = -g(x) + G(x, y, t) \end{cases} \tag{5.68}$$

where $G \in H(\mathcal{R})$.

Theorem 5.28 *Let the functions h, g and $\mathcal{R}$ be continuous and $\mathcal{R}$ be bounded and uniformly continuous w.r.t. $t \in \mathbb{R}$ uniformly w.r.t. (x, y) on compacts from $\mathbb{R}^2$. Suppose that in addition there exist constants $L > N > 0$ and $K > 0$ such that*

(1) $|\mathcal{R}(x, y, t)| \le N$ for all x, y, $t \in \mathbb{R}$;
(2) $g(x)\mathrm{sgn}\, x \ge L$ for $|x| \ge 1$;
(3) $f(x) \ge K$ for $|x| \ge 1$.

Then the system (5.67) is dissipative.

Proof. As well as in the work [270] let us consider the following function $V(x, y) = y^2 - yF(x) + \frac{1}{2}F(x) + 2\int_0^x g(\tau)d\tau$. The derivative of this function taken in virtue of the system (5.68) equals to

$$\dot{V}(x, y, t) = -[y - F(x)]^2 f(x) - g(x)F(x) + [2y - F(x)]G(x, y, t).$$

According to [270] there exists $a > 0$ such that for $x^2 + y^2 \ge a^2$ $\dot{V}_G(x, y, t) < 0$ and $V(x, y) \to +\infty$ for $x^2 + y^2 \to +\infty$. According to Theorem 5.7 the system (5.67) is dissipative. $\qquad\square$

Consider the differential equation of the 3rd order

$$\dddot{x} + a\ddot{x} + b\dot{x} + f(x) = p(x, \dot{x}, \ddot{x}, t)$$

where $a, b \in \mathbb{R}$. Assume $y = ax + \dot{x}$, $z = bx + a\dot{x} + \ddot{x}$. Then this equation is equivalent to the system

$$\begin{cases} \dot{x} = y - ax \\ \dot{y} = z - bx \\ \dot{z} = -f(x) + p(x, y, z, t) \end{cases} \tag{5.69}$$

Along with the system (5.69) consider its H-class

$$\begin{cases} \dot{x} = y - ax \\ \dot{y} = z - bx \\ \dot{z} = -f(x) + g(x, y, z, t) \end{cases} \tag{5.70}$$

where $g \in H(p)$.

Theorem 5.29 *Let the function f and $p(x, y, z, t)$ be continuous and let the following conditions be satisfied:*

1. *$a, b > 0$;*
2. *$0 < \frac{f(x)}{x} < ab$ for $|x| \geq 1$;*
3. *$\lim\limits_{|x| \to +\infty} |f(x)| = +\infty$;*
4. *$\lim\limits_{|x| \to +\infty} |f(x) - abx| = +\infty$;*
5. *p is bounded and uniformly continuous w.r.t. $t \in \mathbb{R}$ uniformly w.r.t. (x, y, t) on compacts from $\mathbb{R}^3$;*
6. *there exists $A > 0$ such that $|p(x, y, z, t)| \leq A$ for all x, y, z and $t \in \mathbb{R}$.*

Then the system (5.69) is dissipative.

Proof.   To prove the formulated statement, following [270] we will consider the function

$$V(x, y, z) = \frac{1}{2}(a^2 x - ay + z)^2 + \frac{1}{2}(z - bx)^2 + \frac{b}{2}y^2 + a \int_a^x [f(\tau) - ab\tau] d\tau.$$

Under the conditions of Theorem 5.29 $V(x, y, z) \to +\infty$ for $x^2 + y^2 + z^2 \to +\infty$ and the derivative of this function taken by virtue of the system (5.70) being to

$$\dot{V}_g(x, y, z, t) = -a(a^2 x - ay + z)^2$$
$$+[abx - f(x)][2(a^2 x - ay + z) - bx][(a^2 - b)x - ay + 2z]g(x, y, z, t).$$

In the same way that in [270] we can show that there exists $a > 0$ such that out of the ball of radius a the function V decreases along the trajectories of (5.70) and, according to Theorem 5.7, the system (5.69) is dissipative. $\square$

    2. Let us give two criterions of convergence of the equation (5.65) that is equivalent to the system

$$\begin{cases} \dot{x} = y \\ \dot{y} = -x - h(x)y + p(t), \end{cases}$$

or, respectively, to the system

$$\begin{cases} \dot{x} = -F(x) + v \\ \dot{v} = -x + p(t) \end{cases} \tag{5.72}$$

where $v = \dot{x} + F(x)$ and $F(x) := \int_0^x h(t)dt$. As usual, along with the system (5.72) we consider its H-class

$$\begin{cases} \dot{x} = -F(x) + v \\ \dot{v} = -x + g(t) \quad (g \in H(p)). \end{cases} \tag{5.73}$$

Theorem 5.30   *If $h(x) > 0$ (excepting maybe the discrete set of points), $p(t)$ is recurrent and $F(+\infty) = +\infty$ (or $F(-\infty) = -\infty$), then the equation (5.65) has a unique bounded on $\mathbb{R}$ uniformly compatible and uniformly globally asymptotically stable solution, i.e. (5.65) is convergent.*

Proof.   Reasoning in the same way that in [272] show that under the conditions of the theorem there exists a closed contour that intersects all the solutions of every system (5.73). According to Theorem 5.6 the system (5.72) is dissipative.

Let us show that the system (5.72) in fact is convergent. Let $\{x_i(t), v_i(t)\}$ $(i = 1, 2)$ be two different solutions of the system (5.73). Construct the function

$$\varphi(t) = |x_1(t) - x_2(t)|^2 + |v_1(t) - v_2(t)|^2$$

and calculate its derivative. We will obtain:

$$\dot{\varphi}(t) = -(x_1(t) - x_2(t))[F(x_1(t)) - F(x_2(t))]. \tag{5.74}$$

Note that if $\dot{\varphi}(t) \equiv 0$ then $x_1(t) \equiv x_2(t)$ and, consequently, $v_1(t) \equiv v_2(t)$. To finish the proof of the theorem is sufficient to refer to Theorem 5.23. $\square$

Theorem 5.31   *If $\psi_1(x) = h(x) - 2D$ $(0 < D < 1)$ and $\psi_2(x) = g(x) - x$ satisfy the conditions:*

(1) $|\psi_1(x)| \le c_1$ for all $x \in \mathbb{R}$;
(2) $|\psi_1(\bar{x}) - \psi_2(x)| \le c_2|\bar{x}) - x|$ for all $\bar{x}, x \in \mathbb{R}$, where $c_1 + c_2 = c < mD = \sqrt{1 - D}D$,

then the system (5.72) has a unique bounded on $\mathbb{R}$ uniformly compatible, uniformly globally asymptotically stable solution.

Proof.   Let

$$\{\varphi(t, x_i, y_i, g)\} = \{(\varphi_1(t, x_i, y_i, g), \varphi_2(t, x_i, y_i, g))$$

$(i = 1, 2, g \in H(f))$ are two solutions of the system (5.73) and

$$\begin{cases} u(t) := \varphi_1(t, x_1, y_1, g) - \varphi_1(t, x_2, y_2, g) \\ v(t) := \varphi_2(t, x_1, y_1, g) - \varphi_2(t, x_2, y_2, g). \end{cases}$$

The functions u and v satisfy the following system of equations

$$\begin{cases} \dot{u} = v \\ \dot{v} = -u - 2Dv - [\psi_2(\varphi_1(t, x_1, y_1, g)) - \psi_2(\varphi_1(t, x_2, y_2, g))] - \\ \quad [\psi_1(\varphi_1(t, x_1, y_1, g))\varphi_2(t, x_1, y_1, g) - \\ \quad \psi_1(\varphi_1(t, x_2, y_2, g))\varphi_2(t, x_2, y_2, g)]. \end{cases}$$

In the same way that in [272] we show that

$$\begin{cases} |u(t)| < (1 + \frac{c_1}{m})\alpha[\exp(-(D - \frac{c}{m})t) + \exp(-Dt)] \\ |v(t)| < (1 + c_1)(1 + \frac{c_1}{m})\alpha[\exp(-(D - \frac{c}{m})t) \\ \quad + \exp(-Dt)] \end{cases} , \qquad (5.75)$$

where

$$\alpha = \frac{1}{m}\sqrt{|u(0)|^2 + 2D|u(0)||v(0)| + |v(0)|^2} \le \frac{1}{m}\sqrt{1 + D}\sqrt{|u(0)|^2 + |v(0)|^2}.$$

From (5.75) follows that

$$\sqrt{|u(t)|^2 + |v(t)|^2} \le \mathcal{N}\exp(-(D - \frac{c}{m})t)\sqrt{|u(0)|^2 + |v(0)|^2},$$

where $\mathcal{N} > 0$ is some constant depending only on c_1, c_2 and D. Now to complete the proof of the theorem is sufficient to refer to Theorem 2.15 and Corollary 2.6.$\Box$

Remark 5.9   *All the statements given in this chapter in the periodical case coincide with previously established results (see [160], [270], [272]).*

5.6   Construction of Lyapunov function for homogeneous systems

Let (X, h, Y) be a locally trivial Banach fibering, $|\cdot| : X \to \mathbb{R}_+$ is a norm on (X, h, Y) coordinated with the metric ρ on X.

Definition 5.9   The autonomous dynamical system $(X, \mathbb{R}_+, \pi)$ is said to be homogeneous of order $m \in \mathbb{R}_+$, if for any $x \in X, t \in \mathbb{R}_+$ and $\lambda > 0$ the equality $\pi(t, \lambda x) = \lambda\pi(\lambda^{m-1}t, x)$ takes place.

Definition 5.10 $\langle(X, \mathbb{T}, \pi), ((Y, \mathbb{T}, \sigma), h\rangle$ is called homogeneous of order $m = 1$ if the autonomous dynamical system $(X, \mathbb{T}, \pi)$ is homogeneous of order $m = 1$.

Theorem 5.32   *For an autonomous homogeneous (of order $m > 1$) dynamical system $(X, \mathbb{R}_+, \pi)$ the following assertions are equivalent:*

1. there exist positive numbers a_i and b_i $(i = 1, 2)$ such that

$$(a_1|x|^{1-m} + b_1 t)^{\frac{1}{1-m}} \le |\pi(t, x)| \le (a_2|x|^{1-m} + b_2 t)^{\frac{1}{1-m}} \qquad (5.76)$$

for all $t \ge 0$ and $x \in X$;

2. *for all $k > m - 1$ there exists a continuous function $V : X \to \mathbb{R}_+$ with the following properties:*

   2.1. *$V(\lambda x) = \lambda^{k-m+1} V(x)$ for all $\lambda \geq 0$ and $x \in X$;*
   2.2. *$\alpha |x|^{k-m+1} \leq V(x) \leq \beta |x|^{k-m+1}$ for all $x \in X$, where α and β are certain positive numbers;*
   2.3. *$V'_\pi(x) = -|x|^k$ for all $x \in X$, where $V'_\pi(x) = \frac{d}{dt} V(\pi(t,x)) \big|_{t=0}$ for $\mathbb{T} = \mathbb{R}_+$ and $V'_\pi(x) = V(\pi(1,x)) - V(x)$ for $\mathbb{T} = \mathbb{Z}_+$.*

Proof. We will show that from 1. follows 2. Let a and b are positive numbers, such that the inequality (5.76) takes place, then for each $k > m - 1$ we define the function $V : X \to \mathbb{R}_+$ by equality

$$V(x) = \int_0^{+\infty} |\pi(t, x)|^k dt. \tag{5.77}$$

First of all we note that by equality (5.77) it is defined correctly the function $V : X \to \mathbb{R}_+$ because the integral, which figures in the second number of equation (5.77) is convergent, moreover it is uniformly convergent w.r.t. x on every bounded set from X. Indeed, since

$$(a_1 |x|^{1-m} + b_1 t)^{\frac{k}{1-m}} \leq |\pi(t, x)|^k \leq (a_2 |x|^{1-m} + b_2 t)^{\frac{k}{1-m}}, \tag{5.78}$$

$$\int_0^{+\infty} |(a|x|^{1-m} + bt)^{\frac{k}{1-m}} dt = \frac{1}{b} \int_{a|x|^{1-m}}^{+\infty} \tau^{\frac{k}{1-m}} d\tau \tag{5.79}$$

and $\frac{k}{1-m} < -1$, then the integral (5.79) is convergent; moreover, the convergence is uniform on every bounded set from X.

We will show that the function V, defined by equality (5.77), is our unknown function. The continuity of V results from the continuity of mapping $\pi : \mathbb{T} \times X \to X$ and uniform convergence of integral (5.79) w.r.t. x on every bounded set from X. We now note that

$$V(\lambda x) = \int_0^{+\infty} |\pi(t, \lambda x)|^k dt = \int_0^{+\infty} \lambda^k |\pi(\lambda^{m-1} t, x)|^k dt$$

$$= \lambda^{k-m+1} \int_0^{+\infty} |\pi(\tau, x)|^k dt = \lambda^{k-m+1} V(x)$$

for all $\lambda > 0$ and $x \in X$. It is not difficult to show that the function V is positive definite. In virtue of (5.78) and (5.79) we have

$$\alpha |x|^{k-m+1} \leq V(x) \leq \beta |x|^{k-m+1} \tag{5.80}$$

for every $x \in X$, where

$$\alpha = \frac{(m-1)a_1^{\frac{k-m+1}{1-m}}}{b_1(k-m+1)} \quad \text{and} \quad \beta = \frac{(m-1)a_2^{\frac{k-m+1}{1-m}}}{b_2(k-m+1)}.$$

From (5.79)-(5.80) results that the function V satisfies and the condition 2.2.

Finally, we note that

$$\frac{d}{dt}V(\pi(t,x)) = -|\pi(t,x)|^k \tag{5.81}$$

and, consequently, $V'_\pi(x) = -|x|^k$ for all $x \in X$.

We will prove now that from condition 2. follows 1. In fact, we denote by $\psi(t) = V(\pi(t,x))$, then in virtue of condition 2.3 we will have

$$\psi'(t) = -|\pi(t,x)|^k \tag{5.82}$$

for all $t \geq 0$.

From the condition 2.2 we have $|\pi(t,x)|^{k-m+1} \geq \frac{1}{\beta}\psi(t)$ and, consequently,

$$\psi'(t) \leq -\frac{1}{\beta^{\frac{k}{k-m+1}}}\psi(t)^{\frac{k}{k-m+1}}$$

for all $t \geq 0$. If $x \neq 0$, then $\psi(t) = V(\pi(t,x)) > 0$ for all $t \geq 0$, therefore

$$V(\pi(t,x)) \leq \left(V^{-\frac{m-1}{k-m+1}}(x) + \frac{m-1}{k-m+1}\frac{1}{\beta^{\frac{k}{k-m+1}}}t\right)^{\frac{1}{1-m}} \tag{5.83}$$

for all $x \in X$ and $t \geq 0$.

From the condition 2.2 and the inequality (5.83) results that $|\pi(t,x)| \leq (a_2|x|^{1-m} + b_2 t)^{\frac{1}{1-m}}$ for all $x \in X$ and $t \geq 0$, where

$$a_2 = (\alpha\beta)^{\frac{m-1}{k-m+1}} \quad \text{and} \quad b_2 = (\alpha)^{\frac{m-1}{k-m+1}}(\beta)^{\frac{k}{k-m+1}}\frac{m-1}{k-m+1}.$$

Analogously can be proved the second inequality. The theorem is completely proved. $\qquad\square$

Corollary 5.10　　*Let $(X,\mathbb{R}_+,\pi)$ be an autonomous homogeneous (of order $m > 1$) dynamical system $(X,\mathbb{R}_+,\pi)$ and the trivial motion of the dynamical system $(X,\mathbb{R}_+,\pi)$ is uniform asymptotic stable, then for every number $k > m-1$ there exists a continuous function $V : X \to \mathbb{R}_+$ which possesses properties 2.1-2.3 from the theorem 5.32.*

Proof.　This assertion directly follows from Theorems 5.32 and 2.35. $\qquad\square$

Theorem 5.33　　*Let a non-autonomous system $\langle(X,\mathbb{T},\pi),((Y,\mathbb{T},\sigma),h\rangle$ be homogeneous of order $m = 1$. Then the conditions 1. and 2. are equivalent:*

1. *there exist positive numbers N_i and ν_i ($i = 1,2$) such that $N_1 e^{-\nu_1 t}|x| \leq |\pi(t,x)| \leq N_2 e^{-\nu_2 t}|x|$ for all $x \in X$ and $t \geq 0$;*

2. *for each $k > 0$ there exists a continuous function $V : X \to \mathbb{R}_+$ satisfying the following conditions :*

   2.1. *$V(\lambda x) = \lambda^k V(x)$ for all $x \in X$ and $\lambda > 0$;*

   2.2. *there exist positive numbers $\alpha \geq 1$ and β so that*

$$\alpha|x|^k \leq V(x) \leq \beta|x|^k \tag{5.84}$$

   for every $x \in X$;

   2.3. *$V'_\pi(x) = -|x|^k$ for all $x \in X$.*

Proof. Let us show that under the conditions of theorem from assumption 1. follows 2. First of all let us show that the function V, defined by equality (5.77) is the unknown, in the case when $\mathbb{T} = \mathbb{R}_+$. We note that by equality (5.77) is correctly defined the function $V : X \to \mathbb{R}_+$, such that

$$N_1^k e^{-\nu_1 k t}|x|^k \leq |\pi(t,x)|^k \leq N_2^k e^{-\nu_2 k t}|x|^k \tag{5.85}$$

for all $t \geq 0$, $x \in X$ and, consequently, the integral in second member of equality (5.77) is convergent uniformly w.r.t. x on every bounded set from X. In particularly, the function V is continuous w.r.t. $x \in X$.

From (5.77) and (5.85) follows that

$$\alpha|x|^k \leq V(x) \leq \beta|x|^k \tag{5.86}$$

for every $x \in X$, where $\alpha = N_1^k(\nu_1 k)^{-1}$ and $\beta = N_2^k(\nu_2 k)^{-1}$. From equality (5.79) we have that $V(\lambda x) = \lambda^k V(x)$ for every $x \in X$ and $\lambda > 0$, and from (5.81) follows the equality $V'_\pi(x) = -|x|^k$.

Let us show that the inverse implication holds too . We now suppose that the conditions 2.1-2.3 of Theorem hold. One denote by $\psi(t) = V(\pi(t,x))$, then the equality (5.82) holds for all $t \geq 0$. From condition 2.2 follows that $|\pi(t,x)|^k \geq \frac{1}{\beta}\psi(t)$ and, consequently,

$$\psi'(t) \leq -\frac{1}{\beta}\psi(t) \tag{5.87}$$

for all $t \geq 0$. From (5.87) we have

$$V(\pi(t,x)) \leq V(x)e^{-\frac{1}{\beta}t} \tag{5.88}$$

for any $t \geq 0$ and $x \in X$. According to (5.86) and (5.88) we have

$$|\pi(t,x)| \leq N_2 e^{-\nu_2 t}|x|$$

for any $t \geq$ and $x \in X$, where $N_2 = (\frac{\beta}{\alpha})^{\frac{1}{k}}$ and $\nu_2 = \frac{1}{\beta k}$. Analogously can be established the second inequality.

If $\mathbb{T} = \mathbb{Z}_+$, then we will define the function $V : X \to \mathbb{R}_+$ by equality

$$V(x) = \sum_{n=1}^{+\infty} |\pi(n,x)|^k. \tag{5.89}$$

The series in second member of equality (5.89) is convergent uniformly w.r.t. x on every bounded subset from X, because under the conditions of Theorem we have

$$|\pi(n,x)|^k \le N^k e^{-\nu k n} |x|^k \tag{5.90}$$

for all $n \in \mathbb{Z}_+$ and $x \in X$ and, consequently, the series (5.89) is majored by geometric series with the denominator $q = e^{-\nu k} < 1$. Thus the function V defined by equality (5.89) is continuous.

From the homogeneity of $(X, \mathbb{T}, \pi)$ of order $m = 1$ and equality (5.89) follows that $V(\lambda x) = \lambda^k V(x)$ for all $\lambda > 0$ and $x \in X$. According to (5.89) and (5.90) we have

$$\alpha |x|^k \le V(x) \le \beta |x|^k \tag{5.91}$$

for any $x \in X$, where $\alpha = 1$ and $\beta = N^k(1 - e^{-\nu k})^{-1}$. Finally, from equality (5.89) follows that $V'_\pi(x) = V(\pi(1,x)) - V(x) = -|x|^k$ for all $x \in X$.

We now will show that from the conditions 2.1–2.3 of Theorem (in the case, when $\mathbb{T} = \mathbb{Z}_+$) follows the estimation (5.84). In fact, we denote by $\psi(n) = V(\pi(n,x))$, then from the conditions 2.2–2.3 we have

$$\Delta\psi(n) = \psi(n+1) - \psi(n) = -|\pi(n,x)|^k \le -\frac{1}{\beta}\psi(n). \tag{5.92}$$

We may assume without loss of generality that $\beta > 1$, then from (5.92) we have

$$V(\pi(n,x)) \le \gamma^n V(x) \tag{5.93}$$

for all $x \in X$ and $n \in \mathbb{Z}_+$, where $\gamma = 1 - \beta^{-1}$. From (5.91) and (5.93) follows the inequality (5.84), if we suppose that $N = (\frac{\beta}{\alpha})^{\frac{1}{k}}$ and $\nu = -\frac{1}{k}ln(1 - \beta^{-1})$. The theorem is proved. $\qquad\square$

Definition 5.11   The trivial section of fiber bundle (X, h, Y) is called globally uniformly asymptotically stable, if

(1) for arbitrary $\varepsilon > 0$ there is $\delta(\varepsilon) > 0$ such that $|\pi(t,x)| < \varepsilon$ for all $t \ge 0$ and $|x| < \delta$;
(2) $\lim\limits_{t \to +\infty} |\pi(t,x)| = 0$ for all $x \in X$, moreover this equality holds uniformly in x on the bounded subsets from X.

Corollary 5.11   *Let $\langle (X, \mathbb{T}, \pi), ((Y, \mathbb{T}, \sigma), h \rangle$ be homogeneous of order $m = 1$. Then the following conditions are equivalent:*

(1) the zero section of fibering (X, h, Y) is uniformly asymptotically stable, i.e.

    (a) for all $\varepsilon > 0$ there exists $\delta(\varepsilon) > 0$ such that $|\pi(t, x)| < \varepsilon$ for all $t \geq 0$ and $|x| < \delta$;

    (b) $\lim\limits_{t \to +\infty} |\pi(t, x)| = 0$ for all $x \in X$, moreover this equality holds uniformly w.r.t. x on every bounded set from X;

(2) there exist positive numbers N and ν such that $|\pi(t, x)| \leq N e^{-\nu t} |x|$ for all $t \geq 0$ and $x \in X$;

(3) for every $k > 0$ there exists a continuous function $V : X \to \mathbb{R}_+$ satisfying the conditions 2.1-2.3 of Theorem 5.33.

Proof. This assertion follows from the Theorems 5.33 and 2.34. $\qquad\qquad\square$

Remark 5.10 *In the case, when the space X is finite dimensional Theorem 5.32 and Corollary 5.10 (for autonomous system with $\mathbb{T} = \mathbb{R}_+$) generalize and refine some results of V.I.Zubov (see, for example [337], Theorems 36 and 37).*

5.7 Differentiable homogeneous systems

Let H be a Hilbert space with scalar product $\langle \cdot, \cdot \rangle$ and the norm $|\cdot| = \sqrt{\langle \cdot, \cdot \rangle}$. We denote by $C(E, B)(C^1(E, B))$ the space of all continuous (differentiable continuous) functions defined on the space E wit values in some Banach space B.

The function $f : E \times \Omega \to F$ is called homogeneous of order k w.r.t. variable $u \in E$ if the equality $f(\lambda x, \omega) = \lambda^k f(x, \omega)$ holds for all $\lambda > 0, x \in E$ and $\omega \in \Omega$.

Lemma 5.6 *The following assertions hold.*

(1) Let Ω be a compact, $f \in C(E \times \Omega, F)$, $f(0, \omega) = 0$ for all $\omega \in \Omega$, $f(\lambda u, \omega) = \lambda^m f(u, \omega)$ (for every $\lambda > 0$ and $(u, \omega) \in E \times \Omega$)). Then there exists $M > 0$ such that $|f(u, \omega)| \leq M|u|^m$ for all $(u, \omega) \in E \times \Omega$.

(2) If the function $f \in C^1(E \times \Omega, F)$ is homogeneous (of order $m \geq 1$), then the function $D_u f(\cdot, \omega) : E \to L(E, F)$ $(\omega \in \Omega)$ will be homogeneous (of order $m-1$), where $L(E, F)$ is the space of all linear continuous operators $A : E \to F$;

(3) The function $f \in C^1(E, F)$ is homogeneous (of order m) if and only if f satisfies the equation $Df(x)x = mf(x)$ for all $x \in E$, where $Df(x)$ is the derivative of Frechet of function $f \in C^1(E)$ in the point x;

Proof. First of all we note that in conditions of Lemma 5.6 there exists $\delta_0 > 0$ such that $|f(u, \omega)| \leq 1$ for all $|u| \leq \delta$ and $\omega \in \Omega$. If we suppose that it is not so, then there exist $\delta_n \to 0, |u_n| < \delta_n$ and $\omega_n \in \Omega$ such that $|f(u_n, \omega_n)| > 1$. Since Ω is compact we may assume that the sequence $\{\omega_n\}$ is convergent. Let $\omega_n \to \omega_0$, then according to continuity of f we have $|f(0,\omega_0)| \geq 1$. On the other hand $f(0,\omega_0) = 0$. The obtained contradiction proves the required assertion.

Thus there exists $\delta > 0$ such that $|f(u,\omega)| \leq 1$ for all $|u| \leq \delta$ and $\omega \in \Omega$ and, consequently, we have

$$|f(u,\omega)| \leq |f(\frac{|u|}{\delta}\frac{u\delta}{|u|},\omega)| = \frac{|u|^m}{\delta^m}|f(\frac{\delta u}{|u|},\omega)| \leq \frac{1}{\delta^m}|u|^m.$$

Let $f \in C^1(E,F)$ and f be homogeneous (of order m), then

$$f(x+h) - f(x) = Df(x)h + r(x,h) \qquad (5.94)$$

and $|r(x,h)| \to 0$ as $|h| \to 0$. We now will replace x by λx in equality (5.94), then we obtain

$$f(\lambda x + h) - f(\lambda x) = Df(\lambda x) + r(\lambda x, h)$$

and, consequently,

$$f(x+u) - f(x) = \lambda^{-m+1}Df(\lambda x)h + \lambda^{-m}r(\lambda x, \lambda u) \qquad (5.95)$$

for all $\lambda > 0$, where $u = \lambda^{-1}h$. From (5.95) follows, that $\lambda^{-m+1}Df(\lambda x) = Df(x)$, i.e. $Df(\lambda x) = \lambda^{m-1}Df(x)$ for any $\lambda > 0$.

Finally, let us prove the third assertion of Lemma. Let $f \in C^1(E,F)$ be a homogeneous function (of order m). Having differentiated the identity $f(\lambda x) = \lambda^m f(x)$ w.r.t. λ and taking $\lambda = 1$ one obtain the identity $Df(x)x = mx$. Let now the identity $Df(x)x = mx$ take place. We denote by $\varphi(\lambda) = \lambda^{-m}f(\lambda x)(\lambda > 0)$. Thus $Df(\lambda x)\lambda x = mf(\lambda x)$, then

$$\varphi'(\lambda) = -m\lambda^{-m+1}f(\lambda x) + \lambda^{-m}Df(\lambda x)x =$$

$$-m\lambda^{-m-1}f(\lambda x) + \lambda^{-m-1}mf(\lambda x) = 0$$

for all $\lambda > 0$ and, consequently, $\varphi(\lambda) = const$ for all $\lambda > 0$. Thus $\varphi(1) = f(x)$, when $\varphi(\lambda) = f(x)$ for any $\lambda > 0$, i.e. $f(\lambda x) = \lambda^m f(x)$ for every $\lambda > 0$ and for all $x \in E$. The lemma is proved. $\qquad\qquad\qquad\qquad\qquad\qquad\qquad\qquad\qquad\qquad\square$

The function $f \in C(E)$ is said to be regular, if for the differential equation

$$x' = f(x) \qquad (5.96)$$

the conditions of existence, uniqueness on $\mathbb{R}_+$ and continuous dependence on initial data are fulfilled, i.e. by equation (5.96) it is generated a semigroup dynamical system $(E,\mathbb{R}_+,\pi)$, where $\pi(t,x)$ is the solution of equation (5.96) with initial condition $\pi(0,x) = x$.

Theorem 5.34 *Let $f \in C'(E, F)$ be regular and homogeneous (of order $m > 1$), then the following conditions are equivalent:*

1. *the zero solution of equation (5.96) is uniform asymptotic stable;*
2. *for all sufficiently large k there exists a continuously differentiable function $V : E \to \mathbb{R}_+$ satisfying the following conditions:*

   2.1. $V(\lambda x) = \lambda^{k-m+1} V(x)$ for all $\lambda \geq 0$ and $x \in E$;

   2.2. there exist positive numbers α and β such that $\alpha |x|^{k-m+1} \leq V(x) \leq \beta |x|^{k-m+1}$ for all $x \in E$;

   2.3. $V'_\pi(x) = DV(x)f(x) = -|x|^k$ for any $x \in E$, where $DV(x)$ is a derivative of Frechet of function V in the point x.

Proof. We suppose that the zero solution of equation (5.96) is uniformly asymptotically stable. Denote by $(E, \mathbb{R}_+, \pi)$ a semigroup dynamical system, generated by equation (5.96). Thus, $f \in C^1(E, F)$ is homogeneous (of order $m > 1$), then according to Theorem 3.4 [337] the equality $\pi(t, \lambda x) = \lambda \pi(\lambda^{m-1} t, x)$ holds for all $x \in E, \lambda \geq 0$ and $t \in \mathbb{R}_+$, i.e. a dynamical system $(E, \mathbb{R}_+, \pi)$ is homogeneous (of order $m > 1$). According to Theorem 5.32 and Corollary 5.10 by equality (5.95) it is defined a continuous function $V : E \to \mathbb{R}_+$, satisfying the condition 2.1–2.3. of Theorem 5.34. Let us show that under conditions of Theorem 5.34 function V will be continuously differentiable and that the equality $V'_\pi(x) = DV(x)x$ holds. To this aim we will formally differentiate the equality (5.77) w.r.t. x, then we will have

$$DV(x)u = \int_0^{+\infty} k|\pi(t, x)|^{k-2} Re\langle D_x \pi(t, x)u, \pi(t, x)\rangle dt. \tag{5.97}$$

Now we will show that the integral figuring in the second hand of formula (5.97), is uniformly convergent w.r.t. x and u on the every ball from E and, consequently, the equality (5.97) really defines a derivative of function V. We note that operator-function $U(t, x) = D_x \pi(t, x)$ satisfies the operational equation

$$X' = \mathcal{A}(t, x)X,$$

where $\mathcal{A}(t, x) = D_x f(\pi(t, x))$, and initial condition $U(0, x) = I$ (I−is identity operator in E). According to Lemma 5.6 the function $\mathcal{A}(t, x) = D_x f(\pi(t, x))$ is homogeneous w.r.t. $\pi(t, x)$ of order $m - 1$ and there exists a number $M > 0$ such that

$$\|\mathcal{A}(t, x)\| \leq M|\pi(t, x)|^{m-1} \tag{5.98}$$

for all $x \in E$ and $t \geq 0$. Since the zero solution of equation (5.96) is uniformly asymptotically stable, in virtue of Corollary 5.10 there exist positive numbers a and b such that

$$|\pi(t, x)| \leq (a|x|^{1-m} + bt)^{\frac{1}{1-m}} \tag{5.99}$$

for any $x \in E$ and $t \geq 0$ and, consequently, from (5.98)– (5.99) we have

$$\|\mathcal{A}(t, x)\| \leq M(a|x|^{1-m} + bt)^{-1}. \tag{5.100}$$

Since

$$\|U(t, x)\| \leq e^{\int_0^t \|\mathcal{A}(\tau, x)\|, d\tau}, \tag{5.101}$$

from (5.100) and (5.101) we have

$$\|U(t, x)\| \leq (a|x|^{1-m} + bt)^{\frac{M}{b}} (a^{-1}|x|^{m-1})^{\frac{M}{b}} \tag{5.102}$$

for every $t \geq 0$ and $x \in E$. By virtue of Inequality (5.102) we obtain

$$|\pi(t, x)|^{k-2} |Re\langle D_x\pi(t, x)u, \pi(t, x)\rangle| \leq |\pi(t, x)|^{k-1} \|D_x\pi(t, x)\| |u| \tag{5.103}$$

$$\leq (a|x|^{1-m} + bt)^{-\frac{k-1}{m-1} + \frac{M}{b}} a^{-\frac{M}{b}} |x|^{(m-1)\frac{m}{b}} |u|$$

for all $t \geq 0$ and $x, u \in E$. From estimation (5.103) follows that the integral in second member of equality (5.97) is uniformly convergent w.r.t. x and u on the each ball from E for sufficiently large k and, consequently, the function V, defined by equality (5.77), under the conditions of Theorem 5.34 is continuously differentiable.

We now note, that

$$V_\pi'(x) = \frac{d}{dt} V(\pi(t, x))\Big|_{t=0} = DV(\pi(t, x))f(\pi(t, x))\Big|_{t=0} = DV(x)f(x)$$

for all $x \in E$. On the other hand, according to Theorem 5.32 $V_\pi'(x) = -|x|^k$. For finishing the proof of the theorem it is sufficient to notice that in conformity with Theorem 5.32 the inverse statement holds, i.e. conditions 2.1–2.3 of Theorem 5.34 imply 1. The theorem is completely proved. □

Now let $f \in C^1(E \times F, E)$ and $\Phi \in C^1(F, F)$. Consider the differential system

$$\begin{cases} u' = f(u, \omega) \\ \omega' = \Phi(\omega) \end{cases} \quad (\omega \in \Omega), \tag{5.104}$$

where Ω is a certain differentiable compact sub-manifold in F, f and Φ are regular functions, i.e. for the differential system (5.104) and equation

$$\omega' = \Phi(\omega) \tag{5.105}$$

are fulfilled the conditions of existence, uniqueness and continuous dependence on initial data on the $\mathbb{R}_+$ for (5.104) and on the $\mathbb{R}$ for (5.105). We denote by $(\Omega, \mathbb{R}, \sigma)$ a group dynamical system, generated by equation (5.105), then the system (5.104) may be written in the form of non-autonomous equation

$$u' = f(u, \omega t), \quad (\omega \in \Omega) \tag{5.106}$$

where $\omega t := \sigma(t, \omega)$. Let $\varphi(t, u, \omega)$ be a solution of equation (5.106) satisfying the initial condition $\varphi(0, u, \omega) = u$, then it is not difficult to see that the triple $\langle E, \varphi, (\Omega, \mathbb{R}, \sigma) \rangle$ is a cocycle over $(\Omega, \mathbb{R}, \sigma)$ with the fiber E. It is easy to check that if a function f is homogeneous of order $m = 1$ w.r.t. $u \in E$, then the equality $\varphi(t, \lambda u, \omega) = \lambda \varphi(t, u, \omega)$, holds for all $t \geq 0$, $\lambda > 0$, $u \in E$ and $\omega \in \Omega$.

Theorem 5.35 *Let the function $f \in C^1(E \times F, F)$ and $\Phi \in C^1(F, F)$ be regular, Ω is compact and the function f is homogeneous (of order $m = 1$) w.r.t. variable $u \in E$, then the following conditions are equivalent:*

1. *the trivial solution of equation (5.106) is uniform exponential stable, i.e. there exist positive numbers N and ν such that*

$$|\varphi(t, u, \omega)| \leq N e^{-\nu t} |u| \tag{5.107}$$

   for all $u \in E, t \geq 0$ and $\omega \in \Omega$;
2. *for all number $k > 1$ there exists a continuously differentiable function $V : E \times \Omega \to \mathbb{R}_+$ satisfying the following conditions:*

   2.1. $V(\lambda u, \omega) = \lambda^k V(u, \omega)$ for all $u \in \Omega$ and $\lambda > 0$;
   2.2. there exist positive numbers α and β such that $\quad \alpha |u|^k \leq V(u, \omega) \leq \beta |u|^k$ for all $u \in E$ and $\omega \in \Omega$;
   2.3.

   $$V'_\pi(u, \omega) := \frac{d}{dt} V(\varphi(t, u, \omega), \omega t)\Big|_{t=0} = D_u V(u, \omega) f(u, \omega) +$$

   $$D_\omega V(u, \omega) \Phi(\omega) = -|u|^k$$

   for all $u \in E$ and $\omega \in \Omega$, where $D_u V$ (respectively $D_\omega V$) is a partial Frechet's derivative of function V w.r.t. variable $u \in E$ (respectively $\omega \in \Omega$).

Proof. Let $\langle (X, \mathbb{R}_+, \pi), (\Omega, \mathbb{R}, \sigma), h \rangle$ be a non-autonomous dynamical system generated by differential system (5.104), where $X = E \times \Omega, h = pr_2$ and $\pi = (\varphi, \sigma)$. The triple (X, h, Ω) is the trivial fiber bundle with the fiber E, moreover the norm of element $x = (u, \omega) \in E \times \Omega$ is defined by equality $|x| = |u|$, where $|u|$ is a norm of element u in the space E. Then under the conditions of Theorem the zero section of non-autonomous dynamical system, constructed above, will satisfy the conditions of Theorem 5.35. Thus to finish the proof of Theorem 5.35 it is sufficient to show that the function $V : E \times \Omega \to \mathbb{R}_+$ from the theorem 5.33 under the conditions of Theorem 5.35 will be continuously differentiable and satisfies the condition

$$V'_\pi(u, \omega) = D_u V(u, \omega) f(u) + D_\omega V(u, \omega) \Phi(\omega)$$

for all $(u, \omega) \in E \times \Omega$.

Let us show that a function $V : E \times \Omega \to \mathbb{R}_+$, defined by equality (5.77), i.e.

$$V(u, \omega) = \int_0^{+\infty} |\varphi(t, u, \omega)|^k dt \tag{5.108}$$

for all $(u, \omega) \in E \times \Omega$ (such that $|\pi(t, x)| = |(\varphi(t, u, \omega), \omega t)| = |\varphi(t, u, \omega)|$) is continuously differentiable. To this aim we will formally differentiate the equality (5.108) w.r.t. variable $u \in E$, thus we obtain

$$D_u V(u, \omega) v = \tag{5.109}$$
$$\int_0^{+\infty} k |\varphi(t, u, \omega)|^{k-2} Re\langle D_u\varphi(t, u, \omega)v, \varphi(t, u, \omega)\rangle dt.$$

Let us show that integral in (5.109) is uniformly convergent w.r.t. $\omega \in \Omega$ and u, v on every bounded set from E and since by formula (5.109) is really defined a derivative of Frechet of function V w.r.t. variable $u \in E$. First of all we note that operator-function $U(t, u, \omega) := D_u\varphi(t, u, \omega)$ satisfies the operational equation

$$X' = \mathcal{B}(t, u, \omega)X, \tag{5.110}$$

where $\mathcal{B}(t, u, \omega) := D_u f(\varphi(t, u, \omega), \omega t)$, and initial condition $U(0, u, \omega) = I$ (I is the identity operator in E). According to Lemma 3.1 function $\mathcal{B}(t, u, \omega) = D_u f(\varphi(t, u, \omega), \omega t)$ is homogeneous w.r.t. $\varphi(t, u, \omega)$ of order $m = 1$ and there exists a number $M > 0$ such that

$$\|\mathcal{B}(t, u, \omega)\| \le M \tag{5.111}$$

for all $(u, \omega) \in E \times \Omega$ and $t \ge 0$. From the inequality (5.111) follows that

$$\|U(t, u, \omega)\| \le e^{\int_0^t \|\mathcal{B}(\tau, u, \omega)\| d\tau} \le e^{Mt}$$

for all $t \ge 0$ and $(u, \omega) \in E \times \Omega$ and, consequently,

$$|\varphi(t, u, \omega)|^{k-2} |Re\langle D_u\varphi(t, u, \omega)v, \varphi(t, u, \omega)\rangle| \tag{5.112}$$
$$|\varphi(t, u, \omega)|^{k-1} \|D_u\varphi(t, u, \omega)\| |v| \le M|v|N^{k-1}e^{-\nu(k-1)t}|u|$$

for any $t \ge 0$, $u, v \in E$ and $\omega \in \Omega$. From the inequality (5.112) follows that for $k > 1$ the integral in the second hand of (5.108) is convergent, moreover uniformly w.r.t. $\omega \in \Omega$ and u, v on the every bounded subset from E. Thus, the function $V(u, \omega)$, defined by formula (5.108) is continuously differentiable w.r.t. variable $u \in E$ and by equality (5.109) it is defined its derivative of Fréchet w.r.t. variable $u \in E$.

Having formally differentiated the equality (5.108) w.r.t. variable $\omega \in F$ we will obtain

$$D_\omega V(u, \omega) w = \tag{5.113}$$
$$\int_0^{+\infty} k |\varphi(t, u, \omega)|^{k-2} Re\langle D_\omega\varphi(t, u, \omega)w, \varphi(t, u, \omega)\rangle dt$$

for all $(u, \omega) \in E \times \Omega$. Let us show that integral in the second member of (5.113) is uniformly convergent w.r.t. $\omega \in \Omega$ and $(u, w) \in E \times F$ on every bounded subset from $E \times F$. First of all we note that from (5.107) we have

$$\varphi(t, u, \omega) = u + \int_0^t f(\varphi(\tau, u, \omega), \omega\tau)d\tau$$

for all $t \geq 0$ and $(u, \omega) \in E \times F$ and, consequently,

$$\begin{aligned} D_\omega\varphi(t, u, \omega) = \int_0^t &D_u f(\varphi(\tau, u, \omega), \omega\tau)D_\omega\varphi(\tau, u, \omega) \\ &+ D_\omega f(\varphi(\tau, u, \omega), \omega\tau)D\Phi(\omega\tau)d\tau. \end{aligned} \tag{5.114}$$

We denote by $\mathcal{V}(t, u, \omega)$ the operator-function $D_\omega\varphi(t, u, \omega)$, then from (5.114) follows that $\mathcal{V}(t, u, \omega)$ satisfies the operational equation

$$Y' = \mathcal{B}(t, u, \omega)Y + \mathcal{F}(t, u, \omega), \tag{5.115}$$

where $\mathcal{F}(t, u, \omega) = D_\omega f(\varphi(t, u, \omega), \omega t)D\Phi(\omega t)$, and initial condition

$$Y(0) = \mathcal{O} \tag{5.116}$$

($\mathcal{O}$ is a null operator, acting from F into F). From the equality (5.115) and the condition (5.116) follows that

$$\mathcal{V}(t, u, \omega) = \int_0^t U(t, \tau, u, \omega)\mathcal{F}(\tau, u, \omega)d\tau \tag{5.117}$$

for all $t \geq 0$ and $(u, \omega) \in E \times \Omega$, where $U(t, \tau, u, \omega)$ is a solution of operational equation (5.110), satisfying the initial condition $U(\tau, \tau, u, \omega) = U(\tau, u, \omega)$, therefore the estimate

$$\|U(t, \tau, u, \omega)\| \leq e^{\int_\tau^t \|\mathcal{B}(\tau, u, \omega)\| d\tau} \leq e^{M(t-\tau)} \tag{5.118}$$

holds for any $t \geq \tau(t, \tau \in \mathbb{R}_+)$ and $(u, \omega) \in E \times \Omega$.

Thus, a function $f(u, \omega)$ is homogeneous w.r.t. u (of order $m = 1$), then $D_\omega f(u, \omega)$ will be the same and, in virtue of Lemma 5.6 there exists a number $M > 0$ such that

$$\|D_\omega f(u, \omega)\| \leq M_1|u|$$

for all $u \in E$ and $\omega \in \Omega$. Thus,

$$\|\mathcal{F}(t, u, \omega)\| = \|D_\omega f(\varphi(t, u, \omega), \omega t)D\Phi(\omega t)| \leq L|\varphi(t, u, \omega)| \tag{5.119}$$

for any $t \geq 0$ and $(u, \omega) \in E \times \Omega$, where $L = M_1 \max\{\|D\Phi(\omega)\| : \omega \in \Omega\}$. From (5.106),(5.117), (5.118) and (5.119) we have

$$|\varphi(t, u, \omega)|^{k-2}|Re\langle D_\omega\varphi(t, u, \omega)w\rangle| \leq |\varphi(t, u, \omega)|^{k-1}\|D_\omega\varphi(t, u, \omega)\||w|$$

$$\leq |\varphi(t, u, \omega)|^{k-1}|w| \int_0^t \|U(t, \tau, u, \omega)\|\|\mathcal{F}(\tau, u, \omega)\|d\tau$$

$$\leq |\varphi(t, u, \omega)|^{k-1}|w| \int_0^t e^{M(t-\tau)}L|\varphi(\tau, u, \omega)|d\tau \leq N^{k-1}e^{-\nu(k-1)t}|u|^{k-1}|w|$$

$$\int_0^t e^{M(t-\tau)}LNe^{-\nu\tau}|u|d\tau = N^k L e^{-\nu(k-1)t}\frac{(e^{Mt} - e^{-\nu t})}{M + \nu}|u|^k|w|$$

for every $t \geq 0, u \in E$ and $\omega \in \Omega$. From this follows, that for sufficiently large k the integral (5.113) is uniformly convergent w.r.t. $\omega \in \Omega$ and $(u, w) \in E \times F$ on the every bounded subset from $E \times F$. Therefore the function $V(u, \omega)$ defined by formula (5.108) is continuously differentiable w.r.t. variable $\omega \in \Omega$ and by formula (5.113) it is defined its Frechet derivative w.r.t. variable $\omega \in \Omega$. So, the function $V : E \times \Omega \to \mathbb{R}_+$ is continuously differentiable w.r.t. variable $u \in E$ and variable $\omega \in \Omega$ both and, consequently, it is continuously differentiable in the sense of Frechet.

Let us note that

$$\frac{d}{dt}V((\varphi(t, u, \omega), \omega t) = D_u V(\varphi(t, u, \omega), \omega t)D_u\varphi(t, u, \omega) +$$

$$D_\omega V(\varphi(t, u, \omega), \omega t)\frac{d}{dt}\omega t = D_u V(\varphi(t, u, \omega), \omega t)f(\varphi(t, u, \omega), \omega t) +$$

$$D_\omega V(\varphi(t, u, \omega), \omega t)\Phi(\omega t)$$

for any $t \geq$ and $(u, \omega) \in E \times \Omega$ and, consequently,

$$V'_\pi(u, \omega) = \frac{d}{dt}V(\varphi(t, u, \omega), \omega t)\Big|_{t=0} = D_u V(u, \omega)f(u, \omega) + D_\omega V(u, \omega)\Phi(u).$$

Thus, under the conditions of Theorem from 1. follows 2.

For finishing the proof of the theorem it is sufficient to remark that according to Theorem 5.33 from the conditions 2.1–2.3 of Theorem follows 1. The theorem is completely proved. $\quad\quad\square$

Remark 5.11 *For finite dimensional systems the theorem 5.34 generalizes and refines the theorem 38 from* [337].

5.8   Global attractors of quasi-homogeneous systems

Let E and F be finite-dimensional Banach spaces. Consider the differential equation

$$x' = f(x), \tag{5.120}$$

where $f \in C(E, E)$ is regular and homogeneous of order $m \geq 1$. Along with equation (5.120) consider also perturbed equation

$$x' = f(x) + F(t, x) \qquad (5.121)$$

with the regular perturbation. Let us remember (see, for example [337], [224] and [227]), that equation (5.121) is called quasi-homogeneous if

$$\limsup_{|x| \to +\infty} \frac{|F(t, x)|}{|x|^m} = c, \qquad (5.122)$$

where c is a certain nonnegative sufficiently small constant. Moreover, the limiting relation (5.122) holds uniformly w.r.t. $t \in \mathbb{R}$.

Along with equation (5.121) consider the family of equations

$$x' = f(x) + G(t, x), \qquad (5.123)$$

where $G \in H(f) := \overline{\{F_\tau : \tau \in \mathbb{R}\}}$ and the bar denotes closure in $C(\mathbb{R} \times E, E)$, $C(\mathbb{R} \times E, E)$ is equipped by topology of convergence on every compact from $\mathbb{R} \times E$. Let us denote by $(Y, \mathbb{R}, \sigma)$ a dynamical system of translations on $Y := H(F)$, by $\varphi(t, u, f + G)$ the solution of equation (5.123), satisfying the condition $\varphi(0, u, f + G) = u$ and $\langle (X, \mathbb{R}_+, \pi), (Y, \mathbb{R}, \sigma), h \rangle$ (i.e. $h := pr_2$ and $\pi := (\varphi, \sigma)$) a non-autonomous dynamical system generated by equation (5.123).

Theorem 5.36 *Let $f \in C^1(E, E)$ be regular, homogeneous of order $m \geq 1$ and the zero solution of equation (5.120) be uniformly asymptotically stable. If $f + F \in C(\mathbb{R} \times E, E)$ is regular, equation (5.121) is quasi-homogeneous, then the following assertions take place:*

(1) a set $I_G = \{u \in E| \sup\{|\varphi(t, u, f + G)| : t \in \mathbb{R}\} < +\infty\}$ is not empty, compact and connected for each $G \in H(F)$;

(2) $\varphi(t, I_G, f + G) = I_{\sigma(t, f+G)}$ for all $t \in \mathbb{R}_+$ and $G \in H(F)$;

(3) a set $I = \bigcup\{I_G | G \in H(F)\}$ is compact and connected;

(4) the equalities

$$\lim_{t \to +\infty} \beta(\varphi(t, M, f + G_{-t}), I_G) = 0$$

and

$$\lim_{t \to +\infty} \beta(\varphi(t, M, f + G), I) = 0$$

take place for all $G \in H(F)$ and bounded subset $M \subseteq E$.

Proof. Let $\langle (X, \mathbb{R}_+, \pi), (Y, \mathbb{R}, \sigma), h \rangle$ be a non-autonomous dynamical system, generated by equation (5.121). Thus, $f \in C^1(E, E)$ and the zero solution of equation

(5.120) is uniformly asymptotically stable, then according to Theorem 5.34 by equality (5.77) it is defined a continuously differentiable function $V : E \to \mathbb{R}_+$, satisfying the conditions 2.1–2.3 of Theorem 5.34. Let us define a function $\mathcal{V} : X \to \mathbb{R}_+$ in the following way: $\mathcal{V}(x) = V(u)$ for all $x = (u, G) \in E \times H(F)$. We note that

$$\mathcal{V}'_{\pi}(x) = \frac{d}{dt} V(\varphi(t, u, f + G))\Big|_{t=0} \tag{5.124}$$
$$= DV(\varphi(t, u, f + G)) \frac{d}{dt} \varphi(t, u, f + G)\Big|_{t=0} = DV(u)(f(u) + G(0, u))$$
$$= DV(u)f(u) + DV(u)G(0, u) = -|u|^k + DV(u)G(0, u).$$

In virtue of condition 2.1 of Theorem 5.34 the function V is homogeneous of order $k - m + 1$, then according to Lemma 5.6 its Frechet derivative $DV(u)$ will be homogeneous of order $k - m$ and, consequently, there exists a number $L > 0$ such that

$$\|DV(u)\| \leq L|u|^{k-m} \tag{5.125}$$

for all $u \in E$. From equality (5.122) follows that for all $\varepsilon > 0$ there exists $r = r(\varepsilon) > 0$ such that

$$|G(t, u)| \leq (c + \varepsilon)|u|^m \tag{5.126}$$

for all $|u| \geq r, t \in \mathbb{R}$ and $G \in H(F)$. From (5.125)–(5.126) we have

$$|DV(u)G(0, u)| \leq L(c + \varepsilon)|u|^k \tag{5.127}$$

and, consequently, from (5.124),(5.127) we will obtain

$$\mathcal{V}'_{\pi}(x) \leq |u|^k(-1 + L(c + \varepsilon)) = -\gamma|x|^k$$

for each $x \in X_r$, where $\gamma = 1 - l(c + \varepsilon) > 0$ since c and ε are a sufficiently small positive numbers. To finish the proof of Theorem it is sufficient to refer to Theorems 2.24- 2.25. $\qquad\square$

Theorem 5.37   *Let $f \in C^1(E, E), \Phi \in C^1(F, F)$ and $f + F \in C^1(E \times F, E)$ be regular, $\Omega \subseteq F$ be a compact invariant set of dynamical system (5.105), the function f be homogeneous (of order $m > 1$) and the zero solution of equation (5.105) be uniformly asymptotically stable. If*

$$|F(u, \omega)| \leq c|u|^m$$

for all $|u| \geq r$ and $\omega \in \Omega$, where r and c are certain positive numbers, and moreover c is sufficiently small, then the following assertions take place:

(1) a set $I_\omega = \{u \in E | \sup\{|\varphi(t, u, \omega)| \; : \; t \in \mathbb{R}\} < +\infty\}$ is not empty, compact and connected for each $\omega \in \Omega$, where $\varphi(t, u, \omega)$ is a solution of equation

$$u' = f(u) + F(u, \omega t)$$

satisfying the initial condition $\varphi(0, u, \omega) = u$;
(2) $\varphi(t, I_\omega, \omega) = I_{\sigma(t, \omega)}$ for all $t \in \mathbb{R}_+$ and $\omega \in \Omega$;
(3) a set $I = \bigcup\{I_\omega \mid \omega \in \Omega\}$ is compact and connected;
(4) the equalities

$$\lim_{t \to +\infty} \beta(\varphi(t, M, \omega_{-t}), I_\omega) = 0 \tag{5.128}$$

and

$$\lim_{t \to +\infty} \beta(\varphi(t, M, \omega), I) = 0 \tag{5.129}$$

take place for all $\omega \in \Omega$ and bounded subset $M \subseteq E$.

Proof. The proof of formulated assertion is carried out using the same scheme as in Theorem 5.36, therefore its proof may be omitted. $\qquad\square$

Theorem 5.38 *Let $f \in C^1(E \times F, E), \Phi \in C^1(F, F)$ and $f + F \in C^1(E \times F, E)$ be regular, $\Omega \subseteq F$ be a compact invariant set of dynamical system (5.105), function f be homogeneous (of order $m = 1$) w.r.t. variable $u \in E$ and the zero solution of equation (5.106) be uniformly asymptotically stable.*
If

$$|F(u, \omega)| \le c|u| \tag{5.130}$$

for all $|u| \ge r$ and $\omega \in \Omega$, where r and c are certain positive numbers, and moreover c is sufficiently small. Then the following assertions take place:

(1) the set $I_\omega = \{u \in E | \sup\{|\varphi(t, u, \omega)| \; : \; t \in \mathbb{R}\} < +\infty\}$ is not empty, compact and connected for each $\omega \in \Omega$, where $\varphi(t, u, \omega)$ is a solution of equation

$$u' = f(u, \omega t) + F(u, \omega t) \tag{5.131}$$

satisfying the initial condition $\varphi(0, u, \omega) = u$;
(2) $\varphi(t, I_\omega, \omega) = I_{\sigma(t, \omega)}$ for all $t \in \mathbb{R}_+$ and $\omega \in \Omega$;
(3) the set $I = \bigcup\{I_\omega | \omega \in \Omega\}$ is compact and connected;
(4) the equalities (5.128) and (5.129) take place for every bounded subset $M \subseteq E$ and $\omega \in \Omega$.

Proof. Let $\langle (X, \mathbb{R}_+, \pi), (Y, \mathbb{R}, \sigma), h \rangle$ be a non-autonomous dynamical system, generated by equation (5.131). Since $f \in C^1(E \times F, E)$ and the zero solution of equation (5.106) is uniformly asymptotically stable, then according to Theorem 5.35 by equality

$$V(u,\omega) = \int_0^{+\infty} |\varphi(t,u,\omega)|^k dt$$

it is defined a continuously differentiable function $V : X = E \times \Omega \to \mathbb{R}_+$, satisfying the conditions 2.1-2.3 of Theorem 5.33. Let us remark that

$$V_\pi'(u,\omega) = \frac{d}{dt} V((\varphi(t,u,\omega),\omega t)\Big|_{t=0} = \{D_u V(\varphi(t,u,\omega),\omega t)\frac{d}{dt}\varphi(t,u,\omega)$$

$$+D_\omega V(\varphi(t,u,\omega),\omega t)\frac{d}{dt}\omega t\}\Big|_{t=0} = D_u V(u,\omega)[f(u,\omega) + F(u,\omega)]$$

$$+D_\omega V(u,\omega)\Phi(\omega) = -|u|^k + D_u V(u,\omega)F(u,\omega). \tag{5.132}$$

Since the function $V(u,\omega)$ is homogeneous of order $k > 1$ w.r.t. variable $u \in E$, then in virtue of Lemma 5.6 there exists a number $L > 0$ such that

$$\|D_u V(u,\omega)\| \le L|u|^{k-1} \tag{5.133}$$

for all $u \in E$ and $\omega \in \Omega$. From (5.130) and (5.133) follows, that

$$|D_u V(u,\omega)F(u,\omega)| \le cL|u|^k \tag{5.134}$$

for each $|u| \ge r$ and $\omega \in \Omega$ and, consequently, according to (5.132), (5.134) we have

$$V_\pi'(u,\omega) \le -|u|^k + cL|u|^k = -\gamma|u|^k$$

for all $\omega \in \Omega$ and $|u| \ge r$, where $\gamma = 1 - cL > 0$ since a number c is sufficiently small. For finishing the proof of the theorem it is sufficiently to refer to Theorem 2.24, 2.25 and 5.3. The theorem is proved. $\qquad\square$

Remark 5.12 *The dissipativity of equation (5.121) is established in* [337] *(theorem 39 and corollary 1).*

Chapter 6

Dissipativity of some classes of equations

6.1 Difference equations

All the results about differential equations, which are presented in previous chapters, for difference equations hold too, because they were formulated and proved for general non-autonomous dynamical systems, both for dynamical systems with continuous time and those with discrete time. Below we will give some of them that we need to study periodical systems with impulse.

Consider a difference equation

$$u(k + 1) = \Phi(k, u(k)) \quad (\Phi \in C(\mathbb{Z} \times E^n, E^n)). \tag{6.1}$$

Denote by $\varphi(\cdot, u, \Phi)$ a solution of the equation (6.1) passing through the point u for $k = 0$. Suppose that Φ is p–periodic in $k \in \mathbb{Z}$, where $p \in \mathbb{Z}$. We set $H(\Phi) := \{\Phi_m = \sigma(\Phi, m) \; : \; 0 \leq m \leq p - 1\}$, and denote by $(Y, \mathbb{Z}, \sigma)$ a dynamical system of translations on $Y := H(\Phi)$ induced by $(C(\mathbb{Z} \times E^n, E^n), \mathbb{Z}, \sigma)$. Let $X := E^n \times Y$ and define a mapping $\pi : X \times \mathbb{Z}_+ \to X$ by the equality $\pi((u, \tilde{\Phi}), k) := (\varphi(k, u, \tilde{\Phi}), \tilde{\Phi}_k)$. From general properties of solutions of the difference equation (6.1) follows that $(X, \mathbb{Z}_+, \pi)$ is a semigroup dynamical system and $\langle (X, \mathbb{Z}_+, \pi), (Y, \mathbb{Z}, \sigma), h \rangle$ $(h := pr_2 : X \to Y)$ is a non-autonomous p–periodic dynamical system. Applying to the above constructed non-autonomous dynamical system the results of section 4.6, we get some assertions for the equation (6.1). Before formulating some statements of this type we will define the notion of dissipativity for the equation (6.1).

Definition 6.1 By analogy with the case of differential equations, the difference equation (6.1) is said to be dissipative, if there is a positive number r such that

$$\limsup_{k \to +\infty} |\varphi(k, v, \tilde{\Phi}| < r$$

for all $v \in E^n$ and $\tilde{\Phi} \in H(\Phi)$.

We define a mapping $\mathbb{P} : E^n \to E^n$ in the following way: $\mathbb{P}(v) := \varphi(p, v, \Phi)$ and denote by (E^n, P) the cascade generated by positive powers of $\mathbb{P}$.

225

Theorem 6.1   *The equation (6.1) is dissipative if and only if the cascade $(E^n, \mathbb{P})$ is dissipative.*

Theorem 6.2   *If the equation (6.1) is dissipative, then:*

1. *there exists $h > 0$ such that for all $a > 0$ there is $k(a) \in \mathbb{Z}_+$ for which $\mathbb{P}^k B[0, a] \subseteq B[0, h]$ for all $k \geq k(a)$;*
2. *the mapping $\mathbb{P}$ has at least one fixed point $u_0 \in E^n$ through which there passes a p–periodic solution of (6.1).*

Theorem 6.3   *If the equation (6.1) is dissipative, then:*

1. *there is a nonempty compact set $I \in E^n$ which possesses the following properties:*

   1.a. $\mathbb{P}(I) = I$;

   1.b. for all $\varepsilon > 0$ there is $\delta(\varepsilon) > 0$ such that $\rho(u, I) < \delta$ implies $\rho(\mathbb{P}^k u, I) < \varepsilon$ for all $k \in \mathbb{Z}_+$;

2. *the set $J := \cup\{J_k := \varphi(k, I, \Phi) : k \in \mathbb{Z}_+\}$ is nonempty, compact and possesses the following properties:*

   2.a. the set J is invariant with respect to (6.1), i.e. $v \in J_m$ implies $\varphi(k, v, \Phi_m) \in J$ for all $k \in \mathbb{Z}_+$;

   2.b. $J_{k+p} = J_k$ for all $k \in \mathbb{Z}_+$;

   2.c. for all $\varepsilon > 0$ there exists $\delta(\varepsilon) > 0$ such that $\rho(\varphi(m, u, \Phi), J_m) < \delta$ implies $\rho(\varphi(k, u, \Phi), J_k) < \varepsilon$ for all $k \geq m$.

Theorem 6.4   *Let the equation (6.1) be dissipative and the function $\Phi : \mathbb{Z} \times \mathbb{C}^n \to \mathbb{C}^n$ be holomorphic in second variable, then the Levinson's center of this equation consists from a single p–periodic trajectory, i.e., the equation (6.1) is convergent.*

Theorems 6.1 – 6.4 directly follow from the results of section 4.6. Concerning Theorem 6.4 it is necessary to add that when the right hand side of (6.1) is holomorphic, then the solution $\varphi(\cdot, z, \Phi)$ $(z \in \mathbb{C}^n)$ is holomorphic too.

Consider the quasi-linear equation

$$u(k + 1) = A(k)u(k) + F(k, u(k)), \tag{6.2}$$

where $A(k)$ is a non-stationary $n \times n$–matrix and the function $F \in C(\mathbb{Z} \times E^n, E^n)$ satisfies some condition of "smallness".

Denote by $U(k, A)$ the Cauchy's matrix for linear equation

$$u(k + 1) = A(k)u(k).$$

Theorem 6.5   *Let $A(k)$ and $F(k, u)$ be p-periodic in $k \in \mathbb{Z}_+$. Suppose that the following conditions hold:*

(1) there are positive numbers M and $q < 1$ such that

$$\|U(k, A)U^{-1}(m, A)\| \leq Mq^{k-m} \qquad (k \geq m); \tag{6.3}$$

(2) $|F(k, u| \leq C + D|u|$ $(C \geq 0,\ 0 \leq D < (1-q)M^{-1})$ for all $u \in E^n$.

Then the equation (6.2) is dissipative.

Proof. Let $\varphi(\cdot, u, \Phi)$ $(\Phi(k, u) := A(k)u + F(k, u))$ be a solution of the equation (6.2) passing through the point $u \in E^n$ for $k = 0$. According to the formula of variation of constants (see, for example, [170]) we have

$$\varphi(k, u, \Phi) = U(k, A)\left(u + \sum_{m=0}^{k-1} U^{-1}(m + 1, A)F(m, \varphi(m, u, \Phi))\right),$$

and, consequently,

$$|\varphi(k, u, \Phi)| \leq Mq^k\left(|u|\right.$$
$$\left. + \sum_{m=0}^{k-1} q^{-m+1}(C + D|\varphi(m, u, \Phi)|)\right). \tag{6.4}$$

We set $u(k) := q^{-k}|\varphi(k, u, \Phi)|$ and, taking into account (6.4), we obtain

$$u(k) \leq M|u| + CMq \sum_{m=0}^{k-1} q^{-m} + DM \sum_{m=0}^{k-1} u(m). \tag{6.5}$$

Denote the right hand side of the equality (6.5) by $v(k)$. Note, that

$$v(k + 1) - v(k) = q^{-k}\frac{CM}{q} + \frac{DM}{q}u(k) \leq \frac{DM}{q}v(k) + \frac{CM}{q}q^{-k},$$

and,hence,

$$v(k + 1) \leq \left(1 + \frac{DM}{q}\right)v(k) + \frac{CM}{q}q^{-k}.$$

From this inequality we obtain

$$v(k) \leq \left(1 + \frac{DM}{q}\right)^{k-1}v(1) + \frac{CM}{q}\frac{1 - q^{k-1}}{1 - q}.$$

Therefore,

$$|\varphi(k, u, \Phi)| \leq (q + DM)^{k-1}qM|u| + \frac{CM}{q - 1}(q^{k-1} - 1), \tag{6.6}$$

because $v(1) = M|u|$. From (6.6) follows that

$$\lim_{k \to +\infty} |\varphi(k, u, \Phi)| \le \frac{CM}{1 - q}.$$

The theorem is proved. $\qquad\square$

Corollary 6.1 *Under the conditions of Theorem 6.5 the equation (6.2) has at least one p-periodic solution.*

Proof. This statement follows from Theorems 6.2 and 6.5. $\qquad\square$

6.2   Equations with impulse

Consider the following differential equation:

$$\dot{u} = f(t, u) \qquad (f \in C(\mathbb{R} \times E^n, E^n)). \tag{6.7}$$

Suppose that $f(t + \tau, u) = f(t, u)$ $(t \in \mathbb{R}, u \in E^n)$ and f is regular, i.e. for the equation

$$\dot{v} = g(t, v) \tag{6.8}$$

the conditions of existence, uniqueness and continuability on $\mathbb{R}_+$ are fulfilled for all $g \in H(f) := \{f_\tau : \tau \in [0, \tau[\}$. Denote by $\varphi(\cdot, v, g)$ a solution of (6.8) passing through the point v for $t = 0$. Note, that

$$\varphi(t + \tau, v, g) = \varphi(t, \varphi(\tau, v, g), g_\tau)$$

for all $t, \tau \in \mathbb{R}_+$, $v \in E^n$ and $g \in H(f)$.

Let $\{t_k\} \subset \mathbb{R}$ $(t_0 = 0)$ and $\{s_k\} \subset E^n$ be such that for certain $p > 0$ $(p \in \mathbb{Z})$ $t_{k+p} - t_k = \tau$ and $s_{k+p} = s_k$ for all $k \in \mathbb{Z}$. It is well known [170] that these conditions are necessary and sufficient for τ–periodicity of the distribution $\sum_{k=-\infty}^{+\infty} s_k \delta_{t_k}$.

Consider a nonlinear τ–periodic equation with impulse

$$\dot{u} = f(t, u) + \sum_{k=-\infty}^{+\infty} s_k \delta_{t_k} = \mathfrak{F}. \tag{6.9}$$

It is known (see, for example, [170]) that, under the conditions above, the equation (6.9) has a unique generalized solution (which is piecewise continuous) passing through the point u when $t = 0$ for every $u \in E^n$. This solution we denote by $\varphi(\cdot, u, \mathfrak{F})$. Thus, $\lim_{t \downarrow 0} \varphi(t, u, \mathfrak{F}) = u$. In addition, on every segment $]t_k, t_{k+1}[$ the equation (6.9) coincides with (6.7). Therefore, the equality

$$\varphi(t, u, \mathfrak{F}) = \varphi(t - t_k, c_k, f_{t_k}) \tag{6.10}$$

holds, where the sequence $\{c_k\} \subset E^n$ satisfies the difference equation

$$c_{k+1} = \varphi(t_{k+1} - t_k, c_k, f_{t_k}) + s_k. \tag{6.11}$$

The equation with impulse (6.9) is said to be dissipative if there exists $r > 0$ such that

$$\limsup_{t \to +\infty} |\varphi(t, v, \widetilde{\mathfrak{F}})| < r \tag{6.12}$$

for all $v \in E^n$ and $\widetilde{\mathfrak{F}} \in H(\mathfrak{F}) := \{\mathfrak{F}_s : s \in [0, \tau[\}$, where $\mathfrak{F}_s$ is an s-translation of the distribution $\mathfrak{F}$.

Remark 6.1 *For periodic equations the condition of dissipativity (6.12) can be simplified because (6.12) is equivalent to the following relation:*

$$\limsup_{t \to +\infty} |\varphi(t, u, \mathfrak{F})| < r. \tag{6.13}$$

Proof. In fact, (6.12) implies (6.13). Conversely. From the equality $\varphi(t + s, u, \mathfrak{F}) = \varphi(t, \varphi(s, u, \mathfrak{F}), \mathfrak{F}_s)$ (if in the point $s \in \mathbb{R}$ the function $\varphi(t, u, \mathfrak{F})$ is discontinuous, then $\varphi(s, u, \mathfrak{F}) := \varphi(s + 0, u, \mathfrak{F})$) for all t, $\tau \in \mathbb{R}_+$. Therefore, we have

$$\varphi(t, v, \widetilde{\mathfrak{F}}) = \varphi(t - \tau + s, \varphi(\tau - s, v, \mathfrak{F}_k), \mathfrak{F})$$

for all $t \geq \tau - s$, where $s \in [0, \tau[$ and $\widetilde{\mathfrak{F}} = \mathfrak{F}_k$. $\square$

Lemma 6.1 *The equation with impulse (6.9) is dissipative if and only if the difference equation (6.11) is dissipative.*

Proof. Let (6.9) be dissipative. Then there exists $r > 0$ such that the condition (6.13) is fulfilled. If $u \in E^n$, then

$$\varphi(k, u, \Phi) = \varphi(t_k, u, \mathfrak{F})$$

and, hence,

$$\limsup_{k \to +\infty} |\varphi(k, u, \Phi)| < r. \tag{6.14}$$

Conversely. Let the equation (6.11) be dissipative. Then there is $r > 0$ such that the condition (6.14) holds. Thus, for all $u \in E^n$ there exists $k_0(u) \in \mathbb{Z}_+$ such that $|\varphi(k, u, \Phi)| < r$ for all $k \geq k_0(u)$. Let $t \geq t_{k_0}$ and $t \in [t_k, t_{k+1}[$ $k \geq k_0)$. By (6.10), $\varphi(t, u, \mathfrak{F}) = \varphi(t - t_k, c_k, f_{t_k})$, where $c_k = \varphi(t_k, u, \mathfrak{F})$. We set

$$R_0 := \max\{|\varphi(s, u, g)| \ : \ s \in [0, \tau], |u| \leq r, \ g \in H(f)\}$$

and note that $|\varphi(t, u, \mathfrak{F})| = |\varphi(t - t_k, c_k, f_{t_k})| \le R_0$ for all $t \ge t_{k_0}$ and, consequently,

$$\limsup_{t \to +\infty} |\varphi(t, u, \mathfrak{F})| \le R_0. \tag{6.16}$$

Moreover, the number R_0 does not depend on $u \in E^n$ in the inequality (6.16). The lemma is proved. $\qquad\square$

Define a mapping $P : E^n \to E^n$ by the equality

$$P(u) := \varphi(\tau, u, \mathfrak{F}) = \varphi(t_p, u, \mathfrak{F}) = \varphi(p, u, \Phi). \tag{6.17}$$

From the equality (6.17) and general properties of solutions of difference equations follows the continuity of the mapping P.

Theorem 6.6   *The equation with impulse (6.9) is dissipative if and only if the cascade (E^n, P), where P is defined by the equality (6.17), is dissipative.*

Proof. This affirmation directly follows from Lemma 6.1 and Theorem 6.1. $\qquad\square$

Corollary 6.2   *If the equation (6.9) is dissipative, then there exists a nonempty compact connected set $I \subset E^n$ possessing the following properties:*

(1) $P(I) = I$, where P is a mapping of E^n onto itself, defined by the formula (6.17);
(2) for an arbitrary $\varepsilon > 0$ there is $\delta(\varepsilon) > 0$ such that $\rho(u, I) < \delta$ implies $\rho(P^k(u), I) < \varepsilon$ for all $k \in \mathbb{Z}_+$;
(3) the equality $\lim\limits_{k \to +\infty} \rho(P^k(u), I) = 0$ holds for all $u \in E^n$.

Proof. Corollary 6.2 follows from Theorems 6.6 and 4.9. $\qquad\square$

Corollary 6.3   *If the equation (6.9) is dissipative, then:*

(1) there is $r > 0$ and for all $a > 0$ there is $k(a) \in \mathbb{Z}_+$ such that $P^k B[0, a] \subseteq B[0, r]$ for all $k \ge k(a)$;
(2) there is $u_0 \in I$ such that $P(u_0) = u_0$ and, consequently, $\varphi(t, u_0, \mathfrak{F})$ is a τ-periodic solution of (6.9).

Consider a quasi-linear equation with impulse

$$\dot{u} = A(t)u + F(t, u) + \sum_{k=-\infty}^{+\infty} s_k \delta_{t_k}. \tag{6.18}$$

The following statement holds.

Theorem 6.7   *Let A and F be τ-periodic in $t \in \mathbb{R}$. The equation (6.18) is dissipative, if the following conditions are fulfilled:*

(1) there are positive numbers N and ν such that

$$\|U(t, A)U^{-1}(s, A)\| \le N e^{-\nu(t-s)},$$

where $U(t, A)$ is a Cauchy matrix for the equation

$$\dot{u} = A(t)u.$$

(2) there exists a sufficiently small positive number ε_0 such that $|F(t, u)| \le C + D|u|$ ($u \in E^n$, $C \ge 0$ and $0 \le D \le \varepsilon_0$).

Proof. Let $f(t, u) := A(t)u + F(t, u)$. According to Lemma 6.1, to prove Theorem 6.7 it is sufficient to show that the difference equation (6.11) is dissipative. Let us rewrite the equation (6.11) as follows:

$$c_{k+1} = \widetilde{A}(k)c_k + \widetilde{F}(k, c_k), \tag{6.19}$$

where $\widetilde{A}(k) := U(t_{k+1} - t_k, A_{t_k})$ and

$$\widetilde{F}(k, u) := \varphi(t_{k+1} - t_k, u, f_{t_k}) + s_k - \widetilde{A}(k)u.$$

Note, that under the conditions of Theorem 6.7 $\widetilde{A}(k)$ and $\widetilde{F}(k, u)$ are p–periodic in $k \in \mathbb{Z}$.

We will show that the trivial solution of the homogeneous equation

$$c_{k+1} = \widetilde{A}(k)c_k \tag{6.20}$$

is uniformly asymptotically stable. In virtue of the periodicity of $\widetilde{A}(k)$ it is sufficient to show that it is asymptotically stable. Note, that

$$U(k, \widetilde{A}) = \prod_{m=0}^{k-1} U(t_{m+1} - t_m, A_{t_m}) = U(t_k - t_{p-1}, A_{t_{p-1}}),$$

therefore, $\|U(k, \widetilde{A})\| \le N e^{-\nu(t_k - t_{p-1})}$. From this fact follows the asymptotic stability of the trivial solution of (6.20).

We will show now that $|\widetilde{F}(k, u)| \le \widetilde{C} + \widetilde{D}(D)|u|$ for all $u \in E^n$, where $\widetilde{C}$ and $\widetilde{D}(D)$ are some positive constants, moreover, $\widetilde{D}(D) \to 0$ as $D \to 0$. Indeed, since $\varphi(t, u, f_l)$ is a solution of the equation

$$\dot{u} = f_l(t, u) = A_l(t)u + F_l(t, u),$$

then $\varphi(t, u, f_l) = U(t, A_l)\left(u + \int_o^t U^{-1}(s, A_l)F_l(, \varphi(s, u, f_l))ds\right)$ and, consequently,

$$\widetilde{F}(k, u) = \int_0^{t_{k+1} - t_k} U(t_{k+1} - t_k, A_{t_k})U^{-1}(s, A_{t_k})F_{t_k}(s, \varphi(s, u, f_{t_k}))ds + s_k. \tag{6.21}$$

Using the same arguments as in Theorem 3.21, we can show that

$$|\varphi(t, u, f_l)| \leq N|u|e^{(-\nu+ND)t} + \frac{CN}{\nu - DN} \tag{6.22}$$

for all $t \geq 0$, $l \in \mathbb{R}$ and $u \in E^n$ $(0 \leq D \leq \nu N^{-1})$. From (6.21) and (6.22) we have

$$|\widetilde{F}(k, u)| \leq \int_0^{t_{k+1}-t_k} N e^{-\nu(t_{k+1}-t_k-s)}(C + D|\varphi(s, u, f_l)|)ds \leq$$

$$N e^{-\nu(t_{k+1}-t_k)} \int_0^{t_{k+1}-t_k} \left(Ce^{\nu s} + De^{\nu s}\left[\frac{CN}{\nu - DN} + N|u|e^{-\nu+ND)s}\right]\right)ds$$

$$= N e^{-\nu(t_{k+1}-t_k)} \left(\left[C + D\frac{CN}{\nu - DN}\right]\frac{e^{\nu(t_{k+1}-t_k)} - 1}{\nu} + \right.$$

$$\left. +N|u|D\frac{e^{ND(t_{k+1}-t_k)} - 1}{ND}\right) = \frac{NC}{\nu - DN}\left(1 - e^{-\nu(t_{k+1}-t_k)}\right) + $$

$$N|u|\left(e^{(ND-\nu)(t_{k+1}-t_k)} - e^{-\nu(t_{k+1}-t_k)}\right) \leq \widetilde{C} + \widetilde{D}(D)|u|,$$

where $\widetilde{C} := NC(\nu - ND)^{-1}$ and

$$\widetilde{D}(D) := N \sup_{0 \leq k \leq p-1}\left[e^{(ND-\nu)(t_{k+1}-t_k)} - e^{-\nu(t_{k+1}-t_k)}\right]. \tag{6.23}$$

From (6.23) follows that $\widetilde{D}(D) \to 0$ as $D \to 0$. Therefore, there is $\varepsilon_0 > 0$, such that $\widetilde{D}(\varepsilon_0) < M^{-1}(1 - q)$ (see Theorem 6.5). According to Theorem 6.5, the difference equation (6.19) is dissipative and, hence, the equation with impulse (6.18) is dissipative too. The theorem is proved. $\qquad\square$

6.3   Convergent periodic equations with impulse

Definition 6.2   Following [137], [270], the equation with impulse (6.9) we will call convergent, if (6.9) has a unique τ–periodic solution $\varphi(t, u, \mathfrak{F})$ which is globally uniformly asymptotically stable.

By analogy, it is possible to define the notion of convergence for the difference equation (6.1). From this definition follows that convergent equations form that part of dissipative equations which has the Levinson's center consisting from a single periodic trajectory (solution) of (6.9).

Lemma 6.2 *The τ-periodic equation with impulse (6.9) is convergent if and only if (6.11) is convergent.*

Proof. Let (6.9) be convergent. Then there is a unique τ-periodic solution $\varphi(t, u_0, \mathfrak{F})$ of (6.9) which is globally uniformly asymptotically stable. Directly from the corresponding definitions follows that $\varphi(t, u_0, \Phi)$ ($\Phi(k, u) := \varphi(t_{k+1} - t_k, u, f_{t_k}) + s_k$) is a unique p-periodic solution of (6.11) and is globally uniformly asymptotically stable.

Conversely. Let (6.11) be convergent and $\varphi(k, u_0, \Phi)$ be its unique p-periodic globally uniformly asymptotically stable solution. Obviously, $\varphi(t, u_0, \mathfrak{F})$ is a unique τ-periodic solution of (6.9). Let $k \in \mathbb{Z}$ and $t \in [t_k, t_{k+1}[$. According to (6.10),

$$|\varphi(t, u, \mathfrak{F}) - \varphi(t, u_0, \mathfrak{F})| = |\varphi(t - t_k, c_k, f_{\tau_k}) - \varphi(t - t_k, c_k^0, f_{t_k})|, \tag{6.24}$$

where $c_k := \varphi(t_k, u, \mathfrak{F})$ and $c_k^0 := \varphi(t_k, u_0, \mathfrak{F})$. The mapping $\varphi : [0, \tau] \times K \times H(f) \to E^n$ ($\varphi : (s, u, g) \to \varphi(s, u, g)$) is uniformly continuous for every compact $K \subset E^n$ and, hence, for all $\varepsilon > 0$ there is $\delta(\varepsilon) > 0$ such that

$$|\varphi(s, u_1, g) - \varphi(s, u_2, g)| < \varepsilon, \tag{6.25}$$

for all $s \in [0, \tau]$, $|u_1 - u_2| < \delta$, $u_1, u_2 \in K$ and $g \in H(f)$.

From (6.24) and (6.25) and the global uniform asymptotic stability of $\varphi(k, u_0, \Phi) = c_k^0$ follows an analogous property for $\varphi(t, u_0, \mathfrak{F})$. The lemma is proved.

Theorem 6.8 *Let $A(k)$ and $F(k, u)$ be p–periodic in $k \in \mathbb{Z}$ and the following conditions be fulfilled:*

(1) there exist positive numbers M and $q < 1$ such that the inequality (6.3) holds;
(2) $|F(k, u_1) - F(k, u_2)| \le L|u_1 - u_2|$ $(0 \le L < M^{-1}(1 - q))$ for all $k \in \mathbb{Z}$ and $u_1, u_2 \in E^n$.

Then the equation (6.2) is convergent.

Proof. Let $\varphi(k, u, \Phi)$ ($\Phi(k, u) := A(k)u + F(k, u)$) be a solution of (6.2) passing through the point u for $k = 0$. In virtue of the formula of variation of constants we have

$$\varphi(k, u, \Phi) = U(k, A)\left(u + \sum_{m=0}^{k-1} U^{-1}(m + 1, A)F(m, \varphi(m, u, \Phi))\right).$$

Consequently,

$$\varphi(k, u_1, \Phi) - \varphi(k, u_2, \Phi) = U(k, A)\Big(u_1 - u_2+$$

$$\sum_{m=1}^{k-1} U^{-1}(m+1, A)[F(m, \varphi(m, u, \Phi)) - F(m, \varphi(m, u_2, \Phi))]\Big).$$

Thus,

$$|\varphi(k, u_1, \Phi) - \varphi(k, u_2, \Phi)| \le Mq^k\Big(|u_1 - u_2|$$

$$+Lq^{-1}\sum_{m=0}^{k-1} q^{-m}|\varphi(k, u_1, \Phi) - \varphi(k, u_2, \Phi)|\Big). \tag{6.26}$$

Let $u(k) := |\varphi(k, u_1, \Phi) - \varphi(k, u_2, \Phi)|q^{-k}$. From (6.26) follows that

$$u(k) \le M\Big(|u_1 - u_2| + Lq^{-1}\sum_{m=0}^{k-1} u(m)\Big). \tag{6.27}$$

Denote by $v(k)$ the right hand side of (6.27). The following inequality

$$v(k+1) - v(k) = LMq^{-1}u(k) \le LMq^{-1}v(k). \tag{6.28}$$

holds. From (6.28) we obtain

$$v(k) \le (1 + LMq^{-1})^{k-1}v(1)$$

and, hence,

$$u(k) \le (1 + LMq^{-1})M|u_1 - u_2|, \tag{6.29}$$

since $v(1) = M|u_1 - u_2|$. From (6.29) we have

$$|\varphi(k, u_1, \Phi) - \Phi(k, u_2, \Phi)| \le (q + LM)^{k-1}qM|u_1 - u_2| \tag{6.30}$$

for all $u_1, u_2 \in E^n$ and $k \in \mathbb{Z}$.

Let us note that under the conditions of Theorem 6.8 we can apply Theorem 6.5 and, consequently, the equation (6.2) admits at least one p-periodic solution $\varphi(k, u_0, \Phi)$ (see also Corollary 6.1). From (6.30) follows the global asymptotic stability of the solution $\varphi(k, u_0, \Phi)$ and, according to the periodicity of (6.2), the solution $\varphi(k, u_0, \Phi)$ will be globally uniformly asymptotically stable. The theorem is proved. $\qquad\square$

Theorem 6.9   *Let $A(t)$ and $F(t, u)$ be τ-periodic in $t \in \mathbb{R}$. The equation (6.18) is convergent, if the following conditions are fulfilled:*

(1) there are positive numbers N and ν such that

$$\|U(t, A)U^{-1}(s, A)\| \leq Ne^{-\nu(t-s)} \quad (t \geq s);$$

(2) there is a sufficiently small positive number ε_0 such that

$$|F(t, u_1) - F(t, u_2)| \leq L|u_1 - u_2|$$

($u_1, u_2 \in E^n$, $t \in \mathbb{R}$ and $0 \leq L \leq \varepsilon_0$).

Proof. According to Lemma 6.2 to prove the convergence of (6.18) it is sufficient to establish that under the conditions of Theorem 6.9 the difference equation (6.11) possesses the same property. As well as in the proof of Theorem 6.7, we will present the equation (6.11) in the form (6.19). Under the conditions of Theorem 6.9 the trivial solution of (6.20) is uniformly asymptotically stable. Further, (6.21) implies that

$$|\widetilde{F}(k, u_1) - \widetilde{F}(k, u_2)| = |\int_0^{t_{k+1}-t_k} U(t_{k+1} - t_k, A_{t_k})U^{-1}(s, A_{t_k}) \times$$

$$[F_{t_k}(s, \varphi(s, u_1, f_{t_k})) - F_{t_k}(s, \varphi(s, u_2, f_{t_k}))]ds \leq \qquad (6.31)$$

$$\int_0^{t_{k+1}-t_k} Ne^{-\nu(t_{k+1}-t_k-s)}L|\varphi(s, u_1, f_{t_k}) - \varphi(s, u_2, f_{t_k})|ds.$$

Reasoning in the same way that in Theorem 5.23 we obtain

$$|\varphi(t, u_1, f_{t_k}) - \varphi(t, u_2, f_{t_k})| \leq N|u_1 - u_2|e^{-(\nu-LN)t} \qquad (6.32)$$

for all $u_1, u_2 \in E^n$ and $t \in \mathbb{R}_+$, where $f_{t_k}(t, u) := A_{t_k}(t)u + F_{t_k}(t, u)$. From (6.31) and (6.32) we have

$$|\widetilde{F}(k, u_1) - \widetilde{F}(k, u_2)| \leq Ne^{-\nu(t_{k+1}-t_k)}L \times$$

$$\int_0^{t_{k+1}-t_k} e^{LN^2 s}N|u_1 - u_2|ds \leq \widetilde{L}(L)|u_1 - u_2|,$$

where

$$\widetilde{L}(L) := \sup_{0 \leq k \leq p-1}\left[e^{(-\nu+LN^2)(t_{k+1}-t_k)} - e^{-\nu(t_{k+1}-t_k)}\right]. \qquad (6.33)$$

From (6.33) follows that $\widetilde{L}(L) \to 0$ for $L \to 0$. Therefore, there exists $\varepsilon_0 > 0$ such that $\widetilde{L}(L) < M^{-1}(1 - q)$ for $0 \leq L \leq \varepsilon_0$. By Theorem 6.8, the equation (6.20) is convergent and to finish the proof of Theorem 6.9 it is sufficient to refer to Lemma 6.2. $\qquad \square$

Remark 6.2   *a. A statement similar to Theorem 6.9 has also been proved in the work* [256].

b. The problem of the convergence and dissipativity of equations with impulse are studied in the works [256], [115], [116], [256].

6.4 Asymptotic stability of linear functional differential equations

Using some ideas and methods developed while studying dissipative dynamical systems we may receive a series of conditions which are equivalent to the asymptotic stability of linear non-autonomous dynamical systems with infinite dimensional phase space. As an application of these result we may receive the corresponding results for linear functional differential equations.

Let (X, h, Y) be a vectorial fibering with the fiber E (E is a Banach space) and $|\cdot| : X \to \mathbb{R}$ be a norm on X, compatible with the distance of X, i.e. $|\cdot|$ is continuous and $|x| = \rho(x, \theta_y)$, where $x \in X_y$, θ_y is the null element of the space X_y and ρ is a distance on X.

Recall that a dynamical system $(X, \mathbb{S}_+, \pi)$ is called locally compact (locally completely continuous) if for any $x \in X$ there are $\delta_x > 0$ and $l_x > 0$ such that $\pi^t B(x, \delta_x)$ $(t \geq l_x)$ is relatively compact.

Theorem 6.10   *Let $(X, \mathbb{S}_+, \pi)$ be locally compact and Y be compact. Then the following conditions are equivalent:*

1. *$\lim\limits_{t \to +\infty} |xt| = 0$ for all $x \in X$;*
2. *all the motions in $(X, \mathbb{S}_+, \pi)$ are relatively compact and $(X, \mathbb{S}_+, \pi)$ does not admit nontrivial compact motions defined on $\mathbb{S}$;*
3. *there are positive numbers N and ν such that $|xt| \leq Ne^{-\nu t}|x|$ for all $x \in X$ and $t \in \mathbb{S}_+$.*

Proof.   From the equality $\lim\limits_{t \to +\infty} |xt| = 0$ follows that Σ_x^+ is relatively compact and $\omega_x \subseteq \Theta := \{\theta_y : y \in J_Y, \text{ where } \theta_y \text{ is the null element of } X_y\}$. Thus, the dynamical system $(X, \mathbb{S}_+, \pi)$ is pointwise dissipative and, according to Theorem 1.10, it is compactly dissipative. Denote by J_X the Levinson's center of the dynamical system $(X, \mathbb{S}_+, \pi)$. We will show that $J_X = \Theta$. Obviously, the set Θ is compact and invariant ($\pi^t \Theta = \Theta$ for all $t \in \mathbb{S}_+$) and, consequently, $\Theta \subseteq J_X$. From the last inclusion follows that $h(J_X) = J_Y$. Now let us show that $J_X = \Theta$. If we suppose that it is not true, then $J_X \setminus \Theta \neq \emptyset$ and, hence, there is $x_0 \in J_X \setminus \Theta$. Since in the set J_X all motions are continuable on $\mathbb{S}$ (see Theorem 1.6), then there exists a continuous mapping $\varphi : \mathbb{S} \to J_X$ such that: $\varphi(0) = x_0$ and $\pi^t \varphi(s) = \varphi(t + s)$ for all $s \in \mathbb{S}$ and $t \in \mathbb{S}_+$. On the other hand, by the linearity of the system $\langle (X, \mathbb{S}_+, \pi), (Y, \mathbb{S}_+, \sigma), h \rangle$ along with the point x_0 all the points λx_0 ($\lambda \in \mathbb{R}$) belong to the set J_X, because J_X is a maximal compact invariant set in X. But the inclusion $\lambda x_0 \in J$ holds for all $\lambda \in \mathbb{R}$ if and only if $x_0 \in \Theta$. The obtained contradiction shows that $J_X = \Theta$.

We will show that in $(X, \mathbb{S}_+, \pi)$ there is no nontrivial compact motions defined on $\mathbb{S}$. In fact, let $x \in X$ be such point that there exists $\varphi : \mathbb{S} \to X$ possessing the following properties: $\varphi(\mathbb{S})$ is relatively compact, $\pi^t \varphi(s) = \varphi(t + s)$ $(t \in \mathbb{S}_+, \ s \in \mathbb{S})$ and $\varphi(0) = x_0$. Since J_X is a maximal compact invariant set, then $\overline{\varphi(\mathbb{S})} \subseteq J_X$ and, in particular, $x_0 \in J_X \subseteq \Theta$. Therefore, $|x_0| = 0$. Thus, we proved that 1. implies 2. The converse implication is evident.

We will show that 1. implies 3. Indeed, since under the conditions of Theorem 6.10 from 1. follows that $(X, \mathbb{S}_+, \pi)$ is pointwise dissipative, then, according to Theorem 1.10, it is locally dissipative. As Y is compact and $(X, \mathbb{S}_+, \pi)$ is locally dissipative, then there is $\delta > 0$ such that

$$\lim_{t \to +\infty} \sup\{|xt| : |x| < \delta\} = 0. \tag{6.34}$$

Taking into consideration (6.34), by standard reasoning (see, for example, [208], [275], and also Theorem 2.38) we may prove that there are N, $\nu > 0$ such that $|xt| \le Ne^{-\nu t}|x|$ for all $x \in X$ and $t \ge 0$. Finally, 3. evidently implies 1. The theorem is proved. $\qquad\square$

Remark 6.3 *a. The condition of local compactness in Theorem 6.10 is essential. To confirm this statement it is sufficient to consider the example 1.8. In this example all motions tend to zero as $t \to +\infty$, but this system is not exponentially stable. It is not difficult to see that the dynamical system from the example 1.8 is not locally compact.*

b. Theorem 6.10 is also true if the space Y is just pseudo-metric.

A very important class of linear non-autonomous dynamical systems with infinite dimensional phase space satisfying the condition of local compactness is the class of linear non-autonomous functional differential equations [175]. Let us recall some notions and denotations from [175]. Let $r > 0$, $C([a, b], \mathbb{R}^n)$ be a Banach space of all continuous functions $\varphi : [a, b] \to \mathbb{R}^n$ equipped with the norm sup. If $[a, b] = [-r, 0]$, then we set $\mathcal{C} := C([-r, 0], \mathbb{R}^n)$. Let $\sigma \in \mathbb{R}$, $A \ge 0$ and $u \in C([\sigma - r, \sigma + A], \mathbb{R}^n)$. We will define $u_t \in \mathcal{C}$ for all $t \in [\sigma, \sigma + A]$ by the equality $u_t(\theta) := u(t + \theta)$, $-r \le \theta \le 0$. Denote by $D = D(\mathcal{C}, \mathbb{R}^n)$ a Banach space of all linear continuous operators acting from $\mathcal{C}$ into $\mathbb{R}^n$ and endowed with operational norm. Consider a linear equation

$$\dot{u} = A(t, u_t), \tag{6.35}$$

where $A : \mathbb{R} \times \mathcal{C} \to \mathbb{R}^n$ is continuous and linear with respect to the second variable, i.e. $A(t, \cdot) \in D$ for all $t \in \mathbb{R}$. Let us set $H(A) := \overline{\{A_s : s \in \mathbb{R}\}}$, where $A_s(t, \cdot) = A(t + s, \cdot)$ and by bar we denote a closure in the compact-open topology on $\mathbb{R} \times \mathcal{C}$.

Along with the equation (6.35) let us consider the family of equations

$$\dot{v} = B(t, v_t), \tag{6.36}$$

where $B \in H(A)$. Let $\varphi(\cdot, \phi, B)$ be a solution of (6.36) passing through the point $\phi \in C$ for $t = 0$ defined for all $t \geq 0$.

Let $Y := H(A)$ and denote by $(Y, \mathbb{R}, \sigma)$ a dynamical system of translations on $H(A)$. Let $X := C \times Y$, $(X, \mathbb{R}_+, \pi)$ be a dynamical system on X defined in the following way: $\pi((\phi, B), \tau) := (\varphi(\tau, \phi, B), B_\tau)$. It is easy to see that the system $\langle (X, \mathbb{R}_+, \sigma), (Y, \mathbb{R}, \sigma), h \rangle$ is linear. We remark a very important property of this system.

The following statement holds.

Lemma 6.3   *Let $H(A)$ be compact. Then for any point $x \in X := C \times H(A)$ there exist a neighborhood U_x of the point x and a positive number $l_x > 0$ such that $\pi^t U_x$ is relatively compact for all $t \geq l_x$, i.e. the dynamical system $(X, \mathbb{R}_+, \pi)$ is locally compact.*

Proof. This assertion follows from Lemmas 2.2.3 and 3.6.1 from [175] and from the compactness of Y. $\square$

Applying to the constructed non-autonomous dynamical system Theorem 6.10 (see also Remark 6.3) and taking into consideration Lemma 6.3, we will obtain the following statement.

Theorem 6.11   *Let $H(A)$ be compact. Then the following statements are equivalent:*

(1) the trivial solution of (6.35) is uniformly exponentially stable, i.e. there are positive numbers N and ν such that $|\varphi(t, \phi, B)| \leq N e^{-\nu t} |\phi|$ for all $B \in H(A)$ and $t \geq 0$;

(2) the trivial solution of (6.36) is uniformly exponentially stable for all $B \in H(A)$.

6.5   Convergence of monotone evolutionary equations

Let H be a Hilbert space. Denote by $L^p_{loc}(\mathbb{R}, H)$ the space of all functions $\varphi : \mathbb{R} \to H$ that are locally integrable with the power p, that is

$$\int_{|t| \leq k} |\varphi(t)|^p dt < +\infty \qquad (p \geq 1, \ k = 0, 1, 2 \ldots).$$

Let us define a topology on $L_{loc}^p(\mathbb{R}, H)$ by the family of semi-norms $|\cdot|_k$:

$$|\varphi|_k := \left(\int\limits_{|t| \le k} |\varphi(t)|^p dt \right)^{\frac{1}{p}} \qquad (k = 0, 1, 2, \ldots)$$

and by $(L_{loc}^p(\mathbb{R}, H), \mathbb{R}, \sigma)$ we denote the dynamical system of translations on $L_{loc}^p(\mathbb{R}, H)$ (see [103]).

Definition 6.3 Recall [30] that an operator $A : D(A) \to H$ $(D(A) \subseteq H)$ is called:

a. monotone, if $Re\langle Au_1 - Au_2, u_1 - u_2 \rangle \ge 0$ for all $u_1, u_2 \in D(A)$;
b. semi-continuous, if the function $\varphi : \mathbb{R} \to \mathbb{R}$ defined by the equality $\varphi(\lambda) := \langle A(u + \lambda v), w \rangle$ is continuous;
c. uniformly monotone, if there exists $\alpha > 0$ such that

$$Re\langle Au - Av, u - v \rangle \ge \alpha |u - v|^2 \tag{6.37}$$

for all $u, v \in D(A)$.

Note, that the family of all monotone operators can be partially ordered with respect to the inclusion of its graphs.

Definition 6.4 A monotone operator is called maximal, if it is maximal among monotone operators.

Consider a differential equation

$$x' + Ax = f(t), \tag{6.38}$$

where $f \in L_{loc}^1(\mathbb{R}, H)$ and A is a maximal monotone operator with the domain $D(A)$. It is known [30] that for all $x_0 \in \overline{D(A)}$ there exists a unique weak solution $\varphi(t, x_0, f)$ of the equation (6.38) satisfying the condition $\varphi(0, x_0, f) = x_0$ and defined on $\mathbb{R}_+$. Let $Y := H^+(f)$ and $(Y, \mathbb{R}_+, \sigma)$ be a dynamical system of translations on Y. We denote by $X := \overline{D(A)} \times Y$ and by $(X, \mathbb{R}_+, \pi)$ a dynamical system on X, where $\pi(t, \langle v, g \rangle) := \langle \varphi(t, v, g), g_t \rangle$. Then the triple $\langle (X, \mathbb{R}_+, \pi), (Y, \mathbb{R}_+, \sigma), h \rangle (h = pr_2 : X \to Y)$ is the non-autonomous dynamical system [187] generated by the equation (6.38). Applying the results obtained in Chapters 1-5 to the constructed in such way non-autonomous dynamical system, we will obtain the following results for the equation (6.38).

Theorem 6.12 *Let $f \in L_{loc}^1(\mathbb{R}, H)$ and $H(f)$ be compact. If A is a maximal monotone operator which is semi-continuous and uniformly monotone, then the equation (6.38) is convergent, i.e. the equation (6.38) admits a unique compact on*

$\mathbb{R}$ *uniformly compatible solution which is globally asymptotically stable. In addition, there exist positive numbers N and ν such that the equation*

$$x' + Ax = g(t) \tag{6.39}$$

holds for all $g \in H(f)$ and admits a unique compact on $\mathbb{R}$ uniformly compatible solution $\varphi(t, v_g, g)$ $(g \in H(f))$ such that

$$|\varphi(t, v, g) - \varphi(t, v_g, g)| \le N e^{\nu t} |v - v_g| \tag{6.40}$$

for all $v \in \overline{D(A)}$ and $t \ge 0$.

Proof. Modifying the results from [187], we obtain that under the conditions of Theorem 6.12 all the solutions of the equation (6.38) are compact on $\mathbb{R}_+$. On the other hand, by the condition (6.37) we have

$$|\varphi(t, x_1, f) - \varphi(t, x_2, f)| \le e^{-\alpha t} |x_1 - x_2| \tag{6.41}$$

for all $t \ge 0$ and $x_1, x_2 \in H$. And, moreover, if $g \in H(f)$, then for solutions of the equation (6.39) the following estimation

$$|\varphi(t, y_1, g) - \varphi(t, y_2, g)| \le e^{-\alpha t} |y_1 - y_2| \tag{6.42}$$

takes place for all $t \ge 0$ and $y_1, y_2 \in H$. The relation (6.42) implies the condition 4. of Theorem 2.15, if we take as $V : X \times X \to \mathbb{R}_+$ the following function $V((v_1, g), (v_2, g)) := |v_1 - v_2|$ $(g \in H(f),\ v_1, v_2 \in \overline{D(A)})$. The theorem is proved$\square$

Corollary 6.4 *Let $f \in L^1_{loc}(\mathbb{R}, H)$ be stationary (τ−periodic, almost periodic, recurrent). Then the equation (6.38) admits a unique stationary (τ−periodic, almost periodic, recurrent) solution wich is globally asymptotically stable.*

Finally, we will give an example of equations of the type (6.38).

Example 6.1   Consider the equation

$$\frac{\partial^2 u}{\partial t^2} = \Delta u - \phi\left(\frac{\partial u}{\partial t}\right) + f(t) \tag{6.43}$$

in the open set $\mathcal{D} \subset \mathbb{R}^n$ with the boundary condition $u|_{\partial \mathcal{D}} = 0$ on the boundary $\partial \mathcal{D}$ of $\mathcal{D}$. Suppose that the function $\phi : \mathbb{R} \to \mathbb{R}$ satisfies the conditions : $\phi(0) = 0$ and $0 < c_1 \le \phi'(\xi) \le c_2$ $(\xi \in \mathbb{R})$. Then the equation (6.43) may be rewritten in the following way:

$$\begin{cases} \frac{\partial u}{\partial t} = v \\ \frac{\partial v}{\partial t} = \Delta u - \phi(v) + f(t) \end{cases} \tag{6.44}$$

We denote by $H := W_0^{1,2}(\Omega) \times L^2(\Omega)$ and define on H a scalar product

$$\langle (u,v), (u^*, v^*) \rangle = \int_\Omega [vv^* + \Delta u \Delta u^* + \lambda u v^* + \lambda u^* v] dx,$$

where λ is a certain positive constant depending only on c_1 and c_2. It is possible to verify (see, for example, [131]) that Theorem 6.12 can be applied to the system (6.44), if the set $H(f) \subset L^1_{loc}(\mathbb{R}, H)$ is compact.

Let $I \subseteq \mathbb{R}$, H be a Hilbert space with scalar product $\langle , \rangle$, $\mathbb{D}(I, \mathbb{R})$ be a space of all infinitely differentiable functions $\varphi : I \to H$ with compact support and $[H]$ be an algebra of all linear bounded operators on H.

Consider the equation

$$\int_\mathbb{R} [\langle u(t), \varphi'(t) \rangle + \langle A(t)u(t), \varphi(t) \rangle + \langle f(t), \varphi(t) \rangle] dt = 0, \qquad (6.45)$$

where $A \in C(\mathbb{R}, [H])$ and $f \in C(\mathbb{R}, H)$. A function $u \in C(I, H)$ is called a solution of the equation (6.45), if the equality (6.45) takes place for all $\varphi \in \mathbb{D}(I, H)$.

Let $x \in H$, $\varphi(t, x, A, f)$ be a solution of the equation (6.45) defined on $\mathbb{R}_+$ and satisfying the condition $\varphi(0, x, A, f) = x$ and

$$\int_\mathbb{R} [\langle u(t), \varphi'(t) \rangle + \langle B(t)u(t), \varphi(t) \rangle + \langle g(t), \varphi(t) \rangle] dt = 0 \qquad (6.46)$$

be a family of equations, where $(B, g) \in H(A, f) := \overline{\{(A_\tau, f_\tau) \mid \tau \in \mathbb{R}\}}$. We will suppose that the operator-function $A \in C(\mathbb{R}, [H])$ is self-adjoint and negatively defined, i.e. $A(t) = -A_1(t) + iA_2(t)$ for all $t \in \mathbb{R}$, where $A_1(t)$ and $A_2(t)$ are self-adjoint and

$$\langle A_1(t)u, u \rangle \geq \alpha |u|^2 \qquad (6.47)$$

for all $t \in \mathbb{R}$ and $u \in H$, where $\alpha > 0$.

Lemma 6.4 [5] *We have*

$$\begin{aligned}
\tfrac{1}{2}\tfrac{d}{dt}|\varphi(t, x, A, f)|^2 = &-\langle A_1(t)\varphi(t, x, A, f), \varphi(t, x, A, f) \rangle \\
&+ Re\langle f(t), \varphi(t, x, A, f) \rangle
\end{aligned} \qquad (6.48)$$

for all $t > 0$.

Lemma 6.5 *The following inequality*

$$|\varphi(t, x, A, f)| \leq |x| + \int_0^t |f(\tau)| d\tau \qquad (6.49)$$

holds for all $t \geq 0$.

Proof. In virtue of the equality (6.48) we have

$$\frac{1}{2}\frac{d}{dt}|\varphi(t,x,A,f)|^2 \le |f(t)||\varphi(t,x,A,f)|.$$

Let $v(t) := |\varphi(t,x,A,f)|^2$. Then $\frac{dv}{dt} \le 2|f(t)|\sqrt{v(t)}$ and, consequently,

$$\sqrt{v(t)} - \sqrt{v(\tau)} \le \int_\tau^t |f(s)|ds.$$

From this follows the inequality (6.49). $\square$

Lemma 6.6   *Let l, r and $\beta > 0, x_0 \in H, A \in C(\mathbb{R},[H])$ and $f \in C(\mathbb{R},[H])$. Then there exists $M = M(f,l,r,\beta,x_0) >$ such that*

$$|\varphi(t,x,B,g) - \varphi(t,x_0,A,f)| \le |x - x_0| +$$

$$M\int_0^t \|B(\tau) - A(\tau)\|d\tau + \int_0^t |g(\tau) - f(\tau)|d\tau \qquad (6.50)$$

for all $t \in [0,l]$ and $x \in B[x_0,r] := \{x \mid x \in H, |x - x_0| \le r\}$, if $|g(t) - f(t)| \le \beta$ and $Re\langle B(t)x,x\rangle \le 0$ for any $t \in [0,l]$ and $x \in H$.

Proof. We denote by $v(t) = \varphi(t,x,B,g) - \varphi(t,x_0,A,f)$. Then

$$\int_\mathbb{R} [\langle v(t),\varphi'(t)\rangle + \langle A(t)v(t),\varphi(t)\rangle +$$

$$\langle B(t) - A(t))v(t),\varphi(t)\rangle + \langle g(t) - f(t),\varphi(t)\rangle]dt = 0$$

for any $\varphi \in \mathbb{D}(\mathbb{R},H)$. By virtue of Lemma 6.4

$$\frac{1}{2}\frac{d}{dt}|v(t)|^2 = Re\langle A(t)v(t),v(t)\rangle$$

$$+Re[\langle(B(t) - A(t))\varphi(t,x,B,f),v(t)\rangle + \langle g(t) - f(t),v(t)\rangle]$$

and, according to Lemma 6.5, we have

$$|v(t)| \le |v(0)| + \int_0^t |(B(\tau) - A(\tau))\varphi(\tau,x,B,g) + g(\tau) - f(\tau)|d\tau$$

$$\le |v(0)| + \int_0^t \|B(\tau) - A(\tau)\||\varphi(\tau,x,B,g)|d\tau + \int_0^t |g(\tau) - f(\tau)|d\tau. \qquad (6.51)$$

On the other hand, according to Lemma 6.5 for $\varphi(t,x,B,g)$ we have

$$|\varphi(t,x,B,g)| \le |x| + \int_0^t |g(\tau)|d\tau \le |x_0| + r + \beta l$$

$$+l \max_{0 \leq t \leq l} |f(t)| := M(f, l, r, \beta, x_0). \tag{6.52}$$

Taking into account the inequalities (6.51) and (6.52), we obtain (6.50). The lemma is proved. $\qquad\square$

Let $\bar{X} = H \times H^+(A, f)$ and denote by X the set of all $\langle u, (B, g) \rangle \in \bar{X}$ such that through the point $u \in H$ there passes a solution $\varphi(t, u, B, g)$ of the equation (6.46) defined on $\mathbb{R}_+$.

Lemma 6.7 *The set $X \subseteq H \times H(A, f)$ is closed in $H \times H(A, f)$.*

Proof. Let $\langle x, (B, g) \rangle \in \bar{X}$. Then there exists a sequence $\langle x_k, (B_k, g_k) \rangle \in X$ such that $x_k \to x$ in the space H, $B_k \to B$ in $C(\mathbb{R}, [H])$ and $g_k \to g$ in $C(\mathbb{R}, H)$. Let l and $\varepsilon > 0$ be such that

$$|x_k - x_m| < \varepsilon, \quad |g_k(t) - g_m(t)| < \varepsilon \quad \text{and} \quad \|B_k(t) - B_m(t)\| \langle \varepsilon \tag{6.53}$$

for all $t \in [0, l]$ and $k, l \geq k_0$. Denote by $r := \sup\{|x_k| \ : \ k \in \mathbb{N}\}$. Then according to Lemma 6.6

$$|\varphi(t, x_k, B_k, g_k) - \varphi(t, x_m, B_m, g_m)| \leq |x_k - x_m| + M \int_0^t \|B_k(\tau) - B_m(\tau)\| d\tau$$

$$+ \int_0^t |g_k(\tau) - g_m(\tau)| d\tau \leq \varepsilon + M\varepsilon l + \varepsilon l \tag{6.54}$$

for all $t \in [0, l]$ and $k, m \geq k_0$, where M is a positive constant which is not dependent on r, l and g. Taking into account that the space $C(\mathbb{R}_+, H)$ is complete and the inequality (6.54), we conclude that the sequence $\{\varphi(t, x_k, B_k, g_k)\}$ is convergent in $C(\mathbb{R}_+, H)$ and, according to the inequality (6.54), $\varphi(t, x, B, g) = \lim_{k \to +\infty} \varphi(t, x_k, B_k, g_k)$. The lemma is proved. $\qquad\square$

Lemma 6.8 *The mapping $\varphi : \mathbb{R}_+ \times X \to H(\varphi : (t, \langle u, B, g \rangle) \to \varphi(t, u, B, g))$ is continuous.*

Proof. Let $t_n \to t, x_k \to x, B_k \to B$ and $g_k \to g$. Then

$$|\varphi(t, x_k, B_k, g_k) - \varphi(t, x, B, g)| \leq |\varphi(t, x_k, B_k, g_k) - \varphi(t_k, x, B, g)|$$

$$+ |\varphi(t_k, x, B, g) - \varphi(t, x, B, g)| \leq \max_{0 \leq t \leq l} |\varphi(t, x_k, B_k, g_k) - \varphi(t, x, B, g)|$$

$$+ |\varphi(t_k, x, B, g) - \varphi(t, x, B, g)|. \tag{6.55}$$

By virtue of the inequality (6.55) and Lemma 6.6 we obtain the necessary assertion. The lemma is proved. $\qquad\square$

Lemma 6.9 *For all $(B, g) \in H^+(A, f)$ and $x_1, x_2 \in H$ we have*

$$|\varphi(t, x_1, B, g) - \varphi(t, x_2, B, g)| \le e^{-\alpha t}|x_1 - x_2|$$

for any $t \in \mathbb{R}_+$.

Proof. If the operator-function $A(t)$ is negatively defined, then every operator-function $B \in H^+(A)$ is negatively defined too and $Re\langle B(t)u, u\rangle \ge \alpha|u|^2$ ($t \in \mathbb{R}, u \in H$), where $\alpha > 0$ is the same constant as the one figuring in (6.47) for the operator-function $A(t)$. Let $\omega(t) := \varphi(t, x_1, B, g) - \varphi(t, x_2, B, g)$. Then, according to Lemma 6.4, we have

$$\frac{1}{2}\frac{d}{dt}|\omega(t)|^2 = Re\langle B(t)\omega(t), \omega(t)\rangle \le -\alpha|\omega(t)|^2$$

and, consequently, $|\omega(t)| \le |\omega(0)|e^{-\alpha t}$ for all $t \in \mathbb{R}_+$. $\square$

Let us define on X a dynamical system in the following way: $\pi(t, x) := \pi(\langle u, (b, g)\rangle, t) = \langle\varphi(t, u, B, g), (B_t, g_t)\rangle$ for all $\langle u, (B, g)\rangle \in X$ and $\mathbb{R}_+$. Let $Y = H(A, f)$ and $(Y, \mathbb{R}, \sigma)$ be a dynamical system of translations on Y and $h = pr_2 : X \to Y$. Then the triple $\langle(X, \mathbb{R}_+, \pi), (Y, \mathbb{R}, \sigma), h\rangle$ is the non-autonomous dynamical system generated by equation (6.45).

Definition 6.5 We will call the equation (6.45) convergent, if it admits a compact solution on $\mathbb{R}$ that is globally asymptotically stable.

According to the results of Chapter 2 (see section 2.5), the equation (6.45) will be convergent if and only if the non-autonomous dynamical system $\langle(X, \mathbb{R}_+, \pi), (Y, \mathbb{R}_+, \sigma), h\rangle$ generated by the equation (6.45) is convergent.

Theorem 6.13 *Let $A \in C(\mathbb{R}, [H]), f \in C(\mathbb{R}, H)$ and $H(A, f)$ be compact. Then the equation (6.45) is convergent.*

Proof. According to the results from [59], [187], all the solutions of the equation (6.45) are compact on $\mathbb{R}_+$ and by virtue of Lemma 6.9 every solution of the equation (6.45) is globally asymptotically stable. $\square$

6.6 Global attractors of non-autonomous Lorenz systems

This section is devoted to the study of non-autonomous Lorenz systems. The problems are formulated and solved in the context of non-autonomous dynamical systems. First, we prove that such systems admit a compact global attractor and characterize its structure. Then, we obtain conditions of convergence of the non-autonomous Lorenz systems, under which all solutions approach a point attractor. Third, we derive a criterion for existence of almost periodic (periodic, quasi-periodic) and recurrent solutions of the systems. Finally, we prove a global averaging principle for non-autonomous Lorenz systems.

The following n-dimensional systems of differential equations are called systems of hydrodynamic type or autonomous Lorenz systems([306]):

$$u_i' = \Sigma_{j,k} b_{ijk} u_j u_k + \Sigma_j a_{ij} u_j + f_i, \ i = 1, 2, ..., n, \tag{6.56}$$

where $\Sigma b_{ijk} u_i u_j u_k$ is identically equal to zero, $\Sigma a_{ij} u_i u_j$ is negative definite, and f_i are constants. The well known three-dimensional Lorenz system for geophysical flows or climate modelling [242] is a special case of this type of systems.

It is known that solutions of (6.56) imbed in some ellipsoid and do not leave it later, i.e. the autonomous system (6.56) is dissipative, and hence admits a compact global attractor. In the vector-matrix form the system (6.56) may be written as:

$$u' = Au + B(u, u) + f, \tag{6.57}$$

where A is a positive definite matrix and $B : H \times H \to H$ (H is a n-dimensional real or complex Euclidean space) is a bilinear form satisfying the identity

$$Re\langle B(u, v), w \rangle = -Re\langle B(u, w), v \rangle$$

for every $u, v, w \in H$.

When f is not a constant vector but a bounded function of time t, it is known that the equation (6.57) also admits a compact global attractor [218].

Our aim is to study the *non-autonomous* version of the equation (6.57). Namely, in this case, the matrix A, the bilinear form B, and the function f all depend on time t. We will consider issues like compact global attractors, convergence, almost periodic (including periodic and quasi-periodic) solutions and recurrent solutions, and averaging principles.

6.6.1 *Non-autonomous Lorenz systems*

Let Ω be a compact metric space, $\mathbb{R} = (-\infty, +\infty), (\Omega, \mathbb{R}, \sigma)$ be a dynamical system on Ω and H be a real or complex Hilbert space. We denote by $L(H)$ $(L^2(H))$ the space of all linear (bilinear) operators on H. When W is some metric space, $C(\Omega, W)$ denotes the space of all continuous functions $f : \Omega \to W$, endowed with the topology of uniform convergence.

Let us consider the non-autonomous Lorenz system

$$u' = A(\omega t)u + B(\omega t)(u, u) + f(\omega t), \quad \omega \in \Omega, \tag{6.58}$$

where $\omega t := \sigma(t, \omega)$, $A \in C(\Omega, L(H))$, $B \in C(\Omega, L^2(H))$ and $f \in C(\Omega, H)$. Note that when the autonomous Lorenz system (6.57) is perturbed by periodic, quasi-periodic, almost periodic or recurrent forces, it can then be written as (6.58). Moreover, we assume that the following conditions are fulfilled:

(1) There exists $\alpha > 0$ such that

$$Re\langle A(\omega)u, u \rangle \leq -\alpha |u|^2 \tag{6.59}$$

for all $\omega \in \Omega$ and $u \in H$, where $|\cdot|$ is a norm in H ;

(2)

$$Re\langle B(\omega)(u, v), w \rangle = -Re\langle B(\omega)(u, w), v \rangle \tag{6.60}$$

for every $u, v, w \in H$ and $\omega \in \Omega$.

Remark 6.4   *a. It follows from (6.60) that*

$$Re\langle B(\omega)(u, v), v) \rangle = 0 \tag{6.61}$$

for every $u, v \in H$ and $\omega \in \Omega$.

   b. From bilinearity and continuity, we obtain

$$|B(\omega)(u, v)| \leq C_B |u||v| \tag{6.62}$$

for all $u, v \in H$ and $\omega \in \Omega$, where $C_B := \sup\{|B(\omega)(u, v)| : \omega \in \Omega,\ u, v \in H,\ |u| \leq 1$ and $|v| \leq 1\}$.

Definition 6.6   We will call the system (6.58) with conditions (6.59) and (6.60) a non-autonomous Lorenz system or a non-autonomous system of hydrodynamic type.

We note that from the conditions (6.60) -(6.62) it follows that

$$|B(\omega)(x_1, x_1) - B(\omega)(x_2, x_2)| \leq C_B(|x_1| + |x_2|)|x_1 - x_2| \tag{6.63}$$

for all $x_1, x_2 \in H$ and $\omega \in \Omega$.

Since the coefficients of (6.58) are locally Lipschitzian with respect to $u \in H$, through every point $x \in H$ passes a unique solution $\varphi(t, x, \omega)$ of equation (6.58) at the initial moment $t = 0$. And this solution is defined on some interval $[0, t_{(x,\omega)})$. Let us note that

$$\begin{aligned}
w'(t) &= 2Re\langle \varphi'(t, x, \omega), \varphi(t, x, \omega) \rangle = 2Re\langle A(\omega t)\varphi(t, x, \omega), \varphi(t, x, \omega) \rangle \\
&\quad + 2Re\langle B(\omega t)(\varphi(t, x, \omega), \varphi(t, x, \omega)), \varphi(t, x, y) \rangle + 2Re\langle f(\omega t), \varphi(t, x, \omega) \rangle \\
&= 2Re\langle A(\omega t)\varphi(t, x, \omega), \varphi(t, x, \omega) \rangle + 2Re\langle f(\omega t), \varphi(t, x, \omega) \rangle \\
&\leq -2\alpha |\varphi(t, x, \omega)|^2 + 2\|f\| |\varphi(t, x, \omega)|,
\end{aligned} \tag{6.64}$$

where $\|f\| := \max\{|f(\omega)| : \omega \in \Omega\}$ and $w(t) := |\varphi(t, x, \omega)|^2$. Then

$$w' \leq -2\alpha w + 2\|f\| w^{\frac{1}{2}}.$$

Thus

$$w(t) \leq v(t)$$

for all $t \in [0, t_{(x,\omega)})$, where $v(t)$ is a solution of the equation

$$v' = -2\alpha v + 2\|f\| v^{\frac{1}{2}},$$

satisfying condition $v(0) = w(0) = |x|^2$. Hence

$$v(t) = [(|x| - \frac{\|f\|}{\alpha})e^{-\alpha t} + \frac{\|f\|}{\alpha}]^2$$

and consequently

$$|\varphi(t, x, \omega)| \leq (|x| - \frac{\|f\|}{\alpha})e^{-\alpha t} + \frac{\|f\|}{\alpha} \tag{6.65}$$

for all $t \in [0, t_{(x,\omega)})$. It follows from the inequality (6.65) that the solution $\varphi(t, x, \omega)$ is bounded and therefore it may be extended to a global solution on $\mathbb{R}_+ = [0, +\infty)$.

Thus we have proved the following theorem.

Theorem 6.14 *Let the conditions (6.59) and (6.60) are fulfilled. Then the following statements hold:*

(i)

$$|\varphi(t, x, \omega)| \leq C(|x|),$$

for all $t \geq 0$, $\omega \in \Omega$ and $x \in H$, where $C(r) = r$ if $r \geq r_0 := \frac{\|f\|}{\alpha}$ and $C(r) = r_0$ if $r \leq r_0$;

(ii)

$$\limsup_{t \to +\infty} \sup\{|\varphi(t, x, \omega)| : |x| \leq r, \omega \in \Omega\} \leq \frac{\|f\|}{\alpha}$$

for every $r > 0$.

The item (i) in this Theorem means that the non-autonomous Lorenz flow is bounded on bounded sets, while the item (ii) implies that the non-autonomous Lorenz system is dissipative, i.e., it admits a bounded absorbing set.

6.6.2 *Non-autonomous dissipative dynamical systems and their attractors*

Let Ω and W be two metric spaces and $(\Omega, \mathbb{R}, \sigma)$ be an autonomous dynamical system on Ω. Let us consider a continuous mapping $\varphi : \mathbb{R}^+ \times W \times \Omega \to W$ satisfying the following conditions:

$$\varphi(0, \cdot, \omega) = id_W \quad \varphi(t + \tau, x, \omega) = \varphi(t, \varphi(\tau, x, \omega), \omega\tau)$$

for all $t, \tau \in \mathbb{R}^+$, $\omega \in \Omega$ and $x \in W$. Here $\omega\tau$ is the short notation for $\sigma_\tau(\omega) := \sigma(\tau, \omega)$. Such a mapping φ (or more explicitly $\langle W, \varphi, (\Omega, \mathbb{R}, \sigma)\rangle$) is called a cocycle over $(\Omega, \mathbb{R}, \sigma)$ with fiber W (see [6, 292]).

Example 6.2 Let E be a Banach space and $C(\mathbb{R} \times E, E)$ be a space of all continuous functions $F : \mathbb{R} \times E \to E$ equipped by the compact–open topology. Let us consider a parameterized differential equation

$$\frac{dx}{dt} = F(\sigma_t\omega, x), \quad \omega \in \Omega$$

on a Banach space E with $\Omega := C(\mathbb{R} \times E, E)$, where $\sigma_t\omega := \sigma(t, \omega)$. We will define $\sigma_t : \Omega \to \Omega$ by $\sigma_t\omega(\cdot, \cdot) = \omega(t + \cdot, \cdot)$ for each $t \in \mathbb{R}$ and interpret $\varphi(t, x, \omega)$ as solution of the initial value problem

$$\frac{d}{dt}x(t) = F(\sigma_t\omega, x(t)), \quad x(0) = x. \tag{6.66}$$

Under appropriate assumptions on $F : \Omega \times E \to E$ (or even $F : \mathbb{R} \times E \to E$ with $\omega(t)$ instead of $\sigma_t\omega$ in (6.66)) to ensure forward existence and uniqueness, then φ is a cocycle on $(C(\mathbb{R} \times E, E), \mathbb{R}, \sigma)$ with fiber E. Note that $(C(\mathbb{R} \times E, E), \mathbb{R}, \sigma)$ is a Bebutov's dynamical system (see for example [32],[102],[292],[300].

Let φ be a cocycle on $(\Omega, \mathbb{R}, \sigma)$ with the fiber E. Then the mapping $\pi : \mathbb{R}^+ \times E \times \Omega \to E \times \Omega$ defined by

$$\pi(t, x, \omega) := (\varphi(t, x, \omega), \sigma_t\omega)$$

for all $t \in \mathbb{R}^+$ and $(x, \omega) \in E \times \Omega$ forms a semi-group $\{\pi(t, \cdot, \cdot)\}_{t \in \mathbb{R}^+}$ of mappings of $X := \Omega \times E$ into itself, thus a semi-dynamical system on the state space X, which is called a skew-product flow [292]. The triplet $\langle (X, \mathbb{R}_+, \pi), (\Omega, \mathbb{R}, \sigma), h\rangle$ (where $h := pr_2 : X \to \Omega$) is a non-autonomous dynamical system (see [33, 102]).

Definition 6.7 Recall that a cocycle φ over $(\Omega, \mathbb{R}, \sigma)$ with the fiber W is called a compact (bounded) dissipative cocycle, if there is a nonempty compact set $K \subseteq W$ such that

$$\limsup_{t \to +\infty}\{\beta(U(t, \omega)M, K) | \omega \in \Omega\} = 0$$

for any $M \in C(W)$ (respectively $M \in B(W)$), where $C(W)(B(W))$ denotes the family of all compact (bounded) subsets of W, β is the semi-distance of Hausdorff and $U(t,\omega) := \varphi(t,\cdot,\omega)$. We can similarly define a compact or bounded dissipative skew-product system.

Lemma 6.10 *Let Ω be a compact metric space and $\langle W,\varphi,(\Omega,\mathbb{R},\sigma)\rangle$ be a cocycle over $(\Omega,\mathbb{R},\sigma)$ with the fiber W. In order for $\langle W,\varphi,(\Omega,\mathbb{R},\sigma)\rangle$ to be compact (bounded) dissipative, it is necessary and sufficient that the skew-product dynamical system $(X,\mathbb{R}_+,\pi)$ is compact (bounded) dissipative.*

Proof. This assertion follows from the corresponding definitions. $\qquad\square$

Definition 6.8 The cocycle $\langle W,\varphi,(Y,\mathbb{R},\sigma\rangle$ is called compact (asymptotically compact) if the associated skew-product dynamical system $(X,\mathbb{R}_+,\pi)$ with $X = W \times Y$ and $\pi = (\varphi,\sigma)$ is compact (respectively asymptotically compact).

Let $(X,\mathbb{R}_+,\pi)$ be compact dissipative and K be a compact set, which attracts all compact subsets of X. Let

$$J = \Omega(K), \tag{6.67}$$

where $\Omega(K) = \bigcap_{t\geq 0}\overline{\bigcup_{\tau\geq t}\pi^\tau K}$. The set J defined by the equality (6.67) does not depend on selection of the attracting set K, and is characterized only by the properties of the dynamical system $(X,\mathbb{R}_+,\pi)$ itself. The set J is called the Levinson's centre of the compact dissipative system $(X,\mathbb{R}_+,\pi)$.

Remark 6.5 *Let $\langle W,\varphi,(\Omega,\mathbb{R},\sigma\rangle$ be a compact dissipative cocycle over $(\Omega,\mathbb{R},\sigma\rangle$ with the fiber W and $\{I_\omega \mid \omega \in \Omega\}$ is its compact global attarctor. Then the skew-product dynamical system $(X,\mathbb{R}_+,\pi)$, generated by cocycle φ, is compact dissipative too and the set $J := \cup\{J_\omega \mid \omega \in \Omega\}$, where $J_\omega := I_\omega \times \{\omega\}$, is the Levinson center of $(X,\mathbb{R}_+,\pi)$.*

Applying the general theorems about non-autonomous dissipative systems to non-autonomous system constructed in the example 6.2, we will obtain series of facts concerning the non-autonomous Lorenz system (6.58). In particular we have the following results.

Theorem 6.15 *Let Ω be a compact metric space, $(\Omega,\mathbb{R},\sigma)$ be a dynamical system on Ω and the conditions (6.59) and (6.60) are fulfilled. If the cocycle φ generated by non-autonomous Lorenz system (6.58) is asymptotically compact, then for every $\omega \in \Omega$, there exists a non-empty compact connected set $I_\omega \subset H$ such that the following conditions hold:*

(1) The set $I = \cup\{I_\omega \; : \; \omega \in \Omega\}$ is compact and connected in H, if the space Ω is connected;

(2)

$$\lim_{t \to +\infty} \sup_{\omega \in \Omega} \beta(U(t, \omega^{-t})M, I) = 0$$

for any bounded set $M \subset H$, where $U(t, \omega) := \varphi(t, \cdot, \omega)$ and β is the semi-distance of Hausdorff;

(3) $U(t, \omega)I_\omega = I_{\omega t}$ for all $t \in \mathbb{R}_+$ and $\omega \in \Omega$;

(4) I_ω consists of those and only those points $x \in H$ through which passes the bounded solutions (on $\mathbb{R}$) of the non-autonomous Lorenz system (6.58).

Proof. This assertion follows from Theorems 6.14, 2.24 and 2.25. $\square$

This theorem states that $\cup\{I_\omega \; : \; \omega \in \Omega\}$ is the compact global attractor of the non-autonomous Lorenz system (6.58) and also characterizes the structure of the sections I_ω of the attractor.

Theorem 6.16   *Under conditions of Theorem 6.15*

$$|\varphi(t, x, \omega)| \leq \frac{\|f\|}{\alpha} \tag{6.68}$$

for all $t \in \mathbb{R}$, $\omega \in \Omega$ and $x \in I_\omega$, where φ is the cocycle generated by Lorenz non-autonomous system (6.58).

Proof. According to Remark 6.5 the set $J := \bigcup\{I_\omega \times \{\omega\} \; : \; \omega \in \Omega\}$ is a Levinson's centre of dynamical system $(X, \mathbb{R}_+, \pi)$ and according to (6.67) for any point $(u_0, y_0) = z \in J$ there exists $t_n \to +\infty, u_n \in H$ and $\omega_n \in \Omega$ such that the sequence $\{u_n\}$ is bounded, $u_0 = \lim_{n \to +\infty} \varphi(t_n, u_n, \omega_n)$ and $\omega_0 = \lim_{n \to +\infty} \omega_n t_n$. From the inequality (6.65), it follows that $|u_0| \leq \frac{\|f\|}{\alpha}$, i.e. $\varphi(t, x, \omega) \in I_{\omega t}$ for all $\omega \in \Omega$ and $t \in \mathbb{R}$, hence $|\varphi(t, x, \omega)| \leq \frac{\|f\|}{\alpha}$ for any $t \in \mathbb{R}$, $x \in I_\omega$ and $\omega \in \Omega$. The theorem is proved. $\square$

Theorem 6.17   *Let φ be the cocycle generated by the Lorenz non-autonomous system (6.58). Under conditions of Theorem 6.15 and further assume that $\alpha^{-2}C_B\|f\| < 1$. Then this cocycle φ is convergent, i.e. for any $\omega \in \Omega$ the set I_ω contains a single point u_ω.*

Proof. Let $\omega \in \Omega$ and $u_1, u_2 \in I_\omega$. We define $\psi(t) := \varphi(t, u_1, \omega) - \varphi(t, u_2, \omega)$ and

$$w(t) := |\varphi(t, u_1, \omega) - \varphi(t, u_2, \omega)|^2.$$

According to Theorem 6.16, the function $w(t)$ is bounded on $\mathbb{R}$. On the other hand, in view of (6.64) and (6.59), we have

$$w'(t) \leq -2\alpha w(t) + 2Re\langle B(\omega t)(\psi(t), \varphi(t, u_2, \omega)), \psi(t)\rangle. \tag{6.69}$$

From the inequalities (6.63), (6.69) and Theorem 6.16, it follows that

$$w'(t) \leq -2\alpha w(t) + 2C_B \frac{\|f\|}{\alpha} w(t).$$

Hence, $w(t) \leq w(0)e^{-2(\alpha - C_B \frac{\|f\|}{\alpha})t}$, i.e.

$$|\varphi(t, u_1, \omega) - \varphi(t, u_2, \omega)| \leq |u_1 - u_2| e^{-(\alpha - \frac{\|f\|}{\alpha} C_B)t}$$

for all $t \geq 0$, $\omega \in \Omega$ and $u_1, u_2 \in I_\omega$. In particular,

$$|u_1 - u_2| \leq |\varphi(t, \varphi(-t, u_1, \sigma(-t, \omega)), \omega) - \tag{6.70}$$
$$\varphi(t, \varphi(-t, u_2, \sigma(-t, \omega)), \omega)| e^{-(\alpha - \frac{\|f\|}{\alpha} C_B)t}$$

for all $t \geq 0$, $\omega \in \Omega$ and $u_1, u_2 \in J_\omega$. Note that $|\varphi(t, u_1, \omega) - \varphi(t, u_2, \omega)|$ is bounded on $\mathbb{R}$. Thus from (6.70) it follows that $u_1 = u_2$, where $\varphi(-t, x, \omega) := u_{\sigma(-t, \omega)}$ for all $x \in I_\omega$, $t \geq 0$ and $\omega \in \Omega$. The theorem is proved. $\qquad\square$

6.6.3 Almost periodic and recurrent solutions of non-autonomous Lorenz systems

In this subsection, we discuss the problem of existence of almost periodic and recurrent solutions of non-autonomous Lorenz systems. Let $\mathbb{T} = \mathbb{R}$ or $\mathbb{R}_+$ and $(X, \mathbb{T}, \pi)$ be a dynamical system.

Definition 6.9 The solution $\varphi(t, x, \omega)$ of non-autonomous Lorenz system (6.58) is called recurrent (almost periodic, quasi-periodic, periodic), if the point $(x, \omega) \in H \times \Omega$ is a recurrent (almost periodic, quasi-periodic, periodic) point of skew-product dynamical system $(X, \mathbb{R}_+, \pi)$ $(X := H \times \Omega$ and $\pi := (\varphi, \sigma))$.

We note (see, for example, [238]),[300] and [302]) that if $\omega \in \Omega$ is a stationary (τ-periodic, almost periodic, quasi periodic, recurrent) point of dynamical system $(\Omega, \mathbb{R}, \sigma)$ and $h : \Omega \to X$ is a homomorphism of dynamical system $(\Omega, \mathbb{R}, \sigma)$ onto $(X, \mathbb{R}_+, \pi)$, then the point $x = h(\omega)$ is a stationary (τ-periodic, almost periodic, quasi periodic, recurrent) point of the system $(X, \mathbb{R}_+, \pi)$.

Let $X := H \times \Omega$ and $\pi := (\varphi, \sigma)$, then mapping $h : \Omega \to X$ is a homomorphism of dynamical system $(\Omega, \mathbb{R}, \sigma)$ onto $(X, \mathbb{R}_+, \pi)$ if and only if $h(\omega) = (u(\omega), \omega)$ for all $\omega \in \Omega$, where $u : \Omega \to H$ is a continuous mapping with the condition that $u(\omega t) = \varphi(t, u(\omega), \omega)$ for all $\omega \in \Omega$ and $t \in \mathbb{R}_+$.

Theorem 6.18 *Let Ω be a compact metric space, the cocycle φ, generated by the non-autonomous Lorenz system (6.58), is asymptotic compact and the conditions (6.59), (6.61)-(6.62) are fulfilled with $\frac{\|f\|C_B}{\alpha^2} < 1$. Then the set I_ω contains a unique* point x_ω ($I_\omega = \{x_\omega\}$) *for every* $\omega \in \Omega$, *the mapping* $u : \Omega \to H$ *defined by* $u(\omega) :=$ x_ω *is continuous and* $u(\omega t) = \varphi(t, u(\omega), \omega)$ *for all* $\omega \in \Omega$ *and* $t \in \mathbb{R}_+$.

Proof. According to Theorem 6.17, it is sufficient to show that the mapping $u : \Omega \to H$ defined above is continuous. Let $\omega \in \Omega$, $\{\omega_n\} \subseteq \Omega$ and $\omega_n \to \omega$. Consider the sequence $\{x_n\} := \{x_{\omega_n}\} \subset I := \bigcup\{I_\omega \mid \omega \in \Omega\}$. Since the set I is compact, then the sequence $\{x_n\}$ is relatively compact. Let x' be a limit point of this sequence, then there is a subsequence $\{x_{k_n}\}$ such that $x_{k_n} \to x'$. Let J be a Levinson center of the skew-product dynamical system $(X, \mathbb{R}_+, \pi)$, generated by the cocycle φ. Note that the point $(x_{k_n}, \omega_{k_n}) \in J_{\omega_{k_n}} := I_{\omega_{k_n}} \times \{\omega_{k_n}\} \subseteq J$ and taking in the consideration that J is compact we obtain that $(x', \omega) \in J$. Thus $(x', \omega) \in J_\omega = I_\omega \times \{\omega\}$ and, consequently, $x' \in I_\omega = \{x_\omega\}$, i.e. the relatively compact sequence $\{x_n\}$ has a unique limit point x_ω. This means that the sequence $\{x_n\}$ converges to x_ω as $n \to +\infty$. The theorem is proved. $\qquad\square$

Corollary 6.5 *Let* Ω *be a compact minimal (almost periodic minimal, quasi-periodic minimal or periodic minimal) set of dynamical system* $(\Omega, \mathbb{R}, \sigma)$. *Then under the conditions of Theorem 6.18, the non-autonomous Lorenz system (6.58) admits a compact global attractor* I, *and for all* $\omega \in \Omega$, *the section* I_ω *of the attractor contains a unique point* x_ω *through which passes a recurrent (almost periodic, quasi-periodic, or periodic) solution of the equation (6.58).*

Let H be a d-dimensional complex Euclidean space, i.e. $H = \mathbb{C}^d$. Denote by $HC(\mathbb{C}^d \times \Omega, \mathbb{C}^d)$ the space of all continuous functions $f : \mathbb{C}^d \times \Omega \to \mathbb{C}^d$ holomorphic in $z \in \mathbb{C}^d$ and equipped with compact-open topology. Consider the differential equation

$$\frac{dz}{dt} = f(z, \sigma_t \omega), \quad (\omega \in \Omega) \tag{6.71}$$

where $f \in HC(\mathbb{C}^d \times \Omega, \mathbb{C}^d)$. Let $\varphi(t, \omega, z)$ be the solution of equation (6.71) passing through the point z at $t = 0$ and defined on $\mathbb{R}^+$. The mapping $\varphi : \mathbb{R}^+ \times \Omega \times \mathbb{C}^d \to \mathbb{C}^d$ has the following properties (see, for example, [47] and [186]):

a) $\varphi(0, z, \omega) = z$ for all $z \in \mathbb{C}^d$.
b) $\varphi(t + \tau, z, \omega) = \varphi(t, \varphi(\tau, z, \omega), \sigma_\tau \omega)$ for all $t, \tau \in \mathbb{R}^+, \omega \in \Omega$ and $z \in \mathbb{C}^d$.
c) Mapping φ is continuous.
d) Mapping $U(t, \omega) := \varphi(t, \cdot, \omega) : \mathbb{C}^d \to \mathbb{C}^d$ is holomorphic for any $t \in \mathbb{R}^+$ and $\omega \in \Omega$.

Definition 6.10 The cocycle $\langle \mathbb{C}^d, \varphi, (\Omega, \mathbb{T}, \theta) \rangle$ is said to be $\mathbb{C}$-analytic if the mapping $U(t, \omega) : \mathbb{C}^d \to \mathbb{C}^d$ is holomorphic for all $t \in \mathbb{R}_+$ and $\omega \in \Omega$.

Example 6.3 Let $(HC(\mathbb{R} \times \mathbb{C}^d, \mathbb{C}^d), \mathbb{R}, \sigma)$ be a dynamical system of translations on $HC(\mathbb{R} \times \mathbb{C}^d, \mathbb{C}^d)$ (Bebutov's dynamical system (see, for example, [102] and [300])). Denote by F the mapping from $\mathbb{C}^d \times HC(\mathbb{R} \times \mathbb{C}^d, \mathbb{C}^d)$ to $\mathbb{C}^d$ defined by equality $F(z, f) := f(0, z)$ for all $z \in \mathbb{C}^d$ and $f \in HC(\mathbb{R} \times \mathbb{C}^d, \mathbb{C}^d)$. Let Ω be the hull $H(f)$ of given function $f \in HC(\mathbb{R} \times \mathbb{C}^d, \mathbb{C}^d)$, that is $\Omega = H(f) := \overline{\{f_\tau | \tau \in \mathbb{R}\}}$, where $f_\tau(t, z) := f(t + \tau, z)$ for all $t, \tau \in \mathbb{R}$ and $z \in \mathbb{C}^d$. Denote the restriction of $(HC(\mathbb{R} \times \mathbb{C}^d, \mathbb{C}^d), \mathbb{R}, \sigma)$ on Ω by $(\Omega, \mathbb{R}, \sigma)$. Then, under appropriate restriction on the given function $f \in HC(\mathbb{R} \times \mathbb{C}^d, \mathbb{C}^d)$, the differential equation $\frac{dz}{dt} = f(z, t) = F(z, \sigma_t f)$ generates a $\mathbb{C}$−analytic cocycle.

Theorem 6.19 *Let $H := \mathbb{C}^d$, Ω be a compact minimal set and the conditions (6.59), (6.61)-(6.62) are fulfilled. Then the non-autonomous Lorenz system admits a compact global attractor $\{I_\omega \mid \omega \in \Omega\}$ and the set I_ω contains a unique point x_ω ($I_\omega = \{x_\omega\}$) for every $\omega \in \Omega$, the mapping $u : \Omega \to H$ defined by equality $u(\omega) := x_\omega$ is continuous and $u(\omega t) = \varphi(t, u(\omega), \omega)$ for all $\omega \in \Omega$ and $t \in \mathbb{R}_+$, where φ is a cocycle generated by the non-autonomous Lorenz system.*

Proof. We note that under the conditions of Theorem 6.19 the right-hand side $f(\omega, z) := A(\omega)z + B(\omega)(z, z) + f(\omega)$ is $\mathbb{C}$-analytic because $D_z f(\omega, z)h = A(\omega)h + B(\omega)(h, z) + B(\omega)(z, h)$ for all $\omega \in \Omega$ and $z \in \mathbb{C}^d$, where $D_z f(\omega, z)$ is a derivative of function $f(\omega, z)$ w.r.t. $z \in \mathbb{C}^d$. Now our statement directly results from Theorems 6.15 and 3.2. The proof is complete. $\qquad\qquad\square$

Corollary 6.6 *Let Ω be a compact minimal (almost periodic minimal, quasi-periodic minimal or periodic minimal) set of dynamical system $(\Omega, \mathbb{R}, \sigma)$. Then under the conditions of Theorem 6.19, the non-autonomous Lorenz system (6.58) admits a compact global attractor I and for all $\omega \in \Omega$, the set I_ω contains a unique point x_ω through which passes a recurrent (almost periodic, quasi-periodic or periodic) solution of equation (6.58).*

6.6.4 *Uniform averaging principle*

Now we consider a uniform averaging principle for a general class of differential equations. In the next subsection, we apply this averaging principle to the non-autonomous Lorenz system (6.58).

Let $C(\mathbb{R} \times H, H)$ be the space of all continuous functions $f : \mathbb{R} \times H \to H$ equipped with compact open topology and let $\mathcal{F} \subseteq C(\mathbb{R} \times H, H)$. In Hilbert space H (with the norm $|\cdot|$ induced by the scalar product) we will consider the family of equations

$$x' = \varepsilon f(t, x), \quad f \in \mathcal{F}, \tag{6.72}$$

containing a small parameter $\varepsilon \in [0, \varepsilon_0]$ $(\varepsilon_o > 0)$.

We assume that on the set $\mathbb{R}_+ \times B[0, r]$, where $B[0, r] := \{x \in H \mid |x| \le r\}$ is a ball of radius $r > 0$ in H, the functions $f \in \mathcal{F}$ are uniformly bounded, i.e. there exists a positive constant M such that

$$|f(t, x)| \le M \tag{6.73}$$

for every $f \in \mathcal{F}$, $t \in \mathbb{R}_+$ and $x \in B[0, r]$, and satisfies the condition of Lipschitz

$$|f(t, x_1) - f(t, x_2)| \le L|x_1 - x_2| \ (x_1, x_2 \in B[0, r]) \tag{6.74}$$

with a constant $L > 0$ does not depending neither on $t \in \mathbb{R}_+$ nor on $f \in \mathcal{F}$.

Furthermore, we assume that the mean value of f is uniform with respect to (w.r.t.) $f \in \mathcal{F}$ and $x \in B[0, r]$

$$f_0(x) := \lim_{T \to +\infty} \frac{1}{T} \int_0^T f(t, x) dt, \tag{6.75}$$

i.e. for every $\varepsilon > 0$ there exists a $l = l(\varepsilon) > 0$ such that

$$\left| \frac{1}{T} \int_0^T f(t, x) dt - f_0(x) \right| < \varepsilon$$

for all $T \ge l(\varepsilon)$, $x \in B[0, r]$ and $f \in \mathcal{F}$, and the function f_0 does not depend on $f \in \mathcal{F}$.

Lemma 6.11   *The condition (6.75) holds if and only if there exists a decreasing continuous function $m : \mathbb{R}_+ \to \mathbb{R}_+$, satisfying the condition $m(t) \to 0$ as $t \to +\infty$, such that*

$$\left| \frac{1}{T} \int_0^T f(t, x) dt - f_0(x) \right| \le m(T) \tag{6.76}$$

for all $T > 0$, $x \in B[0, r]$ and $f \in \mathcal{F}$. The function m does not depend neither on $x \in B[0, r]$ nor on $f \in \mathcal{F}$.

Proof.   Denote by

$$k(T) := \sup_{f \in \mathcal{F}, x \in B[0, r]} \left| \frac{1}{T} \int_0^T f(t, x) dt - f_0(x) \right|.$$

The mapping k possesses the following properties:

(1) $0 \le k(T) \le 2M$, where $M := \sup\{|f(t, x)| \ : \ f \in \mathcal{F}, |x| \le r\}$;
(2) $k(T) \to 0$ as $T \to +\infty$.

Let

$$c_n := \sup_{T \ge n} k(T),$$

then $c_o \geq c_1 \geq \ldots \geq c_n \geq \ldots$ and $c_n \to 0$ as $n \to +\infty$. Define now the function $m : \mathbb{R}_+ \to \mathbb{R}_+$ by the equality

$$m(t) := c_{n-1} + (t - n)(c_n - c_{n-1}) \quad (n \leq t \leq n+1, n = 0, 1, \ldots),$$

where $c_{-1} := c_0 + 1$. The lemma is proved. $\qquad\square$

Lemma 6.12 *Let $\mathcal{F} \subseteq C(\mathbb{R} \times E, E)$ be a family of functions satisfying the condition (6.75), then for every $L > 0$*

$$l(\varepsilon) := \sup\{|\int_0^\tau f(\frac{t}{\varepsilon}, x)dt - f_0(x)| \; : \; 0 \leq \tau \leq L, \; f \in \mathcal{F}, \; |x| \leq r\} \to 0$$

as $\varepsilon \to 0$.

Proof. According to Lemma 6.11 there exists a decreasing continuous function $m : \mathbb{R}_+ \to \mathbb{R}_+$ with the condition $m(t) \to 0$ as $t \to +\infty$ and such that the inequality (6.76) holds. Let $\nu \in (0, 1)$, then

$$l(\varepsilon) \leq \sup\{|\int_0^\tau f(\frac{t}{\varepsilon}, x)dt| \; : \; 0 \leq \tau \leq \varepsilon^\nu, \; f \in \mathcal{F}, \; |x| \leq r\} +$$

$$\sup\{|\int_0^\tau f(\frac{t}{\varepsilon}, x)dt| \; : \; \varepsilon^\nu \leq \tau \leq L, \; f \in \mathcal{F}, \; |x| \leq r\} =$$

$$\sup\{\tau|\frac{\varepsilon}{\tau} \int_0^{\frac{\tau}{\varepsilon}} f(t, x)dt| \; : \; 0 \leq \tau \leq \varepsilon^\nu, \; f \in \mathcal{F}, \; |x| \leq r\} +$$

$$\sup\{\tau|\frac{\varepsilon}{\tau} \int_0^{\frac{\tau}{\varepsilon}} f(t, x)dt| \; : \; \varepsilon^\nu \leq \tau \leq L, \; f \in \mathcal{F}, \; |x| \leq r\} \leq$$

$$m(0)\varepsilon^\nu + Lm(\varepsilon^{\nu-1}) \to 0$$

as $\varepsilon \to 0$. The lemma is proved. $\qquad\square$

Under the assumptions above, it is expedient to consider along with equation (6.72) the *averaged* equation

$$x' = \varepsilon f_o(x).$$

From (6.75) we see that the function f_0 also satisfies the conditions (6.73) and (6.74). Let $\varphi(t, x)$ $(0 \leq t \leq T_0)$ be a solution of equation

$$y' = f_0(y). \tag{6.77}$$

taking values in $B[0, r]$ and passing through the point x at the initial moment $t = 0$. Then, as can easily be seen, the function $\varphi(t, x, \varepsilon) := \varphi(\varepsilon t, x)$ is the solution of equation (6.77) on the interval $0 \leq t \leq \frac{T_0}{\varepsilon}$. We will establish below a connection between $\varphi(t, x, \varepsilon)$ and the solution $\varphi(t, x, f, \varepsilon)$ of equation (6.72) with the initial condition $\varphi(0, x, f, \varepsilon) = x$.

More precisely, we will prove the following assertion.

Theorem 6.20   *Suppose that on $\mathbb{R}_+ \times B[0,r]$, the functions $f \in \mathcal{F}$ satisfy the conditions (6.73)-(6.75). Then for any $\eta > 0$ there exists an $\varepsilon > 0$ $(0 < \varepsilon < \varepsilon_0)$ such that the estimate*

$$|\varphi(t,x,f,\varepsilon) - \varphi(t,x,\varepsilon)| \le \eta \ (0 \le t \le \frac{T_0}{\varepsilon})$$

holds uniformly w.r.t. $f \in \mathcal{F}$ and $x \in B[0,r]$.

Denote by $\mathcal{K}$ the family of all solutions $x : [0, T_0] \to B[0,r]$ of the equation (6.77). Let us prove an auxiliary assertion.

Lemma 6.13   *Let $\mathcal{F} \subseteq C(\mathbb{R} \times H, H)$ be a family of functions satisfying the conditions (6.73)-(6.75). Then the equality*

$$\lim_{\varepsilon \to 0} \int_0^\tau f(\frac{s}{\varepsilon}, x(s))ds = \int_0^\tau f_0(x(s))ds \ \ (0 < \tau \le T_0)$$

holds uniformly w.r.t. $x \in \mathcal{K}$, $\tau \in [0, T_0]$ and $f \in \mathcal{F}$.

Proof.   Observe that

$$\lim_{\varepsilon \to 0} \int_0^\tau f(\frac{\tau}{\varepsilon}, x)d\tau = \tau f_0(x) \tag{6.78}$$

or, equivalently,

$$\lim_{\varepsilon \to 0} \frac{\varepsilon}{\tau} \int_0^{\frac{\tau}{\varepsilon}} f(t,x)dt = f_0(x) \tag{6.79}$$

uniformly w.r.t. $x \in \mathcal{K}$, $\tau \in [0, T_0]$ and $f \in \mathcal{F}$. In fact, according to Lemma 6.12

$$|\frac{\varepsilon}{\tau} \int_0^{\frac{\varepsilon}{\tau}} f(t,x)dt - f_0(x)| \to 0$$

as $\varepsilon \to 0$ uniformly w.r.t. $x \in B[0,r]$, $f \in \mathcal{F}$ and $\tau \in [0, T_0]$. Let us note that the equality (6.79) is equivalent to (6.75). From (6.78) it follows that for any $\tau_1, \tau_2 \in [0, T_0]$ we have

$$\lim_{\varepsilon \to 0} \int_{\tau_1}^{\tau_2} f(\frac{\tau}{\varepsilon}, x)d\tau = \int_{\tau_1}^{\tau_2} f_0(x)d\tau$$

uniformly w.r.t. $x \in B[0,r]$ and $f \in \mathcal{F}$. Hence for any $0 \le \tau_1 < \tau_2 < ...\tau_{n-1} < \tau_n = T_0$, $x_k \in B[0,r]$ $(k = 1, 2, ..., n)$, we conclude that

$$\lim_{\varepsilon \to 0} \sum_1^n \int_{\tau_{k-1}}^{\tau_k} f(\tau, x_k, \varepsilon)d\tau = \sum_1^n \int_{\tau_{k-1}}^{\tau_k} f_0(x_k)d\tau \tag{6.80}$$

uniformly w.r.t. $x_1, x_2, ..., x_n \in B[0,r]$ and $f \in \mathcal{F}$.

If we introduce the step functions $\tilde{x}_n(\tau) := x(\tau_k)$ $(\tau_{k-1} \leq \tau \leq \tau_k;\ \tau_k - \tau_{k-1} = \frac{1}{n};\ k = 1, 2, ..., n$ and $x \in \mathcal{K})$, then from the equality (6.80), we have the following relation

$$\lim_{\varepsilon \to 0} \int_0^\tau f(\frac{s}{\varepsilon}, \tilde{x}_n(s)) ds = \int_0^\tau f_0(\tilde{x}_n(s)) ds. \tag{6.81}$$

Under our assumption the family of functions $\mathcal{K}$ is equicontinuous on $[0, T_0]$ and, consequently,

$$\sup_{x \in \mathcal{K}} \sup_{0 \leq \tau \leq T_0} \|\tilde{x}_n(\tau) - x(\tau)\| \to 0 \tag{6.82}$$

as $n \to +\infty$. Using the condition of Lipschitz (6.74) for the family of functions $\mathcal{F}$ we obtain the estimate

$$|\int_0^\tau f(\frac{s}{\varepsilon}, x(s)) ds - \int_0^\tau f_0(x(s)) ds| \leq \int_0^\tau |f(\frac{s}{\varepsilon}, x(s)) - f(\frac{s}{\varepsilon}, \tilde{x}_n(s))| ds + \tag{6.83}$$

$$|\int_0^\tau [f(\frac{s}{\varepsilon}, \tilde{x}_n(s)) - f_0(\tilde{x}_n(s))] ds| + \int_0^\tau |f_0(x(s)) - f_0(\tilde{x}_n(s))| ds \leq$$

$$2LT_0 \sup_{x \in \mathcal{K}} \sup_{0 \leq \tau \leq T_0} |\tilde{x}_n(\tau) - x(\tau)| + |\int_0^\tau [f(\frac{s}{\varepsilon}, \tilde{x}_n(s)) - f_0(\tilde{x}_n(s))] ds|.$$

From (6.80) - (6.83) immediately we obtain the results in the lemma. $\qquad \square$

Proof. Now we will prove Theorem 6.20. Denote by $\psi(\tau, x, f, \varepsilon)$ (respectively $\bar{\psi}(\tau, x)$) a unique solution of equation

$$x' = f(\frac{\tau}{\varepsilon}, x)$$

(respectively (6.77)) passing through the point $x \in B[0, r]$ at the moment $\tau = 0$ and defined on $[0, \frac{T_0}{\varepsilon}]$. The functions $\psi(\tau, x, f, \varepsilon)$ and $\bar{\psi}(\tau, x)$ satisfy the integral equations

$$\psi(\tau, x, f, \varepsilon) = x + \int_0^\tau f(\frac{s}{\varepsilon}, \psi(s, x, f, \varepsilon)) ds$$

and

$$\bar{\psi}(\tau, x) = x + \int_0^\tau f_0(\bar{\psi}(s, x)) ds,$$

respectively. Using the condition of Lipschitz (6.74), we obtain the estimate

$$|\psi(\tau, x, f, \varepsilon) - \bar{\psi}(\tau, x)| = |\int_0^\tau [f(\frac{s}{\varepsilon}, \psi(s, x, f, \varepsilon)) - f_0(\bar{\psi}(s, x))] ds| \leq$$

$$\int_0^\tau |f(\frac{s}{\varepsilon}, \psi(s, x, f, \varepsilon)) - f(\frac{s}{\varepsilon}, \bar{\psi}(s, x))| ds + |\int_0^\tau [f(\frac{s}{\varepsilon}, \bar{\psi}(s, x)) - f_0(\bar{\psi}(s, x))] ds| \leq$$

$$L \int_0^\tau |\psi(s, x, f, \varepsilon) - \bar{\psi}(s, x)| ds + c(\varepsilon),$$

where

$$c(\varepsilon) := \sup_{0 \le \tau \le T_0, x \in \mathcal{K}} \left| \int_0^\tau [f_0 \frac{s}{\varepsilon}, x(s)) - f_0(x(s))] ds \right|.$$

According to the Gronwall-Bellman inequality (see, for example,[137] or [186]), we can now conclude that

$$|\psi(\tau, x, f, \varepsilon) - \bar{\psi}(\tau, x)| \le \exp(2L\tau) c(\varepsilon)$$

and it remains only to note that by virtue of Lemma 6.13, $c(\varepsilon) \to 0$ as $\varepsilon \to 0$ and

$$|x(t, \varepsilon) - y(\varepsilon t)| = |\psi(\tau, x, f, \varepsilon) - \bar{\psi}(\tau, x)| \le \exp(2L\tau) c(\varepsilon) = \exp(2L\varepsilon t)) c(\varepsilon)$$

for all $t \in [0, \frac{T_0}{\varepsilon}]$. The theorem is thus proved. $\qquad\qquad\square$

In the next subsection, we will also need the following lemma.

Lemma 6.14   *Let $\mathcal{F}$ be a transitive subset of $C(\mathbb{R} \times H, H)$, i.e. there exists a function $g \in \mathcal{F}$ such that $\mathcal{F} = H(g)$, where $H(f)$ if the hull of g. Then the following two assertions are equivalent:*

1. *There exists $f_0 \in C(H, H)$ such that*

$$\lim_{T \to +\infty} \frac{1}{T} \int_0^T f(t, x) dt = f_0(x)$$

   uniformly w.r.t. $f \in \mathcal{F}$ and $x \in B[0, r]$;
2. *There exists $f_0 \in C(H, H)$ such that*

$$\lim_{T \to +\infty} \frac{1}{T} \int_t^{t+T} g(\tau, x) d\tau = f_0(x)$$

   uniformly w.r.t. $t \in \mathbb{R}$ and $x \in B[0, r]$.

Proof. It is evident that 1. implies 2. because $g_t \in \mathcal{F}$ for all $t \in \mathbb{R}$ and, consequently,

$$\frac{1}{T} \int_t^{t+T} g(\tau, x) d\tau = \frac{1}{T} \int_0^T g(t + \tau, x) d\tau \to f_0(x)$$

as $T \to +\infty$ uniformly w.r.t. $t \in \mathbb{R}$ and $x \in B[0, r]$.

Let now $\varepsilon > 0$ and $f \in \mathcal{F} = H(g)$, then there exists a sequence $\{t_n\} \subset \mathbb{R}$ and $L(\varepsilon) > 0$ such that $g_{t_n} \to f$ and

$$\left| \frac{1}{T} \int_0^T g(\tau + t_n, x) d\tau - f_0(x) \right| < \varepsilon \tag{6.84}$$

for all $T > L(\varepsilon)$. Passing to limit as $n \to +\infty$ in the inequality (6.84) we obtain

$$|\frac{1}{T} \int_0^T f(\tau, x)d\tau - f_0(x)| \le \varepsilon$$

for all $T > L(\varepsilon)$. From the latter inequality, the required statement immediately follows. This proves the lemma. $\square$

Remark 6.6 *All the results of this subsection are true for arbitrary Banach space too, not only for Hilbert space.*

6.6.5 *Global averaging principle for the non-autonomous Lorenz systems*

Now we consider a global averaging principle for the non-autonomous Lorenz systems. Let Ω be a compact metric space and $(\Omega, \mathbb{R}, \sigma)$ be a dynamical system on Ω. We consider the "perturbed" non-autonomous Lorenz equation

$$\frac{dx}{dt} = \varepsilon A(\omega t)x + \varepsilon B(\omega t)(x, x) + \varepsilon f(\omega t), \tag{6.85}$$

where $\varepsilon \in [0, \varepsilon_0]$ $(\varepsilon_0 > 0)$ is a small parameter. Suppose that the conditions (6.59)–(6.62) are fulfilled and the following averaging values exist uniformly w.r.t. $\omega \in \Omega$:

$$\overline{A} = \lim_{T \to +\infty} \frac{1}{2T} \int_{-T}^{T} A(\omega t)dt, \tag{6.86}$$

$$\overline{B} = \lim_{T \to +\infty} \frac{1}{2T} \int_{-T}^{T} B(\omega t)dt, \tag{6.87}$$

and

$$\overline{f} = \lim_{T \to +\infty} \frac{1}{2T} \int_{-T}^{T} f(\omega t)dt. \tag{6.88}$$

Remark 6.7 *The conditions* (6.86) − (6.88) *are fulfilled if a dynamical system* $(\Omega, \mathbb{R}, \sigma)$ *is strictly ergodic, i.e. there exists on* Ω *a unique invariant w.r.t.* $(\Omega, \mathbb{R}, \sigma)$ *measure* μ.

Along with equation (6.85), we will also consider the averaged equation

$$\frac{dx}{dt} = \varepsilon\overline{A}x + \varepsilon\overline{B}(x, x) + \varepsilon\overline{f}. \tag{6.89}$$

If we introduce the "slow time" $\tau := \varepsilon t$ ($\varepsilon > 0$), then the equations (6.85) and (6.89) can be written as

$$\frac{dx}{d\tau} = A(\omega\frac{\tau}{\varepsilon})x + B(\omega\frac{\tau}{\varepsilon})(x,x) + f(\omega\frac{\tau}{\varepsilon}) \tag{6.90}$$

and

$$\frac{dx}{d\tau} = \overline{A}x + \overline{B}(x,x) + \overline{f}. \tag{6.91}$$

Remark 6.8 *a. From the conditions (6.61) and (6.87) it follows that*

$$Re\langle \overline{B}(u,v), v \rangle = 0 \tag{6.92}$$

for all $u, v \in H$;

b. From the inequality (6.59) it follows that

$$Re\langle \overline{A}x, x \rangle\rangle \leq -\alpha|x|^2 \tag{6.93}$$

for all $x \in H$.

Theorem 6.21 *Assume the conditions enumerated above are all satisfied. Then for all $T > 0$ and $\rho \geq r_0 := \frac{\|f\|}{\alpha} > 0$, the solution for the non-autonomous Lorenz equation (6.85) approaches the solution of the averaged Lorenz equation (6.89) in the following sense:*

$$\max\{|\varphi(t,x,\omega,\varepsilon) - \overline{\varphi}(t,x,\varepsilon)| : \ 0 \leq t \leq T/\varepsilon, \ |x| \leq \rho, \ and \ \omega \in \Omega\} \to 0 \tag{6.94}$$

as $\varepsilon \to 0$, where $\varphi(t,x,\omega,\varepsilon)$ (respectively $\overline{\varphi}(t,x,\varepsilon)$) is a solution of equation (6.85) (respectively (6.89)), passing through the point x at the initial moment $t = 0$.

Proof. According to Theorem 6.14, we have $|\varphi(t,x,\omega,\varepsilon)| \leq \rho$ and $|\overline{\varphi}(t,x,\varepsilon)| \leq \rho$ for all $t \geq 0$, $|x| \leq \rho, \omega \in \Omega$ and $\varepsilon \in (0,\varepsilon_0]$. If we take $\mathcal{F} := \{f_\omega \mid \omega \in \Omega\} \subset C(\mathbb{R} \times H, H)$, where $f_\omega(t,x) := A(\omega t)x + B(\omega t)(x,x) + f(\omega t)$ for all $t \in \mathbb{R}$ and $x \in H$, then the relation (6.94) follows from Theorem 6.20. This completes the proof. $\square$

Theorem 6.22 *(Global averaging principle for non-autonomous Lorenz systems) Let φ_ε be a cocycle generated by the equation (6.85). Assume the conditions enumerated above are all satisfied. If the cocycle φ_ε($\varepsilon \in [0,\varepsilon_0]$) is asymptotically compact, then the following assertions hold:*

(1) The averaged equation (6.91) admits a compact global attractor $\overline{I} \subset H$;
(2) For every $\varepsilon \in (0,\varepsilon_0]$ the equation (6.85) has a compact global attractor $\{I_\omega^\varepsilon \mid \omega \in \Omega\}$;
(3) The set $I := \cup\{I^\varepsilon \mid \varepsilon \in [0,\varepsilon_0]\}$ is bounded, where $I^0 := \overline{I}$ and $I^\varepsilon := \cup\{I_\omega^\varepsilon \mid \omega \in \Omega\}$;

(4)

$$\lim_{\substack{\varepsilon \to 0 \\ \omega \in \Omega}} \sup \beta(I_\omega^\varepsilon, \overline{I}) = 0$$

and, in particular,

$$\lim_{\varepsilon \to 0} \beta(I^\varepsilon, \overline{I}) = 0.$$

Proof. The first three statements of the theorem follow from Theorems 6.14, 6.15 and Remark 6.5. Now we will prove the fourth statement of the theorem. To this end, we will use the same arguments as in [70, 199]. Let $\lambda > 0$ and $B(\overline{I}, \lambda) = \{x \in H \mid \rho(x, \overline{I}) < \lambda\}$. According to orbital stability of the set $\overline{I}$ (see, for example, [179, Ch.I] or Theorem 1.6), for given λ there exists $\delta = \delta(\lambda) > 0$ (we may consider $\delta(\lambda) < \lambda/2$) such that

$$\overline{\varphi}(t, B(\overline{I}, \delta)) \subset B(\overline{I}, \lambda/2)$$

for all $t \geq 0$. By virtue of boundedness of the set $I = \cup\{I^\varepsilon \mid 0 \leq \varepsilon \leq \varepsilon_0\}$ we may choose $\rho \leq r_0$ such that $I \subset B(0, \rho) = \{x \in H \mid |x| < \rho\}$. Since $\overline{I}$ is a compact global attractor of the system (6.91), then for the closed ball $B[0, \rho] := \{x \in H \mid |x| \leq \rho\}$ and the number $\delta > 0$ there exists $T = T(\rho, \delta) > 0$ such that

$$\overline{\varphi}(t, B[0, \rho]) \subset B(\overline{I}, \delta/2), \quad t \geq T. \tag{6.95}$$

Let $x \in B[0, \rho]$. Then by virtue of Theorem 6.21 for the numbers $\rho \geq r_0$ and $T(\rho, \delta) > 0$ there exists $\mu = \mu(\rho, \delta) > 0$ such that $0 < \varepsilon \leq \mu$, $m(\varepsilon) < \lambda/2$ (see (6.94)), i.e.

$$|\varphi(t, x, \omega, \varepsilon) - \overline{\varphi}(t, x)| < \delta/2 \tag{6.96}$$

for all $x \in B[0, \rho]$, $\omega \in \Omega$, $t \in [0, T/\varepsilon]$ and $0 < \varepsilon \leq \mu$. According to (6.95) we have $\overline{\varphi}(T/\varepsilon, x, \omega, \varepsilon) \in B(\overline{I}, \delta/2)$. Thus, taking into account (6.96), we obtain $\varphi(T/\varepsilon, x, \omega, \varepsilon) \in B(\overline{I}, \delta)$. Let us take the initial point $x_1 := \varphi(T/\varepsilon, x, \omega, \varepsilon)$ and we will repeat for this point the same reasoning as above. Taking into consideration the equality $\varphi(t, x_1, \sigma(T/\varepsilon, \omega), \varepsilon) = \varphi(t + T/\varepsilon, x, \omega, \varepsilon)$, we will have

$$|\varphi(t + T/\varepsilon, x, \omega, \varepsilon) - \overline{\varphi}(t, x_1)| < \delta/2 \tag{6.97}$$

for all $t \in [0, T/\varepsilon]$, $x \in B[0, \rho]$ and $\omega \in \Omega$, where $x_1 = \varphi(T/\varepsilon, x, \omega, \varepsilon)$.

By the inequality (6.97) we obtain again $x_2 := \varphi(2T/\varepsilon, x, \omega, \varepsilon) \in B(\overline{I}, \delta)$ and, consequently,

$$\varphi(t + T/\varepsilon, x, \omega, \varepsilon) \in B(\overline{I}, \lambda/2 + \delta/2) \subset B(\overline{I}, \lambda).$$

If we continue this process and later (by virtue of uniformity w.r.t. $|x| \le \rho$ and $\omega \in \Omega$ of the estimation (6.96) it is possible), we will obtain

$$\varphi(t, x, \omega, \varepsilon) \in B(\overline{I}, \lambda) \tag{6.98}$$

for all $t \ge T/\varepsilon$, $x \in B[0, \rho]$, $\omega \in \Omega$ and $o \le \varepsilon \le \mu$ and, consequently,

$$\varphi(t, x, \sigma(-t, \omega), \varepsilon) \in B(\overline{I}, \lambda)$$

for all $t \ge T/\varepsilon$ and $|x| \le \rho$. Since $I = \cup\{I^\varepsilon \mid 0 \le \varepsilon \le \varepsilon_0\} \subseteq B(0, \rho)$, then according to Theorem 2.24

$$I_\omega^\varepsilon = \bigcap_{t \ge 0} \overline{\bigcup_{\tau \ge t} \varphi(\tau, B[0, \rho], \sigma(-\tau, \omega), \varepsilon)}.$$

Therefore, from (6.98) we have $I_\omega^\varepsilon \subset B(\overline{I}, \lambda)$ for all $\omega \in \Omega$ and $0 < \varepsilon < \mu$. Note that λ is arbitrarily chosen. Hence from the last inclusion we obtain the equality (6.95). The theorem is proved. $\qquad\square$

Chapter 7

Upper semi-continuity of attractors

7.1 Introduction

The problem of upper semi-continuity of global attractors for small perturbations is well studied (see, for example, [179] and references therein) for autonomous and periodical dynamical systems. In the works [40] and [41] this problem was studied for non-autonomous and random dynamical systems.

This chapter is devoted to a systematic study of the problem of upper semi-continuity of compact global attractors and compact pullback attractors of abstract non-autonomous dynamical systems for small perturbations. Several applications of our results are given for different classes of evolutionary equations.

The chapter is organized as follows. In section 2 we study some general properties of maximal compact invariant sets of dynamical systems. In particular, we prove that the compact global attractor and pullback attractor are maximal compact invariant sets (Theorem 7.2).

Section 3 contains the main results about upper semi-continuity of compact global attractors of abstract non-autonomous dynamical systems for small perturbations (Lemmas 7.3, 7.4 and Theorems 7.3, 7.4, 7.5 and 7.6).

In section 4 we give conditions for connectedness and component connectedness of global and pullback attractors (Theorem 7.7).

Section 5 is devoted to an application of our general results obtained in sections 2-4, to the study of different classes of non-autonomous differential equations (quasi-homogeneous systems, monotone systems, non-autonomously perturbed systems, non-autonomous 2D Navier-Stokes equations and quasi-linear functional-differential equations).

7.2 Maximal compact invariant sets

Let W be a complete metric space, $\mathbb{T} = \mathbb{R}$ or $\mathbb{Z}, \Omega$ be a compact metric space, $(\Omega, \mathbb{T}, \sigma)$ be a group dynamical system on Ω and $\langle W, \varphi, (\Omega, \mathbb{T}, \sigma)\rangle$ be a cocycle with fiber W, i.e. the mapping $\varphi : \mathbb{T}_+ \times W \times \Omega \to W$ is continuous and possesses the following properties: $\varphi(0, x, \omega) = x$ and $\varphi(t + \tau, x, \omega) = \varphi(t, \varphi(\tau, x, \omega), \omega t)$, where $\omega t = \sigma(t, \omega)$.

We denote by $X = W \times \Omega, g = pr_1 : X \mapsto W, (X, \mathbb{T}_+, \pi)$ a semi-group dynamical system on X defined by the equality $\pi = (\varphi, \sigma)$, i.e. $\pi^t x = (\varphi(t, u, \omega), \sigma(t, \omega))$ for every $t \in \mathbb{T}_+$ and $x = (u, \omega) \in X = W \times \Omega$. Let $\langle (X, \mathbb{T}_+, \pi), (\Omega, \mathbb{T}, \sigma), h \rangle$ be a non-autonomous dynamical system, where $h = pr_2 : X \mapsto \Omega$.

Definition 7.1   A family $\{I_\omega \mid \omega \in \Omega\}(I_\omega \subset W)$ of nonempty compact subsets of W is called a maximal compact invariant set of cocycle φ, if the following conditions are fulfilled:

(1) $\{I_\omega \mid \omega \in \Omega\}$ is invariant, i.e. $\varphi(t, I_\omega, \omega) = I_{\omega t}$ for every $\omega \in \Omega$ and $t \in \mathbb{T}_+$;
(2) $I = \bigcup\{I_\omega \mid \omega \in \Omega\}$ is relatively compact;
(3) $\{I_\omega \mid \omega \in \Omega\}$ is maximal, i.e. if the family $\{I'_\omega \mid \omega \in \Omega\}$ is relatively compact and invariant, then $I'_\omega \subseteq I_\omega$ for every $\omega \in \Omega$.

Lemma 7.1   *The family $\{I_\omega \mid \omega \in \Omega\}$ is invariant w.r.t. cocycle φ if and only if the set $J = \bigcup\{J_\omega \mid \omega \in \Omega\}(J_\omega := I_\omega \times \{\omega\})$ is invariant with respect to the skew-product dynamical system $(X, \mathbb{T}_+, \pi)$.*

Proof.   Let the family $\{I_\omega \mid \omega \in \Omega\}$ be invariant, $J = \bigcup\{J_\omega | \omega \in \Omega\}$ and $J_\omega = I_\omega \times \{\omega\}$. Then

$$\pi^t J = \bigcup\{\pi^t J_\omega \mid \omega \in \Omega\} = \bigcup\{(\varphi(t, I_\omega, \omega), \omega t) \mid \omega \in \Omega\} \qquad (7.1)$$

$$= \bigcup\{I_{\omega t} \times \{\omega t\} \mid \omega \in \Omega\} = \bigcup\{J_{\omega t} \mid \omega \in \Omega\} = J$$

for all $t \in \mathbb{T}_+$. From the equality (7.1) follows that the family $\{I_\omega \mid \omega \in \Omega\}$ is invariant w.r.t. cocycle φ if and only if a set J is invariant w.r.t. dynamical system $(X, \mathbb{T}_+, \pi)$. $\qquad\square$

Theorem 7.1   *Let the family of sets $\{I_\omega \mid \omega \in \Omega\}$ be maximal, compact and invariant. Then it is closed.*

Proof.   We note that the set $J = \bigcup\{J_\omega \mid \omega \in \Omega\}(J_\omega := I_\omega \times \{\omega\})$ is relatively compact and according to Lemma 7.1 it is invariant. Let $K = \overline{J}$, then K is compact. We will show that K is invariant. If $x \in K$, then there exists $\{x_n\} \subset J$ such that $x = \lim\limits_{n \to +\infty} x_n$. Thus $x_n \in J = \pi^t J$ for all $t \in \mathbb{T}_+$, then for $t \in \mathbb{T}_+$ there exists $\overline{x}_n \in J$ such that $x_n = \pi^t \overline{x}_n$. Since J is relatively compact, then we can suppose that the sequence $\{\overline{x}_n\}$ is convergent. We denote by $\overline{x} = \lim\limits_{n \to +\infty} \overline{x}_n$, then $\overline{x} \in \overline{J}, x = \pi^t \overline{x}$ and, consequently, $x \in \pi^t \overline{J}$ for all $t \in \mathbb{T}_+$, i.e. $\overline{J} = \pi^t \overline{J}$. Let $I' = pr_1 K$, then we have $I' = \bigcup\{I'_\omega \mid \omega \in \Omega\}$, where $I'_\omega = \{u \in W \mid (u, \omega) \in K\}$ and $K_\omega := I'_\omega \times \{\omega\}$.

Since the set K is invariant, then according to Lemma 7.1 the set I' is also invariant w.r.t. cocycle φ. The set I' is compact, because K is compact and $pr_1 : X \mapsto W$ is continuous. According to the maximality of the family $\{I_\omega \mid \omega \in \Omega\}$ we have $I'_\omega \subseteq I_\omega$ for every $\omega \in \Omega$ and, consequently, $I' \subseteq I$. On the other hand $I = pr_1 \overline{J} = I'$ and, consequently, $I' = I$. Thus the set I is compact. The theorem is proved. $\quad\square$

Denote by $C(W)$ the family of all compact subsets of W.

Definition 7.2 Recall that a family $\{I_\omega \mid \omega \in \Omega\}$ ($I_\omega \subset W$) of nonempty compact subsets of W is called

- a compact pullback attractor of the cocycle φ, if the following conditions are fulfilled:

  a. $I = \bigcup\{I_\omega \mid \omega \in \Omega\}$ is relatively compact ;

  b. I is invariant w.r.t. cocycle φ, i.e. $\varphi(t, I_\omega, \omega) = I_{\sigma(t,\omega)}$ for all $t \in \mathbb{T}_+$ and $\omega \in \Omega$;

  c. for every $\omega \in \Omega$ and $K \in C(W)$

  $$\lim_{t \to +\infty} \beta(\varphi(t, K, \omega^{-t}), I_\omega) = 0, \tag{7.2}$$

  where $\beta(A, B) = \sup\{\rho(a, B) : a \in A\}$ is a semi-distance of Hausdorff and $\omega^{-t} := \sigma(-t, \omega)$.

- a compact global attractor, if the following conditions are fulfilled:

  d. a family $\{I_\omega \mid \omega \in \Omega\}$ is compact and invariant;

  f. for every $K \in C(W)$

  $$\lim_{t \to +\infty} \sup_{\omega \in \Omega} \beta(\varphi(t, K, \omega), I) = 0, \tag{7.3}$$

  where $I = \bigcup\{I_\omega \mid \omega \in \Omega\}$.

Theorem 7.2 *A family $\{I_\omega \mid \omega \in \Omega\}$ of nonempty compact subsets of W will be maximal compact invariant set w.r.t. cocycle φ, if one of the following two conditions is fulfilled :*

a. $\{I_\omega \mid \omega \in \Omega\}$ is a compact pullback attractor w.r.t. cocycle φ;
b. $\{I_\omega \mid \omega \in \Omega\}$ is a compact global attractor w.r.t. cocycle φ.

Proof. a. Let the family $\{I_\omega \mid \omega \in \Omega\}$ be a compact pullback attractor. If the family $\{I'_\omega \mid \omega \in \Omega\}$ is a compact and invariant set of cocycle φ, then we have

$$\beta(I'_\omega, I_\omega) = \beta(\varphi(t, I'_{\omega^{-t}}, \omega^{-t}), I_\omega) \le \beta(\varphi(t, K, \omega^{-t}), I_\omega) \to 0$$

as $t \to +\infty$, where $K = \overline{\bigcup\{I'_\omega \mid \omega \in \Omega\}}$, and, consequently, $I'_\omega \subseteq I_\omega$ for every $\omega \in \Omega$, i.e. $\{I_\omega \mid \omega \in \Omega\}$ is maximal.

b. Let the family $\{I_\omega \mid \omega \in \Omega\}$ be a compact global attractor w.r.t. cocycle φ, then according to Theorem 2.24 it is a uniform compact pullback attractor and, consequently, the family $\{I_\omega \mid \omega \in \Omega\}$ is maximal compact invariant set of the cocycle φ. $\square$

Remark 7.1   The family $\{I_\omega \mid \omega \in \Omega\}$ $(I_\omega \subset W)$ is a maximal compact invariant w.r.t. cocycle φ if and only if the set $J = \bigcup \{J_\omega \mid \omega \in \Omega\}$, where $J_\omega = I_\omega \times \{\omega\}$, is a maximal compact invariant in the dynamical system $(X, \mathbb{T}, \pi)$.

7.3   Upper semi-continuity

Lemma 7.2   *Let $\{I_\omega \mid \omega \in \Omega\}$ be a maximal compact invariant set of cocycle φ, then the function $F : \Omega \mapsto C(W)$, defined by equality $F(\omega) := I_\omega$ is upper semi-continuous, i.e. for all $\omega_0 \in \Omega$*

$$\beta(F(\omega_k), F(\omega_0)) \to 0,$$

if $\rho(\omega_k, \omega_0) \to 0$.

Proof.   Let $\omega_0 \in \Omega$, $\omega_k \to \omega_0$ and suppose there exists $\varepsilon_0 > 0$ such that

$$\beta(F(\omega_k), F(\omega_0)) \geq \varepsilon_0.$$

Then there exists $x_k \in I_{\omega_k}$ such that

$$\rho(x_k, I_{\omega_0}) \geq \varepsilon_0. \tag{7.4}$$

As the set I is compact, without loss of generality we can suppose that the sequence $\{x_k\}$ is convergent. Denote by $x = \lim_{k \to +\infty} x_k$, then by virtue of Theorem 7.1 the set $I = \bigcup \{I_\omega \mid \omega \in \Omega\}$ is compact and hence there exists $\omega_0 \in \Omega$ such that $x \in I_{\omega_0} \subset I$.

On the other hand, according to the inequality (7.4) $x \notin I_{\omega_0}$. This contradiction shows that the function F is upper semi-continuous. $\square$

Lemma 7.3   *Let Λ be a compact metric space and $\varphi : \mathbb{T}_+ \times W \times \Lambda \times \Omega \mapsto W$ verify the following conditions:*

1. φ is continuous;

2. for every $\lambda \in \Lambda$ the function $\varphi_\lambda := \varphi(\cdot, \cdot, \lambda, \cdot) : \mathbb{T}_+ \times W \times \Omega \mapsto W$ is a cocycle on Ω with the fiber W;

3. the cocycle φ_λ admits a pullback attractor $\{I_\omega^\lambda \mid \omega \in \Omega\}$ for every $\lambda \in \Lambda$;

4. the set $\bigcup\{I^\lambda \mid \lambda \in \Lambda\}$ is relatively compact, where $I^\lambda = \bigcup\{I_\omega^\lambda \mid \omega \in \Omega\}$,

then the equality

$$\lim_{\lambda \to \lambda_0, \omega \to \omega_0} \beta(I_\omega^\lambda, I_{\omega_0}^{\lambda_0}) = 0 \tag{7.5}$$

holds for every $\lambda_0 \in \Lambda$ and $\omega_0 \in \Omega$ and

$$\lim_{\lambda \to \lambda_0} \beta(I_\lambda, I_{\lambda_0}) = 0 \tag{7.6}$$

for every $\lambda_0 \in \Lambda$.

Proof. Let $Y := \Lambda \times \Omega$ and $\mu : \mathbb{T} \times Y \mapsto Y$ be the mapping defined by the equality $\mu(t, (\lambda, \omega)) := (\lambda, \sigma(t, \omega))$ for every $t \in \mathbb{T}, \lambda \in \Lambda$ and $\omega \in \Omega$. It is clear that the triplet $(Y, \mathbb{T}, \mu)$ is the group dynamical system on Y and $\varphi : \mathbb{T}_+ \times W \times Y \mapsto W$ $(\varphi(t, x, (\lambda, \omega)) := \varphi(t, x, \lambda, \omega))$ is the continuous cocycle on $(Y, \mathbb{T}, \mu)$ with fiber W. Under the conditions of Lemma 7.3 the cocycle φ admits a maximal compact invariant set $\{I_y \mid y \in Y\}$ (where $I_y = I_{(\lambda, \omega)} = I_\omega^\lambda$) because

$$\bigcup \{I_y \mid y \in Y\} = \bigcup \{I_\omega^\lambda \mid \lambda \in \Lambda, \omega \in \Omega\} = \bigcup \{I^\lambda \mid \lambda \in \Lambda\}.$$

According to Lemma 7.2, the function $F : Y \mapsto C(W)$, defined by the equality $F(\lambda, \omega) := I_\omega^\lambda$, is upper semi-continuous and in particular the equality (7.6) holds.

We suppose that the equality (7.6) is not true, then there exist $\varepsilon_0 > 0, \lambda_0 \in \Lambda, \lambda_k \to \lambda_0, \omega_k \in \Omega$ and $x_k \in I_{\omega_k}^{\lambda_k}$ such that

$$\rho(x_k, I_{\lambda_0}) \geq \varepsilon_0. \tag{7.7}$$

Without loss of generality we can suppose that $\omega_k \to \omega_0, x_k \to x_0$ because the sets Ω and $\bigcup \{I_\lambda \mid \lambda \in \Lambda\}$ are compact. According to the inequality (7.7) we have

$$\rho(x_0, I_{\lambda_0}) \geq \varepsilon_0.$$

On the other hand $x_k \in I_{\omega_k}^{\lambda_k}$ and from the equality (7.6) we have

$$x_0 \in I_{\omega_0}^{\lambda_0} \subset I_{\lambda_0}$$

and, consequently,

$$\varepsilon_0 \leq \rho(x_0, I_{\lambda_0}) \leq \beta(I_{\omega_0}^{\lambda_0}, I_{\lambda_0}) = 0.$$

This contradiction shows that the equality (7.6) holds. $\qquad\square$

Corollary 7.1 *Under the conditions of Lemma 7.3 the equality*

$$\lim_{\lambda \to \lambda_0} \beta(I_\omega^\lambda, I_\omega^{\lambda_0}) = 0$$

holds for each $\omega \in \Omega$.

Remark 7.2 The article [41] contains a statement close to Corollary 7.1 in the case when the non-perturbed cocycle φ_{λ_0} is autonomous, i.e. the mapping $\varphi_{\lambda_0} : \mathbb{T}_+ \times W \times \Omega \to W$ does not depend of $\omega \in \Omega$.

Lemma 7.4   *Let the conditions of Lemma 7.3 and the following condition be fulfilled:*

5. *for certain $\lambda_0 \in \Lambda$ the application $F : \Omega \mapsto C(W)$, defined by equality $F(\omega) := I_\omega^{\lambda_0}$ is continuous, i.e. $\alpha(F(\omega), F(\omega_0)) \to 0$ if $\omega \to \omega_0$ for every $\omega_0 \in \Omega$, where α is the full metric of Hausdorff, i.e. $\alpha(A, B) := \max\{\beta(A, B), \beta(B, A)\}$.*

Then

$$\lim_{\lambda \to \lambda_0} \sup_{\omega \in \Omega} \beta(I_\omega^\lambda, I_\omega^{\lambda_0}) = 0. \tag{7.8}$$

Proof. Suppose that the equality (7.8) is not correct, then there exist $\varepsilon_0 > 0, \lambda_k \to \lambda_0, \omega_k \in \Omega$ such that

$$\beta(I_{\omega_k}^{\lambda_k}, I_{\omega_k}^{\lambda_0}) \geq \varepsilon_0. \tag{7.9}$$

On the other hand we have

$$\varepsilon_0 \leq \beta(I_{\omega_k}^{\lambda_k}, I_{\omega_k}^{\lambda_0}) \leq \beta(I_{\omega_k}^{\lambda_k}, I_{\omega_0}^{\lambda_0}) + \beta(I_{\omega_0}^{\lambda_0}, I_{\omega_k}^{\lambda_0})$$
$$\leq \beta(I_{\omega_k}^{\lambda_k}, I_{\omega_0}^{\lambda_0}) + \alpha(I_{\omega_k}^{\lambda_0}, I_{\omega_0}^{\lambda_0}). \tag{7.10}$$

According to Lemma 7.3 (see the equality (7.5))

$$\lim_{k \to +\infty} \beta(I_{\omega_k}^{\lambda_k}, I_{\omega_0}^{\lambda_0}) = 0. \tag{7.11}$$

Under the condition 5. of Lemma 7.4 we have

$$\lim_{k \to +\infty} \alpha(I_{\omega_k}^{\lambda_0}, I_{\omega_0}^{\lambda_0}) = 0. \tag{7.12}$$

From (7.10)-(7.12) passing to the limit as $k \to +\infty$, we obtain $\varepsilon_0 \leq 0$. This contradiction shows that the equality (7.8) holds. $\square$

Definition 7.3   The family of cocycle $\{\varphi_\lambda\}_{\lambda \in \Lambda}$ is called collectively compact dissipative (uniformly collectively compact dissipative), if there exists a nonempty compact set $K \subseteq W$ such that

$$\lim_{t \to +\infty} \sup\{\beta(U_\lambda(t, \omega)M, K)|\omega \in \Omega\} = 0 \qquad \forall \lambda \in \Lambda \tag{7.13}$$

$$(\text{respectively } \lim_{t \to +\infty} \sup\{\beta(U_\lambda(t, \omega)M, K) \mid \omega \in \Omega, \lambda \in \Lambda\} = 0)$$

for all $M \in C(W)$, where $U_\lambda(t, \omega) := \varphi_\lambda(t, \cdot, \omega)$.

Lemma 7.5   *The following conditions are equivalent:*

1. *the family of cocycles $\{\varphi_\lambda\}_{\lambda \in \Lambda}$ is collectively compact dissipative;*
22.a. *every cocycle φ_λ $(\lambda \in \Lambda)$ is compact dissipative;*
  2.b. *the set $\bigcup\{I^\lambda \mid \lambda \in \Lambda\}$ is compact.*

Proof. According to the equality (7.13) every cocycle φ_λ $(\lambda \in \Lambda)$ is compact dissipative and $\bigcup\{I^\lambda \mid \lambda \in \Lambda\} \subseteq K$.

Suppose that the conditions 2.a. and 2.b. hold. Let $K := \overline{\bigcup\{I^\lambda \mid \lambda \in \Lambda\}}$, then the equality (7.13) holds. $\square$

Let $\{\varphi_\lambda\}_{\lambda \in \Lambda}$ be a family of cocycles on $(\Omega, \mathbb{T}, \sigma)$ with fiber W and $\tilde{\Omega} := \Omega \times \Lambda$. On $\tilde{\Omega}$, we define a dynamical system $(\tilde{\Omega}, \mathbb{T}, \tilde{\sigma})$ by equality $\tilde{\sigma}(t, (\omega, \lambda)) := (\sigma(t, \omega), \lambda)$ for all $t \in \mathbb{T}, \omega \in \Omega$ and $\lambda \in \Lambda$. By family of cocycles $\{\varphi_\lambda\}_{\lambda \in \Lambda}$ is generated a cocycle $\tilde{\varphi}$ on $(\tilde{\Omega}, \mathbb{T}, \tilde{\sigma})$ with fiber W, defined in the following way: $\tilde{\varphi}(t, w, (\omega, \lambda)) := \varphi_\lambda(t, w, \omega)$ for all $t \in \mathbb{T}_+, w \in W, \omega \in \Omega$ and $\lambda \in \Lambda$.

Lemma 7.6 *The following conditions are equivalent:*

(1) the family of cocycles $\{\varphi_\lambda\}_{\lambda \in \Lambda}$ is uniformly collectively compact dissipative;
(2) the cocycle $\tilde{\varphi}$ is compact dissipative.

Proof. This assertion follows from the fact that

$$\sup\{\beta(\tilde{U}(t, \tilde{\omega})M, K) \mid \tilde{\omega} \in \tilde{\Omega}\}$$
$$= \sup\{\beta(U_\lambda(t, \omega)M, K) \mid \omega \in \Omega, \ \lambda \in \Lambda\},$$

where $\tilde{U}(t, \tilde{\omega}) := \tilde{\varphi}(t, \cdot, \tilde{\omega})$, and from the corresponding definitions. $\square$

Theorem 7.3 *Let Λ be a compact metric space and $\{\varphi_\lambda\}_{\lambda \in \Lambda}$ be a family of uniformly collectively compact dissipative cocycles on $(\Omega, \mathbb{T}, \sigma)$ with fiber W, then the following statements are true:*

(1) every cocycle φ_λ $(\lambda \in \Lambda)$ is compact dissipative;
(2) the family of compacts $\{I_\omega^\lambda \mid \omega \in \Omega\} = I^\lambda$ is a compact global attractor of cocycle φ_λ, where $I_\omega^\lambda := I_{(\omega, \lambda)}$ and $I := \{I_{(\omega, \lambda)} \mid (\omega, \lambda) \in \tilde{\Omega}\}$ is a Levinson center of cocycle $\tilde{\varphi}$;
(3) the set $\bigcup\{I^\lambda \mid \lambda \in \Lambda\}$ is compact.

Proof. Consider the cocycle $\tilde{\varphi}$ generated by the family of cocycles $\{\varphi_\lambda\}_{\lambda \in \Lambda}$. According to Lemma 7.6 $\tilde{\varphi}$ is compact dissipative and in virtue of the Theorem 2.24 the following assertions take place:

1. $I_{\tilde{\omega}} = \Omega_{\tilde{\omega}}(K) \neq \emptyset$, is compact, $I_{\tilde{\omega}} \subseteq K$ and

$$\lim_{t \to +\infty} \beta(\tilde{U}(t, \tilde{\omega}^{-t})M, I_{\tilde{\omega}}) = 0 \qquad (7.14)$$

for every $\tilde{\omega} \in \tilde{\Omega}$, where

$$\Omega_{\tilde{\omega}}(K) = \bigcap_{t \geq 0} \overline{\bigcup_{\tau \geq t} \tilde{U}(\tau, \tilde{\omega}^{-\tau})K}, \qquad (7.15)$$

$\tilde{\omega}^{-\tau} := \tilde{\sigma}(-\tau, \tilde{\omega})$ and K is a nonempty compact appearing in the equality (7.13);

2. $\tilde{U}(t,\tilde{\omega})I_{\tilde{\omega}} = I_{\tilde{\omega}t}$ for all $\tilde{\omega} \in \tilde{\Omega}$ and $t \in \mathbb{T}_+$;

3. the set $I = \bigcup\{I_{\tilde{\omega}} \mid \tilde{\omega} \in \tilde{\Omega}\}$ is compact.

To finish the proof we note that from the collective compact dissipativity of the family of cocycles $\{\varphi_\lambda\}_{\lambda\in\Lambda}$ it follows that every cocycle φ_λ will be compact dissipative. Let $\{I_\omega^\lambda \mid \omega \in \Omega\} = I^\lambda$ be a Levinson center of the cocycle φ_λ, then according to Theorem 2.24

$$I_\omega^\lambda = \bigcap_{t \geq 0} \overline{\bigcup_{\tau \geq t} U_\lambda(\tau, \omega^{-\tau})K}. \tag{7.16}$$

From (7.15) and (7.16) it follows that $I_\omega^\lambda = \Omega_{\tilde{\omega}}(K) = I_{\tilde{\omega}}$ and, consequently, $I^\lambda = \bigcup\{I_\omega^\lambda \mid \omega \in \Omega\} \subseteq \bigcup\{I_\omega^\lambda \mid \omega \in \Omega,\ \lambda \in \Lambda\} = I$ for all $\lambda \in \Lambda$. Thus $\bigcup\{I^\lambda \mid \lambda \in \Lambda\} \subseteq I$ and, consequently, it is compact. $\qquad\square$

Definition 7.4   The family $\{(X,\mathbb{T}_+,\pi_\lambda)\}_{\lambda\in\Lambda}$ of autonomous dynamical systems is called collectively (uniformly collectively) asymptotic compact if for every bounded positive invariant set $M \subseteq X$ (i.e. $\pi_\lambda(t,M) \subseteq M$ for all $t \in \mathbb{T}_+$ and $\lambda \in \Lambda$) there exists a nonempty compact K such that

$$\lim_{t\to+\infty} \beta(\pi_\lambda^t M, K) = 0 \qquad \forall\, \lambda \in \Lambda \tag{7.17}$$

$$(\lim_{t\to+\infty} \sup_{\lambda\in\Lambda} \beta(\pi_\lambda^t M, K) = 0).$$

Definition 7.5   The bounded set $K \subset X$ is called absorbing (uniformly absorbing) for the family $\{(X,\mathbb{T}_+,\pi_\lambda)\}_{\lambda\in\Lambda}$ of autonomous dynamical systems if for any bounded subset $B \subset X$ there exists a number $L = L(\lambda, B) > 0$ $(L = L(B) > 0)$ such that $\pi_\lambda^t B \subseteq K$ for all $t \geq L(\lambda, B)$ $(t \geq L(B))$ and $\lambda \in \Lambda$.

Theorem 7.4   *Let Λ be a complete metric space. If the family $\{(X,\mathbb{T}_+,\pi_\lambda)\}_{\lambda\in\Lambda}$ of autonomous dynamical systems admits an absorbing bounded set $K \subset X$ and is collectively asymptotic compact, then $\{(X,\mathbb{T}_+,\pi_\lambda)\}_{\lambda\in\Lambda}$ admits a global compact attractor, i.e. there exists a nonempty compact set $K \subset X$ such that*

$$\lim_{t\to+\infty} \beta(\pi_\lambda^t B, K) = 0 \tag{7.18}$$

for all $\lambda \in \Lambda$ and bounded $B \subset X$.

Proof.   Let the family $\{(X,\mathbb{T}_+,\pi_\lambda)\}_{\lambda\in\Lambda}$ of autonomous dynamical systems be collectively asymptotic compact and a bounded M be its absorbing set. According to Theorem 1.5 the nonempty set $K = \Omega(M)$ is compact and the equality (7.18) holds. The theorem is proved. $\qquad\square$

Theorem 7.5 *Let Λ be a complete compact metric space. If the family $\{(X, \mathbb{T}_+, \pi_\lambda)\}_{\lambda \in \Lambda}$ of autonomous dynamical systems admits a uniformly absorbing bounded set $K \subset X$ and it is uniformly collectively asymptotic compact, then $\{(X, \mathbb{T}_+, \pi_\lambda)\}_{\lambda \in \Lambda}$ admits a uniform compact global attractor, i.e. there exists a nonempty compact set $K \subset X$ such that*

$$\lim_{t \to +\infty} \sup_{\lambda \in \Lambda} \beta(\pi_\lambda^t B, K) = 0 \qquad (7.19)$$

for all bounded $B \subset X$.

Proof. Consider the autonomous dynamical system $(\tilde{X}, \mathbb{T}_+, \tilde{\pi})$ on $\tilde{X} := X \times \Lambda$ defined by equality $\tilde{\pi}(t, (x, \lambda)) := (\pi_\lambda(t, x), \lambda)$ for all $t \in \mathbb{T}_+, x \in X$ and $\lambda \in \Lambda$. We note that under the conditions of Theorem 7.5 the bounded set $K \times \Lambda$ is absorbing for dynamical system $(\tilde{X}, \mathbb{T}_+, \tilde{\pi})$ if the set K is uniformly absorbing for the family $\{(X, \mathbb{T}_+, \pi_\lambda)\}_{\lambda \in \Lambda}$ and $(\tilde{X}, \mathbb{T}_+, \tilde{\pi})$ is asymptotically compact. According to Theorem 1.24 the dynamical system $(\tilde{X}, \mathbb{T}_+, \tilde{\pi})$ admits a compact global attractor $\tilde{K} \subset \tilde{X} := X \times \Lambda$. To finish the proof it is sufficient to note that the set $K := pr_1 \tilde{K} \subset X$ is compact and

$$\sup_{\lambda \in \Lambda} \beta(\pi_\lambda^t B, K) \le \beta(\tilde{\pi}_\lambda^t B, K_0) \to 0$$

as $t \to +\infty$, where $K_0 := K \times \Lambda \supset \tilde{K}$, for all bounded subset $B \subset X$. $\qquad \square$

Let φ be a cocycle on $(\Omega, \mathbb{T}, \sigma)$ with fiber W and $(X, \mathbb{T}_+, \pi)$ be a skew-product dynamical system, where $X := W \times \Omega$ and $\pi(t, (w, \omega)) := (\varphi(t, w, \omega), \omega t)$ for all $t \in \mathbb{T}_+, w \in W$ and $\omega \in \Omega$.

Definition 7.6 The cocycle φ is called asymptotically compact (a family of cocycles $\{\varphi_\lambda\}_{\lambda \in \Lambda}$ is called collectively asymptotically compact) if a skew-product dynamical system $(X, \mathbb{T}_+, \pi)$ (a family of skew-product dynamical systems $(X, \mathbb{T}_+, \pi_\lambda)_{\lambda \in \Lambda}$) is asymptotically compact.

Theorem 7.6 *Let Ω and Λ be compact metric spaces, W be a Banach space and $\{\varphi_\lambda\}_{\lambda \in \Lambda}$ be a family of cocycles on $(\Omega, \mathbb{T}, \sigma)$ with fiber W. If there exist $r > 0$ and the function $V_\lambda : W \times \Omega \to \mathbb{R}_+$ for all $\lambda \in \Lambda$, with the following properties:*

(1) the family of cocycles $\{\varphi_\lambda\}_{\lambda \in \Lambda}$ is collectively asymptotically compact;

(2) the family of functions $\{V_\lambda\}_{\lambda \in \Lambda}$ is collectively bounded on bounded sets and for every $c \in \mathbb{R}_+$ the sets $\{x \in X_r \mid V_\lambda(x) \le c\}$ uniformly bounded;

(3) $V_\lambda'(w, \omega) \le -c(|w|)$ for all $w \in W_r = \{w \in W \mid |w| \ge r\}, \omega \in \Omega$ and $\lambda \in \Lambda$, where $c : \mathbb{R}_+ \to \mathbb{R}_+$ is positive on $[r, +\infty)$, $V_\lambda'(w, \omega) := \limsup_{t \to 0+} t^{-1}[V_\lambda(\varphi_\lambda(t, w, \omega), \omega t) - V_\lambda(w, \omega)]$ if $\mathbb{T} = \mathbb{R}_+$ and $V_\lambda'(w, \omega) := V_\lambda(\varphi_\lambda(1, w, \omega), \sigma(1, \omega)) - V_\lambda(w, \omega)$ if $\mathbb{T} = \mathbb{Z}_+$.

Then every cocycle φ_λ ($\lambda \in \Lambda$) admits a uniform compact global attractor I^λ ($\lambda \in \Lambda$) and the set $\bigcup\{I^\lambda \mid \lambda \in \Lambda\}$ is compact.

Proof.   Let $X := W \times \Omega$ and $(X, \mathbb{T}, \pi_\lambda)$ be a skew-product dynamical system, generated by the cocycle φ_λ , then (X, h, Ω) , where $h := pr_2 : X \to \Omega$, is a trivial fiber bundle with fiber W. Under the conditions of Theorem 7.6 and according to Theorem 5.2 the non-autonomous dynamical system $\langle (X, \mathbb{T}_+, \pi_\lambda), (\Omega, \mathbb{T}, \sigma), h \rangle$ admits a compact global attractor J^λ and according to Theorem 2.24 the cocycle φ_λ admits a compact global attractor $I^\lambda = \{I^\lambda_\omega \mid \omega \in \Omega\}$, where $I^\lambda_\omega := pr_1 J^\lambda_\omega$ and $J^\lambda_\omega = pr_2^{-1}(\omega) \bigcap J^\lambda$.

Let $\tilde{\Omega} := \Omega \times \Lambda$, $(\tilde{\Omega}, \mathbb{T}, \tilde{\sigma})$ be a dynamical system on $\tilde{\Omega}$ defined by the equality $\tilde{\sigma}(t, (\omega, \lambda)) := (\sigma(t, \omega), \lambda)$ (for all $t \in \mathbb{T}, \omega \in \Omega$ and $\lambda \in \Lambda$), $\tilde{X} := W \times \tilde{\Omega}$ and $(\tilde{X}, \mathbb{T}_+, \tilde{\pi})$ be an autonomous dynamical system defined by equality $\tilde{\pi}(t, (w, \tilde{\omega})) := (\pi_\lambda(t, w), (\omega t, \lambda))$ for all $\tilde{\omega} := (\omega, \lambda) \in \tilde{\Omega} := \Omega \times \Lambda$. Note that the triplet $(\tilde{X}, h, \Omega)$, where $h := pr_2 : \tilde{X} \to \tilde{\Omega}$, is a trivial fiber bundle with fiber $W, \langle (\tilde{X}, \mathbb{T}_+, \tilde{\pi}), (\tilde{\Omega}, \mathbb{T}, \tilde{\sigma}), h \rangle$ is a non-autonomous dynamical system. The function $\tilde{V} : \tilde{X}_r := W_r \times \tilde{\Omega} \to \mathbb{R}_+$, defined by the equality $\tilde{V}(\tilde{x}) := V_\lambda(w, \omega)$ for all $\tilde{x} := (w, (\omega, \lambda)) \in \tilde{X}_r$ under the conditions of Theorem 7.6, all the conditions of Theorem 5.3 hold and, consequently, the dynamical system $(\tilde{X}, \mathbb{T}_+, \tilde{\pi})$ admits a compact global attractor. To finish the proof of the theorem it is sufficiently to note that if the dynamical system $(\tilde{X}, \mathbb{T}_+, \tilde{\pi})$ admits a compact global attractor $\tilde{J}$, then the family of cocycles $\{\varphi_\lambda\}_{\lambda \in \Lambda}$ is uniformly collectively compact dissipative and according to Theorem 7.3 the set $I = \bigcup\{I^\lambda \mid \lambda \in \Lambda\}$ is compact, where $I^\lambda = \{I^\lambda_\omega \mid \omega \in \Omega\}$ is the compact global attractor of cocycle φ_λ. The theorem is proved. $\square$

7.4   Connectedness

Recall that the space W possesses the property (S) if for every compact $K \in C(W)$ there exists a compact connected set $V \in C(W)$ such that $K \subseteq V$.

If $M \subseteq W$, for each $\omega \in \Omega$, we write

$$\Omega_\omega(M) = \bigcap_{t \geq 0} \overline{\bigcup_{\tau \geq t} \varphi(\tau, M, \omega^{-\tau})} .$$

Lemma 7.7   *Suppose that the cocycle φ admits a compact pullback attractor $\{I_\omega \mid \omega \in \Omega\}$, then the following assertions hold:*

a. *$\emptyset \neq \Omega_\omega(M) \subseteq I_\omega$ for every $M \in C(W)$ and $\omega \in \Omega$;*

b. *the family $\{\Omega_\omega(M) \mid \omega \in \Omega\}$ is compact and invariant w.r.t. cocycle φ for every $M \in C(W)$;*

c. if $I = \bigcup\{I_\omega \mid \omega \in \Omega\} \subseteq M$, then the following inclusion $I_\omega \subseteq \Omega_\omega(M)$ holds for every $\omega \in \Omega$.

Proof. The first and second assertions follow from the definition of pullback attractor and from the equalities (2.83)-(2.84).

Let I be a subset of M, then

$$I_\omega = \varphi(t, I_{\omega-t}, \omega^{-t}) \subseteq \varphi(t, I, \omega^{-t}) \subseteq \varphi(t, M, \omega^{-t}) \tag{7.20}$$

and according to the equality (2.83) we have $I_\omega \subseteq \Omega_\omega(M)$. $\qquad\square$

Theorem 7.7 *Let W possess the property (S) and let the cocycle φ admit a compact pullback attractor $\{I_\omega \mid \omega \in \Omega\}$, then:*

(1) the set I_ω is connected for every $\omega \in \Omega$;

(2) if the space Ω is connected, then the set $I = \bigcup\{I_\omega | \omega \in \Omega\}$ also is connected.

Proof. 1. Since the equality (7.2) holds and the space W possesses the property (S), then there exists a connected compact $V \in C(W)$ such that $I \subseteq V$ and

$$\lim_{t \to +\infty} \beta(\varphi(t, V, \omega^{-t}), I_\omega) = 0, \tag{7.21}$$

for every $\omega \in \Omega$. We shall show that the set I_ω is connected. If we suppose that it is not true, then there are $A_1, A_2 \neq \emptyset$, closes and $A_1 \bigsqcup A_2 = I_\omega$. Let $0 < \varepsilon_0 < d(A_1, A_2)$ and $L = L(\varepsilon_0) > 0$ be such that

$$\beta(\varphi(t, V, \omega^{-t}), I_\omega) < \frac{\varepsilon_0}{3} \tag{7.22}$$

for all $t \geq L(\varepsilon_0)$. We note that the set $\varphi(t, V, \omega^{-t})$ is connected and according to the inclusion (7.20) and the inequality (7.22) the following condition

$$\varphi(t, V, \omega^{-t}) \bigcap (W \setminus [B(A_1, \tfrac{\varepsilon_0}{3}) \bigsqcup B(A_2, \tfrac{\varepsilon_0}{3})]) \neq \emptyset$$

is fulfilled for every $t \geq L(\varepsilon_0)$ and $\omega \in \Omega$, where $B(A, \varepsilon) := \{u \in W \mid \rho(u, A) < \varepsilon\}$. Then there exists $t_n \to +\infty$ and $u_n \in W$ such that

$$u_n \in \varphi(t_n, V, \omega^{-t_n}) \bigcap (W \setminus [B(A_1, \tfrac{\varepsilon_0}{3}) \bigsqcup B(A_2, \tfrac{\varepsilon_0}{3})]). \tag{7.23}$$

According to the equality (7.21) it is possible to suppose that the sequence $\{u_n\}$ is convergent. We denote by $u := \lim_{n \to +\infty} u_n$, then from Lemma 2.15 follows that $u \in \Omega_\omega(V)$. Since $I \subseteq V$, then according to Lemma 2.15 we have $u \in \Omega_\omega(V) \subseteq I_\omega \subseteq I$. On the other hand according to (7.23) we have $u \notin B(A_1, \tfrac{\varepsilon_0}{3}) \bigsqcup B(A_2, \tfrac{\varepsilon_0}{3})$. This contradiction shows that the set I_ω is connected.

2. Let the space Ω be connected. According to Lemma 7.3 the function $F : \Omega \mapsto C(W)$, defined by equality $F(\omega) = I_\omega$ is upper semi-continuous and from the

corollary 1.8.13 [83] (see also Lemma 3.1 [148]) follows that the set $I = \bigcup\{I_\omega \mid \omega \in \Omega\} = F(\Omega)$ is connected. $\qquad\square$

Corollary 7.2   *Let W be a metric space with the property (S) and let the cocycle φ admit a compact global attractor $\{I_\omega \mid \omega \in \Omega\}$, then:*

(1) the set I_ω is connected for every $\omega \in \Omega$;

(2) if the space Ω is connected, then the set $I = \bigcup\{I_\omega \mid \omega \in \Omega\}$ also is connected.

Proof.   This assertion follows from Theorems 7.2, 7.7 and Lemma 7.2. $\qquad\square$

7.5   Applications

7.5.1   *Quasi-homogeneous systems*

Let E and G be two finite dimensional spaces.

Definition 7.7   The function $f \in C(E \times G, E)$ is called [94, 102] homogeneous of order m with respect to variable $u \in E$ if the equality $f(\lambda u, \omega) = \lambda^m f(u, \omega)$ holds for all $\lambda \geq 0, u \in E$ and $\omega \in G$.

Theorem 7.8   *Let $f \in C^1(E), \Phi \in C^1(G), \Omega \subseteq G$ be a compact invariant set of dynamical system*

$$\omega' = \Phi(\omega), \tag{7.24}$$

the function f be homogeneous (of order $m > 1$) and a zero solution of equation

$$u' = f(u) \tag{7.25}$$

be uniformly asymptotically stable. If $F \in C^1(E \times G, E)$ and

$$|F(u, \omega)| \leq c|u|^m$$

for all $|u| \geq r$ and $\omega \in \Omega$, where r and c are certain positive numbers, then there exists a positive number λ_0 such that for all $\lambda \in \Lambda = [-\lambda_0, \lambda_0]$ the following holds:

(1) a set $I_\omega^\lambda := \{u \in E \mid \sup\{|\varphi_\lambda(t, u, \omega)| : t \in \mathbb{R}\} < +\infty\}$ is not empty, compact and connected for each $\omega \in \Omega$, where $\varphi_\lambda(t, u, \omega)$ is a unique solution of equation $u' = f(u) + \lambda F(u, \omega t)$ satisfying the initial condition $\varphi_\lambda(0, u, \omega) = u$;

(2) $\varphi_\lambda(t, I_\omega^\lambda, \omega) = I_{\sigma(t,\omega)}^\lambda$ for all $t \in \mathbb{R}_+$ and $\omega \in \Omega$;

(3) the set $I^\lambda = \bigcup\{I_\omega^\lambda \mid \omega \in \Omega\}$ is compact and connected;

(4) the equalities

$$\lim_{t \to +\infty} \beta(\varphi_\lambda(t, M, \omega_{-t}), I_\omega^\lambda) = 0 \tag{7.26}$$

and

$$\lim_{t \to +\infty} \beta(\varphi_\lambda(t, M, \omega), I^\lambda) = 0$$

take place for all $\lambda \in \Lambda, \omega \in \Omega$ and bounded subset $M \subseteq E$.

(5) the set $\bigcup\{I^\lambda \mid \lambda \in \Lambda\}$ is compact;

(6) the equality

$$\lim_{\lambda \to 0} \sup_{\omega \in \Omega} \beta(I^\lambda_\omega, 0) = 0$$

holds.

Proof. Under the condition of Theorem 7.8 according to Theorem 5.34 by the equality

$$V(u) = \int_0^{+\infty} |\pi(t, u)|^k dt,$$

where $\pi(t, u)$ is a solution of equation (7.25) with condition $\pi(0, u) = u$, is defined a continuously differentiable function $V : E \to \mathbb{R}_+$, verifying the following conditions:

a. $V(\mu u) = \mu^{k-m+1} V(u)$ for all $\mu \geq 0$ and $u \in E$;

b. there exist positive numbers α and β such that $\alpha|u|^{k-m+1} \leq V(u) \leq \beta|u|^{k-m+1}$ for all $u \in E$;

c. $V'(u) = DV(u)f(u) = -|u|^k$ for all $u \in E$, where $DV(u)$ is a derivative of Frechet of function V in the point u.

Let us define a function $\mathfrak{V} : X \to \mathbb{R}_+$ $(X := E \times \Omega)$ in the following way: $\mathfrak{V}(u, \omega) := V(u)$ for all $(u, \omega) \in X$. Note that

$$\mathfrak{V}'(u, \omega) := \frac{d}{dt} V(\varphi_\lambda(t, u, \omega))|_{t=0} = -|u|^k + DV(u)\lambda F(u, \omega)$$

and there exists $\lambda_0 > 0$ such that the inequality

$$\mathfrak{V}'(u, \omega) \leq -\nu|u|^k$$

holds for all $\omega \in \Omega$ and $|u| \geq r$, where $\nu = 1 - \lambda_0 cL > 0$ (see the proof of Theorem 4.3 [102]).

For finishing the proof of the theorem it is sufficient to refer to Theorem 7.6 and Lemma 7.4. $\qquad \square$

Theorem 7.9 *Let $f \in C^1(E \times G, E), \Phi \in C^1(F), \Omega \subseteq G$ be a compact invariant set of dynamical system (7.24), the function f be homogeneous (of order $m = 1$) w.r.t. variable $u \in E$ and a zero solution of equation*

$$u' = f(u, \omega t) \quad (\omega \in \Omega) \tag{7.27}$$

be uniformly asymptotically stable. If $|F(u, \omega)| \le c|u|$ for all $|u| \ge r$ and $\omega \in \Omega$, where r and c are certain positive numbers, then there exists a positive number λ_0 such that for all $\lambda \in \Lambda = [-\lambda_0, \lambda_0]$ the following assertions take place:

(1) a set $I_\omega^\lambda := \{u \in E \mid \sup\{|\varphi_\lambda(t, u, \omega)| : t \in \mathbb{R}\} < +\infty\}$ is not empty, compact and connected for each $\omega \in \Omega$, where $\varphi_\lambda(t, u, \omega)$ is a unique solution of equation

$$u' = f(u, \omega t) + \lambda F(u, \omega t)$$

verifying the initial condition $\varphi_\lambda(0, u, \omega) = u$;

(2) $\varphi_\lambda(t, I_\omega^\lambda, \omega) = I_{\sigma(t,\omega)}^\lambda$ for all $t \in \mathbb{R}_+$ and $\omega \in \Omega$;

(3) a set $I^\lambda := \bigcup\{I_\omega^\lambda \mid \omega \in \Omega\}$ is compact and connected;

(4) the equalities

$$\lim_{t \to +\infty} \beta(\varphi_\lambda(t, M, \omega_{-t}), I_\omega^\lambda) = 0$$

and

$$\lim_{t \to +\infty} \beta(\varphi_\lambda(t, M, \omega), I^\lambda) = 0 \tag{7.28}$$

take place for all $\lambda \in \Lambda, \omega \in \Omega$ and bounded subset $M \subseteq E$.

(5) the set $\bigcup\{I^\lambda \mid \lambda \in \Lambda\}$ is compact;

(6) the following equality holds:

$$\lim_{\lambda \to 0} \sup_{\omega \in \Omega} \beta(I_\omega^\lambda, 0) = 0 \, .$$

Proof. The proof of this assertion is similar to the proof of the theorem 7.8. $\quad\square$

7.5.2 *Monotone systems*

Let $f \in C(E \times \Omega, E)$ be a function satisfying

$$\mathrm{Re}\langle f(u_1, \omega) - f(u_2, \omega), u_1 - u_2 \rangle \le -k|u_1 - u_2|^\alpha \tag{7.29}$$

for all $\omega \in \Omega$ and $u_1, u_2 \in E$ ($k > 0$ and $\alpha \ge 2$).

Theorem 7.10 [316, 92] *If the function f verifies the condition (7.29), then the following statements are true:*

(1) the set $I_\omega = \{u \in E \mid \sup_{t \in \mathbb{R}} |\varphi(t, u, \omega)| < +\infty\}$ contains a single point $\{\varphi(\omega)\}$ for all $\omega \in \Omega$, where $\varphi(t, u, \omega)$ is a solution of equation (7.26) with condition $\varphi(0, u, \omega) = u$;

(2) the inequalities

$$|\varphi(t, u, \omega) - \varphi(\omega t)| \le e^{-kt}|u - \varphi(u)| \quad (\text{if } \alpha = 2),$$

$$|\varphi(t, u, \omega) - \varphi(\omega t)| \le (|u - \varphi(u)|^{2-\alpha} + (\alpha - 2)t)^{\frac{1}{2-\alpha}} \quad (if \ \alpha > 2)$$

hold for all $t \ge 0, u \in E$ and $\omega \in \Omega$;

(3) the function $\gamma : \Omega \to E$, defined by equality $\gamma(\omega) = I_\omega$ is continuous and $\gamma(\omega t) := \varphi(t, \gamma(\omega), \omega)$ for all $t \ge 0, u \in E$ and $\omega \in \Omega$.

Theorem 7.11 *Let $f \in C(E \times \Omega, E)$ be a function verifying the condition (7.29) and $F \in C(E \times \Omega, E)$ be a function with the condition*

$$Re\langle F(u_1, \omega) - F(u_2, \omega), u_1 - u_2 \rangle \le L|u_1 - u_2|^\alpha \tag{7.30}$$

for all $u_1, u_2 \in E$ and $\omega \in \Omega$, where L is some positive number.

Then there exists a positive number λ_0 such that for all $|\lambda| \le \lambda_0$ the following hold:

(1) the set $I_\omega^\lambda := \{u \in E \mid \sup\limits_{t \in \mathbb{R}} |\varphi_\lambda(t, u, \omega)| < +\infty\}$ contains a single point $\{\varphi_\lambda(\omega)\}$ for each $\omega \in \Omega$, where $\varphi_\lambda(t, u, \omega)$ is a unique solution of the equation

$$u' = f(u, \omega t) + \lambda F(u, \omega t) \quad (\omega \in \Omega) \tag{7.31}$$

satisfying the initial condition $\varphi_\lambda(0, u, \omega) = u$;

(2) the function $\gamma_\lambda : \Omega \to E$ defined by equality $\gamma_\lambda(\omega) = I_\omega^\lambda$ is continuous and $\gamma_\lambda(\omega t) = \varphi_\lambda(t, \gamma_\lambda(\omega), \omega)$ for all $t \ge 0, u \in E$ and $\omega \in \Omega$.

(3)

$$\lim_{\lambda \to 0} \sup_{\omega \in \Omega} |\gamma_\lambda(\omega) - \gamma_0(\omega)| = 0.$$

Proof. Let $g := f + \lambda F$, then from (7.29)–(7.32) follows that

$$Re\langle g(u_1, \omega) - g(u_2, \omega), u_1 - u_2 \rangle \le (-k + L|\lambda|)|u_1 - u_2|^\alpha \tag{7.32}$$

for all $u_1, u_2 \in E$ and $\omega \in \Omega$. From (7.32) follows that there exists $\lambda_0 > 0$ such that $-k + L|\lambda| \le -k + L\lambda_0 < 0$ for all $|\lambda| \le \lambda_0$ and according to Theorem 7.10 the assertions 1. and 2. of the theorem are true.

It is clear that for $|\lambda| \le \lambda_0$ the cocycle φ_λ generated by the equation (7.31) admits a compact global attractor $I^\lambda = \{\gamma_\lambda(\omega) \mid \omega \in \Omega\}$.

Now we will show that the set $\bigcup\{I^\lambda \mid \lambda \in \Lambda = [-\lambda_0, \lambda_0]\}$ is compact. Let $V(u) := \frac{1}{2}|u|^2$, then

$$V'(u) = \frac{d}{dt}V(\varphi_\lambda(t, u, \omega))|_{t=0} = Re\langle g(u, \omega), u \rangle \tag{7.33}$$

$$= Re\langle g(u, \omega) - g(0, \omega), u \rangle + Re\langle g(0, \omega), u \rangle \le (-k + L|\lambda_0|)|u|^2 + C|u|$$

$$= |u|^2(-k + L|\lambda_0| + \frac{C}{|u|^{\alpha-1}}),$$

where $C := \max\{|g(0,\omega)| \; : \; \omega \in \Omega, \; \lambda \in \Lambda\}$. From the equality (7.33) follows that there exists $r > 0$ such that for all $|u| \geq r$

$$V'(u) \leq -\nu|u|^2, \tag{7.34}$$

where $\nu = k - L|\lambda_0| - \frac{C}{r^{\alpha-1}} > 0$. Now to finish the proof of Theorem 7.10 it is sufficient to refer to Theorem 7.6. The theorem is proved. $\qquad\square$

7.5.3   *Quasi-linear systems*

Consider a non-autonomous quasi-linear system

$$u' = A(\omega t)u + \lambda f(u, \omega t) \quad (\omega \in \Omega)$$

on E. Denote by $[E]$ the space of all linear continuous operators acting onto E and equipped with the operational norm.

Theorem 7.12   *Let $A \in C(\Omega, [E]), f \in C(E \times \Omega, E)$ and let the following conditions be fulfilled:*

(1) there exists a positive constant α_0 such that $\mathrm{Re}\langle A(\omega)u, u \rangle \leq -\alpha_0|u|^2$ for all $u \in E$ and $\omega \in \Omega$;

(2) for any $r > 0$ there exists a positive constant $L(r)$ such that

$$|f(u_1, \omega) - f(u_2, \omega)| \leq L|u_1 - u_2|$$

for all $u_1, u_2 \in B[0, r] := \{u \in E \mid |u| \leq r\}$ and $\omega \in \Omega$.

Then there exists a positive constant λ_0 such that for all $\lambda \in \Lambda := [-\lambda_0, \lambda_0]$ the following statements are true:

(1) the set $I_\omega^\lambda := \{u \in E \mid \sup\{|\varphi_\lambda(t, u, \omega)| : t \in \mathbb{R}\} < +\infty\}$ is not empty, compact and connected for each $\omega \in \Omega$, where $\varphi_\lambda(t, u, \omega)$ there is a unique solution of equation

$$u' = A(\omega t)u + \lambda F(u, \omega t)$$

satisfying the initial condition $\varphi_\lambda(0, u, \omega) = u$;

(2) $\varphi_\lambda(t, I_\omega^\lambda, \omega) = I_{\sigma(t,\omega)}^\lambda$ for all $t \in \mathbb{R}_+$ and $\omega \in \Omega$;

(3) the set $I^\lambda := \bigcup\{I_\omega^\lambda \mid \omega \in \Omega\}$ is compact and connected;

(4) the equalities

$$\lim_{t \to +\infty} \beta(\varphi_\lambda(t, M, \omega_{-t}), I_\omega^\lambda) = 0,$$

$$\lim_{t \to +\infty} \beta(\varphi_\lambda(t, M, \omega), I^\lambda) = 0$$

hold for all $\lambda \in \Lambda, \omega \in \Omega$ and bounded subset $M \subseteq E$.

(5) the set $\bigcup\{I^\lambda \mid \lambda \in \Lambda\}$ is compact;
(6) the following equality is true

$$\lim_{\lambda \to 0} \sup_{\omega \in \Omega} \beta(I_\omega^\lambda, 0) = 0.$$

Proof. Let λ_0 be a positive number such that $\nu = \alpha_0 - \lambda_0 \alpha > 0$, then the function $F_\lambda(u, \omega) := A(\omega)u + \lambda f(u, \omega)$ verifies the condition

$$\text{Re}\langle F_\lambda(u, \omega), u \rangle \leq -\nu |u|^2 + \lambda_0 \beta \tag{7.35}$$

for all $|\lambda| \leq \lambda_0, \omega \in \Omega$ and $u \in E$.

From the inequality (7.35) follows (see, for example, [97, p.11]) that the inequality

$$|\varphi_\lambda(t, u, \omega)|^2 \leq |u|^2 e^{-2\nu t} + \frac{\lambda_0 \beta}{\nu}(1 - e^{-2\nu t}) \tag{7.36}$$

holds for all $t \in \mathbb{R}_+, \lambda \in \Lambda$ and $(u, \omega) \in E \times \Omega$ and, consequently, the family of cocycles $\{\varphi_\lambda\}_{\lambda \in \Lambda}$ admits a bounded absorbing set. Now to finish the proof of the theorem it is sufficient to refer to Theorems 7.3, 7.5 and Lemma 7.4. $\quad\square$

7.5.4 *Non-autonomously perturbed systems*

Theorem 7.13 *Suppose that $f \in C(E)$ is uniformly Lipschitzian and an autonomous system (7.24) has a global attractor I. Furthermore suppose that $F \in C(E \times \Omega, E)$ is uniformly Lipschitz in $u \in E$ and it is uniformly bounded on $E \times \Omega$, i.e. $\sup\{|F(u, \omega)| : (u, \omega) \in E \times \Omega\} = K < +\infty$. Then there exists a positive number $\lambda_0 > 0$ such that for all $\lambda \in \Lambda = [-\lambda_0, \lambda_0]$ the following are true:*

(1) the set $I_\omega^\lambda := \{u \in E \mid \sup\{|\varphi_\lambda(t, u, \omega)| : t \in \mathbb{R}\} < +\infty\}$ is not empty, compact and connected for each $\omega \in \Omega$, where $\varphi_\lambda(t, u, \omega)$ there is a unique solution of equation $u' = f(u) + \lambda F(u, \omega t)$ satisfying the initial condition $\varphi_\lambda(0, u, \omega) = u$;
(2) $\varphi_\lambda(t, I_\omega^\lambda, \omega) = I_{\sigma(t,\omega)}^\lambda$ for all $t \in \mathbb{R}_+$ and $\omega \in \Omega$;
(3) the set $I^\lambda := \bigcup\{I_\omega^\lambda \mid \omega \in \Omega\}$ is compact and connected;
(4) the equalities

$$\lim_{t \to +\infty} \beta(\varphi_\lambda(t, M, \omega_{-t}), I_\omega^\lambda) = 0$$

and

$$\lim_{t \to +\infty} \beta(\varphi_\lambda(t, M, \omega), I^\lambda) = 0$$

take place for all $\lambda \in \Lambda, \omega \in \Omega$ and bounded subset $M \subseteq E$.
(5) the set $\bigcup\{I^\lambda \mid \lambda \in \Lambda\}$ is compact;

(6) the following equality is true

$$\lim_{\lambda \to 0} \sup_{\omega \in \Omega} \beta(I_\omega^\lambda, I) = 0 .$$

Proof. According to Theorem 22.5 from [327] (see also [217, 220]), under its conditions there exists a continuous function $V : E \setminus I \to \mathbb{R}_+$ with the following properties:

a. V is uniformly Lipschitz in E, i.e. there exists a constant $L > 0$ such that $|V(u_1) - V(u_2)| \le L|u_1 - u_2|$ for all $u_1, u_2 \in E$.

b. there exist continuous strictly increasing functions $a, b : \mathbb{R}_+ \to \mathbb{R}_+$ with $a(0) = b(0) = 0$ and $0 < a(r) < b(r)$ for all $r > 0$ such that $a(\beta(u, I)) \le V(u) \le b(\beta(u, I))$ for all $u \in E$, where $\beta(u, I) := \sup\{\rho(u, x) \mid x \in I\}$.

c. there exists a constant $c > 0$ such that $V'(u) \le -cV(u)$ for all $u \in E \setminus I$, where $V'(u) := \limsup_{t \to 0+} t^{-1}[V(\pi(t, u)) - V(u)]$ and $\pi(t, u)$ is a unique solution of equation (7.25) with initial condition $\pi(0, u) = u$.

Define a function $\mathfrak{V} : X \to \mathbb{R}_+$ ($X := E \times \Omega$) in the following way: $\mathfrak{V}(x) := V(u)$ for all $x = (u, \omega) \in X$. Note that

$$\mathfrak{V}'(u, \omega) = \limsup_{t \to 0+} V(\varphi_\lambda(t, u, \omega))|_{t=0} \le LKr - c\mathfrak{V}(u, \omega)$$

(see [220, p.11]) for all $u \in E$. Then there exist $\lambda_0 > 0$ and $r_0 > 0$ such that

$$\mathfrak{V}'(u, \omega) \le -LK\lambda_0$$

for all $|u| \ge r_0$ and $\omega \in \Omega$.

To finish the proof of the theorem it is sufficient to refer to Theorem 7.6 and Lemma 7.4. $\square$

Remark 7.3   Similar theorem was proved in [220, Th.4.1] for the pullback attractors of non-autonomously perturbed systems.

7.5.5   *Non-autonomous 2D Navier-Stokes equations*

Let $G \subset \mathbb{R}^2$ be a bounded domain with C^2 smooth boundary,

$$V := \{u \in (\mathring{W}_2^1(G))^2, \ \mathrm{div}\, u(x) = 0\}, \ H := \overline{V}^{(L_2(G))^2},$$

V' be the dual space of V, $(\mathring{W}_2^1(G))^2$ denotes the Sobolev space of functions having two components, and let π be the orthogonal projector from $(L_2(G))^2$ onto H. The operator $F(u, v) := \pi(u, \nabla)v$ has values in V'.

Let Ω be a compact complete metric space, $(\Omega, \mathbb{R}, \sigma)$ be a dynamical system on Ω, $\mathcal{F} \in C(V \times \Omega, V)$ and satisfy the the following conditions:

(i) $|\mathcal{F}(u_1, \omega) - \mathcal{F}(u_2, \omega)| \le L|u_1 - u_2|$ for all $u_1, u_2 \in V$ and $\omega \in \Omega$;

(ii) $\operatorname{Re}\langle \mathcal{F}(u, \omega), u \rangle \le M|u|^2 + N$ for all $u \in V$ and $\omega \in \Omega$, where L, M and N are some positive constants.

Consider the perturbed $2D$ Navier-Stokes equation

$$u' + \nu Au + B(u, u) = \mathcal{F}(u, \omega t) \quad (\omega \in \Omega) \tag{7.37}$$

on H, where $B : V \times V \to V'$ is a bilinear continuous map and A is the extension of $-\pi\nabla$ with zero boundary conditions on V and $\nu > 0$. In particular, there exists $\lambda_1 > 0$ such that

$$\langle Au, u \rangle \ge |u|_V^2 \ge \lambda_1 |u|_H^2$$

for any $u \in V$. According to [286],[311] by equation (7.37) is generated a cocycle $\varphi(t, u, \omega)$ on $(\Omega, \mathbb{R}, \sigma)$ with fiber H, where $\varphi(t, u, \omega)$ is a unique solution of equation (7.37) with the condition $\varphi(0, u, \omega) = u$.

Lemma 7.8 *Under the conditions (i) and (ii) the following assertions holds:*

(1) for any $T > 0$, $\nu > 0$, $\omega \in \Omega$ and any $u \in H$ the equation (7.34) has a unique solution $\varphi(t, u, \omega)$ with path in $C([0, T], H)$;

(2) the energy inequality holds

$$\frac{d}{dt}|\varphi(t, u, \omega)|_H^2 + \nu\lambda_1\varphi(t, u, \omega)|_H^2 \le \frac{|\mathcal{F}(0, \omega t)|_H^2}{\nu\lambda_1} + 2L|\varphi(t, u, \omega)|_H^2 \tag{7.38}$$

for all $t \in [0, T], u \in H$ and $\omega \in \Omega$;

(3) the mapping $\varphi : \mathbb{R}_+ \times H \times \Omega \to H$ is continuous.

Proof. The assertions 1. and 2. follow from [111] (see also Lemma 3.1 [286]).

Now we will prove that the mapping $\varphi : \mathbb{R}_+ \times H \times \Omega \to H$ is continuous. Let $t_0 \in \mathbb{R}_+, u_0 \in H$ and $\omega_0 \in \Omega$, then we have

$$|\varphi(t, u, \omega) - \varphi(t_0, u_0, \omega_0)|_H \tag{7.39}$$

$$\le |\varphi(t, u, \omega) - \varphi(t, u_0, \omega_0)|_H + |\varphi(t, u_0, \omega_0) - \varphi(t_0, u_0, \omega_0)|_H.$$

Denote by $w(t) := \varphi(t, u, \omega) - \varphi(t, u_0, \omega_0)$ and $f(t) := \mathcal{F}(\varphi(t, u, \omega), \omega t) - \mathcal{F}(\varphi(t, u_0, \omega_0), \omega_0 t)$, then the function $w(t)$ verifies the following equation

$$\frac{dw}{dt} + \nu Aw + B(w, w) + B(w, u_1) + B(u_1, w) = f(t), \tag{7.40}$$

where $u_1 := \varphi(t, u_0, \omega_0)$. Using the well-known identity $\langle B(u, v), v \rangle = 0 \; (\forall v \in H)$, where $\langle \cdot, \cdot \rangle$ is the scalar product in H, we obtain

$$\frac{1}{2}\frac{d}{dt}|w|_H^2 = \langle \dot{w}, w \rangle = \langle -\nu Aw - B(w,w) - B(w,u_1) - B(u_1,w) \tag{7.41}$$
$$+ f(t), w \rangle = -\nu \langle Aw, w \rangle - \langle B(w,w), w \rangle - \langle B(w,u_1), w \rangle - \langle B(u_1,w), w \rangle$$
$$+ \langle f(t), w \rangle = -\nu \langle Aw, w \rangle - \langle B(w,u_1), w \rangle + \langle f(t), w \rangle.$$

Bearing in mind the inequality $|u|_{L_4}^2 \le |w|_H |w|_V$ (see [232]) we obtain

$$|\langle B(w,u_1), w \rangle| \le |w|_{L_4}^2 |u_1|_V$$
$$\le |w|_H |w|_V |u_1|_V \le \frac{\nu}{2}|w|_V^2 + \frac{1}{2\nu}|w|_H^2 |u_1|_V^2.$$

Taking into account that $|(f,w)| \le |f|_H |w|_H$, we get from (7.41)

$$\frac{1}{2}\frac{d}{dt}|w|_H^2 \le -\nu\lambda_1 |w|_V^2 + \frac{\nu\lambda_1}{2}|w|_V^2 + \frac{1}{2\nu\lambda_1}|w|_H^2 |u_1|_V^2 + |f|_H |w|_H. \tag{7.42}$$

We remark that

$$|f(t)|_H = |\mathcal{F}(\varphi(t,u,\omega),\omega t) - \mathcal{F}(\varphi(t,u_0,\omega_0),\omega_0 t)|_H \tag{7.43}$$

$$\le L|\varphi(t,u_0,\omega_0) - \varphi(t,u_0,\omega_0)| + |\mathcal{F}(\varphi(t,u_0,\omega_0),\omega t) - \mathcal{F}(\varphi(t,u_0,\omega_0),\omega_0 t)|$$

and, consequently, from (7.41)-(7.43), we obtain

$$\frac{1}{2}\frac{d}{dt}|w|_H^2 \le \left(\frac{1}{2\nu}|u_1|_V^2 + L + \frac{1}{2}\right)|w|_H^2 + \frac{|f|^2}{2}. \tag{7.44}$$

From this differential inequality we deduce that

$$|w(t)|_H^2 \le \exp\left(\int_0^t \left(\frac{1}{2\nu}|\varphi(\tau,u_0,\omega_0)|_V^2 + L + \frac{1}{2}\right)d\tau\right)|u - u_0|_H^2$$
$$+ \int_0^t \exp\left(-\int_0^\tau \left(\frac{1}{2\nu}|\varphi(s,u_0,\omega_0)|_V^2 + L + \frac{1}{2}\right)ds\right) \tag{7.45}$$
$$\times \frac{1}{2}|\mathcal{F}(\varphi(\tau,u_0,\omega_0),\omega\tau) - \mathcal{F}(\varphi(\tau,u_0,\omega_0),\omega_0\tau)|d\tau.$$

Since $\mathcal{F} \in C(H \times \Omega, H)$, then

$$\max_{0 \le t \le T}|\mathcal{F}(\varphi(t,u_0,\omega_0),\omega t) - \mathcal{F}(\varphi(t,u_0,\omega_0),\omega_0 t)| \to 0$$

as $\omega \to \omega_0$ and, consequently, from (7.45) we obtain

$$\max_{0 \le t \le T}|w(t)| \to 0. \tag{7.46}$$

From (7.39) and (7.46) we obtain the continuity of mapping φ. The lemma is proved.

$\square$

Corollary 7.3 *Under the conditions (i) and (ii) there exists a positive number* $L_0 < \frac{\nu\lambda_1}{2}$ *such that if* $L < L_0$, *then the following inequality*

$$|\varphi(t,u,\omega)|^2 \le e^{(-\nu\lambda_1+2L_0)t}|u|^2 + \frac{|f|^2}{\nu\lambda_1(-2L_0+\nu\lambda_1)}$$

holds for all $t \ge 0$, $u \in H$ *and* $\omega \in \Omega$, *where* $|f| := \max_{\omega \in \Omega}|\mathcal{F}(0,\omega)|$.

This assertion follows from the second assertion of Lemma 7.8.

Theorem 7.14 *There exists a positive number* $L_0 > 0$ *such that the cocycle* φ *generated by (7.37) admits a compact global attractor, if* $L \le L_0$.

Proof. According to Lemma 3.1 [286] there exists $L_0 > 0$ (for example $L_0 < \frac{\nu\lambda_1}{2}$) such that the cocycle φ admits a bounded absorbing set if $L < L_0$. On the other hand the cocycle φ is compact, i.e. the mapping $\varphi(t,\cdot,\cdot) : V \times \Omega \to V$ is completely continuous for all $t > 0$. To finish the proof of the theorem it is sufficient to refer to Theorem 2.24. $\qquad\square$

Theorem 7.15 *Under the conditions (i) and (ii) there exists a positive number* λ_0 *such that the following is true:*

(1) *the set* $I_\omega^\lambda := \{u \in H \mid \sup_{t \in \mathbb{R}}|\varphi_\lambda(t,u,\omega)| < +\infty\}$ *is not empty, compact and connected for all* $\omega \in \Omega$ *and* $\lambda \in \Lambda := [-\lambda_0, -\lambda_0]$, *where* $h \in H$ *and* $\varphi_\lambda(t,u,\omega)$ *is a unique solution of the equation*

$$u' + \nu Au + F(u,u) + h = \lambda\mathcal{F}(u,\omega t) \quad (\omega \in \Omega) \tag{7.47}$$

satisfying the initial condition $\varphi_\lambda(0,u,\omega) = u$;

(2) $\varphi_\lambda(t, I_\omega^\lambda, \omega) = I_{\sigma(t,\omega)}^\lambda$ *for all* $\lambda \in \Lambda$, $t \in \mathbb{R}_+$ *and* $\omega \in \Omega$;

(3) *the set* $I^\lambda = \bigcup\{I_\omega^\lambda \mid \omega \in \Omega\}$ *is compact and connected;*

(4) *the equalities*

$$\lim_{t \to +\infty} \beta(\varphi_\lambda(t, M, \omega_{-t}), I_\omega^\lambda) = 0$$

and

$$\lim_{t \to +\infty} \beta(\varphi_\lambda(t, M, \omega), I^\lambda) = 0$$

take place for all $\lambda \in \Lambda$, $\omega \in \Omega$ *and bounded subset* $M \subseteq E$.

(5) *the set* $\bigcup\{I^\lambda \mid \lambda \in \Lambda\}$ *is compact and connected;*

(6) *the equality*

$$\lim_{\lambda \to 0} \sup_{\omega \in \Omega} \beta(I_\omega^\lambda, I) = 0$$

holds, where I is a Levinson center for the equation

$$u' + \nu Au + F(u, u) + h = 0.$$

Proof. Let $\tilde{F}(u, \omega t) := -h + \lambda \mathcal{F}(u, \omega t)$ and $\lambda_0 < \frac{\nu \lambda_1}{2L}$, then for the equation

$$u' + \nu Au + F(u, u) = \lambda \tilde{F}(u, \omega t) \quad (\omega \in \Omega)$$

the conditions of Theorem 7.14 are fulfilled. Let φ_λ be a cocycle generated by equation (7.43), then according to Corollary 7.3 the family of cocycle $\{\varphi_\lambda\}_{\lambda \in \Lambda}$ admits a collectively absorbing bounded set. Since the imbedding V into H is compact, to finish the proof of the theorem it is sufficient to refer to Theorem 7.6 and Lemma 7.4. The theorem is proved. $\square$

7.5.6 *Quasi-linear functional-differential equations*

Let $r > 0$, $C([a, b], \mathbb{R}^n)$ be the Banach space of continuous functions $\nu : [a, b] \to \mathbb{R}^n$ with the sup-norm . If $[a, b] := [-r, 0]$, then suppose $\mathcal{C} := C([-r, 0], \mathbb{R}^n)$. Let $\sigma \in \mathbb{R}, A \geq 0$ and $u \in C([\sigma - r, \sigma + A], \mathbb{R}^n)$. For any $t \in [\sigma, \sigma + A]$ define $u_t \in \mathcal{C}$ by the equality $u_t(\theta) = u(t + \theta), -r \leq \theta \leq 0$. Let us define by $C(\Omega \times \mathcal{C}, \mathbb{R}^n)$ the space of all continuous functions $f : \Omega \times \mathcal{C} \to \mathbb{R}^n$, with compact-open topology and let $(\Omega, \mathbb{R}, \sigma)$ be a dynamical system on the compact metric space Ω. Consider the equation

$$u' = f(\omega t, u_t) \qquad (\omega \in \Omega), \tag{7.48}$$

where $f \in C(\Omega \times \mathcal{C}, \mathbb{R}^n)$. We will suppose that the function f is regular, that is for any $\omega \in \Omega$ and $u \in \mathcal{C}$ the equation (7.48) has a unique solution $\varphi(t, u, \omega)$ which is defined on $\mathbb{R}_+ = [0, +\infty)$. Let $X := \mathcal{C} \times \Omega$, and $\pi : X \times \mathbb{R}_+ \to X$ be a dynamical system on X defined by the following rule: $\pi(\tau, (u, \omega)) := (\varphi_\tau(u, \omega), \omega \tau)$, then the triplet $\langle (X, \mathbb{R}_+, \pi), (\Omega, \mathbb{R}, \sigma), h \rangle$ $(h := pr_2 : X \to \Omega)$ is a non-autonomous dynamical system, where $\varphi_\tau(u, \omega)(\theta) := \varphi(\tau + \theta, u, \omega)$.

From the general properties of solutions of (7.48) (see, for example [179]), we have the following statement.

Theorem 7.16 *The following statements are true:*

(1) The non-autonomous dynamical system $\langle (X, \mathbb{R}_+, \pi), (\Omega, \mathbb{R}, \sigma), h \rangle$ generated by equation (7.45) is conditionally completely continuous;

(2) Let Ω be compact and the function $f : \Omega \times \mathcal{C} \to \mathbb{R}^n$ be bounded on $\Omega \times B$ for any bounded set $B \subset \mathcal{C}$, then the non-autonomous dynamical system generated by the equation (7.48) is conditionally completely continuous (in particular, it is asymptotically compact).

Denote by $[\mathcal{C}]$ the space of all linear continuous operators acting onto $\mathcal{C}$ and equipped with the operational norm.

Theorem 7.17 [78] *Let $A \in C(\Omega, [\mathcal{C}]), f \in C(\mathcal{C} \times \Omega, E)$ and let the following conditions be fulfilled:*

(1) a zero solution of equation

$$u' = A(\omega t)u_t \tag{7.49}$$

is uniformly asymptotically stable, i.e. there exist positive numbers N and ν such that $|\varphi_0(t, u, \omega)| \le Ne^{-\nu t}|u|$ for all $t \ge 0$, $u \in \mathcal{C}$ and $\omega \in \Omega$, where $\varphi_0(t, u, \omega)$ is a solution of equation (7.49) with condition that $\varphi_0(0, u, \omega) = u$;
(2) there exists a positive constant L such that

$$|f(u_1, \omega) - f(u_2, \omega)| \le L|u_1 - u_2|$$

for all $u_1, u_2 \in \mathcal{C}$ and $\omega \in \Omega$.

Then there exists a positive constant $\varepsilon_0(\varepsilon_0 < \frac{\nu}{N})$ such that

$$|\varphi(t, u, \omega)| \le \frac{NM}{\nu - NL} + (N|u| - \frac{NM}{\nu - NL})e^{-(\nu - NL)t}$$

for all $t \ge 0, L \in (0, \varepsilon_0)$, $u \in \mathcal{C}$ and $\omega \in \Omega$, where $\varphi(t, u, \omega)$ is a unique solution of the equation

$$u' = A(\omega t)u_t + f(u_t, \omega t)$$

with the condition $\varphi(0, u, \omega) = u$ and $M := \max\{|f(0, \omega)| : \omega \in \Omega\}$.

Consider a non-autonomous quasi-linear system

$$u' = A(\omega t)u_t + \lambda f(u_t, \omega t) \quad (\omega \in \Omega) \tag{7.50}$$

on $\mathcal{C}$.

Theorem 7.18 *Let $f \in C(\mathcal{C} \times \Omega, E)$ and let the inequality*

$$|f(u_1, \omega) - f(u_2, \omega)| \le L|u_1 - u_2|$$

take place for all $u_1, u_2 \in \mathcal{C}$ and $\omega \in \Omega$, where L is some positive number.
Then there exists a positive number λ_0 such that for all $\lambda \in \Lambda := [-\lambda_0, \lambda_0]$ the following statements are true:

(1) the set $I_\omega^\lambda := \{u \in \mathcal{C} \mid \sup\{|\varphi_\lambda(t, u, \omega)| : t \in \mathbb{R}\} < +\infty\}$ is not empty, compact and connected for each $\omega \in \Omega$, where $\varphi_\lambda(t, u, \omega)$ there is a unique solution of equation (7.50) satisfying the initial condition $\varphi_\lambda(0, u, \omega) = u$;
(2) $\varphi_\lambda(t, I_\omega^\lambda, \omega) = I_{\sigma(t, \omega)}^\lambda$ for all $t \in \mathbb{R}_+$ and $\omega \in \Omega$;

(3) the set $I^\lambda = \bigcup\{I_\omega^\lambda \mid \omega \in \Omega\}$ is compact and connected;

(4) the equalities

$$\lim_{t \to +\infty} \beta(\varphi_\lambda(t, M, \omega_{-t}), I_\omega^\lambda) = 0,$$

$$\lim_{t \to +\infty} \beta(\varphi_\lambda(t, M, \omega), I^\lambda) = 0$$

hold for all $\lambda \in \Lambda, \omega \in \Omega$ and bounded subset $M \subseteq E$.

(5) the set $\bigcup\{I^\lambda \mid \lambda \in \Lambda\}$ is compact;

(6) the following equality holds

$$\lim_{\lambda \to 0} \sup_{\omega \in \Omega} \beta(I_\omega^\lambda, 0) = 0.$$

Proof. Let λ_0 be a positive number such that $\nu = \lambda_0 L < \nu/N$, then the function $F_\lambda(u, \omega) := A(\omega)u + \lambda f(u, \omega)$ satisfies the condition

$$|F_\lambda(u, \omega)| \leq \nu|u| + M \tag{7.51}$$

(with $M := \max_{\omega \in \Omega} |f(0, \omega)|$) for all $|\lambda| \leq \lambda_0$, $\omega \in \Omega$ and $u \in C$. From the inequality (7.51) and Theorem 7.16 follows that the family of cocycles $\{\varphi_\lambda\}_{\lambda \in \Lambda}$ admits a bounded absorbing set. Now to finish the proof of theorem it is sufficient to refer to Theorems 7.3, 7.5, 7.15 and Lemma 7.4. $\qquad\square$

Chapter 8

The relationship between pullback, forward and global attractors

Non-autonomous dynamical systems can often be formulated in terms of a cocycle mapping for the dynamics in the state space that is driven by an autonomous dynamical system in what is called a parameter space. Traditionally the driving system is topological and the resulting cartesian product system forms an autonomous semi–dynamical system that is known as a skew–product flow. Results on global attractors for autonomous semi–dynamical systems can thus be adapted to such non-autonomous dynamical systems via the associated skew–product flow [64, 75, 88, 83, 111, 185, 292, 318].

A new type of attractor, called a pullback attractor, was proposed and investigated for non-autonomous or random dynamical systems [97, 126, 153, 262, 284]. Essentially, it consists of a parametrized family of nonempty compact subsets of the state space that are mapped onto each other by the cocycle mapping as the parameter is changed by the underlying driving system. Pull back attraction to a component subset for a fixed parameter value is achieved by starting progressively earlier in time, that is, at parameter values that are carried forward to the fixed value. The deeper reason for this procedure is that a cocycle can be interpreted as a mapping between the fibers of a fiber bundle where the image fiber is fixed. The kernels of a global attractor of the skew–product flows considered in [111] are very similar. This differs from the more conventional forward convergence where the parameter value of the limiting object also evolves with time, in which case the parametrized family could be called a forward attractor.

Pullback attractors and forward attractors can, of course, be defined for non-autonomous dynamical systems with a topological driving system [218, 220, 221]. In fact, when the driving system is the shift operator on the real line, forward attraction to a time varying solution, say, is the same as the attraction in Lyapunov asymptotic stability. The situation of a compact parameter space is dynamically more interesting as the associated skew–product flow may then have a global attractor. The relationship between the global attractor of the skew–product system and the pullback and forward attractors of the cocycle system is investigated in

287

this chapter. We also note that forward attractors are stronger than global attractors if we suppose a compact set of non-autonomous perturbations. An example is presented in which the cartesian product of the component subsets of a pullback attractor is not a global attractor of the skew–product flow. This set is, however, a maximal compact invariant subset of the skew–product flow. By a generalization of some stability results of Zubov [336] it is asymptotically stable. Thus a pullback attractor always generates a local attractor of the skew–product system, but this need not be a global attractor. If, however, the pullback attractor generates a global attractor in the skew–product flow and if, in addition, its component subsets depend lower continuously on the parameter, then the pullback attractor is also a forward attractor. Several examples illustrating these results are presented in the final section.

8.1   Pullback, forward and global attractors

A general non-autonomous dynamical system is defined here in terms of a cocycle mapping φ on a state space W that is driven by an autonomous dynamical system σ acting on a base space Ω, which will be called the parameter space. In particular, let W be a complete metric space, let Ω be a compact metric space and let $\mathbb{T}$, the time set, be either $\mathbb{R}$ or $\mathbb{Z}$.

Let $(\Omega, \mathbb{T}, \sigma)$ be a dynamical system on the metric space Ω. Recall that the triple $\langle U, \varphi, (\Omega, \mathbb{T}, \sigma) \rangle$ is called a cocycle (or non-autonomous dynamical system) on the state space W, if the mapping $\varphi : \mathbb{T}_+ \times W \times \Omega \to \Omega$ is continuous, $\varphi(0, u, \omega) = u$ and $\varphi(t * \tau, u, \omega) = \varphi(t, \varphi(\tau, u, \omega), \omega\tau)$ fora all $u \in W$ and $t, \tau \in \mathbb{T}_+$.

Let X be the cartesian product of W and Ω. Then the mapping $\pi : \mathbb{T}_+ \times X \to X$ defined by

$$\pi(t, (u, \omega)) := (\varphi(t, u, \omega), \sigma_t \omega)$$

forms a semi–group on X over $\mathbb{T}_+$ [290, 291].

The autonomous semi–dynamical system $(X, \mathbb{T}_+, \pi) = (W \times \Omega, \mathbb{T}_+, (\varphi, \sigma))$ is called the skew–product dynamical system associated with the cocycle $\langle W, \varphi, (\Omega, \mathbb{T}, \sigma) \rangle$.

For example, let W be a Banach space and let the space $C = C(\mathbb{R} \times W, W)$ of continuous functions $f : \mathbb{R} \times W \to W$ be equipped with the compact open topology. Consider the autonomous dynamical system $(C, \mathbb{R}, \sigma)$, where σ is the shift operator on C defined by $\sigma_t f(\cdot, \cdot) := f(\cdot + t, \cdot)$ for all $t \in \mathbb{T}$. Let Ω be the hull $H(f)$ of a given functions $f \in C$, that is,

$$\Omega = H(f) := \overline{\bigcup_{t \in \mathbb{R}} \{f(\cdot + t, \cdot)\}},$$

and denote the restriction of $(C, \mathbb{R}, \sigma)$ to Ω by $(\Omega, \mathbb{R}, \sigma)$. Let $F : \Omega \times W \to W$ be the continuous mapping defined by $F(\omega, u) := \omega(0, u)$ for $\omega \in \Omega$ and $u \in W$. Then, under appropriate restrictions on the given function $f \in C$ (see Sell [290, 291]) defining Ω, the differential equation

$$u' = \omega(t, u) = F(\sigma_t \omega, u) \tag{8.1}$$

generates a cocycle $\langle W, \varphi, (\Omega, \mathbb{R}, \sigma) \rangle$, where $\varphi(t, u, \omega)$ is the solution of (8.1) with the initial value u at time $t = 0$.

Let β denote the Hausdorff semi-distance between two nonempty sets of a metric space Y, that is,

$$\beta(A, B) = \sup_{a \in A} \inf_{b \in B} \rho(A, B),$$

and let $\mathcal{D}(W)$ be $C(W)$ or $B(W)$, where $C(W)$ $(B(W))$ is a classes of sets containing the compact subsets (the bounded subsets) of the metric space W.

The definition of a global attractor for an autonomous semi–dynamical system $(X, \mathbb{T}^+, \pi)$ is well known. Specifically, a nonempty compact subset $\mathcal{A}$ of X which is π-invariant, that is, satisfies

$$\pi(t, \mathcal{A}) = \mathcal{A} \quad \text{for all } t \in \mathbb{T}_+, \tag{8.2}$$

is called a global attractor for $(X, \mathbb{T}^+, \pi)$ with respect to $\mathcal{D}(X)$ if

$$\lim_{t \to \infty} \rho(\pi(t, D), \mathcal{A}) = 0 \tag{8.3}$$

for every $D \in \mathcal{D}(X)$.

Conditions for the existence of such global attractors and examples can be found in $[20, 88, 179, 314, 318]$. Of course, semi–dynamical systems need not be a skew–product systems, but when they are, the following definition will be used.

Suppose that the skew–product dynamical system $(X, \mathbb{T}_+, \pi) = (W \times \Omega, \mathbb{T}_+, (\varphi, \sigma))$ has a global attractor $\mathcal{A}$. Then we will call the set $\mathcal{A}$ the global attractor with respect to $\mathcal{D}(W)$ of the associated cocycle $\langle W, \varphi, (\Omega, \mathbb{T}, \sigma) \rangle$.

Definition 8.1 The global attractor $\mathcal{A}$ with respect to $\mathcal{D}(W)$ of the skew–product dynamical system $(X, \mathbb{T}_+, \pi) = (W \times \Omega, \mathbb{T}_+, (\varphi, \sigma))$ will be called the global attractor with respect to $\mathcal{D}(W)$ of the associated cocycle $\langle W, \varphi, (\Omega, \mathbb{T}, \sigma) \rangle$.

Other types of attractors, in particular pullback attractors, that consist of a family of nonempty compact subsets of the state space of the cocycle mapping have been proposed for non-autonomous or random dynamical systems $[126, 127, 218, 284, 285]$.

Definition 8.2   Let $I = \{I_\omega \mid \omega \in \Omega\}$ be a family of nonempty compact sets of W for which $\bigcup_{\omega \in \Omega} I_\omega)$ is relatively compact and let I be φ–invariant with respect to a cocycle $\langle W, \varphi, (\Omega, \mathbb{T}, \sigma) \rangle$, that is, satisfies

$$\varphi(t, I_\omega, \omega) = I_{\sigma_t \omega} \qquad \text{for all } t \in \mathbb{T}^+,\ \omega \in \Omega. \tag{8.4}$$

Definition 8.3   The family I is called a pullback attractor of $\langle W, \varphi, (\Omega, \mathbb{T}, \sigma) \rangle$ with respect to $\mathcal{D}(U)$ if

$$\lim_{t \to \infty} \beta(\varphi(t, D, \sigma_{-t}\omega), I_\omega) = 0 \tag{8.5}$$

for any $D \in \mathcal{D}(W)$ and $\omega \in \Omega$, or a uniform pullback attractor if the convergence (8.5) is uniform in $\omega \in \Omega$, that is, if

$$\lim_{t \to \infty} \sup_{\omega \in \Omega} \beta(\varphi(t, D, \sigma_{-t}\omega), I_\omega) = 0.$$

Definition 8.4   The family I is called a forward attractor if the forward convergence

$$\lim_{t \to \infty} \beta(\varphi(t, D, \omega), I_{\sigma_t \omega}) = 0$$

holds instead of the pullback convergence (8.5), or a uniform forward attractor if this forward convergence is uniform in $\omega \in \Omega$, that is, if

$$\lim_{t \to \infty} \sup_{\omega \in \Omega} \beta(\varphi(t, D, \omega), A(\sigma_t \omega)) = 0.$$

The main task of this chapter is to investigate connections of different types of attractors.

It follows directly from the definition that a pullback attractor is unique. Obviously, any uniform pullback attractor is also a uniform forward attractor, and vice versa.

If I is a forward attractor for the cocycle $\langle W, \varphi, (\Omega, \mathbb{T}, \sigma) \rangle$, then ([88] Lemma 4.2) the subset

$$J = \bigcup_{\omega \in \Omega} \left(I_\omega \times \{\omega\} \right) \tag{8.6}$$

of X is the global attractor for the skew–product dynamical system $(X, \mathbb{T}_+, \pi)$. A weaker result holds when I is a pullback attractor. The inverse property is not true in general.

Although we could formulate for weaker assumptions we will restrict ourselves to the case that $\bigcup_{\omega \in \Omega} I_\omega$ is compact and $\mathcal{D}(W)$ consists of compact sets.

Our considerations can be embedded into the more general theory of pullback attractors with a domain of attraction $\mathcal{D}$ consisting of family of sets $D = \{D(\omega)\}_{\omega \in \Omega}$ such that $\bigcup_{\omega \in \Omega} D(\omega)$ is relatively compact in W, see [285].

The following existence result for pullback attractors is adapted from [127, 211].

Theorem 8.1 *Let $\langle W, \varphi, (\Omega, \mathbb{T}, \sigma) \rangle$, with Ω compact be a cocycle and suppose that there exists a family of nonempty sets $C = \{C(\omega)\}_{\omega \in \Omega}$, $\bigcup_{\omega \in \Omega} C(\omega)$ relatively compact such that*

$$\lim_{t \to \infty} \beta(\varphi(t, D, \sigma_{-t}\omega), C(\omega)) = 0$$

for any bounded subset D of W and any $\omega \in \Omega$. Then there exists a pullback attractor.

A related result is given by Theorem 2.24: if the skew–product system $(X, \mathbb{T}_+, \pi)$ has a global attractor $\mathcal{A}$, then the cocycle $\langle W, \varphi, (\Omega, \mathbb{T}, \sigma) \rangle$ has a pullback attractor.

We now continue to derive properties of pullback attractors an skew–product dynamical systems.

Lemma 8.1 *If I is a pullback attractor of a cocycle $\langle W, \varphi, (\Omega, \mathbb{T}, \sigma) \rangle$, where Ω is compact, then the subset $\mathcal{A}$ of X defined by (8.6) is the maximal π-invariant compact set of the associated skew–product dynamical system $(X, \mathbb{T}_+, \pi)$.*

Proof. The π-invariance follows from the φ-invariance of I via

$$\pi(t, J) = \bigcup_{\omega \in \Omega} (\varphi(t, I_\omega, \omega), \sigma_t\omega) = \bigcup_{\omega \in \Omega} (I_{\sigma_t\omega}, \sigma_t\omega) = J.$$

Now $J \subset \bigcup_{\omega \in \Omega} I_\omega \times \Omega$, where Ω is compact and $\bigcup_{\omega \in \Omega} I_\omega$ is relatively compact, so J is relatively compact. Hence $\mathcal{B} := \bar{J}$ is compact, from which it follows that

$$B(\omega) := \{u : (u, \omega) \in \mathcal{B}\}$$

is a compact set in W for each $\omega \in \Omega$ and that the set

$$\bigcup_{\omega \in \Omega} B(\omega) \subset \mathrm{pr}_1 \mathcal{B}$$

is relatively compact. On the other hand, $\mathcal{B}$ is π-invariant since

$$\pi(t, \mathcal{B}) = \pi(t, \bar{J}) = \overline{\pi(t, J)} = \bar{J} = \mathcal{B}$$

for the continuous mapping $\pi(t, \cdot)$. In addition, $\varphi(t, B(\omega), \omega) = B(\sigma_t\omega)$ holds, that is, the $B(\omega)$ are φ-invariant, since

$$\pi(t, \mathcal{B}) = \bigcup_{p \in P} (\varphi(t, B(p), p), \sigma_t p) = \mathcal{B} = \bigcup_{p \in P} (B(\sigma_t p), \sigma_t p)$$

and $\sigma_t\omega = \sigma_t\hat{\omega}$ implies that $\omega = \hat{\omega}$ for the homeomorphism σ_t. The construction shows $B(\omega) \supset I_\omega$. By the φ-invariance of the $B(\omega)$ and the pullback attraction property it follows then that $B(\omega) = I_\omega$ such that $\mathcal{A} = \mathcal{B}$. Hence $\mathcal{A}$ is compact.

To prove that the compact invariant set J is maximal, let J' be any other compact invariant set the of skew–product dynamical system $(X, \mathbb{T}_+, \pi)$. Then $J' = \{I'(\omega)\}_{\omega \in \Omega}$ is a family of compact φ–invariant subsets of W and by pullback attraction

$$\beta(I'_\omega, I\omega) = \beta(\varphi(t, I'_{(\sigma_{-t}\omega)}, \sigma_{-t}\omega), I_\omega)$$
$$\leq \beta(\varphi(t, K, \sigma_{-t}\omega), I_\omega) \to 0$$

as $t \to +\infty$, where $K := \overline{bigcup_{\omega \in \Omega} I'_\omega}$ is compact. Hence $I'_\omega \subseteq I_\omega$ for every $\omega \in \Omega$, i.e. $J' \subseteq J$, which means J is maximal for $(X, \mathbb{T}_+, \pi)$. $\qquad\square$

Definition 8.5   A set–valued mapping $M = \{M(\omega)\}_{\omega \in \Omega}$ for $M(\omega) \in W$ is called upper semi–continuous if

$$\lim_{\omega \to \omega_0} \beta(M(\omega), M(\omega_0)) = 0 \quad \text{for any } \omega_0 \in \Omega.$$

We call such a set–valued mapping lower semi–continuous if

$$\lim_{\omega \to \omega_0} \beta(M(\omega_0), M(\omega)) = 0 \quad \text{for any } \omega_0 \in \Omega.$$

Definition 8.6 M is called continuous if it is both upper and lower semi–continuous.

Note that a set–valued mapping M is upper semi–continuous if and only if the graph of is closed in $W \times \Omega$.

It follows straightforwardly from Lemma 8.1:

Corollary 8.1   *The set valued mapping $\omega \to I_\omega$ formed with the components sets of a pullback attractor $I = \{I_\omega \mid \omega \in \Omega\}$ of a cocycle $\langle W, \varphi, (\Omega, \mathbb{T}, \sigma) \rangle$, where Ω is compact, is upper semi–continuous.*

The following example shows that, in general, a pullback attractor need not also be a forward attractor nor form a global attractor of the associated skew–product dynamical system.

Example 8.1   Let f be the function on $\mathbb{R}$ defined by

$$f(t) = -\left(\frac{1+t}{1+t^2}\right)^2, \qquad t \in \mathbb{R},$$

and let $(\Omega, \mathbb{R}, \sigma)$ be the autonomous dynamical system $\Omega := H(f)$, the hull of f in $C(\mathbb{R}, \mathbb{R})$, with the shift operator σ. Note that

$$\Omega = H(f) = \bigcup_{h \in \mathbb{R}} \{f(\cdot + h)\} \cup \{0\}.$$

Finally, let E be the evaluation functional on $C(\mathbb{R}, \mathbb{R})$, that is $E(\omega) := \omega(0) \in \mathbb{R}$.

Lemma 8.2 *The functional*

$$\gamma(\omega) = -\int_0^\infty e^{-\tau} E(\sigma_\tau \omega) \, d\tau = -\int_0^\infty e^{-\tau} \omega(\tau) \, d\tau$$

is well defined and continuous on Ω, and the function of $t \in \mathbb{R}$ given by

$$\gamma(\sigma_t \omega) = -e^t \int_t^\infty e^{-\tau} \omega(\tau) \, d\tau = \begin{cases} \dfrac{1}{1 + (t+h)^2} & : \omega = \sigma_h f \\[2mm] 0 & : \omega = 0 \end{cases}$$

is the unique solution of the differential equation

$$x' = x + E(\sigma_t \omega) = x + \omega(t)$$

that exists and is bounded for all $t \in \mathbb{R}$.

Proof. The proof is by straightforward calculation, so will be omitted. $\qquad\square$

Consider now the non-autonomous differential equation

$$u' = g(\sigma_t \omega, u), \tag{8.7}$$

where

$$g(\omega, u) := \begin{cases} -u - E(\omega)u^2 & : 0 \le u\gamma(\omega) \le 1,\ \omega \neq 0 \\[2mm] -\dfrac{1}{\gamma(\omega)}\left(1 + \dfrac{E(\omega)}{\gamma(\omega)}\right) & : 1 < u\,\gamma(\omega),\ \omega \neq 0 \\[2mm] -u & : 0 \le u,\ \omega = 0 \end{cases} \quad\cdot$$

It is easily seen that this equation has a unique solution passing through any point $u \in W = \mathbb{R}^+$ at time $t = 0$ defined on $\mathbb{R}$. These solutions define a cocycle mapping

$$\varphi(t, u_0, \omega) = \begin{cases} \dfrac{u_0}{e^t(1 - u_0\gamma(\omega)) + u_0\gamma(\sigma_t\omega)} & : 0 \le u_0\gamma(\omega) \le 1,\ \omega \neq 0 \\[2mm] u_0 + \dfrac{1}{\gamma(\sigma_t\omega)} - \dfrac{1}{\gamma(\omega)} & : 1 < u_0\,\gamma(\omega),\ \omega \neq 0 \\[2mm] e^{-t}u_0 & : 0 \le u_0,\ \omega = 0 \end{cases} \quad\cdot \tag{8.8}$$

According to the construction, the cocycle mapping φ admits as its only invariant sets $I_\omega = \{0\}$ for $\omega \in \Omega$. To see that the $I_\omega = \{0\}$ form a pullback attractor, observe that

$$\varphi(t, u_0, \sigma_{-t}\omega) = \begin{cases} \dfrac{u_0}{e^t(1 - u_0\gamma(\sigma_{-t}\omega)) + u_0\gamma(\omega)} & : 0 \le u_0\gamma(\omega) \le 1,\ \omega \neq 0 \\[2mm] u_0 + \dfrac{1}{\gamma(\omega)} - \dfrac{1}{\gamma(\sigma_{-t}\omega)} & : 1 < u_0\,\gamma(\omega),\ \omega \neq 0 \\[2mm] e^{-t}u_0 & : 0 \le u_0,\ \omega = 0 \end{cases} \quad\cdot$$

In particular, note that $t \to \gamma(\sigma_t \omega)^{-1}$ is a solution of the differential equation (8.7). Since $\gamma(\sigma_{-t}\omega)^{-1}$ tends to $+\infty$ sub-exponentially fast for $t \to \infty$, it follows that

$$\varphi(t, u, \sigma_{-t}\omega) \le \frac{1}{2} L\, e^{-\frac{1}{2}t}$$

for any $u \in [0, L]$ for any $L \ge 0$ and $\omega \in \Omega$ provided t is sufficiently large. Consequently $I = \{I_\omega \mid \omega \in \Omega\}$ with $I_\omega = \{0\}$ for all $\omega \in \Omega$ is a pullback attractor for φ. In view of (8.8), the stable set $W^s(J) := \{x \in X \mid \lim_{t \to +\infty} \rho(\pi^t x, J) = 0\}$ of J, that is, the set of all points in X that are attracted to J by π, is given by

$$W^s(J) = \{(u, \omega) : \omega \in \Omega,\ u \ge 0,\ u\,\gamma(\omega) < 1\} \neq X.$$

Hence the cocycle mapping φ in this example admits a pullback attractor that is neither a forward attractor for φ nor a global attractor of the associated skew–product flow.

Other examples for different kinds of attractors are given by Scheutzow [283] for random dynamical systems generated by one dimensional stochastic differential equations. However, these considerations are based on the theory of Markov processes.

8.2   Asymptotic stability in α-condensing semi–dynamical systems

To continue to investigate general relations between pull back attractors and skew–product flows we have to derive some results from the general stability theory. We start with some definitions.

Let $(X, \mathbb{T}_+, \pi)$ be a semi–dynamical system. The ω– limit set of a set M is defined to be

$$\omega(M) = \bigcap_{\tau \ge 0} \overline{\bigcup_{t \ge \tau} \pi(t, M)}.$$

Definition 8.7   A set M is called Lyapunov stable if for any $\varepsilon > 0$ there exists an $\delta > 0$ such that $\pi(t, B(M, \delta)) \subseteq B(U, \varepsilon)$ for $t \ge 0$.

Definition 8.8 M is called a local attractor if there exists a neighborhood $B(M, \gamma)$ ($\gamma > 0$) of M such that $B(M, \gamma) \subseteq W^s(M)$.

Definition 8.9   A set M which is Lyapunov stable and a local attractor is called asymptotically stable.

Note that any asymptotically stable compact set M also attracts compact sets contained in $\mathcal{U}(M)$.

Definition 8.10 Recall that a π–invariant compact set M is said to be locally maximal if there exists a number $\delta > 0$ such that any π–invariant compact set contained in the open δ-neighborhood $B(M, \delta)$ of M is in fact contained in M.

Definition 8.11 An autonomous semi–dynamical system $(X, \mathbb{T}_+, \pi)$ is called [179] α-condensing if $\pi(t, B)$ is bounded and

$$\alpha(\pi(t, B)) < \alpha(B)$$

for all $t > 0$ for any bounded set B of X with $\alpha(B) > 0$.

Remark 8.1 *In the book [179] there are a lot class of dynamical systems which possesses with property. For example: every d.s. on the finite-dimensional space, every d.s. with compact $\pi^t(t > 0)$ or $\pi^t = m(t) + r(t)(m(t) : X \to X$ is compact for every $t > 0$ and $r(t)x \to 0$ as $t \to +\infty$ uniformly w.r.t. x on every bounded set X).*

Theorem 8.2 *Let M be a locally maximal compact set for an α-condensing semi–dynamical system $(X, \mathbb{T}_+, \pi)$. Then M is Lyapunov stable if and only if there exists a $\delta > 0$ such that*

$$\boldsymbol{\alpha}_{\gamma_x} \cap M = \emptyset$$

for any entire trajectory γ_x through any $x \in B(M, \delta) \setminus M$.

Proof. A proof of the necessity direction was given by Zubov in [336] Theorem 7 for a locally compact space X. This proof remains also true for a non locally compact space under consideration here. Really, let M be a compact invariant set for (X, T_+, π) stable in the positive direction. If we suppose that this assertion is not true, then there exist $x \notin M, \gamma_x$ and $\tau_n \to -\infty$ such that $\rho(\gamma_x(\tau_n), M) \to 0$ as $n \to \infty$. Let $0 < \varepsilon < \rho(x, M)$ and $\delta(\varepsilon) > 0$ the corresponding positive number from stability of set M, then for sufficiently large n we have $\rho(\gamma_x(\tau_n), M) < \delta(\varepsilon)$ and, consequently, $\rho(\pi^t \gamma_x(\tau_n), M) < \varepsilon$ for all $t \geq 0$. In particularly for $t = -\tau_n$ we have $\rho(x, M) = \rho(\pi^{-\tau_n} \gamma_x(\tau_n), M) < \varepsilon$. The obtained contradiction prove our assertion.

For the sufficiency direction, consider first the case $\mathbb{T}_+ = \mathbb{Z}_+$ and let $B(M, \delta_0)$ be a neighborhood such that M is locally maximal in $B(M, \delta_0)$. Suppose that M is not Lyapunov stable, but that the other condition of the theorem holds. Then there exist an $\varepsilon_0 > 0$ and sequences $\delta_n \to 0$, $x_n \in B(M, \delta_n)$, $k_n \to \infty$ such that $\pi(k, x_n) \in B(M, \varepsilon_0)$ for $0 \leq k \leq k_n - 1$ and $\pi(k_n, x_n) \notin B(M, \varepsilon_0)$. This ε_0 has to be chosen sufficiently small such that

$$\beta(\pi(1, B(M, \varepsilon_0)), M) < \frac{\delta_0}{2}.$$

Define $A = \{x_n\}$ and $B = \cup_{n \in \mathbb{N}} \{\pi(k, x_n) \mid 0 \leq k \leq k_n - 1\}$. Then $\alpha(A) = 0(\alpha$ is the measure of non-compactness of Kuratowskii) since A is relatively compact.

In addition, $\pi(1, B) \subseteq B(M, \delta_0)$, so $\pi(1, B)$ is bounded. Suppose that B is not relatively compact, so $\alpha(B) > 0$. It follows by the properties of the measure of non-compactness for the non relatively compact set B that

$$\alpha(B) = \alpha(A \cup \pi(1, B) \cap B) \leq \max(\alpha(A), \pi(1, B)) = \alpha(\pi(1, B)) < \alpha(B)$$

which is a contradiction. This shows that B is relatively compact. Now $\tilde{\gamma}_{\tilde{x}}$ is an entire trajectory of the discrete–time semi–dynamical system above with $\tilde{\gamma}_{\tilde{x}}(0) = \tilde{x}$ and $\overline{\tilde{\gamma}_{\tilde{x}}(\mathbb{Z}_-)} \subset \bar{B}$. Thus the α limit set $\boldsymbol{\alpha}_{\tilde{\gamma}_{\tilde{x}}}$ is nonempty, compact and invariant. In addition, $\boldsymbol{\alpha}_{\tilde{\gamma}_{\tilde{x}}} \subset B(M, \varepsilon_0)$, hence $\boldsymbol{\alpha}_{\tilde{\gamma}_{\tilde{x}}} \subset M$ because M is a locally maximal invariant compact set. On the other hand, $\tilde{\gamma}_{\tilde{x}}(0) = \tilde{x} \in B(M, \varepsilon_0) \setminus M$, so $\boldsymbol{\alpha}_{\tilde{\gamma}_{\tilde{x}}} \cap M = \emptyset$ holds by the assumptions. This contradiction proves the sufficiency of the condition in the discrete–time case.

Now let $\mathbb{T}_+ = \mathbb{R}_+$ and suppose that $\boldsymbol{\alpha}_{\gamma_x} \cap M = \emptyset$ where $x \notin M$ holds for the continuous–time system. Then it also holds for the restricted discrete–time system generated by $\pi_1 := \pi(1, \cdot)$ because any entire trajectory γ_x of the restricted discrete–time system can be extended to an entire trajectory of the continuous–time system via

$$\gamma_x(t) = \pi(\tau, \gamma_x(n)), \quad n \in \mathbb{Z}, \quad t = n + \tau, \quad 0 < \tau < 1.$$

Consequently, the set M is Lyapunov stable with respect to the restricted discrete–time dynamical system generated by π_1. Since M is compact, for every $\varepsilon > 0$ there exists a $\mu > 0$ such that

$$\rho(\pi(t, x), M) < \varepsilon \quad \text{for all } t \in [0, 1], \ x \in B(M, \mu).$$

In view of the first part of the proof above, there is a $\delta > 0$ such that

$$\rho(\pi(n, x), M) < \min(\mu, \varepsilon) \quad \text{for } x \in B(M, \delta) \quad \text{for } n \in \mathbb{Z}_+.$$

The Lyapunov stability of M for the continuous dynamical system $(X, \mathbb{R}_+, \pi)$ then follows from the semi–group property of π. $\qquad\square$

The next lemma will be needed to formulate the second main theorem of this section. Asymptotic stability here means (locally) Lyapunov stability and attracting.

Lemma 8.3   *Let M be a compact subset of X that is positively invariant for a semi–dynamical system $(X, \mathbb{T}_+, \pi)$. Then M is asymptotically stable if and only if $\omega(M)$ is locally maximal and asymptotically stable.*

Proof. Suppose that M is asymptotically stable. Then there exists a closed positively invariant bounded neighborhood C of M contained in its stable set $W^s(M)$. The mapping π can be restricted to the complete metric space C to form a semi–dynamical system $(C, \mathbb{T}^+, \pi)$. Since M is a locally attracting set it attracts compact subsets of C. The assertion then follows by Theorems 2.4.2 and 3.4.2 in [179] because $\omega(M) = \bigcap_{t \in \mathbb{T}^+} \pi(t, M)$.

Suppose instead that $\omega(M)$ is asymptotically stable and locally maximal. Since M is compact, $\omega(M) = \bigcap_{t \geq 0} \pi(t, M)$. Hence there exist $\eta > 0$ and $\tau \in \mathbb{T}^+$ such that

$$\pi(\tau, M) \subset B(\omega(M), \eta) \subset W^s(\omega(M)).$$

Now $\pi^{-1}(\tau, B(\omega(M), \eta))$, where π^{-1} denotes the pre–image of $\pi(\tau, \cdot)$ for fixed τ, is an open neighborhood of M and $\pi(\tau, \pi^{-1}(\tau, B(\omega(M), \eta))) \subset W^s(\omega(M))$. Hence for any $x \in \pi^{-1}(\tau, B(\omega(M), \eta) \subset W^s(\omega(M))$ we have that $\pi(t, x)$ tends to $\omega(M)$ as $t \to \infty$, from which it follows that $\pi(t, x)$ also tends to M because $M \supset \Omega(M)$.

Then, if M were not Lyapunov stable, there would exist $\varepsilon_0 > 0$, $\delta_n \to 0$, $x_n \in B(M, \delta_n)$ and $t_n \to \infty$ such that

$$\rho(\pi(t_n, x_n), M) \geq \varepsilon_0. \tag{8.9}$$

For sufficiently large n_0, the set $\overline{\{x_n\}}_{n \geq n_0}$ would then be contained in the pre–image $\pi^{-1}(1, B(\omega(M), \eta))$. Since M is compact, so is the set $\overline{\{x_n\}}_{n \geq n_0}$. This set would thus be attracted by $\omega(M) \subset M$, which contradicts (8.9) . $\qquad\square$

Lemma 8.4 *Let M be a compact subset of X that is a positively invariant set for an asymptotically compact semi–dynamical system $(X, \mathbb{T}_+, \pi)$. Then the set M is asymptotically stable if and only if $\omega(M)$ is locally maximal and Lyapunov stable.*

Proof. The necessity follows by Lemma 8.3. Suppose instead that $\omega(M)$ is locally maximal and Lyapunov stable. Then for any $\varepsilon > 0$ there exists a $\delta > 0$ such that

$$\pi(t, B(\omega(M), \delta) \subset B(\omega(M), \varepsilon) \qquad \text{for all } t \geq 0.$$

By the assumption of asymptotic compactness, $\omega(B(\omega(M), \delta))$ is nonempty and compact with

$$\lim_{t \to \infty} \beta(\pi(t, B(\omega(M), \delta)), \omega(B(\omega(M), \delta))) = 0$$

(see [179] Corollary 2.2.4.). Since $\omega(M)$ is locally maximal, $\omega(B(\omega(M), \delta)) \subset \omega(M)$ for sufficiently small $\delta > 0$, which means that $\omega(M)$ is asymptotically stable. The conclusion then follows by Lemma 8.3. $\qquad\square$

Corollary 8.2 *Let $(X, \mathbb{T}_+, \pi)$ be asymptotically compact and let M be a compact π–invariant set. Then M is asymptotically stable if and only if M is locally maximal and Lyapunov stable.*

Proof. Indeed, $M = \omega(M)$ here, so just apply Lemma 8.4. $\qquad\square$

The next theorem is a generalization to infinite dimensional spaces and α-condensing systems of Theorem 8 of Zubov [336] characterizing the asymptotic stability of a compact set.

Theorem 8.3 *Let* $(X, \mathbb{T}_+, \pi)$ *be an* α-*condensing semi–dynamical system and let* $M \subset X$ *be a compact invariant set. Then the set* M *is asymptotically stable if and only if*

(i) M *is locally maximal, and*

(ii) there exists a $\delta > 0$ *such that* $\boldsymbol{\alpha}_{\gamma_x} \cap M = \emptyset$ *for any entire trajectory* γ_x *through any* $x \in B(M, \delta) \setminus M$.

Proof. By Lemma 2.3.5 in [179] any α-condensing semi–dynamical system is asymptotically compact, so the assertion follows easily from Theorem 8.2 and Corollary 8.2. $\qquad\square$

Definition 8.12 A cocycle $\langle W, \varphi, (\Omega, \mathbb{T}, \sigma)\rangle$ will be called α-condensing if the set $\varphi(t, B, \Omega)$ is bounded and

$$\alpha(\varphi(t, B, \Omega)) < \alpha(B)$$

for all $t > 0$ for any bounded subset B of W with $\alpha(B) > 0$.

Lemma 8.5 *Suppose that the cocycle* $\langle W, \varphi, (\Omega, \mathbb{T}, \sigma)\rangle$ *is* α-*condensing. Then the associated skew–product flow* $(X, \mathbb{T}_+, \pi)$ *is also* α-*condensing.*

Proof. Let $M := \bigcup_{\omega \in \Omega}(M_\omega \times \{\omega\})$ be a bounded set in X. Then M can be covered by finitely many balls $M_i \subset X$, $i = 1, \cdots, n$, of largest radius $\alpha(M) + \varepsilon$ for an arbitrary $\varepsilon > 0$. The sets $\mathrm{pr}_1 M_i \subset W$, $i = 1, \cdots, n$, cover $\mathrm{pr}_1 M$. The sets M_i are balls so $\alpha(\mathrm{pr}_1 M_i) = \alpha(M_i) < \alpha(M) + \varepsilon$ for $i = 1, \cdots, n$. It is easily seen that

$$\pi(t, M) = \bigcup_{\omega \in \Omega} \{\pi(t, (M_\omega, \omega))\} =$$

$$\bigcup_{\omega \in \Omega} \{(\varphi(t, M_\omega, \omega), \sigma_t \omega)\} \subset \varphi(t, \mathrm{pr}_1 M, \Omega) \times \Omega.$$

Since φ is α-condensing, the set $\varphi(t, \mathrm{pr}_1 M, \Omega)$ is bounded. Hence

$$\alpha(\pi(t, M)) \leq \alpha(\varphi(t, \mathrm{pr}_1 M, \Omega) \times \Omega) \tag{8.10}$$

$$\leq \alpha(\varphi(t, \mathrm{pr}_1 M, \Omega)) < \alpha(\mathrm{pr}_1 M) \leq \alpha(M) \quad \text{for each } t > 0.$$

The second inequality above is true by the compactness of Ω. Indeed, Ω can be covered by finitely many open balls Ω_i of arbitrarily small radius. Hence

$$\alpha(\varphi(t, \mathrm{pr}_1 M, \Omega) \times \Omega) \leq \max_i \alpha(\varphi(t, \mathrm{pr}_1 M, \Omega) \times \Omega_i) \leq \alpha(\varphi(t, \mathrm{pr}_1 M, \Omega)) + \varepsilon$$

for arbitrarily small $\varepsilon > 0$. The conclusion of Lemma follows by (8.10). $\qquad\square$

8.3 Uniform pullback attractors and global attractors

It was seen earlier that the set $\cup_{\omega\in\Omega}(I_\omega \times \{\omega\}) \subset X$ which was defined in terms of the pullback attractor $I = \{I_\omega\}_{\omega\in\Omega}$ of a cocycle $\langle W, \varphi, (\Omega, \mathbb{T}, \sigma)\rangle$ is the maximal π-invariant compact subset of the associated skew–product system $(X, \mathbb{T}_+, \pi)$, but need not be a global attractor. However, this set is always a attractor under the additional assumption that the cocycle φ is α-condensing.

Theorem 8.4 *Let Ω be a compact space, $\langle W, \varphi, (\Omega, \mathbb{T}, \sigma)\rangle$ be an α-condensing cocycle with a pullback attractor $I = \{I_\omega\}_{\omega\in\Omega}$ and define $J = \cup_{\omega\in\Omega}(I_\omega \times \{\omega\})$. Then*

(i) the α-limit set $\boldsymbol{\alpha}_{\gamma_x}$ of any entire trajectory γ_x passing through $x \in X \setminus J$ is empty.

(ii) J is asymptotically stable with respect to π.

Proof. Suppose that there exists an entire trajectory γ_x through $x = (u, \omega) \in X \setminus J$ such that $\boldsymbol{\alpha}_{\gamma_x} \neq \emptyset$. Then there exists a subsequence $-\tau_n \to \infty$ such that $\gamma_x(\tau_n)$ converges to a point in $\boldsymbol{\alpha}_{\gamma_x}$. The set $K = \mathrm{pr}_1 \overline{\cup_{n\in\mathbb{N}} \gamma_x(\tau_n)}$ is compact since $\overline{\cup_{n\in\mathbb{N}} \gamma_x(\tau_n)}$ is compact. Also $I = \{I_\omega \mid \omega \in \Omega\}$ is a pullback attractor, so

$$\lim_{n\to\infty} \beta(\varphi(-\tau_n, K, \sigma_{\tau_n}\omega), I_\omega) = 0$$

from which it follows that $u \in I_\omega$. Hence $(u, \omega) \in J$, which is a contradiction. This proves the first assertion.

By Lemma 8.5 $(X, \mathbb{T}^+, \pi)$ is α-condensing. According to Lemma 8.1 J is a maximal compact invariant set of $(X, \mathbb{T}^+, \pi)$ since I is a pullback attractor of the cocycle φ. The second assertion then follows from Theorem 8.4 and from the first assertion of this theorem. $\qquad\square$

Remark 8.2 *(i) The skew–product system in the example in Section 2 has only a* local *attractor associated with the pullback attractor.*

(ii) If in addition to the assumptions of Theorem 8.4 the stable set $W^s(J)$ of J satisfies $W^s(J) = X$, then J is in fact a global attractor (see Theorem 1.13).

Theorem 8.5 *Suppose that Ω compact, $\langle W, \varphi, (\Omega, \mathbb{T}, \sigma)\rangle$ is a cocycle with a pullback attractor $I = \{I_\omega \mid \omega \in \Omega\}$ and suppose that $W^s(J) = X$, where $J = \cup_{\omega\in\Omega}(I_\omega \times \{\omega\})$.*

If the mapping $\omega \to I_\omega$ is lower semi–continuous, then I is a uniform pullback attractor, hence a uniform forward attractor.

Proof. Suppose that the uniform convergence

$$\lim_{t\to\infty} \sup_{\omega\in\Omega} \beta(\varphi(t,D,\omega), I_{\sigma_t\omega}) = 0$$

is not true for some $D \in C(W)$. Then there exist $\varepsilon_0 > 0$, a set $D_0 \in C(W)$ and sequences $t_n \to \infty$, $\omega_n \in \Omega$ and $u_n \in D_0$ such that:

$$\rho(\varphi(t_n, u_n, \omega_n), I_{\sigma_{t_n}\omega}) \geq \varepsilon_0. \tag{8.11}$$

Now Ω is compact and J is a global attractor by Remark 8.2 (ii), so it can be supposed that the sequences $\{\varphi(t_n, u_n, \omega_n)\}$ and $\{\sigma_{t_n}\omega\}$ are convergent. Let $\bar{u} = \lim_{n\to\infty} \varphi(t_n, u_n, \omega_n)$ and $\bar{\omega} = \lim_{n\to\infty} \sigma_{t_n}\omega_n$. Then $\bar{u} \in I_{\bar{\omega}}$ because $\bar{x} = (\bar{u}, \bar{\omega}) \in J$. On the other hand, by (8.11),

$$\begin{aligned}
\varepsilon_0 &\leq \beta(\varphi(t_n, \omega_n, x_n), I_{\sigma_{t_n}\omega_n}) \\
&\leq \beta(\varphi(t_n, \omega_n, x_n), I_{\bar{\omega}}) + \beta(I_{\bar{\omega}}, I_{\sigma_{t_n}\omega_n}).
\end{aligned}$$

By the lower semi–continuity of $\omega \to I_\omega$ it follows then that $\bar{u} \notin A(\bar{\omega})$, which is a contradiction. $\qquad\square$

Remark 8.3   *The example 8.1 shows that Theorem 8.5 is in general not true without the assumption that $W^s(J) = X$. In view of Corollary 8.1, the set valued mapping $\omega \to I_\omega$ will, in fact, then be continuous here.*

8.4   Examples of uniform pullback attractors

Several examples illustrating the application of the above results, in particular of Theorem 8.5, are now presented. More complicated examples will be discussed in Chapter 12 (section 12.5).

8.4.1   *Periodic driving systems*

Consider a periodical dynamical system $(\Omega, \mathbb{T}, \sigma)$, that is, for which there exists a minimal positive number T such that $\sigma_T \omega = \omega$ for any $\omega \in \Omega$.

Theorem 8.6   *Suppose that a non-autonomous α-condensing dynamical system $\langle U, \varphi, (\Omega, \mathbb{R}, \sigma)\rangle$ with a periodical dynamical system $(\Omega, \mathbb{R}, \sigma)$ has a pullback attractor $I = \{I_\omega\}_{\omega\in\Omega}$. Then IJ is a uniform forward attractor for $\langle W, \varphi, (\Omega, \mathbb{R}, \sigma)\rangle$.*

Proof.   Consider a sequence $\omega_n \to \omega$. By the periodicity of the driving system there exists a sequence $\tau_n \in [0, T]$ such that $\omega_n = \sigma_{\tau_n}\omega$. By compactness, there is a convergent subsequence (indexed here for convenience like the full one) $\tau_n \to \tau \in$

$[0, T]$. Hence

$$\omega = \lim_{n \to \infty} \omega_n = \lim_{n \to \infty} \sigma_{\tau_n} \omega = \sigma_\tau \omega$$

which means $\tau = 0$ or T. Suppose that $\tau = T$. Then

$$\lim_{n \to \infty} \beta(I_\omega, I_{\omega_n}) = \lim_{n \to \infty} \beta(I_\omega, \varphi(\tau_n, I_\omega, \omega))$$
$$= \beta(I_\omega, \varphi(T, I_\omega, \omega)) = \beta(I_\omega, A(\sigma_T \omega)) = 0$$

since φ is continuous and $I_{\omega_n} = I_{\sigma_{\tau_n} \omega} = \varphi(\tau_n, I_\omega, \omega)$ by the φ–invariance of J. Hence the set valued $\omega \to I_\omega$ is lower semi–continuous.

Now $\varphi(nT, u_0, \omega) = \varphi(nT, u_0, \sigma_{-nT} \omega)$ since $\omega = \sigma_{-nT} \omega$ by the periodicity of the driving system $(\Omega, \mathbb{R}, \sigma)$. Hence from pullback convergence

$$\lim_{n \to \infty} \rho(\varphi(nT, u_0, \omega), I_\omega) = \lim_{n \to \infty} \rho(\varphi(nT, u_0, \sigma_{-nT} \omega), I_\omega) = 0$$

for any $(u_0, \omega) \in U \times P$. On the other hand

$$\sup_{s \in [0,T]} \rho(\varphi(s + nT, u_0, \omega), A(\sigma_{s+nT} \omega))$$
$$= \sup_{s \in [0,T]} \rho(\varphi(s, \varphi(nT, u_0, \omega), \sigma_{nT} \omega), \varphi(s, A(\sigma_{nT} \omega), \sigma_{nT} \omega)) = 0$$

by the cocycle property of φ and the φ–invariance of J. Hence

$$\lim_{t \to \infty} \rho((\varphi(t, u_0, \omega), \sigma_t \omega), J) = 0,$$

where $J = \cup_{\omega \in \Omega} (I_\omega \times \{\omega\})$. This shows that $W^s(J) = X$. The result then follows by Theorem 8.5. $\qquad\square$

Consider the 2–dimensional Navier-Stokes equation in the operator form

$$\frac{du}{dt} + \nu Au + B(u) = f(t), \quad u(0) = u_0 \in H, \tag{8.12}$$

which can be interpreted as an evolution equation on the rigged space $V \subset H \subset V'$, where V and H are certain Banach spaces. In particular, here $W = H$, which is in fact a Hilbert space, for the phase space. Then, from [311, Chapter 3],

Lemma 8.6 *The 2–dimensional Navier-Stokes equation (8.12) has a unique solution $u(\cdot, u_0, f)$ in $C(0, T; H)$ for each initial condition $u_0 \in H$ and forcing term $f \in C(0, T; H)$ for every $T > 0$. Moreover, $u(t, u_0, f)$ depends continuously on (t, u_0, f) as a mapping from $\mathbb{R}^+ \times H \times C(\mathbb{R}, H)$ to H.*

Now suppose that f is a periodic function in $C(\mathbb{R}, H)$ and define $\sigma_t f(\cdot) := f(\cdot + t)$. Then $\Omega := \bigcup_{t \in \mathbb{R}} \sigma_t f$ is a compact subset of $C(\mathbb{R}, H)$. By Lemma 8.6 the mapping $(t, u_0, \omega) \to \varphi(t, u_0, \omega)$ from $\mathbb{R}^+ \times H \times C(\mathbb{R}, H) \to H$ defined by $\varphi(t, u_0, \omega) = u(t, u_0, \omega)$ is continuous and forms a cocycle mapping with respect to σ on Ω.

By [311] Theorem III.3.10 the mapping φ is completely continuous and hence α-condensing.

Lemma 8.7   *The cocycle $\langle H, \varphi, (\Omega, \mathbb{R}, \sigma) \rangle$ generated by the Navier-Stokes equation (8.12) with periodic forcing term in $C(\mathbb{R}, H)$ has a pullback attractor.*

Proof. The solution of the Navier-Stokes equation satisfies an energy inequality

$$|u(t)|_H^2 + \lambda_1 \nu \int_0^t |u(\tau)|_H^2 \, d\tau \leq |u_0|_H^2 + \frac{1}{\nu} \int_0^t |w(\tau)|_{V'}^2 \, d\tau$$

where λ_1 is the smallest eigenvalue of A. It follows that the balls $B[0, R(\omega)]$ in H with center zero and square radius

$$R^2(\omega) = \frac{1}{\nu} \int_{-\infty}^0 e^{\nu \lambda_1 \tau} |w(\tau)|_{V'}^2 \, d\tau$$

is a pullback attracting family of sets in the sense of Theorem 8.1. In particular, $C(\omega) := \varphi(1, B[0, R(\sigma_{-1}\omega)], \sigma_{-1}\omega)$ satisfies all of the required properties of Theorem 8.1 because $\varphi(1, \cdot, \omega)$ is completely continuous. $\square$

This theorem and Theorem 8.6 give

Theorem 8.7   *The cocycle $\langle H, \varphi, (\Omega, \mathbb{R}, \sigma) \rangle$ generated by the Navier-Stokes equation (8.12) with periodic forcing term in $C(\mathbb{R}, H)$ has a uniform pullback attractor, which is also a uniform forward attractor.*

Remark 8.4   *See [149] for a related result involving a different type of non-autonomous attractor.*

8.4.2   *Pullback attractors with singleton component sets*

Now pullback attractors with singleton component sets, that is with

$$I_\omega = \{a(\omega)\}, \qquad a(\omega) \in W,$$

will be considered.

Lemma 8.8   *Let $\langle W, \varphi, (\Omega, \mathbb{T}, \sigma) \rangle$ be a cocycle, $J = \{I_\omega\}_{\omega \in \Omega}$ be a pullback attractor with singleton component sets. Then the mapping $\omega \to I_\omega$ is continuous.*

Proof. This follows from Corollary 8.1 since the upper semi–continuity of a set valued mapping $\omega \to I_\omega$ reduces to continuity when the I_ω are single point sets. $\square$

It follows straightforwardly from this lemma and Theorem 8.5 that

Theorem 8.8 *Suppose that Ω is compact and the cocycle $\langle W, \varphi, (\Omega, \mathbb{T}, \sigma)\rangle$ has a pullback attractor $I = \{I_\omega\}_{\omega \in \Omega}$ with singleton component sets which generates a global attractor $J = \cup_{\omega \in \Omega} I_\omega \times \{\omega\}$. Then J is a uniform pullback attractor and, hence, also a uniform forward attractor.*

The previous theorem can be applied to differential equations on a Hilbert space $(H, \langle \cdot, \cdot \rangle)$ of the form

$$u' = F(\sigma_t \omega, u) \tag{8.13}$$

where $F \in C(\Omega \times H, H)$ is uniformly dissipative, that is, there exist $\nu \geq 2$, $\alpha > 0$

$$\langle F(\omega, u_1) - F(\omega, u_2), u_1 - u_2 \rangle \leq -\alpha |u_1 - u_2|^\nu \tag{8.14}$$

for any u_1, $u_2 \in H$ and $\omega \in \Omega$.

Theorem 8.9 [92] *The cocycle $\langle H, \varphi, (\Omega, \mathbb{T}, \sigma)\rangle$ generated by (8.13) has a uniform pullback attractor that consists of singleton component subsets.*

For example, the equation

$$u' = F(\sigma_t \omega, x) = -u|u| + g(\sigma_t \omega)$$

with $g \in C(\Omega, \mathbb{R})$ satisfies

$$\langle W_1 - u_2, F(\sigma_t \omega, u_1) - F(\sigma_t \omega, u_2) \rangle \leq -\frac{1}{2} |u_1 - u_2|^2 (|u_1| + |u_2|) \leq -\frac{1}{2} |u_1 - u_2|^3,$$

which is condition (8.14) with $\alpha = \frac{1}{2}$ and $\nu = 3$.

The above considerations apply also to nonlinear non-autonomous partial differential equations with a uniform dissipative structure, such as the dissipative quasi–geostrophic equations

$$u_t + J(\psi, u) + \beta \psi_x = \nu \Delta u - r u + f(x, y, t) , \tag{8.15}$$

with relative vorticity $u(x, y, t) = \Delta \psi(x, y, t)$, where $J(f, g) = f_x g_y - f_y g_x$ is the Jacobian operator. This equation can be supplemented by homogeneous Dirichlet boundary conditions for both ψ and u

$$\psi(x, y, t) = 0, \qquad u(x, y, t) = 0 \quad \text{on } \partial D , \tag{8.16}$$

and an initial condition,

$$u(x, y, 0) = u_0(x, y) \quad \text{on } D,$$

where D is an arbitrary bounded planar domain with area $|D|$ and piecewise smooth boundary. Let W be the Hilbert space $L_2(D)$ with norm $|\cdot|$.

Theorem 8.10   *Assume that*

$$\frac{r}{2} + \frac{\pi\nu}{|D|} > \frac{1}{2}\beta\left(\frac{|D|}{\pi} + 1\right)$$

and that the wind forcing $f(x, y, t)$ is temporally periodic (quasi-periodic, almost periodic, recurrent) with its $L^2(D)$–norm bounded uniformly in time $t \in \mathbb{R}$ by

$$\|f(\cdot, \cdot, t)\| \le \sqrt{\frac{\pi r}{3|D|}}\left[\frac{r}{2} + \frac{\pi\nu}{|D|} - \frac{1}{2}\beta\left(\frac{|D|}{\pi} + 1\right)\right]^{\frac{3}{2}}.$$

Then the dissipative quasigeostrophic model (8.15)–(8.16) has a unique temporally periodic (quasi-periodic, almost periodic, recurrent) solution that exists for all time $t \in \mathbb{R}$.

The proof in [141] (for the almost periodic case) involves explicitly constructing a uniform pullback and forward absorbing ball in $L^2(D)$ for the vorticity ω, hence implying the existence of a uniform pullback attractor as well as a global attractor for the associate skew–product flow system for which the component sets are singleton sets. The parameter set Ω here is the hull of the forcing term f in $L^2(D)$ and a completely continuous cocycle $\varphi(t, u_0, \omega) = \omega(t, u_0, \omega)$ with respect to the shift operator σ on Ω that is continuous in all variables.

8.4.3   *Distal dynamical systems*

A function $\gamma_{(u,\omega)} : \mathbb{R} \to W$ represents an entire trajectory $\gamma_{(u,\omega)}$ of a cocycle $\langle W, \varphi, (\Omega, \mathbb{T}, \sigma)\rangle$ if $\gamma_{(u,\omega)}(0) = u \in W$ and $\varphi(t, \gamma_{(u,\omega)}(\tau), \sigma_\tau\omega) = \gamma_{(u,\omega)}(t + \tau)$ for $t \ge 0$ and $\tau \in \mathbb{R}$.

Definition 8.13   A cocycle is called distal on $\mathbb{T}_-$ if

$$\inf_{t \in \mathbb{T}_-} \rho\left(\gamma_{(u_1,\omega)}(t), \gamma_{(u_2,\omega)}(t)\right) > 0$$

for any entire trajectories $\gamma_{(u_1,\omega)}$ and $\gamma_{(u_2,\omega)}$ with $u_1 \ne u_2 \in W$ and any $\omega \in \Omega$.

Definition 8.14   A cocycle φ is said to be uniformly Lyapunov stable if for any $\varepsilon > 0$ there exists a $\delta = \delta(\varepsilon) > 0$ such that

$$\rho\left(\varphi(t, u_1, \omega), \varphi(t, u_2, \omega)\right) < \delta$$

for all u_1, $u_2 \in W$ with $\rho(u_1, u_2) < \varepsilon$, $\omega \in \Omega$ and $t \ge 0$.

Recall that an autonomous dynamical system $(\Omega, \mathbb{T}, \sigma)$ is called minimal if Ω does not contain proper compact subsets which are σ–invariant.

The following lemma is due to Furstenberg [156] (see also [32, Chapter 3] or [238, Chapter 7] Proposition 4).

Lemma 8.9 *Suppose that a cocycle $\langle W, \varphi, (\Omega, \mathbb{T}, \sigma) \rangle$ is distal on $\mathbb{T}_-$ and that $(\Omega, \mathbb{T}, \sigma)$ is minimal. In addition suppose that a compact subset J of X is $\pi-$invariant with respect to the skew–product system $(X, \mathbb{T}_+, \pi)$. Then the mapping $\omega \to I_\omega := \{u \in W : (u, \omega) \in J\}$ is continuous.*

The following theorem gives the existence of uniform forward attractors.

Theorem 8.11 *Suppose that the cocycle $\langle W, \varphi, (\Omega, \mathbb{T}, \sigma) \rangle$ is uniformly Lyapunov stable and that the skew–product system $(X, \mathbb{T}_+, \pi)$ has a global attractor $J = \cup_{\omega \in \Omega} I_\omega \times \{\omega\}$. Then $I = \{I_\omega\}_{\omega \in \Omega}$ is a uniform forward attractor for $\langle W, \varphi, (\Omega, \mathbb{T}, \sigma) \rangle$.*

Proof. Suppose that the cocycle $\langle W, \varphi, (\Omega, \mathbb{T}, \sigma) \rangle$ is not distal. Then there is a $\omega_0 \in \Omega$, a sequence $t_n \to \infty$ and entire trajectories $\gamma_{(u_1, \omega_0)}$, $\gamma_{(u_2, \omega_0)}$ with $u_1 \neq u_2$ such that

$$\lim_{n \to \infty} \rho \left(\gamma_{(u_1, \omega_0)}(-t_n), \gamma_{(u_2, \omega_0)}(-t_n) \right) = 0.$$

Let $\varepsilon = \rho(u_1, u_2) > 0$ and choose $\delta = \delta(\varepsilon) > 0$ from the uniformly Lyapunov stability property. Then

$$\rho \left(\gamma_{(u_1, \omega_0)}(-t_n), \gamma_{(u_2, \omega_0)}(-t_n) \right) < \delta$$

for sufficiently large n. Hence

$$\rho \left(\varphi(t, \gamma_{(u_1, \omega_0)}(-t_n), \sigma_{-t_n} \omega_0), \varphi(t, \gamma_{(u_2, \omega_0)}(-t_n), \sigma_{-t_n} \omega_0) \right) < \varepsilon$$

for $t \geq 0$ and, in particular,

$$\varepsilon = \rho(u_1, u_2) = \rho(\varphi(t_n, \gamma_{(u_1, \omega_0)}(-t_n), \sigma_{-t_n} \omega_0),$$
$$\varphi(t_n, \gamma_{(u_2, \omega_0)}(-t_n), \sigma_{-t_n} \omega_0)) < \varepsilon$$

for $t = t_n$, which is a contradiction. The cocycle is thus distal, so $\omega \to I_\omega$ is continuous by Lemma 8.9. The result then follows from Theorem 8.5 since $\{I_\omega \mid \omega \in \Omega\}$ generates a pullback attractor. $\square$

This theorem will now be applied to the non-autonomous differential equation (8.13) on a Hilbert space H, which is assumed to generate a cocycle φ that is continuous on $\mathbb{T}_+ \times \Omega \times H$ and asymptotically compact.

Theorem 8.12 *Suppose that $F \in C(H \times \Omega, H)$ satisfies the dissipativity conditions*

$$\langle F(u_1, \omega) - F(u_2, \omega), u_1 - u_2 \rangle \leq 0 \tag{8.17}$$

$$\langle F(u, \omega), u \rangle \leq -\mu(|u|) \tag{8.18}$$

for u_1, u_2, $u \in H$ and $\omega \in \Omega$, where $\mu : [R, \infty) \to \mathbb{R}_+ \setminus \{0\}$. Suppose also that (8.13) generates a cocycle φ that is continuous and asymptotically compact. Finally, suppose that $(\Omega, \mathbb{T}, \sigma)$ is a minimal dynamical system.

Then the cocycle $\langle H, \varphi, (\Omega, \mathbb{T}_+, \sigma) \rangle$ has a uniform pullback attractor.

Proof. It follows by the chain rule applied to $|u|^2$ for a solution of (8.13) that $|\varphi(t, u, \omega)| < |u|$ for $|u| > R$, $t > 0$ and $\omega \in \Omega$. Hence, according to Theorem 5.5, the non-autonomous dynamical system $\langle (X, \mathbb{T}_+, \pi), (\Omega, \mathbb{T}_+, \sigma), h \rangle$ (where $X := H \times \Omega, h := pr_2$ and $\pi := (\varphi, \sigma)$) has a global attractor. On the other hand, by (8.17),

$$|\varphi(t, u_1, \omega) - \varphi(t, u_2, \omega)| \le |u_1 - u_2|$$

for $t \ge 0$, $\omega \in \Omega$ and u_1, $u_2 \in H$. Theorem 8.11 then gives the result. $\qquad\square$

The above theorem holds for a differential equation (8.13) on $H = \mathbb{R}$ with

$$F(\omega, u) = \begin{cases} -(u + 1) + g(\omega) : u < -1 \\ \qquad g(\omega) \qquad : |u| \le 1 \\ -(u - 1) + g(\omega) : u > 1. \end{cases}$$

where $g \in C(\Omega, \mathbb{R})$.

Chapter 9

Pullback attractors of $\mathbb{C}$-analytic systems

One of the most studied classes of nonlinear ODEs is the class of $\mathbb{C}$ - analytic differential equations, i.e. the equations

$$\frac{dz}{dt} = f(t, z), \qquad (9.1)$$

where the right hand side f is a holomorphic function with respect to complex variable $z \in \mathbb{C}^d$. Let $\varphi(t, f, z)$ be a unique solution of equation (9.1) with initial condition $\varphi(0, f, z) = z$ and be defined on $\mathbb{R}_+$. In virtue of fundamental theory of ODEs with holomorphic right hand side (see, for example [122] and [186]) the mapping φ possesses the following properties:

1. $\varphi(0, f, z) = z$.
2. $\varphi(t + \tau, f, z) = \varphi(t, f_\tau, \varphi(\tau, f, z))$ for every $t, \tau \in \mathbb{R}^+$ and $z \in \mathbb{C}^d$, where f_τ is a τ- translation of function f.
3. φ is continuous.
4. $\varphi(t, f, \cdot) : \mathbb{C}^d \to \mathbb{C}^d$ is holomorphic for every t and f.

The properties 1.-4. will be the basis of our research of abstract $\mathbb{C}$−analytic non-autonomous dynamical system.

The dissipative periodic equation (9.1) was studied by I.L.Zinchenko [334] and he proved that in this case the equation (9.1) admits a unique periodic globally uniformly asymptotically stable solution. This result was generalized for almost periodic equations (9.1) by D.N.Cheban [62] and [84] (see also the Chapter 3). He studied this problem within the framework of general $\mathbb{C}$- analytic non-autonomous dynamical systems.

In this chapter we study the structure of global pullback attractors of general $\mathbb{C}$- analytic cocycles with noncompact base (in terminology of equation (9.1): the right hand side f is unbounded with respect to time $t \in \mathbb{R}$).

This chapter is organized as follows. In section 1 we introduce the class of $\mathbb{C}$−analytic cocycle.

In section 2 we establish some general facts about non-autonomous dynamical systems. We introduce the semigroup E_ω^+, E_ω^- and E_ω acting on the fiber X_ω of stratification (X, h, Ω). These semigroups are sub-semigroups of Ellis semigroup (in the case of compact base Ω) and play an important role in the study of non-autonomous dynamical system.

Section 3 is devoted to positively uniformly stable cocycles. For this class of cocycles we prove that on every compact invariant set the corresponding cocycle can be prolonged uniquely in the negative direction.

In section 4 we study the structure of compact global pullback attractor of $\mathbb{C}-$ analytic cocycles with compact base. The main result in this section is Theorem 9.4 which states that for considered class of cocycles the pullback attractor $\{I_\omega \mid \omega \in \Omega\}$ is trivial, i.e. the section I_ω contains a single point.

Section 5 is devoted to study of the uniform dissipative cocycles with noncompact base. For this class of cocycles we prove the triviality of its global pullback attractor (see Theorem 9.7).

In section 6 we introduce the class of cocycles possessing the property of dissipativity (non-uniform) with non-compact base. The main result in this section is Theorem 9.8 which describes the structure of compact pullback attractor of mentioned class of cocycles. In particular its triviality is proved.

Section 7 is devoted to application of our general results, obtained in sections 3-6 to study of differential equations (ODEs, Caratheodory equations with almost periodic coefficients, almost periodic ODEs with impulse).

This paper is organized as follows. In section 1 we introduce the class of $\mathbb{C}-$ analytic cocycle.

In section 2 we establish some general facts about non-autonomous dynamical systems. We introduce the semigroup E_ω^+, E_ω^- and E_ω acting on the fiber X_ω of stratification (X, h, Ω). These semigroups are sub-semigroups of Ellis semigroup (in the case of compact base Ω) and play an important role in the study of non-autonomous dynamical system.

Section 3 is devoted to positively uniformly stable cocycles. For this class of cocycles we prove that on every compact invariant set the corresponding cocycle can be prolonged uniquely in the negative direction.

In section 4 we study the structure of compact global pullback attractor of $\mathbb{C}-$ analytic cocycles with compact base. The main result in this section is Theorem 9.4 which states that for considered class of cocycles the pullback attractor $\{I_\omega \mid \omega \in \Omega\}$ is trivial, i.e. the section I_ω contains a single point.

Section 5 is devoted to study of the uniform dissipative cocycles with noncompact base. For this class of cocycles we prove the triviality of its global pullback attractor (see Theorem 9.7).

In section 6 we introduce the class of cocycles possessing the property of dissipativity (non-uniform) with non-compact base. The main result in this section is

Theorem 9.8 which describes the structure of compact pullback attractor of mentioned class of cocycles. In particular its triviality is proved.

Section 7 is devoted to application of our general results, obtained in sections 3-6 to study of differential equations (ODEs, Caratheodory equations with almost periodic coefficients, almost periodic ODEs with impulse).

9.1 $\mathbb{C}$-analytic cocycles

Let Ω be a complete metric space, let $\mathbb{T}$, the time set, be either $\mathbb{R}$ or $\mathbb{Z}$, $\mathbb{T}_+ = \{t \in \mathbb{T} \mid \ t \geq 0\}$ $(\mathbb{T}_- = \{t \in \mathbb{T} \mid \ t \leq 0\})$, let $(\Omega, \mathbb{T}, \sigma)$ be an autonomous two-sided dynamical system on Ω and E^d be a d-dimensional real $(\mathbb{R}^d)$ or complex $(\mathbb{C}^d)$ Euclidean space with the norm $|\cdot|$.

Denote by $HC(\mathbb{C}^d \times \Omega, \mathbb{C}^d)$ the space of all the continuous functions $f : \mathbb{C}^d \times \Omega \to \mathbb{C}^d$ holomorphic in $z \in \mathbb{C}^d$ and equipped by compact-open topology. Consider the differential equation

$$\frac{dz}{dt} = f(z, \sigma_t \omega), \quad (\omega \in \Omega) \tag{9.2}$$

where $f \in HC(\mathbb{C}^d \times \Omega, \mathbb{C}^d)$. Let $\varphi(t, z, \omega)$ be the solution of equation (9.2) passing through the point z for $t = 0$ and defined on $\mathbb{R}_+$. The mapping $\varphi : \mathbb{R}_+ \times \mathbb{C}^d \times \Omega \to \mathbb{C}^d$ has the following properties (see, for example, [122] and [186]):

a) $\varphi(0, z, \omega) = z$ for all $z \in \mathbb{C}^d$.

b) $\varphi(t + \tau, z, \omega) = \varphi(t, \varphi(\tau, z, \omega), \sigma_\tau \omega)$ for all $t, \tau \in \mathbb{R}_+, \omega \in \Omega$ and $z \in \mathbb{C}^d$.

c) the mapping φ is continuous.

d) the mapping $\varphi(t, \omega) := \varphi(t, \cdot, \omega) : \mathbb{C}^d \to \mathbb{C}^d$ is holomorphic for any $t \in \mathbb{R}_+$ and $\omega \in \Omega$.

Definition 9.1   The cocycle $\langle \mathbb{C}^d, \varphi, (\Omega, \mathbb{T}, \sigma) \rangle$ is called $\mathbb{C}$-analytic if the mapping $\varphi(t, \omega) : \mathbb{C}^d \to \mathbb{C}^d$ is holomorphic for all $t \in \mathbb{T}_+$ and $\omega \in \Omega$.

Example 9.1   Let $(HC(\mathbb{R} \times \mathbb{C}^d, \mathbb{C}^d), \mathbb{R}, \sigma)$ be a dynamical system of translations on $HC(\mathbb{R} \times \mathbb{C}^d, \mathbb{C}^d)$ (Bebutov's dynamical system (see, for example, [300])). Denote by F the mapping from $\mathbb{C}^d \times HC(\mathbb{R} \times \mathbb{C}^d, \mathbb{C}^d)$ to $\mathbb{C}^d$ defined by equality $F(z, f) := f(0, z)$ for all $z \in \mathbb{C}^d$ and $f \in HC(\mathbb{R} \times \mathbb{C}^d, \mathbb{C}^d)$. Let Ω be the hull $H(f)$ of a given function $f \in HC(\mathbb{R} \times \mathbb{C}^d, \mathbb{C}^d)$, that is $\Omega = H(f) := \overline{\{f_\tau \mid \tau \in \mathbb{R}\}}$, where $f_\tau(t, z) := f(t + \tau, z)$ for all $t, \tau \in \mathbb{R}$ and $z \in \mathbb{C}^d$. Denote the restriction of $(HC(\mathbb{R} \times \mathbb{C}^d, \mathbb{C}^d), \mathbb{R}, \sigma)$ on Ω by $(\Omega, \mathbb{R}, \sigma)$. Then, under appropriate restriction on the given function $f \in HC(\mathbb{R} \times \mathbb{C}^d, \mathbb{C}^d)$ defining Ω, the differential equation $\frac{dz}{dt} = f(t, z) = F(z, \sigma_t f)$ generates a $\mathbb{C}$−analytic cocycle.

Remark 9.1 *Analogously as above every difference equation with holomorphic right hand side generates a C-analytic cocycle with discrete time $\mathbb{Z}_+$.*

9.2　Some general facts about non-autonomous dynamical systems

Definition 9.2　The point $\omega \in \Omega$ is called (see, for example, [302] and [304]) positively (negatively) stable in the sense of Poisson if there exists a sequence $t_n \to +\infty$ ($t_n \to -\infty$ respectively) such that $\sigma_{t_n}\omega \to \omega$. If the point ω is Poisson stable in both directions, in this case it is called Poisson stable.

Denote by $\mathfrak{N}_\omega = \{\{t_n\} \mid \sigma_{t_n}\omega \to \omega\}$, $\mathfrak{N}_\omega^+ := \{\{t_n\} \in \mathfrak{N}_\omega \mid t_n \to +\infty\}$ and $\mathfrak{N}_\omega^- := \{\{t_n\} \in \mathfrak{N}_\omega | t_n \to -\infty\}$.

Definition 9.3　(Conditional compactness). Let (X, h, Ω) be a fiber space, i.e. X and Ω be two metric spaces and $h : X \to \Omega$ be a homomorphism from X into Ω. The subset $M \subseteq X$ is said to be conditionally relatively compact, if the pre-image $h^{-1}(\Omega') \bigcap M$ of every relatively compact subset $\Omega' \subseteq \Omega$ is a relatively compact subset of X, in particularly $M_\omega := h^{-1}(\omega) \bigcap M$ is relatively compact for every ω. The set M is called conditionally compact if it is closed and conditionally relatively compact.

Example 9.2　Let K be a compact space, $X := K \times \Omega$, $h = pr_2 : X \to \Omega$, then the triplet (X, h, Ω) be a fiber space, the space X is conditionally compact, but not compact.

Let $\langle (X, \mathbb{T}_+, \pi), (\Omega, \mathbb{T}, \sigma), h \rangle$ be a non-autonomous dynamical system and $\omega \in \Omega$ be a positively Poisson stable point. Denote by

$$E_\omega^+ := \{\xi \mid \ \exists\{t_n\} \in \mathfrak{N}_\omega^+ \ \ \text{such that} \ \ \pi^{t_n}|_{X_\omega} \to \xi\},$$

where $X_\omega := \{x \in X \mid \ h(x) = \omega\}$ and $\to$ means the pointwise convergence.

Lemma 9.1　*Let $\omega \in \Omega$ be a positively Poisson stable point, $\langle (X, \mathbb{T}^+, \pi), (\Omega, \mathbb{T}, \sigma), h \rangle$ be a non-autonomous dynamical system and X be a conditionally compact space, then E_ω^+ is a nonempty compact sub-semigroup of the semigroup $X_\omega^{X_\omega}$ (w.r.t. composition of mappings).*

Proof.　Let $\{t_n\} \in \mathfrak{N}_\omega^+$, then $\sigma_{t_n}\omega \to \omega$ and, consequently, the set

$$Q := \overline{\bigcup \{\pi^{t_n}(X_\omega) | n \in \mathbb{N}\}}$$

is compact, because X is conditionally compact. Thus $\{\pi^{t_n}|_{X_\omega}\} \subseteq Q^{X_\omega}$ and according to Tyhonov's Theorem this sequence is relatively compact. Let ξ be a limit point of $\{\pi^{t_n}|_{X_\omega}\}$, then $\xi \in E_\omega^+$ and, consequently, $E_\omega^+ \neq \emptyset$.

We note that $E_\omega^+ \subseteq X_\omega^{X_\omega}$ and, consequently, E_ω^+ is relatively compact. Let now $\xi_1, \xi_2 \in E_\omega^+$, we will prove that $\xi_1 \cdot \xi_2 \in E_\omega^+$. Since $\xi_1, \xi_2 \in E_\omega^+$, then there are two sequences $\{t_n^i\} \in \mathfrak{N}_\omega^+$ $(i = 1, 2)$ such that

$$\pi^{t_n^i}|_{X_\omega} \to \xi_i \quad (i = 1, 2).$$

Denote by $\xi := \xi_1 \cdot \xi_2 \in X_\omega^{X_\omega} \subseteq Q^{X_\omega}$, then we have $\pi^{t_n} \cdot \xi_2 \to \xi_1 \cdot \xi_2 = \xi$ as $n \to +\infty$. Let $U_\xi \subset X_\omega^{X_\omega}$ be an arbitrary open neighborhood of point ξ in $X_\omega^{X_\omega}$, then from relation (9.3) results that there exists a number $n_1(\xi) \in \mathbb{N}$ such that $\pi^{t_n} \cdot \xi_2 \in U_\xi$ for all $n \geq n_1(\xi)$. Now we fix $n \geq n_1(\xi)$, then there exist an open neighborhood $U_{\pi^{t_n^1} \cdot \xi_2} \subset U_\xi$ of point $\pi^{t_n^1} \cdot \xi_2 \in Q^{X_\omega}$ and a number $m_n \in \mathbb{N}$ such that

$$\pi^{t_n^1} \cdot \pi^{t_m^2}|_{X_\omega} \in U_{\pi^{t_n^1} \cdot \xi_2}$$

for any $n \geq n_1(\xi)$ and $m \geq m_n(\xi)$ and, consequently,

$$\pi^{t_n^1} \cdot \pi^{t_m^2}|_{X_\omega} \in U_\xi$$

for any $n \geq n_1(\xi)$ and $m \geq m_n(\xi)$. Thus from sequence $\{\pi^{t_n^1 + t_m^2}|_{X_\omega}\}$ it is possible to extract a subsequence $\{\pi^{t_{n_k}^1 + t_{m_k}^2}|_{X_\omega}\}$ $(t_{n_k}^1 + t_{m_k}^2 \to +\infty)$ such that $\pi^{t_{n_k}^1 + t_{m_k}^2}|_{X_\omega} \to \xi$ and, consequently, $\xi = \xi_1 \cdot \xi_2 \in E_\omega^+$. The Lemma is proved. $\qquad\square$

Corollary 9.1 *Let $\omega \in \Omega$ be a negatively Poisson stable point, $\langle(X, \mathbb{T}, \pi),$ $(\Omega, \mathbb{T}, \sigma), h\rangle$ be a two-sided non-autonomous dynamical system and X be a conditionally compact space, then $E_\omega^- = \{\xi|\ \exists\{t_n\} \in \mathfrak{N}_\omega^-$ such that $\pi^{t_n}|_{X_\omega} \to \xi\}$ is a nonempty compact sub-semigroup of semigroup $X_\omega^{X_\omega}$.*

Proof. This assertion follows from Lemma 9.1. $\qquad\square$

Lemma 9.2 *Let $\omega \in \Omega$ be a two-sided Poisson stable point, $\langle(X, \mathbb{T}, \pi), (\Omega, \mathbb{T}, \sigma), h\rangle$ be a two-sided non-autonomous dynamical system and X be a conditionally compact space, then $E_\omega = \{\xi|\ \exists\{t_n\} \in \mathfrak{N}_\omega$ such that $\pi^{t_n}|_{X_\omega} \to \xi\}$ is a nonempty compact sub-semigroup of the semigroup $X_\omega^{X_\omega}$.*

Proof. This assertion can be proved using the same type of arguments as well as in the proof of Lemma 9.1 and therefore we omit the details. $\qquad\square$

Corollary 9.2 *Under the conditions of Lemma 9.2 E_ω^+ and E_ω^- are two nonempty sub-semigroups of the semigroup E_ω.*

Lemma 9.3 *Under the conditions of Lemma 9.2 the following assertions hold:*

(1) if $\xi_1 \in E_\omega^-$ and $\xi_2 \in E_\omega^+$, then $\xi_1 \cdot \xi_2 \in E_\omega^- \bigcap E_\omega^+$.
(2) $E_\omega^- \bigcap E_\omega^+$ is a sub-semigroup of the semigroup E_ω^-, E_ω^+ and E_ω.

(3) $E_\omega^- \cdot E_\omega \subseteq E_\omega^-$ *and* $E_\omega^+ \cdot E_\omega \subseteq E_\omega^+$, *where* $A_1 \cdot A_2 := \{\xi_1 \cdot \xi_2 | \xi_i \in A_i \quad (i = 1, 2)\}$ *and* $A_i \subseteq E_\omega$.

(4) *if at least one of the sub-semigroups* E_ω^- *or* E_ω^+ *is a group, then* $E_\omega^- = E_\omega^+ = E_\omega$.

Proof. Let $\xi_1 \in E_\omega^-$ and $\xi_2 \in E_\omega^+$, then there are $t_n^1 \to -\infty$ and $t_n^2 \to +\infty$ such that $\sigma_{t_n^i} \omega \to \omega$ and $\pi^{t_n^i}|_{X_\omega} \to \xi_i (i = 1, 2)$. Using the same type arguments as well as in the proof of Lemma 9.1 we may choose the subsequence $\{t_{n_k}^1 + t_{m_k}^2\} \subset \{t_n^1 + t_m^2\}$ with the following properties: a) $t_{n_k}^1 + t_{m_k}^2 \geq k$ or $t_{n_k}^1 + t_{m_k}^2 \leq -k$ and b) $\pi^{t_{n_k}^1 + t_{m_k}^2} \to \xi_1 \cdot \xi_2$, i.e. $\xi_1 \cdot \xi_2 \in E_\omega^+ \bigcap E_\omega^-$.

The second statement follows from the first one.

Let $\xi_1 \in E_\omega^+$ (E_ω^-, respectively) and $\xi_2 \in E_\omega$, then there exist two sequences $t_n^1 \to +\infty$ (or $-\infty$, respectively) and t_n^2 such that $\pi^{t_n^i} \to \xi_i \quad (i = 1, 2)$. Then we may choose a subsequence $\{t_{n_k}^1 + t_{m_k}^2\}$ with the following properties:

$$a) t_{n_k}^1 + t_{m_k}^2 \geq k \quad (\leq -k, \text{ respectively}) \quad \text{and} \quad b) \pi^{t_{n_k}^1 + t_{m_k}^2} \to \xi_1 \cdot \xi_2,$$

and consequently, $\xi_1 \cdot \xi_2 \in E_\omega^+$ (E_ω^-, respectively).

Finally, let E_ω^- be a subgroup of the semigroup E_ω. According to the third statement of Lemma 9.3 $E_\omega^- \cdot E_\omega \subseteq E_\omega^-$. Since E_ω^- is a nonempty compact invariant set w.r.t. E_ω, then in E_ω^- exists a compact minimal subset $I \subset E_\omega^-$, i.e. $I \neq \emptyset$, compact and $u \cdot E_\omega = I$ for every $u \in I$. Let now $u \in I$ be an idempotent element of right ideal I of semigroup E_ω, then u is an unit element $(u(x) = x \quad \forall x \in X_\omega)$ of I because $I \subseteq E_\omega^-$ and E_ω^- according to conditions of Lemma 9.3 is a subgroup of the semigroup E_ω. Thus we have $E_\omega = u \cdot E_\omega = I \subseteq E_\omega^-$ and, consequently, $E_\omega^- = E_\omega$. Analogously $E_\omega^+ = E_\omega$. The theorem is proved. $\quad\square$

Lemma 9.4 *Let* $\omega \in \Omega$ *be a two-sided Poisson stable point,* $\langle (X, \mathbb{T}, \pi), (\Omega, \mathbb{T}, \sigma), h \rangle$ *be a two-sided non-autonomous dynamical system and* X *be a conditionally compact space and*

$$\inf_{n \in \mathbb{N}} \rho(x_1 t_n, x_2 t_n) > 0 \tag{9.3}$$

for all $\{t_n\} \in \mathfrak{N}_\omega^-$ *and* $x_1, x_2 \in X_\omega \quad (x_1 \neq x_2)$, *then* E_ω^- *is a subgroup of the semigroup* E_ω.

Proof. Indeed, if $u \in E_\omega^-$ is an arbitrary idempotent element of E_ω^-, then $u^2 = u$ and there exists a sequence $\{t_n\} \in \mathfrak{N}_\omega^-$ such that $\pi^{t_n} \to u$. According to (9.3) we have $u(x_1) \neq u(x_2)$ for all $x_1 \neq x_2 \quad (x_1, x_2 \in X_\omega)$. On the other hand $u^2(x) = u(x)$ for all $x \in X_\omega$ and, consequently, $u(x) = x$ for all $x \in X_\omega$. Thus every idempotent of semigroup E_ω^- is an unit element of E_ω (in particular E_ω^-) and, consequently, E_ω^- is a group (see, for example [32]). $\quad\square$

Lemma 9.5 *Let $\omega \in \Omega$ be a two-sided Poisson stable point, $\langle (X, \mathbb{T}, \pi), (\Omega, \mathbb{T}, \sigma), h \rangle$ be a two-sided non-autonomous dynamical system and X be a conditionally compact space and the condition (9.3) holds for all $\{t_n\} \in \mathfrak{N}_\omega^-$ and $x_1, x_2 \in X_\omega (x_1 \neq x_2)$, then inequality (9.3) is fulfilled for any $\{t_n\} \in \mathfrak{N}_\omega^+$ and $x_1, x_2 \in X_\omega (x_1 \neq x_2)$.*

Proof. According to Lemma 9.3 under the conditions of Lemma 9.5 we have $E_\omega^- = E_\omega^+ = E_\omega$ and E_ω is a group. Suppose that for some sequence $\{t_n\} \in \mathfrak{N}_\omega^+$

$$\inf_{n \in \mathbb{N}} \rho(x_1 t_n, x_2 t_n) = 0. \tag{9.4}$$

Since $\sigma_{t_n} \omega \to \omega$ and the space X is conditionally compact, the sequence $\{\pi^{t_n}|_{X_\omega}\}$ is relatively compact in Q^{X_ω}, where $Q := \overline{\bigcup \{\pi^{t_n}(X_\omega) | n \in \mathbb{N}\}}$ and, consequently, we may assume that it is convergent. Let $\xi := \lim_{n \to +\infty} \pi^{t_n}|_{X_\omega}$, then from equality (9.4) results that $\xi(x_1) = \xi(x_2) \ (x_1 \neq x_2)$, but $\xi \in E_\omega$ and E_ω is a group and, consequently, ξ is a one-to-one mapping. The obtained contradiction proves our assertion. $\qquad \square$

Lemma 9.6 *Let $\omega \in \Omega$ be a positively Poisson stable point, $\langle (X, \mathbb{T}_+, \pi), (\Omega, \mathbb{T}, \sigma), h \rangle$ be a non-autonomous dynamical system, generated by cocycle φ , $pr_1(\bigcup_{t \geq 0} \pi^t X_\omega)$ be relatively compact and*

$$A_\omega(X_\omega) := (\bigcap \overline{\bigcup_{t \geq 0} \pi^\tau X_\omega}) \bigcap X_\omega,$$

then for any $x \in A_\omega(X_\omega)$ there exists an entire trajectory of dynamical system $(X, \mathbb{T}_+, \pi)$ passing through point x for $t = 0$ and $pr_1(\gamma(\mathbb{T}))$ $(\gamma(\mathbb{T}) := \{\gamma(t) | t \in \mathbb{T}\})$ is relatively compact.

Proof. Let $x \in A_\omega(X_\omega)$, then there are $\{t_n\} \in \mathfrak{N}_\omega$ and $x_n \in X_\omega$ such that $x = \lim_{n \to \infty} \pi^{t_n} x_n$, $\sigma_{t_n} \omega \to \omega$ and $t_n \to +\infty$. We consider the sequence $\{\gamma_n\} \subset C(\mathbb{T}, M)$, where $M := \overline{\bigcup_{t \geq 0} \pi^t X_\omega}$, defined by equality

$$\gamma_n(t) = \pi^{t+t_n} x_n, \quad \text{if} \quad t \geq -t_n \quad \text{and} \quad \gamma_n(t) = x_n \quad \text{for} \quad t \leq t_n.$$

Now we will prove that the sequence $\{\gamma_n\}$ is equicontinuous on every segment $[-l, l] \subset \mathbb{T}$. If we suppose that it is not true, then there exist $\varepsilon_0, l_0 > 0, t_n^i \in [-l_0, l_0]$ and $\delta_n \to 0$ $(\delta_n > 0)$ such that

$$|t_n^1 - t_n^2| \leq \delta_n \quad \text{and} \quad \rho(\gamma_n(t_n^1), \gamma_n(t_n^2)) \geq \varepsilon_0. \tag{9.5}$$

We may suppose that $t_n^i \to t_0$ $(i = 1, 2)$. From (9.5) we obtain

$$\varepsilon_0 \leq \rho(\gamma_n(t_n^1), \gamma_n(t_n^2)) = \rho(\pi^{t_n^1 + l_0}(\pi^{t_n - l_0} x_n), \pi^{t_n^2 + l_0}(\pi^{t_n - l_0} x_n)) \tag{9.6}$$

for sufficiently large n $(t_n \geq l_0)$. Note that the sequence $\{\pi^{t_n - l_0} x_n\}$ is relatively compact and $h(\pi^{t_n - l_0} x_n) = \sigma_{t_n - l_0} h(x_n) = \sigma_{t_n - l_0} \omega \to \sigma_{-l_0} \omega$. Let $\bar{x} = \lim\limits_{n \to \infty} \pi^{t_n - l_0} x_n$, then passing to limit in the inequality (9.6) we obtain $\varepsilon_0 \leq 0$. The obtained contradiction proves our assertion.

Now taking into account the conditional compactness of set K we can affirm that $\{\gamma_n\}$ is a relatively compact sequence of $C(\mathbb{T}, M)$. Let γ be a limit point of sequence $\{\gamma_n\}$, then there exists a subsequence $\{\gamma_{k_n}\}$ such that $\gamma(t) = \lim\limits_{n \to \infty} \gamma_{k_n}(t)$ uniformly on every segment $[-l, l] \subset \mathbb{T}$. In particular $\gamma \in C(\mathbb{T}, M)$. We note that $\pi^t \gamma(s) = \lim\limits_{n \to \infty} \pi^t \gamma_{k_n}(s) = \lim\limits_{n \to \infty} \gamma_{k_n}(s + t) = \gamma(s + t)$ for all $t \in \mathbb{T}_+$ and $s \in \mathbb{T}$. Finally, we see that $\gamma(0) = \lim\limits_{n \to \infty} \gamma_{k_n}(0) = \lim\limits_{n \to \infty} \pi^{t_{k_n}} x_{k_n} = x$, i.e. γ is an entire trajectory of dynamical system $(X, \mathbb{T}_+, \pi)$ passing through point x. The Lemma is completely proved. $\square$

9.3   Positively uniformly stable cocycles

Let E^d be a d–dimensional real ($\mathbb{R}^d$) or complex ($\mathbb{C}^d$) Euclidean space with the norm $|\cdot|, \rho$ be the distance generated by this norm, Ω be a metric space and the triplet $\langle E^d, \varphi, (\Omega, \mathbb{T}, \sigma) \rangle$ be a cocycle on the state space E^d.

Theorem 9.1   *Let $\langle E^d, \varphi, (\Omega, \mathbb{T}, \sigma) \rangle$ be a cocycle with the following properties:*

(1) It admits a conditionally relatively compact invariant set $\{I_\omega \mid \omega \in \Omega\}$ (i.e. $\bigcup\{I_\omega \mid \omega \in \Omega'\}$ is relatively compact subset of E^d for any relatively compact subset Ω' of Ω).

(2) The cocycle φ is positively uniformly stable on $\{I_\omega \mid \omega \in \Omega\}$.

Then all motions on $J := \bigcup\{J_\omega \mid \omega \in \Omega\}$ ($J_\omega := I_\omega \times \{\omega\}$) may be continued uniquely to the left and define on J a two-sided dynamical system $(J, \mathbb{T}, \pi)$, i.e. the skew-product system $(X, \mathbb{T}_+, \pi)$ generates on J a two-sided dynamical system $(J, \mathbb{T}, \pi)$.

Proof. First step: we will prove that the set $J \subset X$ is distal in the negative direction w.r.t. the non-autonomous dynamical system $\langle (X, \mathbb{T}_+, \pi), (\Omega, \mathbb{T}, \sigma), h \rangle$, i.e. for all $\omega \in \Omega$ and $u_1, u_2 \in I_\omega (u_1 \neq u_2)$ the following inequality holds

$$\inf_{t \leq 0} \rho(\gamma_1(t), \gamma_2(t)) > 0 \tag{9.7}$$

for all $\gamma_i \in \Phi_{(u_i, \omega)} (i = 1, 2)$, where by $\Phi_{(u, \omega)}$ it is denoted the family of all the entire trajectories of $(X, \mathbb{T}_+, \pi)$ passing through point (u, ω) and belonging to J. If it is not true, then there exist $\omega_0 \in \Omega, u_i^0 \in I_{\omega_0}$ $(u_1^0 \neq u_2^0), \gamma_i^0 \in \Phi_{(u_i^0, \omega_0)}$ $(i = 1, 2)$ and

$-t_n \to -\infty$ such that

$$\rho(\gamma_1^0(-t_n), \gamma_2^0(-t_n)) \to 0 \qquad (9.8)$$

as $n \to \infty$. Let $\varepsilon := \rho(u_1^0, u_2^0) > 0$ and $\delta = \delta(\varepsilon) > 0$ be chosen from positively uniformly stability of cocycle φ on family of compact subsets $\{I_\omega \mid \omega \in \Omega\}$, then for sufficiently large n from (9.8) we have $\rho(\gamma_1^0(-t_n), \gamma_2^0(-t_n)) < \delta$ and, consequently, $\varepsilon = \rho(u_1^0, u_2^0) = \rho(\pi^{t_n}\gamma_1^0(-t_n), \pi^{t_n}\gamma_2^0(-t_n)) < \varepsilon$. The obtained contradiction proves our assertion.

Second step: we will prove that for any $\omega \in \Omega$ and $u \in I_\omega$ the set $\Phi_{(u,\omega)}$ contains only one entire trajectory of $(X, \mathbb{T}_+, \pi)$ belonging to J. Let $\Phi := \bigcup\{\Phi_{(u,\omega)} \mid (u,\omega) \in J\} \subset C(\mathbb{T}, X)$, where $C(\mathbb{T}, X)$ is a space of all the continuous functions $f : \mathbb{T} \to X$ equipped with compact-open topology and $(C(\mathbb{T}, X), \mathbb{T}, \lambda)$ is Bebutov's dynamical system (dynamical system of translations (see, for example, [292, 300])). It is easy to verify that Φ is a closed and invariant subset of dynamical system $(C(\mathbb{T}, X), \mathbb{T}, \lambda)$ and, consequently, induces on the set Φ the dynamical system $(\Phi, \mathbb{T}, \lambda)$. Let H be a mapping from Φ into Ω, defined by equality $H(\gamma) := h(\gamma(0))$, then it is possible to verify (see [83]) that the triplet $\langle (\Phi, \mathbb{T}, \lambda), (\Omega, \mathbb{T}, \sigma), H \rangle$ is a non-autonomous dynamical system. Now we will show that this non-autonomous dynamical system is distal on the negative direction, i.e.

$$\inf_{t \le 0} \rho(\gamma_1^t, \gamma_2^t) > 0$$

for all $\gamma_1, \gamma_2 \in H^{-1}(\omega)$ $(\gamma_1 \ne \gamma_2)$ and $\omega \in \Omega$. Indeed, otherwise there exist $\omega_0, \gamma_1, \gamma_2 \in H^{-1}(\omega_0)$ $(\gamma_1 \ne \gamma_2)$ and $t_n \to +\infty$ such that $\rho(\gamma_1^{-t_n}, \gamma_2^{-t_n}) \to 0$ (where $\gamma^\tau := \sigma(\gamma, \tau)$, i.e. $\gamma^\tau(s) := \gamma(\tau + t)$ for all $s \in \mathbb{T}$) as $n \to \infty$ and, consequently,

$$|\gamma_1(-t_n) - \gamma_2(-t_n)| \le \rho(\gamma_1^{-t_n}, \gamma_2^{-t_n}) \to 0. \qquad (9.9)$$

Since $\gamma_1 \ne \gamma_2$, then there exists $t_0 \in \mathbb{T}$ such that $\gamma_1(t_0) \ne \gamma_2(t_0)$. Let $\tilde{\gamma}_i(t) := \gamma_i(t + t_0)$ for all $t \in \mathbb{T}$, then $\tilde{\gamma}_i \in \Phi_{\omega_0}$ and from inequality (9.9) we have

$$|\tilde{\gamma}_1(-t_n) - \tilde{\gamma}_2(-t_n)| \to 0. \qquad (9.10)$$

as $n \to \infty$, $-t_n - t_0 \to -\infty$. Thus we found $\omega_0 := h(\gamma_i(t_0))$ and $u_i := pr_1\gamma_i(t_0)$ $(i = 1, 2)$, $u_1, u_2 \in I_{\omega_0}$ $(u_1 \ne u_2)$ and the entire trajectories $\tilde{\gamma}_i \in \Phi_{(\omega, u_i)}(i = 1, 2)$ such that $\tilde{\gamma}_1$ and $\tilde{\gamma}_2$ are proximal (see (9.10)). But (9.10) and (9.7) are contradictory. Thus the negative distality of the non-autonomous dynamical system $\langle (\Phi, \mathbb{T}, \sigma), (\Omega, \mathbb{T}, \sigma), H \rangle$ is proved.

Now we can prove that for any $\omega \in \Omega$ and $u \in I_\omega$ the set $\Phi_{(\omega, u)}$ contains a unique entire trajectory. In fact, if it is not true, then there exists $(\omega_0, u_0) \in \Omega \times E^d$ and two different trajectories $\gamma_1, \gamma_2 \in \Phi_{(\omega_0, u_0)}(\gamma_1 \ne \gamma_2)$. In virtue of above γ_1 and

γ_2 are negatively distal with respect to $\langle (\Phi, \mathbb{T}, \sigma), (\Omega, \mathbb{T}, \sigma), H \rangle$, i.e.

$$\alpha(\gamma_1, \gamma_2) := \inf_{t \leq 0} \rho(\gamma_1^t, \gamma_2^t) > 0$$

and, consequently, $\rho(\gamma_1(t), \gamma_2(t)) \geq \alpha(\gamma_1, \gamma_2) > 0$ for all $t \geq 0$. In particular $\gamma_1(0) \neq \gamma_2(0)$. The obtained contradiction proves our statement.

Third step: let now $\tilde{\pi}$ be a mapping from $\mathbb{T} \times J$ into J defined by equality

$$\tilde{\pi}(t, x) = \pi(t, x) \quad \text{if} \quad t \leq 0 \quad \text{and} \quad \gamma_x(t) \quad \text{if} \quad t < 0$$

for all $x \in J$, where γ_x is a unique entire trajectory of the dynamical system $(X, \mathbb{T}_+, \pi)$ passing through point x and belonging to J. To prove that $(J, \mathbb{T}, \tilde{\pi})$ is a two-sided dynamical system on J it is sufficient to verify the continuity of the mapping $\tilde{\pi}$. Let $x \in J$, $t \in \mathbb{T}-, x_n \to x$ and $t_n \to t$, then there is a $l_0 > 0$ such that $t_n \in [-l_0, l_0]$ and, consequently,

$$\rho(\tilde{\pi}(t_n, x_n), \tilde{\pi}(t, x)) = \rho(\pi^{t_n + l_0} \gamma_{x_n}(-l_0), \pi^{t + l_0} \gamma_x(-l_0)) \leq \qquad (9.11)$$

$$\rho(\pi^{t_n + l_0} \gamma_{x_n}(-l_0), \pi^{t_n + l_0} \gamma_x(-l_0)) + \rho(\pi^{t_n + l_0} \gamma_x(-l_0), \pi^{t + l_0} \gamma_x(-l_0)).$$

Reasoning as in the proof of Lemma 9.2 it is possible to establish that the sequence $\{\gamma_{x_n}\}$ is relatively compact in $C(\mathbb{T}, J)$ and that every limit point of this sequence $\gamma \in \Phi$ and $\gamma(0) = x$. Taking into account the result of the second step we claim that $\gamma_{x_n} \to \gamma_x$ uniformly on every segment $[-l, l] \subset \mathbb{T}(l > 0)$. In particular, $\gamma_{x_n}(-l_0) \to \gamma_x(-l_0)$. Passing now to limit in inequality (9.11) when $n \to \infty$ we obtain the continuity of mapping $\tilde{\pi}$ in the point (t, x). The theorem is completely proved. $\qquad \square$

Remark 9.2   *Theorem 9.1 is true and in the case if we replace the condition 2. by the following: for arbitrary $\varepsilon > 0$ there exist two positive numbers $\delta(\varepsilon)$ and $L(\varepsilon)$ such that*

$$\rho(\varphi(t, \omega, u_1), \varphi(t, \omega, u_2)) < \varepsilon \qquad (9.12)$$

for all $\omega \in \Omega, t \geq L(\varepsilon)$ and $u_1, u_2 \in I_\omega$ with condition $\rho(u_1, u_2) < \delta$.

9.4   The compact global pullback attractors of $\mathbb{C}$-analytic cocycles with compact base

In this section we suppose that $\langle \mathbb{C}^d, \varphi, (\Omega, \mathbb{T}, \sigma) \rangle$ is a $\mathbb{C}-$analytic cocycle and Ω is a compact space.

Theorem 9.2   *Let $\langle \mathbb{C}^d, \varphi, (\Omega, \mathbb{T}, \sigma) \rangle$ be a $\mathbb{C}-$analytic cocycle admitting a compact global pullback attractor $\{I_\omega \mid \omega \in \Omega\}$, then:*

(1) *The compact invariant set $J = \bigcup\{J_\omega \mid \omega \in \Omega\}$ of the skew-product dynamical system $(X, \mathbb{T}_+, \pi)$ $(X := \mathbb{C}^d \times \Omega,\ \pi := (\varphi, \sigma))$ is asymptotically stable.*

(2) *There exists a positive number δ_0 such that the cocycle φ is positively uniformly stable on the compact set $\mathfrak{B}[I, \delta] := \bigcup\{B[I_\omega, \delta] \mid \omega \in \Omega\}$, where $B[I_\omega, \delta] := \{z \in \mathbb{C}^d \mid \rho(z, I_\omega) \leq \delta\}$, for all $0 < \delta < \delta_0$.*

(3) *The skew-product dynamical system $(X, \mathbb{T}_+, \pi)$ generates on J a group dynamical system $(J, \mathbb{T}, \pi)$.*

Proof.   Denote by $X := \mathbb{C}^d \times \Omega$ and by $(X, \mathbb{T}_+, \pi)$ the skew-product dynamical system. Then under the conditions of the theorem the set $J = \bigcup\{J_\omega | \omega \in \Omega\}$ is a nonempty compact invariant set and according to Theorem 4.1 [100] is asymptotically stable with respect to $(X, \mathbb{T}_+, \pi)$. In particular there exists a $\delta_0 > 0$ such that the set $B[J, \delta_0] := \{x \in X \mid \rho(x, J) \leq \delta_0\}$ is positively invariant. Since Ω is compact and $\pi^t(u, \omega) := (\varphi(t, u, \omega), \sigma_t \omega)$, then there exists a positive number $C = C(\delta_0)$ such that $|\varphi(t, u, \omega)| \leq C$ for all $\omega \in \Omega$ and $u \in B[I_\omega, \delta_0]$. Taking into account the connectedness of set I_ω (see, for example [126] and also Theorem 2.25) according to Cauchy's Theorem for all $\delta < \delta_0$ there exists a positive number $L(\delta)$ such that

$$|\varphi(t, \omega, u_1) - \varphi(t, \omega, u_2)| \leq L(\delta)|u_1 - u_2| \tag{9.13}$$

for all $\omega \in \Omega, t \in \mathbb{R}^+$ and $u_1, u_2 \in B[I_\omega, \delta]$. It is easy to see that from inequality (9.13) results the positively uniformly stability of set $B[I, \delta]$ for every $0 < \delta < \delta_0$. Particularly the set $I := \bigcup\{I_\omega \mid \omega \in \Omega\}$ will be positively uniformly stable and to finish the proof of Theorem it is sufficiently to apply Theorem 9.1 to our situation for the skew-product system $(X, \mathbb{T}_+, \pi)$. The Theorem is completely proved. $\square$

Definition 9.4   The cocycle $\langle \mathbb{E}^d, \varphi, (\Omega, \mathbb{T}, \sigma) \rangle$ is called linear (see, for example, [6],[33] and [292]) if the mapping $\varphi(t, \omega) : E^d \to E^d$ is linear for every $t \in \mathbb{T}_+$ and $\omega \in \Omega$.

Theorem 9.3   *Let $\langle \mathbb{C}^d, \varphi, (\Omega, \mathbb{T}, \sigma) \rangle$ be a linear cocycle, then the following conditions are equivalent:*

(1) 1. $\lim\limits_{t \to +\infty} |\varphi(t, \omega, u)| = 0$ *for all $u \in E^d$ and $\omega \in \Omega$.*

(2) 2. *There exist positive numbers N, ν such that $|\varphi(t, \omega, u)| \leq N \exp(-\nu t)|u|$ for all $t \in \mathbb{T}_+, \omega \in \Omega$ and $u \in E^d$.*

Proof.   This statement follows from Theorems 1.10 and 2.38. $\square$

Theorem 9.4   *Let $\langle \mathbb{C}^d, \varphi, (\Omega, \mathbb{T}, \sigma) \rangle$ be a $\mathbb{C}-$analytic cocycle admitting a compact pullback attractor $\{I_\omega \mid \ \omega \in \Omega\}$, and let every point $\omega \in \Omega$ be positively Poisson stable. Then the following assertions hold:*

(1) For every $\omega \in \Omega$ the set I_ω consists of a unique point $\nu(\omega)$.

(2) $\nu(\sigma_t\omega) = \varphi(t, \nu(\omega), \omega)$ for all $\omega \in \Omega$ and $t \in \mathbb{T}_+$.

(3) The mapping $\omega \to \gamma(\omega)$ is continuous, where $\gamma := (\nu, Id_\Omega)$.

(4) Every point $\gamma(\omega)$ is positively Poisson stable.

(5) The continuous invariant section ν is uniformly asymptotically stable, i.e.

     (a) for arbitrary $\varepsilon > 0$ there exists $\delta(\varepsilon) > 0$ such that $\rho(z, \nu(\omega)) < \delta$ implies $\rho(\varphi(t, z, \omega), \nu(\sigma_t\omega)) < \varepsilon$ for all $t \geq 0$ and $\omega \in \Omega$.

     (b) There exists $\delta_0 > 0$ such that

$$\lim_{t \to +\infty} \rho(\varphi(t, z, \omega), \nu(\sigma_t\omega)) = 0$$

for all $\omega \in \Omega$ and z with the condition $\rho(z, \nu(\omega)) \leq \delta_0$.

Proof. Under the conditions of Theorem 9.3 there exists a positive number δ_0 such that the set $M := \mathfrak{B}[J, \delta_0]$ is a compact and positively invariant set of skew-product system $(X, \mathbb{T}_+, \pi)$ (see the proof of Theorem 9.2), where $\mathfrak{B}[I, \delta_0] := \bigcup\{B[I_\omega, \delta_0] \times \{\omega\} \mid \omega \in \Omega\}$. Denote by $E = E(M, \mathbb{T}_+, \pi)$ the Ellis semigroup of the dynamical system $(M, \mathbb{T}_+, \pi)$, $E_\omega := \{\xi \in E \mid \xi M_\omega \subseteq M_\omega\}$, where $M_\omega := \{(u, \omega) \mid (u, \omega) \in M\}$ and $E_\omega^+ := \{\xi \in E_\omega \mid \exists \{t_n\} \in \mathfrak{N}_\omega^+ \text{ such that } \pi^{t_n}|_{M_\omega} \to \xi\}$. According to Theorem 9.2 and Lemma 9.1 E_ω^+ is a nonempty compact sub-semigroup of the Ellis semigroup E. Note that every mapping $\xi \in E_\omega^+$, which maps $B[I_\omega, \delta_0]$ into I_ω, is holomorphic because, according to Theorem 9.2, the convergence $\pi^{t_n}|_{M_\omega} \to \xi$ is uniform on M_ω. Consider an idempotent $v \in E_\omega^+$, then $v(v(u)) = v(u)$ for all $u \in M_\omega$ and, consequently, $v(p) = p$ for every $p \in v(M_\omega) = v(I_\omega)$. Since v is holomorphic and $v(I_\omega)$ is a compact connected set, then [167] $v(I_\omega)$ contains only one point $\nu(\omega)$. On the other hand we have $v(v(u)) = v(u)$ for all $u \in M_\omega$, i.e. $v(\nu(\omega)) = v(u)$. Thus there exists a sequence $t_n \to +\infty$ such that

$$|\varphi(t_n, u, \omega) - \varphi(t_n, \nu(u), \omega)| = 0 \tag{9.14}$$

for all $u \in M_\omega$. Taking into account the positively uniformly stability of cocycle φ from (9.14) we obtain the equality

$$|\varphi(t, u, \omega) - \varphi(t, \nu(u), \omega)| = 0 \tag{9.15}$$

for all $u \in B(I_\omega, \delta_0) := \{u \in \mathbb{C}^d \mid \rho(u, I_\omega) < \delta_0\}$. Now we will prove that $I_\omega = \{\nu(\omega)\}$ for every $\omega \in \Omega$. Let $0 < \delta < \delta_0, u \in I_\omega$ and $h \in \mathbb{C}^d$ with condition $|h| < \delta$, then according to equality (9.15) we have

$$\lim_{t \to +\infty} \sup_{|h| \leq \delta} |\varphi(t, \omega, u + h) - \varphi(t, \omega, u)| = 0 \tag{9.16}$$

for all $\omega \in \Omega$ and $u \in I_\omega$. In virtue of Cauchy's formula (see [42] and also [85])

$$U(t, (u, \omega))w = \tag{9.17}$$

$$\frac{1}{(2\pi i)^d} \int\limits_{|v_1|=\frac{\delta}{2}} \cdots \int\limits_{|v_d|=\frac{\delta}{2}} \frac{\varphi(t,\omega,u+v) - \varphi(t,\omega,u)}{v_1 \cdot \ldots \cdot v_d} \sum_{k=1}^{d} \frac{w_k}{v_k} dv_1 \cdot \ldots \cdot dv_d,$$

where $U(t,(u,\omega)) := \frac{\partial \varphi(t,\omega,u)}{\partial u}$ for all $(u,\omega) \in M$ and $t \in \mathbb{T}_+$. From (9.16) and (9.17) it follows that $\lim\limits_{t \to +\infty} \|U(t,(u,\omega))\| = 0$ for all $(u,\omega) \in M$. According to Theorem 9.3 there exist positive numbers N and α such that

$$\|U(t,(u,\omega))\| \leq N \exp(-\alpha t) \tag{9.18}$$

for any $(u,\omega) \in M$. Let now $u_1, u_2 \in I_\omega$ and $\psi : [0,1] \to B(I_\omega,\delta)$ be a continuously differentiable function with properties: $\psi(0) = u_1$ and $\psi(1) = u_2$. Consider the function $\Delta(s) := \varphi(t,\psi(s),\omega)$, then according to Lagrange's formula we have

$$\Delta(1) - \Delta(0) = \Delta'(\tau), \tag{9.19}$$

where $0 < \tau < 1$. Hence from (9.18) and (9.19) we have

$$|\varphi(t,u_1,\omega) - \varphi(t,u_2,\omega)| \leq N_1 \exp(-\alpha t)|u_1 - u_2| \tag{9.20}$$

for all $t \in \mathbb{T}_+, \omega \in \Omega$ and $u_1, u_2 \in M_\omega$, where $N_1 := N \cdot m$ and $m := \max\limits_{0 \leq s \leq 1} |\psi'(s)|$.

To finish the proof of theorem it is sufficient to remark that according to Theorem 9.1 on set J there is defined a two-sided dynamical system $(J,\mathbb{T},\pi)$ and, in particular, through every point $u \in I_\omega$ passes a unique entire trajectory of cocycle φ, i.e. the function $\varphi(t,\omega,u)$ $(u \in I_\omega$ and $\omega \in \Omega)$ is defined on $\mathbb{T}$. If $u_1 \neq u_2$ $(u_1, u_2 \in I_\omega)$, then from (9.20) follows that

$$|\varphi(-t,\omega,u_1) - \varphi(-t,\omega,u_2)| \geq N_1 \exp(\alpha t)|u_1 - u_2|$$

for all $t \in \mathbb{T}_+$. But the trajectories $\varphi(t,\omega,u_i) \in I_{\sigma_t \omega}(i = 1,2; t \in \mathbb{T})$ are bounded on $\mathbb{T}$. The obtained contradiction proves our assertion. Thus we have $I_\omega = \{\nu(\omega)\}$.

Now it is easy to see that the mapping $\omega \to \gamma(\omega)$ is continuous, where $\gamma = (\nu, Id_\Omega)$, and $\pi^t \gamma(\omega) = \gamma(\sigma_t \omega)$ for all $\omega \in \Omega$ and $t \in \mathbb{T}_+$ and, consequently, $\nu(\sigma_t \omega) = \varphi(t,\omega,\nu(\omega))$ for all $\omega \in \Omega$ and $t \in \mathbb{T}_+$.

Next we will note that every point $\gamma(\omega)$ is positively Poisson stable. Indeed, let $\{t_n\} \in \mathfrak{N}_\omega$, then $\pi^{t_n} \gamma(\omega) = \gamma(\sigma_{t_n} \omega) \to \gamma(\omega)$ and, consequently, $\{t_n\} \in \mathfrak{N}_{\gamma(\omega)}$. Finally, the uniformly asymptotically stability of continuous and invariant section ν results from Theorem 9.2. The theorem is completely proved. $\quad\square$

9.5 The uniform dissipative cocycles with noncompact base

Let Ω be a complete metric space (generally speaking noncompact), $\langle E^d, \varphi, (\Omega, \mathbb{T}, \sigma) \rangle$ be a cocycle on the state space E^d and $(X, \mathbb{T}_+, \pi)$ be the corresponding skew-product dynamical system, where $X := E^d \times \Omega$ and $\pi := (\varphi, \sigma)$.

Definition 9.5   The cocycle $\langle E^d, \varphi, (\Omega, \mathbb{T}, \sigma) \rangle$ is said to be dissipative if for any $\omega \in \Omega$ there is a positive number r_ω such that

$$\limsup_{t \to +\infty} |\varphi(t, \omega, u)| < r_\omega$$

for all $\omega \in \Omega$ and $u \in E^d$, i.e. for all $u \in E^d$ and $\omega \in \Omega$ there exists a positive number $L(\omega, u)$ such that $|\varphi(t, \omega, u)| < r_\omega$ for all $t \geq L(\omega, u)$.

Definition 9.6   The cocycle $\langle E^d, \varphi, (\Omega, \mathbb{T}, \sigma) \rangle$ is said to be uniformly dissipative if there exists a positive number r (r is not depend upon $\omega \in \Omega$) such that for any $R > 0$ there is a positive number $L(R)$ such that $|\varphi(t, \omega, u)| < r$ for all $\omega \in \Omega$ and $|u| \leq R$ and $t \geq L(R)$.

Theorem 9.5   *([100, 127, 217]) Let $\langle E^d, \varphi, (\Omega, \mathbb{T}, \sigma) \rangle$ be an uniformly dissipative cocycle, then it admits a compact global pullback attractor $\{I_\omega \mid \omega \in \Omega\}$ with $|u| \leq r$ for all $u \in I_\omega$ and $\omega \in \Omega$, where r is the positive number in definition 9.6.*

Theorem 9.6   *Let $\langle \mathbb{C}^d, \varphi, (\Omega, \mathbb{T}, \sigma) \rangle$ be a $\mathbb{C}-$analytic uniformly dissipative cocycle, then the following statements hold:*

i) *The cocycle φ admits a compact global pullback attractor $\{I_\omega \mid \omega \in \Omega\}$ with $|u| \leq r$ for all $u \in I_\omega$ and $\omega \in \Omega$, where r is the positive number in condition (9.23).*

ii) *For any $R > 0$ there exist positive constants $C = C(R)$ and $L(R)$ such that*

$$\rho(\varphi(t, \omega, u_1), \varphi(t, \omega, u_2)) \leq C\rho(u_1, u_2) \tag{9.21}$$

for all $t \geq L(R), \omega \in \Omega$ and $u_1, u_2 \in \mathbb{C}^d$ with the condition $|u_i| \leq R$ $(i = 1, 2)$.

iii) *For arbitrary $\varepsilon > 0$ there exist $L(\varepsilon) > 0$ and $\delta(\varepsilon) > 0$ such that*

$$|\varphi(t, \omega, u + h) - \varphi(t, \omega, u))| < \varepsilon$$

for all $t \geq L(\varepsilon), u \in I_\omega, \omega \in \Omega$ and $|h| < \delta$.

iv) *The set J of $(X, \mathbb{T}_+, \pi)$ is negatively distal, i.e.*

$$\inf_{t \leq 0} \rho(\gamma_1(t), \gamma_2(t)) > 0,$$

where γ_i (i=1,2) is a entire trajectory passing through point $(u_i, \omega) \in J, u_1 \neq u_2$ and $\gamma_i(\mathbb{S}) \subseteq J$.

v) *On the set J there is defined a two-sided dynamical system $(J, \mathbb{T}, \pi)$ generated by skew-product system $(X, \mathbb{T}_+, \pi)$.*

Proof.   The first assertion of theorem results from Theorem 9.5. Let now $R > 0$ and $R' > R$, according to uniformly dissipativity of cocycle φ there exists $L(R') > 0$ such that $|\varphi(t, \omega, u)| < r$ for all $t \geq L(R'), \omega \in \Omega$ and $|u| \leq R'$. In virtue of Cauchy's formula for $R < R'$ there is a constant $C(R) > 0$ such that $|\frac{\partial \varphi}{\partial u}(t, u, \omega)| \leq C(R)$ for all $t \geq L(R), \omega \in \Omega$ and $|u| \leq R$ and, consequently the inequality (9.21) holds.

The third assertion we will prove by method of contradiction. If it is not true, then there exist $\varepsilon_0 > 0, \delta_n \to 0 \quad (\delta_n > 0), |h_n| < \delta_n \quad (h_n \in \mathbb{C}^d), t_n \geq n, \quad \omega_n \in \Omega$ and $u_n \in I_{\omega_n}$ such that

$$|\varphi(t_n, \omega_n, u_n + h_n) - \varphi(t_n, \omega_n, u_n))| \geq \varepsilon_0$$

Let now $R > r$ and $C(R), L(R)$ be positive constants figuring in the inequality (9.24), then we have the following inequality

$$\varepsilon_0 \leq |\varphi(t_n, \omega_n, u_n + h_n) - \varphi(t_n, \omega_n, u_n))| \leq C(R)|h_n| \leq C(R)\delta_n. \tag{9.22}$$

Passing to limit in the inequality (9.22) as $n \to \infty$ we obtain $\varepsilon_0 \leq 0$. The obtained contradiction proves our assertion.

The fourth and fifth statements follow from Theorem 9.1 (see also Remark 9.2) because from condition iii) results that for arbitrary $\varepsilon > 0$ there exist two positive constants $\delta(\varepsilon)$ and $L(\varepsilon)$ satisfying the inequality (9.12). The Theorem is completely proved. $\qquad\square$

Theorem 9.7 *Let $\langle \mathbb{C}^d, \varphi, (\Omega, \mathbb{T}, \sigma) \rangle$ be a $\mathbb{C}-$analytic uniformly dissipative cocycle and every point $\omega \in \Omega$ be positively Poisson stable, then:*

1. *The set I_ω consists of only one point $\nu(\omega)$ for every ω.*
2. *The mapping $\omega \to \gamma(\omega)$ is continuous, where $\gamma := (\nu, Id_\Omega)$.*
3. *$\nu(\sigma_t \omega) = \varphi(t, \nu(\omega), \omega)$ for all $\omega \in \Omega$ and $t \in \mathbb{T}_+$.*
4. *$\lim\limits_{t \to +\infty} \rho(\varphi(t, \sigma_{-t}\omega)z, \nu(\omega)) = 0$ for every $\omega \in \Omega$ uniformly with respect to z in compact subsets of $\mathbb{C}^d$.*
5. *Every point $\gamma(\omega)$ is positively Poisson stable.*
6. *$\lim\limits_{t \to +\infty} \rho(\varphi(t, z, \omega), \nu(\sigma_t \omega)) = 0$ for all $\omega \in \Omega$ and $z \in \mathbb{C}^d$, i.e. every positive semi trajectory $\varphi(t, \omega, z)$ is asymptotically Poisson stable in positive direction.*

Proof. Let $\langle \mathbb{C}^d, \varphi, (\Omega, \mathbb{T}, \sigma) \rangle$ be a $\mathbb{C}-$analytic uniformly dissipative cocycle, then according to Theorem 9.6 this cocycle has the properties i)-v). Let $\{I_\omega \mid \omega \in \Omega\}$ be the compact global pullback attractor of cocycle φ and let $(X, \mathbb{T}_+, \pi)$ be the skew-product dynamical system. Denote by

$$E_\omega^+ := \{\xi| \quad \exists\{t_n\} \in \mathfrak{N}_\omega^+, \ \pi^{t_n}|_{X_\omega} \to \xi\}.$$

Since the cocycle φ possesses the property ii), then the pointwise convergence $\pi^{t_n}|_{X_\omega} \to \xi$ coincides with uniform convergence on every compact subset of $X_\omega := \mathbb{C}^d \times \{\omega\}$ and, consequently, every mapping $\xi \in E_\omega^+$ is holomorphic. As well as in Lemma 9.1 it is possible to show that E_ω^+ is a nonempty compact semigroup w.r.t. composition of mappings. Consider the idempotent element v of semigroup

E_ω^+. We will show that $v(X_\omega) \subseteq I_\omega$. Indeed, $v \in E_\omega^+$ and, consequently, there exists a sequence $\{t_n\} \in \mathfrak{N}_\omega^+$ such that $v = \lim\limits_{n\to\infty} \pi^{t_n}|_{X_\omega}$. Let $\bar{x} \in v(X_\omega)$, i.e. $\bar{x} = v(x)$ for some $x \in X_\omega$. This means that $\bar{x} = \lim\limits_{n\to\infty} \pi^{t_n} x$. According to Lemma 9.6 there exists an entire trajectory γ of the skew-product system $(X, \mathbb{T}_+, \pi)$ passing through the point $\bar{x}$ for $t = 0$ and $\gamma(\mathbb{T}) := \{\gamma(t) \mid t \in \mathbb{T}\}$ is conditionally relatively compact. Taking into account that J is a maximal invariant set of $(X, \mathbb{T}_+, \pi)$ with relatively compact $pr_1 J$ we have $\bar{x} \in J_\omega$, i.e. $v(X_\omega) \subseteq J_\omega$. Since $X_\omega = \mathbb{C}^d \times \{\omega\}$ and v is holomorphic by virtue of Liouville's Theorem the holomorphic function v is a constant, i.e. there exists $\gamma(\omega) \in J_\omega$ such that $v(X_\omega) = \{\gamma(\omega)\}$. We note that $v^2 = v$ and, consequently, $v(v(x)) = v(x)$ for all $x \in X_\omega$, i.e. $v(\gamma(\omega)) = v(x)$. Thus, there exists a sequence $t_n \to +\infty$ such that

$$\lim_{t\to+\infty} \rho(\pi^{t_n}\gamma(\omega), \pi^{t_n} x) = 0. \tag{9.23}$$

Taking into consideration the property ii) of cocycle φ we obtain from (9.23) the equality

$$\lim_{n\to\infty} \rho(\varphi(t, z, \omega), \varphi(t, \nu(\omega), \omega)) = 0 \tag{9.24}$$

for all $z \in \mathbb{C}^d$, where $\gamma := (\nu, Id_\Omega)$.

Now we will show that there exists $\delta_0 > 0$ such that for arbitrary $\varepsilon > 0$ there is $L(\varepsilon) > 0$ with the property

$$|\varphi(t, \omega, u + h) - \varphi(t, \omega, u)| < \varepsilon \tag{9.25}$$

for all $(u, \omega) \in J$, $t \geq L(\varepsilon)$ and uniformly w.r.t. $|h| \leq \delta_0$. If it is not true, then there are $\delta \to +0, \varepsilon_0 > 0, |h_n| \leq \delta_n, \omega_n \in \Omega, u_n \in I_{\omega_n}$ and $t_n \geq n$ such that

$$|\varphi(t_n, \omega_n, u_n + h_n) - \varphi(t_n, \omega_n, u_n)| \geq \varepsilon_0.$$

On the other hand, according to property ii), there exists $C(R) > 0$ $(R > \sup\limits_{n\in\mathbb{N}} \delta_n)$ such that for sufficiently large n we have

$$\varepsilon_0 \leq |\varphi(t_n, \omega_n, u_n + h_n) - \varphi(t_n, \omega_n, u_n)| \leq C(R)\delta_n. \tag{9.26}$$

Taking into account that $\delta_n \to 0$ from (9.26) it follows that $\varepsilon_0 \leq 0$. The obtained contradiction prove our assertion.

From equality (9.17) and inequality (9.25) it follows that

$$\lim_{t\to+\infty} \|U(t, (u, \omega))\| = 0 \tag{9.27}$$

uniformly with respect to $(u, \omega) \in \mathfrak{B}[J, \delta_0]$. Denote by

$$m(t) := \sup\{\|U(t, (u, \omega))\| \quad |(u, \omega) \in \mathfrak{B}[J, \delta_0]\},$$

then

a) $m(t) \to 0$ as $t \to +\infty$.

b) $\exists L > 0$ such that m is bounded on $[L, +\infty)$.

c) $m(t + \tau) \le m(t)m(\tau)$ for all $t, \tau \ge L$.

From a)-c) it follows (see, for example, [91]) that there exist $N, \alpha > 0$ such that $m(t) \le N \exp(-\alpha t)$ for all $t \ge L$ and, consequently, from (9.27) we have

$$\|U(t, (u, \omega))\| \le N \exp(-\alpha t)$$

for all $(u, \omega) \in \mathfrak{B}[J, \delta_0]$ and $t \ge L$. Using the same arguments as well as as in the proof of Theorem 9.4 we conclude that $I_\omega = \{\nu(\omega)\}$ for all $\omega \in \Omega$.

Now we will prove that the mapping $\omega \to \gamma(\omega)$ is continuous. Let $\omega_n \to \omega$ and consider the sequence $\{\gamma(\omega_n)\} \subset J$. Since J is conditionally compact, this sequence is relatively compact. Let $\bar{x}$ be a limit point of $\{\gamma(\omega_n)\}$, then it is easy to see that $\bar{x} \in J_\omega = \{\gamma(\omega)\}$ and, consequently, $\gamma(\omega)$ is a unique limit point of relatively compact sequence $\{\gamma(\omega_n)\}$. Hence $\gamma(\omega_n) \to \gamma(\omega)$.

The equality $\nu(\sigma_t\omega) = \varphi(t, \nu(\omega), \omega)$ follows from invariance of J and from equality $J_\omega = \{\gamma(\omega)\}$ for all $\omega \in \Omega$, taking into account that $\nu(\omega) = pr_1\gamma(\omega)$.

The equality 4. follows from equality $J_\omega = \{\gamma(\omega)\} = \{(\nu(\omega), \omega)\}$ and from the fact that $\{\nu(\omega) \mid \omega \in \Omega\}$ is a compact global pullback attractor of cocycle φ.

The stability in the sense of Poisson in the positive direction of point $\gamma(\omega)$ follows from the continuity of γ and the equality $\pi^t\gamma(\omega) = \gamma(\sigma_t\omega)$ for all $t \in \mathbb{T}_+$ and $\omega \in \Omega$.

The sixth assertion follows from (9.24) and the equality $\varphi(t, \nu(\omega), \omega) = \nu(\sigma_t\omega)$ for all $t \in \mathbb{T}_+$ and $\omega \in \Omega$. The Theorem is completely proved. $\qquad\square$

9.6 The compact and local dissipative cocycles with noncompact base

Definition 9.7 The cocycle $\langle E^d, \varphi, (\Omega, \mathbb{T}, \sigma) \rangle$ is said to be compact dissipative if for any nonempty compact $\Omega' \subset \Omega$ there is a positive number $r_{\Omega'}$ such that for arbitrary $R > 0$ there exists a positive number $L(R, \Omega')$ with the following property

$$|\varphi(t, \omega, u)| < r_{\Omega'} \tag{9.28}$$

for all $\omega \in \Omega', |u| \le R$ and $t \ge L(R, \Omega')$.

Definition 9.8 The cocycle $\langle E^d, \varphi, (\Omega, \mathbb{T}, \sigma) \rangle$ is said to be locally dissipative if for any $\omega \in \Omega$ and $R > 0$ there are positive numbers r_ω, δ_ω and $L(R, \omega)$ such that

$$|\varphi(t, \tilde{\omega}, u)| < r_{\Omega'} \tag{9.29}$$

for all $\tilde{\omega} \in B(\omega, \delta_\omega) := \{\tilde{\omega} \in \Omega \mid \rho(\tilde{\omega}, \omega) < \delta_\omega\}, |u| \le R$ and $t \ge L(R, \omega)$.

Lemma 9.7 *Every locally dissipative cocycle φ is compact dissipative.*

Proof. Suppose that φ is locally dissipative and let $\Omega' \subset \Omega$ be a nonempty compact set. According to locally dissipativity of φ for every $\omega \in \Omega$, and $R > 0$ there exist $r_\omega, L(R, \omega) > 0$ and $\delta_\omega > 0$ such that the inequality (9.29) holds. Considering the open covering $\bigcup \{ B(\omega, \delta_\omega) \mid \omega \in \Omega' \}$ of compact set Ω', we may extract the finite sub covering $\bigcup \{ B(\omega_i, \delta_{\omega_i}) \mid i = \overline{1,k} \}$. Let $L(R, \Omega') := \max \{ L(R, \omega_i) \mid i = \overline{1,k} \}$, then it is clear that inequality (9.28) holds for all $\omega \in \Omega', |u| \le R$ and $t \ge L(R, \Omega)$. The Lemma is proved. $\qquad\square$

Remark 9.3 *Compact dissipativity, generally speaking, does not imply locally dissipativity.*

Lemma 9.8 *Let $\langle E^d, \varphi, (\Omega, \mathbb{T}, \sigma) \rangle$ be a compact dissipative $\mathbb{C}-$analytic cocycle. Then for any nonempty compact $\Omega' \subset \Omega$ and $R > 0$ there exist $L(\Omega', R)$ and $C = C(\Omega', R) > 0$ such that*

$$\rho(\varphi(t, u_1, \omega), \varphi(t, u_2, \omega)) \le C(\Omega', R)\rho(u_1, u_2) \tag{9.30}$$

for any $\omega \in \Omega', |u_i| \le R$ $(i = 1, 2)$ and $t \ge L(\Omega', R)$.

Proof. Let $R > 0$ and $R' > R$, then according to the compact dissipativity of cocycle φ for nonempty compact $\Omega' \subset \Omega$ and $R' > 0$ there exist $r_{\Omega'} > 0$ and $L(\Omega', R') > 0$ such that $|\varphi(t, u, \omega)| < r_{\Omega'}$ for all $\omega \in \Omega, |u| \le R'$ and $t \ge L(\Omega', R')$. In view of Cauchy's formula for $R < R'$ there exists a constant $C = C(R, \Omega) > 0$ such that $|\frac{\partial \varphi}{\partial u}(t, u, \omega)| \le C(R, \Omega')$ for all $t \ge L(R', \Omega'), \omega \in \Omega'$ and $|u| \le R$ and, consequently, the inequality (9.30) holds for $|u_1|, |u_2| \le R, \omega \in \Omega'$ and $t \ge L(\Omega', R) := \inf \{ L(\Omega', R') \mid R' > R \}$. $\qquad\square$

Lemma 9.9 *Let $\langle E^d, \varphi, (\Omega, \mathbb{T}, \sigma) \rangle$ be a compact dissipative $\mathbb{C}-$analytic cocycle, and γ_1, γ_2 are two entire bounded trajectories passing through point (u_1, ω) and (u_2, ω) for $t = 0$ respectively, then the following assertions hold:*

(1) If $\omega \in \Omega$ is negatively Poisson stable and $\{t_n\} \in \mathfrak{N}_\omega^-$, then

$$\inf_{n \in \mathbb{N}} \rho(\gamma_1(t_n), \gamma_2(t_n)) > 0 \tag{9.31}$$

if $u_1 \ne u_2$.

(2) If $\omega \in \Omega$ is positively Poisson stable and $\{t_n\} \in \mathfrak{N}_\omega^+$, then the equality

$$\inf_{n \in \mathbb{N}} \rho(\varphi(t_n, \omega, u_1), \varphi(t_n, \omega, u_2)) = 0 \tag{9.32}$$

implies

$$\lim_{t \to +\infty} \rho(\varphi(t, \omega, u_1), \varphi(t, \omega, u_2)) = 0. \tag{9.33}$$

Proof. Let $\omega \in \Omega$ be a negative Poisson stable point, $\{t_n\} \in \mathfrak{N}_\omega^-$ and $u_1 \neq u_2$. If the equality (9.31) is not true, then $\rho(\gamma_1(t_n), \gamma_2(t_n)) \to 0$ as $n \to \infty$. Denote by $R :=$ $\sup_{t \in \mathbb{T}} \max\{|\gamma_1(t)|, |\gamma_2(t)|\} > 0, \Omega' := \overline{\{\sigma_{t_n}\omega \mid n \in \mathbb{N}\}} \subset \Omega$ and let $C(\Omega', R), L(\Omega', R)$ be the corresponding constants figuring in the inequality (9.30), then we obtain

$$\rho(u_1, u_2) = \rho(\varphi(-t_n, \gamma_1(t_n), \sigma_{t_n}\omega), \varphi(-t_n, \gamma_2(t_n), \sigma_{t_n}\omega)) \qquad (9.34)$$
$$\leq C(\Omega', R)\rho(\gamma_1(t_n), \gamma_2(t_n))$$

for sufficiently large n $(-t_n \geq L(\Omega', R))$. Passing to limit in the inequality (9.34) as $n \to \infty$ we have $\rho(u_1, u_2) \leq 0$, but $u_1 \neq u_2$. The obtained contradiction prove the first statement of Lemma 9.9 .

Let now $\omega \in \Omega$ be a positive Poisson stable point and $\{t_n\} \in \mathfrak{N}_\omega^+$ such that the inequality (9.33) holds, then for arbitrary $\varepsilon > 0$ there exists $n_0 = n_0(\varepsilon)$ such that

$$\rho(\varphi(t, u_1, \omega), \varphi(t, u_2, \omega)) < \frac{\varepsilon}{2C(\Omega', R)} \qquad (9.35)$$

and, consequently, according to Lemma 9.8 we obtain

$$\rho(\varphi(t, u_1, \omega), \varphi(t, u_2, \omega)) = \rho(\varphi(t - t_n, \varphi(t_n, u_1, \omega), \sigma_{t_n}\omega), \qquad (9.36)$$
$$\varphi(t - t_n, \varphi(t_n, u_2, \omega), \sigma_{t_n}\omega) \leq C(\Omega', R)\rho(\varphi(t_n, u_1, \omega), \varphi(t_n, u_2, \omega))$$

for all $t \geq t_n + L(\Omega', R)$, where $R := \sup\{\max\{|\varphi(t, u_1, \omega)|, |\varphi(t, u_2, \omega)|\} \mid t \in \mathbb{T}_+\}$. Denote by $L(\varepsilon) := t_{n_0(\varepsilon)} + L(\Omega, R)$, then from inequalities (9.35) and (9.36) we obtain

$$\rho(\varphi(t, u_1, \omega), \varphi(t, u_2, \omega)) < \varepsilon$$

for all $t \geq L(\varepsilon)$ and, consequently, (9.33) holds. The lemma is proved. $\qquad\square$

Theorem 9.8 *Let $\langle E^d, \varphi, (\Omega, \mathbb{T}, \sigma)\rangle$ be a compact dissipative $\mathbb{C}-$analytic cocycle admitting a compact pullback attractor $\{I_\omega \mid \omega \in \Omega\}$ and every point $\omega \in \Omega$ be two-sided Poisson stable, then the following assertions hold:*

1. *The set I_ω consists of only one point $\nu(\omega)$, i.e. $I_\omega = \{\nu(\omega)\}$ for every $\omega \in \Omega$.*
2. *The mapping $\omega \to \gamma(\omega)$ is continuous, where $\gamma = (\nu, Id_\Omega)$.*
3. *$\nu(\sigma_t\omega) = \varphi(t, \nu(\omega), \omega)$ for all $\omega \in \Omega$ and $t \in T^+$.*
4. *The point $\gamma(\omega)$ is Poisson's stable for all $\omega \in \Omega$.*
5. *$\lim_{t \to +\infty} \beta(\varphi(t, \sigma_{-t}\omega)K, \nu(\omega)) = 0$ for all compact subsets K of $\mathbb{C}^d$, where $\beta(A, B)$ is the semi-distance of Hausdorff between A and B.*
6. *$\lim_{t \to +\infty} |\varphi(t, \omega, z) - \nu(\sigma_t\omega)| = 0$ for all $\omega \in \Omega$ and $z \in \mathbb{C}^d$.*

Proof. To prove the first assertion of Theorem 9.8 we consider the non-autonomous dynamical system $\langle(\Phi, \mathbb{T}, \lambda), (\Omega, \mathbb{T}, \sigma), H\rangle$ constructed in the proof of Theorem 9.1. Using the same type of argument as in Theorem 9.1 and taking into consideration the Lemma 9.9 we may state that the system $\langle(\Phi, \mathbb{T}, \lambda), (\Omega, \mathbb{T}, \sigma), H\rangle$ possesses the following properties:

a) Φ is conditionally compact invariant set.

b) for every $\omega \in \Omega, \gamma_1, \gamma_2 \in \varphi_\omega(\gamma_1 \neq \gamma_2)$ and $\{t_n\} \in \mathfrak{N}_\omega^-$ holds

$$\inf_{n\in\mathbb{N}} \rho(\gamma_1^{t_n}, \gamma_2^{t_n}) > 0,$$

where γ^t is a t-translation of γ, i.e. $\gamma^t(s) := \gamma(t + s)$ for all $s \in \mathbb{T}$.

Then according to Lemmas 9.3–9.5 we have that $E_\omega^- = E_\omega^+ = E_\omega$ is a group. Particularly there are two sequences $t_n^1 \to +\infty$ and $t_n^2 \to -\infty$ such that

$$\lim_{n\to\infty} \gamma^{t_n^i} = \gamma \quad (i = 1, 2). \tag{9.37}$$

for all $\gamma \in \Phi_\omega$, i.e. every entire trajectory γ of global pullback attractor $\{I_\omega \mid \omega \in \Omega\}$ is two-sided Poisson stable. On the other hand according to Lemma 9.9 we have

$$\lim_{t\to+\infty} \rho(\gamma_1(t), \gamma_2(t)) = 0. \tag{9.38}$$

From (9.37) and (9.38) we obtain

$$\rho(\gamma_1(t), \gamma_2(t)) = \lim_{n\to\infty} \rho(\gamma_1(t + t_n^1), \gamma_2(t + t_n^1)) = 0$$

for all $t \in \mathbb{T}$, i.e. $\gamma_1 = \gamma_2$. Thus there exists a unique entire trajectory $\tilde{\gamma}_\omega \in \Phi_\omega$, i.e. $\Phi_\omega = \{\tilde{\gamma}_\omega\}$ and, consequently, $I_\omega = \{\tilde{\gamma}_\omega(0)\} := \{\gamma(\omega)\}$, where $\gamma(\omega) := \tilde{\gamma}_\omega(0)$.

The proof of item 2.– 6. of Theorem 9.8 uses the same type of arguments as in Theorem 9.7. The Theorem is proved. $\qquad\square$

9.7   Applications

9.7.1   *ODEs*

Consider the differential equation

$$\frac{dz}{dt} = f(t, z), \tag{9.39}$$

where $f \in CH(\mathbb{R} \times \mathbb{C}^d, \mathbb{C}^d)$ and the family of equations

$$\frac{dz}{dt} = g(t, z), \tag{9.40}$$

where $g \in H(f) := \overline{\{f_\tau \mid \tau \in \mathbb{R}\}}$ and f_τ is a τ-translation of function f w.r.t. variable t, i.e. $f_\tau(t, z) := f(t + \tau, z)$ for all $t \in \mathbb{R}$ and $z \in \mathbb{C}^d$. Denote by $\varphi(t, z, g)$ the solution of equation (9.40) with the initial condition $\varphi(0, z, g) = z$, then φ is a $\mathbb{C}$-analytic cocycle on $\mathbb{C}^d$.

Definition 9.9 The equation (9.39) is called dissipative if there exists a positive number r such that $\limsup\limits_{t\to+\infty}|\varphi(t,z,g)| < r$ for all $z \in \mathbb{C}^d$ and $g \in H(f)$.

Definition 9.10 The function $f \in CH(\mathbb{R} \times \mathbb{C}^d, \mathbb{C}^d)$ is called positively (negatively) Poisson stable in $t \in \mathbb{R}$ uniformly w.r.t. z on compact subsets of $\mathbb{C}^d$ [292],[300] if there exists $t_n \to +\infty$ ($t_n \to -\infty$, respectively) such that $f(t+t_n, z) \to f(t,z)$ as $n \to \infty$ uniformly on every compact subsets of $\mathbb{R} \times \mathbb{C}^d$.

Theorem 9.9 *Suppose that the following conditions hold:*

(1) The set $H(f) \subset CH(\mathbb{R} \times \mathbb{C}^d, \mathbb{C}^d)$ is compact.
(2) Every function $g \in H(f)$ is positively Poisson stable (in this case function f is called [300],[304] *quasi recurrent).*
(3) The equation (9.39) is dissipative.

Then every equation (9.40) admits a unique bounded on $\mathbb{R}$ and positively Poisson stable solution. This solution is globally uniformly asymptotically stable.

Proof. We consider the dynamical system of translations (Bebutov's system [292],[300]) $(CH(\mathbb{R} \times \mathbb{C}^d, \mathbb{C}^d), \mathbb{R}, \sigma)$. Since $H(f)$ is invariant and closed subset of $CH(\mathbb{R} \times \mathbb{C}^d, \mathbb{C}^d)$, then on set $H(f)$ it is induced a dynamical system $(H(f), \mathbb{R}, \sigma)$. Let $\Omega := H(f)$, then on space $\mathbb{C}^d$ it is defined a $\mathbb{C}$-analytic cocycle $\langle \mathbb{C}^d, \varphi, (\Omega, \mathbb{R}, \sigma)\rangle$, generated by equation (9.39). According to the general properties of equation (9.39) with holomorphic right hand side f, the cocycle φ will be $\mathbb{C}$-analytic (see, for example, [122]). To finish the proof of this theorem it is sufficient to apply Theorem 9.4. $\square$

Definition 9.11 (Bohr's almost periodic function) The function $f \in CH(\mathbb{R} \times \mathbb{C}^d, \mathbb{C}^d)$ is called almost periodic (in the sense of Bohr) in $t \in \mathbb{R}$ uniformly w.r.t. z on compact subsets of $\mathbb{C}^d$ [238, 292, 300] if for every $\varepsilon > 0$ and nonempty compact subset $K \subset \mathbb{C}^d$ the set

$$\mathfrak{T}(\varepsilon, f, K) := \{\tau \in \mathbb{R} | \max_{z \in K} |f(t+\tau, z) - f(t,z)| < \varepsilon\}$$

is relatively dense on $\mathbb{R}$, i.e. there is a number $l = l(\varepsilon, f, K) > 0$ such that

$$\mathfrak{T}(\varepsilon, f, K) \bigcap [a, a+l] \neq \emptyset$$

for all $a \in \mathbb{R}$.

Definition 9.12 The equation (9.39) is called pullback dissipative if for every $g \in H(f)$ there exists a positive number r_g such that for all $R > 0$

$$\limsup_{t\to+\infty} |\varphi(t, g_{-t}, z)| < r_g$$

uniformly w.r.t. $|z| \leq R$.

Theorem 9.10   *Let $f \in CH(\mathbb{R} \times \mathbb{C}^d, \mathbb{C}^d)$ be an almost periodic function in $t \in \mathbb{R}$ uniformly w.r.t. z on compact subsets of $\mathbb{C}^d$ and the equation (9.39) be pullback dissipative, then every equation (9.40) admits a unique bounded solution $\nu_g(t)$, which is almost periodic and satisfies the following conditions:*

(1) $\nu_g(t)$ is uniformly asymptotically stable (locally).
(2) $\displaystyle \lim_{t \to +\infty} \sup_{|z| \leq R} |\varphi(t, \omega_{-t})z - \nu_g(0)| = 0.$

Proof.   The proof of this theorem uses the same type of argument as the proof of Theorem 9.9 and is based on the Theorem 9.3. $\square$

9.7.2   *Caratheodory differential equations*

Consider now the equation (9.39) with right hand side f satisfying the conditions of Caratheodory (see, for example,[292]) and holomorphic w.r.t. variable $z \in \mathbb{C}^d$. The space of all the Carateodory functions we denote by $\mathfrak{C}H(\mathbb{R} \times \mathbb{C}^d, \mathbb{C}^d)$. The topology on this space is defined by family of semi-norm (see [292])

$$p_{n,m}(f) := \int_{-n}^{n} \max_{|z| \leq m} |f(t,z)| dt.$$

This space is metrizable and on $\mathfrak{C}H(\mathbb{R} \times \mathbb{C}^d, \mathbb{C}^d)$ the dynamical system of translations $(\mathfrak{C}H(\mathbb{R} \times \mathbb{C}^d, \mathbb{C}^d), \mathbb{R}, \sigma)$ can be defined.

Using the standard arguments for ODEs (see,for example, [122] and [186]) one can prove that every equation (9.40) admits a unique solution $\varphi(t, z, g)$ with initial condition $\varphi(0, z, g) = z$ and supplementary the mapping $\varphi(t, g) := \varphi(t, \cdot, g) : \mathbb{C}^d \to \mathbb{C}^d$ is holomorphic. Thus if the solutions $\varphi(t, z, g)$ are defined on $\mathbb{R}_+$, the mapping $\varphi : \mathbb{R}_+ \times \mathbb{C}^d \times H(f) \to \mathbb{C}^d$ defines a $\mathbb{C}-$analytic cocycle on $\mathbb{C}^d$ with the base $H(f)$, where $H(f) := \overline{\{f_\tau | \tau \in \mathbb{R}\}}$ and the bar denotes the closure in the space $\mathfrak{C}H(\mathbb{R} \times \mathbb{C}^d, \mathbb{C}^d)$. Hence we may apply the general results from sections 1.–6. to cocycle φ, generated by equation (9.39) with Caratheodory's right hand side, and we will obtain some results for this type of equations. For example the following assertion holds.

Theorem 9.11   *Let $f \in \mathfrak{C}H(\mathbb{R} \times \mathbb{C}^d, \mathbb{C}^d)$ be an almost periodic function in $t \in \mathbb{R}$ (in the sense of Stepanov [238]) uniformly w.r.t. z on compact subsets of $\mathbb{C}^d$, i.e. for every $\varepsilon > 0$ and compact subset $K \subset \mathbb{C}^d$ the set*

$$\mathfrak{T}(\varepsilon, f, K) := \{\tau \in \mathbb{R} | \int_0^1 \max_{x \in K} |f(t + \tau + s, z) - f(t + s, z)| ds < \varepsilon\}$$

is relatively dense on $\mathbb{R}$*. Suppose that the equation (9.39) is dissipative, then every equation (9.40) admits a unique almost periodic (in the sense of Bohr) solution* $\nu_g(t)$ *which is globally uniformly asymptotically stable.*

9.7.3 *ODEs with impulses*

Let $\{t_n\}_{n\in\mathbb{Z}}$ be a sequence of real numbers, $\inf\{t_{n+1}-t_n|\quad n\in\mathbb{Z}\}>0$, $p:\mathbb{R}\to\mathbb{C}^d$ be a continuously differentiable function on every interval (t_n,t_{n+1}), continuous to the right in every point $t=t_n$, almost periodic in the sense of Stepanov and

$$p'(t) = \sum_{n\in\mathbb{Z}} s_n\delta_{t_n},$$

where $s_n := p(t_n+0) - p(t_n-0)$ (i.e. the function p is piecewise constant). More information about the function described above can be found in the books [170] and [279].

Consider the equation with impulses

$$\frac{dz}{dt} = f(t,z) + \sum_{n\in\mathbb{Z}} s_n\delta_{t_n} \tag{9.41}$$

or equivalently

$$\frac{dz}{dt} = f(t,z) + p'(t). \tag{9.42}$$

At the same time we consider the family of equations

$$\frac{dz}{dt} = g(t,z) + q'(t), \tag{9.43}$$

where $(g,q)\in H(f,p) := \overline{\{(f_\tau,p_\tau)|\tau\in\mathbb{R}\}}$ and by the bar we denote the closure in the product-space $CH(\mathbb{R}\times\mathbb{C}^d,\mathbb{C}^d)\times\mathfrak{C}(\mathbb{R},\mathbb{C}^d)$.

Denote by $\varphi(t,z,g,q)$ the unique solution of equation (9.43) (see [170] and [279]) satisfying the initial condition $\varphi(0,z,g,q)=z$. This solution is continuous on every interval (t_n,t_{n+1}) and continuous to the right in every point $t=t_n$ (see [170] and [279]).

Definition 9.13 The equation (9.41) is called dissipative if there exists a number $r>0$ such that $\limsup\limits_{t\to+\infty}|\varphi(t,z,g,q)|<r$ for every $(g,q)\in H(f,p)$ and $z\in\mathbb{C}^d$.

Using the transformation $w:=z+q(t)$ we can transform the equation (9.43) into the equation

$$\frac{dw}{dt} = g(t,w+q(t)). \tag{9.44}$$

Remark 9.4   *Denote by $\tilde{\varphi}(t, z, g, q)$ the cocycle defined by the family of equations (9.44), then it is clear that the cocycle φ generated by (9.42) is dissipative if and only if the cocycle $\tilde{\varphi}$ generated by (9.44) is dissipative.*

Theorem 9.12   *Let $f \in CH(\mathbb{R} \times \mathbb{C}^d, \mathbb{C}^d)$ be a Bohr's almost periodic function in $t \in \mathbb{R}$ uniformly with respect to z on every compacts subsets of $\mathbb{C}^d$, locally Lipshitz in z uniformly w.r.t. $t \in \mathbb{T}$ and $p \in \mathfrak{C}(\mathbb{R}, \mathbb{C}^d)$ be a Stepanov almost periodic function bounded on $\mathbb{R}$. If the equation (9.42) is dissipative, then for every $(g, q) \in H(f, p)$ the equation (9.43) admits a unique Stepanov almost periodic solution and this solution is globally uniformly asymptotically stable .*

Proof.   Let $\varphi(t, z, g, q)$ be the cocycle generated by equation (9.42) and let $\tilde{\varphi}(t, w, g, q)$ be the cocycle generated by equation (9.44). Then we have the following equality

$$\varphi(t, z, g, q) = q(t) + \tilde{\varphi}(t, z - q(0), g, q). \tag{9.45}$$

Under the conditions of Theorem 9.12 the cocycle $\tilde{\varphi}$ is dissipative, $\mathbb{C}-$analytic and the right hand side of equation (9.44) under the conditions of Theorem 9.12 is Stepanov almost periodic in $t \in \mathbb{R}$ uniformly on every compact of $\mathbb{C}^d$ w.r.t. z. To finish the proof of this theorem we apply the Theorem 9.11 to the equation (9.44) and take into consideration the relation (9.45). The theorem is proved. $\square$

Chapter 10

Pullback attractors under discretization

A defining characteristic of an autonomous dynamical system is its dependence on the time that has elapsed only and not on the absolute time itself. Consequently, limiting objects, such as attractors, actually exist for all time as invariant sets under the evolution of the autonomous system. Although such concepts can also be used for general non-autonomous systems, where the absolute starting time is as important as the time elapsed since starting, they are often too restrictive and exclude many interesting types of dynamical behaviour. A simple example is that of an asymptotically stable solution that is neither constant nor periodic. What are the limiting or attracting points here? What are the corresponding invariant sets, and, important for numerical considerations, how can one assure convergence to a particular point in such an invariant set? The forwards running convergence of an asymptotically stable solution is of little direct use in constructing the limiting solution since this solution may itself be changing with increasing time. An alternative is to use *pullback convergence,* that is to hold fixed an absolute time instant and to consider the limiting values at this time instant of other solutions that start progressively early in absolute time. The limiting sets now depend on absolute time and are invariant under the evolution of the system, that is, are carried forward onto each other as time increases. This idea was introduced several years ago in the context of random dynamical systems [126, 153, 154, 284], which are intrinsically non-autonomous, but had been used already in the 1960s by M.Krasnoselskii [223] to establish the existence of solutions of deterministic systems that are bounded over the entire time axis. It has also been applied recently [217, 219] to investigate variable time-step (hence non-autonomous) numerical approximations of global attractors of autonomous systems governed by dissipative ordinary differential equations.

Another key idea in [126, 153, 154, 284] is to formulate the non-autonomous dynamics on $\mathbb{R}^d$ in terms of a cocycle mapping φ that is driven by an underlying autonomous system σ on some parameter set Ω. At its simplest, Ω is just the absolute time set $\mathbb{R}$ and σ is the shift operator that essentially resets the starting time to the current absolute time value. More useful is to consider for Ω a function space of admissible vector fields as proposed by G. R. Sell [292] or as a probability sample space as in [126, 153, 154, 284], where the current parameter value takes the role of absolute time and is adjusted by σ with the passage of time. The advantage here is, in the first case at least, that the parameter space can be topologized (often as a compact space) and the product system (σ, φ) is an *autonomous* semi–dynamical system known as *skew product flow* on the new product state space $\Omega \times \mathbb{R}^d$. The extensive theory of autonomous dynamical systems can then be applied to such skew product flows, in particular concepts such as invariant sets, limit sets and attractors, but just how these manifest themselves in terms of the original dynamics on the original state space $\mathbb{R}^d$ and what relationship, if any, they have with pullback convergence need to be clarified.

In this chapter we investigate the effect of time discretization on the pullback attractor of a non-autonomous ordinary differential equation for which the vector fields depend on a parameter that varies in time rather than depending directly on time itself. The parameter space is assumed to be compact so the skew product flow formalism as well as cocycle formalism also applies and the vector fields have a strong dissipative structure that implies the existence of a compact set that absorbs all compact sets under the resulting non-autonomous dynamics. The numerical scheme considered is a general 1–step scheme such as the Euler scheme with variable time-steps. Our main result is to show that the numerical scheme interpreted as a discrete time non-autonomous dynamical system, hence discrete time cocycle mapping and skew product flow on an extended parameter space, also possesses a cocycle attractor and that its component subsets converge upper semi–continuously to those of the cocycle attractor of the original system governed by the differential equation. This is a non-autonomous analogue of a result of P. E. Kloeden and J. Lorenz [216] on the discretization of an autonomous attractor; see also [179, 308]. We will also see that the corresponding skew product flow systems have global attractors with the cocycle attractor component sets as their cross-sectional sets in the original state space $\mathbb{R}^d$. Finally, we investigate the periodicity and almost periodicity of the discretized pullback attractor when the parameter dynamics in the ordinary differential equation is periodic or almost periodic and the pullback attractor consists of singleton valued component sets, i.e. the pullback attractor is a single trajectory.

The chapter is organized as follows. Pullback attractors, cocycles and skew product flows are defined in Section 1 and a theorem is stated, summarizing results from the literature on the relationship between pullback attractors and global attractors of skew product flows. The class of non-autonomous differential equations and the corresponding variable time-step 1–step schemes to be considered are introduced in Section 2 and their cocycle formalism is then established in Section 3. The main result, Theorem 10.2, is formulated and proved in Section 4. Section 5 is devoted to periodic and almost periodic behaviour when the pullback attractor is a single trajectory. Finally, the section 6 contains the proof of a lemma used earlier, which compares the cocycle mappings of the original continuous time system and of the discrete time numerical systems.

10.1 Non-autonomous dynamical systems and pullback attractors

Consider an autonomous dynamical system on a metric space Ω described by a group $\sigma = \{\sigma_t\}_{t\in\mathbb{T}}$ of mappings of Ω into itself, where the time set $\mathbb{T}$ is either $\mathbb{Z}$ (discrete time) or $\mathbb{R}$ (continuous time).

Let W be a complete metric space and consider a continuous mapping $\varphi :$ $\mathbb{T}_+ \times W \times \Omega \to W$ satisfying the properties

$$\varphi(0, \cdot, \omega) = \mathrm{id}_W, \qquad \varphi(\tau + t, u, \omega) = \varphi(\tau, \varphi(t, u, \omega), \sigma_t\omega)$$

for all $t, \tau \in \mathbb{T}_+$, $\omega \in \Omega$ and $u \in W$. The mapping φ is called a (continuous) cocycle on W with respect to σ. Then the mapping $\pi : \mathbb{T}_+ \times W \times \Omega \to W \times \Omega$ defined by

$$\pi(t, u, \omega) := (\varphi(t, u, \omega), \sigma_t\omega)$$

for all $t \in \mathbb{T}_+$, $(u, \omega) \in \Omega \times W$ forms an autonomous semi–dynamical system on the state space $W \times \Omega$, i.e. the set of mappings $\{\pi(t, \cdot, \cdot)\}_{t\in\mathbb{T}_+}$ of $W \times \Omega$ into itself is a semi–group, which is called a continuous skew product flow [292].

The usual concept of a global attractor for the autonomous semi–dynamical system π on the state space $\Omega \times W$ can be used here. It is the maximal nonempty compact subset $\mathcal{A}$ of $\Omega \times W$ which is π–invariant, that is

$$\pi(t, \mathcal{A}) = \mathcal{A} \quad \text{for all} \quad t \in \mathbb{T}_+,$$

and attracts all compact subsets of $\Omega \times W$, that is

$$\lim_{t\to\infty} \beta\left(\pi(t, \mathcal{D}), \mathcal{A}\right) = 0 \qquad \text{for all} \quad \mathcal{D} \in C(W \times \Omega),$$

where $C(W \times \Omega)$ is the space of all nonempty compact subsets of $\Omega \times W$ and β is the Hausdorff semi-distance on $C(W \times \Omega)$.

Another type of attractor, called a pullback attractor, consists of subsets of the original state space W, which is advantageous for discretizations of the non-autonomous system.

Theorem 10.1 *Let φ be a continuous cocycle on W with respect to a group σ of continuous mappings on Ω and let $\pi = (\varphi, \sigma)$ be the corresponding skew product flow on $W \times \Omega$. In addition, suppose that there is a nonempty compact subset B of W and for every $D \in C(W)$ there exists a $T(D) \in \mathbb{T}_+$, which is independent f $\omega \in$*

Ω, *such that*

$$\varphi(t, D, \omega) \subset B \quad \text{for all} \quad t > T(D). \tag{10.1}$$

Then

1. *there exists a unique pullback attractor $I = \{I_\omega \mid \omega \in \Omega\}$ of the cocycle φ on W, where*

$$I_\omega = \bigcap_{\tau \in \mathbb{T}_+} \overline{\bigcup_{t > \tau} \varphi(t, B, \sigma_{-t}\omega)}. \tag{10.2}$$

2. *there exists a global compact attractor $\mathcal{A}$ of the autonomous semi–dynamical system π on $W \times \Omega$, where*

$$\mathcal{A} = \bigcap_{\tau \in \mathbb{T}_+} \overline{\bigcup_{t > \tau} \pi(t, B \times \Omega)}.$$

3.

$$J = \bigcup_{\omega \in \Omega} I_\omega \times \{\omega\}.$$

See Crauel and Flandoli [126] and Schmalfuß [284] for the proof of Assertion 1. and Cheban and Fakeeh [83] and Hale [179] for the proof of Assertion 2. Assertion 3. has been proved by Cheban [88].

Remark 10.1 *Assertion 1. true remains under weaker conditions. For instance, the sets D in the absorbing condition (10.1) could be parameter dependent, the parameter space Ω need not be compact nor the mappings σ_t continuous (which is the situation in random dynamical systems, see Arnold [6]). Note that the validity of Assertion 3. for non-uniform absorbing times, a situation which occurs in important applications, remains open. See [97] for a systematic investigation of the relationship between the pullback and forwards attractors of the cocycle system and the global attractor of the associated skew product flow.*

10.2   Non-autonomous quasi-linear differential equation

We consider a non-autonomous quasi-linear differential equation

$$\dot{u} = A(\omega)u + f(u, \omega t) \tag{10.3}$$

on $\mathbb{R}^d$ where $\omega \in \Omega$, on which there exists a group of mappings $\sigma_t : \Omega \to \Omega$ for all $t \in \mathbb{R}$, where $\omega t := \omega t$. A solution $x(t) = \varphi(t, u_0, \omega)$ of (10.3) with initial value $x(0)$

$= u_0$ satisfies the equation

$$\frac{d\varphi}{dt}(t, u_0, \omega) = A(\sigma_t\omega)\varphi(t, u_0, \omega) + f(\sigma_t\omega, \varphi(t, u_0, \omega)).$$

Our assumptions are

D1. Ω is a compact metric space and $(t, \omega) \mapsto \sigma_t\omega$ is continuous.

D2. $\omega \mapsto A(\omega)$ is continuous and satisfies

$$(A(\omega)u, u) \leq -\alpha(\omega)\,|u|^2$$

for all $(u, \omega) \in \mathbb{R}^d \times \Omega$.

D3. $(u, \omega) \mapsto f(u, \omega)$ is continuous, is locally Lipschitz in u uniformly in $\omega \in \Omega$ and satisfies

$$(f(u, \omega), u) \leq a(\omega)\,|u|^2 + c(\omega)$$

for all $(u, \omega) \in \mathbb{R}^d \times \Omega$.

D4. $\alpha(\omega) > 0$, $\alpha(\omega) - a(\omega) \geq \alpha_0 > 0$ and $c(\omega) \leq c_0 < \infty$ for all $\omega \in \Omega$.

Remark 10.2 . *The mapping f in **D3** denotes the remaining part of the differential equation. It may also contain linear terms that has not been included in linear part with the matrix $A(\omega)$. Though seemingly superfluous, it is sometimes convenient to distinguish the matrix operator $A(\omega)$ in this way (especially in infinite dimensional generalizations, which are not considered here).*

Remark 10.3 *By the above continuity and the compactness of Ω we have the finite uniform upper bounds*

$$A_0 := \sup_{\omega \in \Omega} \|A(\omega)\|, \qquad F_R := \sup_{\omega \in \Omega, |u| \leq R} |F(u, \omega)|.$$

We also consider a variable time-step one-step explicit numerical scheme corresponding to the differential equation (10.3), such as the Euler scheme, which we write as

$$u_{n+1} = u_n + h_n F(h_n, \sigma_{t_n}\omega, u_n) \tag{10.4}$$

where $t_0 = 0$ and $t_n = \sum_{j=0}^{n-1} h_j$, $t_{-n} = -\sum_{j=1}^{n} h_{-j}$ for $n \geq 1$ for $\{h_j\}_{n \in \mathbb{Z}}$ a given two sided sequence of positive terms and $F : [0, 1] \times \Omega \times \mathbb{R}^d \to \mathbb{R}^d$ is the increment function. We make the following assumptions:

N1. F is continuous in all of its variables and locally Lipschitz in u uniformly in $(h, p) \in [0, 1] \times \Omega$;

N2. The numerical scheme (10.4) satisfies a local discretization error estimate of the form

$$|\varphi(h_0, u_0, \omega) - u_1| \leq h_0 \mu_R(h_0), \qquad |u_0| \leq R,$$

for each $R > 0$, where $\mu_R(h) > 0$ for $h > 0$ and $\mu_R(h) \to 0$ as $h \to 0+$.

N3. F satisfies the consistency condition

$$F(0, u, \omega) = A(\omega)u + f(u, \omega)$$

for all $(u, \omega) \in \mathbb{R}^d \times \Omega$;

N4. F satisfies the Lipschitz consistency condition

$$|(F(h, u, \omega) - A(\omega)u - f(u, \omega)) -$$
$$(F(h, v, \omega) - A(\omega)v - f(v, \omega))| \leq \bar{\mu}_R(h)|u - v|$$

for all $|u|$, $|v| \leq R$ uniformly in $\omega \in \Omega$, where $\bar{\mu}_R(h) > 0$ for $h > 0$ and $\bar{\mu}_R(h) \to 0$ as $h \to 0+$.

For example, $F(h, u, \omega) \equiv A(\omega)u + f(u, \omega)$ for the Euler scheme applied to the differential equation (10.3). Note that for one-step order schemes such as the Euler and Runge–Kutta schemes, $\mu(h)$ is typically of the form $K_R h^p$ for some integer $p \geq 1$. In our case here this would require the differentiability of F in ω as well as u and of σ_t in t, which is too restrictive for certain applications.

Our main result (see Theorem 10.2 below) is that non-autonomous dynamical systems generated by the differential equation (10.3) and the numerical scheme (10.4) both have pullback and global attractors, and that the numerical attractors converge upper semi–continuously to the corresponding attractors of the differential equation as the step size goes to zero. We will apply Theorem 10.1 to establish the existence of such attractors, but first we need to show in what sense the numerical scheme (10.4) generates a discrete time cocycle mapping and skew product flow.

An example

We consider the 3–dimensional Lorenz system with time dependent coefficients

$$\dot{u}_1 = -p(t)u_1 + p(t)u_2$$
$$\dot{u}_2 = r(t)u_1 - u_2 - u_1 u_3$$
$$\dot{u}_3 = -b(t)u_3 + u_1 u_2$$

(see Temam [314], Chapter I.2.3). Assuming that p and r are differentiable functions of $\mathbb{R}$ into itself, we can rewrite this system after the transformation $u_3 := u_3 - r(t) - p(t)$ in the form (10.3) with the matrix

$$A(\omega) = \begin{pmatrix} -p(0) & p(0) & 0 \\ -p(0) & -1 & 0 \\ 0 & 0 & -1 \end{pmatrix}$$

and the mapping

$$f(u,\omega) = \begin{pmatrix} 0 \\ -u_1 u_3 \\ u_1 u_2 - \dot{\sigma}(0) - \dot{r}(0) - b(0)(r(0) + \sigma(0)) - (b(0) - 1)u_3 \end{pmatrix},$$

where $u = (u_1, u_2, u_3)$ and $\omega := \omega(\cdot) = (b(\cdot), r(\cdot), \sigma(\cdot)) \in C(\mathbb{R}, \mathbb{R}) \times C^1(\mathbb{R}, \mathbb{R}) \times C^1(\mathbb{R}, \mathbb{R})$.

Suppose that b is almost periodic in $C(\mathbb{R}, \mathbb{R})$ and that σ, r are almost periodic in $C^1(\mathbb{R}, \mathbb{R})$ with

$$0 < \sigma_{\min} := \inf_{t \in R} \sigma(t), \qquad 1 < \inf_{t \in R} b(t).$$

Then define σ_t for each $t \in \mathbb{R}$ by $\sigma_t p(\cdot) := \omega(\cdot + t)$. Finally, define the parameter set Ω to be the closed hull [292]

$$P := \overline{\bigcup_{t \in \mathbb{R}} \sigma_t(b(\cdot), r(\cdot), p(\cdot))},$$

which is a compact subset of $C(\mathbb{R}, \mathbb{R}) \times C^1(\mathbb{R}, \mathbb{R}) \times C^1(\mathbb{R}, \mathbb{R})$.

We need to check that the assumptions **D1**–**D4** are satisfied by the differential equation with this matrix A and function f. The first assumption **D1** follows straighforwardly and the second **D2** holds with $\alpha(\omega) \equiv \alpha_0 := \min(\sigma_{\min}, 1) > 0$ with the assumptions on p and r ensuing the continuity of A. These assumptions on p and r also ensure the continuity of f needed in **D3** and the estimate is given by

$$(f(u,\omega), u) = (-\dot{p}(0) - \dot{r}(0))u_3 - b(0)(r(0) + p(0))u_3 - (b(0) - 1)u_3^2$$

$$\leq \frac{(\dot{p}(0) + \dot{r}(0))^2}{2(b(0) - 1)} + \frac{b^2(0)(r(0) + p(0))^2}{2(b(0) - 1)} =: c(\omega)$$

with $a(\omega) \equiv 0$. Finally, it is obvious from these definitions of the constants that **D4** is also satisfied, in particular with $c_0 := \sup_{\omega \in \Omega} c(\omega) < \infty$.

The Euler scheme for this differential equation also satisfies assumptions **N1–N4**. Since $F(h, u, \omega) \equiv A(\omega)u + f(u, \omega)$ here, assumptions **N3** and **N4** hold trivially, while assumption **N1** follows from **D1–D3** above. Assumption **N2** also follows from **D1–D3** with the proof being almost the same as in the first part of the proof in the Appendix. Note that if b is also assumed to be continuously differentiable, then we have the usual second order local discretization error here. One can show that assumptions **N1–N4** are also satisfied by higher order schemes such as Runge–Kutta schemes, but the details are not as straightforward or clean as for the Euler scheme.

10.3   Cocycle property

The solution $\varphi(t, u_0, \omega)$ of the differential equation (10.3) satisfies assumptions **N1–N4** the initial condition

$$\varphi(0, u_0, \omega) = u_0 \quad \text{for all} \quad (u_0, \omega) \in \mathbb{R}^d \times \Omega$$

and the cocycle property

$$\varphi(s + t, u_0, \omega) = \varphi(s, \varphi(t, u_0, \omega), \sigma_t p) \quad \text{for all} \quad s, t \in \mathbb{R}_+, \quad (u_0, \omega) \in \mathbb{R}^d \times \Omega$$

with respect to the autonomous dynamical system generated by the group $\{\sigma_t\}_{t \in \mathbb{R}}$ on Ω. (Existence of such solutions for all $t \in \mathbb{R}_+$ is assured by that of an absorbing set to be established in the proof of Theorem 10.2 below). By our assumptions the mapping $(t, u, \omega) \mapsto \varphi(t, u, \sigma_t \omega)$ is continuous. Moreover, the mapping $\varphi := (\varphi, \sigma)$ defined on $\mathbb{R}_+ \times \mathbb{R}^d \times \Omega$

$$\varphi(t, u, \omega) := (\varphi(t, u, \omega), \sigma_t \omega) \quad \text{for all} \quad (t, u_0, \omega) \in \mathbb{R}_+ \times \mathbb{R}^d \times \Omega$$

generates an autonomous semi-dynamical system, that is a skew product flow, on the state space $X := \mathbb{R}^d \times \Omega$.

The situation is somewhat more complicated for the discrete time system generated by the numerical scheme with variable time steps. For this we will restrict the choice of admissible step-size sequences. For each $\delta > 0$, we define $\mathcal{H}^\delta$ to be the set of all two sided sequences $\{h_n\}_{n \in \mathbb{Z}}$ satisfying

$$\frac{1}{2}\delta \leq h_n \leq \delta \tag{10.5}$$

for each $n \in \mathbb{Z}$ (the particular factor $1/2$ here is chosen just for convenience). The set $\mathcal{H}^\delta$ is compact metric space with the metric

$$\rho_{\mathcal{H}^\delta}\left(\mathbf{h}^{(1)}, \mathbf{h}^{(2)}\right) = \sum_{n=-\infty}^{\infty} 2^{-|n|} \left|h_n^{(1)} - h_n^{(2)}\right|.$$

We then consider the shift operator $\tilde{\sigma} : \mathcal{H}^\delta \to \mathcal{H}^\delta$ defined by $\tilde{\sigma}\mathbf{h} = \tilde{\sigma}\{h_n\}_{n\in\mathbb{Z}} := \{h_{n+1}\}_{n\in\mathbb{Z}}$. The operator $\tilde{\sigma}$ is a homeomorphism with respect to the above metric on $\mathcal{H}^\delta$ and its group of iterates $\{\tilde{\sigma}_n\}_{n\in\mathbb{Z}}$ forms a discrete time autonomous dynamical system on the compact metric space $(\mathcal{H}^\delta, \rho_{\mathcal{H}^\delta})$. Finally, for a given sequence $\{h_n\}_{n\in\mathbb{Z}}$ we set $t_0 = 0$ and define $t_n = t_n(\mathbf{h}) := \sum_{j=0}^{n-1} h_j$ and $t_{-n} = t_{-n}(\mathbf{h}) := -\sum_{j=1}^{n} h_{-j}$ for $n \geq 1$.

Now we introduce the parameter space $\mathcal{Q}^\delta := \mathcal{H}^\delta \times \Omega$ for a fixed $\delta > 0$ and we use the following lemma to introduce a discrete time autonomous dynamical system $\Theta = \{\Theta_n\}_{n\in\mathbb{Z}}$ on $\mathcal{Q}^\delta$.

Lemma 10.1 *The mappings $\Theta_n : \mathcal{Q}^\delta \to \mathcal{Q}^\delta$, $n \in \mathbb{Z}$, generated by iteration of*

$$\Theta_0 := \mathrm{id}_{\mathcal{Q}^\delta}, \qquad \Theta_1(\mathbf{h}, p) := (\tilde{\sigma}_1\mathbf{h}, \sigma_{h_0}p), \qquad \Theta_{-1}(\mathbf{h}, p) := (\tilde{\sigma}_{-1}\mathbf{h}, \sigma_{-h_{-1}}p)$$

form a group of continuous mappings on $\mathcal{H}^\delta \times P$.

Proof. We first show that $\Theta_1\Theta_{-1} = \Theta_{-1}\Theta_1 = \Theta_0 = \mathrm{id}_{\mathcal{Q}^\delta}$. Indeed we have

$$\Theta_1\Theta_{-1}(\mathbf{h}, p) = \Theta_1\left(\tilde{\sigma}_{-1}\mathbf{h}, \sigma_{h_{-1}}p\right) = \left(\tilde{\sigma}_1\tilde{\sigma}_{-1}\mathbf{h}, \sigma_{h_{-1}}\sigma_{-h_{-1}}p\right) = (\mathbf{h}, p)$$

$$\Theta_{-1}\Theta_1(\mathbf{h}, p) = \Theta_{-1}\left(\tilde{\sigma}_1\mathbf{h}, \sigma_{h_0}p\right) = \left(\tilde{\sigma}_{-1}\tilde{\sigma}_1\mathbf{h}, \sigma_{-h_0}\sigma_{h_0}p\right) = (\mathbf{h}, p).$$

The continuity of the Θ_n mappings follows from the facts that $(t, p) \mapsto \sigma_t p$ is continuous, $\tilde{\sigma}$ is a homeomorphism and the composition and cartesian products of continuous mappings are continuous. $\square$

We define a mapping $\psi : \mathbb{Z}_+ \times \mathbb{R}^d \times \mathcal{Q}^\delta \to \mathbb{R}^d$ by

$$\psi(0, u_0, q) := u_0, \qquad \psi(n, u_0, q) = \psi(n, u_0, (\mathbf{h}, p)) := u_n \qquad n \geq 1,$$

where u_n is the nth iterate of the numerical scheme (10.4) with initial value $u_0 \in \mathbb{R}^d$, initial parameter $\omega \in \Omega$ and step-size sequence $\mathbf{h} \in \mathcal{H}^\delta$. These mappings are continuous on $\mathbb{R}^d \times \mathcal{Q}^\delta$. They also satisfy a cocycle property with respect to Θ. $\square$

Lemma 10.2 *ψ is a discrete time cocycle over $(\mathcal{Q}^\delta, \mathbb{Z}, \Theta)$ $\mathbb{R}^d$ with fiber $\mathbb{R}^d$.*

Proof. We write the numerical scheme (10.4) for a given $(\mathbf{h}, p)$ and u_0 as

$$u_{n+1} = u_n + h_n F\left(h_n, \sigma_{t_n(\mathbf{h})}p, u_n\right), \qquad n \in \mathbb{Z}_+,$$

where $t_0(\mathbf{h}) = 0$ and $t_n(\mathbf{h}) = \sum_{j=0}^{n-1} h_j$ for $n \geq 1$. Hence, in terms of the ψ mapping we have

$$\psi(n + 1, u_0, (\mathbf{h}, p)) = \psi(n, u_0, (\mathbf{h}, p)) + h_n F(h_n, \sigma_{t_n(\mathbf{h})}p, \psi(n, u_0, (\mathbf{h}, p)))$$

in general and

$$\psi(1, u_0, (\mathbf{h}, p)) = u_0 + h_0 F\left(h_0, \sigma_{t_0(\mathbf{h})}p, u_0\right)$$

for $n = 0$ since by definition $\psi(0, u_0, (\mathbf{h}, p)) = u_0$, which gives the identity property. The cocycle property is established as follows. Let $n \geq 0$. Then

$$\psi(1 + n, u_0, q) = \psi(1 + n, u_0, (\mathbf{h}, p))$$

$$= \psi(n, u_0, (\mathbf{h}, p)) + h_n F\left(h_n, \sigma_{t_n(\mathbf{h})}p, \psi(n, u_0, (\mathbf{h}, p))\right)$$

$$= \psi(n, u_0, (\mathbf{h}, p)) + (\tilde{\sigma}_n\mathbf{h})_0 F\left((\tilde{\sigma}_n\mathbf{h})_0, \sigma_{t_0(\tilde{\sigma}_n\mathbf{h})}p, \psi(n, u_0, (\mathbf{h}, p))\right)$$

$$= \psi\left(1, p, \Theta_n(\mathbf{h}), \psi(n, u_0, (\mathbf{h}, p))\right) = \psi\left(1, \psi(n, u_0, q), \Theta_n q\right),$$

that is $u_{1+n} = = \psi\left(1, u_n, \Theta_n q\right)$. Iterating this n times, we obtain

$$u_n = \psi\left(1, u_0, \Theta_{n-1}q\right) \circ \cdots \circ \psi\left(1, u_0, \Theta_0 q\right).$$

Similarly with $m \geq 2$ we have

$$\psi(m + n, u_0, q) = u_{m+n} = \psi\left(1, \cdot, \Theta_{m+n-1}q\right) \circ \cdots \circ \psi\left(1, \cdot, \Theta_n q\right) \circ$$

$$\circ\psi\left(1, \cdot, \Theta_{n-1}q\right) \circ \cdots \circ \psi\left(1, u_0, \Theta_0 q\right)$$

$$= \psi\left(1, \cdot, \Theta_{m+n-1}q\right) \circ \cdots \circ \psi\left(1, u_n, \Theta_n q\right)$$

$$= \psi\left(1, \cdot, \Theta_{m-1}\Theta_n q\right) \circ \cdots \circ \psi\left(1, u_n, \Theta_0\Theta_n q\right)$$

$$= \psi\left(m, u_n, \Theta_n q\right) = \psi\left(m, \psi\left(n, q, u_0\right), \Theta_n q\right),$$

which is the desired cocycle property.

Finally we define $\Psi := (\psi, \Theta)$ and observe that the mappings $\Psi(n, \cdot\cdot)$ are continuous on $\mathbb{R}^d \times \mathcal{Q}^\delta$ for $n \in \mathbb{Z}_+$. $\qquad\square$

Remark 10.4 *The mappings ψ, Θ and Ψ are defined in the same way for each $\delta > 0$, so we do not index them with δ.*

From the cocycle and group properties we obtain

Lemma 10.3 $\Psi = (\psi, \Theta)$ *is a discrete time autonomous semi–dynamical system on the state space $\mathbb{R}^d \times \mathcal{Q}^\delta$.*

10.4 Main result

Our main result is the to establish the existence of pullback attractors for the non-autonomous dynamical systems (NDS) generated by the differential equation and numerical scheme and to show that the components of the numerical pullback attractor are upper semi–continuous in their parameter and converge upper semi–continuously to the corresponding components of the differential equation's pullback attractor. Global attractors also exist for the corresponding skew product flows and converge upper semi–continuously in an appropriate sense.

Theorem 10.2 *Let Assumptions **D1–D4** and **N1–N3** hold. Then the continuous time NDS (φ, σ) generated by the differential equation (10.3) has a pullback attractor $I = \{I_\omega \mid \omega \in \Omega\}$ and the discrete time NDS (Ψ, Θ) generated by the numerical scheme (10.4) has a pullback attractor $I^\delta = \{I_q^\delta \mid q \in \mathcal{Q}^\delta\}$, provided the maximal step-size δ is sufficiently small, such that the set-valued mappings $\omega \mapsto I_\omega$ and $(p, \mathbf{h})$ $\mapsto I_{(p,\mathbf{h})}^\delta$ are upper semi–continuous with respect to the Hausdorff semi-distance and satisfy*

$$\lim_{\delta \to 0+} \sup_{\mathbf{h} \in \mathcal{H}^\delta} \beta \left(I_{(\omega,\mathbf{h})}^\delta, I_\omega \right) = 0 \qquad \text{for each} \quad \omega \in \Omega.$$

Moreover, the corresponding skew product flows have global attractors J and J^δ, respectively, of the form

$$J = \bigcup_{\omega \in \Omega} I_\omega \times \{\omega\}, \qquad J^\delta = \bigcup_{q \in \mathcal{Q}^\delta} I_q^\delta \times \{q\},$$

which satisfy

$$\lim_{\delta \to 0+} \beta \left(\mathrm{Pr}_{\mathbb{R}^d \times \Omega} J^\delta, J \right) = 0.$$

Proof. The proof of the convergence assertions in Theorem 10.2 will follow immediately from an application of Theorem 10.1 after it has been shown that the ball $B[0; R_0]$ in $\mathbb{R}^d$ with centre 0 and radius $R_0 := 3\sqrt{c_0/\alpha_0}$ is a forwards absorbing set uniformly in all parameters for both the continuous time and discrete time cocycle systems under consideration. The convergence assertions require additional work.

10.4.1 *Existence of an absorbing set*

Write $x(t)$ for the solution $\varphi(t, u_0, \omega)$ of (10.3), so

$$\frac{dx}{dt}(t) = A(\sigma_t \omega)x(t) + f(\sigma_t \omega, x(t)).$$

The following estimate will also be used to construct an absorbing set.

$$\frac{d}{dt}|x(t)|^2 = 2\left(\frac{dx}{dt}(t), x(t)\right) = 2\left(A(\sigma_t p)x(t) + f(\sigma_t p, x(t)), x(t)\right) \tag{10.6}$$

$$= 2\left(A(\sigma_t p)x(t)\right) + 2\left(f(\sigma_t p, x(t)), x(t)\right)$$

$$\leq -2\alpha(\sigma_t p)\,|x(t)|^2 + 2a(\sigma_t p)\,|x(t)|^2 + 2c(\sigma_t p)$$

$$\leq -2\left(\alpha(\sigma_t p) - a(\sigma_t p)\right)|x(t)|^2 + 2c_0 \leq -2\alpha_0\,|x(t)|^2 + 2c_0$$

and so

$$|x(t)|^2 \leq |u_0|^2 e^{-2\alpha_0 t} + \frac{c_0}{\alpha_0}\left(1 - e^{-2\alpha_0 t}\right).$$

This implies that the ball $B[0; R_0]$ with radius $R_0 = 3\sqrt{c_0/\alpha_0}$ is a forwards absorbing and positively invariant set for all solutions of the differential equation (10.3) uniformly in $\omega \in \Omega$. Note for later purposes that the ball $B[0; 2R_0/3]$ is also positively invariant for the differential equation.

The proof that $B[0; R_0]$ is a uniform forwards absorbing set for the numerical scheme (10.4) is more complicated. First we show that the inequality (10.6) implies

$$|x(t)| \leq |u_0|e^{-\alpha_0 t} + \sqrt{\frac{c_0}{\alpha_0}}\left(1 - e^{-\alpha_0 t}\right). \tag{10.7}$$

as long as $|x(t)| \geq \frac{1}{3}R_0 = \sqrt{\frac{c_0}{\alpha_0}}$. To see this note that (10.6) can be rewritten as

$$\frac{d}{dt}|x(t)|^2 \leq -2\alpha_0\,|x(t)|^2 + 2c_0 \leq -2\alpha_0\,|x(t)|^2 + 2\frac{c_0}{\sqrt{\frac{c_0}{\alpha_0}}}\,|x(t)|$$

and hence as

$$\frac{d}{dt}|x(t)| \leq -\alpha_0\,|x(t)| + \sqrt{c_0\alpha_0}$$

as long as $|x(t)| \geq \frac{1}{3}R_0$.

We note also by continuity that there exists a $T = T(c_0, \alpha_0) > 0$ such that $|x(t)| = |\varphi(t, u_0, \omega)| \geq \frac{1}{3}R_0$ for all $t \in [0, T]$, u_0 with $|u_0| \geq \frac{2}{3}R_0$ and $\omega \in \Omega$.

We fix an $R \gg R_0$ and let K_R be the constant in the local discretization error estimate for the ball $B[0; R]$. Then from (10.7) with $x(h) = \varphi(h, u_0, \omega)$ and

Assumption **N2** we have

$$|u_1| \leq |\varphi(h, u_0, \omega)| + |\varphi(h, u_0, \omega) - u_1|$$

$$\leq |u_0|e^{-\alpha_0 h} + \sqrt{\frac{c_0}{\alpha_0}} \left(1 - e^{-\alpha_0 h}\right) + K_R h \mu_R(h) \tag{10.8}$$

for $u_0 \in A[R, 2R_0/3] := B[0; R] \setminus B[0; 2R_0/3]$ and $h \in (0, T]$. Now let $\delta_0 \in (0, 1] \cap (0, T]$ be such that

$$K_R h \mu_R(h) \leq \frac{1}{3} R_0, \qquad \frac{K_R h \mu_R(h)}{1 - e^{-\alpha_0 h}} \leq R - \frac{1}{3} R_0$$

for $h \in (0, \delta_0]$. Then from (10.8) for $u_0 \in A[R, 2R_0/3]$ and $h \in (0, \delta_0]$ we have

$$|u_1| \leq |u_0|e^{-\alpha_0 h} + \sqrt{\frac{c_0}{\alpha_0}} \left(1 - e^{-\alpha_0 h}\right) + K_R h \mu_R(h)$$

$$\leq Re^{-\alpha_0 h} + \frac{1}{3} R_0 \left(1 - e^{-\alpha_0 h}\right) + \left(R - \frac{1}{3} R_0\right) \left(1 - e^{-\alpha_0 h}\right) \leq R,$$

so $u_1 \in B[0; R]$. In addition, for $u_0 \in B[0; 2R_0/3]$ we have $x(h) \in B[0; 2R_0/3]$, so

$$|u_1| \leq |\varphi(h, u_0, \omega)| + |\varphi(h, u_0, \omega) - u_1| \leq \frac{2}{3} R_0 + K_R h \mu_R(h) \leq \frac{2}{3} R_0 + \frac{1}{3} R_0 = R_0$$

for $h \in (0, \delta_0]$, so $u_1 \in B[0; R]$ here too. Hence, the ball $B[0; R]$ is positively invariant for the numerical scheme for $u_0 \in B[0; R]$ and $h \in (0, \delta_0]$. We can thus apply the inequality (10.8) iteratively as long as the $u_n \in A[R, 2R_0/3]$, that is we have

$$|u_{n+1}| \leq |u_n|e^{-\alpha_0 h} + \sqrt{\frac{c_0}{\alpha_0}} \left(1 - e^{-\alpha_0 h}\right) + K_R h \mu_R(h) \tag{10.9}$$

when $u_0, u_1, \ldots, u_n \in A[R, 2R_0/3]$ and $h \in (0, \delta_0]$.

Now further restrict the step-size so that

$$\frac{K_R h \mu_R(h)}{1 - e^{-\alpha_0 h}} \leq \frac{1}{2} R_0 - \frac{1}{3} R_0 = \frac{1}{2} R_0 - \frac{1}{2} \sqrt{\frac{c_0}{\alpha_0}}$$

for all $h \in (0, \delta_1]$, where $\delta_1 \in (0, \delta_0)$. Using a similar argument as above for the ball $B[0; R]$, we can show that the ball $B[0; R_0]$ is positively invariant for the numerical scheme with step-sizes $h \in (0, \delta_1]$.

To show that the ball $B[0; R_0]$ is absorbing, we further restrict the step-size so that

$$\frac{1}{2}\left(1 + e^{-\alpha_0 h}\right) \le e^{-\frac{1}{4}\alpha_0 h}$$

for all $h \in (0, \delta_2]$, where $\delta_2 \in (0, \delta_1]$. Then, if $u_0 \in A[R, R_0]$, from inequality (10.9) we have

$$|u_1| \le |u_0| e^{-\alpha_0 h} + \sqrt{\frac{c_0}{\alpha_0}}\left(1 - e^{-\alpha_0 h}\right) + K_R h \mu_R(h)$$

$$\le |u_0| e^{-\alpha_0 h} + \sqrt{\frac{c_0}{\alpha_0}}\left(1 - e^{-\alpha_0 h}\right) + \left(\frac{1}{2} R_0 - \sqrt{\frac{c_0}{\alpha_0}}\right)\left(1 - e^{-\alpha_0 h}\right)$$

$$\le |u_0| e^{-\alpha_0 h} + \frac{1}{2}|u_0|\left(1 - e^{-\alpha_0 h}\right)$$

$$\le \frac{1}{2}\left(1 + e^{-\alpha_0 h}\right)|u_0| \le e^{-\frac{1}{4}\alpha_0 h}|u_0|.$$

In particular, when $u_0, u_1, \ldots, u_j \in B[0; R] \setminus B[0; R_0]$ we can iterate the last inequality to obtain

$$|u_j| \le e^{-j\frac{1}{8}\alpha_0 \delta}|u_0|$$

if we use variable step-sizes with $\delta/2 \le h_j \le \delta$ with $\delta \in (0, \delta_2]$. Obviously, there exists a finite integer $J_R(u_0)$ such that $|u_j| \le R_0$ for all $j \ge J_R(u_0)$, that is the ball $B[0; R_0]$ is absorbing for the numerical scheme for all step-size sequences $\mathbf{h} \in \mathcal{H}^\delta$ with $\delta \in (0, \delta_2]$. Note that this holds uniformly in $\omega \in \Omega$.

10.4.2   *Upper semi–continuity of the pullback attractor component sets*

Let $\psi(n, x, q)$ denote the numerical trajectory and Θ the shift operator on $\mathcal{Q}^\delta$. The mappings $q \mapsto \Theta q$ and $(x, q) \mapsto \psi(n, x, q)$ for each integer positive n are continuous.

The absorbing set $B = B[0; R_0]$ is compact absorbing set and forwards invariant uniformly in $q \in \mathcal{Q}^\delta$. Hence, by the cocycle property, the compact sets $\psi(n, B, \Theta_{-n}q)$ are nested with increasing n and the pullback attractor has component sets defined by

$$I_q^\delta := \bigcap_{n \ge 0} \psi(n, B, \Theta_{-n}q)$$

This means, in particular, that $\beta(\psi(n, B, \Theta_{-n}q), I_q^\delta) \to 0$ as $n \to \infty$. Let $\varepsilon > 0$ and pick $n_0 = n_0(\varepsilon, q) > 0$ so that

$$\psi(n_0, B, \Theta_{-n_0}q) \subset B\left(I_q^\delta, \varepsilon\right),$$

where $B\left(I_q^\delta, \varepsilon\right)$ is the ball of radius ε about I_q^δ.

Now the compact set-valued mappings $q \mapsto \psi(n, B, \Theta_{-n}q)$ are continuous in q with respect to the Hausdorff semi-distance for each fixed n. Fix $\varepsilon > 0$ and pick $n_0 = n_0(\varepsilon, \bar{q})$ from above. Then there exists $\delta(\varepsilon, \bar{q}) = \delta(\varepsilon, n_0(\varepsilon, \bar{q})) > 0$ such that

$$\beta\left(\psi(n_0, B, \Theta_{-n_0}q), \psi(n_0, B, \Theta_{-n_0}\bar{q})\right) < \varepsilon$$

for all $q \in \mathcal{Q}^\delta$ with $\rho(q, \bar{q}) < \delta(\varepsilon, \bar{q})$. In particular,

$$\psi(n_0, B, \Theta_{-n_0}q) \subset B\left(\psi(n_0, B, \Theta_{-n_0}\bar{q}), \varepsilon\right)$$

for all $q \in \mathcal{Q}^\delta$ with $\rho_{\mathcal{Q}^\delta}(q, \bar{q}) < \delta(\varepsilon, \bar{q})$. Hence we have

$$I_q^\delta \subset \psi(n_0, B, \Theta_{-n_0}q) \subset B\left(\psi(n_0, B, \Theta_{-n_0}\bar{q}), \varepsilon\right)$$
$$\subset B\left(B\left(I_{\bar{q}}^\delta, \varepsilon\right), \varepsilon\right) = B\left(I_{\bar{q}}^\delta, 2\varepsilon\right)$$

that is $I_q^\delta \subset B\left(I_{\bar{q}}^\delta, 2\varepsilon\right)$ or equivalently $\beta\left(I_q^\delta, I_{\bar{q}}^\delta\right) < 2\varepsilon$ for all $q \in \mathcal{Q}^\delta$ with $\rho(q, \bar{q}) < \delta(\varepsilon, \bar{q})$. This means the mapping $q \mapsto I_q^\delta$ is upper semi–continuous.

The proof for the mapping $\omega \mapsto A_\omega$ is essentially the same.

10.4.3 *Upper semi–continuous convergence of the discretized pullback attractors*

We will now prove the upper semi–continuous convergence of the discretized pullback attractor component sets to their continuous time counterparts. For the proof we need the following lemma on the convergence of the numerical trajectories to the corresponding continuous time trajectory with convergence of the maximum step size to zero. Its proof is given in the appendix.

Lemma 10.4 *For fixed $t > 0$ and $\mathbf{h} = \{h_n\}_{n \in \mathbb{Z}} \in \mathcal{H}^\delta$ for some $\delta > 0$, let $N(t, \mathbf{h})$ be the positive integer such that*

$$h_{-1} + h_{-2} + \cdots + h_{-N(t,\mathbf{h})} \leq t < h_{-1} + h_{-2} + \cdots + h_{-N(t,\mathbf{h})-1}$$

and consider a sequence (of step-size sequences) with $\mathbf{h}^m \in \mathcal{H}^{\delta_m}$*, where* $\delta_m \to 0$ *as* $m \to \infty$*. Then*

$$\psi\left(N(t, \mathbf{h}^m), u_m, \Theta_{-N(t,\mathbf{h}^m)}(\mathbf{h}^m, \omega_m)\right) \to \varphi(t, u_0, \sigma_{-t}\omega) \quad as \quad m \to \infty$$

for any sequence $u_m \to u_0 \in \mathbb{R}^d$ *and* $\omega_m \to \omega \in \Omega$.

We suppose that the upper semi–continuous convergence assertion of Theorem 10.2 is not true. Then there exists an $\varepsilon_0 > 0$ and sub-sequences (for convenience we use the original index) $\mathbf{h}^m$ in $\mathcal{H}^{\delta_m}$ and $a_m \in I^{\delta_m}_{(\omega,\mathbf{h}^m)}$ for $m \in \mathbb{N}$ such that

$$\mathrm{dist}\left(a_m, I_{(\omega,\mathbf{0})}\right) \geq \varepsilon_0. \tag{10.10}$$

Note that the component sets $I_{(\omega,\mathbf{h})}$ and A_q^δ are contained in the ball $B[0; R_0]$ where $R_0 = 3\sqrt{c_0/\alpha_0}$ for each $\omega \in \Omega$. Then $a_m \in A^{\delta_m}_{(p,\mathbf{h}^m)} \subset B[0; R_0]$ for each for $m \in \mathbb{N}$ and $B[0; R_0]$ is compact, so there exists a convergent subsequence (again we use the original index) $a_m \to a_* \in B[0; R_0]$. Thus we have

$$\mathrm{dist}\left(a_*, I_{(\omega,\mathbf{0})}\right) \geq \varepsilon_0.$$

Now choose $t > 0$ sufficiently large so that by pullback attraction

$$\mathrm{dist}\left(\varphi\left(t, \sigma_{-t}p, B[0; R_0]\right), I_{(\omega,\mathbf{0})}\right) < \frac{1}{2}\varepsilon_0. \tag{10.11}$$

By the invariance property of a pullback attractor there exist $b_m \in A^{\delta_m}_{\Theta_{-N(t,\mathbf{h}^m)}(\omega,\mathbf{h}^m)}$ such that

$$\psi\left(N(t, \mathbf{h}^m), \Theta_{-N(t,\mathbf{h}^m)}(p, \mathbf{h}^m), b_m\right) = a_m. \tag{10.12}$$

Since the $A^{\delta_m}_{\Theta_{-N(t,\mathbf{h}^m)}(\omega,\mathbf{h}^m)} \subset B[0; R_0]$ for each for $m \in \mathbb{N}$, there exists a convergent subsequence (once again we use the original index) $b_m \to b_* \in B[0; R_0]$. By (10.12) and Lemma 10.4 we have

$$\varphi\left(t, \sigma_{-t}\omega, b_*\right) = a_*,$$

which contradicts (10.10) with respect (10.11). This contradiction proves the upper semi–continuous convergence of the numerical pullback attractor components.

10.4.4   *Upper semi–continuous convergence of the discretized global attractors*

Let $J \subseteq \mathbb{R}^d \times \Omega$ be the global attractor of the continuous time skew product flow dynamical system $\pi(t, u, \omega) = (\varphi(t, u, \omega), \sigma_t\omega)$ and $J^\delta \subset \mathcal{H}^\delta \times \mathbb{R}^d \times \Omega$ the global attractor of the discrete time semi–dynamical system $\pi^\delta(n, \mathbf{h}, u, \omega) =$

$(\Theta_n(\mathbf{h}, \omega), \psi^\delta(n, x, (\mathbf{h}, \omega))$ based on the numerical scheme, where we include the superscript δ on π^δ and ψ^δ for emphasis. Then J^δ converges to $\mathcal{A}$ as $\delta \to 0$ uniformly in the sense that

$$\lim_{\delta \to 0} \sup_{(\mathbf{h}^\delta, x^\delta, \omega^\delta) \in J^\delta} \beta\left(\left(p^\delta, x^\delta\right), J\right) = 0. \tag{10.13}$$

Suppose that (10.13) is not true. Then there exist sequences $\delta_n \to 0$ and $\left(\mathbf{h}^{\delta_n}, x^{\delta_n}, \omega^{\delta_n}\right) \subset J^{\delta_n}$ and an $\varepsilon_0 > 0$ such that

$$\beta\left(\left(x^{\delta_n}, \omega^{\delta_n}\right), \mathcal{A}\right) \geq \varepsilon_0.$$

Let B be a absorbing compact set in $\mathbb{R}^d$ that is independent of δ. Since $J^\delta \subset \mathcal{H}^\delta \times B \times \Omega$ and $\mathcal{H}^\delta \times B \times \Omega$ is compact, we can select a subsequence (we use the same index for convenience) such that $\omega^{\delta_n} \to \omega^0$ and $x^{\delta_n} \to x^0$ as $n \to \infty$. Hence

$$\beta\left(\left(x^0, \omega^0\right), J\right) \geq \varepsilon_0. \tag{10.14}$$

On the other hand, since B is compact and the absorption is uniform in $\omega \in \Omega$, there exists a $t > 0$ such that

$$\beta\left(\pi(t, B, \Omega), J\right) < \frac{1}{2}\varepsilon_0.$$

In addition, by invariance, $\pi^\delta(j, J^\delta) = J^\delta$ for all $j \in \mathbb{N}$. Now choose for j the integer $N(t, \mathbf{h}^{\delta_n})$, where $N(t, \mathbf{h})$ is defined in Lemma 10.4. Then we can find a $\left(\hat{\mathbf{h}}^{\delta_n}, \hat{\omega}^{\delta_n}, \hat{x}^{\delta_n}\right) \in \mathcal{H}^\delta \times B \times \Omega$ such that

$$\pi^{\delta_n}\left(N(t, \mathbf{h}^{\delta_n}), \hat{\mathbf{h}}^{\delta_n}, \hat{x}^{\delta_n}\right), \hat{\omega}^{\delta_n} = \left(\mathbf{h}^{\delta_n}, x^{\delta_n}\right), \omega^{\delta_n};$$

indeed we can define $\left(\hat{\mathbf{h}}^{\delta_n}, \hat{\omega}^{\delta_n}\right)$ by $\Theta_{-N(t, \mathbf{h}^{\delta_n})}(\mathbf{h}^{\delta_n}, \omega^{\delta_n})$. By a similar compactness argument, there is subsequence of this subsequence (again we use the original index) such that $\left(\hat{x}^{\delta_n}, \hat{p}^{\delta_n}\right) \to \left(\hat{x}^0, \hat{\omega}^0\right) \in B \times \Omega$. By Lemma 10.4 we then have

$$x^{\delta_n} = \psi^{\delta_n}\left(N(t, \mathbf{h}^{\delta_n}), \left(\hat{\mathbf{h}}^{\delta_n}, \hat{x}^{\delta_n}, \hat{\omega}^{\delta_n}\right)\right) \to \varphi\left(t, \hat{x}^0, \hat{\omega}^0\right) = x^0$$

while

$$\Theta_{N(t, \mathbf{h}^{\delta_n})}\left(\hat{\mathbf{h}}^{\delta_n}, \hat{\omega}^{\delta_n}\right) = \left(\mathbf{h}^{\delta_n}, \omega^{\delta_n}\right) \to \left(\mathbf{0}, \omega^0\right) = \left(\mathbf{0}, \sigma_t \hat{\omega}^0\right).$$

Combining these results we have

$$\pi^{\delta_n}\left(N(t, \mathbf{h}^{\delta_n}), \hat{\mathbf{h}}^{\delta_n}, \hat{x}^{\delta_n}, \hat{\omega}^{\delta_n}\right) \to \left(\mathbf{0}, \pi\left(t, \hat{x}^0, \hat{\omega}^0\right)\right) = \left(\mathbf{0}, x^0, \omega^0\right)$$

and hence $\beta\left(\left(x^0, \omega^0\right), J\right) < \frac{1}{2}\varepsilon_0$, which contradicts (10.14). Hence the original assertion (10.13) must be true. This completes the proof of the main theorem, Theorem 10.2. $\qquad\square$

10.5   Singleton set-valued pullback attractor case

Let $u_1(t)$ and $u_2(t)$ be two solutions of the differential equation (10.3) with the same initial parameter ω but different initial values in the positively absorbing ball $B[0; R_0]$ and write

$$\Delta(t) = u_1(t) - u_2(t), \qquad \Delta_{f,\omega}(t) = f(\sigma_t\omega, u_1(t)) - f(\sigma_t\omega, u_2(t)).$$

Let L_0 be the local Lipschitz constant of f in $B[0; R_0]$, which by assumption is uniform in $\omega \in \Omega$, so

$$|f(\omega, u_1(t)) - f(\omega, u_2(t))| \leq L_0 \, |u_1(t) - u_2(t)|$$

or $|\Delta_{f,\omega}(t)| \leq L_0|\Delta(t)|$ in $B[0; R_0]$. We assume that $L_0 \leq \alpha_0/2$. Then similarly to earlier (but now we do not use the inner product inequality on the f)

$$\frac{d}{dt}|\Delta(t)|^2 = 2\left(\frac{d}{dt}\Delta(t), \Delta(t)\right)$$

$$= 2\left(A(\sigma_t\omega)\Delta(t), \Delta(t)\right) + 2\left(\Delta_{f,\omega}(t), \Delta(t)\right)$$

$$\leq -2\alpha_0\,|\Delta(t)|^2 + 2|\Delta_{f,\omega}(t)||\Delta(t)|$$

$$\leq -2\alpha_0\,|\Delta(t)|^2 + 2L_0|\Delta(t)|^2$$

$$\leq -2\left(\alpha_0 - L_0\right)|\Delta(t)|^2 \leq -\alpha_0\,|\Delta(t)|^2$$

and so

$$|\Delta(t)| \leq |\Delta(0)|e^{-(\alpha_0/2)t},$$

which means the solution operator $u \mapsto \phi(t, \omega, u)$ is a contraction mapping on the ball $B[0; R_0]$ for each $t > 0$ and $\omega \in \Omega$.

Now consider the numerical scheme. Let

$$y_1 = u_1 + hF(h, \omega, u_1), \qquad y_2 = u_2 + hF(h, \omega, u_2)$$

where $h \in [\delta/2, \delta]$ and $u_1, u_2 \in B[0; R_0]$. Write

$$\Delta x = u_1 - u_2, \qquad \Delta y = y_1 - y_2, \qquad \Delta_F(h) = F(h, \omega, u_1) - F(h, \omega, u_2),$$

so $|\Delta F| \leq L_{00}|\Delta x|$ in $B[0; R_0]$, where L_{00} is the local Lipschitz constant of F in u on $B[0; R_0]$, uniformly in $(h, \omega) \in [0, 1] \times \Omega$. We assume the Lipschitz consistency condition **N4** here, so

$$|\Delta_F(h) - A\Delta x - \Delta_{f,\omega}(h)| \leq \bar{\mu}_R(h)|\Delta x|$$

where we omit the parameter ω for convenience. Thus we have

$$|\Delta y|^2 = (\Delta y, \Delta y) = (\Delta x + h\Delta_F(h), \Delta x + h\Delta_F(h))$$

$$= (\Delta x, \Delta x) + 2h(\Delta_F(h), \Delta x) + h^2(\Delta_F(h), \Delta_F(h))$$

$$= |\Delta x|^2 + 2h(A\Delta x + \Delta_{f,\omega}(h), \Delta x) + h^2|\Delta_F(h)|^2$$
$$+ 2h(\Delta_F(h) - A\Delta x - \Delta_{f,\omega}(h), \Delta x)$$

$$\leq |\Delta x|^2 - 2h\alpha_0|\Delta x|^2 + 2h|\Delta_{f,\omega}(h)|\,|\Delta x| + h^2 L_{00}^2|\Delta x|^2$$
$$+ 2h|\Delta_F(h) - A\Delta x - \Delta_{f,\omega}(h)||\Delta x|$$

$$\leq |\Delta x|^2(1 - 2h\alpha_0) + 2hL_0\,|\Delta x|^2 + h^2 L_{00}^2|\Delta x|^2$$
$$+ 2h\bar{\mu}_R(h)|\Delta x|^2$$

$$\leq |\Delta x|^2\left(1 - 2h(\alpha_0 - L_0) + h^2 L_{00}^2 + 2h\bar{\mu}_R(h)\right)$$

$$\leq |\Delta x|^2\left(1 - h\alpha_0 + h^2 L_{00}^2 + 2h\bar{\mu}_R(h)\right),$$

where we have used the assumption that $L_0 \leq \alpha_0/2$. By further restricting h from above we can assure that

$$|\Delta y|^2 \leq |\Delta x|^2\left(1 - h\alpha_0/2\right) \leq |\Delta x|^2\gamma(\delta\alpha_0/4)$$

for $h \in [\delta/2, \delta]$ and δ sufficiently small. This means that the numerical solution satisfies the contractive condition

$$\left|\psi(n, u_0, (\bar{h}, \omega)) - \psi(n, \bar{x}_0, (\bar{h}, \omega))\right| \leq |u_0 - \bar{x}_0|\gamma_0^n$$

for all $u_0, \bar{x}_0 \in B[0; R_0]$, $\omega \in \Omega$ and step-size sequence $\bar{h} \in \mathcal{H}^\delta$, where $\gamma_0 := \sqrt{\gamma(\delta\alpha_0/4)}$.

From the Contraction Mapping Principal we conclude that the original and numerical pullback attractors each consist of a single trajectory. The continuity of their component elements with respect to the parameter follows from Theorem 10.2 and the fact that upper semi–continuity there reduces to continuity for the singleton set-valued mappings.

Theorem 10.3 *Let the Assumptions* **D1–D4** *and* **N1–N4** *hold and suppose that $2L_0 \leq \alpha_0$ and that δ is sufficiently small. Then the pullback attractors of Theorem 10.2 consist of singleton component sets, that is $I_\omega = \{a^*(\omega)\}$ and $I_{(\mathbf{h},\omega)}^\delta = \{a_\delta^*(\mathbf{h}, \omega)\}$, where the mappings $\omega \mapsto a^*(\omega)$ and $(\mathbf{h}, \omega) \mapsto a_\delta^*(\mathbf{h}, \omega)$ are continuous.*

These singleton valued pullback attractor–trajectories inherit the periodicity or almost periodicity of the differential equation and of the differential equation and step-size sequence, respectively. This is formulated in the following theorem, the proof of which will be presented in the remainder of this section. The periodic case is straightforward, while the almost periodic case is considerably more complicated and requires the introduction of appropriate definitions and a number of auxiliary results.

Recall that the set $A \subset P$ is called minimal with respect to a dynamical system $(\Omega, \mathbb{R}, \sigma)$ if it is nonempty, closed and invariant and if no proper subset of A has these properties.

Theorem 10.4   *Suppose that the assumptions of Theorem 10.3 hold and that Ω is minimal. Then the singleton valued pullback attractor–trajectory $I_\omega = \{a^*(\omega)\}$ is periodic (resp., almost periodic) if $\omega \in \Omega$ is periodic (resp., almost periodic), whereas the numerical singleton valued pullback attractor–trajectory $I^\delta_{(\mathbf{h},\omega)} = \{a^*_\delta(\mathbf{h},\omega)\}$ is periodic (resp., almost periodic) if $q = (\mathbf{h},\omega) \in \mathcal{Q}^\delta$ is periodic (resp., almost periodic).*

Definition 10.1   A sequence $\mathbf{h} = \{h_n\}_{n \in \mathbb{Z}}$ is m–periodic if $h_{n+m} = h_n$ for all $n \in \mathbb{Z}$ or, equivalently, if $\tilde{\sigma}_m \mathbf{h} = \mathbf{h}$, where m is the smallest integer for which these equalities hold.

Recall that we have defined a time sequence $\{t_n(\mathbf{h})\}_{n \in \mathbb{Z}}$ by $t_0(\mathbf{h}) = 0$, $t_n(\mathbf{h}) := \sum_{j=0}^{n-1} h_j$ and $t_{-n}(\mathbf{h}) := -\sum_{j=1}^{n} h_{-j}$ for $n \geq 1$ corresponding to a given sequence $\mathbf{h} = \{h_n\}_{n \in \mathbb{Z}}$.

Lemma 10.5   *Let $\mathbf{h} \in \mathcal{H}^\delta$ be m–periodic and let $\omega \in \Omega$ be τ–periodic with respect to σ, that is with $\sigma_\tau \omega = \omega$ where $\tau \in \mathbb{R}^+$. Then the point $(\mathbf{h},\omega) \in \mathcal{Q}^\delta = \mathcal{H}^\delta \times \Omega$ is periodic with respect to $\Theta = (\tilde{\sigma}, \sigma)$ if and only if $t_m(\mathbf{h})/\tau$ is rational.*

Proof.   Suppose that $t_m(\mathbf{h})/\tau = k/l$ for some $k, l \in \mathbb{N}$. Then $l t_m(\mathbf{h}) = k\tau$ and $\Theta_{lm}(\mathbf{h},\omega) = (\mathbf{h}, \sigma_{l t_m(\mathbf{h})}\omega) = (\mathbf{h}, \sigma_{k\tau}\omega) = (\mathbf{h},\omega)$. On the other hand, suppose that $\Theta_k(\mathbf{h},\omega) = (\mathbf{h},\omega)$ for some $k \in \mathbb{N}$. Then $(\tilde{\sigma}_k \mathbf{h}, \sigma_{t_k(\mathbf{h})}\omega) = (\mathbf{h},\omega)$, which implies that $k = l_1 m$ and $t_k(\mathbf{h}) = l_1 t_m(\mathbf{h}) = l_1 t_m(\mathbf{h}) = l_2 \tau$ where $l_1, l_2 \in \mathbb{N}$. Hence $t_m(\mathbf{h})/\tau = l_1/l_2$. $\square$

The following result can be found in [32, 304].

Theorem 10.5   *Let $(X, \mathbb{T}, \pi)$ be a dynamical system on a compact metric space (X, ρ). Then a point $x \in X$ is almost periodic if and only if for every $\varepsilon > 0$ there* *exists a $\delta = \delta(\varepsilon) > 0$ such that*

$$\rho\left(\pi(t+t_1,x), \pi(t+t_2,x)\right) < \varepsilon, \qquad \text{for all } t \in \mathbb{T},$$

whenever $\rho\left(\pi(t_1,x), \pi(t_2,x)\right) < \delta$.

Definition 10.2 A sequence $\{c_n\}$ in $\mathbb{R}$ is said to be *almost periodic* if the function $\varphi : \mathbb{Z} \to \mathbb{R}$ defined by $\varphi(n) := c_n$ for $n \in \mathbb{Z}$ is almost periodic.

Definition 10.3 A sequence $\{\tau_n\}$ in $\mathbb{R}$ will be called *regular* if it has the form

$$\tau_n = an + c_n, \qquad \text{for all } n \in \mathbb{Z},$$

where $a \in \mathbb{R}$ is a constant and $\{c_n\}$ is an almost periodic sequence; see Samoilenko and Trofimchuk [280, 281].

Theorem 10.6 *Suppose that the time sequence $\{t_n(\mathbf{h})\}_{n\in\mathbb{Z}}$ corresponding to $\mathbf{h} \in \mathcal{H}^\delta$ is regular. In addition, suppose that dynamical system $(\Omega, \mathbb{R}, \sigma)$ is minimal and almost periodic, that is Ω is minimal and every point $\omega \in \Omega$ is almost periodic. Then the point $(\mathbf{h}, \omega) \in \mathcal{Q}^\delta$ is almost periodic for the dynamical system $(\mathcal{Q}^\delta, \mathbb{Z}, \Theta)$.*

Proof. Since the point $\omega \in \Omega$ is almost periodic for the dynamical system $(\Omega, \mathbb{R}, \sigma)$, then by Theorem 10.5 the point $\omega \in \Omega$ will be almost periodic relatively to the discrete time dynamical system $(\Omega, \mathbb{Z}, \sigma^{(a)})$, where $\sigma^{(a)} = \{\sigma_{an}\}_{n\in\mathbb{Z}}$ and $t_n(\mathbf{h}) = an + c_n$ is the regularity representation of $\{t_n(\mathbf{h})\}_{n\in\mathbb{Z}}$. We apply Theorem 10.5 to the dynamical system $(\Omega, \mathbb{R}, \sigma)$. Given $\varepsilon > 0$, let $\delta(\varepsilon) \in (0, \varepsilon/3)$ be such that

$$\rho(\sigma_t p_1, \sigma_t p_2) < \frac{\varepsilon}{3} \tag{10.15}$$

for every $t \in \mathbb{R}$ and $\omega_1, \omega_2 \in \Omega$ with $\rho(\omega_1, \omega_2) < \delta$. Then we use uniform continuity on the compact space Ω: given the above $\delta(\varepsilon) > 0$, let $\gamma(\varepsilon) \in (0, \delta(\varepsilon))$ be such that

$$\rho(\sigma_s \omega, \omega) < \delta \tag{10.16}$$

for every $\omega \in \Omega$ and $s \in \mathbb{R}$ with $|s| \leq \gamma$.

Now for this $\gamma(\varepsilon) > 0$ we denote by $\mathcal{M}_{\gamma(\varepsilon)}$ the relatively dense subset of $\mathbb{Z}$ subset for which

$$\rho(\tilde{\sigma}_{n+m}\mathbf{h}, \tilde{\sigma}_n\mathbf{h}) < \gamma(\varepsilon), \quad |c_{n+m} - c_n| < \gamma(\varepsilon), \quad \rho(\sigma_{a(n+m)}\omega, \sigma_{an}\omega) < \gamma(\varepsilon) \tag{10.17}$$

for all $m \in \mathcal{M}_\varepsilon$, $n \in \mathbb{Z}$ and $\omega \in \Omega$. From (10.15)–(10.17) we have

$$\rho(\Theta_{n+m}(\mathbf{h}, \omega), \Theta_n(\mathbf{h}, \omega)) = \rho(\tilde{\sigma}_{n+m}\mathbf{h}, \tilde{\sigma}_n\mathbf{h}) + \rho(\sigma_{t_{n+m}(\mathbf{h})}\omega, \sigma_{t_n(\mathbf{h})}\omega)$$

$$= \rho(\tilde{\sigma}_{n+m}\mathbf{h}, \tilde{\sigma}_n\mathbf{h}) + \rho(\sigma_{a(n+m)+c_{n+m}}\omega, \sigma_{an+c_n}\omega) < \rho(\tilde{\sigma}_{n+m}\mathbf{h}, \tilde{\sigma}_n\mathbf{h})$$

$$+\rho(\sigma_{a(n+m)}(\sigma_{c_{n+m}}\omega), \sigma_{a(n+m)}(\sigma_{c_n}\omega)) \quad + \rho(\sigma_{a(n+m)}(\sigma_{c_n}\omega), \sigma_{an}(\sigma_{c_n}\omega))$$

$$< \gamma(\varepsilon) + \frac{\varepsilon}{3} + \gamma(\varepsilon) < \frac{\varepsilon}{3} + \frac{\varepsilon}{3} + \frac{\varepsilon}{3} = \varepsilon$$

for all $n \in \mathbb{Z}$ and $m \in \mathcal{M}_{\gamma(\varepsilon)}$. Hence the point $(\mathbf{h}, p) \in \mathcal{Q}^\delta$ is almost periodic for the dynamical system $(\mathcal{Q}^\delta, \mathbb{Z}, \Theta)$. $\qquad\square$

Corollary 10.1 *Let* $\mathbf{h} \in \mathcal{H}^\delta$ *be* m*–periodic and let* $\omega \in \Omega$ *be almost periodic for the dynamical system* $(\Omega, \mathbb{R}, \sigma)$. *Then the point* $(\mathbf{h}, \omega) \in \mathcal{Q}^\delta$ *is almost periodic for the dynamical system* $(\mathcal{Q}^\delta, \mathbb{Z}, \Theta)$. *In particular, if* $t_m(\mathbf{h})/\tau$ *is irrational, then point* $(\mathbf{h}, p)$ *is almost periodic, but not periodic.*

As the final step in our proof of Theorem 10.4, we need the following lemma, which we prove directly here noting that the result also follows from Theorems 1 and 2 in [276].

Lemma 10.6 *Suppose that the assumptions of Theorem 10.3 hold and let* $\{a_\delta^*(q)\}_{q \in \mathcal{Q}^\delta}$ *denote the singleton valued pullback attractor for the numerical scheme (10.4). Then the function* $n \mapsto a_\delta^*(\Theta_n q)$ *for* $n \in \mathbb{Z}$ *is periodic (resp., almost periodic) if the point* q *is periodic (resp., almost periodic) for the dynamical system* $(\mathcal{Q}^\delta, \mathbb{Z}, \Theta)$.

Proof. Let $q \in \mathcal{Q}^\delta$ be m–periodic, that is $\Theta_m q = q$. Then $a_\delta^*(\Theta_{n+m} q) = a_\delta^*(\Theta_n \Theta_m q)$ $= a_\delta^*(\Theta_n q)$ for every $n \in \mathbb{Z}$. Hence $n \mapsto a_\delta^*(\Theta_n q)$ is periodic.

The function $a_\delta^* : \mathcal{Q}^\delta \to \mathbb{R}^d$ defined by $q \mapsto a_\delta^*(q)$ for each $q \in \mathcal{Q}^\delta$ is continuous, hence uniformly continuous, on the compact space $\mathcal{Q}^\delta$. That is, for every $\varepsilon > 0$ there exists a $\delta(\varepsilon) > 0$ such that $|a_\delta^*(q_1) - a_\delta^*(q_2)| < \varepsilon$ whenever $\rho_{\mathcal{Q}^\delta}(q_1, q_2) < \delta$. Now let the point q be almost periodic and for $\delta = \delta(\varepsilon) > 0$ denote by M_δ the relatively dense subset of $\mathbb{Z}$ such that $\rho(\Theta_{n+m} q, \Theta_n q) < \delta$ for all $m \in M_\delta$ and $n \in \mathbb{Z}$. From this and the uniform continuity we have

$$|a_\delta^*(\Theta_{n+m} q) - a_\delta^*(\Theta_n q)| < \varepsilon$$

for all $n \in \mathbb{Z}$ and $m \in M_{\delta(\varepsilon)}$. Hence $n \mapsto a_\delta^*(\Theta_n q)$ is almost periodic. $\qquad\square$

In conclusion, we can restate the assertions of Theorem 10.4 in more detail as follows.

Corollary 10.2 *Suppose that the assumptions of Theorem 10.3 hold and that Ω is minimal. In addition, let $\{a_\delta^*(\mathbf{h}, \omega)\}_{(\mathbf{h},\omega)\in\mathcal{Q}^\delta}$ denote the singleton valued pullback attractor for the numerical scheme (10.4).*

1. Let $\mathbf{h} \in \mathcal{H}^\delta$ be m–periodic and $\omega \in \Omega$ be τ–periodic. Then $n \mapsto a_\delta^*(\Theta_n(\mathbf{h}, \omega)\}$ is periodic if $t_m(\mathbf{h})/\tau$ is rational and almost periodic if $t_m(\mathbf{h})/\tau$ is irrational.

2. Let the time sequence $\{t_n(\mathbf{h})\}_{n\in\mathbb{Z}}$ corresponding to $\mathbf{h} \in \mathcal{H}^\delta$ be regular and let the point $\omega \in \Omega$ be almost periodic. Then $n \mapsto a_\delta^*(\Theta_n(\mathbf{h}, \omega)\}$ is almost periodic.

10.6 Appendix: Proof of Lemma 10.4

Consider an autonomous dynamical system on a compact metric space Ω described by a group $\sigma = \{\sigma_t\}_{t\in\mathbb{R}}$ of mappings of Ω into itself such that the mapping $(t, \omega) \mapsto \sigma_t \omega$ is continuous. Consider also an ordinary differential equation

$$\dot{u} = g(u, \omega t) \quad (\omega \in \Omega)$$

on $\mathbb{R}^d$ with a unique solution $\varphi(t, u_0, \omega)$ satisfying the initial value problem

$$\frac{d}{dt}\varphi(t, u_0, \omega) = g(\varphi(t, u_0, \omega), \sigma_t \omega), \qquad x(0, u_0, \omega) = u_0.$$

We assume that $(u, \omega) \mapsto g(u, \omega)$ is continuous on $\mathbb{R}^d \times \Omega$ and locally Lipschitz in u uniformly in ω, that is for each $R > 0$ there exists an L_R such that

$$|g(x, \omega) - g(y, \omega)| \le L_R|x - y|, \qquad \forall x, y \in B[0; R].$$

In particular, then for a sequence of times t_n and time steps $h_n = t_{n+1} - t_n$ this gives in integral equation form

$$\varphi(t_{n+1}, u_0, \omega) = \varphi(t_n, u_0, \omega) + \int_{t_n}^{t_{n+1}} g(\sigma_\tau \omega, \varphi(\tau, u_0, \omega)) \, d\tau.$$

In future we just write $x(t)$ for this solution. By the Mean Value Theorem there exists $\tau_n \in [0, 1]$ such that

$$\varphi(t_{n+1}, u_0, \omega) = \varphi(t_n, u_0, \omega) + x(t_{n+1}) = x(t_n) +$$
$$h_n \, g(\sigma_{t_n + \tau_n h_n} \omega, \varphi(t_n + \tau_n h_n, u_0, \omega)).$$

The corresponding higher order scheme solution is

$$u_{n+1} = u_n + h_n \, F(h_n, \sigma_{t_n} \omega, u_n),$$

where the increment function $F(h, u, \omega)$ is continuous and satisfies the consistency condition

$$F(0, u, \omega) = g(u, \omega), \qquad \forall\, x, \omega.$$

Let $x(t) := \varphi(t, u_0, \omega)$, then

$$x(t_{n+1}) - u_{n+1} = x(t_n) - u_n +$$
$$h_n\left[g(x(t_n + \tau_n h_n), \sigma_{t_n + \tau_n h_n}\omega) - F(h_n, \sigma_{t_n}\omega, u_n)\right]$$

so the global discretization error $E_n := |x(t_n) - u_n|$ is estimated by

$$E_{n+1} \le E_n + h_n\left|g(x(t_n + \tau_n h_n), \sigma_{t_n + \tau_n h_n}\omega) - F(h_n, \sigma_{t_n}\omega, u_n)\right|$$

$$\le E_n + h_n\left|g(x(t_n + \tau_n h_n), \sigma_{t_n + \tau_n h_n}\omega) - g(x(t_n), \sigma_{t_n + \tau_n h_n}\omega)\right|$$
$$+ h_n\left|g(x(t_n), \sigma_{t_n + \tau_n h_n}\omega) - g(x(t_n), \sigma_{t_n}\omega)\right|$$
$$+ h_n\left|g(x(t_n), \sigma_{t_n}\omega) - g(u_n, \sigma_{t_n}\omega)\right|$$
$$+ h_n\left|g(u_n, \sigma_{t_n}\omega) - F(h_n, \sigma_{t_n}\omega, u_n)\right|$$

$$\le E_n + h_n L_R\left|x(t_n + \tau_n h_n) - x(t_n)\right| + h_n \omega_g(\Delta_n\omega; R)$$
$$+ h_n L_R\left|x(t_n) - u_n\right| + h_n \omega_F(h_n\omega; R)$$

$$= (1 + h_n L_R)E_n + h_n L_R\left|\int_{t_n}^{t_n + \tau_n h_n} g(x(s), \sigma_s\omega)\, ds\right|$$
$$+ h_n \omega_g(\Delta_n\omega; R) + h_n \omega_F(h_n\omega; R)$$

$$\le (1 + h_n L_R)E_n + h_n^2 L_R M_R + h_n \omega_g(\Delta_n\omega; R) + h_n \omega_F(h_n\omega; R)$$

where $M_R := \max_{\omega\in\Omega, x\in B[0;R]} |g(u, \omega)|$ and $\omega_g(\delta; R)$ is the modulus of continuity of $g(\sigma.\omega, x)$ uniformly in $\omega \in \Omega$ and $u \in B[0; R]$ and $\omega_F(h_n\omega; R)$ is the modulus of continuity of $F(\cdot, u, \omega)$ uniformly in $\omega \in \Omega$ and $u \in B[0; R]$ that is

$$\omega_g(\delta; R) := \sup_{0\le t\le\delta}\ \sup_{\omega\in\Omega,\ x\in B[0;R]} |g(\sigma_t\omega, x) - g(\omega, x)|$$

and

$$\omega_F(\delta; R) := \sup_{0\le h\le\delta}\ \sup_{\omega\in\Omega,\ x\in B[0;R]} |F(h, u, \omega) - F(0, u, \omega)|.$$

Here $\omega_g(\delta; R) \to 0$ and $\omega_F(\delta; R) \to 0$ as $\delta \to 0$.

Now we consider an interval $[0, T]$ and restrict to step-sizes $h_n \in [\delta/2, \delta]$ for some $\delta > 0$. Note then that $t_{n+1} = \sum_{j=0}^{n} h_j$ then satisfies $n\delta/2 \le t_n \le n\delta$ with $t_n \le T$, which means $n\delta \le 2T$ for these choices of n. The above difference inequality thus satisfies

$$E_{n+1} \le (1 + L_R\delta)E_n + \delta\left(\omega_g(\delta; R) + \omega_F(\delta; R) + L_R M_R \delta\right)$$

and hence with $E_0 = 0$ yields

$$E_n \le \delta\left(\omega_g(\delta; R) + \omega_F(\delta; R) + L_R M_R \delta\right) \frac{(1 + L_R\delta)^n - 1}{(1 + L_R\delta) - 1}$$

$$\le \left(\omega_g(\delta; R) + \omega_F(\delta; R) + L_R M_R \delta\right) \frac{1}{L_R} e^{L_R n\delta}$$

$$\le \left(\omega_g(\delta; R) + \omega_F(\delta; R) + L_R M_R \delta\right) \frac{1}{L_R} e^{2L_R T},$$

that is

$$|x(t_n) - u_n| \le \left(\omega_g(\delta; R) + \omega_F(\delta; R) + L_R M_R \delta\right) \frac{1}{L_R} e^{2L_R T}.$$

Hence for $t \in (t_n, t_{n+1})$, we have

$$|x(t) - u_n| \le |x(t) - x(t_n)| + |x(t_n) - u_n| \le$$

$$\left| \int_{t_n}^{t} g(\sigma_s\omega, x(s))\, ds \right| + \left(\omega_g(\delta; R) + \omega_F(\delta; R) + L_R M_R \delta\right) \frac{1}{L_R} e^{2L_R T} \le$$

$$\delta M_R + \left(\omega_g(\delta; R) + \omega_F(\delta; R) + L_R M_R \delta\right) \frac{1}{L_R} e^{2L_R T}.$$

Let us now consider variable parameters and initial values. let $\omega_j \to \omega$ in Ω and $u_{0j} \to u_0$ in $\mathbb{R}^d$. Let $\varphi(t, u_0, \omega)$ and $u_n(u_0, \omega)$, etc, denote the corresponding solutions. By continuity in initial conditions and parameters uniformly on a compact time interval $[0, T]$, we have

$$\varphi(t, u_{0j}, \omega_j) \to \varphi(t, u_0, \omega),$$

as $j \to \infty$ for $t \in [0, T]$.

Combining all of these partial results for $t \in (t_n, t_{n+1})$ we obtain

$$|u_n(u_{0j}, \omega_j) - x(t, u_0, \omega)| \le |u_n(u_{0j}, \omega_j) - x(t, u_{0j}, \omega_j)| +$$

$$|x(t, u_{0j}, \omega_j) - x(t, u_0, \omega)| \le \delta M_R + ((\omega_g(\delta; R) + \omega_F(\delta; R) +$$

$$L_R M_R \delta) \frac{1}{L_R} e^{2L_R T} + |x(t, u_{0j}, \omega_j) - x(t, u_0, \omega)|$$

which converges to zero as the maximum step-size δ converges to zero and j tends to ∞.

Chapter 11

Global attractors of non-autonomous Navier-Stokes equations

This chapter is devoted to the study of non-autonomous Navier-Stokes equations. It is proved that such systems admit compact global attractors. This problem is formulated and solved in the terms of general non-autonomous dynamical systems. We give conditions of convergence of non-autonomous Navier-Stokes equations . A test of existence of almost periodic (quasi periodic, recurrent, pseudo recurrent) solutions of non-autonomous Navier-Stokes equations is given. We prove the global averaging principle for non-autonomous Navier-Stokes equations.

We consider the two-dimensional Navier-Stokes system

$$u' + q(t) \sum_{i=1}^{2} u_i \partial_i u = \nu \Delta u - \nabla p + \phi(t)$$
$$div\ u = 0, \quad u|_{\partial D} = 0, \tag{11.1}$$

where D is an open bounded set with boundary $\partial D \in C^2$. This equation can be written in the following form

$$u' + Au + B(t)(u, u) = f(t) \tag{11.2}$$

on the corresponding Sobolev's space E, where $-A$ is a Stokes operator, $B(t)$ is a bilinear form satisfying the identity

$$Re\langle B(t)(u, v), w \rangle = -Re\langle B(t)(u, w), v \rangle \tag{11.3}$$

for all $t \in \mathbb{R}$ and $u, v, w \in E$, and f is forcing term.

In the work $[112],[140],[199],[200]$ there is studied a non-stationary equation (11.2), when f is a function of time $t \in \mathbb{R}$. It is shown that the equation with compact f (in particularly, almost periodic) admits a compact global attractor and also for small nonlinear (bilinear) term it was proved the existence a unique almost periodic (quasi periodic, periodic) solution of equation (11.2) if the forcing term f is almost periodic (quasi periodic, periodic).

The aim of the chapter is to study the equation (11.2) in the case, when the the bilinear operator B, and the function f are non-stationary. The conditions under

357

which a non-stationary equation of type (11.2) admits a compact global attractor are indicated.

The theorem of "partial" averaging on finite interval for ordinary differential equations it was proved in the work [183]. The works [140],[199] and [200] are devoted to generalization of method of averaging for dissipative partial differential equations. We prove the theorem of "partial" averaging for non-autonomous Navier-Stokes equation (11.2) (i.e. the bilinear form and forcing term are non-stationaries).

This chapter is organized as follows:

In Section 1 we introduce a class of non-autonomous Navier-Stokes equations and establish its dissipativity (Theorem 11.4).

In Section 2 we prove that non-autonomous Navier-Stokes equations admit a compact global attractor (Theorem 11.6).

Section 3 is devoted to study of the problem of existence of almost periodic (quasi periodic, recurrent, pseudo recurrent) solutions of non-autonomous Navier-Stokes equations (Corollaries 11.3) and we give the conditions of convergence of this equations (Theorem 11.8).

In Section 4 we prove the uniform averaging principle for the non-autonomous Navier-Stokes equations on the finite segment (Theorem 11.9).

Section 5 is devoted to prove the global averaging principle for non-autonomous Navier-Stokes equations on the semi-axis (Theorems 11.10,11.11 and 11.12).

11.1   Non-autonomous Navier-Stokes equations

Some results from the theory of semigroups of linear operators [188] and PDEs [199], [294],[314] are collected below.

A closed operator A with domain $D(A)$ that is dense in a Banach space X is called a sectorial operator if for some $a \in \mathbb{R}$ and $\varphi \in (0, \frac{\pi}{2})$ the sector

$$S_{a,\varphi} := \{\lambda \in \mathbb{C}, \pi \geq |arg(\lambda - a)| \geq \varphi\} \tag{11.4}$$

is contained in the resolvent set and for $\lambda \in S_{a,\varphi}$

$$\|(\lambda I - A)^{-1}\|_{X \to X} \leq \frac{c}{|\lambda - a| + 1}. \tag{11.5}$$

For a sectorial operator A the analytic semigroup of linear bounded operators in X is defined and denoted by e^{-At}, $t \geq 0$.

Let A be a sectorial operator with $Re\sigma(A) > 0$. For $\alpha \in (0, 1)$ we define fractional powers of A as follows:

$$A^{\alpha} := (A^{-\alpha})^{-1}, \text{ where } A^{-\alpha} := \frac{1}{\Gamma(\alpha)} \int_0^{\infty} t^{\alpha-1} e^{-At} dt.$$

The corresponding domains $D(A^\alpha)$ are Banach spaces with norm given by

$$|\cdot|_\alpha := |\cdot|_{D(A^\alpha)} = |A^\alpha \cdot|.$$

Theorem 11.1 *The following estimates are valid:*

(1)

$$\|e^{-At}\|_{X\to X} \le Ce^{-at}, \quad t \ge 0, \tag{11.6}$$

(2)

$$\|A^\alpha e^{-At}\|_{X\to X} \le C_\alpha t^{-\alpha}e^{-at}, \quad t > 0. \tag{11.7}$$

Let Ω be a compact metric space, $\mathbb{R} = (-\infty, +\infty), (\Omega, \mathbb{R}, \sigma)$ be a dynamical system on Ω, $\mathcal{E}$ be a real or complex Hilbert space, $L(\mathcal{E})$ be the space of all linear continuous operators on $\mathcal{E}$, $L^2(\mathcal{E})$ be the space of all bilinear continuous operators $B : \mathcal{E} \times \mathcal{E} \to \mathcal{F}$ and $C(\Omega, W)$ be a space of all continuous functions $f : \Omega \to W$ (W is some metric space), endowed with the topology of uniform convergence. Let us consider the equation

$$u' + Au + B(\omega t)(u, u) = f(\omega t), \tag{11.8}$$

$(\omega \in \Omega)$ where $\omega t := \sigma(t, \omega), B \in C(\Omega, L^2(\mathcal{E})), f \in C(\Omega, \mathcal{E})$ and A is a linear operator.

Below we will use some notions, denotations and results from [200]. Let Hilbert spaces E, F, X satisfy $E \subset F$; $E, F, X \subset \mathcal{E}$, each embedding being dense and continuous.

Operator A. We further suppose that the linear operator A is densely defined in $\mathcal{E}$ and such that the linear equation

$$u' + Au = 0 \tag{11.9}$$

generates the c_0-semigroup of linear bounded operators

$$e^{-At} : \mathcal{E} \to \mathcal{E}, \quad \varphi(t, x) := e^{-At}x,$$

which for $t > 0$ can be extended to the linear bounded operators from F to E satisfying the following estimates

$$\|e^{-At}\|_{E\to E} \le Ke^{-at}, \tag{11.10}$$

$$\|e^{-At}\|_{F\to E} \le Kt^{-\alpha_1}e^{-at}, \quad 0 \le \alpha_1 < 1, \tag{11.11}$$

$$\|Ae^{-At}\|_{F\to E} \le Kt^{-\alpha_2}e^{-at}, \quad 0 \le \alpha_2 < 2. \tag{11.12}$$

We also suppose that the following condition is satisfied

$$Ae^{At} = e^{At}A, \tag{11.13}$$

in the sense of $L(F, E) := \{A : F \to E \mid A \text{ is linear and bounded }\}$ equipped with the operationel norm.

Bilinear operator B. Denote by $L^2(E, F)$ the space of all bilinear continuous operators $B : E \times E \to F$ with the norm

$$\|B\| := \sup\{|B(u, v)|_F \; : \; |u|_E \leq 1, |v|_E \leq 1\}.$$

Let $C(\Omega, L^2(E, F))$ be a space of all continuous mappings $B : \Omega \to L^2(E, F)$ and

$$C_B := \sup\{|B(\omega)(u, v)|_F \; : \; \omega \in \Omega, \; |u|_E \leq 1, |v|_E \leq 1\},$$

then the mapping $F : \Omega \times E \to F$ $(F(\omega, u) := B(\omega)(u, u))$ satisfies the following inequality

$$|B(\omega)(u_1, u_1) - B(\omega)(u_2, u_2)|_F \leq C_B(|u_1|_E + |u_2|_E)|u_1 - u_2|_E \tag{11.14}$$

for all $u_1, u_2 \in E$.

From the inequality (11.14) follows that on every ball $B[0, R] := \{u \in E \; : \; |u|_E \leq R\}$ we have

$$|B(\omega)(u_1, u_1) - B(\omega)(u_2, u_2)|_F \leq 2C_B R|u_1 - u_2|_E \tag{11.15}$$

for all $u_1, u_2 \in E$.

Remark 11.1 *The space of all the bilinear continuous operators $C(\Omega, L^2(E, F))$ is a Banach space with the norm $\|B\| := C_B$.*

Function f. The external force $f : \Omega \to X$ is continuous, i.e. $f \in C(\Omega, X)$.

Operators e^{-At}. The operators e^{-At} $(t > 0)$ can be extended to the linear bounded operators from X to E satisfying the estimates

$$\|e^{-At}\|_{X \to E} \leq Kt^{-\beta_1}e^{-at}, \; 0 \leq \beta_1 < 1, \tag{11.16}$$

$$\|Ae^{-At}\|_{X \to E} \leq Kt^{-\beta_2}e^{-at}, \; 0 \leq \beta_2 < 2, \tag{11.17}$$

and the equation (11.13), this time in the sense of $L(X, E)$.

We suppose that the following conditions are fulfilled:

(1) there exists $\alpha > 0$ such that

$$Re\langle Au, u \rangle \geq \alpha|u|_E^2 \tag{11.18}$$

for all $u \in E$; ;

(2)

$$Re\langle B(\omega)(u,v),w\rangle = -Re\langle B(\omega)(u,w),v\rangle \tag{11.19}$$

for every $u,v,w \in E$ and $\omega \in \Omega$.

Remark 11.2   *a. It follows from (11.19) that*

$$Re\langle B(\omega)(u,v),v)\rangle = 0 \tag{11.20}$$

for every $u,v \in E$ and $\omega \in \Omega$.
   b.

$$|B(\omega)(u,v)|_F \le C_B|u|_E|v|_E \tag{11.21}$$

for all $u,v \in E$ and $\omega \in \Omega$, where $C_B := \sup\{|B(\omega)(u,v)|_F : \omega \in \Omega,\ u,v \in E,\ |u|_E \le 1,\ and\ |v|_E \le 1\}$.

The equation (11.8) with conditions (11.18) and (11.19) is called a non-autonomous Navier-Stokes equation. We will consider the mild solutions of the equation (11.8), i.e. $u \in C([0,T],E)$ and satisfy the following integral equation

$$u(t) = e^{-At}x + \int_0^t e^{-A(t-s)}(-B(\omega s)(u(s),u(s)) + f(\omega s))ds. \tag{11.22}$$

Theorem 11.2   *Let $x_0 \in E$, $r > 0$ and the conditions (11.10),(11.11) and (11.18) are fulfilled, then there exist positive numbers $\delta = \delta(x_0,r)$ and $T = T(x_0,r)$ such that the equation (11.22) admits a unique solution $\varphi(t,x,\omega)$ ($x \in B[x_0,\delta] := \{x \in E \mid |x - x_0| \le \delta\}$) defined on the interval $[0,T]$ with the conditions: $\varphi(0,x,\omega) = x$, $|\varphi(t,x,\omega) - x_0| \le r$ for all $t \in [0,T]$ and the mapping $\varphi : [0,T] \times B[x_0,\delta] \times \Omega \to E$ $((t,x,\omega) \to \varphi(t,x,\omega))$ is continuous.*

Proof.   Let $x_0 \in E$, $r > 0$, $\delta > 0$ and $T > 0$. We consider a space $C_{x_0,r,\delta,T}$ of all continuous functions $\psi : [0,T] \times B[x_0,\delta] \times \Omega \to B[x_0,r]$ equipped with the distance

$$d(\psi_1,\psi_2) := \sup\{|\psi_1(t,x,\omega) - \psi(t,x,\omega)|_E : 0 \le t \le T, x \in B[x_0,\delta], \omega \in \Omega\}$$

is a complete metric space.
   We define the operator Φ acting onto $C_{x_0,r,\delta,T}$ by the equality

$$(\Phi\psi)(t,x,\omega) = e^{-At}x + \int_0^t e^{-A(t-s)}(-B(\omega s)(\psi(s,x,\omega),\psi(s,x,\omega)) + f(\omega s))ds.$$

There exist $\delta_1 = \delta_1(x_0, r) > 0$ and $T_1 = T_1(x_0, r) > 0$ such that $\Phi C_{x_0, r, \delta, T} \subseteq C_{x_0, r, \delta, T}$ for all $\delta \in (0, \delta_1]$ and $T \in (0, T_1]$. In fact,

$$|(\Phi\psi)(t, x, \omega) - x_0|_E \le |e^{-At}x - x_0|_E +$$

$$|\int_0^t e^{-A(t-s)} B(\omega s)(\psi(s, x, \omega), \psi(s, x, \omega)))ds|_E + |\int_0^t e^{-A(t-s)} f(\omega s)ds|_E \le$$

$$m(\delta, T) + \int_0^t Ke^{-a(t-s)}(t-s)^{-\alpha_1}|\psi(s, x, \omega)|_E^2 ds +$$

$$\int_0^t Ke^{-a(t-s)}(t-s)^{-\beta_1}\|f\|ds \le m(\delta, T) + K(|x_0|_E + r)^2 \frac{T^{1-\alpha_1}}{1-\alpha_1}$$

$$+K\|f\|\frac{T^{1-\beta_1}}{1-\beta_1} := d_1(x_0, r, \delta, T) \to 0$$

as $\delta + T \to 0$, where $m(\delta, T) := \sup\{|e^{-tA}x - x_0|_E \ : \ t \in [0, T], x \in B[x_0, r]\}$ and $\|f\| := \sup\{|f(\omega)|_X \ : \ \omega \in \Omega\}$. Thus there exist $\delta_1 = \delta_1(x_0, r) > 0$ and $T_1 = T_1(x_0, r) > 0$ such that $d_1(x_0, r, \delta, T) \le r$ for all $\delta \in (0, \delta_1]$ and $T \in (0, T_1]$.

Let now $\psi_1, \psi_2 \in C_{x_0, r, \delta, T}$, then

$$|(\Phi\psi_1)(t, x, \omega)) - (\Phi\psi_2)(t, x, \omega))|_E =$$

$$|\int_0^t [B(\omega s)(\psi_1(s, x, \omega), \psi_1(s, x, \omega)) - B(\omega s)(\psi_2(s, x, \omega), \psi_2(s, x, \omega))]|_E \le$$

$$2C_B(|x_0|_E + r)Td(\psi_1, \psi_2)$$

and, consequently, $d(\Phi\psi_1, \Phi\psi_2) \le L(x_0, r, T)d(\psi_1, \psi_2)$, where $L(x_0, r, T) := 2C_B(|x_0| + r)T \to 0$ as $T \to 0$. Thus there exists $T_2 = T_2(x_0, r) > 0$ such that $L(x_0, r, T) < 1$ for all $T \in (0, T_2]$. Denote by $\delta(x_0, r) := \delta_1(x_0, r)$ and $T(x_0, r) := min(T_1(x_0, r), T_2(x_0, r))$, then the mapping $\Phi : C_{x_0, r, \delta, T} \to C_{x_0, r, \delta, T}$ is a contraction and, consequently, there exists a unique function $\varphi \in C_{x_0, r, \delta, T}$ satisfying the equation (11.22) on the interval $[0, T]$. The theorem is proved. $\square$

Remark 11.3 *The theorem 11.2 is true and for the equation*

$$u' + Au = \mathcal{F}(\omega t, u)$$

if the continuous function $\mathcal{F} : \Omega \times E \to F$ satisfies the following conditions:

(1)

$$\sup\{|\mathcal{F}(\omega, 0)|_E \ : \ \omega \in \Omega\} < \infty$$

(Ω ,generally speaking, is not compact);
(2) F is locally Lipschitz, i.e. for every $r > 0$ there exists $L(r) > 0$ such that

$$|\mathcal{F}(\omega, u_1) - \mathcal{F}(\omega, u_2)|_F \le L(r)|u_1 - u_2|_E$$

for all $u_1, u_2 \in E$ with condition: $|u_i|_E \le r \ (i = 1, 2)$.

Theorem 11.3 *Let $\mathcal{K}$ be a family of solutions of equation (11.22) satisfying the following condition: there exists a positive constant M such that $|x(t)|_{\mathcal{D}(A)} \leq M$ for all $t \in \mathbb{R}_+$ ($|x|_{\mathcal{D}(A)} := |Ax|_E$). If there exists $\tilde{C}_B > 0$ such that*

$$|B(\omega)(u,v)|_F \leq \tilde{C}_B |u|_{\mathcal{D}(A)} |v|_{\mathcal{D}(A)}$$

for all $u, v \in \mathcal{D}(A)$, then this family of functions is uniform equicontinuous on $\mathbb{R}_+$, i.e. for every $\varepsilon > 0$ there exists $\delta(\varepsilon) > 0$ such that $|t_1 - t_2| < \delta$ implies $|x(t_1) - x(t_2)| < \varepsilon$ for all $t_1, t_2 \in \mathbb{R}_+$ and $x \in \mathcal{K}$.

Proof. Let $\psi \in \mathcal{K}$ and $x := \psi(0)$, then $\psi(t) = \varphi(t, x, \omega)$ for all $t \in \mathbb{R}_+$ and we have

$$|\varphi(t, x, \omega) - x|_E \leq |e^{-At}x - x|_E + |\int_0^t e^{-A(t-\tau)}(-B(\omega s)(\varphi(s, x, \omega),$$

$$\varphi(s, x, \omega)) + f(\omega s))ds|_E \leq \int_0^t e^{-as}s^{-\alpha_1}|x|_{\mathcal{D}(A)}ds +$$

$$\int_0^t e^{-as}(t-s)^{-\alpha_1}C_b|\varphi(s, x, \omega)|^2_{\mathcal{D}(A)}ds + \int_0^t e^{-a(t-s)}(t-s)^{-\beta_1}\|f\|ds \leq$$

$$\frac{t^{1-\alpha_1}}{1-\alpha_1}M + C_B M^2 \frac{t^{1-\alpha_1}}{1-\alpha_1} + \|f\|\frac{t^{1-\beta_1}}{1-\beta_1}.$$

From the last inequality we obtain

$$\sup\{|\varphi(t, x, \omega) - x|_E \mid |x|_{\mathcal{D}(A)} \leq M, \ \omega \in \Omega\} \to 0$$

as $t \to 0$ and, consequently,

$$|\varphi(t_2, x, \omega) - \varphi(t_1, x, \omega)|_E = |\varphi(t_2 - t_1, \varphi(t_1, x, \omega), \omega t_1) - \varphi(t_1, x, \omega)|_E \leq$$

$$\sup\{|\varphi(t_2 - t_1, x, \omega) - x|_E \ : \ |x|_{\mathcal{D}(A)} \leq M, \ \omega \in \Omega\} \to 0$$

as $t_2 - t_1 \to 0$. The theorem is proved. $\qquad\square$

Example 11.1 We consider the two-dimensional Navier-Stokes system

$$u' + q(t)\sum_{i=1}^{2} u_i \partial_i u = \nu\Delta u - \nabla p + \phi(t) \tag{11.23}$$

$$div \ u = 0, \quad u|_{\partial D} = 0,$$

where D is an open bounded set with smooth boundary $\partial D \in C^2$.

The functional setting of the problem is well known [228],[311] . We denote by H and V the closure of the linear space $\{u \in C_0^\infty(D)^2, \ div \ u = 0\}$ in $L^2(D)^2$ and $H_0^1(D)^2$, respectively. Denote by P the corresponding orthogonal projection $P : L_2(D)^2 \to H$. We further set

$$A := -\nu P\Delta, \ B(t)(u,v) := q(t)P(\sum_{i=1}^{2} u_i\partial_i v).$$

The Stokes operator A is self-adjoint positive with domain $\mathcal{D}(A)$ dense in H. The inverse operator is compact. We define the Hilbert spaces $\mathcal{D}(A^\alpha)$, $\alpha \in (0,1]$ as the domains of the powers of A in the standard way. Furthermore, $V := \mathcal{D}(A^{1/2})$, and $|u|_{\mathcal{D}(A^{1/2})} = |\nabla u|$.

Applying P we write (11.23) as the evolution equation of the following form

$$u' + Au + \mathcal{B}(t)(u,u) = \mathcal{F}(t), \quad \mathcal{F}(t) := P\phi(t). \tag{11.24}$$

Let $\mathcal{F} \in C(\mathbb{R},H)$ ($X := H$) and $\mathcal{B} \in C(\mathbb{R},L^2(H,\mathcal{D}(A^{-\delta})))$ ($F := \mathcal{D}(A^{-\delta})$). Denote by $Y := C(\mathbb{R},H) \times C(\mathbb{R},L^2(H,\mathcal{D}(A^{-\delta})))$ and $(Y,\mathbb{R},\sigma)$ a dynamical system of translations (Bebutov's dynamical system, see for example, [300],[302] and [292]). Let $\Omega := H(\mathcal{B},\mathcal{F}) = \overline{\{(\mathcal{B}_\tau,\mathcal{F}_\tau)|\ \tau \in \mathbb{R}\}}$, where $\mathcal{B}_\tau(t) := \mathcal{B}(t+\tau)$ (respectively $\mathcal{F}_\tau(t) := \mathcal{F}(t+\tau)$) for all $t \in \mathbb{R}$, by bar we denote a closure in the compact-open topology and $(\Omega,\mathbb{R},\sigma)$ be a dynamical system of translations on Ω.

Along with the equation (11.24) we consider its H-class

$$u' + Au + \tilde{\mathcal{B}}(t)(u,u) = \tilde{\mathcal{F}}(t), \tag{11.25}$$

where $(\tilde{\mathcal{B}},\tilde{\mathcal{F}}) \in H(\mathcal{B},\mathcal{F})$. Let $B : \Omega \to L^2(H,\mathcal{D}(A^{-\delta}))$ (respectively $f : \Omega \to H$) be a mapping defined by equality

$$B(\omega) = B(\tilde{\mathcal{B}},\tilde{\mathcal{F}}) := \tilde{\mathcal{B}}(0)\ (f(\omega) = f(\tilde{\mathcal{B}},\tilde{\mathcal{F}}) := \tilde{\mathcal{F}}(0)),$$

where $\omega = (\tilde{\mathcal{B}},\tilde{\mathcal{F}}) \in \Omega$, then the equation (11.24) and its H-class can be written in the form (11.21).

We now set in the notation above $E := \mathcal{D}(A^{1/2})$, $X := H$, $F := \mathcal{D}(A^{-\delta})$ and see that (11.10)-(11.12), (11.18) and (11.16)-(11.17) are valid with $\alpha_1 = 1/2 + \delta$, $\beta_1 = 1/2$, $\beta_2 = 3/2$.

According to Theorem 11.2 through every point $x \in H$ passes a unique solution $\varphi(t,x,\omega)$ of equation (11.8) at the initial moment $t = 0$. And this solution is defined on some interval $[0,t_{(x,\omega)})$. Let us note, that

$$\begin{aligned}
w'(t) &= 2Re\langle \varphi'(t,x,\omega),\varphi(t,x,\omega)\rangle = 2Re\langle A(\omega t)\varphi(t,x,\omega),\varphi(t,x,\omega)\rangle \\
&\quad +2Re\langle B(\omega t)(\varphi(t,x,\omega),\varphi(t,x,\omega)),\varphi(t,x,y)\rangle + 2Re\langle f(\omega t),\varphi(t,x,\omega)\rangle \\
&= 2Re\langle A(\omega t)\varphi(t,x,\omega),\varphi(t,x,\omega)\rangle + 2Re\langle f(\omega t),\varphi(t,x,\omega)\rangle \\
&\leq -2\alpha|\varphi(t,x,\omega)|_E^2 + 2\|f\||\varphi(t,x,\omega)|_E,
\end{aligned} \tag{11.26}$$

where $\|f\| := \max\{|f(\omega)|_X : \omega \in \Omega\}$ and $w(t) = |\varphi(t,x,\omega)|_E^2$. Then

$$w' \leq -2\alpha w + 2\|f\|w^{\frac{1}{2}} \tag{11.27}$$

and consequently

$$w(t) \leq v(t) \tag{11.28}$$

for all $t \in [0, t_{(x,\omega)})$, where $v(t)$ is an upper solution of equation

$$v' = -2\alpha v + 2\|f\|v^{\frac{1}{2}}, \tag{11.29}$$

satisfying condition $v(0) = w(0) = |x|^2$. Thus

$$v(t) = [(|x|_E - \frac{\|f\|}{\alpha})e^{-\alpha t} + \frac{\|f\|}{\alpha}]^2 \tag{11.30}$$

and consequently

$$|\varphi(t,x,\omega)|_E \leq (|x|_E - \frac{\|f\|}{\alpha})e^{-\alpha t} + \frac{\|f\|}{\alpha} \tag{11.31}$$

for all $t \in [0, t_{(x,\omega)})$. It follows from the inequality (11.27) that solution $\varphi(t,x,\omega)$ is bounded and therefore it may be prolonged on $\mathbb{R}_+ = [0, +\infty)$.

Thus we have proved the following theorem.

Theorem 11.4 *Let the conditions (11.18) and (11.19) are fulfilled. Then the following statements hold:*

(i) Every solution $\varphi(t,x,\omega)$ of non-autonomous Navier-Stokes equation (11.8) is bounded and therefore it may be prolonged on $\mathbb{R}_+$.

(ii)

$$|\varphi(t,x,\omega)|_E \leq C(|x|_E), \tag{11.32}$$

for all $t \geq 0$, $\omega \in \Omega$ and $x \in E$, where $C(r) = r$ if $r \geq r_0 := \frac{\|f\|}{\alpha}$ and $C(r) = r_0$ if $r \leq r_0$;

(iii)

$$\limsup_{t \to +\infty} \sup\{|\varphi(t,x,\omega)|_E \ : \ |x|_E \leq r, \omega \in \Omega\} \leq \frac{\|f\|}{\alpha} \tag{11.33}$$

for every $r > 0$.

Lemma 11.1 *Under the conditions of Theorem 11.4 we have*

$$\int_t^{t+l} |\varphi(\tau,x,\omega)|_E^2 d\tau \leq \frac{r^2}{2\alpha} + \frac{r}{\alpha}l\|f\| := M(r) \tag{11.34}$$

for all $t \geq 0$ and $r \geq r_0$.

Proof. From the equality (11.27) after integration in t between t and $t+l$ we obtain

$$2\alpha \int_t^{t+l} |\varphi(\tau,x,\omega)|_E^2 d\tau \leq |\varphi(t,x,\omega)|_E^2 + 2rl\|f\| \tag{11.35}$$

and,consequently,

$$\int_t^{t+l} |\varphi(\tau, x, \omega)|_E^2 d\tau \le \frac{r^2}{2\alpha} + \frac{r}{\alpha} l \|f\| := M(r). \tag{11.36}$$

$\square$

Lemma 11.2 *([314, Ch.3]) (The Uniform Gronwall Lemma). Let g, h, y, be three positive locally integrable functions on $]t_0, \infty[$ such that y' is locally integrable on $]t_0, \infty[$, and which satisfy*

$$y' \le gy + h \text{ for } t \ge t_0,$$
$$\int_t^{t+l} g(s)ds \le a_1, \quad \int_t^{t+l} h(s)ds \le a_2, \quad \int_t^{t+l} y(s)ds \le a_3 \text{ for } t \ge t_0,$$

where l, a_1, a_2, a_3, are positive constants. Then

$$y(t + l) \le (\frac{a_3}{l} + a_2)e^{a_1} \ \forall \ t \ge t_0.$$

Theorem 11.5 *Under the conditions of Theorem 11.4 if*

$$|\langle B(\omega)(u, v), w \rangle| \le C|u|^{1/2}|Au|^{1/2}|v|_{1/2}|w| \tag{11.37}$$
$$\forall \ u \in \mathcal{D}(A), \ v \in V, \ w \in H,$$

then

$$|\varphi(t, x, \omega)|_{\mathcal{D}(A)} \le K(r) \ \forall |x| \le r \ (r \ge r_0) \tag{11.38}$$

for all $t \ge 0$ and $\omega \in \Omega$, where $K(r)$ is some positive constant depending only on r.

Proof. Since

$$\langle A\varphi(t, x, \omega), \varphi(t, x, \omega) \rangle = \frac{1}{2} \frac{d}{dt} |\varphi(t, x, \omega)|_E^2 \tag{11.39}$$

by taking the scalar product of (11.8) with Au we find

$$\frac{1}{2} \frac{d}{dt} |\varphi(t, x, \omega)|_E^2 + |A\varphi(t, x, \omega)|_E^2 + \tag{11.40}$$
$$\langle B(\varphi(t, x, \omega), \varphi(t, x, \omega)), A\varphi(t, x, \omega) \rangle = \langle f(\omega t), A\varphi(t, x, \omega) \rangle.$$

Taking into account the inequality

$$|\langle f(\omega t), A\varphi(t, x, \omega) \rangle| \le |f(\omega t)|_E |A\varphi(t, x, \omega)|_E \le$$
$$\frac{1}{4}|A\varphi(t, x, \omega)|^2 + \|f\|^2 \tag{11.41}$$

and using (11.39) and the Young inequality we obtain

$$
\begin{aligned}
&|\langle B(\omega)(\varphi(t,x,\omega),\varphi(t,x,\omega)),A\varphi(t,x,\omega)\rangle| \leq \\
&c_1|\varphi(t,x,\omega)|^{1/2}\|\varphi(t,x,\omega)\|\,|A\varphi(t,x,\omega)|^{3/2} \leq \\
&\frac{1}{4}|A\varphi(t,x,\omega)|^2 + c_1'|\varphi(t,x,\omega)|^2\|\varphi(t,x,\omega)\|^4.
\end{aligned}
\tag{11.42}
$$

Hence

$$
\frac{1}{2}\frac{d}{dt}\|\varphi(t,x,\omega)\|^2 + |A\varphi(t,x,\omega)|^2 \leq \|f\|^2 + \frac{1}{2}|A\varphi(t,x,\omega)|^2 +
$$
$$
c_1'|\varphi(t,x,\omega)|^2\|\varphi(t,x,\omega)\|^4.
\tag{11.43}
$$

From this inequality according to Gronwal lemma we can prove that $|\varphi(t,x,\omega)|_{\mathcal{D}(A)}$ is uniformly (w.r.t. x and ω) bounded on interval $[0,l]$. Applying the uniform Gronwal lemma with g,h,y replaced by

$$
c_1'|\varphi(t,x,\omega)|^2\|\varphi(t,x,\omega)\|^2 \;,\; \|f\|^2 \;,\; \|\varphi(t,x,\omega)\|^2
\tag{11.44}
$$

we obtain that $\|\varphi(t,x,\omega)\|^2$ is bounded on $[l,\infty[$ and, consequently, it is bounded on $[0,\infty[$ uniformly w.r.t. $\|x\| \leq r$ and $\omega \in \Omega$. The theorem is proved. $\square$

11.2 Attractors of non-autonomous dynamical systems

Lemma 11.3 *The non-autonomous Navier-Stokes equation (11.8) generates a cocycle φ (or more explicit $\langle E,\varphi,(\Omega,\mathbb{R},\sigma)\rangle$), where $\varphi(t,x,\omega)$ is a unique solution of equation (11.8) defined on $\mathbb{R}_+$ with the initial condition $\varphi(0,x,\omega)=x$.*

Proof. In fact, according to Theorems 11.2 the mapping $\varphi : \times E \times \Omega \to E$ $((t,x,\omega) \to \varphi(t,x,\omega))$ is continuous and in view of uniqueness of solution $\varphi(t,x,\omega)$ we have the following identity: $\varphi(t+\tau,x,\omega) = \varphi(t,\varphi(\tau,x,\omega),\omega\tau)$ for all $t,\tau \in \mathbb{R}_+, x \in E$ and $\omega \in \Omega$, where $\omega\tau := \sigma(\tau,\omega)$. $\square$

Example 11.2 Let E be a Banach space and $C(\mathbb{R} \times E,E)$ be a space of all continuous functions $F : \mathbb{R} \times E \to E$ equipped with the compact-open topology. Let us consider a parameterized differential equation

$$
\frac{dx}{dt} + Ax = F(\sigma_t\omega,x) \quad (\omega \in \Omega)
$$

on a Banach space E with $\Omega := C(\mathbb{R} \times E,E)$, where $\sigma_t\omega := \sigma(t,\omega)$ and the linear operator A is densely defined in E and such that the linear equation

$$
u' + Au = 0
$$

generates the c_0-semigroup of linear bounded operators

$$e^{-At} : E \to E, \quad \varphi(t, x) := e^{-At}x.$$

We will define $\sigma_t : \Omega \to \Omega$ by $\sigma_t \omega(\cdot, \cdot) = \omega(t + \cdot, \cdot)$ for each $t \in \mathbb{R}$ and interpret $\varphi(t, x, \omega)$ as mild solution of the initial value problem

$$\frac{d}{dt}x(t) + Ax = F(\sigma_t\omega, x(t)), \quad x(0) = x. \tag{11.45}$$

Under appropriate assumptions on $F : \Omega \times E \to E$ (or even $F : \mathbb{R} \times E \to E$ with $\omega(t)$ instead of $\sigma_t\omega$ in (11.45)) to ensure forwards existence and uniqueness, then φ is a cocycle on $(C(\mathbb{R} \times E, E), \mathbb{R}, \sigma)$ with fiber E, where $(C(\mathbb{R} \times E, E), \mathbb{R}, \sigma)$ is a Bebutov's dynamical system (see for example [32],[102],[292]) [300].

Definition 11.1   Recall that the cocycle $\langle W, \varphi, (Y, \mathbb{R}, \sigma) \rangle$ is called compact (asymptotically compact) if the skew-product dynamical system $(X, \mathbb{R}_+, \pi)$ $(X = W \times Y, \pi = (\varphi, \sigma))$ is compact (respectively asymptotic compact).

Let $(X, \mathbb{R}_+, \pi)$ be compact dissipative and K be a compact set, which attracts all compact subsets of X. Suppose

$$J = \Omega(K), \tag{11.46}$$

where $\Omega(K) = \bigcap_{t \geq 0} \overline{\bigcup_{\tau \geq t} \pi^\tau K}$. The set J is called the Levinson's center of the compact dissipative system $(X, \mathbb{R}_+, \pi)$.

Applying general theorems about non-autonomous dissipative systems (see Chapter 2) to non-autonomous system constructed in the example 11.2, we will obtain series of facts concerning the equation (11.8). In particular, from Theorems 11.4, 2.19 and 2.20 follows the theorem below.

Theorem 11.6   *Let Ω be a compact metric space, $(\Omega, \mathbb{R}, \sigma)$ be a dynamical system on Ω and the conditions (11.18) and (11.19) are fulfilled. If the cocycle φ generated by non-autonomous Navier-Stokes equation is asymptotically compact, then for every $\omega \in \Omega$ there exists a non-empty compact and connected $I_\omega \subset E$ such that the following conditions hold:*

(1) the set $I = \cup\{I_\omega \;:\; \omega \in \Omega\}$ is compact and connected in E;
(2) I is connected if Ω is connected;
(3)

$$\lim_{t \to +\infty} \sup_{\omega \in \Omega} \beta(U(t, \omega^{-t})M, I) = 0$$

for any bounded set $M \subset E$, where $U(t, \omega) = \varphi(t, \cdot, \omega)$ and β is the semi-distance of Hausdorff;
(4) $U(t, \omega)I_\omega = I_{\omega t}$ for all $t \in \mathbb{R}_+$ and $\omega \in \Omega$;

(5) I_ω consists of those and only those points $x \in E$ through which passes the solution of the equation (11.8) bounded on $\mathbb{R}$.

Theorem 11.7 *Under conditions of Theorem 11.6*

$$|\varphi(t, x, \omega)| \leq \frac{\|f\|}{\alpha}$$

for all $t \in \mathbb{R}$, $\omega \in \Omega$ and $x \in I_\omega$, where φ is a cocycle, generated by non-autonomous Navier-Stokes equation.

Proof. The set $J = \bigcup \{I_\omega \times \{\omega\} : \omega \in \Omega\}$ is a Levinson's center of dynamical system $(X, \mathbb{R}_+, \pi)$ and according to (11.46) for any point $(u_0, y_0) = z \in J$ there exists $t_n \to +\infty, u_n \in E$ and $\omega_n \in \Omega$ such that the sequence $\{u_n\}$ is bounded, $u_0 = \lim\limits_{n \to +\infty} \varphi(t_n, u_n, \omega_n)$ and $\omega_0 = \lim\limits_{n \to +\infty} \omega_n t_n$. From the inequality (11.27) follows that $|u_0| \leq \frac{\|f\|}{\alpha}$, i.e. $\varphi(t, x, \omega) \in I_{\omega t}$ for all $\omega \in \Omega$ and $t \geq 0$, hence $|\varphi(t, x, \omega)| \leq \frac{\|f\|}{\alpha}$ for any $t \in \mathbb{R}, x \in I_\omega$ and $\omega \in \Omega$. The theorem is proved. $\qquad\Box$

11.3 Almost periodic and recurrent solutions of non-autonomous Navier-Stokes equations

Definition 11.2 An autonomous dynamical system $(\Omega, \mathbb{T}, \sigma)$ is said to be pseudo recurrent if the following conditions are fulfilled:

a) Ω is compact;

b) $(\Omega, \mathbb{T}, \sigma)$ is transitive, i.e. there exists a point $\omega_0 \in \Omega$ such that $\Omega = \overline{\{\omega_0 t \mid t \in \mathbb{T}\}}$;

c) every point $\omega \in \Omega$ is stable in the sense of Poisson, i.e.

$$\mathfrak{N}_\omega = \{\{t_n\} \mid \omega t_n \to \omega \text{ and } |t_n| \to +\infty\} \neq \emptyset.$$

Lemma 11.4 *([105]) Let $\langle (X, \mathbb{T}_1, \pi), (\Omega, \mathbb{T}_2, \sigma), h \rangle$ be a non-autonomous dynamical system and the following conditions are fulfilled:*

1) $(\Omega, \mathbb{T}_2, \sigma)$ is pseudo recurrent;

2) $\gamma \in C(\Omega, X)$ is an invariant section of the homomorphism $h : X \to \Omega$, i.e. $h(\gamma(\omega)) = \omega$ and $\gamma(\sigma(t, \omega)) = \pi(t, \gamma(\omega))$ for all $\omega \in \Omega$ and $t \in \mathbb{T}_2$.

Then the autonomous dynamical system $(\gamma(\Omega), \mathbb{T}_2, \pi)$ is pseudo recurrent too.

Definition 11.3 The solution $\varphi(t, x, \omega)$ of non-autonomous Navier-Stokes equation (11.8) is called recurrent (pseudo recurrent, almost periodic, quasi periodic), if the point $(x, \omega) \in H \times \Omega$ is a recurrent (pseudo recurrent, almost periodic, quasi periodic) point of skew-product dynamical system $(X, \mathbb{R}_+, \pi)$ $(X = H \times \Omega$ and $\pi = (\varphi, \sigma))$.

We note (see, for example, [238] and [300],[302]) that if $\omega \in \Omega$ is a stationary (τ-periodic, almost periodic, quasi periodic, recurrent) point of dynamical system $(\Omega, \mathbb{R}, \sigma)$ and $h : \Omega \to X$ is a homomorphism of dynamical system $(\Omega, \mathbb{R}, \sigma)$ onto $(X, \mathbb{R}_+, \pi)$, then the point $x = h(\omega)$ is a stationary (τ-periodic, almost periodic, quasi periodic, recurrent) point of the system $(X, \mathbb{R}_+, \pi)$.

Let $X := H \times \Omega$ and $\pi := (\varphi, \sigma)$, then mapping $h : \Omega \to X$ is a homomorphism of dynamical system $(\Omega, \mathbb{R}, \sigma)$ onto $(X, \mathbb{R}_+, \pi)$ if and only if $h(\omega) = (u(\omega), \omega)$ for all $\omega \in \Omega$, where $u : \Omega \to H$ is a continuous mapping with the condition that $u(\omega t) = \varphi(t, u(\omega), \omega)$ for all $\omega \in \Omega$ and $t \in \mathbb{R}_+$.

The following affirmations hold:

Lemma 11.5   *Let Ω be a compact metric space, A be a linear operator densely defined in E such that the equation*

$$x' + Ax = 0$$

generates a c_0-semigroup $\{U(t)\}_{t \geq 0}$. If the condition (11.18) is fulfilled, then

$$\|U(t)\| \leq \exp(-\alpha t)$$

for all $t \in \mathbb{R}_+$, where $U(t)$ is a Cauchy operator of equation (11.48).

Proof. Let $\varphi(t, x) := U(t)x$, then according to the inequality 11.18 we have

$$\frac{d}{dt}|\varphi(t, x)|^2 \leq -2\alpha|\varphi(t, x)|^2$$

and, consequently,$|\varphi(t, x)| \leq \exp(-\alpha t)|x|$ for all $x \in H$ and $t \in \mathbb{R}_+$. Thus we have $|U(t)x| \leq \exp(-\alpha t)|x|$, therefore $\|U(t)\| \leq \exp(-\alpha t)$ for all $t \in \mathbb{R}_+$. $\qquad\square$

Lemma 11.6   *Suppose that the condition (11.18) is fulfilled. Then for every function $f \in C(\Omega, H)$ there exists a unique function $\gamma \in C(\Omega, H)$ defined by equality*

$$\gamma(\omega) = \int_{-\infty}^{0} U(-\tau)f(\omega\tau)d\tau$$

such that

$$\gamma(\omega t) = \varphi(t, \gamma(\omega), \omega) \tag{11.48}$$

for every $\omega \in \Omega$ and $t \in \mathbb{R}_+$, where $\varphi(t, x, \omega)$ is a solution of equation

$$u' = Au + f(\omega t)$$

with the initial condition $\varphi(0, x, \omega) = x$ *and the following inequality*

$$\|\gamma\| \leq \frac{1}{\alpha}\|f\|$$

takes place.

Proof. The formulated statement results from Lemma 11.5 and Proposition 7.33 from [47]. $\qquad\square$

Lemma 11.7 *Let Ω be a compact metric space, the cocycle φ, generated by the non-autonomous Navier-Stokes equation (11.8) and $\alpha^{-2}\|f\|C_B < 1$, then the following inequality*

$$|\varphi(t, x_1, \omega) - \varphi(t, x_2, \omega)| \leq e^{-(\alpha - C_B \frac{\|f\|}{\alpha})t}|x_1 - x_2|$$

$$(x_1, x_2 \in B[0, r_0], r_0 := \frac{\|f\|}{\alpha}, \omega \in \Omega \ and \ t \in \mathbb{R}_+)$$

takes place.

Proof. Let $r_0 := \frac{\|f\|}{\alpha}$ and $x_1, x_2 \in B[0, r_0] := \{x \in E \ : \ |x| \leq r_0\}$. According to Theorem 11.4 we have $|\varphi(t, x_i, \omega)| \leq r_0$ for all $t \geq 0, \omega \in \Omega$ and $i = 1, 2$. Denote by $\psi(t) := \varphi(t, x_1, \omega) - \varphi(t, x_2, \omega)$, then we obtain

$$\frac{d}{dt}|\psi(t)|^2 = 2Re\langle A\psi(t), \psi(t)\rangle + 2Re\langle B(\omega t)(\psi(t), \varphi(t, x_2, \omega)), \psi(t)\rangle \leq$$

$$-2\alpha|\psi(t)|^2 + 2C_B|\varphi(t, x_2, \omega)||\psi(t)|^2 \leq$$

$$-2\alpha|\psi(t)|^2 + 2C_B\frac{\|f\|}{\alpha}|\psi(t)|^2 = -2(\alpha - C_B\frac{\|f\|}{\alpha})|\psi(t)|^2$$

and, consequently,

$$|\psi(t)|^2 \leq e^{-2(\alpha - C_B\frac{\|f\|}{\alpha})t}|\psi(0)|^2.$$

Thus we have

$$|\varphi(t, x_1, \omega) - \varphi(t, x_2, \omega)| \leq e^{-(\alpha - C_B\frac{\|f\|}{\alpha})t}|x_1 - x_2|$$

$$(x_1, x_2 \in B[0, r_0], \omega \in \Omega \ and \ t \in \mathbb{R}_+)$$

for all $x_1, x_2 \in B[0, r_0], \omega \in \Omega$ and $t \in \mathbb{R}_+$. The Lemma is proved. $\qquad\square$

Theorem 11.8 *Let Ω be a compact metric space, φ be the cocycle, generated by the non-autonomous Navier-Stokes equation (11.8) and $\frac{\|f\|C_B}{\alpha^2} < 1$. Then there exists a function $\gamma \in C(\Omega, B[0, r_0])$ such that:*

a.

$$\gamma(\omega t) = \varphi(t, \gamma(\omega), \omega) \tag{11.49}$$

for every $\omega \in \Omega$ and $t \in \mathbb{R}_+$, where $\varphi(t, x, \omega)$ is a solution of equation (11.8) with the initial condition $\varphi(0, x, \omega) = x$;

b.

$$\|\gamma\| \le \frac{\|f\|}{\alpha};$$

$$(11.50)$$

c.

$$|\varphi(t, x, \omega) - \gamma(\omega t)| \le e^{-(\alpha - C_B \frac{\|f\|}{\alpha})t}|x - \gamma(\omega)| \qquad (11.51)$$

for all $x \in E$, $\omega \in \Omega$ and $t \in \mathbb{R}_+$, where $\|\gamma\| := \sup\{|\gamma(\omega)| : \omega \in \Omega\}$.

Proof. Let $\Gamma := C(\Omega, B[0, r_0])$ $(C(\Omega, E))$ be a space all the continuous functions $f : \Omega \to B[0, r_0]$ (respectively $f : \Omega \to E$) equipped with the distance

$$d(f_1, f_2) = \max\{|f_1(\omega) - f_2(\omega)| : \omega \in \Omega\} .$$

Then (Γ, d) (respectively $(C(\Omega, E), d)$) is a complete metric space.

Let $t \in \mathbb{R}_+$. We define the mapping $S^t : \Gamma \to C(\Omega, E)$ by the equality

$$(S^t \nu)(\omega) := U(t, \omega^{-t})\nu(\omega^{-t})$$

for all $\omega \in \Omega$, where $\omega^{-t} := \sigma(-t, \omega)$ and $U(t, \omega) := \varphi(t, \cdot, \omega)$. According to Theorem 11.4 we have $S^t(\Gamma) \subseteq \Gamma$ for all $t \in \mathbb{R}_+$. It easy to see that the family of mappings $\{S^t \mid t \in \mathbb{R}_+\}$ possesses the following properties:

(1)

$$S^0 = Id_\Gamma$$

and

(2)

$$S^{t+\tau} = S^t S^\tau$$

for all $t, \tau \in \mathbb{R}_+$.

Thus $\{S^t \mid t \in \mathbb{R}_+\}$ forms a commutative semigroup with identity element. Now we will show that the mapping $S^t (t > 0)$ is a contraction. In fact, let $\nu_1, \nu_2 \in \Gamma$, then we have

$$(S^t \nu_1)(\omega) - (S^t \nu_2)(\omega) = U(t, \omega^{-t})\nu_1(\omega^{-t}) - U(t, \omega^{-t})\nu_2(\omega^{-t}). \qquad (11.52)$$

From the lemma 11.7 and the equality (11.52) it follows that

$$d(S^t \nu_1, S^t \nu_2) \le e^{-(\alpha - C_B \frac{\|f\|}{\alpha})t} d(\nu_1, \nu_2)$$

for all $t \in \mathbb{R}_+$ and, consequently, there exists a unique common fixed point $\gamma \in \Gamma$, i.e. $S^t \gamma = \gamma$ for all $t \in \mathbb{R}_+$. In particularly

$$U(t, \omega^{-t})\gamma(\omega^{-t}) = \gamma(\omega)$$

for all $t \in \mathbb{R}_+$ and $\omega \in \Omega$. From this equality follows that

$$\gamma(\omega t) = U(t, \omega)\gamma(\omega) = \varphi(t, \gamma(\omega), \omega)$$

for all $t \in \mathbb{R}_+$ and $\omega \in \Omega$.

Let $x \in E, \varphi(t, x, \omega)$ be a unique solution of equation (11.8) with the initial condition $\varphi(0, x, \omega) = x$ and $\gamma \in \Gamma$ the function with the property (11.49). Denote by $\psi(t) := \varphi(t, x, \omega) - \gamma(\omega t)$, then we have

$$\frac{d}{dt}|\psi(t)|^2 = 2Re\langle A\psi(t), \psi(t)\rangle + 2Re\langle B(\omega t)(\psi(t), \gamma(\omega t)), \psi(t)\rangle \leq$$
$$-2\alpha|\psi(t)|^2 + 2C_B|\gamma(\omega t)||\psi(t)|^2 \leq -2\alpha|\psi(t)|^2 +$$
$$2C_B\frac{\|f\|}{\alpha}|\psi(t)|^2 = -2(\alpha - C_B\frac{\|f\|}{\alpha})|\psi(t)|^2$$

and, consequently,

$$|\psi(t)|^2 \leq e^{-2(\alpha - C_B\frac{\|f\|}{\alpha})t}|\psi(0)|^2.$$

Thus we have

$$|\varphi(t, x, \omega) - \gamma(\omega t)| \leq e^{-(\alpha - C_B\frac{\|f\|}{\alpha})t}|x - \gamma(\omega)|$$

for all $x \in E$, $\omega \in \Omega$ and $t \in \mathbb{R}_+$. The theorem is proved. $\square$

Corollary 11.1 *Under the conditions of Theorem 11.8 there exists a unique function $\gamma \in C(\Omega, E)$ such that*

$$\gamma(\omega t) = \varphi(t, \gamma(\omega), \omega) \tag{11.53}$$

for all $t \in \mathbb{R}_+$ and $\omega \in \Omega$.

Proof. Let $\overline{\gamma} \in C(\Omega, E)$ be a function satisfying the equality (11.53) and $\gamma \in \Gamma = C(\Omega, B[0, r_0])$ the function from Theorem 11.8. Since $\overline{\gamma}(\omega t) = \varphi(t, \overline{\gamma}(\omega), \omega)$ for all $t \in \mathbb{R}_+$ and $\omega \in \Omega$, then according to the inequality (11.51) we have

$$|\overline{\gamma}(\omega t) - \gamma(\omega t)| \leq e^{-(\alpha - C_B\frac{\|f\|}{\alpha})t}|\overline{\gamma}(\omega) - \gamma(\omega)| \tag{11.54}$$

for all $t \in \mathbb{R}_+$ and $\omega \in \Omega$. In particularly, from (11.54) we obtain

$$|\overline{\gamma}(\omega) - \gamma(\omega)| \leq | \leq e^{-(\alpha - C_B\frac{\|f\|}{\alpha})t}\|\overline{\gamma} - \gamma\| \tag{11.55}$$

for all $t \in \mathbb{R}_+$ and $\omega \in \Omega$, where $\omega^{-t} := \sigma(-t, \omega)$ and $\|\overline{\gamma} - \gamma\| := \max\{|\overline{\gamma}(\omega) - \gamma(\omega)| : \omega \in \Omega\}$. Passing to the limit in the inequality (11.55) we obtain $\overline{\gamma}(\omega) = \gamma(\omega)$ for all $\omega \in \Omega$. $\qquad\square$

Corollary 11.2   *Under the conditions of Theorem 11.8 the equation (11.8) admits a compact global attractor $\{I_\omega \ : \ \omega \in \Omega\}$ and $I_\omega = \{\gamma(\omega)\}$ for all $\omega \in \Omega$, where $\gamma \in \Gamma$ is a function from Theorem 11.8.*

Corollary 11.3   *Let Ω be a compact minimal set containing only the periodic (quasi periodic, almost periodic, recurrent, pseudo recurrent) motions, then under conditions of Theorem 11.8 the non-autonomous Navier-Stokes equation (11.8)) admits a unique periodic (quasi periodic, almost periodic, recurrent, pseudo recurrent) solution $\gamma(\omega t)$ and every other solution of this equation is asymptotic periodic (asymptotic quasi periodic, asymptotic almost periodic, asymptotic recurrent, asymptotic pseudo recurrent).*

Proof.   Let $\gamma \in \Gamma$ be a function from Theorem 11.8, then according this theorem we have $\varphi(t, \gamma(\omega), \omega) = \gamma(\omega t)$ for all $t \in \mathbb{R}_+$ and $\omega \in \Omega$ and, consequently, the solution $\varphi(t, \gamma(\omega), \omega)$ is periodic (quasi periodic, almost periodic, recurrent). Let $\varphi(t, x, \omega)$ be a arbitrary solution of equation (11.8), then taking into consideration the inequality (11.51) we conclude that $\varphi(t, x, \omega)$ is asymptotic periodic (asymptotic quasi periodic, asymptotic almost periodic, asymptotic recurrent, asymptotic pseudo recurrent). $\qquad\square$

11.4   Uniform averaging for a finite interval

We shall be dealing with the non-autonomous Navier-Stokes equation

$$u' + \varepsilon A u + \varepsilon B(\omega t)(u, u) = \varepsilon f(\omega t), \tag{11.56}$$

where $\varepsilon \in [0, \varepsilon_0]$, A is linear and $B(\omega)$ is a bilinear operator, f is a forcing term.

Existence of partial averaged.   Below we will suppose that $B(\omega) = B_0(\omega) + B_1(\omega)$ ($B_0, B_1 \in C(\Omega, L^2(E, F))$) for all $\omega \in \Omega$ and the average of $B_1(\omega)$ is equal to 0, that is,

$$\lim_{t \to \infty} \frac{1}{t} \int_0^t B_1(\omega \tau) d\tau = 0 \tag{11.57}$$

uniformly with respect to $\omega \in \Omega$.

Remark 11.4   *1. The condition (11.57) is fulfilled if a dynamical system $(\Omega, \mathbb{R}, \sigma)$ is strictly ergodic, i.e. on Ω exists a unique invariant w.r.t. $(\Omega, \mathbb{R}, \sigma)$ measure μ.*

2. *According to Lemma 5.1 from* [104] *the equality (11.57) takes place if and only if there exists a positive continuous on* $\mathbb{R}_+$ *function* k *with* $\lim\limits_{t\to\infty} k(t) = 0$ *such that*

$$\|\frac{1}{t}\int_0^t B_1(\omega\tau)d\tau\|_{L^2(E,F)} \le k(t) \tag{11.58}$$

for all $\omega \in \Omega$ *and* $t \in \mathbb{R}_+$.

The bilinear operator $B_0 \in C(\Omega, L^2(E,F))$ satisfies the condition

$$Re\langle B_0(\omega)(u,v), w\rangle = -Re\langle B_0(\omega)(u,w), v\rangle \tag{11.59}$$

for all $u, v \in E$ and $w \in F$.

The forcing term $f(\omega) = f_0(\omega) + f_1(\omega)$ for all $\omega \in \Omega$ ($f_0, f_1 \in C(\Omega, X)$) and the function f_1 has the average which is equal to 0, i.e. there exists a positive continuous on $\mathbb{R}_+$ function k_1 with $\lim\limits_{t\to\infty} k_1(t) = 0$ such that

$$|\frac{1}{t}\int_0^t k_1(\omega\tau)d\tau|_X \le k_1(t) \tag{11.60}$$

for all $\omega \in \Omega$ and $t \in \mathbb{R}_+$.

Along with equation (11.56) we consider also the partial averaged equation

$$u' + \varepsilon A x + \varepsilon B_0(\omega t)(u,u) = \varepsilon f_0(\omega t). \tag{11.61}$$

If we introduce the "slow time" $\tau := \varepsilon t$ ($\varepsilon > 0$), then the equations (11.56) and (11.61) can be written in the following way

$$u' + Au + B(\omega\frac{\tau}{\varepsilon})(u,u) = f(\omega\frac{\tau}{\varepsilon}) \tag{11.62}$$

and

$$\bar{u}' + A\bar{u} + B_0(\omega\frac{\tau}{\varepsilon})(\bar{u},\bar{u}) = f_0(\omega\frac{\tau}{\varepsilon}). \tag{11.63}$$

We will consider the mild solutions $u(t)$ and $\bar{u}(t)$ of the equations (11.62) and (11.63), i.e. $u, \bar{u} \in C([0,T], E)$ and satisfy the following integral equations

$$u(\tau) = e^{-A\tau}x + \int_0^\tau e^{-A(\tau-s)}(-B(\omega\frac{s}{\varepsilon})(u(s), u(s)) + f(\omega\frac{s}{\varepsilon}))ds, \tag{11.64}$$

and

$$\bar{u}(\tau) = e^{-A\tau}x + \int_0^\tau e^{-A(\tau-s)}(-B_0(\omega\frac{s}{\varepsilon})(\bar{u}(s), \bar{u}(s)) + f_0(\omega\frac{s}{\varepsilon}))ds. \tag{11.65}$$

Denote by $\varphi(\tau, x, \omega, \varepsilon)$ ($\bar{\varphi}(\tau, x, \omega, \varepsilon)$) a unique solution of equation (11.64) (respectively (11.65)). According to Theorem 11.4 the cocycle $\varphi(\cdot, \cdot, \cdot, \varepsilon)$ ($\bar{\varphi}(\cdot, \cdot, \cdot, \varepsilon)$), generated by equation (11.64) (respectively (11.65)), has an absorbing ball $B[0, R_0]$ ($B[0, \bar{R}_0]$) in E, where $R_0 := \frac{\|f\|}{\alpha}$ ($\bar{R}_0 := \frac{\|f_0\|}{\alpha}$). This means that for every ball $B[0, R]$ (respectively $B[0, \bar{R}]$) there exists a positive number $L = L(R)$ (respectively $\bar{L} = \bar{L}(\bar{R})$) such that

$$U(t, \omega, \varepsilon)B[0, R] \subseteq B[0, R_0] \tag{11.66}$$

$$(\bar{U}(t, \omega, \varepsilon)B[0, \bar{R}] \subseteq B[0, \bar{R}_0]) \tag{11.67}$$

for all $t \geq L$ $(t \geq \bar{L})$, $\varepsilon \in [0, \varepsilon_0]$ and $\omega \in \Omega$, where $U(t, \omega, \varepsilon) := \varphi(t, \cdot, \omega, \varepsilon)$ $\bar{U}(t, \omega, \varepsilon) := \bar{\varphi}(t, \cdot, \omega, \varepsilon)$.

According to Theorem 11.4 the cocycle $\varphi(\cdot, \cdot, \cdot, \varepsilon)$ $(\bar{\varphi}(\cdot, \cdot, \cdot, \varepsilon))$ is uniformly bounded for $t \geq 0$, that is, for every ball $B[0, R_1]$ $(B[0, \bar{R}_1])$ there exists a ball $B[0, R_2]$ $(B[0, \bar{R}_2])$ such that

$$U(t, \omega, \varepsilon)B[0, R_1] \subseteq B[0, R_2] \tag{11.68}$$

$$(\bar{U}(t, \omega, \varepsilon)B[0, \bar{R}_1] \subseteq B[0, \bar{R}_2]) \tag{11.69}$$

for all $t \geq 0$, $\varepsilon \in [0, \varepsilon_0]$ and $\omega \in \Omega$.

Let $C(\mathbb{R} \times E, E)$ be the space of all continuous functions $f : \mathbb{R} \times E \to E$ equipped with compact open topology and let $\mathcal{F} \subseteq C(\mathbb{R} \times E, E)$.

Theorem 11.9   *Let $L > 0$ be arbitrary but fixed. If $\varphi(0, x, \omega, \varepsilon) = \bar{\varphi}(0, x, \omega, \varepsilon) = x \in B[0, \overline{R}_0]$, that is, the initial points coincide and belong to the absorbing ball of equation (11.63) and the condition (11.32) is fulfilled, then the following relation takes place*

$$\sup\{|\varphi(t, x, \omega, \varepsilon) - \bar{\varphi}(t, x, \omega, \varepsilon)|_E \ : \ 0 \leq t \leq L, \ |x|_E \leq \overline{R}_0, \ \omega \in \Omega\} \to 0 \tag{11.70}$$

as $\varepsilon \to 0$.

Proof. The proof below goes along the same lines as the proofs of the corresponding results from [140],[199] and [200]. We set $v(t) := \varphi(t, x, \omega, \varepsilon) - \bar{\varphi}(t, x, \omega, \varepsilon)$. Subtracting the equation (11.64) from the equation (11.65), we obtain

$$v(t) = \int_0^t e^{(t-s)A}(-B(\omega\frac{s}{\varepsilon})(v(s), \varphi(s, x, \omega, \varepsilon)) -$$

$$B(\omega\frac{s}{\varepsilon})(\bar{\varphi}(s, x, \omega, \varepsilon), v(s)))ds - \int_0^t e^{(t-s)A}B_1(\omega\frac{s}{\varepsilon})(\bar{\varphi}(s, x, \omega, \varepsilon),$$

$$\bar{\varphi}(s, x, \omega, \varepsilon))ds + \int_0^t e^{(t-s)A}f_1(\omega s)ds \tag{11.71}$$

According to Theorem 11.4 $|\varphi(t,x,\omega,\varepsilon)|, |\bar{\varphi}(t,x,\omega,\varepsilon)| \leq r_0$ for all $t \geq 0$, where $r_0 := \max\{\frac{\|f\|}{\alpha}, \frac{\|f_1\|}{\alpha}\}$. In view of (11.71) $v(t)$ satisfies the inequality

$$|v(t)|_E \leq |\int_0^t e^{(t-s)A}(B(\omega\frac{s}{\varepsilon})(v(s),\varphi(s,x,\omega,\varepsilon)) +$$

$$B(\omega\frac{s}{\varepsilon})(\bar{\varphi}(s,x,\omega,\varepsilon),v(s)))ds|_E + |\int_0^t e^{(t-s)A}B_1(\omega\frac{s}{\varepsilon})(\bar{\varphi}(s,x,\omega,\varepsilon),$$

$$\bar{\varphi}(s,x,\omega,\varepsilon))ds|_E + |\int_0^t e^{(t-s)A}f_1(\omega\frac{s}{\varepsilon})ds|_E , \tag{11.72}$$

for all $t \in [0,L]$. By (11.11) and (11.12) we see that the first term on the right-hand side of (11.72) is less than

$$2r_0 K C_B \int_0^t e^{-a(t-s)}(t-s)^{\alpha_1}|v(s)|_E ds . \tag{11.73}$$

We now show that the sum of the second and third terms in (11.72) tends to 0 as $\varepsilon \to 0$ uniformly w.r.t. $t \in [0,L], |x| \leq R_0$ and $\omega \in \Omega$. Before estimating the second term, we must integrate it by parts. We have

$$\int_0^t e^{-(t-s)A}B_1(\omega\frac{s}{\varepsilon})(\bar{\varphi}(s,x,\omega),\bar{\varphi}(s,x,\omega)ds =$$

$$e^{-tA}\int_0^t B_1(\omega\frac{s}{\varepsilon})(\bar{\varphi}(s,x,\omega),\bar{\varphi}(s,x,\omega)ds + \tag{11.74}$$

$$\int_0^t Ae^{-(t-s)A}\int_0^s B_1(\omega\frac{\tau}{\varepsilon})(\bar{\varphi}(\tau,x,\omega),\bar{\varphi}(\tau,x,\omega)d\tau ds.$$

Hence

$$|\int_0^t e^{-(t-s)A}B_1(\omega\frac{s}{\varepsilon})(\bar{\varphi}(s,x,\omega),\bar{\varphi}(s,x,\omega)ds|_E \leq$$

$$|e^{-tA}\int_0^t B_1(\omega\frac{s}{\varepsilon})(\bar{\varphi}(s,x,\omega),\bar{\varphi}(s,x,\omega)ds|_E + \tag{11.75}$$

$$|\int_0^t Ae^{-(t-s)A}\int_0^s B_1(\omega\frac{\tau}{\varepsilon})(\bar{\varphi}(\tau,x,\omega),\bar{\varphi}(\tau,x,\omega)d\tau ds \leq$$

$$e^{-aL}k_2(\varepsilon) + \frac{L^{1-\alpha_1}}{1-\alpha_1}k_2(\varepsilon) := c(\varepsilon),$$

where

$$k_2(\varepsilon) := \sup\{|\int_0^t B_1(\omega\frac{s}{\varepsilon})(\bar{\varphi}(s,x,\omega),\bar{\varphi}(s,x,\omega)ds|_E :$$

$$\omega \in \Omega, \ |x|_E \leq R_0, \ 0 \leq t \leq L\}. \tag{11.76}$$

According to Lemma 6.13 $k_2(\varepsilon) \to 0$ as $\varepsilon \to 0$, because according to Theorems 11.5 and 11.3 the family of functions $\mathcal{K} := \{\bar{\varphi}(\cdot, x, \omega) \ : \ |x|_E \leq R_0, \ \omega \in \Omega\}$ is equicontinuous on $\mathbb{R}_+$.

We begin with the third term. Integrating by part in s and taking into account the inequalities (11.11)-(11.12), (11.16)-(11.17) and (11.59) we find

$$|\int_0^t e^{(t-s)A} f_1(\omega \frac{s}{\varepsilon}) ds|_E = |e^{At} \int_0^t f_1(\omega \frac{s}{\varepsilon}) ds +$$

$$\int_0^t Ae^{(t-s)A} \int_t^s f_1(\omega \frac{\tau}{\varepsilon}) d\tau ds|_E \leq \|e^{A\tau}\|_{X \to E} |\int_0^t f_1(\omega \frac{s}{\varepsilon}) ds|_X +$$

$$\int_0^t \|Ae^{A(t-s)}\|_{X \to E} |\int_t^s f_1(\omega \frac{s}{\varepsilon}) ds|_X$$

$$\leq K t^{1-\beta_1} e^{-at} k_1(\frac{t}{\varepsilon}) + \int_0^t K(t-s)^{1-\beta_2} e^{-a(t-s)} k_1(\frac{t-s}{\varepsilon}) ds. \qquad (11.77)$$

Let $\alpha \in [0,1), \nu \in (0,1)$ and $\beta \in [0,2)$. Since

$$t^\alpha k_1(\frac{t}{\varepsilon}) \leq \sup_{0 \leq t \leq \varepsilon^\nu} t^\alpha k_1(\frac{t}{\varepsilon}) + \sup_{\varepsilon^\nu \leq t \leq L} t^\alpha k_1(t) \leq$$

$$\varepsilon^{\alpha\nu} k_1(0) + L^\alpha k_1(\varepsilon^{\nu-1}) \to 0 \text{ as } \varepsilon \to 0$$

and

$$\int_0^t s^{1-\beta} k_1(\frac{s}{\varepsilon}) ds = \int_0^{\varepsilon^\nu} s^{1-\beta} k_1(\frac{s}{\varepsilon}) ds + \int_{\varepsilon^\nu}^t s^{1-\beta} k_1(\frac{s}{\varepsilon}) ds \qquad (11.78)$$

$$\leq k_1(0) \frac{\varepsilon^{\nu(2-\beta)}}{2-\beta} + k(\varepsilon^{\nu-1}) \frac{(t^{2-\beta} - \varepsilon^{\nu(2-\beta)})}{2-\beta}$$

$$\leq k_1(0) \frac{\varepsilon^{\nu(2-\beta)}}{2-\beta} + k(\varepsilon^{\nu-1}) \frac{(L^{2-\beta} - \varepsilon^{\nu(2-\beta)})}{2-\beta} \text{ as } \varepsilon \to 0$$

uniformly w.r.t. $t \in [0, L]$ and $\omega \in \Omega$, then

$$\sup_{0 \leq t \leq L} |\int_0^t e^{(t-s)A} f_1(\omega \frac{s}{\varepsilon}) ds|_E \to 0 \text{ as } \varepsilon \to 0. \qquad (11.79)$$

Thus,

$$|v(t)|_E \leq C(\varepsilon) + D \int_0^t (t-s)^{-\alpha_1} |v(s)|_E ds \qquad (11.80)$$

for all $t \in [0, L]$, where $C(\varepsilon) := Ke^{-aL}(L^{1-\beta_1} + L^{1-\beta_2}) k_1(\frac{L}{\varepsilon}) + c(\varepsilon) \to 0$ as $\varepsilon \to 0$ and $D := 2R_0 C_B K$.

We now use the known inequality [188, Ch.7]. If

$$u(t) \leq a + b \int_0^t (t-s)^{\beta-1} u(s) ds, \ 0 < \beta \leq 1, \qquad (11.81)$$

then

$$u(t) \le aG_\beta([b\Gamma(\beta)]^{1/\beta}t), \tag{11.82}$$

where $G_\alpha(x)$ is a monotone function, while $\Gamma(\beta)$ is a gamma function.

In our case we have

$$|v(t)|_E \le C(\varepsilon)G_\beta([b\Gamma(\beta)]^{1/\beta}t) \le \tag{11.83}$$
$$C(\varepsilon)G_\beta([b\Gamma(\beta)]^{1/\beta}L) := d(\varepsilon) \to 0 \ (\beta := 1 - \alpha_1 \in (0,1])$$

as $\varepsilon \to 0$ uniformly w.r.t. $\omega \in \Omega$, $x \in B[0, R_0]$ and $t \in [0, L]$ for every $L > 0$. The theorem is proved. $\qquad\square$

11.5 The global averaging principle for Navier-Stokes equations

Let Ω be a compact metric space, $(\Omega, \mathbb{R}, \sigma)$ be a dynamical system on Ω, E be a Banach space and $\langle E, \varphi, (\Omega, \mathbb{R}, \sigma)\rangle$ be a cocycle on $(\Omega, \mathbb{R}, \sigma)$ with fiber E.

Definition 11.4 A family of nonempty compact sets $\{ I_\omega \mid \omega \in \Omega\}$ $(I_\omega \subset E)$ is called a local attractor (local forward attractor) if the followings conditions are fulfilled:

(1)

$$I = \bigcup\{I_\omega \ : \ \omega \in \Omega\}$$

   is compact;

(2)

$$\varphi_{\lambda_0}(t, I_\omega^{\lambda_0}, \omega) = I_{\sigma(t,\omega)}^{\lambda_0}$$

   for all $t \in \mathbb{R}_+$ and $\omega \in \Omega$;

(3) there exists $R_0 > 0$ such that $I \subset B(0, R_0) := \{x \in E|\ |x| < R_0\}$ and

$$\lim_{t\to\infty} \sup_{\omega\in\Omega} \beta(\varphi(t, B[0, R_0], \omega), I) = 0$$

(respectively $\lim_{t\to\infty} \sup_{\omega\in\Omega} \beta(\varphi(t, B[0, R_0], \omega), I_{\omega t}) = 0$)

Theorem 11.10 *Let Λ be a compact metric space, E be a banach space and φ_λ ($\lambda \in \Lambda$) be a cocycle on $(\Omega, \mathbb{R}, \sigma)$ with fiber E. Suppose that the following conditions are fulfilled:*

(1) the cocycle ϕ_{λ_0} admits a local forward attractor,

(2) the following relation takes place

$$m_L(\lambda) := \sup\{|\varphi_\lambda(t,x,\omega) - \varphi_{\lambda_0}(t,x,\omega)| \; :$$
$$0 \le t \le L, \omega \in \Omega, |x| \le R_0\} \to 0 \qquad (11.84)$$

as $\lambda \to \lambda_0$ for every positive number L;
(3) every cocycle φ_λ is asymptotically compact.

Then the next statements are valid:

a. *there exists a positive number μ such that for all $\lambda \in B[\lambda_0,\mu] := \{\lambda \in \Lambda \; : \; \rho(\lambda,\lambda_0) \le \mu\}$ the cocycle φ_λ admits in $B[0,R_0]$ a forward attractor $\{I_\omega \; : \; \omega \in \Omega\}$;*

b.

$$\sup_{\omega \in \Omega} \beta(I_\omega^\lambda, I_\omega^{\lambda_0}) \to 0$$

as $\lambda \to \lambda_0$.

Proof. Let $\rho > 0$ be an arbitrary small number such that $B[I^{\lambda_0},\rho] \subset B[0,R_0]$. We choose $L = L(\frac{\rho}{3})$ according to the condition

$$\varphi_{\lambda_0}(t, B[0,R_0],\omega) \subset B[I_\omega^{\lambda_0}, \frac{\rho}{3}]$$

for all $\omega \in \Omega$ and $t \ge L(\frac{\rho}{3})$. Now we choose $\varepsilon_0 = \varepsilon_0(L)$ so that $m(\lambda) < \frac{\rho}{3}$ for all $\lambda \in B[\lambda_0,\varepsilon_0]$.

Let $t_1 := L$, then we have $\varphi_{\lambda_0}(t_1,x,\omega) \in B[I_{\omega t_1}^{\lambda_0}, \frac{\rho}{3}]$ and $\varphi_\lambda(t_1,x,\omega) \in B[0,R_0]$. We take the point $x_1 := \varphi_\lambda(t_1,x,\omega)$ as the initial point and we consider $\varphi_\lambda(t,x_1,\omega t_1)$ on the segment $[0,L]$,

$$\varphi_{\lambda_0}(t,x_1,\omega t_1); \; \varphi_\lambda(t,x_1,\omega t_1) = \varphi_\lambda(t,\varphi_\lambda(t_1,x,\omega),\omega t_1) = \varphi_\lambda(t+t_1,x,\omega).$$

On this segment $\varphi_\lambda(t,x_1,\omega t_1)$ and $\varphi_{\lambda_0}(t,x_1,\omega t_1)$ will diverge by the value less than $\frac{\rho}{3}$. Since $\varphi_{\lambda_0}(t,x_1,\omega t_1) \in B[I_{\omega t_1}^{\lambda_0}, \frac{\rho}{3}]$, we get $\varphi_\lambda(2t_1,x,\omega) \in B[I_{\omega 2t_1}, \frac{2\rho}{3}]$.

If we take the point $x_2 := \varphi_\lambda(2t_1,x,\omega)$ as the initial one, then we see that the situation is similar to that occurred at the previous step.

Repeating this process, we arrive at a conclusion that $\varphi_\lambda(t,x,\omega) \in B[I_{\omega t}^{\lambda_0}, \rho] \subset B(0,R_0)$ for all $t \ge L(\frac{\rho}{3})$ and $\omega \in \Omega$. Since the cocycle φ_λ is asymptotic compact then it admits a forward attractor $\{I_\lambda \; : \; \omega \in \Omega\}$ such that $I^\lambda := \bigcup\{I_\omega^\lambda \; : \; \omega \in \Omega\} \subseteq B[I^{\lambda_0},\rho]$ and, consequently, $\beta(I^\lambda,I^{\lambda_0}) \to 0$ as $\lambda \to \lambda_0$.

Below we proved the inclusion $\varphi_\lambda(t,B[0,R_0],\omega) \subseteq B[I_{\omega t}^{\lambda_0}, \frac{\rho}{3}]$ for all $t \ge L$ and $\omega \in \Omega$ and, consequently, we obtain

$$\varphi_\lambda(t,B[0,R_0],\omega_t) \subseteq B[I_\omega^{\lambda_0}, \frac{\rho}{3}] \qquad (11.85)$$

for all $t \geq L$ and $\omega \in \Omega$. Taking onto consideration that

$$I_\omega^\lambda = \bigcap_{t \geq 0} \overline{\bigcup_{\tau \geq t} \varphi_\lambda(\tau, B[0, R_0], \sigma(-\tau, \omega))} \tag{11.86}$$

from (11.85) and (11.86) it follows that $I_\omega^\lambda \subseteq B[I_\omega^{\lambda_0}, \rho]$ for all $\omega \in \Omega$ and $\lambda \in B[\lambda_0, \varepsilon_0]$ and, consequently, $\sup_{\omega \in \Omega} \beta(I_\omega^\lambda, I_\omega^{\lambda_0}) \to 0$ as $\lambda \to \lambda_0$. The theorem is proved. $\qquad\square$

Remark 11.5   *1. The second condition of Theorem 11.10 is fulfilled, for example, if the space E is finite-dimensional and the mapping $\varphi : \mathbb{R}_+ \times E \times \Omega \times \Lambda \to E$, defined by the equality $\varphi(t, x, \omega, \lambda) := \varphi_\lambda(t, x, \omega)$, is continuous.*

In fact, if we suppose that it is not true, then there exist $L_0 > 0$, $\lambda_k \to \lambda_0$, $x_k \in B[0, R_0]$, $t_l \in [0, L_0]$ and $\omega_k \in \Omega$ such that

$$|\varphi_{\lambda_k}(t_k, x_k, \omega_k) - \varphi_{\lambda_0}(t_k, x_k, \omega_k)| \geq \varepsilon_0 > 0. \tag{11.87}$$

Since the sets $B[0, R_0], \Omega$ and $[0, L_0]$ are compacts, we can suppose that the sequences $\{x_k\}$, $\{t_k\}$ and $\{\omega_k\}$ are convergent. Denote by $t_0 := \lim_{k \to \infty} t_k$, $x_0 := \lim_{k \to \infty} x_k$ and $\omega_0 := \lim_{k \to \infty} \omega_k$. Passing to limit in the equality (11.87) and taking into account the continuity of the mapping φ we obtain $0 \geq \varepsilon_0$. The obtained contradiction prove our statement.

2. Under the conditions of Theorem 11.10 if we suppose that the cocycle φ_{λ_0} admits a compact global forward attractor $\{I_\omega^{\lambda_0} : \omega \in \Omega\}$, i.e.

$$\lim_{t \to \infty} \sup_{\omega \in \Omega} \beta(\varphi_{\lambda_0}(t, B[0, R], \omega), I_{\omega t}) = 0$$

for every $R > 0$, then should be naturally to hope that for the λ sufficiently close to λ_0 the cocycle φ_λ also will admits a compact global forward attractor $\{I_\omega^\lambda : \omega \in \Omega\}$ in the small neighborhood of I^{λ_0}. Unfortunately, generally speaking, this assertion is not true.

In fact, let φ_0 be a cocycle (dynamical system) generated by the equation $x' = -x$ and φ_λ be a cocycle generated by the equation $x' = -x + \lambda x^3$ $(\lambda > 0)$. It is clear that the cocycle φ_0 (φ_λ) admits a compact global attractor $I^0 = \{0\}$ $(I^\lambda = [-\lambda^{-1/2}, \lambda^{-1/2}])$. In the small neighborhood of the attractor $I^0 = \{0\}$ the cocycle φ_λ (for small λ) admits a local (but not global) attractor $I^\lambda = \{0\}$.

Theorem 11.11   *Let Λ be a compact metric space, $(\Omega, \mathbb{R}, \sigma)$ be a dynamical system on the compact metric space Ω, E be a Banach space and φ_λ $(\lambda \in \Lambda)$ be a cocycle on $(\Omega, \mathbb{R}, \sigma)$ with fiber E. Suppose that the following conditions are fulfilled:*

(1) the cocycle φ_{λ_0} admits a compact global forward attractor;
(2) the following relation takes place

$$m_L(\lambda) := \sup\{|\varphi_\lambda(t, x, \omega) - \varphi_{\lambda_0}(t, x, \omega)| \ :$$

$$0 \le t \le L, \ \omega \in \Omega, \ |x| \le R_0\} \to 0$$

as $\lambda \to \lambda_0$ for every positive number L;

(3) every cocycle φ_λ admits a compact global attractor $\{I_\omega^\lambda \ : \ \omega \in \Omega\}$;

(4) the set $I := \bigcup\{I^\lambda \ : \ \lambda \in \Lambda\}$ is bounded in E.

Then the following equality

$$\lim_{\lambda \to \lambda_0} \sup_{\omega \in \Omega} \beta(I_\omega^\lambda, I_\omega^{\lambda_0}) = 0$$

is fulfilled and, in particularly,

$$\lim_{\lambda \to \lambda_0} \beta(I^\lambda, I^{\lambda_0}) = 0.$$

Proof. Suppose that the conditions of the theorem are fulfilled. According to the condition 4. there exists a positive number R_0 such that $I \subset B(0, R_0)$. Reasoning as in Theorem 11.10 for all $\rho > 0$ we will find a $L = L(\frac{\rho}{3}) > 0$ and $\delta_0 = \delta_0(\rho) > 0$ such that

$$\varphi_\lambda(t, I_\omega^\lambda, \omega) \subseteq B[I_{\omega t}^{\lambda_0}, \rho]$$

for all $t \ge L$ and $\omega \in \Omega$ and, consequently,

$$I_\omega^\lambda = \varphi_\lambda(t, I_{\sigma(-t,\omega)}, \sigma(-t, \omega)) \subseteq B[I_\omega^{\lambda_0}, \rho]$$

for all $\omega \in \Omega$ and $\rho(\lambda, \lambda_0) < \delta_0$. The theorem is proved. $\qquad\square$

Theorem 11.12 *Let $\varepsilon \in (0, \varepsilon_0)$, Ω be compact and connected and φ_ε ($\bar{\varphi}_\varepsilon$) be a cocycle generated by the equation (11.56) (respectively by the equation (11.61)).*
Suppose that the following conditions are fulfilled:

(1) $B(\omega) := B_0(\omega) + B_1(\omega) \ (\omega \in \Omega)$, $B_0, B_1 \in C(\Omega, L^2(E, F))$;

(2) *the bilinear forms B and B_0 satisfy the condition (11.19);*

(3) *the average of $B_1(\omega)$ is equal to 0, i.e.* $\lim_{t \to \infty} \frac{1}{t} \int_0^t B_1(\omega s) ds = 0$ *uniformly w.r.t.* $\omega \in \Omega$;

(4) *the bilinear form B_0 satisfies the condition (11.32);*

(5) *the cocycles φ_ε and $\bar{\varphi}_\varepsilon$ ($\varepsilon \in (0, \varepsilon_0]$)) are asymptotically compact .*

Then the following statements are true:

a. *for every $\varepsilon \in (0, \varepsilon_0]$) and $\omega \in \Omega$ the set $I_\omega^\varepsilon := \{x \in E \ : \ \text{the solution } \varphi_\varepsilon(t, x, \omega)$ of equation (11.57) is defined and bounded on $\mathbb{R}\}$ (respectively $\bar{I}_\omega^\varepsilon := \{x \in E \mid \text{the solution } \bar{\varphi}_\varepsilon(t, x, \omega)$ of equation (11.61) is defined and bounded on $\mathbb{R}\}$) is nonempty, compact and connected;*

b. *the cocycle φ_ε ($\bar{\varphi}_\varepsilon$) admits a compact global attractor $\{I_\omega^\varepsilon \ : \ \omega \in \Omega\}$ (respectively $\{\bar{I}_\omega^\varepsilon \ : \ \omega \in \Omega\}$);*

c. *the set I^ε (respectively $\bar{I}^\varepsilon$) is compact and connected;*

d. *the set $I := \bigcup\{I^\varepsilon \ : \ \varepsilon \in [0,\varepsilon_0]\}$ (respectively $\bar{I} := \bigcup\{\bar{I}^\varepsilon \ : \ \varepsilon \in [0,\varepsilon_0]\}$, where $\bar{I}^\varepsilon := \bigcup\{\bar{I}^\varepsilon_\omega \ : \ \omega \in \Omega\}$) is compact, where $I^\varepsilon := \bigcup\{I^\varepsilon_\omega \ : \ \omega \in \Omega\}$, $I^0 = \bar{I}^0 := \bigcup\{\bar{I}^0_\omega \ : \ \omega \in \Omega\}$ and $\{\bar{I}^0_\omega \ : \ \omega \in \Omega\}$ is a compact global attractor of equation (11.61), when $\varepsilon = 1$;*

e. $\lim\limits_{\varepsilon \to 0} \beta(I^\varepsilon, \bar{I}^0) = 0$, *where β is a semi-distance of Hausdorff;*

f. *If a dynamical system $(\Omega, \mathbb{R}, \sigma)$ is periodic, i.e. there exists $\omega_0 \in \Omega$ such that $\omega_0\tau = \omega_0$ and $\Omega = \{\omega_0 t \ : \ t \in [0,\tau)\}$, then*

$$\lim_{\varepsilon \to 0} \sup_{\omega \in \Omega}\{\beta(I^\varepsilon_\omega, I^0_\omega)\} = 0.$$

Proof. Let $\varepsilon \in (0,\varepsilon_0)$, Ω be compact and connected and φ_ε ($\bar{\varphi}_\varepsilon$) be a cocycle generated by the equation (11.56) (respectively by the equation (11.61)), then we have

$$\varphi_\varepsilon(t,x,\omega) = \varphi(\varepsilon t, x, \omega, \varepsilon) \tag{11.88}$$

$$(\text{respectively } \bar{\varphi}_\varepsilon(t,x,\omega) = \bar{\varphi}(\varepsilon t, x, \omega, \varepsilon))$$

for all $t \in \mathbb{R}_+, x \in E$ and $\omega \in \Omega$, where $\varphi(\cdot,\cdot,\cdot,\varepsilon)$ (respectively $\bar{\varphi}(\cdot,\cdot,\cdot,\varepsilon)$) is a cocycle generated by the equation (11.62) (respectively (11.63)). From the equality (11.88) it follows that $\{I^\varepsilon_\omega \ : \ \omega \in \Omega\}$ (respectively $\{\bar{I}^\varepsilon_\omega \ : \ \omega \in \Omega\}$,) is a compact global attractor of the equation (11.62) (respectively (11.63)). Now to finish the proof of theorem it is sufficient to apply Theorems 11.9, 11.11, 7.7 and Lemmas 7.3, 7.4. The theorem is proved. $\square$

Chapter 12

Global attractors of V-monotone dynamical systems

Differential equations with monotone right hand side make one of the most studied classes of nonlinear equations (see, for example, [30], [185], [240], [316], [333] and the literature quoted there).

The problem of existence of almost periodic solutions of monotone nonlinear almost periodic equation has been studied by many authors (see [113], [114], [184], [189], [238], [316], [333] and others).

The purpose of this chapter is to study the global attractors of general V- monotone non-autonomous dynamical systems and their applications to different classes of differential equations (ODEs, ODEs with impulse, some classes of evolutionary partial differential equations).

For autonomous equations an analogous problem was studied before (see, for example, [20], [43],[314]), but for non-autonomous dynamical systems this problem is considered for the first time in our book.

12.1 Global attractors of V-monotone NDS

Lemma 12.1 *Let $(X, \mathbb{S}_+, \pi)$ be a compactly k-dissipative dynamical system with the Levinson's center J. Then the dynamical system $(X \times X, \mathbb{S}_+, \pi \times \pi)$ (where $\pi \times \pi(t, (x_1, x_2)) := (\pi(t, x_1), \pi(t, x_2))$ for all $t \in \mathbb{S}$ and $x_1, x_2 \in X$) is also compactly k-dissipative and $J \times J$ is its Levinson's center.*

Proof. Let $(X, \mathbb{S}_+, \pi)$ be a compactly k-dissipative dynamical system. Then there exists a nonempty compact $K \subseteq X$ such that

$$\lim_{t \to +\infty} \beta(\pi^t M, K) = 0 \tag{12.1}$$

for every $M \in C(X)$. Let $K' := K \times K$ and $M' \in C(X \times X)$. Then from the equality (12.1) follows that

$$\lim_{t \to +\infty} \beta((\pi \times \pi)^t M', K') = 0$$

385

and, consequently, $(X \times X, \mathbb{S}_+, \pi \times \pi)$ is compactly k-dissipative and the set $J' :=$ $\Omega(K')$ is its Levinson's center. We note that $\Omega(K \times K) \subseteq \Omega(K) \times \Omega(K)$ and, consequently $J' \subseteq J \times J$, because $J = \Omega(K)$. On the other hand, the set $J \times J$ is a compact invariant set for the dynamical system $(X \times X, \mathbb{S}_+, \pi \times \pi)$ and, consequently, $J \times J \subseteq J'$ because J' is a maximal compact invariant set in $(X \times X, \mathbb{S}_+, \pi \times \pi)$. The Lemma is proved. $\qquad\square$

Let Ω be a compact topological space , (E, h, Ω) be locally trivial Banach stratification [29] and $|\cdot|$ be a norm on (E, h, Ω) co-ordinate with the metric ρ on E (that is $\rho(x_1, x_2) = |x_1 - x_2|$ for any $x_1, x_2 \in X$ such that $h(x_1) = h(x_2)$).

Recall that a non-autonomous dynamical system $\langle (X, \mathbb{T}_+, \pi), (\Omega, \mathbb{S}, \sigma), h \rangle$ is said to be uniformly stable in positive direction on compacts of X [62], if for arbitrary $\varepsilon > 0$ and $K \subseteq X$ there is $\delta = \delta(\varepsilon, K) > 0$ such that the inequality $\rho(x_1, x_2) < \delta$ $(h(x_1) = h(x_2))$ implies that $\rho(\pi^t x_1, \pi^t x_2) < \varepsilon$ for $t \in \mathbb{T}^+$.

Denote by $X \dot\times X = \{(x_1, x_2) \in X \times X \mid h(x_1) = h(x_2)\}$.

Definition 12.1　The non-autonomous dynamical system $\langle (X, \mathbb{S}_+, \pi), (\Omega, \mathbb{S}, \sigma), h \rangle$ is called (see [113], [114] and [333], [238]) V-monotone, if there exists a function $V : X \dot\times X \to \mathbb{R}_+$ with the following properties:

a. V is continuous.
b. V is positively defined, i.e. $V(x_1, x_2) = 0$ if and only if $x_1 = x_2$.
c. $V(x_1 t, x_2 t) \le V(x_1, x_2)$ for all $(x_1, x_2) \in X \dot\times X$ and $t \in \mathbb{S}_+$.

Theorem 12.1　*Let $\langle (X, \mathbb{S}_+, \pi), (\Omega, \mathbb{S}, \sigma), h \rangle$ be V-monotone and compactly dissipative, then it is uniformly stable in positive direction on compacts from X.*

Proof.　Let $\langle (X, \mathbb{S}_+, \pi), (\Omega, \mathbb{S}, \sigma), h \rangle$ be a V-monotone non-autonomous dynamical system and let it be not uniformly stable in positive direction on compacts from X. Then there are $\varepsilon_0 > 0$, a sequence $\{t_n\} \subseteq \mathbb{S}_+$ ($t_n \to +\infty$ as $n \to +\infty$), a sequence $\delta_n \to 0$ ($\delta_n > 0$), a compact $K_0 \subseteq X$ and sequences $\{x_n^i\} \subseteq K_0$ ($i = 1, 2$) such that

$$\rho(x_n^1, x_n^2) < \delta_n \quad \text{and} \quad \rho(x_n^1 t_n, x_n^2 t_n) \ge \varepsilon_0 \qquad (12.2)$$

for all $n \in \mathbb{N}$. Since the dynamical system $\langle (X, \mathbb{S}_+, \pi), (\Omega, \mathbb{S}, \sigma), h \rangle$ is compactly dissipative, then we may suppose without loss of generality that the sequences $\{x_n^i\}$ ($i = 1, 2$) and $\{x_n^i t_n\}$ ($i = 1, 2$) are convergent. We denote by $x^i = \lim_{n \to +\infty} x_n^i$ ($i = 1, 2$) and $\bar{x}^i = \lim_{n \to +\infty} x_n^i t_n$ ($i = 1, 2$). According to the inequality (12.2) we obtain $x^1 = x^2$ and $\bar{x}^1 \ne \bar{x}^2$. On the other hand, in view of V-monotonicity of $\langle (X, \mathbb{S}_+, \pi), (\Omega, \mathbb{S}, \sigma), h \rangle$, we have

$$V(x_n^1 t_n, x_n^2 t_n) \le V(x_n^1, x_n^2) \qquad (12.3)$$

for all $n \in \mathbb{N}$. Passing to limit in (12.3) as $n \to +\infty$, we obtain the equality $\bar{x}^1 = \bar{x}^2$ which contradicts to the inequality (12.2). This contradiction proves Theorem 12.1.

$\square$

Corollary 12.1 *Let $\langle (X, \mathbb{S}_+, \pi), (\Omega, \mathbb{S}, \sigma), h \rangle$ be a V-monotone compactly dissipative non-autonomous dynamical system and Ω be minimal. Then:*

1. *J is uniformly orbitally stable in positive direction, i.e. for $\varepsilon > 0$ there is $\delta(\varepsilon) > 0$ such that the inequality $\rho(x, J_{h(x)}) < \delta$ implies that $\rho(\pi^t x, J_{h(\pi^t x)}) < \varepsilon$ for $t \geq 0$;*
2. *J is an attractor of compact sets from X, i.e. for $\varepsilon > 0$ and a compact $K \subseteq X$ there is $L(\varepsilon, K) > 0$ such that $\pi^t K_\omega \subseteq \tilde{B}(J_{\sigma_t \omega}, \varepsilon)$ for $\omega \in \Omega$ and $t \geq L(\varepsilon, K)$, where $\tilde{B}(M, \varepsilon) := \{ x \in X \mid \rho(x, M_{h(x)}) < \varepsilon \}$;*
3. *all motions on J can be continued to the left and J is bilaterally distal;*
4. *$J_\omega := X_\omega \bigcap J$ for $\omega \in \Omega$ and is a connected set if X_ω is connected, and for distinct ω_1 and ω_2 the sets J_{ω_1} and J_{ω_2} are homeomorphic;*
5. *J is formed of recurrent trajectories, and every two arbitrary points $x_1, x_2 \in J_\omega$ ($\omega \in \Omega$) are mutually recurrent.*

Theorem 12.2 *Let $\langle (X, \mathbb{S}_+, \pi), (\Omega, \mathbb{S}, \sigma), h \rangle$ be a V-monotone compactly dissipative non-autonomous dynamical system, Ω be a minimal set and J be its Levinson's center. Then*

$$V(x_1 t, x_2 t) = V(x_1, x_2) \tag{12.4}$$

for all $x_1, x_2 \in J$ such that $h(x_1) = h(x_2)$.

Proof. Let $x_1, x_2 \in J$ be such that $h(x_1) = h(x_2)$. Then according to Corollary 12.1 $x_1, x_2 \in J_\omega$ ($h(x_1) = h(x_2) = \omega \in \Omega$) are mutually recurrent and, consequently, the function $\phi \in C(\mathbb{S}, \mathbb{R})$ defined by the equality $\phi(t) := V(x_1 t, x_2 t)$ ($t \in \mathbb{S}$) is recurrent. On the other hand, since $\langle (X, \mathbb{S}_+, \pi), (\Omega, \mathbb{S}, \sigma), h \rangle$ is V-monotone, we obtain $\phi(t_2) \leq \phi(t_1)$ for all $t_2 \geq t_1$ ($t_1, t_2 \in \mathbb{S}$). As the function ϕ is recurrent and monotone, then it is a constant. In fact, if we suppose that there exists $t_2 > t_1$ such that $\phi(t_2) < \phi(t_1)$, then in virtue of the recurrence of the function ϕ there exists a sequence $\{s_n\}$ such that $s_n \to +\infty$ as $n \to +\infty$ and $\phi(s_n) \to \phi(t_1)$. On the other hand, $\phi(s_n) < \phi(t_2)$ for a sufficiently large n and, consequently, $\phi(t_1) \leq \phi(t_2) < \varphi(t_1)$. The obtained contradiction proves Theorem 12.2. $\square$

Corollary 12.2 *Under the conditions of Theorem 12.2, if a non-autonomous dynamical system $\langle (X, \mathbb{S}_+, \pi), (\Omega, \mathbb{S}, \sigma), h \rangle$ is strictly monotone, i.e. $V(x_1 t, x_2 t) < V(x_1, x_2)$ for all $t > 0$ and $(x_1, x_2) \in X \dot{\times} X$ ($x_1 \neq x_2$), then $J_\omega := J \bigcap X_\omega$ consists of a single point for all $\omega \in \Omega$.*

Theorem 12.3   *Let $\langle (X, \mathbb{S}_+, \pi), (\Omega, \mathbb{S}, \sigma), h \rangle$ be a V-monotone compactly dissipative non-autonomous dynamical system with the compact minimal base Ω and J be its Levinson's center. Then for every point $x \in X_y$ there exists a unique recurrent point $p \in J_\omega$ such that*

$$\lim_{t \to +\infty} \rho(xt, pt) = 0, \tag{12.5}$$

i.e. every trajectory of this system is asymptotically recurrent.

Proof. Let $\omega \in \Omega$ and $x \in X_\omega$ be an arbitrary point. Since $\langle (X, \mathbb{S}_+, \pi), (\Omega, \mathbb{S}, \sigma), h \rangle$ is compactly dissipative, then the semi-trajectory $\{xt \mid t \in \mathbb{S}_+\}$ is relatively compact. According to Lemma 22.29 from [32], there exists a recurrent point $p \in \omega_x \cap X_\omega$ and a sequence $t_n \to +\infty$ such that

$$\lim_{n \to +\infty} \rho(xt_n, pt_n) = 0. \tag{12.6}$$

According to Theorem 12.1, the non-autonomous dynamical system $\langle (X, \mathbb{S}_+, \pi),$ $(\Omega, \mathbb{T}, \sigma), h \rangle$ is uniformly stable in positive direction on compact $K := \overline{\{xt \mid t \in \mathbb{S}^+ \}}$. Let $\varepsilon > 0$ and $\delta = \delta(\varepsilon, K) > 0$ be chosen out of the uniform stability of K. Then from equality (12.5) we have $\rho(xt_n, pt_n) < \delta(\varepsilon, K)$ for all sufficiently large n and, consequently,

$$\rho(x(t_n + t), p(t_n + t)) < \varepsilon$$

for all $t \geq 0$ or $\rho(xt, pt) < \varepsilon$ for every $t \geq t_n$, i.e. the equality (12.5) holds.

Thus, we found a recurrent point $p \in \omega_x \cap X_\omega$ satisfying the equality (12.5). It is clear that $p \in J_\omega$. Now we will prove that point p with the property (12.5) is unique. Let us suppose that $p_1, p_2 \in J_\omega$ and

$$\lim_{t \to +\infty} \rho(xt, p_i t) = 0 \tag{12.7}$$

for $i = 1, 2$. Then according to Corollary 12.1 the points p_1 and p_2 are mutually recurrent. In particular, there exists a sequence $t_n \to +\infty$ such that

$$\lim_{n \to +\infty} p_i t_n = p_i \ (i = 1, 2). \tag{12.8}$$

On the other hand, from the equality (12.7) we have

$$\lim_{n \to +\infty} \rho(p_1 t_n, p_2 t_n) = 0. \tag{12.9}$$

From (12.8) and (12.9) follows that $p_1 = p_2$. The theorem is completely proved. $\square$

Corollary 12.3   *Under the conditions of Theorem 12.3 the following assertions hold:*

(1) ω-limit set (which is denoted as ω_x) of every point $x \in X$ is a compact minimal set.

(2) if $x_1, x_2 \in X_\omega$ ($\omega \in \Omega$), then $\omega_{x_1} = \omega_{x_2}$ or $\omega_{x_1} \bigcap \omega_{x_2} = \emptyset$.

12.2 On the structure of Levinson center of V-monotone NDS

Definition 12.2 (X, ρ) is called [189] a metric space with segments, if for any $x_1, x_2 \in X$ and $\alpha \in [0, 1]$ the intersection of $B[x_1, \alpha r]$ (a closed ball with its center at x and radius αr, where $r = \rho(x_1, x_2)$) and $B[x_2, (1 - \alpha)r]$ consists of a unique element $S(\alpha, x_1, x_2)$.

Definition 12.3 A metric space (X, ρ) is called [189] strictly-convex, if (X, ρ) is a metric space with segments and for any $x_1, x_2, x_3 \in X$, $x_2 \neq x_3$ and $\alpha \in (0, 1)$ there holds the inequality $\rho(x_1, S(a, x_2, x_3)) < \max\{\rho(x_1, x_2), \rho(x_1, x_3)\}$.

Definition 12.4 Let X be a strictly metric-convex space. A subset M of X is said to be metric-convex, if $S(\alpha, x_1, x_2) \in M$ for any $\alpha \in (0, 1)$ and $x_1, x_2 \in M$.

We note that every convex closed subset X of the Hilbert space H equipped with the metric $\rho(x_1, x_2) = |x_1 - x_2|$ is strictly metric-convex.

Let $x \in X$. Denote by Φ_x the family of all entire trajectories of the dynamical system $(X, \mathbb{S}_+, \pi)$ passing through the point x for $t = 0$, i.e. $\gamma \in \Phi_x$ if and only if $\gamma : \mathbb{S} \to X$ is a continuous mapping with the properties: $\gamma(0) = x$ and $\pi^t \gamma(\tau) = \gamma(t + \tau)$ for all $t \in \mathbb{S}_+$ and $\tau \in \mathbb{S}$.

Theorem 12.4 *Let $\langle (X, \mathbb{S}_+, \pi), (\Omega, \mathbb{S}, \Theta), h \rangle$ be a V-monotone compactly dissipative non-autonomous dynamical system, J be its Levinson center and the following conditions be held:*

1. *$V(x_1, x_2) = V(x_2, x_1)$ for all $(x_1, x_2) \in X \dot\times X$.*
2. *$V(x_1, x_2) \leq V(x_1, x_3) + V(x_3, x_2)$ for all $x_1, x_2, x_3 \in X$ with the condition $h(x_1) = h(x_2) = h(x_3)$.*
3. *the space (X_ω, V_ω) is strictly metric-convex for all $\omega \in \Omega$, where $X_\omega := h^{-1}(\omega) = \{x \in X | h(x) = \omega\}$ ($\omega \in \Omega$) and $V_\omega := V|_{X_\omega \times X_\omega}$.*

If $\gamma_{x_i} \in \Phi_{x_i}$ ($i = 1, 2$) and $x_1, x_2 \in J_\omega$ ($\omega \in \Omega$), then the function $\gamma : \mathbb{S} \to X$ ($\gamma(t) = S(\alpha, \gamma_{x_1}(t), \gamma_{x_2}(t))$ for all $t \in \mathbb{S}$) defines an entire trajectory of the dynamical system $(X, \mathbb{S}_+, \pi)$.

Proof. Let $\omega \in \Omega$, $\alpha \in [0, 1]$ and $x_1, x_2 \in J_\omega$. We denote by $x := S(\alpha, x_1, x_2)$. Let $\gamma_{x_i} \in \Phi_{x_i}$ ($i = 1, 2$). Consider the function $\gamma : \mathbb{S} \to J$ defined by the equality

$$\gamma(t) := S(\alpha, \gamma_{x_1}(t), \gamma_{x_2}(t)) \tag{12.10}$$

for all $t \in \mathbb{S}$. We will show that γ is an entire trajectory of $(X, \mathbb{S}_+, \pi)$ with the condition $\gamma(0) = x$. In fact, according to Theorem 12.2

$$V(\gamma_1(t), \gamma_2(t)) = V(x_1, x_2) = d$$

for all $t \in \mathbb{S}$. Since $V(\gamma_1(t), \gamma(t)) = \alpha d$ for all $t \in \mathbb{S}$ and $\langle (X, \mathbb{S}_+, \pi), (Y, \mathbb{S}, \Theta), h \rangle$ is V-monotone, we have

$$V(\gamma_1(t + \tau), \pi^t \gamma(\tau)) = V(\pi^t \gamma_1(\tau), \pi^t \gamma(\tau)) \le V(\gamma_1(\tau), \gamma(\tau)) \le \alpha d$$

and

$$V(\gamma_2(t + \tau), \pi^t \gamma(\tau)) = V(\pi^t \gamma_2(\tau), \pi^t \gamma(\tau)) \le V(\gamma_2(\tau), \gamma(\tau)) \le (1 - \alpha)d$$

and, consequently,

$$\pi^t \gamma(\tau) \in S(\alpha, \gamma_1(t + \tau), \gamma_2(t + \tau)).$$

So, $\pi^t \gamma(\tau) = \gamma(t + \tau)$ for all $\tau \in \mathbb{S}$ and $t \in \mathbb{S}_+$. Now we will prove that the function γ is continuous. It is clear that γ is continuous on $\mathbb{S}_+$. Let $t_0 \in \mathbb{S}$, $t_0 \le 0$ and $t = t_0 + h$ ($|h| < \delta$, $\delta > 0$). Then we have

$$\rho(\gamma(t_0 + h), \gamma(t_0)) = \rho(\pi^{t_0 + |t_0| + \delta + h} \gamma(-|t_0| - \delta),$$
$$\pi^{t_0 + |t_0| + \delta} \gamma(-|t_0| - \delta)). \tag{12.11}$$

Passing to limit in (12.11) as $h \to 0$, we obtain $\lim\limits_{h \to 0} \gamma(t_0 + h) = \gamma(t_0)$. $\qquad\square$

We denote by $\mathcal{K} = \{a \in C(\mathbb{S}_+, \mathbb{R}_+) \mid a(0) = 0, \ a \text{ is strictly increasing}\}$.

Theorem 12.5   *Under the conditions of Theorem 12.4, if, in addition, a non-autonomous dynamical system $\langle (X, \mathbb{S}_+, \pi), (\Omega, \mathbb{S}, \Theta), h \rangle$ is boundedly k-dissipative and there exists a function $a \in \mathcal{K}$ with the property $\lim\limits_{t \to +\infty} a(t) = +\infty$ such that $a(\rho(x_1, x_2)) \le V(x_1, x_2)$ for all $(x_1, x_2) \in X \dot{\times} X$, then J_ω will be metric-convex for all $\omega \in \Omega$, where $J_\omega := J \bigcap X_\omega$ and J is the Levinson's center of $(X, \mathbb{S}_+, \pi)$.*

Proof.   Let $\omega \in \Omega$, $x_1, x_2 \in J_\omega, \alpha \in (0, 1)$ and $x := S(\alpha, x_1, x_2)$. We will show that $x \in J_\omega$. Since J is invariant, there exist entire trajectories γ_1 and γ_2 such that $\gamma_i(0) = x_i$ ($i = 1, 2$) and $\gamma_i(t) \in J_{\theta_t \omega}$ for all $t \in \mathbb{S}$. Consider the function $\gamma : \mathbb{S} \to J$ defined by the equality (12.10). According to Theorem 12.4, the mapping γ defines an antire trajectory of the dynamical system $(X, \mathbb{S}_+, \pi)$. We note that the trajectory γ is bounded. In fact, by the construction of γ we have $V(\gamma_1(t), \gamma(t)) \le \alpha d$, and according to the conditions of Theorem 12.5 we obtain $a(\rho(\gamma(t), \gamma_1(t))) \le V(\gamma(t), \gamma_1(t)) \le \alpha d$. and, consequently, $\rho(\gamma(t), \gamma_1(t)) \le a^{-1}(\alpha d)$ for all $t \in \mathbb{S}$. Thus, the trajectory γ is bounded and, taking into account that the set $\gamma(\mathbb{S})$ is invariant and the Levinson's center J attracts every bounded set from X, we have $x \in J_\omega$. The theorem is proved. $\qquad\square$

12.3   Almost periodic solutions of *V*-monotone systems

Let (X, ρ) be a metric space.

Definition 12.5   Recall that a function $\varphi : \mathbb{S} \to X$ is called almost periodic (in the sense of Bohr), if for every $\varepsilon > 0$ there exists a relatively dense subset A_ε of $\mathbb{R}$ such that

$$\rho(\varphi(t + \tau), \varphi(t)) < \varepsilon$$

for all $t \in \mathbb{R}$ and $\tau \in A_\varepsilon$.

Lemma 12.2   [238] *Let Ω be a compact minimal set and $M \subseteq X$ be a compact invariant set of $(X, \mathbb{S}_+, \pi)$. If the non-autonomous dynamical system $\langle (X, \mathbb{S}_+, \pi), (\Omega, \mathbb{S}, \sigma), h \rangle$ is distal on M in negative direction, then the mapping $\omega \longmapsto M_\omega := M \cap X_\omega$ is continuous with respect to the Hausdorff metric.*

Lemma 12.3   *Let $M \subseteq X$ be a compact invariant set of $(X, \mathbb{S}_+, \pi)$. If the non-autonomous dynamical system $\langle (X, \mathbb{S}_+, \pi), (\Omega, \mathbb{S}, \sigma), h \rangle$ is uniformly stable in positive direction on compacts from X, then $\langle (X, \mathbb{S}_+, \pi), (\Omega, \mathbb{S}, \sigma), h \rangle$ is distal in negative direction on the invariant set M.*

Proof.   We will prove that under the conditions of Lemma 12.3 a compact invariant set M is distal in negative direction. Really, if we suppose the contrary, then there exist $\omega_0 \in \Omega, t_n \geq 0$, $x_1 \neq x_2$ $(x_1, x_2 \in M$ and $h(x_1) = h(x_2))$ and $\gamma_{x_i} \in \Phi_{x_i}$ $(i = 1, 2)$ such that

$$\rho(\gamma_{x_1}(-t_n), \gamma_{x_2}(-t_n)) \to 0$$

as $n \to +\infty$. Let $\varepsilon_0 = \rho(x_1, x_2) > 0$ and $\delta_0 = \delta(\varepsilon_0, M) > 0$ is chosen from the condition of the uniform stability in positive direction of $\langle (X, \mathbb{S}_+, \pi), (\Omega, \mathbb{S}, \sigma), h \rangle$ on M. Then for sufficiently large n we have

$$\rho(\gamma_{x_1}(-t_n), \gamma_{x_2}(-t_n)) < \delta_0$$

and, consequently,

$$\rho(\pi^t \gamma_{x_1}(-t_n), \pi^t \gamma_{x_2}(-t_n)) < \varepsilon_0$$

for all $t \geq 0$. In particular, for $t = -t_n$ we have

$$\rho(x_1, x_2) = \rho(\pi^{t_n} \gamma_{x_1}(-t_n), \pi^{t_n} \gamma_{x_2}(-t_n)) < \varepsilon_0 = \rho(x_1, x_2).$$

The obtained contradiction proves the necessary statement. $\square$

Corollary 12.4   *Under the conditions of Lemma 12.3, if Ω is a compact minimal set, then the mapping $\omega \longmapsto J_\omega$ is continuous with respect to the Hausdorff metric.*

This assertion follows from Lemmas 12.2 and 12.3.

Lemma 12.4 *Let (M, ρ) be a compact strictly metric-convex space and E be a compact sub-semigroup of isometrics of the semigroup M^M (i.e. $E \subseteq M^M$ and $\rho(\xi x_1, \xi x_2) = \rho(x_1, x_2)$ for all $x_1, x_2 \in M$). Then there exists a common fixed point $\bar{x} \in M$ of E, i.e. $\xi(\bar{x}) = \bar{x}$ for all $\xi \in E$.*

Proof. Let x_0 be an arbitrary point of M. We denote by $\ell(x) = \sup\limits_{\xi \in E} \rho(\xi(x), x_0)$ and

$$\ell_0 = \inf_{x \in M} \ell(x). \tag{12.12}$$

Let $\{x_n\} \subseteq M$ be a minimizing sequence for (12.12), i.e.

$$\ell_0 \leq \ell(x_n) \leq \ell_0 + \frac{1}{n} \tag{12.13}$$

for all $n \in \mathbb{N}$. Since the set M is compact, we may suppose that the sequence $\{x_n\}$ is convergent. Let $x' = \lim\limits_{n \to +\infty} x_n$. Then passing to limit in the inequality (12.13), we obtain $\ell(x') \leq \ell_0$. On the other hand, $\ell(x) \geq \ell_0$ for all $x \in M$ and, consequently, $\ell(x') = \ell_0$. We put $M' = \{x \in M \mid \ell(x) = \ell_0\}$. Then $M' \neq \emptyset$. The set M' is invariant with respect to the semigroup E, i.e. $\xi(M') \subseteq M'$ for all $\xi \in E$. In fact, let $\eta \in E$ and $x' \in M'$. Then

$$\ell(\eta(x')) = \sup_{\eta \in E} \rho(\xi(\eta(x')), x_0) \leq \sup_{\xi \in E} \rho(\xi(x'), x_0) = \ell(x') = \ell_0$$

and, consequently, $\ell(\eta(x')) = \ell_0$. We will show now that the set M' contains a single point. Really, if we suppose the contrary, then there exist $x_1, x_2 \in M'$ ($x_1 \neq x_2$). We consider $x = S(\frac{1}{2}, x_1, x_2)$, which is an element from M, because M is strictly metric-convex. Since the semigroup E is compact, then under the conditions of Lemma 12.4 there exists $\xi \in E$ such that $\ell_0 \leq \ell(x) = \rho(\xi(x), x_0)$. On the other hand, according to the strict metric-convexity of M we have $\rho(\xi(x), x_0) < \max\{\rho(\xi(x_1), x_0), \rho(\xi(x_2), x_0)\} \leq \ell_0$ and, consequently, $\ell(x) < \ell_0$. The obtained contradiction proves that M' contains a unique point $\bar{x}$. Taking into account the invariance of the set M' with respect to the semigroup E, we have $\xi(\bar{x}) = \bar{x}$ for all $\xi \in E$. The Lemma is proved. $\square$

Theorem 12.6 *Let $\langle (X, \mathbb{S}_+, \pi), (\Omega, \mathbb{S}, \sigma), h \rangle$ be a V-monotone bounded k-dissipative NDS, J be its Levinson's center and the following conditions be held:*

1. $V(x_1, x_2) = V(x_2, x_1)$ for all $(x_1, x_2) \in X \dot{\times} X$.

2. $V(x_1, x_2) \leq V(x_1, x_3) + V(x_3, x_2)$ for all $x_1, x_2, x_3 \in X$ with the condition $h(x_1) = h(x_2) = h(x_3)$.

3. the space (X_ω, V_ω) is strictly metric-convex for all $\omega \in \Omega$, where $X_\omega := h^{-1}(\omega) := \{x \in X \mid h(x) = \omega\}$ $(\omega \in \Omega)$ and $V_\omega := V|_{X_\omega \times X_\omega}$.

Then the set-valued mapping $\omega \to J_\omega$ admits at least one continuous invariant section, i.e. there exists a continuous mapping $\nu : \Omega \to J$ with the properties: $h(\nu(\omega)) = \omega$ and $\nu(\sigma(t, y)) = \pi(t, \nu(\omega))$ for all $t \in \mathbb{S}$ and $\omega \in \Omega$.

Proof. According to Corollary 12.1, under the conditions of Theorem 12.6 the semi-group dynamical system $(X, \mathbb{S}_+, \pi)$ defines on J a group dynamical system $(J, \mathbb{S}, \pi)$. Let $\omega_0 \in \Omega$ be an arbitrary point of Ω , J be the Levinson's center of the dynamical system $(X, \mathbb{S}_+, \pi)$ and $E = E(J, \mathbb{S}, \pi)$ be the Ellis semigroup of the dynamical system $(J, \mathbb{S}, \pi)$, i.e. $E(J, \mathbb{S}, \pi) = \overline{\{\pi^t \mid t \in \mathbb{S}\}}$, where by bar we denote the close in J^J (J^J is equipped with the topology of Tihonoff). We denote by $E_{\omega_0} := \{\xi \in E \mid \xi(J_{\omega_0}) \subseteq J_{\omega_0}\}$. Then under the conditions of Theorem 12.2, $E_{\omega_0} \neq \emptyset$ is a compact sub-semigroup of Ellis of the semigroup E. According to Theorem 12.2, we have $V(\xi(x_1), \xi(x_2)) = V(x_1, x_2)$ for all $(x_1, x_2) \in J_{\omega_0} \times J_{\omega_0}$ and consequently, under the conditions of Theorem 12.6 we have a strictly metric-convex (with respect to the complete metric $V_{\omega_0} := V|J_{\omega_0} \times J_{\omega_0}$) compact set J_{ω_0} and a compact semigroup of isometrics E_ω acting on J_{ω_0}. Applying Lemma 12.4, we obtain a common fixed point $\bar{x}_{\omega_0} \in J_{\omega_0}$. We denote by $\Sigma := H(\bar{x}_{\omega_0}) = \overline{\{\bar{x}_{\omega_0} t \mid t \in \mathbb{S}\}}$. It is clear that Σ is a compactly invariant set of $(J, \mathbb{S}, \pi)$. Obviously, $\Sigma_{\omega_0} := \Sigma \bigcap J_{\omega_0} = \{x_{\omega_0}\}$. We will prove that $\Sigma_\omega := \Sigma \bigcap J_\omega$ contains a single point $\bar{x}_\omega$. It is evident that $\Sigma_\omega \neq \emptyset$ for all $\omega \in \Omega$, because Ω and Σ are compactly invariant sets and Ω is minimal. Now we will prove that Σ_ω contains exactly one point. If we suppose the contrary, then there exist $\omega \in \Omega$ and $x_1, x_2 \in \Sigma_\omega$ such that $x_1 \neq x_2$. Since the set Ω is minimal, then there exists a sequence $\{t_n\} \to -\infty$ such that $\omega t_n \to \omega_0$ as $n \to +\infty$. Taking into consideration the compactness of Σ, we may suppose that the sequences $\{x_i t_n\}$ $(i = 1, 2)$ are convergent. We denote by $x_i' = \lim_{n \to +\infty} x_i t_n$. It is clear that $x_i' \in \Sigma_{\omega_0}$ and, consequently, $x_1' = x_2'$. On the other hand, according to Corollary 12.3 the dynamical system $(J, \mathbb{S}, \pi)$ is distal in negative direction and, consequently, $x_1' \neq x_2'$. The obtained contradiction proves our statement. Thus, we have a compactly invariant set $\Sigma \subseteq J$ with the following property $\Sigma_\omega = \Sigma \bigcap J_\omega = \{x_\omega\}$ for all $\omega \in \Omega$. Now we define a mapping $\nu : \Omega \to J$ by the following equality: $\nu(\omega) = x_\omega$ for all $\omega \in \Omega$. It is easy to verify that ν is a continuous and invariant section of the set-valued mapping h^{-1}. The theorem is proved. $\qquad\square$

Corollary 12.5 *Under the conditions of Theorem 12.6 the Levinson's center of the dynamical system $(X, \mathbb{S}_+, \pi)$ contains at least one stationary (τ ($\tau > 0$) - periodic, quasi-periodic, almost periodic) point, if the minimal set Ω consists of a stationary (τ ($\tau > 0$) - periodic, quasi-periodic, almost periodic) point.*

Remark 12.1   *a. In the proof of Lemma 12.4 we used the well known Favard method of minimax (see, for example, [151] and [238]).*

b. In the second part of the proof of Theorem 12.6 we used some ideas of Zhikov [238] (Ch. VII) and Shcherbakov [302] (Ch. III).

12.4   Pullback attractors of V-monotone NDS

Let E be a Banach space and $\langle E, \varphi, (\Omega, \mathbb{S}, \sigma), h \rangle$ be a cocycle on the state space E. Denote by $X := E \times \Omega$, $(X, \mathbb{S}_+, \pi)$ a skew-product dynamical system, i.e. $\pi = (\varphi, \sigma)$ and $\langle (X, \mathbb{S}_+, \pi), (\Omega, \mathbb{S}, \sigma), h \rangle$ is a non-autonomous dynamical system generated by the cocycle φ, where $h := pr_2 : X \to \Omega$.

Definition 12.6   The cocycle φ is called V-monotone (see [113], [114], [238], [333]), if there exists a continuous function $\mathcal{V} : E \times E \times \Omega \to \mathbb{R}^+$ with the following properties:

a. $\mathcal{V}(u_1, u_2, \omega) \geq 0$ for all $\omega \in \Omega$ and $u_1, u_2 \in \Omega$.
b. $\mathcal{V}(u_1, u_2, \omega) = 0$ if and only if $u_1 = u_2$.
c. $\mathcal{V}(\varphi(t, \omega, u_1), \varphi(t, \omega, u_2), \sigma_t \omega) \leq \mathcal{V}(u_1, u_2, \omega)$ for all $u_1, u_2 \in E, \omega \in \Omega$ and $t \in \mathbb{S}_+$.

Lemma 12.5   *The cocycle φ is $V-$monotone if and only if the non-autonomous dynamical system $\langle (X, \mathbb{S}_+, \pi), (\Omega, \mathbb{S}, \sigma), h \rangle$ generated by the cocycle φ is V-monotone.*

Proof.   Let φ be a V-monotone cocycle. Then we define a mapping $V : X \dot\times X \to \mathbb{R}^+$ by the equality

$$V((\omega, u_1), (\omega, u_2)) := \mathcal{V}(u_1, u_2, \omega) \qquad (12.14)$$

for all $(\omega, u_1), (\omega, u_2) \in X \dot\times X$, where $\mathcal{V} : E \times E \times Y \to \mathbb{R}^+$ is the function figuring in the definition of V-monotone cocycle φ and $X \dot\times X := \{ (x_1, x_2) \in X \times X \mid h(x_1) = h(x_2) \}$.

From the equality (12.14) and a.-c. follows that V possesses the following properties:

i) $V(x_1, x_2) \geq 0$ for all $(x_1, x_2) \in X \dot\times X$ and $V(x_1, x_2) = 0$ if and only if $x_1 = x_2$.
ii) V is continuous.
iii) $V(x_1 t, x_2 t) \leq V(x_1, x_2)$ for all $(x_1, x_2) \in X \dot\times X$ and $t \in \mathbb{S}_+$.

The inverse statement is proved by analogy. $\square$

Theorem 12.7   *Let φ be a $V-$monotone cocycle admitting a compact pullback attractor $\{I_\omega \mid \omega \in \Omega \}$ and let exist a function $a \in \mathcal{K}$ with the property $a(r) \to +\infty$*

as $r \to +\infty$ such that $\mathcal{V}(u_1, u_2, \omega) \geq a(\rho(u_1, u_2))$ *for all* $u_1, u_2 \in E$ *and* $\omega \in \Omega$. *If the NDS* $\langle (X, \mathbb{S}_+, \pi), (\Omega, \mathbb{S}, \sigma), h \rangle$ *generated by the cocycle* φ *is* α-*condensing, then it admits a compact global attractor* J *and* $J_\omega = I_\omega \times \{\omega\}$ *for every* $\omega \in \Omega$, *where* $J_\omega := h^{-1} \bigcap J$.

Proof. First step: we will prove that under the conditions of Theorem 12.7 all trajectory of the dynamical system $(X, \mathbb{S}_+, \pi)$ are positively compact. Really, let $x \in X$ ($x := (u, \omega)$) and $q := (p, \omega) \in J_\omega := I_\omega \times \{\omega\}$ be an arbitrary point. Then in virtue of the $V-$monotonicity of the cocycle φ we have

$$\mathcal{V}(\varphi(t, u, \omega), \varphi(t, p, \omega), \sigma_t \omega) \leq \mathcal{V}(u, p, \omega) \tag{12.15}$$

for all $t \in \mathbb{S}_+$. From the inequality (12.15) results that $\{\varphi(t, \omega, u) \mid t \in \mathbb{S}_+ \}$ is bounded. If we suppose that it is not true, then there exists a sequence $t_n \to +\infty$ such that

$$\lim_{n \to +\infty} \rho(\varphi(t_n, u, \omega), \varphi(t_n, p, \omega)) = +\infty. \tag{12.16}$$

Under the conditions of Theorem 12.7 from (12.15) and (12.16) follows

$$\mathcal{V}(u, p, \omega) \geq \mathcal{V}(\varphi(t_n, u, \omega), \varphi(t_n, p, \omega), \sigma_{t_n} \omega) \geq$$

$$a(\rho((\varphi(t_n, u, \omega), \varphi(t_n, p, \omega)) \to +\infty$$

as $n \to +\infty$. The obtained contradiction proves our assertion.

Taking into consideration the asymptotic compactness of the dynamical system $(X, \mathbb{S}_+, \pi)$ and the boundedness of the positively invariant set $M := \{(\varphi(t, \omega, u), \sigma_t \omega) \mid t \in \mathbb{S}_+ \}$, we conclude that it is relatively compact. Thus, under the conditions of Theorem 12.7 every positive semi-trajectory is relatively compact.

Second step: since φ admits a compact global pullback attractor, then according to Theorems 8.4 and 7.2 the set J is asymptotically stable and it is the maximal compact invariant set of X. We will show that $W^s(J) = X$. Really, let $x \in X$ be an arbitrary point. Then its ω-limit set ω_x is nonempty, compact and invariant, because $\{xt \mid t \in \mathbb{S}_+ \}$ is relatively compact. Taking into account the maximality of J, we have $\omega_x \subseteq J$ for all $x \in X$ and, consequently, $W^s(J) = X$, i.e. J is the compact global attractor of $(X, \mathbb{S}_+, \pi)$. The theorem is proved. $\square$

12.5 Applications

12.5.1 *Finite-dimensional systems*

Denote by $\mathbb{R}^n$ a real $n-$dimensional Euclidean space with the scalar product $\langle , \rangle$ and the norm $|\cdot|$ generated by the scalar product. Let $[\mathbb{R}^n]$ be a space of all the linear mappings $A : \mathbb{R}^n \to \mathbb{R}^n$ equipped with the operational norm.

Theorem 12.8 *Let Ω be a compact minimal set, $F \in C(\Omega \times \mathbb{R}^n, \mathbb{R}^n)$, $W \in C(\Omega, [\mathbb{R}^n])$ and the following conditions be held:*

1. *the matrix-function W is positively defined, i.e. $\langle W(\omega)u, u\rangle \in \mathbb{R}$ for all $\omega \in \Omega$, $u \in \mathbb{R}^n$ and there exists a positive constant a such that $\langle W(\omega)u, u\rangle \geq a|u|^2$ for all $\omega \in \Omega$ and $u \in \mathbb{R}^n$.*

2. *the function $t \to W(\sigma_t\omega)$ is differentiable for every $\omega \in \Omega$ and $\dot{W}(\omega) \in C(\Omega, [\mathbb{R}^n])$, where $\dot{W}(\omega) := \frac{d}{dt}W(\sigma_t\omega)|_{t=0}$.*

3. *$\langle \dot{W}(\omega)(u-v) + W(\omega)(F(\omega, u) - F(\omega, v)), u-v\rangle \leq 0$ for all $\omega \in \Omega$ and $u, v \in \mathbb{R}^n$.*

4. *there exist a positive constant r and a function $c : [r, +\infty) \to (0, +\infty)$ such that $\langle \dot{W}(\omega)u + W(\omega)F(\omega, u), u\rangle \leq -c(|u|)$ for all $|u| > r$.*

Then the equation

$$u' = F(u, \sigma_t\omega) \qquad (12.17)$$

generates a cocycle φ on $\mathbb{R}^n$ which admits a compact global attractor $I = \{I_\omega \mid \omega \in \Omega\}$ with the following properties:

a. *I_ω is a nonempty, compact and convex subset of $\mathbb{R}^n$ for every $\omega \in \Omega$.*

b. *$I = \bigcup\{I_\omega \mid \omega \in \Omega\}$ is connected.*

c. *the mapping $\omega \to I_\omega$ is continuous with respect to the metric of Hausdorff.*

d. *$I = \{I_\omega \mid \omega \in \Omega\}$ is invariant, i.e. $\varphi(t, \omega, I_\omega) = I_{\sigma_t\omega}$ for all $\omega \in \Omega$ and $t \in \mathbb{R}_+$.*

e. *$\lim\limits_{t\to+\infty} \beta(\varphi(t, \sigma_t\omega)M, I_\omega) = 0$ for all $M \in C(\mathbb{R}^n)$ and $\omega \in \Omega$.*

f. *$\limsup\limits_{t\to+\infty}\{\beta(\varphi(t, M, \sigma_t\omega), I) \mid \omega \in \Omega\} = 0$ for any $M \in C(\mathbb{R}^n)$, where $I = \bigcup\{I_\omega \mid \omega \in \Omega\}$.*

g. *$I = \{I_\omega \mid \omega \in \Omega\}$ is a uniform forward attractor ,i.e.*

$$\lim_{t\to+\infty} \sup_{\omega\in\Omega} \beta(\varphi(t, \omega)M, I_{\sigma_t\omega}) = 0$$

for any $M \in C(\mathbb{R}^n)$.

h. *the equation (12.17) admits at least one stationary ($\tau-$periodic, quasi-periodic, almost periodic) solution, if the point $\omega \in \Omega$ is stationary ($\tau-$periodic, quasi-periodic, almost periodic).*

Proof. Since the function $F \in C(\Omega \times \mathbb{R}^n, \mathbb{R}^n)$, then, according to the theorem of Peano (see, for example, [186]), the equation (12.17) admits at least one solution $u(t)$ ($t \in [0, t_u), t_u > 0$) with the condition $u(0) = x$ for every $x \in \mathbb{R}^n$. We will show that under the conditions of Theorem 12.8 this solution is unique. In fact, let $u_i(t)$ ($i = 1, 2$) be two solutions of the equation (12.17) defined on $[0, t_0)$ with the condition $u_i(0) = x$ ($i = 1, 2$). We consider the function $\langle W(\omega t)u_1(t), u_2(t)\rangle$.

According to the conditions 1. and 3. of Theorem 12.8 we have

$$a|u_1(t) - u_2(t)|^2 \leq 0$$

for all $t \in [0, t_0)$ and, consequently, $u_1(t) = u_2(t)$ for all $t \in [0, t_0)$.

Now we will prove that every solution of the equation (12.17) is defined on $\mathbb{R}_+$. Let $u \in \mathbb{R}^n$ and $\varphi(t, u, \omega)$ be the unique solution of the equation (12.17) defined on $[0, t_{(u,\omega)})$. To prove that $t_{(u,\omega)} = +\infty$ it is sufficient to show that the solution $\varphi(t, u, \omega)$ is bounded on $[0, t_{(u,\omega)})$. We denote by $b := \max_{\omega \in \Omega} \|W(\omega)\|$ and $T_1(u, \omega) = \{t \in [0, t_{(u,\omega)}) \mid |\varphi(t, u, \omega)| \leq r\}$ and $T_2(u, \omega) = [0, t_{(u,\omega)}) \setminus T_1(u, \omega)$. It is clear that the set $T_2(u, \omega)$ is open and, consequently, $T_2(u, \omega) = \bigcup_\alpha \{(t_\alpha, t_\beta) \mid \beta = \beta(\alpha)\}$. For all $t \in T_2(u, \omega)$ there exists α such that $t \in (t_\alpha, t_\beta)$, $|\varphi(t_\alpha, u, \omega)| = |\varphi(t_\beta, u, \omega)|$ and $|\varphi(t, u, \omega)| > r$. Consequently,

$$a|\varphi(t, u, \omega)|^2 \leq \langle W(\sigma_t\omega)\varphi(t, u, \omega), \varphi(t, u, \omega)\rangle \leq \qquad (12.18)$$

$$\langle W(\omega t_\alpha)\varphi(t_\alpha, u, \omega), \varphi(t_\alpha, u, \omega)\rangle \leq br^2.$$

From the inequality (12.18) follows that $|\varphi(t, u, \omega)| \leq \sqrt{\frac{b}{a}}r$ and, consequently, we obtain $\sup\{|\varphi(t, u, \omega)| \mid t \in [0, t_{(u,\omega)})\} \leq r_0 := \max\{r, \sqrt{\frac{b}{a}}r\}$. Thus, the equation (12.17) defines a cocycle φ on $\mathbb{R}^n$.

Let $X = \mathbb{R}^n \times \Omega$, $(X, \mathbb{R}_+, \pi)$ be a skew-product dynamical system and $\langle(X, \mathbb{R}_+, \pi), (\Omega, \mathbb{R}, \sigma), h\rangle$ be the non-autonomous dynamical system generated by the equation (12.17). Denote by $\mathcal{V} : X \to \mathbb{R}^+$ the function defined by the equality $\mathcal{V}(u, \omega) := \langle W(\omega)u, u\rangle$ for all $(u, \omega) \in X := \mathbb{R}^n \times \Omega$. If $|\varphi(t, u, \omega)| > r$ for all $t \in (t_1, t_2) \subset \mathbb{R}_+$, then

$$\frac{d}{dt}\mathcal{V}(\sigma_t\omega, \varphi(t, u, \omega)) = \langle \dot{W}(\sigma_t\omega)\varphi(t, u, \omega), \varphi(t, u, \omega)\rangle +$$

$$\langle W(\sigma_\omega)F(\sigma_t\omega, \varphi(t, u, \omega)), \varphi(t, u, \omega)\rangle \leq -c(|\varphi(t, u, \omega)|) < 0.$$

In view of Theorem 5.3 the non-autonomous dynamical system $\langle(X, \mathbb{R}_+, \pi), (\Omega, \mathbb{R}, \sigma), h\rangle$ admits a compact global attractor.

Let $V : X \dot{\times} X \to \mathbb{R}^+$ be the function defined by the equality $V(\omega, u_1, u_2) := \langle W(\omega)(u_1 - u_2), u_1 - u_2\rangle$. Then

$$\frac{d}{dt}V(\sigma_t\omega, \varphi(t, u_1, \omega), \varphi(t, u_2, \omega)) = \langle \dot{W}(\sigma_t\omega)(\varphi(t, u_1, \omega) -$$

$$\varphi(t, u_2, \omega)), \varphi(t, u_1, \omega) - \varphi(t, u_2, \omega)\rangle +$$

$$\langle W(\sigma_t\omega)(F(\sigma\omega, \varphi(t, u_1, \omega)) -$$

$$F(\sigma\omega, \varphi(t, u_2, \omega))), \varphi(t, u_1, \omega) - \varphi(t, u_2, \omega)\rangle \leq 0$$

and, consequently, $V(\sigma_t\omega, \varphi(t, u_1, \omega), \varphi(t, u_2, \omega)) \leq V(\omega, u_1, u_2)$ for all $\omega \in \Omega$, $u_1, u_2 \in \mathbb{R}^n$ and $t \in \mathbb{R}_+$. It is easy to verify that the function V satisfies all the conditions of Theorems 12.4, 12.5 and the statements a.-h. follow from Corollary 12.1, Theorems 12.4, 12.5 and Lemma 12.4. The theorem is proved. $\square$

Example 12.1   As an example that illustrates this theorem we can consider the following equation

$$u' = g(u) + f(\sigma_t\omega),$$

where $f \in C(\Omega, \mathbb{R})$ and

$$g(u) = \begin{cases} (u+1)^2 & , u < -1 \\ 0 & , |u| \leq 1 \\ -(u-1)^2) & , u > 1. \end{cases}$$

Example 12.2   Let us consider the equation

$$x'' + p(x)x' + ax = f(\sigma_t\omega),$$

where $p \in C(\mathbb{R}, \mathbb{R})$, $f \in C(\Omega, \mathbb{R})$ and a is a positive number. Denote by $y := x' + F(x)$, where $F(x) := \int_0^x p(s)ds$. Then we obtain the system

$$\begin{cases} x' = y - F(x) \\ y' = -ax + f(\sigma_t\omega). \end{cases} \tag{12.19}$$

Theorem 12.9   *Suppose the following conditions hold:*

1. $p(x) \geq 0$ for all $x \in \mathbb{R}$.
2. there exist positive numbers r and k such that $p(x) \geq k$ for all $|x| \geq r$.

Then the non-autonomous dynamical system generated by (12.19) is compactly dissipative and $V-$monotone.

Proof.   Let $X := \Omega \times \mathbb{R}$ and $\langle (X, \mathbb{R}_+, \pi), (\Omega, \mathbb{R}, \sigma), h \rangle$ be the non-autonomous dynamical system generated by (12.19). We define the function $V : X \to \mathbb{R}^+$ by the equality

$$V(\omega, x, y) := y^2 - yF(x) + \frac{1}{2}F^2(x) + ax^2.$$

Then

$$\frac{d}{dt}V(\pi^t(\omega, x, y))|_{t=0} = -p(x)[y - F(x)]^2 - axF(x) + (2y - F(x))f(\omega).$$

According to [270] (see the proof of Theorem 12.1.2), there exists $R > 0$ such that $\frac{d}{dt}V(\pi^t(\omega,x,y))|_{t=0} < 0$ for all $x^2 + y^2 \geq R^2$ and $V(\omega,x,y) \to +\infty$ as $x^2 + y^2 \to +\infty$. In view of Theorem 5.3 the non-autonomous dynamical system $\langle(X,\mathbb{R}_+,\pi),(\Omega,\mathbb{R},\sigma),h\rangle$ is compactly dissipative.

Let $V : X \dot{\times} X \to \mathbb{R}^+$ be the function defined by the equality $V((u_1,\omega),(u_2,\omega)) := \langle(u_1 - u_2), u_1 - u_2\rangle$. Then

$$\frac{d}{dt}V(\sigma_t\omega, \varphi(t,u_1,\omega), \varphi(t,u_2,\omega)) = -(x_1(t) - x_2(t))[F(x_1(t)) - F(x_2(t))] \leq 0$$

for all $t \in \mathbb{R}$, where $x_i(t) = pr_1\varphi(t,u_i,\omega)$ $(i = 1,2)$, and, consequently, $V(\pi^t(u_1,\omega),\pi^t(u_2,\omega)) \leq V((u_1,\omega),(u_2,\omega))$ for all $t \in \mathbb{R}^+$. To finish the proof it is sufficient to refer to Theorem 12.8. The theorem is proved. $\qquad\square$

12.5.2 *Caratheodory's differential equations*

Let us consider now the equation (12.17) with the right hand side f satisfying the conditions of Caratheodory (see, for example,[292]). The space of all Caratheodory's functions we denote by $\mathfrak{C}(\mathbb{R} \times \mathbb{R}^n, \mathbb{R}^n)$. Topology on this space is defined by the family of semi-norms (see [292])

$$d_{k,m}(f) := \int_{-k}^{k} \max_{|x|\leq m} |f(t,x)|dt.$$

This space is metrizable, and on $\mathfrak{C}(\mathbb{R} \times \mathbb{R}^n, \mathbb{R}^n)$ there can be defined a dynamical system of translations $(\mathfrak{C}(\mathbb{R} \times \mathbb{R}^n, \mathbb{R}^n), \mathbb{R}, \sigma)$.

We consider the equation

$$\frac{dx}{dt} = f(t,x), \qquad\qquad (12.20)$$

where $f \in \mathfrak{C}(\mathbb{R} \times \mathbb{R}^n, \mathbb{R}^n)$, and the family of equations

$$\frac{dx}{dt} = g(t,x), \qquad\qquad (12.21)$$

where $g \in H(f) := \overline{\{f_\tau \mid \tau \in \mathbb{R}\,\}}$, and f_τ is a $\tau-$translation of the function f w.r.t. the variable t, i.e. $f_\tau(t,x) := f(t + \tau, x)$ for all $t \in \mathbb{R}$ and $x \in \mathbb{R}^n$ and by bar is denoted the closure in the space $\mathfrak{C}(\mathbb{R} \times \mathbb{R}^n, \mathbb{R}^n)$. Denote by $\varphi(t,x,g)$ the solution of equation the (12.21) with the initial condition $\varphi(0,g,x) = x$. Then φ is a cocycle on $\mathbb{R}^n$ (see, for example [292]) with the base $H(f)$. Hence, we may apply the general results from section 1-4 to the cocycle φ generated by the equation (12.20) with a Caratheodory's right hand side and obtain some results for this type of equations.

For instance, the following assertion holds.

Theorem 12.10 *Let $f \in \mathfrak{C}(\mathbb{R} \times \mathbb{R}^n, \mathbb{R}^n)$ be an almost periodic function in $t \in \mathbb{R}$ (in the sense of Stepanoff [238]) uniformly w.r.t. x on compacts from $\mathbb{R}^n$, i.e. for* *every $\varepsilon > 0$ and compact $K \subset \mathbb{R}^n$ the set*

$$\mathfrak{T}(\varepsilon, f, K) := \{\tau \in \mathbb{R} \mid \int_0^1 \max_{x \in K} |f(t + \tau + s, x) - f(t + s, x)| ds < \varepsilon \}$$

is relatively dense on $\mathbb{R}$. Suppose that

1. *$\langle f(t, x_1) - f(t, x_2), x_1 - x_2 \rangle \leq 0$ for all $t \in \mathbb{R}$ and $x_1, x_2 \in \mathbb{R}^n$.*
2. *there exists a positive constant r and a function $c : [r, +\infty) \to (0, +\infty)$ such that $\langle f(t, u), u \rangle \leq -c(|u|)$ for all $|u| > r$.*

Then on $\mathbb{R}^n$ the equation (12.20) generates a cocycle φ which admits a compact global attractor $I = \{I_g \mid g \in H(f) \}$ with the following properties:

a. *I_g is a nonempty, compact and convex subset of $\mathbb{R}^n$ for every $g \in H(f)$.*
b. *$I = \bigcup \{I_g \mid g \in H(f) \}$ is connected.*
c. *the mapping $g \to I_g$ is continuous with respect to the metric of Hausdorff.*
d. *$I = \{I_g \mid g \in H(f)\}$ is invariant, i.e. $\varphi(t, g, I_g) = I_{\sigma_t g}$ for all $g \in H(f)$ and $t \in \mathbb{R}_+$.*
e. *$\lim_{t \to +\infty} \beta(\varphi(t, \sigma_{-t} g) M, I_g) = 0$ for all $M \in C(\mathbb{R}^n)$ and $g \in H(f)$.*
f. *$\lim_{t \to +\infty} \sup \{\beta(\varphi(t, \sigma_{-t} g) M, I) \mid g \in H(f) \} = 0$ for any $M \in C(\mathbb{R}^n)$, where $I = \bigcup \{I_g \mid g \in H(f) \}$.*
g. *$I = \{I_g \mid g \in H(f) \}$ is a uniform forward attractor, i.e.*

$$\lim_{t \to +\infty} \sup_{g \in H(f)} \beta(\varphi(t, g) M, I_{\sigma_t g}) = 0$$

for any $M \in C(\mathbb{R}^n)$.
h. *the equation (12.20) admits at least one stationary ($\tau-$periodic, quasi-periodic, almost periodic) solution, if the function $f \in \mathfrak{C}(\mathbb{R} \times \mathbb{R}^n, \mathbb{R}^n)$ is stationary ($\tau-$periodic, quasi-periodic, almost periodic) in $t \in \mathbb{R}$ uniformly w.r.t. x on compacts from $\mathbb{R}^n$.*

Proof. This theorem is proved using the same arguments that we used in the proof of Theorem 12.8. $\qquad \square$

12.5.3 *ODEs with impulse*

Let $\{t_k\}_{k \in \mathbb{Z}}$ be a two-sided sequence of real numbers, $p : \mathbb{R} \to \mathbb{R}^n$ be a continuously differentiable on every interval (t_k, t_{k+1}) function, continuous to the right in every point $t = t_k$, bounded on $\mathbb{R}$, almost periodic in the sense of Stepanoff and

$$p'(t) = \sum_{k \in \mathbb{Z}} s_k \delta_{t_k},$$

where $s_k := p(t_k + 0) - p(t_k - 0)$.

Consider the equation with impulse

$$\frac{dx}{dt} = f(t,x) + \sum_{k\in\mathbb{Z}} s_k \delta_{t_k}$$

or, what is equivalent,

$$\frac{dx}{dt} = f(t,x) + p'(t) \tag{12.22}$$

and paralelly consider the family of equations

$$\frac{dx}{dt} = g(t,x) + q'(t), \tag{12.23}$$

where $(g,q) \in H(f,p) := \overline{\{(f_\tau, p_\tau) | \tau \in \mathbb{R}\}}$ and by bar we denote the closure in the product-space $\mathfrak{C}(\mathbb{R} \times \mathbb{R}^n, \mathbb{R}^n) \times \mathfrak{C}(\mathbb{R}, \mathbb{R}^n)$.

Denote by $\varphi(t,x,g,q)$ the unique solution of the equation (12.23) (see [170] and [279]) satisfying the initial condition $\varphi(0,x,g,q) = x$. This solution is continuous on every interval (t_k, t_{k+1}) and continuous to the right in every point $t = t_k$ (see [170] and [279]).

By the transformation

$$x := y + q(t) \tag{12.24}$$

we can bring the equation (12.23) to the equation

$$\frac{dy}{dt} = g(t, y + q(t)). \tag{12.25}$$

Theorem 12.11 *Let $f \in C(\mathbb{R} \times \mathbb{R}^n, \mathbb{R}^n)$ be a Bohr's almost periodic function in $t \in \mathbb{R}$ uniformly with respect to x on every compact from $\mathbb{R}^n$ and $p \in \mathfrak{C}(\mathbb{R}, \mathbb{R}^n)$ be a Stepanoff's almost periodic function. Suppose that $\langle f(t,x_1) - f(t,x_2), x_1 - x_2 \rangle \le 0$ for all $t \in \mathbb{R}$ and that there exist positive numbers α, L_1, L_2 and r such that*

$$\langle f(t,x), x \rangle \le -L_1 |x|^{\alpha+1} \quad and \quad |f(t,x)| \le L_2 |x|^\alpha$$

for all $t \in \mathbb{R}$ and $|x| > r$ $(x \in \mathbb{R}^n)$.

Then the equation (12.22) admits a compact global attractor $\{I_{(g,q)} \mid (g,q) \in H(f,p)\}$ with the following properties:

a. *$I_{(g,q)}$ is a nonempty, compact and convex subset of $\mathbb{R}^n$ for every $(g,q) \in H(f,p)$.*

b. *$I = \bigcup\{I_{(g,q)} \mid (g,q) \in H(f,p)\}$ is disconnected.*

c. *$I = \{I_{(g,q)} \mid (g,q) \in H(f,p)\}$ is invariant, i.e. $\varphi(t, I_{(g,q)}, g, q) = I_{\sigma_t(g,q)}$ for all $(g,q) \in H(f,p)$ and $t \in \mathbb{R}_+$.*

d. *$\lim\limits_{t\to+\infty} \beta(\varphi(t, \sigma_t(g,q))M, I_{(g,q)}) = 0$ for all $M \in C(\mathbb{R}^n)$ and $(g,q) \in H(f,g)$.*

e. *$\lim\limits_{t\to+\infty} \sup\{\beta(\varphi(t, \sigma_{-t}(g,q))M, I) \mid (g,q) \in H(f,p)\} = 0$ for any $M \in C(\mathbb{R}^n)$, where $I = \bigcup\{I_{(g,q)} \mid (g,q) \in H(f,p)\}$.*

f. $I = \{I_g \mid g \in H(f)\}$ is a uniform forward attractor ,i.e.

$$\lim_{t \to +\infty} \sup_{(g,q) \in H(f,p)} \beta(\varphi(t,g,q)M, I_{\sigma_t(g,q)}) = 0$$

for any $M \in C(\mathbb{R}^n)$.

g. the equation (12.22) admits at least one stationary (τ - periodic, quasi-periodic, almost periodic in the sense of Stepanoff) solution, if the function $(f,p) \in \mathfrak{C}(\mathbb{R} \times \mathbb{R}^n, \mathbb{R}^n) \times \mathfrak{C}(\mathbb{R}, \mathbb{R}^n)$ is stationary (τ - periodic, quasi-periodic, almost periodic in $t \in \mathbb{R}$) uniformly w.r.t. x on compacts of $\mathbb{R}^n$.

Proof. Let $\varphi(t,x,g,q)$ be the cocycle generated by the family of equations (12.23) and $\tilde{\varphi}(t,y,g,q)$ be the cocycle generated by the family of equations (12.25). Then we have the following equality

$$\varphi(t,x,g,q) = q(t) + \tilde{\varphi}(t,x - q(0),g,q)). \tag{12.26}$$

We will show that it is possible to apply Theorem12.10 to the equation

$$\frac{dy}{dt} = f(t, y + p(t)).$$

Really,

$$\langle f(t, y_1 + p(t)) - f(t, y_2 + p(t)), y_1 - y_2 \rangle$$

for all $t \in \mathbb{R}$ and $y_1, y_2 \in \mathbb{R}^n$, and

$$\langle f(t, y + p(t)), y \rangle = \langle f(t, y + p(t)), y + p(t) \rangle - \langle f(t, y + p(t)), p(t) \rangle \le \tag{12.27}$$

$$-L_1 |y + p(t)|^{\alpha+1} + L_2 \|p\| |y + p(t)|^{\alpha}$$

for all $t \in \mathbb{R}$ and $|y + p(t)| > r$, where $\|p\| := \sup\{|p(t)| \mid t \in \mathbb{R}\}$. Taking into account the fact that the function p is bounded on $\mathbb{R}$, from (12.27) we obtain the existence of positive numbers R (that sufficiently large) and L_1^0, L_2^0 such that

$$\langle f(t, y + p(t)), y \rangle \le -L_1^0 |y|^{\alpha+1} \text{ and } \langle f(t, y + p(t)) \rangle \le L_2^0 |y|^{\alpha}$$

for all $t \in \mathbb{R}$ and $|y| > R$. To finish the proof of the theorem it is sufficient to apply Theorem 12.10 and take into consideration the relations (12.24) and (12.26). The theorem is proved. $\square$

12.5.4   *Evolution equations with monotone operators*

Let H be a real Hilbert space with the inner product $\langle , \rangle, |\cdot| = \sqrt{\langle , \rangle}$ and E be a reflexive Banach space contained in H algebraically and topologically. Furthermore, let E be dense in H, and here H can be identified with a subspace of the dual E' of E and $\langle , \rangle$ can be extended by continuity to $E' \times E$.

We consider the initial value problem

$$u'(t) + Au(t) = f(\sigma_t \omega) \tag{12.28}$$

$$u(0) = u, \tag{12.29}$$

where $A : E \to E'$ is bounded (generally nonlinear),

$$|Au|_{E'} \leq C|u|_E^{p-1} + K, u \in E, p > 1,$$

coercive,

$$\langle Au, u \rangle \geq a|u|_E^p, u \in E, a > 0,$$

monotone,

$$\langle Au_1 - Au_2, u_1 - u_2 \rangle \geq 0, u_1, u_2 \in E,$$

and semi-continuous (see [240]).

A nonlinear "elliptic" operator

$$Au = -\sum_{i=1}^{n} \frac{\partial}{\partial x_i} \phi(\frac{\partial u}{\partial x_i}) \quad \text{in } D \subset \mathbb{R}^n$$

$$u = 0 \text{ on } \partial D,$$

where D is a bounded domain in $\mathbb{R}^n$, $\phi(\cdot)$, is an increasing function satisfying

$$\phi|_{[-1,1]} = 0, \ c|\xi|^p \leq \sum_{i=1}^{n} \xi_i \phi(\xi_i) \leq C|\xi|^p \ (\text{ for all } |\xi| \geq 2),$$

and provides an example with $H = L^2(D), E = W_0^{1,p}(D), \ E' = W^{-1,p'}(D), \ p' = \frac{p}{p-1}$.

The following result is established in [240] (Ch.2 and Ch.4). If $x \in H$ and $f \in C(\Omega, E')$, $p' = \frac{p}{p-1}$, then there exists a unique solution $\varphi \in C(\mathbb{R}_+, H)$ of (12.28) – (12.29).

We denote by $\varphi(\cdot, u, \omega)$ the unique solution of (12.28) and (12.29). According to [187], $\varphi(\cdot, u, \omega)$ is a continuous cocycle on H.

Theorem 12.12 *Suppose that the operator A satisfies the conditions above and the cocycle φ generated by the equation (12.28) is asymptotically compact. Then the equation (12.28) admits a compact global attractor $I = \{I_\omega \mid \omega \in \Omega\}$ possessing the following properties:*

a. I_ω is a nonempty, compact and convex subset of H for every $\omega \in \Omega$.

b. $I = \bigcup\{I_\omega \mid \omega \in \Omega\}$ is connected.

c. the mapping $\omega \to I_\omega$ is continuous with respect to the metric of Hausdorff.

d. $I = \{I_\omega \mid \omega \in \Omega \}$ *is invariant, i.e.* $\varphi(t, I_\omega, \omega) = I_{\sigma_t \omega}$ *for all* $\omega \in \Omega$ *and* $t \in \mathbb{R}_+$.

e. $\lim\limits_{t \to +\infty} \beta(\varphi(t, \sigma_{-t}\omega)M, I_\omega) = 0$ *for all* $M \in C(H)$ *and* $\omega \in \Omega$.

f. $\lim\limits_{t \to +\infty} \sup\{\beta(\varphi(t, \sigma_t\omega)M, I)\mid \omega \in \Omega \} = 0$ *for any* $M \in C(H)$, *where* $I = \bigcup\{I_\omega \mid \omega \in \Omega \}$.

g. $I = \{I_\omega \mid \omega \in \Omega \}$ *is a uniform forward attractor ,i.e.*

$$\lim_{t \to +\infty} \sup_{\omega \in \Omega} \beta(\varphi(t, \omega)M, I_{\sigma_t \omega}) = 0$$

for any $M \in C(H)$.

h. *the equation (12.28) admits at least one stationary (* τ *- periodic, quasi-periodic, almost periodic) solution, if the point* $\omega \in \Omega$ *is stationary (* τ *- periodic, quasi-periodic, almost periodic).*

Proof. Let $X := H \times \Omega$, $(X, \mathbb{R}_+, \pi)$ be a skew-product dynamical system and $\langle (X, \mathbb{R}_+, \pi), (\Omega, \mathbb{R}, \sigma), h \rangle$ be a non-autonomous dynamical system generated by the equation (12.28). Denote by $\mathcal{V} \to \mathbb{R}^+$ the function defined by the equality $\mathcal{V}(\omega, u) := \langle u, u \rangle$ for all $(u, \omega) \in X = H \times \Omega$. If $|\varphi(t, u, \omega)| > r := (\frac{\|f\|}{a})^{\frac{1}{p-1}}$ (where $\|f\| := \max\{|f(\omega)| \mid \omega \in \Omega \}$) for all $t \in (t_1, t_2) \subset \mathbb{R}_+$, then

$$\frac{d}{dt}\mathcal{V}(\sigma_t\omega, \varphi(t, u, \omega)) = \langle A(\varphi(t, u, \omega)), \varphi(t, u, \omega) \rangle$$
$$\leq -a|\varphi(t, u, \omega)|^p + \|f\||\varphi(t, u, \omega)| < 0.$$

In view of Theorem 5.3 the non-autonomous dynamical system $\langle (X, \mathbb{R}_+, \pi), (\Omega, \mathbb{R}, \sigma), h \rangle$ admits a compact global attractor.

Let $V : X \dot\times X \to \mathbb{R}^+$ be a function defined by the equality $V(\omega, u_1, u_2) := \langle u_1 - u_2, u_1 - u_2 \rangle$. Then

$$\frac{d}{dt}V(\sigma_t\omega, \varphi(t, u_1, \omega), \varphi(t, u_2, \omega))$$
$$= 2\langle \frac{d}{dt}(\varphi(t, u_1, \omega) - \varphi(t, u_2, \omega)), \varphi(t, u_1, \omega) - \varphi(t, u_2, \omega) \rangle$$
$$= 2\langle A(\varphi(t, \omega, u_1)) - A(\varphi(t, u_2, \omega)), \varphi(t, u_1, \omega) - \varphi(t, u_2, \omega) \rangle \leq 0$$

and, consequently, $V(\sigma_t\omega, \varphi(t, u_1, \omega), \varphi(t, u_2, \omega)) \leq V(\omega, u_1, u_2)$ for all $\omega \in \Omega$, $u_1, u_2 \in H$ and $t \in \mathbb{R}_+$. It is easy to verify that the function V satisfies all the conditions of Theorems 12.4, 12.5 and the statements a.-h. follow from Theorems 12.4, 12.5, 12.6 and Corollary 12.5. The theorem is proved. $\square$

Remark 12.2 *If the injection of E into H is compact, then the cocycle φ generated by the equation (12.28), evidently, is asymptotically compact.*

Example 12.3 A typical example of the equation of the type (12.28) is the equation

$$\frac{\partial}{\partial t}u = \sum_{i=1}^{n}\frac{\partial}{\partial x_i}\phi\left(\frac{\partial u}{\partial x_i}\right) + f(\sigma_t\omega), \ u|_{\partial D} = 0 \qquad (12.30)$$

with "nonlinear Laplacian" $Au = \sum_{i=1}^{n}\frac{\partial}{\partial x_i}\phi\left(\frac{\partial u}{\partial x_i}\right)$, where $\phi(\cdot)$ is an increasing function satisfying the condition

$$c|\xi|^p \le \sum_{i=1}^{n}\xi_i\phi(\xi_i) \le C|\xi|^p$$

for all $|\xi| \ge 2$ and $\phi|_{[-1,1]} = 0$, provides an example with $H = L^2(D), E = W_0^{1,p}(D), \ E' = W^{-1,p'}(D), p' = \frac{p}{p-1}$. It is possible to verify (see, for example, [240],[30] and [20]) that the "nonlinear Laplacian" satisfies all the conditions of Theorem 12.12 and, consequently, (12.30) admits a compact global attractors with the properties a.-h. We note that the attractor of the equation (12.30) is not trivial, i.e. the set I_ω is not single point set, at least for certain $\omega \in \Omega$.

Remark 12.3 *If the operator* $A := \sum_{i=1}^{n}\frac{\partial}{\partial x_i}\phi\left(\frac{\partial u}{\partial x_i}\right)$ *is uniformly elliptic, i.e.* $c|\xi|^p \le \sum_{i=1}^{n}\xi_i\phi(\xi_i) \le C|\xi|^p$ *(for all* $\xi \in \mathbb{R}^n$*), then the set* I_ω *is the single point set for all* $\omega \in \Omega$ *(for autonomous systems see* [314], *Ch.III), because in this case the non-autonomous dynamical system generated by the equation (12.30) is strictly monotone.*

Chapter 13

Linear almost periodic dynamical systems

13.1 Bounded motions of linear systems

Assume that X and Y are complete metric spaces, $\mathbb{R}$ ($\mathbb{Z}$) is the group of real numbers (integers), $\mathbb{S} = \mathbb{R}$ or $\mathbb{S} = \mathbb{Z}, \mathbb{S}_+ = \{s \in \mathbb{S} \mid s \geq 0\}, \mathbb{S}_- = \{s \in \mathbb{S} \mid s \leq 0\}$, and $\mathbb{T} = \mathbb{S}_+, \mathbb{S}_-$ or $\mathbb{S}$. Let $(X, \mathbb{T}, \pi)$ $((Y, \mathbb{S}, \sigma))$ be a semigroup (group) dynamical system on $X(Y)$.

Recall that a triple $\langle (X, \mathbb{T}, \pi), (Y, \mathbb{S}, \sigma), h \rangle$, where h is a homomorphism of $(X, \mathbb{T}, \pi)$ onto $(Y, \mathbb{S}, \sigma)$, is called [69] a non-autonomous dynamical system.

Definition 13.1 $\langle (X, \mathbb{T}, \pi), (Y, \mathbb{S}, \sigma), h \rangle$ is said [32, 333] to be distal on $\mathbb{S}_+$ in the fiber $X_y := \{x \in X \mid h(x) = y\}$ if $\inf\{\rho(x_1 t, x_2 t) \mid t \in \mathbb{S}_+\} > 0$ for all $x_1, x_2 \in X_y, x_1 \neq x_2$.

For group non-autonomous dynamical systems the distalness on $\mathbb{S}_-$ and $\mathbb{S}$ on the fiber X_y can be defined likewise.

Definition 13.2 A non-autonomous system is said to be distal on $\mathbb{S}_+(\mathbb{S}_-, \mathbb{S})$ if it is distal in every fiber $X_y, y \in Y$.

Assume that $(X_i, \mathbb{T}, \pi_i)$ is a dynamical system on $X_i, i = 1, \ldots, k$; let $X := X_1 \times \ldots \times X_k$, and let $\pi := (\pi_1, \ldots, k) : X \times \mathbb{T} \to X$ be defined by the formula

$$\pi(x, t) := (\pi_1(x_1, t), \ldots, \pi_k(x_k, t))$$

for all $t \in \mathbb{T}$ and $x := (x_1, \ldots, x_k) \in X$.

Definition 13.3 The dynamical system $(X, \mathbb{T}, \pi)$, where $X := X_1 \times \ldots \times X_k$ and $\pi := (\pi_1, \ldots, \pi_k)$, is called the direct product of the dynamical systems $(X_i, \mathbb{T}, \pi_i), i = 1, \ldots, k$ and denoted by $(X_1, \mathbb{T}, \pi_1) \times, \ldots, (X_k, T, \pi_k)$.

If $X_i = X, i = 1, \ldots, k$, and $\pi_i = \pi, i = 1, \ldots, k$, then

$$(X, \mathbb{T}, \pi) \times (X, \mathbb{T}, \pi) \times \ldots \times (X, \mathbb{T}, \pi) := (X^k, \mathbb{T}, \pi).$$

407

The direct product of group dynamical systems is defined likewise.

Definition 13.4   The points $x_1, ..., x_k \in X$ are said to be jointly recurrent if the point $(x_1, ..., x_k) \in X^k$ is recurrent in the dynamical system $(X^k, \mathbb{T}, \pi)$.

Lemma 13.1   [32, 333]. *The following assertions hold.*

(1) Assume that X is compact and $(Y, \mathbb{S}, \sigma)$ is minimal. If the group non-autonomous dynamical system $\langle (X, \mathbb{S}, \pi), (Y, \mathbb{S}, \sigma), h \rangle$ is distal on $\mathbb{S}_+(\mathbb{S}_-)$, then it is distal on $\mathbb{S}$.

(2) Assume that X is compact, $(Y, \mathbb{S}, \sigma)$ is minimal, and $y \in Y$. Then the following conditions are equivalent :

    (a) the group non-autonomous system $\langle (X, \mathbb{S}, \pi), (Y, \mathbb{S}, \sigma), h \rangle$ is distal on $\mathbb{S}$ in the fiber X_y;

    (b) for any points $x_1, ..., x_k \in X$, where k is any positive integer ≥ 2, the point $(x_1, ..., x_k) \in X^k$ is recurrent in $(X^k, \mathbb{S}, \pi)$.

Let $C(\mathbb{S}, X)$ be the space of all continuous maps $\varphi : \mathbb{S} \to X$ equipped with the compact-open topology and let $(C(\mathbb{S}, X), \mathbb{S}, \sigma)$ be the dynamical system of translations (shifts) on $C(\mathbb{S}, X)$. Let d be a metric on $C(\mathbb{S}, X)$ consistent with its topology (for example, the Bebutov metric).

Lemma 13.2   [282]. *Let $(X, \mathbb{S}_+, \pi)$ be a semigroup dynamical system and assume that for any $t \in \mathbb{S}_+$ the map $\pi^t : X \to X$ is a homeomorphism and $\tilde{\pi}$ is the map of $X \times \mathbb{S}$ to X defined by the equality*

$$\tilde{\pi}(x, t) := \begin{cases} \pi(x, t), & (x, t) \in X \times \mathbb{S}_+, \\ (\pi^{-t})^{-1}(x), & (x, t) \in X \times \mathbb{S}_-. \end{cases}$$

Then the triple $(X, \mathbb{S}, \tilde{\pi})$ is a group dynamical system.

Lemma 13.3   [32]. *Assume that $\langle (X, \mathbb{S}_+, \pi), (Y, \mathbb{S}, \sigma), h \rangle$ is a non-autonomous dynamical system, Y is compact, $(Y, \mathbb{S}, \sigma)$ is minimal, and $x_0 \in X_y$ has a relatively compact semi-trajectory $\{x_0 t \mid t \in \mathbb{S}_+\}$. Then one can find a recurrent point $x \in \omega_{x_0}$ $(x \in X_y)$ and a sequence $t_n \to +\infty$ such that $\rho(x_0 t_n, x t_n) \to 0$ as $n \to +\infty$.*

Recall that a dynamical system $(X, \mathbb{S}, \pi)$ is said to be asymptotically compact if for any bounded positively invariant set $B \subseteq X$ there is a non-empty compact set $K \subseteq X$ such that

$$\lim_{t \to +\infty} \sup\{\rho(xt, K) \mid x \in B\} = 0.$$

Remark 13.1   *(i) Assume that $x \in X$ is such that $\{xt \mid t \in \mathbb{S}_+\}$ is bounded and $(X, \mathbb{S}_+, \pi)$ is asymptotically compact. Then $\{xt \mid t \in \mathbb{S}_+\}$ is relatively compact.*

(ii) Let $M \subseteq X$ be bounded and invariant. Then M is relatively compact if the dynamical system $(X, \mathbb{S}_+, \pi)$ is asymptotically compact. In particular, if $x \in X$ and $\gamma \in \Phi_x$ is such that $\gamma(\mathbb{S})$ is bounded, then $\gamma(\mathbb{S})$ is relatively compact, where Φ_x is a set of all entire trajectories of $(X, \mathbb{S}_+, \pi)$ passing through x at $t = 0$.

If $X := E \times Y, \pi := (\varphi, \sigma)$, that is, $\pi((u, y), t) := (\varphi(t, x, y), \sigma(t, y))$ for all $(u, y) \in E \times Y$ and $t \in S$, then the non-autonomous dynamical system $\langle (X, \mathbb{T}, \pi), (Y, \mathbb{S}, \sigma), h \rangle$, where $h := pr_2 : X \to Y$, is called [275] a skew product over $(Y, \mathbb{S}, \sigma)$ with the fiber E.

Let (X, h, Y) be a locally trivial Banach fiber bundle [28].

Recall that a non-autonomous dynamical system $\langle (X, \mathbb{T}, \pi), (Y, \mathbb{S}, \sigma), h \rangle$ is said [275], [33] to be linear if the map $\pi^t : X_y \to X_{yt}$ is linear for every $t \in \mathbb{T}$ and $y \in Y$.

If $\langle (X, \mathbb{T}, \pi), (Y, \mathbb{S}, \sigma), h \rangle$ is a skew product over $(Y, \mathbb{S}, \sigma)$ with the fiber E, then it is linear if and only if E is a Banach space and the map $\varphi(t, \cdot, y) : E \to E$ is linear for every $y \in Y$ and $t \in \mathbb{T}$.

Throughout the rest of this section we assume that Y is compact, the dynamical system $(Y, \mathbb{S}, \sigma)$ is minimal, $X = E \times Y, E$ is a Banach space with the norm $|\cdot|$, the non-autonomous dynamical system $\langle (X, \mathbb{T}, \pi), (Y, \mathbb{S}, \sigma), h \rangle$ is linear, $\pi = (\varphi, \sigma)$, and $h = pr_2$.

Let $F \subseteq E \times Y$ be a closed vectorial subset of the trivial fiber bundle $(E \times Y, pr_2, Y)$ that is positively invariant relative to $(X, \mathbb{T}, \pi)$. We put

$$\mathbb{B}^+ = \{(x, y) \in F \mid \sup |\varphi(t, x, y)| \; : \; t \in \mathbb{S}_+ < +\infty\}.$$

The set $\mathbb{B}^-$ is defined likewise. If $\langle (X, \mathbb{S}_+, \pi), (Y, \mathbb{S}, \sigma), h \rangle$ is a semigroup non-autonomous dynamical system, then $\mathbb{B}$ is the set of all points of F with the following property: there is an entire trajectory of the dynamical system $(F, \mathbb{S}_+, \pi)$ bounded on $\mathbb{S}$ that passes through this point. We put $\mathbb{B}_y^+ := \mathbb{B}^+ \bigcap X_y$ and $\mathbb{B}_y := \mathbb{B} \bigcap X_y, y \in Y$.

Theorem 13.1   *The following conditions are equivalent :*

(i) there is an $M > 0$ such that

$$|\varphi(t, x, y)| \le M|x| \tag{13.1}$$

for all $(x, y) \in \mathbb{B}^+(\mathbb{B}^-, \mathbb{B})$ and $t \in \mathbb{S}_+(\mathbb{S}_-, \mathbb{S})$;
(ii) $\mathbb{B}^+(\mathbb{B}^-, \mathbb{B})$ is closed in F.

Proof.   We prove this theorem in the case when $\mathbb{T} = \mathbb{S}_+$. In the case when $\mathbb{T} = \mathbb{S}_-$ or $\mathbb{T} = \mathbb{S}$ it can be proved in a similar way. We claim that (i) implies (ii). Assume that $(x, y) \in \mathbb{B}^+$. Then there is an $(x_n, y_n) \in \mathbb{B}^+$ such that $(x_n, y_n) \to (x, y)$ as $n \to +\infty$. By condition (i), the inequality

$$|\varphi(t, x_n, y_n)| \le M|x_n| \tag{13.2}$$

is valid for all $n = 1, 2, \ldots$ and $t \in \mathbb{S}_+$. Passing to the limit in (13.2) as $n \to +\infty$, we obtain that $|\varphi(t, x, y)| \leq M|x|$ for all $t \in \mathbb{S}_+$, that is, $(x, y) \in \mathbb{B}^+$.

Now we claim that (ii) implies (i). Let $\mathbb{B}^+$ be closed and let $y \in Y$. We put

$$d(y) = \sup\{|\varphi(t, x, y)| \ : \ t \in \mathbb{S}_+, (x, y) \in \mathbb{B}^+, |x| \leq 1\}. \tag{13.3}$$

We claim that the function $d : Y \to \mathbb{R}_+$ defined by formula (13.3) is lower semi-continuous. Assume the contrary. Then there are $\varepsilon > 0, y \in Y$, and $y_n \to y$ such that

$$\lim_{n \to +\infty} d(y_n) = d(y) - \varepsilon. \tag{13.4}$$

Formula (13.3) implies that there are $|x_n| \leq 1$ and $\{t_n\} \subseteq \mathbb{S}_+((x_n, y) \in \mathbb{B}^+)$ such that

$$d(y) = \lim_{n \to +\infty} |\varphi(t_n, x_n, y)|.$$

Therefore, for $\varepsilon > 0$ one can find a k such that the inequality

$$||\varphi(t_n, x_n, y)| - d(y)| < \frac{\varepsilon}{4} \tag{13.5}$$

is valid for all $n \geq k$. Since the map $\varphi(t_k, x_k, \cdot) : Y \to X$ is continuous, there is an $n = n(k)$ such that

$$|\varphi(t_k, x_k, y_n) - \varphi(t_k, x_k, y)| < \frac{\varepsilon}{4} \tag{13.6}$$

for all $n \geq n(k)$. Inequalities (13.5) and (13.6) imply that

$$|d(y) - |(t_k, x_k, y_n)|| < \frac{\varepsilon}{4} \tag{13.7}$$

for all $n \geq n(k)$. Hence,

$$d(y) - d(y_n) \leq \frac{\varepsilon}{2} \tag{13.8}$$

for all $n \geq n(k)$. On the other hand, formula (13.4) implies that

$$d(y) - d(y_n) \geq 3\frac{\varepsilon}{4} \tag{13.9}$$

if n is sufficiently large.

Inequality (13.9) contradicts (13.8). This contradiction proves that $d : Y \to \mathbb{R}_+$ is lower semi-continuous. Hence, this function has a set of points of continuity $D \subseteq Y$ of the type G_δ. Let $p \in D$, then there exists positive numbers δ_p and M_p such that $d(y) \leq M_p$ for all

$$y \in B[p, \delta_p] := \{y \in Y \mid \rho(y, p) \leq \delta_p\} \subseteq Y.$$

Since Y is minimal, there are negative numbers $t_1, \ldots, t_m$ such that $Y = \bigcup_{i=1}^{m} \sigma(t_i, B[p, \delta_p])$ (see [238, p.134]). We put $L := \max\{|t_i| \ : \ i = 1, \ldots, m\}$.

We claim that the family of operators $\{\pi^t \mid t \in [0, L]\}$ is uniformly continuous, that is, for any $\varepsilon > 0$ there is a $\delta(\varepsilon) > 0$ such that $|x| \leq \delta$ implies that $|xt| \leq \varepsilon$ for all $t \in [0, L]$. Assume the contrary. Then there are $\varepsilon_0 > 0, \delta_n \downarrow 0, |x_n| < \delta_n, y_n \in Y$, and $t_n \in [0, L]$ such that

$$|\varphi(t_n, x_n, y_n)| \geq \varepsilon_0. \tag{13.10}$$

Since Y and $[0, L]$ are compact, we can assume that the sequences $\{y_n\}$ and $\{t_n\}$ are convergent. Put $y_0 = \lim_{n \to +\infty} y_n$ and $t_0 := \lim_{n \to +\infty} t_n$. Passing to the limit in (13.10) as $n \to +\infty$, we obtain $0 \geq \varepsilon_0$. The last inequality contradicts the choice of ε_0. This contradiction proves the above assertion.

If $\alpha > 0$ is such that $|\varphi(t, x, y)| \leq 1$ for all $t \in [0, L], |x| \leq \alpha$, and $y \in Y$, then

$$|\varphi(t, x, y)| \leq \alpha^{-1}|x| \tag{13.11}$$

for all $t \in [0, L], x \in E$, and $y \in Y$. Assume that $q \in Y, y \in B[p, \delta_p]$, and t_i are such that $q = yt_i$. Then

$$|\varphi(t, x, q)| = |\varphi(t, x, yt_i)| = |\varphi(t + t_i, \varphi(-t_i, x, yt_i), y)| =$$

$$|\varphi(-t_i, x, yt_i)||\varphi(t + t_i, \frac{\varphi(-t_i, x, yt_i)}{|\varphi(-t_i, x, yt_i)|}, y)|. \tag{13.12}$$

The set $\mathbb{B}^+$ is positively invariant and contains (x, q). Therefore, $\mathbb{B}^+$ contains $\pi^{-t_i}(x, q)$, where $\pi^{-t_i}(x, q) = (\varphi(-t_i, x, q), \sigma(-t_i, q))$. Hence, $\pi^{-t_i}(x, q) = (\varphi(-t_i, x, q), \sigma(-t_i, q)) \in \mathbb{B}_y^+$ and

$$|\varphi(t + t_i, \frac{\varphi(-t_i, x, yt_i)}{|\varphi(-t_i, x, yt_i)|, y)|}, y)| \leq M_p \tag{13.13}$$

for all $t \geq L$. On the other hand, inequality (13.11) implies that

$$|\varphi(-t_i, x, yt_i)| \leq \alpha^{-1}|x| \tag{13.14}$$

for all $x \in E$. Formulas (13.12)-(13.14) imply that $|\varphi(t, x, q)| \leq M|x|$ for all $t \in \mathbb{S}_+$ and $x \in E$, where $M = \alpha^{-1} \max\{1, M_p\}$. This completes the proof of the theorem.

Lemma 13.4 *Assume that $\langle (X, \mathbb{S}_+, \pi), (Y, \mathbb{S}, \sigma), h \rangle$ is a linear non-autonomous dynamical system, $(X, \mathbb{S}_+, \pi)$ is asymptotic compact, and there is an $M > 0$ such that*

$$|\gamma(t)| \leq M|x| \tag{13.15}$$

for all $\gamma \in \Phi_{(x,y)}, (x, y) \in \mathbb{B}$, and $t \in \mathbb{S}$. Then the set $\Phi_0 := \bigcup\{\Phi_{(x,y)} \mid (x, y) \in \mathbb{B}, |x| \leq 1\}$ is relatively compact in $C(\mathbb{S}, X)$.

Proof. Consider the set $K := \{\gamma(t) \mid t \in \mathbb{S}, \gamma \in \Phi_0\} \subseteq \mathbb{B}$. It is obvious that K is invariant. By (13.15), it is bounded. Since $(X, \mathbb{S}_+, \pi)$ is asymptotic compact, K is relatively compact (see remark 13.1). We claim that the family of functions $\Phi_0 \subseteq C(\mathbb{S}, X)$ is equicontinuous. Assume the contrary. Then there are $\varepsilon_0 > 0, \delta_n \downarrow 0, \{t_n^i\}(i = 1, 2)$, and $\{\gamma_n\} \subseteq \Phi_0$ such that $|t_n^1 - t_n^2| < \delta_n$ and

$$|\gamma_n(t_n^1) - \gamma_n(t_n^2)| \geq \varepsilon_0. \tag{13.16}$$

Without loss of generality we can assume that $t^2 - n > t_n^1$. Then inequality (13.16) implies that

$$|\pi^{\tau_n} x_n - x_n| \geq \varepsilon_0 \tag{13.17}$$

for all $n = 1, 2, ...$, where $\tau_n := t_n^2 - t_n^1$ and $x_n := \gamma(t_n^1)$. Since $\{x_n\} \subseteq K$, we can assume that the sequence $\{x_n\}$ converges. Let $x_0 = \lim\limits_{n \to +\infty} x_n$. Passing to the limit in inequality (13.17) as $n \to +\infty$, we obtain the inequality $0 \geq \varepsilon_0$, which contradicts the choice of ε_0. To complete the proof of the lemma, it is sufficient to apply the Arzela–Ascoli- theorem. $\qquad\square$

Theorem 13.2 *Assume that $\langle (X, \mathbb{S}_+, \pi), (Y, \mathbb{S}, \sigma), h \rangle$ is a linear non-autonomous dynamical system, $(X, \mathbb{S}_+, \pi)$ is asymptotic compact, and there is an $M > 0$ such that inequality (13.15) is valid for all $\gamma \in \Phi_{(x,y)}, (x, y) \in \mathbb{B}$, and $t \in S$. Then the following assertions hold:*

(i) Any two different entire trajectories γ_1 and γ_2 $(h(\gamma_1(0)) = h(\gamma_2(0)))$ are jointly recurrent;

(ii) For any $(x, y) \in \mathbb{B}$ the set $\Phi_{(x,y)}$ consists of a single entire recurrent trajectory;

(iii) $\mathbb{B}$ is closed in F;

(iv) $(X, \mathbb{S}_+, \pi)$ induces a group dynamical system $(\mathbb{B}, \mathbb{S}, \pi)$ on $\mathbb{B}$;

(v) For any $y \in Y$ the set $\mathbb{B}_y$ is finite-dimensional and $\dim \mathbb{B}_y$ does not depend on $y \in Y$.

Proof. Assume that $(x, y) \in \mathbb{B}$, and let $\gamma \in \Phi_{(x,y)}$ be bounded on $\mathbb{S}$. By Lemma 13.4, the set $H(\gamma) := \overline{\{\gamma_\tau \mid \tau \in \mathbb{S}\}}$ is compact in $C(\mathbb{S}, X)$, since $\gamma_\tau \in \Phi_{\gamma(\tau)}$, where γ_τ is the shift of γ by τ and the bar denotes closure in $C(\mathbb{S}, X)$. Consider the group non-autonomous dynamical system $\langle (H(\gamma), \mathbb{S}, \lambda), (Y, \mathbb{S}, \sigma), h \rangle$, where $(H(\gamma), \mathbb{S}, \lambda)$ is the dynamical system of shifts on $H(\gamma)$ induced by the Bebutov's system $(C(\mathbb{S}, X), \mathbb{S}, \sigma)$ and $\mu : H(\gamma) \to Y$ is the map defined by the equality $\mu(\psi) = h(\psi(0))$. Under the conditions of Theorem 13.2 (see also inequality (13.15)) the non-autonomous dynamical system $\langle (H(\gamma), \mathbb{S}, \lambda), (Y, \mathbb{S}, \sigma), \mu \rangle$ is negatively distal, that is, $\inf\{d\lambda(t, \gamma_1), \lambda(t, \gamma_2)) : t \in \mathbb{S}_-\} > 0$ for any $\gamma_1, \gamma_2 \in H(\gamma)$ such that $\gamma_1 \neq \gamma_2$ and $\mu(\psi_1) = \mu(\psi_2)$. By Lemma 1 in [238, p.104], $\langle (H(\gamma), \mathbb{S}, \lambda), (Y, \mathbb{S}, \sigma), h \rangle$ is distal on $\mathbb{S}$. Therefore, γ_1 and γ_2 are jointly recurrent. In particular, they are recurrent. Moreover, by Lemma 13.1, any two entire trajectories $\gamma_1 \in \Phi_{(x,y)}$ and $\gamma_2 \in \Phi_{(x,y)}$ are jointly recurrent.

We claim that for any $(x, y) \in \mathbb{B}$ the set $\Phi_{(x,y)}$ consists of a single entire recurrent trajectory. Assume the contrary. Then there are $\gamma_1, \gamma_2 \in \Phi_{(x,y)}$ such that $\gamma_1 \neq \gamma_2$. Putting $\gamma(t) := \gamma_1(t) - \gamma_2(t)$, we obtain a recurrent function $\gamma \neq 0$ such that $\gamma(t) = 0$ for all $t \in \mathbb{S}_+$, which is impossible.

Now let $(x, y) \in \bar{\mathbb{B}}$. Then there is an $(x_n, y_n) \in \mathbb{B}$ such that $(x_n, y_n) \to (x, y)$. Let γ_n be the (unique) entire trajectory of $(F, \mathbb{S}_+, \pi)$ bounded on $\mathbb{S}$ and satisfying the condition $\gamma_n(0) = (x_n, y_n)$. By Lemma 13.4, we can assume that the sequence $\{\gamma_n\}$ converges in $C(\mathbb{S}, X)$. Inequality (13.15) implies that $\gamma = \lim\limits_{n \to +\infty} \gamma_n$ is an entire trajectory of $(F, \mathbb{S}_+, \pi)$ bounded on $\mathbb{S}$. Moreover, $\gamma(0) = (x, y)$. Thus, $\gamma \in \Phi_{(x,y)}$, whence $(x, y) \in \mathbb{B}$.

Since $\mathbb{B}$ is closed and invariant, $(X, \mathbb{S}_+, \pi)$ induces a dynamical system $(B, \mathbb{S}_+, \pi)$ on $\mathbb{B}$, and $\pi^t \mathbb{B} = \mathbb{B}$ for all $t \in \mathbb{S}_+$. By assertion (ii) of the theorem, $\pi^t : \mathbb{B} \to \mathbb{B}$ $(t \in \mathbb{S}_+)$ is a one-to-one map and $(\pi^t)^{-1}(b) = \gamma_b(-t)$ for all $t \in \mathbb{S}_+$ and $b \in \mathbb{B}$, where $\{\gamma_b\} = \Phi_b$. Lemma 13.4 implies that $\pi^t : \mathbb{B} \to \mathbb{B}$ is a homeomorphism. To prove assertion (iv), it is sufficient to apply Lemma 13.2.

We now prove the last assertion of the theorem. Since $(X, \mathbb{S}_+, \pi)$ is asymptotic compact, $K := \{(x, y) \mid (x, y) \in \mathbb{B}, |x| \leq 1, y \in Y\}$ is compact. Therefore, every linear subspace $\mathbb{B}_y$ of $X_y := E \times \{y\}$ is finite-dimensional. Let $y \in Y$ and let $x_1, ..., x_k \in \mathbb{B}_y$ be a basis in $\mathbb{B}_y$. Then Lemma 13.1 implies that the points $x_1, ..., x_k$ are jointly recurrent. Let q be an arbitrary point of Y. Since Y is minimal, there is a sequence $\{t_n\} \subset \mathbb{S}$ such that $yt_n \to q$ and $\pi(t_n, x_i) \to \xi_i$ $(i = 1, ..., k)$ as $n \to +\infty$, where $\xi_1, ..., \xi_k \in \mathbb{B}_q$ and the points $\xi_1, ..., \xi_k$ are jointly recurrent.

We claim that $\xi_1, ..., \xi_k$ are linearly independent. Assume the contrary. Then there are constants $c_1, ..., c_k$ such that $c_1 \xi_1 + ... + c_k \xi_k = 0$ and $\sum_{i=1}^{k} |c_i| \neq 0$. Since $(\xi_1, ..., \xi_k) \in H(x_1, ..., x_k)$ and $(x_1, ..., x_k)$ is recurrent, there is a $\{\tau_n\} \subset \mathbb{S}$ such that $q\tau_n \to y$ and $\pi(\xi_i, \tau_n) \to x_i$ $(i = 1, ..., k)$ as $n \to +\infty$. Therefore,

$$c_1 x_1 + ... + c_k x_k = \lim\limits_{n \to +\infty} \pi(\tau_n, c_1 \xi_1 + ... + c_k \xi_k) = 0.$$

The last relation contradicts the choice of $x_1, ..., x_k$. Thus, $\dim \mathbb{B}_q \geq \dim \mathbb{B}_y$ for all $q \in Y$. Since Y is minimal, the reverse inequality also holds. Hence, $\dim \mathbb{B}_q = dim \mathbb{B}_y$ for all $q \in Y$, which completes the proof of the theorem. $\square$

Theorem 13.3 *Let $\langle (X, \mathbb{S}_+, \pi), (Y, \mathbb{S}, \sigma), h \rangle$ be a linear non-autonomous dynamical system and assume that $(X, \mathbb{S}_+, \pi)$ is asymptotic compact. Then the following conditions are equivalent :*

(i) there is an $M > 0$ such that (13.15) is valid for all $\gamma \in \Phi_{(x,y)}, (x, y) \in \mathbb{B}$, and $t \in \mathbb{S}$;

(ii) $\mathbb{B}$ is closed in F.

Proof. By Theorem 13.2, the theorem will be proved if we prove that (ii) implies (i). By Theorem 13.1, there is an $M > 0$ such that (13.1) is valid for all $(x, y) \in \mathbb{B}$ and $t \in \mathbb{S}_+$. Since $\mathbb{B}$ is invariant, for any $(x, y) \in B$ and $\gamma \in \Phi_{(x,y)}$ the inequality

$$|\gamma(t)| \geq M^{-1}|x| \tag{13.18}$$

is valid for all $t \in \mathbb{S}_-$. Repeating the arguments used in Lemma 13.4, we can show that $H(\gamma) := \overline{\{\gamma_\tau \mid \tau \in \mathbb{S}\}}$ is compact in $C(\mathbb{S}, X)$. Consider the group dynamical system $\langle (H(\gamma), \mathbb{S}, \lambda), (Y, \mathbb{S}, \sigma), \mu \rangle$ (see the proof of Theorem 13.2). Inequality (13.18) implies that the non-autonomous system $(H(\gamma), \mathbb{S}, \lambda), (Y, \mathbb{S}, \sigma), \mu \rangle$, is distal on $\mathbb{S}_-$. By Lemma 13.1, γ is recurrent. Therefore,

$$\sup\{|\gamma(t)| \ : \ t \in \mathbb{S}\} = \sup\{|\gamma(t)| \ : \ t \in \mathbb{S}_+\} \leq M|x|$$

for all $\gamma \in \Phi_{(x,y)}$ and $(x, y) \in \mathbb{B}$. Hence, condition (i) holds and the theorem is proved. $\qquad\qquad\square$

Theorem 13.4   *Let $\langle (X, \mathbb{S}_+, \pi), (Y, \mathbb{S}, \sigma), h \rangle$ be a linear non-autonomous dynamical system. Assume that $(X, \mathbb{S}_+, \pi)$ is asymptotic compact and $\mathbb{B}^+$ is closed. Then*

(i) $\mathbb{B}$ is closed, and

(ii) for any $(x, y) \in \mathbb{B}^+$ there is a recurrent point $(x, y) \in \mathbb{B}$ such that
$$\lim_{t \to +\infty} |\varphi(t, x_0, y) - \varphi(t, x, y)| = 0.$$

Proof.   Assume that $\mathbb{B}^+$ is closed. By Theorem 13.1, there is an $M > 0$ such that (13.1) is valid for all $(x, y) \in \mathbb{B}^+$ and $t \in \mathbb{S}_+$. Assume that $(x, y) \in \mathbb{B} \subseteq \mathbb{B}^+$. Then inequality (13.1) implies that (13.18) is valid for all $t \in \mathbb{S}_-$ and $\gamma \in \Phi_{(x,y)}$. Repeating the arguments used in the proof of Theorem 13.3, we obtain (13.15) for all $\gamma \in \Phi_{(x,y)}, (x, y) \in \mathbb{B}$ and $t \in \mathbb{S}$. To complete the proof of the first assertion of the theorem, it is sufficient to apply Theorem 13.2.

Now we prove the second assertion of the theorem. Let $(x_0, y) \in \mathbb{B}^+$. Since $(X, \mathbb{S}_+, \pi)$ is asymptotic compact, the semi-trajectory $\{\pi^t(x_0, y) \mid t \in \mathbb{S}_+\}$ of the point (x_0, y) is relatively compact. Therefore, $\omega_{(x_0, y)} \neq \emptyset$ is compact and invariant. By Lemma 13.3, one can find a recurrent point $(x, y) \in \omega_{(x_0, y)}$ and a sequence $t_n \to +\infty$ such that

$$\lim_{n \to +\infty} |\varphi(t_n, x_0, y) - \varphi(t_n, x, y)| = 0. \tag{13.19}$$

Inequality (13.15) implies that

$$|\varphi(t, x_0, y) - \varphi(t, x, y)| \leq M|\varphi(t_n, x_0, y) - \varphi(t_n, x, y)| \tag{13.20}$$

for all $t \geq t_n$. Formulas (13.19) and (13.20) imply the desired assertion, which completes the proof of the theorem. $\qquad\qquad\square$

Remark 13.2 *The second assertion of Theorem 13.3 remains true even if we do not assume that $\mathbb{B}^+$ is closed.*

We conclude this section with a condition under which a linear non-autonomous system is asymptotic compact.

Let $P : X \to X$ be a projection of the vector bundle, that is, $P_y := P|_{X_y}$ projection in X_y for every $y \in Y$. Then P is said to be completely continuous if $P(M)$ is relatively compact for any bounded set $M \subseteq X$.

Lemma 13.5 *Let $\langle (X, \mathbb{S}_+, \pi), (Y, \mathbb{S}, \sigma), h \rangle$ be a linear non-autonomous dynamical system. Assume that the maps $\pi^t := \pi(t, \cdot) : X \to X$ can be represented as sums $\pi(t, x) := \pi_1(t, x) + \pi_2(t, x)$ for all $t \in \mathbb{S}_+$ and $x \in X$, and that the following conditions hold.*

(i) $|\pi_1(t, x)| \le m(t, r)$ for all $t \in \mathbb{S}_+$ and $x \in X$, where $m : \mathbb{R}_+^2 \to R_+$ and for every $r \ge 0$ the function $m(t, r)$ tends to zero as $t \to +\infty$.

(ii) The maps $\pi_2(t, \cdot) : X \to X$ $(t > 0)$ are conditionally completely continuous, that is, $\pi_2(t, A)$ is relatively compact for any $t > 0$ and any bounded positively invariant set $A \subseteq X$.

Then the dynamical system $(X, \mathbb{S}_+, \pi)$ is asymptotically compact.

Proof. Let $A \subseteq X$ be a bounded positively invariant set. Then $A = \Sigma^+(A) := \{\pi^t A \mid t \in \mathbb{S}_+\}$. Since Y is compact and (X, h, Y) is local trivial, there is an $r > 0$ such that $A \subset \{x \in X \ : \ |x| \le r\}$. We claim that for any $\{x_k\} \subset A$ and $t_n \to +\infty$ the sequence $\{x_k t_k\}$ is relatively compact. We can cover $M := \{x_k t_k\}$ with a finite ε-net for any $\varepsilon > 0$. Assume that $\varepsilon > 0$ and $l > 0$ are such that $m(l, r) < \frac{\varepsilon}{2}$. We represent M as the union $M_1 \bigcup M_2$, where $M_1 := \{x_k t_k\}_{k=1}^{k_1}, M_2 := \{x_k t_k\}_{k=k_1+1}^{\infty}$ and $k_1 := \max\{k \mid t_k \le l\}$. The set M_2 is a subset of $\pi^l(\Sigma^+(A))$ whose elements can be represented in the form $\pi_1(t, x) + \pi_2(t, x)$ $(x \in \Sigma^+(A))$. Since $\pi_2(l, \Sigma^+(A))$ is relatively compact, we can cover it with a finite $(\frac{\varepsilon}{2})$-net. For any $y \in \pi_1(l, \Sigma^+(A))$ there is an $x \in \Sigma^+(A)$ such that $y = \pi_1(l, x)$ and $|y| = |\pi_1(l, x)| \le m(l, r) < \frac{\varepsilon}{2}$. Therefore, the zero section Θ of the vector bundle (X, h, Y) is an $(\frac{\varepsilon}{2})$-net of $\pi_1(l, \Sigma^+(A))$. Since Y is compact and (X, h, Y) is local trivial, the zero section Θ is compact. Hence, M_2 and M are covered by the ε-net $\theta \bigcup M_1$. Since Θ is compact and the space Y is complete, $M = \{x_k t_k\}_{k=1}^{\infty}$ is relatively compact. We complete the proof by applying Lemma 1.3. $\square$

Corollary 13.1 *Let $\langle (X, \mathbb{S}_+, \pi), (Y, \mathbb{S}, \sigma), h \rangle$ be a linear non-autonomous dynamical system and let $P : X \to X$ be a completely continuous projection. Assume that there are positive numbers N and ν such that $|\pi^t Q(x)| \le N e^{-\nu t} |x|$ for all $x \in X$ and $t \in \mathbb{S}_+$, where $Q : X \to X$ and $Q_y := Q|_{X_y} = I_y - P_y$ for all $y \in Y$ $(I_y := id_{X_y})$. Then $(X, \mathbb{S}_+, \pi)$ is asymptotic compact.*

Proof. To deduce this corollary from Lemma 13.5, it is sufficient to observe that $\pi(t,x) = \pi_1(t,x) + \pi_2(t,x)$ for all $x \in X$ and $t \in \mathbb{S}_+$, where $\pi_1(t,x) := \pi^t Q(x)$ and $\pi_2(t,x) := \pi^t P(x)$. Under the hypotheses of Corollary 13.1, for every $t > 0$ the map $\pi_2(t,\cdot) := \pi^t P$ is completely continuous and $|\pi_1(t,x)| \le Ne^{-\nu t}|x|$. Hence, Lemma 13.5 is applicable. $\qquad\square$

Remark 13.3   *In the proofs of Lemma 13.5 and Corollary 13.1 we used only the compactness of Y. Hence, these statements are also valid in the case when the dynamical system $(Y, \mathbb{S}, \sigma)$ is not minimal.*

13.2   Bounded solutions of linear equations

Let $[E]$ be the Banach space of all bounded linear operators that act on a Banach space E equipped with the operator norm. Let Λ be a complete metric space of closed linear operators that act on E (for example, $\Lambda = [E]$ or $\Lambda = \{A_0 + B \mid B \in [E]\}$, where A_0 is a closed operator that acts on E). Let $C(R, \Lambda)$ be the space of all continuous operator-valued functions $A : R \to \Lambda$ equipped with the compact-open topology and let $(C(\mathbb{R}, \Lambda), \mathbb{R}, \sigma)$ be the dynamical system of shifts on $C(\mathbb{R}, \Lambda)$.

Linear ordinary differential equations. Let $\Lambda = [E]$ and consider the differential equation

$$u' = A(t)u, \tag{13.21}$$

where $A \in C(\mathbb{R}, [E])$. Consider the H-class of equation (13.21), that is, the family of equations

$$v' = B(t)v, \tag{13.22}$$

with $B \in H(A) := \overline{\{A_\tau \mid \tau \in \mathbb{R}\}}$, $A_\tau(t) = A(t+\tau)$, and $t \in \mathbb{R}$, where the bar denotes closure in $C(\mathbb{R}, [E])$. Let $\varphi(t, v, B)$ be the solution of equation (13.22) that satisfies the condition $\varphi(0, v, B) = v$.

We put $Y := H(A)$ and denote the dynamical system of shifts on $H(A)$ by $(Y, \mathbb{R}, \sigma)$. We put $X := E \times Y$ and define a dynamical system on X by setting $\pi(t, (v, B)) := (\varphi(t, v, B), B_t)$ for all $(v, B) \in E \times Y$ and $t \in \mathbb{R}$. Then $\langle (X, \mathbb{R}, \pi), (Y, \mathbb{R}, \sigma), h \rangle$ is a linear group non-autonomous dynamical system, where $h := pr_2 : X \to Y$. Applying the results of section 13.1 to this system, we obtain the following assertions.

Lemma 13.6   [51, 60]

(i) The map $(t, u, A) \mapsto \varphi(t, u, A)$ of $\mathbb{R} \times E \times C(\mathbb{R}, [E])$ to E is continuous, and

(ii) the map $A \mapsto U(\cdot, A)$ of $C(\mathbb{R}, [E])$ to $C(\mathbb{R}, [E])$ is continuous, where $U(t, A)$ is the Cauchy's operator [132] of equation (13.21).

Theorem 13.5 *Assume that $A \in C(\mathbb{R}, [E])$ is recurrent (that is, $H(A)$ is a compact minimal set of $(C(\mathbb{R}, [E]), \mathbb{R}, \sigma)$). Then the following conditions are equivalent:*

(a) the set

$$\mathbb{B}^+(\mathbb{B}^-, \mathbb{B}) = \{(v, \mathbb{B}) \in E \times H(A) \mid \sup_{t \in \mathbb{R}_+ \ (\mathbb{R}_-, \mathbb{R})} |\varphi(t, v, B)| < +\infty\} \quad (13.23)$$

is closed in $E \times H(A)$, and

(b) there is a positive number M such that

$$|\varphi(t, v, B)| \le M|v| \quad (13.24)$$

for all $t \in \mathbb{R}_+(\mathbb{R}_-, \mathbb{R})$ and $(v, B) \in \mathbb{B}^+(\mathbb{B}^-, \mathbb{B})$.

Corollary 13.2 *Let $A \in C(\mathbb{R}, [E])$ be recurrent. Then the following assertions are equivalent :*

(i) all solutions of all equations (13.22) are bounded on $\mathbb{R}_+(\mathbb{R}^-, \mathbb{R})$, and

(ii) there is an $M > 0$ such that (13.24) is valid for all $v \in E, B \in H(A)$, and $t \in \mathbb{R}_+(\mathbb{R}_-, \mathbb{R})$.

Theorem 13.6 *Assume that $A \in C(\mathbb{R}, [E])$ is recurrent, the linear non-autonomous dynamical system generated by equation (13.21) is asymptotic compact, and all solutions of all equations (13.22) are bounded on $\mathbb{R}_+$. Then*

(i) there is an $M > 0$ such that (13.24) is valid for all $t \in \mathbb{R}_+, v \in E$ and $B \in H(A)$,

(ii) the set $\mathbb{B}$ defined by formula (13.23) is closed in $E \times H(A)$,

(iii) all solutions of all equations (13.22) bounded on $\mathbb{R}$ are recurrent,

(iv) for any $B \in H(A)$ equation (13.22) has only a finite number n_B of solutions that are linearly independent and bounded on $\mathbb{R}$, and $n_B = n_A$ for all $B \in H(A)$,

(v) for any $v_0 \in E$ and $B \in H(A)$ there is a $(v, B) \in \mathbb{B}$ such that

$$\lim_{t \to +\infty} |\varphi(t, v_0, B) - \varphi(t, v, B)| = 0,$$

that is, any solution of any equation (13.22) is asymptotic recurrent.

We now formulate some sufficient conditions for the asymptotic compactness of the linear non-autonomous dynamical system generated by equation (13.21).

Theorem 13.7 *Let $A \in C(\mathbb{R}, [E]), A(t) = A_1(t) + A_2(t)$ for all $t \in \mathbb{R}$, and assume that $H(A_i), i = 1, 2$, are compact and the following conditions hold.*

(i) The zero solution of the equation

$$u' = A_1(t)u \quad (13.25)$$

is uniformly asymptotically stable, that is, there are positive numbers N and ν such that

$$\|U(t, A_1)U^{-1}(\tau, A_1)\| \le Ne^{-\nu(t-\tau)} \tag{13.26}$$

for all $t \ge \tau$ ($t, \tau \in \mathbb{R}$), where $U(t, A_1)$ is the Cauchy's operator of the equation (13.25).

(ii) *The family of operators $\{A_2(t) \mid t > 0\}$ is uniformly completely continuous, that is, for any bounded set $A \in E$ the set $\{A_2(t)A \mid t > 0\}$ is relatively compact.*

Then the linear non-autonomous dynamical system generated by equation (13.21) is asymptotic compact.

Proof. Let $B \in H(A)$. Then there are $\{t_n\} \subset \mathbb{R}$ and $B_i \in H(A_i)$, $i = 1, 2$, such that $B(t) = B_1(t) + B_2(t)$ and $B_i(t) = \lim_{n \to +\infty} A_i(t + t_n)$. Note that

$$\varphi(t, v, B) = U(t, B_1)v + \int_0^t U(t, B_1)U^-1(\tau, B_1)B_2(\tau)\varphi(\tau, v, B)d\tau. \tag{13.27}$$

By Lemma 13.6,

$$U(t, B_i) = \lim_{n \to +\infty} U(t, A_{it_n}), \ \ A_{it_n}(t) := A_i(t + t_n),$$

and the equality

$$U(t, A_{1t_n})U^-1(\tau, A_{1t_n}) = U(t + t_n, A_1)U^-1(\tau + t_n, A_1)$$

and inequality (13.26) imply that

$$\|U(t, B_1)U^-1(\tau, B_1)\| \le Ne^{-\nu(t-\tau)} \tag{13.28}$$

for all $t \ge \tau$ and $B_1 \in H(A_1)$. By Lemma 13.5, Theorem 13.7 will be proved if we can prove that the set

$$\left\{ \int_0^t U(t, B_1)U^-1(\tau, B_1)B_2(\tau)\varphi(\tau, v, B)d\tau \mid (v, B) \in \mathcal{A} \right\}$$

is relatively compact for every $t > 0$ and every bounded positively invariant set $\mathcal{A} \subseteq E \times Y$. We put

$$K_A := \overline{\{B_2(t)\varphi(t, v, B) \mid t \in \mathbb{R}_+, (v, B) \in \mathcal{A}\}}.$$

Then

$$\int_0^t U(t, B_1)U^-1(\tau, B_1)B_2(\tau)\varphi(\tau, v, B)d\tau \tag{13.29}$$

$$\in t \cdot \overline{conv}\{U(t, B_1)U^-1(\tau, B_1)w \mid 0 \le \tau \le t, \ B_1 \in\in H(A_1), w \in K_A\}.$$

Since $H(A_1), H(A)$, and K_A are compact sets, formula (13.29), condition (ii) of Theorem 13.7, and Lemma 13.6 imply that $\bigcup\{U(t, B_1)U^{-1}(\tau, B_1)w \mid 0 \leq \tau \leq t, B_1 \in H(A_1), w \in K_A\}$ is compact, which completes the proof of the theorem. $\square$

Theorem 13.8 *Let $H(A)$ be compact and assume that there is a finite-dimensional projection $P \in [E]$ such that*

(i) the family of projections $\{P(t) \mid t \in \mathbb{R}\}$, where $P(t) := U(t, A)PU^{-1}(t, A)$, is relatively compact in $[E]$, and

(ii) there are positive numbers N and ν such that

$$\|U(t, A)QU^{-1}(\tau, A)\| \leq Ne^{-\nu(t-\tau)}$$

for all $t \geq \tau$, where $Q := I - P$.

Then the linear non-autonomous dynamical system generated by equation (13.21) is asymptotically compact.

Proof. Since the family of projections $\{P(t) \mid t \in \mathbb{R}\}$ is relatively compact, we can assume that the sequence $\{P(t_n)\}$ converges. Let $P(B) := \lim\limits_{n \to +\infty} P(t_n)$. We claim that the family $\mathbb{H} := \overline{\{P(t) \mid t \in \mathbb{R}\}}$ is uniformly completely continuous, where the bar denotes closure in $[E]$. Indeed, let $\mathcal{A}$ be a bounded subset of $E, \{x_n\} \subseteq \{QA \mid Q \in \mathbb{H}\}$, and $\varepsilon_n \downarrow 0$. Then there are $t_n \in \mathbb{R}$ and $v_n \in \mathcal{A}$ such that $\rho(x_n, P(t_n)v_n) \leq \varepsilon_n$. Since the sequence $\{P(t_n)\}$ is relatively compact, we can assume that it converges. Let $L := \lim\limits_{n \to +\infty} P(t_n)$. Then L is completely continuous, which implies that the sequence $\{x_n'\} := \{Lv_n\}$ is relatively compact. Note that

$$\rho(x_n, x_n') \leq \rho(x_n, P(t_n)v_n) + \rho(P(t_n)v_n, Lv_n) \leq \varepsilon_n + \|P(t_n) - L\|\|v_n\|,$$

which implies that $\rho(x_n, x_n') \to 0$ as $n \to +\infty$. Hence, $\{x_n\}$ is relatively compact.

Assume that $B \in H(A)$ and $\{t_n\} \subset \mathbb{R}$ are such that

$$B = \lim\limits_{n \to +\infty} A_{t_n}, \quad P(B) := \lim\limits_{n \to +\infty} P(A_{t_n}),$$

where $P(A_{t_n}) := U(t_n, A)PU^{-1}(t_n, A)$. The assertions proved above imply that the family $\{P(B) \mid B \in H(A)\}$ is uniformly completely continuous. Note that $Q(B) = \lim\limits_{n \to +\infty} Q(A_{t_n})$, where $Q(B) := I - P(B)$ and $Q(A_{t_n}) := I - P(A_{t_n})$. Moreover, condition (ii) of Theorem 13.8 implies that

$$\|U(t, A_{t_n})Q(A)U^{-1}(\tau, A_{t_n})\| \leq Ne^{-\nu(t-\tau)} \tag{13.30}$$

for all $t \geq \tau$. Passing to the limit in (13.30) as $n \to +\infty$ and taking Lemma 13.6 into account, we obtain that

$$\|U(t, B)Q(B)U^{-1}(\tau, B)\| \leq Ne^{-\nu(t-\tau)}$$

for all $t \geq \tau$ and $B \in H(A)$. We complete the proof of the theorem by observing that $U(t, B)Q(B) + U(t, B)P(B) = U(t, B)$ and applying Lemma 13.5. $\square$

Linear functional-differential equations. Let $r > 0, C([a, b], \mathbb{R}^n)$ be the Banach space of all continuous functions $\varphi : [a, b] \to \mathbb{R}^n$ with the norm sup. For $[a, b] := [-r, 0]$ we put $\mathcal{C} := C([-r, 0], \mathbb{R}^n)$. Let $c \in \mathbb{R}, a \geq 0$, and $u \in C([c - r, c + a], \mathbb{R}^n)$. We define $u_t \in \mathcal{C}$ for any $t \in [c, c + a]$ by the relation $u_t(\theta) := u(t + \theta), -r \geq \theta \geq 0$. Let $\mathfrak{A} = \mathfrak{A}(\mathcal{C}, \mathbb{R}^n)$ be the Banach space of all linear operators that act from $\mathcal{C} \to \mathbb{R}^n$ equipped with the operator norm, let $C(\mathbb{R}, \mathfrak{A})$ be the space of all operator-valued functions $A : \mathbb{R} \to \mathfrak{A}$ with the compact-open topology, and let $(C(\mathbb{R}, \mathfrak{A}), \mathbb{R}, \sigma)$ be the dynamical system of shifts on $C(\mathbb{R}, A)$. Let $H(\mathfrak{A}) := \overline{\{\mathfrak{A}_\tau \mid \tau \in \mathbb{R}\}}$, where $\mathfrak{A}_\tau$ is the shift of the operator-valued function $\mathfrak{A}$ by τ and the bar denotes closure in $C(\mathbb{R}, \mathfrak{A})$.

Consider the linear functional-differential equation with delay

$$u' = \mathfrak{A}(t)u_t \tag{13.31}$$

along with the family of equations

$$v' = \mathfrak{B}(t)v_t, \tag{13.32}$$

where $\mathfrak{B} \in H(\mathfrak{A})$. Let $\varphi(t, v, \mathfrak{B})$ be the solution of equation (13.32) satisfying the condition $\varphi(0, v, \mathfrak{B}) = v$ and defined for all $t \geq 0$. Let $Y := H(\mathfrak{A})$ and denote the dynamical system of shifts on $H(\mathfrak{A})$ by $(Y, \mathbb{R}, \sigma)$. Let $X := \mathcal{C} \times Y$ and let $\pi := (\varphi, \sigma)$ be the dynamical system on X defined by the equality $\pi(\tau, (v, \mathfrak{B})) := (\varphi(\tau, v, \mathfrak{B}), \mathfrak{B}_\tau)$. The non-autonomous system $\langle (X, \mathbb{R}_+, \pi), (Y, \mathbb{R},), h \rangle (h := pr_2 : X \to Y)$ is linear.

Lemma 13.7   *Let $H(A)$ be compact in $C(\mathbb{R}, \mathfrak{A})$. Then the linear non-autonomous dynamical system $\langle (X, \mathbb{R}_+, \pi), (Y, \mathbb{R}, \sigma), h \rangle$ generated by equation (13.31) is completely continuous, that is, for any bounded set $A \subseteq X$ there is an $l = l(A) > 0$ such that $\pi^l A$ is relatively compact.*

Proof. This follows from general properties of solutions of linear functional- differential equations with delay (see, for example, [175], Lemmas 2.2.3 and 3.6.1) since $Y = H(\mathfrak{A})$ is compact. $\square$

Applying the results obtained in section 13.1 to the linear non-autonomous dynamical system generated by equation (13.31), we obtain the following assertions.

Theorem 13.9   *Let $A \in C(\mathbb{R}, \mathfrak{A})$ be recurrent. Then the following conditions are equivalent :*

(i) all solutions of all equations (13.32) are bounded on $\mathbb{R}_+$,

(ii) there is a positive number M such that $|\varphi(t, v, B)| \le M|v|$ for all $t \ge 0, v \in C$, and $B \in H(A)$.

Theorem 13.10 *Let $A \in C(\mathbb{R}, \mathfrak{A})$ be recurrent, and assume that all solutions of all equations (13.32) are bounded on $\mathbb{R}_+$. Then*

(i) the set of all the solutions of all equations (13.32) that are bounded on $\mathbb{R}$ is closed in $C(R, C) \times H(A)$,

(ii) all the solutions of all equations (13.32) that are bounded on $\mathbb{R}$ are recurrent,

(iii) for any $B \in H(A)$ equation (13.32) has only a finite number n_B of solutions that are linearly independent and bounded on $\mathbb{R}$, and $n_B = n_A$ for all $B \in H(A)$,

(iv) all solutions of all equations (13.32) are asymptotic recurrent.

Now consider the neutral functional-differential equation

$$\frac{d}{dt}Du_t = A(t)u_t, \tag{13.33}$$

where $A \in C(\mathbb{R}, \mathfrak{A})$ and the operator $D \in \mathfrak{A}$ is atomic at zero [175, p.67]. Like (13.31), equation (13.33) generates a linear non-autonomous dynamical system $\langle(X, \mathbb{R}_+, \pi), (Y, \mathbb{R}, \sigma), h\rangle$, where $X := C \times Y$, $Y := H(A)$, and $\pi := (\varphi, \sigma)$.

Lemma 13.8 *Let $H(A)$ be compact, and assume that the operator D is stable, that is, the zero solution of the homogeneous difference equation $Dy_t = 0$ is uniformly asymptotically stable [175]. Then the linear non-autonomous dynamical system $(X, \mathbb{R}_+, \pi), (Y, \mathbb{R}, \sigma), h\rangle$ generated by equation (13.33) is asymptotically compact.*

Proof. This follows from Theorem 12.3.2 and Lemma 12.4.1 in [175], since $Y = H(A)$ is compact. $\square$

Therefore, Theorems 13.9 and 13.10 hold for equations (13.33).

Linear partial differential equations. Consider the differential equation (13.21) with unbounded coefficients. Let $A \in C(\mathbb{R}, \Lambda)$, where Λ is a complete metric space of linear closed operators that act on E (for example, $\Lambda := \{A_0 + B \mid B \in [E]\}$, where A_0 is a closed operator that acts on E). Consider the H-class (13.22) of equation (13.21), where $B \in H(A)$. We assume that the following conditions are fulfilled for equation (13.21) and its H-class:

(i) for any $v \in E$ and $B \in H(A)$ equation (13.22) has precisely one solution $\varphi(0, v, B)$ that is defined on $\mathbb{R}_+$ and satisfies the condition $\varphi(0, v, B) = v$;

(ii) the map $\varphi : (t, v, B) \to \varphi(t, v, B)$ is continuous in the topology of $\mathbb{R}_+ \times E \times C(\mathbb{R}, \mathfrak{A})$;

(iii) for every $t \in \mathbb{R}_+$ the map $U(t, \cdot) : H(A) \to [E]$ is continuous, where $U(t, B)$ is the Cauchy's operator of equation (13.22), that is, $U(t, B)v := \varphi(t, v, B)$ for all $t \in \mathbb{R}_+$ and $v \in E$.

Under the above assumptions equation (13.21) generates a linear non-autonomous dynamical system to which the results obtained in section 13.1 can be applied. Therefore, Theorems 13.5 and 13.6 hold in this case.

In conclusion we consider a partial differential equation that satisfies conditions (i)-(iii).

Example 13.1 A closed linear operator $A : D(A) \to E$ with dense domain $D(A)$ is said [188] to be sectorial if one can find a $\phi \in (0, \frac{\pi}{2})$, an $M \geq 1$, and a real number a such that the sector

$$S_{a,\phi} := \{\lambda \mid |\arg(\lambda - a)| \leq \pi, \lambda \neq a\}$$

lies in the resolvent set $\rho(A)$ of A and $\|(\lambda I - A)^{-1}\| \leq M|\lambda - a|^{-1}$ for all $\lambda \in S_{a,\phi}$. An important class of sectorial operators is formed by elliptic operators [188], [202].

Consider the differential equation

$$u' = (A_1 + A_2(t))u, \tag{13.34}$$

where A_1 is a sectorial operator that does not depend on $t \in \mathbb{R}$, and $A_2 \in C(\mathbb{R}, [E])$.

The results of [238], [188] imply that equation (13.34) satisfies conditions (i)-(iii). Therefore, analogies of Theorems 13.5 and 13.6 hold for equation (13.34).

Remark 13.4 *Statements similar to Theorems 13.5 and 13.6 hold for difference equations and can be deduced from the results of section 13.1 by applying these results to linear non-autonomous dynamical systems with discrete time generated by the corresponding difference equations.*

13.3 Finite-dimensional systems

Throughout this section we assume that the Banach space E is finite-dimensional and its norm $|\cdot|$ is induced by the scalar product $\langle \cdot, \cdot \rangle$, that is, $|\cdot|^2 := \langle \cdot, \cdot \rangle$. We propose several conditions for equations (13.21) in a finite-dimensional space that are equivalent to the uniform bistability of the zero solution of equation (13.21), and prove that the uniform Lyapunov stability of the zero solution of equation (13.21) implies that there is a frame of solutions of equation (13.21) bounded on $\mathbb{R}$ whose Gram determinant is separated from zero.

Let $x_1, ..., x_k$ be a set of vectors in E. Let us recall [23] that the Gram determinant $\Gamma(x_1, ..., x_k)$ of the vectors $x_1, ..., x_k$ is defined to be the determinant

$|\langle x_i, x_j \rangle|^k_{i,j=1}$. The Gram determinant of the vectors $x_1, ..., x_k$ is no-negative, and equals zero only if the vectors $x_1, ..., x_k$ are linearly dependent.

Theorem 13.11 *The following assertions are equivalent:*

(i) there is an $M > 0$ such that (13.23) is valid for all $t \in \mathbb{R}$ and $(x, y) \in \mathbb{B}$;

(ii) $\mathbb{B}$ is closed;

(iii) B is a subbundle of F, that is, B is closed and there is a k such that $dim\mathbb{B}_y = k$ for all $y \in Y$;

(iv) all the motions in F that are non-trivial and bounded on $\mathbb{R}$ are separated from zero, that is, $\inf\{ |\varphi(t, x, y)| : t \in \mathbb{R}\} > 0$ for any $(x, y) \in \mathbb{B}$ such that $x \neq 0$;

(v) one can find a $y_0 \in Y$ and a basis $\xi_1, ..., \xi_k \in \mathbb{B}_{y_0}$ such that

$$\inf_{t \in \mathbb{R}} \Gamma(\xi_1, ..., \xi_k; t) := \alpha > 0, \tag{13.35}$$

where $\xi_i = (x_i, y_0), i = 1, ..., k$ and $\Gamma(\xi_1, ..., \xi_k; t)$ is the Gram determinant of the vectors $\pi(t, \xi_i) \in E \times Y, i = 1, ..., k$.

Proof. Assertions (i) and (ii) are equivalent by Theorem 13.1. By Theorems 13.1 and 13.2, (ii) implies (iii) (the reverse implication is obvious). The equivalence of conditions (iii) and (iv) follows from [33], Theorem 8.22.

Assume that (iv) is fulfilled, let $y_0 \in Y$, and let $\xi_1, ..., \xi_k \in \mathbb{B}_y$ be a basis in $\mathbb{B}$. We claim that (13.35) holds. Assume the contrary. Then one can find a $\{t_n\} \subset \mathbb{R}$ such that $|t_n| \to +\infty$ and $\Gamma(\xi_1, ..., \xi_k; t_n) \to 0$ as $n \to +\infty$. Since all non-zero motions in F are separated from zero, Lemma 13.1 implies that $\xi_1, ..., \xi_k$ and y are jointly recurrent. Without loss of generality, we can assume that the sequences $\{\pi(t_n, \xi_i)\}, i = 1, ..., k$ and $\{\sigma(t_n, y)\}$ are convergent.

Let

$$\eta_i := \lim_{n \to +\infty} \pi(t_n, \xi_i), \quad i = 1, ..., k, \quad q := \lim_{n \to +\infty} \sigma(t_n, y_0).$$

Then

$$\Gamma(\eta_1, ..., \eta_k) = \lim_{n \to +\infty} \Gamma(\pi(t_n, \xi_1), ..., \pi(t_n, \xi_k)) = 0.$$

Hence, $\eta_1, ..., \eta_k$ are linearly dependent. Repeating the above argument, we deduce from the last fact that $\xi_1, ..., \xi_k$ are linearly dependent, which contradicts their choice. Hence, (iv) implies (v).

We claim that (v) implies (iv). First, we prove that for any $q \in Y$ one can find a basis $\eta_1, ..., \eta_k$ such that $\Gamma(\eta_1, ..., \eta_k; t) \geq \alpha > 0$. Indeed, since Y is minimal, there is a sequence $\{t_n\} \subset \mathbb{R}$ such that $\sigma(t_n, y_0) \to q$. Since $\xi_1, ..., k \in \mathbb{B}_{y_0}$, we can assume that the sequences $\{\pi(t_n, \xi_i)\}, i = 1, ..., k$ are convergent. Let $\eta_i := \lim_{n \to +\infty} \pi(t_n, \xi_i)$.

Then $\eta_1, ..., \eta_k \in \mathbb{B}_q$ and

$$\Gamma(\eta_1, ..., \eta_k; t) = \lim_{n \to +\infty} \Gamma(\xi_1, ..., \xi_k; t + t_n) \qquad (13.36)$$

for all $t \in \mathbb{R}$. Therefore, $\eta_1, ..., \eta_k$ are linearly independent. Hence, $n_q := \dim\mathbb{B}_q \geq n_{y_0} := \dim\mathbb{B}_{y_0}$. Since Y is minimal, the reverse inequality also holds. Therefore, $n_q = n_{y_0}$ for all $q \in Y$. This implies that $\eta_1, ..., \eta_k$ is a basis in $\mathbb{B}_q$. Thus, we have shown that for any $q \in Y$ there is a basis $\eta_1, ..., \eta_k \in \mathbb{B}_q$ that satisfies condition (13.36). For any $q \in Y$ and $(x, q) \in \mathbb{B}_q$ we have $\inf\{\,|\varphi(t, x, q)| \,:\, t \in \mathbb{R}\} > 0$. Indeed, if we assume the contrary, then there are $(x_0, q) \in \mathbb{B}_q$, $x_0 \neq 0$, and $|t_n| \to +\infty$ such that

$$|\varphi(t_n, x_0, y)| \to 0 \qquad (13.37)$$

as $n \to +\infty$. Let $\eta_1', ..., \eta_k'$ be a basis in $\mathbb{B}_q$, and let $\eta_1' = (x_0, q)$. Then (13.37) implies that

$$\Gamma(\eta_1' t_n, ..., \eta_k' t_n) \to 0. \qquad (13.38)$$

Since $\eta_1, ..., \eta_k$ and $, \eta_1', ..., \eta_k'$ are two bases in $\mathbb{B}_q$, there is a non-degenerate linear transformation that transforms the first basis into the second, and vice versa. Formula (13.38) implies that a similar relation holds for the basis $\eta_1, ..., \eta_k$, which contradicts inequality (13.36). This contradiction completes the proof of the implication (v)$\Longrightarrow$ (iv). The theorem is proved. $\qquad\square$

Applying Theorem 13.11 to the linear non-autonomous dynamical system generated by equation (13.21), we obtain the following assertion.

Theorem 13.12   *Let $A \in C(\mathbb{R}, [E])$ be recurrent. Then the following conditions are equivalent:*

(a) the set $\mathbb{B} := \{(u, B) \in E \times H(A) \mid \sup_{t \in \mathbb{R}} |\varphi(t, u, B)| < +\infty\}$ is closed in $E \times H(A)$;

(b) there is a positive number M such that

$$|\varphi(t, u, B)| \leq M|u| \qquad (13.39)$$

for all $t \in \mathbb{R}$ and $(u, B) \in \mathbb{B}$;

(c) $\mathbb{B}$ is closed and all fibres $\mathbb{B}_B$ have the same dimension, that is, all the equations (13.22) have the same number of solutions that are linearly independent and bounded on $\mathbb{R}$;

(d) all non-trivial solutions of all equations in the H-class (13.22) bounded on $\mathbb{R}$ are separated from zero, that is,

$$\inf_{t \in \mathbb{R}} |\varphi(t, u, B)| > 0 \qquad (13.40)$$

for all $(u, B) \in \mathbb{B}, u \neq 0$;

(e) *there is a basis* $\varphi_1, ..., \varphi_k(k = dim\mathbb{B}_A$ *and* $\mathbb{B}_A := \{(u, A) \mid (u, A) \in \mathbb{B}\})$ *that consists of solutions of equation (13.21) bounded on* $\mathbb{R}$ *and satisfies the condition* $\Gamma(\varphi_1, ..., \varphi_k; t) > 0$ *for all* $t \in \mathbb{R}$, *where* $\Gamma(\varphi_1, ..., k; t) := |\langle \varphi_i(t), \varphi_j(t) \rangle|_{i,j=1}^{n}$ *is the Gram determinant of* $\varphi_1, ..., \varphi_k$.

Corollary 13.3 *Let* $A \in C(\mathbb{R}, [E])$ *be recurrent. Then the following conditions are equivalent:*

(i) *all solutions of all equations from the H-class (13.22) are bounded on* $\mathbb{R}$;

(ii) *there is an* $M > 0$ *such that* $|\varphi(t, u, B)| \le M|u|$ *for all* $t \in \mathbb{R}$ *and* $u \in E$;

(iii) *all non-zero solutions of all equations from the H-class (13.22) are bounded on* $\mathbb{R}$ *and separated from zero, that is, (13.40) is valid for all* $(u, B) \in \mathbb{B}$ *such that* $u \ne 0$;

(iv) *there are positive numbers* C *and such that* $\|U(t, A)\| \le C$ *for all* $t \in \mathbb{R}$ *and* $\inf\{|det\ U(t, A)| : t \in \mathbb{R}\} = \alpha$, *where* $U(t, A)$ *is the Cauchy operator of equation (13.21).*

Proof. This follows from Theorem 13.12, since $\Gamma(\varphi_1, ..., \varphi_n; t) = |det\ U(t, A)|^2$, where $\varphi_1, ..., \varphi_n$ are the column-vectors of the matrix $U(t, A)$. $\square$

Remark 13.5 *The equivalence of conditions (i) and (ii) was established in* [39], [204]. *The implication (iv)$\Longrightarrow$ (i) sharpens a result in* [204].

Theorem 13.13 *Assume that all solutions of all equations from the H-class (13.22) are bounded on* $\mathbb{R}_+$. *Then there is an* $M > 0$ *such that* $|\varphi(t, u, B)| \le M|u|$ *for all* $t \in \mathbb{R}$ *and* $(u, B) \in \mathbb{B}$, *that is,* $\mathbb{B}$ *is closed.*

Proof. This follows from Theorems 13.1 and 13.4. $\square$

In conclusion we consider examples that illustrate the above results.

Example 13.2 Let $a \in C(\mathbb{R}, \mathbb{R})$ be the Bohr almost periodic function defined by the equality

$$a(t) := \sum_{k=0}^{\infty} \frac{1}{(2k+1)^{3/2}} \sin \frac{t}{2k+1}, \tag{13.41}$$

and let

$$h(t) := \int_0^t a(s)ds = \sum_{k=0}^{\infty} \frac{2}{(2k+1)^{1/2}} \sin^2 \frac{t}{2(2k+1)}.$$

Note that $a(t + t_n) \to -a(t)$ uniformly on $\mathbb{R}$, where $t_n := (2n + 1)!!$. Therefore, $-a \in H(a) := \overline{\{a_\tau \mid \tau \in \mathbb{R}\}}$. Using the inequality $|\sin t| \ge \frac{1}{2}|t|$ with $|t| \le 1$, we obtain that

$$h(t) = \sum_{k=0}^{\infty} \frac{1}{(2k+1)^{1/2}} \sin^2 \frac{t}{2(2k+1)} \geq \sum_{k \geq \frac{1}{2}(\frac{|t|}{2}-1)} \frac{t^2}{8} \frac{1}{(2k+1)^{5/2}}$$

$$\geq \frac{t^2}{8} \int_{|s| \geq \frac{1}{2}(\frac{|t|}{2}-1)} \frac{ds}{(2s+1)^{5/2}} = \frac{t^2 2^{3/2}}{24|t|^{3/2}} = \frac{1}{6\sqrt{2}} |t|^{1/2} \to +\infty$$

as $|t| \to +\infty$. This implies that the module of all non-zero solutions of the equation

$$x' = a(t)x \tag{13.42}$$

tend to $+\infty$ as $|t| \to +\infty$, whereas those of the equation

$$y' = b(t)y, \tag{13.43}$$

with $b := -a \in H(a)$ tend to zero.

The above example is a slight modification of the well-known example of Favard (see [137, p.435] or [150]). Our case differs from Favard's example in that the solutions of equation (13.43) are not only bounded on $\mathbb{R}$, but they tend to zero as $|t| \to +\infty$. Thus, a non-zero solution of equation (13.43) is asymptotically stable, but the zero solution of equation (13.42) is not, even though $a \in H(b)$.

Example 13.3   Assume that $a \in C(\mathbb{R}, \mathbb{R})$ is defined by the formula $a(t) := -1 + \sin t^{\frac{1}{3}}$, $t \in \mathbb{R}$. For equation (13.42) the sets

$$\mathbb{B}^+ := \{(x,b) \mid x \in \mathbb{R}, \ b \in H(a)\} = \mathbb{R} \times H(a),$$
$$\mathbb{B} := \{(0,b) \mid b \in H(a)\} \cup \{(x,\theta) \mid x \in \mathbb{R}\},$$

where $\theta \in C(\mathbb{R}, \mathbb{R})$ is the function identically equal to zero, are closed. Thus, the recurrence of A in Theorem 13.12 (Theorem 13.13) is a sufficient condition, but this condition is not necessary.

13.4   Relationship between different types of stability

In 1962 W. Hahn [168] posed the problem of whether asymptotic stability implies uniform stability for linear equation

$$x' = A(t)x \qquad (x \in \mathbb{R}^n) \tag{13.44}$$

with almost periodic coefficients. C. C. Conley and R. K. Miller [123] gave a negative answer to this by constructing a scalar equation $x' = a(t)x$ with the property that every solution $\varphi(t, x, a) \to 0$ as $t \to +\infty$, but the null solution is not uniformly stable (see also [89]). From the results of R. J. Sacker and G. R. Sell [275] and I. U. Bronshteyn [33, p.141] follows, that for linear system (13.44) with recurrent (in particular, almost periodic) matrix from asymptotic stability of null solution of system (13.44) and all system

$$x' = B(t)x, \qquad (13.45)$$

where $B \in H(A) := \overline{\{A_\tau : \tau \in \mathbb{R}\}}$, A_τ is the translation of matrix A on τ and by bar is denoted the closure in the topology of uniform convergence uniform on every compact from $\mathbb{R}$, follows the uniform stability of null solution of system (13.44). Finally we note that from results of author [65] follows the validity of above mentioned result for arbitrary system (13.44) with compact matrix (i.e. when $H(A)$ is compact). Below we study the relationship between the asymptotic stability and uniform stability of null solution of system (13.44) in the arbitrary Banach space.

Our main result is that for linear system (13.44) with recurrent coefficients in the arbitrary Banach space the following statement takes place: if the null solution of equation (13.44) and all the equation (13.45) are asymptotically stable, then the null solution of equation (13.44) is uniformly stable.

From the theorem of Banach-Steinhauss follows that point dissipativity and compact dissipativity are equivalent for autonomous linear system. One example of linear autonomous dynamical system which is compact dissipative, but is not local dissipative, is constructed in the section 1.6 (see example1.8).

Theorem 13.14 *Let $\langle (X, \mathbb{S}_+, \pi), (Y, \mathbb{S}, \sigma), h \rangle$ be a linear non-autonomous dynamical system and the following conditions hold:*

1. Y is compact and minimal (i.e. $Y = H(y) := \overline{\{yt : t \in \mathbb{S}\}}$ for all $y \in Y$);
2. for any $x \in X$ there exists $C_x \geq 0$ such that

$$|xt| \leq C_x \qquad (13.46)$$

for all $t \in \mathbb{S}_+$;
3. the mapping $y \mapsto \|\pi_y^t\|$ is continuous, where $\|\pi_y^t\|$ is a norm of linear operator $\pi_y^t := \pi^t|_{X_y}$, for every $t \in \mathbb{S}_+$.

Then there exists $M \geq 0$ such that the inequality

$$|\pi^t x| \leq M|x| \qquad (13.47)$$

takes place for all $t \in \mathbb{S}_+$ and $x \in X$.

Proof. From condition 2. and theorem of Banach-Steinhauss follows the uniform boundedness of the family of linear operators $\{\pi_y^t : t \in \mathbb{S}_+\}$ for every $y \in Y$, i.e. for any $y \in Y$ there exists $M_y \geq 0$ such that $\|\pi_y^t\| \leq M_y$ for all $t \in \mathbb{S}_+$. We put

$$d(y) := \sup\{\|\pi_y^t\| : t \in \mathbb{S}_+\} \qquad (13.48)$$

and claim that the function $d : Y \to \mathbb{R}_+$, defined by equality (13.48) be lower-semicontinuous, i.e. $\liminf\limits_{y_n \to y} d(y_n) \geq d(y)$ for all $y \in Y$ and $\{y_n\} \to y$. Suppose that it is not true, then there exist $y \in Y, \{y_n\}$ and $\varepsilon > 0$ such that

$$\liminf_{y_n \to y} d(y_n) = d(y) - \varepsilon. \tag{13.49}$$

From the equality (13.48) follows $d(y) = \lim\limits_{n \to +\infty} \|\pi_y^{t_n}\|$ for some sequence $\{t_n\} \subseteq \mathbb{S}_+$ and, consequently, there exists k such that

$$\left| \|\pi_y^{t_n}\| - d(y) \right| < \frac{\varepsilon}{4} \tag{13.50}$$

for all $n \geq k$. According to continuity of mapping $y \mapsto \|\pi_y^t\|$ there exists $n(k)$ such that

$$\left| \|\pi_{y_n}^{t_k}\| - \|\pi_y^{t_k}\| \right| < \frac{\varepsilon}{4} \tag{13.51}$$

for all $n \geq n(k)$. From (13.50) and (13.51) follows

$$\left| d(y) - \|\pi_{y_n}^{t_k}\| \right| < \frac{\varepsilon}{2} \tag{13.52}$$

for all $n \geq n(k)$. From inequality (13.52) results

$$|d(y) - d(y_n)| \leq \frac{\varepsilon}{2} \tag{13.53}$$

for all $n \geq n(k)$. Inequality (13.53) contradicts (13.49). This contradiction proves that $d : Y \to \mathbb{R}_+$ is lower semi-continuous. Hence, this function has a set of points of continuity $D \subset Y$ of the type G_δ. Let $p \in D$, then there exist a positive numbers δ_p and M_p such that $d(y) \leq M_p$ for all $y \in B[p, \delta_p] = \{y \in Y \mid \rho(y, p) \leq \delta_p\} \subset Y$.

Since Y is minimal, there are negative numbers $t_1, t_2, ..., t_m$ such that $Y = \bigcup_{i=1}^{m} \sigma(S[p, \delta_p], t_i)$ (see [238, p.134]). We put $L := \max\{t_i | i = 1, 2, ..., m\}$. Assume that $m \in Y, y \in B[p, \delta_p]$ and t_i are such that $m = yt_i$. Then

$$|xt| = |\pi_y^{t+t_i}(\pi_{yt_i}^{-t_i}(x))| \leq M_p C |x| \tag{13.54}$$

for all $x \in X$ and $t \geq L$, where

$$C := \max\{\max\{\|\pi_y^{-t_i}\| \mid y \in Y\}, i = 1, 2, ..., m\}.$$

We claim that the family of operators $\{\pi^t : t \in [0, L]\}$ is uniformly continuous, that is, for any $\varepsilon > 0$ there is a $\delta(\varepsilon) > 0$ such that $|x| \leq \delta$ implies $|xt| \leq \varepsilon$ for all $t \in [0, L]$. Assume the contrary. Then there are $\varepsilon_0 > 0, \delta_n \to 0$ $(\delta_n > 0), |x_n| < \delta_n$ and $t_n \in [0, L]$ such that

$$|x_n t_n| \geq \varepsilon_0. \tag{13.55}$$

Since (X, h, Y) is locally trivial Banach fiber bundle and Y is compact, then the zero section $\Theta = \{\theta_y : y \in Y\}$ of (X, h, Y) is compact and, consequently, we can assume that the sequences $\{x_n\}$ and $\{t_n\}$ are convergent. Put $x_0 = \lim\limits_{n\to+\infty} x_n$ and $t_0 = \lim\limits_{n\to+\infty} t_n$, then $x_0 = \theta_{y_0}$ $(y_0 = h(x_0))$. Passing to the limit in (13.55) as $n \to +\infty$, we obtain $0 = |x_0 t_0| \geq \varepsilon_0$. The last inequality contradicts the choice of ε_0. This contradiction proves the above assertion. If $\gamma > 0$ is such that $|\pi^t x| \leq 1$ for all $|x| \leq \gamma$ and $t \in [0, L]$, then

$$|xt| \leq \frac{1}{\gamma}|x| \tag{13.4.12}$$

for all $t \in [0, L]$ and $x \in X$. We put $M := \max\{\gamma^{-1}, M_p C\}$, then from (13.54) and (13.4.12) follows the inequality (13.47) for all $t \geq 0$ and $x \in X$. The theorem is proved. $\qquad\square$

Remark 13.6 *a. If the fiber bundle (X, h, Y) is finite-dimensional, then the condition 3. of Theorem 13.14 holds.*

b. Let $X := E \times Y$, where E is a Banach space and $\pi := (\varphi, \sigma)$, i.e. $\pi^t x :=$ $(\varphi(t, u, y), \sigma^t y)$ for all $t \in \mathbb{S}_+$ and $x := (u, y) \in X = E \times Y$. Then the condition 3. of Theorem 13.14 holds, if for every $t \in \mathbb{S}_+$ the mapping $U(t, \cdot) : Y \to [E]$ is continuous, where $U(t, y)u = \varphi(t, u, y)$ for any $(t, u, y) \in \mathbb{S}_+ \times E \times Y$ and $[E]$ is a Banach space of all continuous operators acting onto E and equipped with the operational norm.

Theorem 13.15 *Let $\langle (X, \mathbb{S}_+, \pi), (Y, \mathbb{S}, \sigma), h \rangle$ be a linear non-autonomous dynamical system, Y be a compact minimal set and the mapping $y \mapsto \|\pi_y^t\|$ be continuous for each $t \in \mathbb{S}_+$. Then from point dissipativity of $\langle (X, \mathbb{S}_+, \pi), (Y, \mathbb{S}, \sigma), h \rangle$ follows its compact dissipativity.*

Proof. Assume that the conditions of Theorem 13.15 are fulfilled and the non-autonomous dynamical system $\langle (X, \mathbb{S}_+, \pi), (Y, \mathbb{S}, \sigma), h \rangle$ is point dissipative, then according to Theorem 2.36 for every $x \in X$ there exists a constant $C_x \geq 0$ such that the inequality (13.46) takes place for all $x \in X$ and $t \in \mathbb{S}_+$. Next to complete the proof of Theorem 13.15, it is sufficient to refer to Theorems 13.14 and 2.37. $\qquad\square$

Linear ordinary differential equations.

Let Λ be a complete metric space of linear operators that act on Banach space E and $C(\mathbb{R}, \Lambda)$ be a space of all continuous operator-functions $A : \mathbb{R} \to \Lambda$ equipped with open-compact topology and $(C(\mathbb{R}, \Lambda), \mathbb{R}, \sigma)$ be a dynamical system of shifts on $C(\mathbb{R}, \Lambda)$.

Let $\Lambda = [E]$ and consider the linear differential equation

$$u' = A(t)u \quad , \tag{13.56}$$

where $A \in C(\mathbb{R}, \Lambda)$. Along with equation (13.56), we shall also consider its $H-$class, that is, the family of equations

$$v' = B(t)v \quad , \tag{13.57}$$

where $B \in H(A) := \overline{\{A_\tau : \tau \in \mathbb{R}\}}, A_\tau(t) = \mathcal{A}(t + \tau)$ $(t \in \mathbb{R})$ and the bar denotes closure in $C(\mathbb{R}, \Lambda)$. Let $\varphi(t, u, \mathcal{B})$ be the solution of equation (13.57) that satisfies the condition $\varphi(0, v, B) = v$. We put $Y := H(A)$ and denote the dynamical system of shifts on $H(A)$ by $(Y, \mathbb{R}, \sigma)$, then the triple $\langle (X, \mathbb{R}_+, \pi), (Y, \mathbb{R}, \sigma), h \rangle$ is a linear non-autonomous dynamical system, where $X := E \times Y, \pi := (\varphi, \sigma)$ (i.e. $\pi(\tau, (v, \mathcal{B})) := (\varphi(\tau, v, B), B_\tau)$ and $h := pr_2 : X \to Y$. Applying Theorem 13.15 to this system, we obtain the following assertion.

Theorem 13.16   *Let $A \in C(\mathbb{R}, \Lambda)$ be recurrent (i.e. $H(A)$ is compact minimal set of $(C(\mathbb{R}, \Lambda), \mathbb{R}, \sigma)$) and the zero solution of equation (13.56) and all equation (13.57) are asymptotically stable, i.e. $\lim\limits_{t \to +\infty} |\varphi(t, v, B)| = 0$ for all $v \in E$ and $B \in H(A)$. Then the zero solution of equation (13.56) is uniformly stable, i.e. there exists $M \geq 0$ such that $|\varphi(t, v, B)| \leq M|v|$ for all $t \geq 0, v \in E$ and $B \in H(A)$.*

Proof.   According to Lemma 2 [60] the mapping $B \mapsto \varphi(t, \cdot, B)$ from $H(A)$ into $[E]$ is continuous for all $t \in \mathbb{R}$. To finish the proof of the theorem it suffices to refer to Theorem 13.14.
$\square$

Linear Partial differential equations.   Let Λ be some complete metric space of linear closed operators acting into Banach space E (for example $\Lambda = \{A_0 + B | B \in [E]\}$, where A_0 is a closed operator that acts on E). We assume that the following conditions are fulfilled for equation (13.56) and its $H-$ class (13.57):

a. for any $v \in E$ and $B \in H(A)$ equation (13.57) has exactly one solution that is defined on $\mathbb{R}_+$ and satisfies the condition $\varphi(0, v, B) = v$;

b. the mapping $\varphi : (t, v, B) \to \varphi(t, v, B)$ is continuous in the topology of $\mathbb{R}_+ \times E \times C(\mathbb{R}; \Lambda)$;

c. for every $t \in \mathbb{R}_+$ the mapping $U(t, \cdot) : H(A) \to [E]$ is continuous, where $U(t, \cdot)$ is the Cauchy's operator of equation (13.57), i.e. $U(t, B)v := \varphi(t, v, B)$ $(t \in \mathbb{R}_+, v \in E$ and $\in H(A)$).

Under the above assumptions the equation (13.56) generates a linear non-autonomous dynamical system $\langle (X, \mathbb{R}_+, \pi), (Y, \mathbb{R}, \sigma), h \rangle$, where $X = E \times Y, \pi = (\varphi, \sigma)$ and $h := pr_2 : X \to Y$. Applying Theorem 13.15 to this system, we will obtain the analogue of Theorem 13.16 for different classes of partial differential equations.

We will consider an example of partial differential equation which satisfies the above conditions a.-c. Let $\mathcal{H}$ be a Hilbert space with a scalar product $\langle \cdot, \cdot \rangle = |\cdot|^2, \mathcal{D}(\mathbb{R}_+, \mathcal{H})$ be a set of all infinite differentiable and finite into $\mathbb{R}_+$ functions with values into $\mathcal{H}$.

Denote by $(C(\mathbb{R}, [\mathcal{H}]), \mathbb{R}, \sigma)$ a dynamical system of shifts on $C(\mathbb{R}, [\mathcal{H}])$. Consider the equation

$$\int_{\mathbb{R}_+} [\langle u(t), \varphi'(t) \rangle + \langle A(t)u(t), \varphi(t) \rangle] dt = 0, \qquad (13.58)$$

along with the family of equations

$$\int_{\mathbb{R}_+} [\langle u(t), \varphi'(t) \rangle + \langle B(t)u(t), \varphi(t) \rangle] dt = 0, \qquad (13.59)$$

where $B \in H(A) := \overline{\{A_\tau | \tau \in \mathbb{R}\}}, A_\tau(t) := (t + \tau)$ and the bar denotes closure in $C(\mathbb{R}, [\mathcal{H}])$.

The function $u \in C(\mathbb{R}_+, \mathcal{H})$ is called a solution of equation (13.58), if (13.58) takes place for all $\varphi \in \mathcal{D}(\mathbb{R}_+, \mathcal{H})$.

Assume that the operator $A(t)$ is self-adjoint. Let $(H(A), \mathbb{R}, \sigma)$ be a dynamical system of shifts on $H(A), \varphi(t, v, B)$ be a solution of equation (13.59) with condition $\varphi(0, v, B) = v, \tilde{X} = \mathcal{H} \times H(\mathcal{A})$, X be a set of all the points $\langle u, B \rangle \in \tilde{X}$ such that through point $u \in \mathcal{H}$ passes a solution $\varphi(t, u, A)$ of equation (13.58) defined on $\mathbb{R}_+$. According to Lemma 2.21 [92] the set X is closed in $\tilde{X}$. In virtue of Lemma 2.22 [92] the triple $(X, \mathbb{R}_+, \pi)$ is a dynamical system on X (where $\pi := (\varphi, \sigma)$) and $\langle (X, \mathbb{R}_+, \pi), (Y, \mathbb{R}, \sigma), h \rangle$ is a linear non-autonomous dynamical system, where $h := pr_2 : X \to Y = H(A)$. Applying the results from [5] it is possible to show that for every t the mapping $B \mapsto U(t, \mathcal{B})$ (where $U(t, B)v := \varphi(t, v, B)$) from $H(A)$ into $[\mathcal{H}]$ is continuous and, consequently, for this system is applicable Theorem 13.14. Thus the following assertion takes place.

Theorem 13.17 *Let $A \in C(\mathbb{R}, [\mathcal{H}])$ be recurrent and the zero solution of equation (13.56) and all equation (13.57) are asymptotically stable, i.e. $\lim_{t \to +\infty} |\varphi(t, v, B)| = 0$ for all $v \in E$ and $B \in H(A)$. Then the zero solution of equation (13.56) is uniformly stable, i.e. there exists $M \geq 0$ such that $|\varphi(t, v, B)| \leq M|v|$ for all $t \geq 0, v \in \mathcal{H}$ and $B \in H(A)$.*

We will give the example of limit problem reducing to equation of type (13.58). Let Ω be a bounded domain in $\mathbb{R}^n, \Gamma$ be frontier of $\Omega, Q = \mathbb{R}_+ \times \Omega$ and $S = \mathbb{R}_+ \times \Gamma$. Consider in Q the first initial limit problem for equation

$$\frac{\partial u}{\partial t} = L(t)u \qquad (u|_{t=0} = \varphi, u|_S = 0), \qquad (13.60)$$

where

$$L(t)u := \sum_{i,j=1}^{n} \frac{\partial}{\partial x_i}\left(a_{ij}(t, x)\frac{\partial u}{\partial x_j}\right) - a(t, x)u$$

The operator $A(t)$ according to theorem of Riesz is defined by equality

$$\langle A(t)u, \varphi \rangle = - \int_\Omega [\sum_{i,j=1}^n a_{ij}(t,x)\frac{\partial u}{\partial x_j}\frac{\partial \varphi}{\partial x_i} + a(t,x)u\varphi]dx.$$

If $a_{ij}(t,x) = a_{ji}(t,x)$ and the functions $a_{ij}(t,x)$ and $a(t,x)$ are recurrent (almost periodic) with respect to $t \in \mathbb{R}$ uniformly with respect to $x \in \Omega$, then for equation (13.60) it is applicable Theorem 13.17, if $\mathcal{H} = \dot{W}_2^1(\Omega)$

Linear functional-differential equations. Denote by $\mathfrak{A} = \mathfrak{A}(C, \mathbb{R}^n)$ a Banach space of all linear continuous operators acting from $C := C([-r, 0], \mathbb{R}^n)$ into $\mathbb{R}^n$, equipped by operational norm. Consider the equation

$$u' = A(t)u_t \quad , \tag{13.61}$$

where $A \in C(\mathbb{R}, \mathfrak{A})$. We put $H(A) := \overline{\{A_\tau : \tau \in \mathbb{R}\}}, A_\tau(t) := A(t + \tau)$ and the bar denotes closure in the topology of uniform convergence on every compact from $\mathbb{R}$.

Along with equation (13.61) we also consider the family of equations

$$u' = B(t)u_t \quad , \tag{13.62}$$

where $B \in H(A)$. Let $\varphi_t(v, B)$ be a solution of equation (13.62) with condition $\varphi_0(v, B) = v$, defined on $\mathbb{R}_+$. We put $Y := H(A)$ and denote by $(Y, \mathbb{R}, \sigma)$ a dynamical system of shifts on $H(A)$. Let $X := C \times Y$ and $\pi := (\varphi, \sigma)$ a dynamical system on X, defined by equality $\pi(t, (v, B)) := (\varphi_\tau(v, B), B_\tau)$. A non-autonomous dynamical system $\langle (X, \mathbb{R}_+, \pi), (Y, \mathbb{R}, \sigma), h \rangle$ $(h := pr_2 : X \to Y)$ is linear. The following assertion takes place.

Lemma 13.9 *Let $H(A)$ be compact in $C(\mathbb{R}, \mathfrak{A})$, then the non-autonomous dynamical system $\langle (X, \mathbb{R}_+, \pi), (Y, \mathbb{R}, \sigma), h \rangle$ generated by equation (13.61) is completely continuous.*

Proof. Let $B \subset C = C[-r, 0]$ be a bounded set and $t \geq r$. According to the continuity of mapping $\varphi : \mathbb{R}_+ \times C \times H(A) \mapsto C$ and compactness of $H(A)$ there exists a positive number M such that $|\varphi_\tau(v, B)| \leq M$ and $|B(\tau)\varphi_\tau(v, B)| \leq M$ for all $\tau \in [0, t], B \in H(A)$ and $v \in B$, and, consequently, $|\dot{\varphi}(\tau, v, B)| \leq M$ for all $\tau \in [0, t], B \in H(A)$ and $v \in B$, i.e. the family of functions $\{\varphi_t(v, B) : B \in H(A), v \in B\}$ (for $t \geq r$) is uniformly continuous on $[-r, 0]$. Therefore this family of functions is relatively compact. The Lemma is proved. $\square$

Theorem 13.18 *Let $H(A)$ be compact. Then the following assertion are equivalent:*

(1) for any $B \in H(A)$ the zero solution of equation (13.62) is asymptotically stable, i.e. $\lim_{t \to +\infty} |\varphi_t(v, B)| = 0$ for all $v \in C$ and $B \in H(A)$;

(2) *the zero solution of equation (13.61) is uniformly asymptotically stable, i.e. there are the positive numbers N and ν such that $|\varphi_t(v,B)| \le Ne^{-\nu t}|v|$ for all $t \ge 0, v \in C$ and $B \in H(A)$.*

Proof. Let $\langle (X,\mathbb{R}_+,\pi),(Y,\mathbb{R},\sigma),h \rangle$ be a linear non-autonomous dynamical system, generated by equation (13.61). According to Lemma 13.9 this system is completely continuous and to finish the proof it is sufficiently to refer to Theorems 2.38 and 6.10. $\square$

Consider the neutral functional differential equation

$$\frac{d}{dt}Dx_t = A(t)x_t \quad , \tag{13.63}$$

where $A \in C(\mathbb{R},\mathfrak{A})$ and $D \in \mathfrak{A}$ is non-atomic at zero operator [175, p.67]. As well as in the case of equation (13.61), the equation (13.63) generates a linear dynamical system $\langle (X,\mathbb{R}_+,\pi),(Y,\mathbb{R},\sigma),h \rangle$, where $X := C \times Y, Y := H(A)$ and $\pi := (\varphi,\sigma)$. The following statement takes place.

Lemma 13.10 *Let $H(A)$ be compact and the operator D is stable, i.e. the zero solution of homogeneous difference equation $Dy_t = 0$ is uniformly asymptotically stable. Then a linear non-autonomous dynamical system $\langle (X,\mathbb{R}_+,\pi),(Y,\mathbb{R},\sigma),h \rangle$, generated by equation (13.63), is asymptotically compact.*

Proof. According to [179] (see, p.119, formula (5.18)) the mapping $\varphi_t(\cdot,B) : C \to C$ can be written as

$$\varphi_t(\cdot,B) = S_t(\cdot) + U_t(\cdot,B)$$

for all $B \in H(A)$, where $U_t(\cdot,B)$ is conditionally completely continuous for $t \ge r$ and there exist constants $N > 0, \nu > 0$ such that $\|S_t\| \le Ne^{-\nu t}(t \ge 0)$. To finish the proof of Lemma 13.10 it is sufficiently to refer to Theorem 2.22. $\square$

Theorem 13.19 *Let $A \in C(\mathbb{R},\mathfrak{A})$ be recurrent (i.e. $H(A)$ is compact minimal in the dynamical system of shifts $(C(\mathbb{R},\mathfrak{A}),\mathbb{R},\sigma)$) and D is stable, then the following assertions are equivalent:*

(1) the zero solution of equation (13.61) and all equation

$$\frac{d}{dt}Dx_t = B(t)x_t \quad , \tag{13.64}$$

where $B \in H(A)$, are asymptotically stable, i.e. $\displaystyle\lim_{t \to +\infty} |\varphi_t(v,B)| = 0$ for all $v \in C$ and $B \in H(A)$ $(\varphi_t(v,B)$ is a solution of equation (13.64) with condition $\varphi_0(v,B) = v)$;

(2) the zero solution of equation (13.63) is uniformly exponentially stable, i.e. there are a positive numbers N and ν such that $|\varphi_t(v, B)| \le N e^{-\nu t} |v|$ for all $t \ge 0$, $v \in C$ and $B \in H(A)$.

Proof. Let $\langle (X, \mathbb{R}_+, \pi), (Y, \mathbb{R}, \sigma), h \rangle$ be a linear non-autonomous dynamical system, generated by equation (13.63). According to Lemma 13.10 this system is asymptotically compact. To finish the proof of Theorem 13.19 it is sufficiently to refer to Theorems 2.38 and 1.25. The theorem is proved. $\qquad\square$

13.5   Linear α-condensing systems

Let $A(t)$ be a continuous $n \times n$ matrix-function and $H(A)$ be the family of all matrix-functions $B = \lim\limits_{n \to +\infty} A_{t_n}$, where $\{t_n\} \subset \mathbb{R}$, $A_{t_n}(t) = A(t_n + t)$ and the convergence $A_{t_n} \to B$ is uniform on every compact subset of $\mathbb{R}$. The following result is well known.

Theorem 13.20   [275, 33, 65] *Let A be a bounded and uniformly continuous matrix-function on $\mathbb{R}$, then the following conditions are equivalent:*

1. *The trivial solution of equation*

$$x' = A(t)x \tag{13.65}$$

   is uniformly exponentially stable.
2. *The trivial solution of equation (13.65) is uniformly asymptotically stable.*
3. *The trivial solution of equation (13.65) and every equation*

$$y' = B(t)y \quad (B \in H(A)) \tag{13.66}$$

   is asymptotically stable.

For equations in infinite-dimensional spaces conditions 1., 2., and 3. are not equivalent; see example 1.8 and also [81, 121, 249]. However, in the general infinite-dimensional case condition 1. implies condition 2., and condition 2. implies condition 3.

Linear non-autonomous dynamical systems satisfying one of the conditions 1., or 2., or 3. are studied in [81]; see also Chapter 2 (section 2.11). We will show that if the operator corresponding to the the Cauchy's problem for (13.65) satisfies some compactness condition, then condition 3 implies condition 1 (see Theorems 13.23 and 13.24).

For recurrent (almost periodic) systems this result is made precise in Theorems 13.25 and 13.26. Applications of this result to different classes of linear evolution equations (ordinary linear differential equations in a Banach space, retarded and neutral functional differential equations, some classes of evolution partial differential equations) are given.

Assume that X and Y are complete metric spaces, $\mathbb{R}$ is the set of real numbers, $\mathbb{Z}$ is the set of integer numbers, $\mathbb{S} = \mathbb{R}$ or $\mathbb{Z}$, $\mathbb{S}_+ = \{t \in \mathbb{S} : t \geq 0\}$ and $\mathbb{S}_- = \{t \in \mathbb{S} | t \leq 0\}$.

Recall that a dynamical system $(X, \mathbb{S}_+, \pi)$ is said to be conditionally β-condensing [179] if there exists $t_0 > 0$ such that $\beta(\pi^{t_0} B) < \beta(B)$ for all bounded sets B in X with $\beta(B) > 0$. The dynamical system $(X, \mathbb{S}_+, \pi)$ is said to be β-condensing if it is conditionally β-condensing and the set $\pi^{t_0} B$ is bounded for all bounded sets $B \subseteq X$.

According to Lemma 2.3.5 in [179, p.15] and Lemma 3.3 in [82] the conditional condensing dynamical system $(X, \mathbb{S}_+, \pi)$ is asymptotically compact.

Let E be a metric space, $X := E \times Y$, $A \subset X$, and $A_y := \{x \in A : pr_2 x = y\}$. Then $A = \cup\{A_y : y \in Y\}$. Let $\tilde{A}_y := pr_1 A_y$ and $\tilde{A} = \cup\{\tilde{A}_y : y \in Y\}$. Note that if the space Y is compact, then a set $A \subset X$ is bounded in X if and only if the set $\tilde{A}$ is bounded in E.

Lemma 13.11   *The equality $\alpha(A) = \alpha(\tilde{A})$ takes place for all bounded sets $A \subset X$, where $\alpha(A)$ and $\alpha(\tilde{A})$ are the Kuratowskii's measure of non-compactness for the sets $A \subset X$ and $\tilde{A} \subset E$.*

Proof. Let $\varepsilon > 0$ and A be a bounded subset in X, then there are sets $A_1, A_2, \ldots, A_n$ such that $A = \cup\{A_i : i = 1, 2, \ldots, n\}$ and $\operatorname{diam} A_i < \alpha(A) + \varepsilon$. Note that $\tilde{A} = \cup\{\tilde{A}_i : i = 1, 2, \ldots, n\}$ and $\operatorname{diam} \tilde{A}_i \leq \operatorname{diam} A_i < \alpha(A) + \varepsilon$, and consequently, $\alpha(\tilde{A}) \leq \alpha(A)$.

Let ε be a positive constant, A be a bounded set in X, $\tilde{A} = \cup\{\tilde{A}_k : k = 1, 2, \ldots, m\}$ and $\operatorname{diam} \tilde{A}_k < \alpha(\tilde{A}) + \varepsilon$. Since Y is compact, there are sets $Y_1, Y_2, \ldots, Y_\ell$ such that $Y_1 \cup Y_2 \cup \cdots \cup Y_\ell = Y$ and $\operatorname{diam} Y_j < \varepsilon$ $(j = 1, 2, \ldots \ell)$. Let $A_i = pr_1^{-1}(\tilde{A}_i) \cap A$, and

$$A_{ij} = pr_2^{-1}(pr_2(pr_1^{-1}(\tilde{A}_i) \cap A) \cap Y_j) \cap A_i.$$

Note that $A_{ij} \subseteq \tilde{A}_i \times Y_j$, and that

$$\operatorname{diam} A_{ij} \leq \operatorname{diam} \tilde{A}_i + \operatorname{diam} Y_j < \alpha(\tilde{A}) + \varepsilon + \varepsilon = \alpha(A) + 2\varepsilon.$$

Since $A = \cup\{A_{ij} : i = 1, 2, \ldots, n, j = 1, 2, \ldots, \ell\}$, it follows that $\alpha(A) \leq \alpha(\tilde{A})$ and $\alpha(A) = \alpha(\tilde{A})$. which concludes the present proof. $\square$

Definition 13.5   A cocycle $\langle E, \varphi, (Y,, \sigma) \rangle$ is called conditionally α-condensing if there exists $t_0 > 0$ such that for any bounded set $B \subseteq E$ the inequality $\alpha(\varphi(t_0, B, Y)) < \alpha(B)$ holds if $\alpha(B) > 0$. The cocycle φ is called α-condensing if it is a conditional α-condensing cocycle and the set $\varphi(t_0, B, Y) = \cup\{\varphi(t_0, u, Y)|u \in B, y \in Y\}$ is bounded for all bounded set $B \subseteq E$.

Definition 13.6 A cocycle φ is called conditional α-contraction of order $k \in [0,1)$, if there exists $t_0 > 0$ such that for any bounded set $B \subseteq E$ for which $\varphi(t_0, B, Y) = \cup\{\varphi(t_0, u, Y)|u \in B, y \in Y\}$ is bounded the inequality $\alpha(\varphi(t_0, B, Y)) \leq k\alpha(B)$ holds. The cocycle φ is called α-contraction if it is a conditional α-contraction cocycle and the set $\varphi(t_0, B, Y) = \cup\{\varphi(t_0, u, Y)|u \in B, y \in Y\}$ is bounded for all bounded sets $B \subseteq E$.

Lemma 13.12 *Let Y be compact and the cocycle φ be α-condensing. Then the skew-product dynamical system $(X, \mathbb{S}_+, \pi)$, generated by the cocycle φ, is α-condensing too.*

Proof. Let $A \subset X$ be a bounded subset, $t_0 > 0$ and $\alpha(A) > 0$, then

$$\pi(t_0, A) = \cup\{\pi(t_0, A_y|y \in Y\}$$
$$= \cup\{(\varphi(t_0, A_y, y), yt)|y \in Y\} \subseteq \varphi(t_0, \tilde{A}, Y) \times Y. \tag{13.67}$$

Since A is bounded, $\tilde{A}$ is also bounded in E and according to the condition of the lemma the set $\varphi(t_0, \tilde{A}, Y)$ is bounded and, consequently, $\pi(t_0, A)$ is bounded. According to Lemma 13.11 and (13.67) we have

$$\alpha(\pi(t_0, A)) = \alpha(\cup\{(\varphi(t_0, A_y, y), yt_0)|y \in Y\}) \leq \alpha(\varphi(t_0, \tilde{A}, Y)) < \alpha(\tilde{A}) = \alpha(A).$$

The lemma is proved. $\square$

Theorem 13.21 *Let E be a Banach space, φ be a cocycle on $(Y, \mathbb{S}, \sigma)$ with fiber E and the following conditions be fulfilled:*

(1) $\varphi(t, u, y) = \psi(t, u, y) + \gamma(t, u, y)$ for all $t \in \mathbb{S}_+, u \in E$ and $y \in Y$.
(2) There exists a function $m : \mathbb{R}_+^2 \to \mathbb{R}_+$ satisfying the condition $m(t, r) \to 0$ as $t \to +\infty$ (for every $r > 0$) such that $|\psi(t, u_1, y) - \psi(t, u_2, y)| \leq m(t, r)|u_1 - u_2|$ for all $t \in \mathbb{S}_+, u_1, u_2 \in B[0, r]$ and $y \in Y$.
(3) $\gamma(t, A, Y)$ is compact for all bounded $A \subset X$ and $t > 0$.

Then the cocycle φ is an α-contraction.

Proof. Let $\varepsilon > 0$ and A be a bounded set in E, then there are sets $A_1, A_2, \ldots, A_n$ such that $A = \cup\{A_i : i = 1, 2, \ldots, n\}$ and diam $A_i < \alpha(A) + \varepsilon$ for $i = 1, 2, \ldots, n$. Since Y is compact, then there are a sets $Y_1, Y_2, \ldots, Y_m$ such that $Y_1 \cup Y_2 \cup \cdots \cup Y_m = Y$ with condition diam $Y_j < \varepsilon$ for all $j = 1, 2, \ldots, m$.

Let $r := \operatorname{diam} A$ and t_0 be a positive number such that $m(t_0, r) < 1$. We note that

$$\varphi(t_0, A, Y) \subseteq \psi(t_0, A, Y) + \gamma(t_0, A, Y)$$
$$= \cup\{\psi(t_0, A_i, Y_j)|i = 1, 2, \dots, n; j = 1, 2, .., m\} + \gamma(t_0, A, Y). \qquad (13.68)$$

According to the conditions of Theorem 13.21, $\alpha(\gamma(t_0, A, Y)) = 0$ and

$$\operatorname{diam} \psi(t_0, A_i, y) \le m(t_0, r) \operatorname{diam} A_i$$

for all $y \in Y$. Thus we have

$$|\psi(t_0, u_1, y_1) - \psi(t_0, u_2, y_2)| \le |\psi(t_0, u_1, y_1) - \psi(t_0, u_2, y_1)|$$
$$+|\psi(t_0, u_2, y_1) - \psi(t_0, u_2, y_2)| \qquad (13.69)$$

and, consequently,

$$\operatorname{diam} \psi(t_0, A_i, y) \le m(t_0, r) \operatorname{diam} A_i + \operatorname{diam} \psi(t_0, u_2, Y_j) \quad \text{for all} \quad y \in Y_j . \quad (13.70)$$

Since Y is compact, from (13.69)-(13.70) follows the inequality

$$\operatorname{diam} \psi(t_0, A_i, Y_j) \le m(t_0, r) \operatorname{diam} A_i \le m(t_0, r)(\alpha(A) + \varepsilon)$$

and, consequently, $\alpha(\varphi(t_0, A, Y)) \le m(t_0, r)\alpha(A)$. The theorem is proved. $\qquad \square$

13.6 Exponential stable systems

Theorem 13.22 *Let* $\langle(X, \mathbb{S}_+, \pi), (Y, \mathbb{S}, \sigma), h\rangle$ *be a linear non-autonomous dynamical system, Y be a compact set. Then the following conditions are equivalent:*

1. *The non-autonomous dynamical system $\langle(X, \mathbb{S}_+, \pi), (Y, \mathbb{S}, \sigma), h\rangle$ is uniformly exponentially stable, i.e. there exist two positive constants N and ν such that $|\pi(t, x)| \le Ne^{-\nu t}|x|$ for all $t \in \mathbb{S}_+$ and $x \in X$.*
2. $\|\pi^t\| \to 0$ *as $t \to +\infty$, where $\|\pi^t\| = \sup\{|\pi^t x| : x \in X, |x| \le 1\}$.*
3. *The non-autonomous dynamical system $\langle(X, \mathbb{S}_+, \pi), (Y, \mathbb{S}, \sigma), h\rangle$ is locally dissipative.*

Proof. According to Theorem 2.38, conditions 1. and 3. are equivalent. Now we will prove that the conditions 1. and 2. are equivalent. It is clear that from 1. follows 2. According to condition 2. there exists $L > 0$ such that

$$\|\pi^t\| \le 1 \qquad (13.71)$$

for all $t \ge L$. We claim that the family of operators $\{\pi^t : t \in [0, L]\}$ is uniformly continuous, that is, for any $\varepsilon > 0$ there is a $\delta(\varepsilon) > 0$ such that $|x| \le \delta$ implies

$|xt| \leq \varepsilon$ for all $t \in [0, L]$. On the contrary, assume that there are $\varepsilon_0 > 0$, $\delta_n > 0$ with $\delta_n \to 0$, $|x_n| < \delta_n$ and $t_n \in [0, L]$ such that

$$|x_n t_n| \geq \varepsilon_0. \tag{13.72}$$

Since (X, h, Y) is a locally trivial Banach fiber bundle and Y is compact, the zero section $\Theta := \{\theta_y : y \in Y, \theta_y \in X_y, |\theta_y| = 0\}$ of (X, h, Y) is compact and, consequently, we can assume that the sequences $\{x_n\}$ and $\{t_n\}$ are convergent. Put $x_0 = \lim\limits_{n \to +\infty} x_n$ and $t_0 = \lim\limits_{n \to +\infty} t_n$, then $x_0 = \theta_{y_0}$ $(y_0 = h(x_0))$. Passing to the limit in (13.72) as $n \to +\infty$, we obtain $0 = |x_0 t_0| \geq \varepsilon_0$. This last inequality contradicts the choice of ε_0, and hence proves the above assertion. If $\gamma > 0$ is such that $|\pi^t x| \leq 1$ for all $|x| \leq \gamma$ and $t \in [0, L]$, then

$$|xt| \leq \frac{1}{\gamma} |x| \tag{13.73}$$

for all $t \in [0, L]$ and $x \in X$. We put $M := \max\{\gamma^{-1}, 1\}$, then from (13.71) and (13.73) follows

$$\|\pi^t\| \leq M \tag{13.74}$$

for all $t \geq 0$ and $x \in X$. Consider the function $m(t) := \|\pi^t\|$. We note that $m(t+\tau) \leq m(t)m(\tau)$ for all $t, \tau \in \mathbb{S}_+$ and $m(t) \leq M$ for all $t \in \mathbb{S}_+$ and $m(t) \to 0$ as $t \to +\infty$. According to Lemma 2.19 there exist positive numbers N and ν such that $m(t) \leq Ne^{-\nu t}$ for all $t \in \mathbb{S}_+$. Thus $|\pi(t, x)| \leq \|\pi^t\||x| \leq Ne^{-\nu t}|x|$ for all $t \in \mathbb{S}_+$ and $x \in X$. The theorem is proved. $\qquad\square$

Let $\mathbb{B} := \{x \in X : \exists \gamma \in \Phi_x \text{ such that } \sup\limits_{t \in \mathbb{S}} |\gamma(t)| < +\infty\}$.

Theorem 13.23 *Let $\langle (X, \mathbb{S}_+, \pi), (Y, \mathbb{S}, \sigma), h \rangle$ be a linear non-autonomous dynamical system, Y be compact and $(X, \mathbb{S}_+, \pi)$ be conditionally α-condensing. Then the following assertions are equivalent:*

(1) The non-autonomous dynamical system $\langle (X, \mathbb{S}_+, \pi), (Y, \mathbb{S}, \sigma), h \rangle$ is point dissipative and this system doesn't admit non-trivial bounded trajectories on $\mathbb{S}$, i.e. $\mathbb{B} \subseteq \Theta = \{\theta_y : y \in Y, \theta_y \in X_y, |\theta_y| = 0\}$.

(2) The non-autonomous dynamical system $\langle (X, \mathbb{S}_+, \pi), (Y, \mathbb{S}, \sigma), h \rangle$ is uniformly exponentially stable.

Proof. Denote by $\Theta = \{\theta_y : y \in Y, \theta_y \in X_y, |\theta_y| = 0\}$ the zero section of the vector fibering (X, h, Y). Since (X, h, Y) is locally trivial and Y is compact, the zero section Θ is compact and an invariant set of the dynamical system $(X, \mathbb{S}_+, \pi)$. Taking into account that the dynamical system $(X, \mathbb{S}_+, \pi)$ is conditionally α-condensing, according to Theorem 2.4.8 [179] the set Θ is orbitally stable and in particular there exists a positive constant N such that $|xt| \leq N|x|$ for all $t \in \mathbb{S}_+$ and $x \in X$. By virtue of Theorem 2.38 the dynamical system $(X, \mathbb{S}_+, \pi)$ is compact dissipative and according to Theorem 2.38 $(X, \mathbb{S}_+, \pi)$ is local dissipative. It follows from Theorem 13.22 that $(X, \mathbb{S}_+, \pi)$ is uniformly exponentially stable.

Let now the non-autonomous dynamical system $\langle (X, \mathbb{S}_+, \pi), (Y, \mathbb{S}, \sigma), h \rangle$ be uniformly exponentially stable, then according to Theorem 13.22 it is locally dissipative. Let J be its Levinson's centre (i.e. maximal compact invariant set of dynamical system $(X, \mathbb{S}_+, \pi)$) . We note that according to the linearity of non-autonomous dynamical system $\langle (X, \mathbb{S}_+, \pi), (Y, \mathbb{S}, \sigma), h \rangle$ we have $J = \Theta$. Let φ be a entire bounded trajectory of dynamical system $(X, \mathbb{S}_+, \pi)$. Since the non-autonomous dynamical system $\langle (X, \mathbb{S}_+, \pi), (Y, \mathbb{S}, \sigma), h \rangle$ is conditionally α-condensing, in particularly, it is asymptotically compact and the set $M = \varphi(\mathbb{S})$ is relatively compact. In fact, the set M is invariant $\Omega(M) = \overline{M}$ and in view of Lemma 3.3 [82] the set M is relatively compact. We note that $\varphi(\mathbb{S}) \subseteq J = \Theta$ because J is the maximal compact invariant set of $(X, \mathbb{S}_+, \pi)$. The theorem is proved. $\qquad\square$

Remark 13.7 *Theorem A in [277] implies a version of Theorem 13.23 under slightly stronger assumptions.*

Theorem 13.24 *Let $\langle (X, \mathbb{S}_+, \pi), (Y, \mathbb{S}, \sigma), h \rangle$ be a linear non-autonomous dynamical system, Y be compact and $(X, \mathbb{S}_+, \pi)$ be completely continuous, i.e. for any bounded set $A \subseteq X$ there exists a positive number ℓ such that $\pi^\ell(A)$ is relatively compact. Then the following assertions are equivalent:*

1. *The non-autonomous dynamical system $\langle (X, \mathbb{S}_+, \pi), (Y, \mathbb{S}, \sigma), h \rangle$ is uniformly exponentially stable.*

2. $\lim\limits_{t \to +\infty} |\pi^t x| = 0$ *for all $x \in X$.*

Proof. It is clear that condition 1 implies 2. Now we will show that condition 1 follows from 2. According to Theorem 2.38 the non-autonomous dynamical system $\langle (X, \mathbb{S}_+, \pi), (Y, \mathbb{S}, \sigma), h \rangle$ is point dissipative. Since the dynamical system $(X, \mathbb{S}_+, \pi)$ is completely continuous, by virtue of Theorem 2.38 the dynamical system $(X, \mathbb{S}_+, \pi)$ is locally dissipative. To prove the theorem it is sufficient to refer to Theorem 13.22 $\square$

13.7 Linear system with a minimal base

In this section we study a linear system $\langle (X, \mathbb{S}_+, \pi), (Y, \mathbb{S}, \sigma), h \rangle$ with compact minimal base $(Y, \mathbb{S}, \sigma)$.

Theorem 13.25 *Let $\langle (X, \mathbb{S}_+, \pi), (Y, \mathbb{S}, \sigma), h \rangle$ be a linear non-autonomous dynamical system and the following conditions hold:*

(1) Y is compact and minimal.

(2) The dynamical system $(X, \mathbb{S}_+, \pi)$ is asymptotically compact.

(3) The mapping $y \mapsto \|\pi_y^t\|$ is continuous, where $\|\pi_y^t\|$ is the norm of the linear operator $\pi_y^t := \pi^t|_{X_y}$, for every $t \in \mathbb{S}_+$ or $(X, \mathbb{S}_+, \pi)$ is a skew-product dynamical system.

Then the non-autonomous dynamical system $\langle (X, \mathbb{S}_+, \pi), (Y, \mathbb{S}, \sigma), h \rangle$ is uniformly exponentially stable if and only if

$$\lim_{t \to +\infty} |\pi^t x| = 0 \tag{13.75}$$

for all $x \in X$.

Proof. It is clear that from uniform exponential stability of the non-autonomous dynamical system $\langle (X, \mathbb{S}_+, \pi), (Y, \mathbb{S}, \sigma), h \rangle$ follows the equality (13.75).

From condition (13.75) and minimality of $(Y, \mathbb{S}, \sigma)$ by virtue of Theorem 13.14 it follows that for the non-autonomous dynamical system $\langle (X, \mathbb{S}_+, \pi), (Y, \mathbb{S}, \sigma), h \rangle$ the inequality

$$|\pi(t, x)| \le M|x| \tag{13.76}$$

holds for all $t \in \mathbb{S}_+$ and $x \in X$. According to Theorem 2.38 this system is compact dissipative. Since the dynamical system $(X, \mathbb{S}_+, \pi)$ is asymptotically compact, to finish the proof of the Theorem it is sufficient to remark that according to Theorem 2.13 [86] every compact dissipative and asymptotically compact dynamical system is local dissipative. Thus a linear non-autonomous dynamical system $\langle (X, \mathbb{S}_+, \pi), (Y, \mathbb{S}, \sigma), h \rangle$ is local dissipative and, consequently, it is uniform exponentially stable. The theorem is proved. $\qquad\square$

Theorem 13.26   *Suppose that the following conditions are satisfied:*

(1) A dynamical system $(Y, \mathbb{S}, \sigma)$ is compact and minimal.

(2) A linear non-autonomous dynamical system $\langle (X, \mathbb{S}_+, \pi), (Y, \mathbb{S}, \sigma), h \rangle$ is generated by cocycle φ (i.e. $X = E \times Y$, $\pi = (\varphi, \sigma)$ and $h = pr_2 : X \mapsto Y$).

(3) The dynamical system $(X, \mathbb{S}_+, \pi)$ is conditionally α-condensing.

(4) There exists a positive number M such that $|\varphi(t, u, y)| \le M|u|$ for all $t \in Y$ and $t \in \mathbb{S}_+$.

Then there are two vectorial positively invariant sub-fiberings (X^0, h, Y) and (X^s, h, Y) of (X, h, Y) such that:

a. $X_y = X_y^0 + X_y^s$ and $X_y^0 \cap X_y^s = \theta_y$ for all $y \in Y$, where $\theta_y = (0, y) \in X = E \times Y$ and 0 is the zero in the Banach space E.

b. The vector sub-fibering (X^0, h, Y) is finite dimensional, invariant (i.e. $\pi^t X^0 = X^0$ for all $t \in \mathbb{S}_+$) and every trajectory of the dynamical system $(X, \mathbb{S}_+, \pi)$ belonging to X^0 is recurrent.

c. *There exist two positive numbers N and ν such that $|\varphi(t, u, y)| \leq Ne^{-\nu t}|u|$ for all $(u, y) \in X^s$ and $t \in \mathbb{S}_+$.*

Proof. Let $X^0 := \mathbb{B}$, then according to Theorem 13.2, statement b. holds. Denote by P_y the projection of $X_y := h^{-1}(y)$ to $\mathbb{B}_y := \mathbb{B} \cap h^{-1}(y)$, then $P_y(u, y) = (\mathcal{P}(y)u, y)$ for all $u \in E, \mathcal{P}^2(y) = \mathcal{P}(y)$ and the mapping $\mathcal{P} : Y \to [E]$ ($y \mapsto \mathcal{P}(y)$) is continuous, where by $[E]$ denotes the set of all linear continuous operators acting on E. Now we set $X^s_y := \mathcal{Q}(y)X_y$ and $X^s := \cup\{X^s_y : y \in Y\}$, where $\mathcal{Q}(y) := Id_E - \mathcal{P}(y)$. We will show that X^s is closed in X. In fact, let $\{x_n\} = \{(u_n, y_n)\} \subseteq X^s$ and $x_0 = (u_0, y_0) = \lim_{n \to \infty} x_n$. Note that $P_{y_0}(x_0) = (\mathcal{P}(y_0)u_0, y_0) = (\lim_{n \to \infty} \mathcal{P}(y_n)u_n, y_0) = (0, y_0) = \theta_{y_0}$ and, consequently, $x_0 \in X^s_{y_0} \subseteq X^s$.

Let $(X^s, \mathbb{S}_+, \pi)$ be the dynamical system induced by $(X, \mathbb{S}_+, \pi)$. It is clear that under the conditions of Theorem 13.26 the dynamical system $(X^s, \mathbb{S}_+, \pi)$ is asymptotically compact and every positive semi-trajectory is relatively compact and, consequently, $\lim_{t \to \infty} |\pi(t, x)| = 0$ for all $x \in X^s$ because the dynamical system $(X^s, \mathbb{S}_+, \pi)$ doesn't have a non-trivial entire trajectory bounded on $\mathbb{S}$. In fact, if we suppose that it is not true, then there exist $x_0 = (u_0, y_0)$ and $t_n \to +\infty$ such that: $|u_0| \neq 0, \lim_{n \to +\infty} \pi(t_n, x) = x_0$ and through point x_0 pass a non-trivial entire trajectory bounded on $\mathbb{S}$. This contradiction proves the necessary assertion. Thus we can apply Theorem 2.38 according which there exist two positive constants N and ν such that $|\varphi(t, u, y)| \leq Ne^{-\nu t}|u|$ for all $(u, y) \in X^s$ and $t \in \mathbb{S}_+$. The theorem is proved. $\square$

13.8 Some classes of uniformly exponentially stable equations

Let Λ be a complete metric space of linear operators that act on a Banach space E and $C(\mathbb{R}, \Lambda)$ be the space of all continuous operator-functions $A : \mathbb{R} \to \Lambda$ equipped with the open-compact topology and $(C(\mathbb{R}, \Lambda), \mathbb{R}, \sigma)$ be the dynamical system of shifts on $C(\mathbb{R}, \Lambda)$.

Linear ordinary differential equations. Let $\Lambda = [E]$ and consider the linear differential equation

$$u' = A(t)u , \tag{13.77}$$

where $A \in C(\mathbb{R}, \Lambda)$. Along with equation (13.77), we shall also consider its H-class, that is, the family of equations

$$v' = B(t)v , \tag{13.78}$$

where $B \in H(A) := \overline{\{A_\tau : \tau \in \mathbb{R}\}}$, $A_\tau(t) = A(t + \tau)$ $(t \in \mathbb{R})$, and the bar denotes closure in $C(\mathbb{R}, \Lambda)$. Let $\varphi(t, u, B)$ be the solution of equation (13.78) that satisfies the condition $\varphi(0, u, B) = u$. We put $Y := H(A)$ and denote the dynamical system of shifts on $H(A)$ by $(Y, \mathbb{R}, \sigma)$. Then the triple $\langle (X, \mathbb{R}_+, \pi), (Y, \mathbb{R}, \sigma), h \rangle$ is a linear non-autonomous dynamical system, where $X := E \times Y$, $\pi := (\varphi, \sigma)$; i.e., $\pi((v, B), \tau) := (\varphi(\tau, v, B), B_\tau)$ and $h := pr_2 : X \to Y)$.

Lemma 13.13   [51, 60]

(i) *The mapping $(t, u, A) \mapsto \varphi(t, u, A)$ of $\mathbb{R} \times E \times C(\mathbb{R}, [E])$ to E is continuous, and*

(ii) *the mapping $A \mapsto U(\cdot, A)$ of $C(\mathbb{R}, [E])$ to $C(\mathbb{R}, [E])$ is continuous, where $U(\cdot, A)$ is the Cauchy's operator [132] of equation (13.77).*

Theorem 13.27   *Let $A \in C(\mathbb{R}, \Lambda)$ be compact (i.e. $H(A)$ is a compact set of $(C(\mathbb{R}, \Lambda), \mathbb{R}, \sigma)$), then the following conditions are equivalent:*

1. *The trivial solution of equation (13.77) is uniformly exponentially stable, i.e. there exist positive numbers N and ν such that $\|U(t, A)U(\tau, A)^{-1}\| \leq Ne^{-(t-\tau)}$ for all $t \geq \tau$.*
2. *There exist positive numbers N and ν such that $\|U(t, B)U(\tau, B)^{-1}\| \leq Ne^{-(t-\tau)}$ for all $t \geq \tau$ and $B \in H(A)$.*
3. $\limsup\limits_{t \to +\infty} \{\|U(t, B)\| : B \in H(A)\} = 0.$

Proof.   Applying Theorem 13.22 to the non-autonomous system $\langle (X, \mathbb{R}_+, \pi), (Y, \mathbb{R}, \sigma), h \rangle$, generated by equation (13.77), we obtain the equivalence of conditions 2. and 3. According to Lemma 3 [60] conditions 1. and 2. are equivalent. The theorem is proved. $\square$

Theorem 13.28   *Let $A \in C(\mathbb{R}, \Lambda)$ be recurrent with respect to $t \in \mathbb{S}$ (i.e. $H(A)$ is a compact and minimal set of $(C(\mathbb{R}, \Lambda), \mathbb{R}, \sigma)$), the non-autonomous system $\langle (X, \mathbb{R}_+, \pi), (Y, \mathbb{R}, \sigma), h \rangle$ generated by equation (13.77) is asymptotically compact. Then the following conditions are equivalent:*

1. *The trivial solution of equation (13.77) is uniformly exponentially stable, i.e. there exist positive numbers N and ν such that $\|U(t, A)U(\tau, A)^{-1}\| \leq Ne^{-(t-\tau)}$ for all $t \geq \tau$.*
2. $\limsup\limits_{t \to +\infty} |\varphi(t, u, B)| = 0$ *for every $u \in E$ and $B \in H(A)$.*

Proof.   According to Lemma 13.13 the mapping $U(t, \cdot) : [E] \to [E]$ is continuous and, consequently, the mapping $B \mapsto \|U(t, B)\|$ is also continuous for every $t \in \mathbb{S}$. Now applying Theorem 13.22 to non-autonomous system $\langle (X, \mathbb{R}_+, \pi), (Y, \mathbb{R}, \sigma), h \rangle$ generated by equation (13.77), we obtain the equivalence of conditions 1. and 2. The theorem is proved. $\square$

We now formulate some sufficient conditions for the $\alpha-$ condensedness (in particular, asymptotic compactness) of the linear non-autonomous dynamical system generated by equation (13.77).

Theorem 13.29 *Let $A \in C(\mathbb{R}, [E]), A(t) = A_1(t) + A_2(t)$ for all $t \in \mathbb{R}$, and assume that $H(A_i)$ ($i = 1, 2$) are compact and the following conditions hold:*

(i) The zero solution of the equation

$$u' = A_1(t)u \tag{13.79}$$

is uniformly asymptotically stable, that is, there are positive numbers N and ν such that

$$\|U(t, A_1)U^{-1}(\tau, A_1)\| \le Ne^{-\nu(t-\tau)} \tag{13.80}$$

for all $t \ge \tau$ $(t, \tau \in \mathbb{R})$, where $U(t, A_1)$ is the Cauchy's operator of equation (13.79).

(ii) The family of operators $\{A_2(t) : t > 0\}$ is uniformly completely continuous, that is, for any bounded set $A \subset E$ the set $\{A_2(t)A : t > 0\}$ is relatively compact.

Then the linear non-autonomous dynamical system generated by equation (13.77) is an α-contraction.

Proof. First of all we note that the set $\varphi(t, A, Y)$ is bounded for every $t > 0$ and bounded set $A \subseteq E$. Let $B \in H(A)$. Then there are $\{t_n\} \subset \mathbb{S}$ such that $B(t) = B_1(t) + B_2(t)$ and $B_i(t) = \lim\limits_{t \to +\infty} A_i(t + t_n)$. Note that

$$\varphi(t, v, B) = U(t, B_1)v + \int_0^t U(t, B_1)U^{-1}(\tau, B_1)B_2(\tau)\varphi(\tau, v, B)d\tau.$$

By Lemma 13.13,

$$U(t, B_i) = \lim\limits_{t \to +\infty} U(t, A_{it_n}), \quad A_{it_n}(t) = A_i(t + t_n),$$

and the equality

$$U(t, A_{1t_n})U^{-1}(\tau, A_{1t_n}) = U(t + t_n, A_1)U^{-1}(\tau + t_n, A_1)$$

and inequality (13.80) imply that

$$\|U(t, B_1)U^{-1}(\tau, B_1)\| \le Ne^{-\nu(t-\tau)} \tag{13.81}$$

for all $t \ge \tau$ and $B_1 \in H(A_1)$. Theorem 13.29 will be proved if we can prove that the set

$$\left\{ \int_0^t U(t, B_1)U^{-1}(\tau, B_1)B_2(\tau)\varphi(\tau, v, B)d\tau : (v, B) \in A \right\}$$

is relatively compact for every $t > 0$ and every bounded positively invariant set $A \subseteq E \times Y$. We put

$$K_A^t = \overline{\{B_2(\tau)\varphi(\tau, v, B) : \tau \in [0, t], (v, B) \in A\}}$$

and we note that the set K_A^t is compact. Really, the set $\varphi([0, t], A) = \cup\{\varphi(\tau, v, B) : \tau \in [0, t], (v, B) \in A\}$ is bounded because $|\varphi(\tau, v, B)| \leq e^{Mt} r$ for all $\tau \in [0, t]$ and $(v, B) \in A$, where $r = \sup\{|v| : \exists B \in H(A), \text{ such that } (v, B) \in A\}$ and $M = \sup\{\|A(t)\| : t \in \mathbb{S}\}$. Then

$$\int_0^t U(t, B_1)U^{-1}(\tau, B_1)B_2(\tau)\varphi(\tau, v, B)d\tau \tag{13.82}$$

$$\in t \, \overline{\text{conv}}\{U(t, B_1)U^{-1}(\tau, B_1)w : 0 \leq \tau \leq t, B_1 \in H(A_1), w \in K_A^t\}.$$

Since $H(A_1), H(A)$, and K_A^t are compact sets, then formula (13.82), condition (ii) of Theorem 13.29, and Lemma 13.13 imply that $\{U(t, B_1)U^{-1}(\tau, B_1)w : 0 \leq \tau \leq t, B_1 \in H(A_1), w \in K_A^t\}$ is compact, which completes the proof of the theorem. $\square$

Theorem 13.30 *Let $H(A)$ be compact and assume that there is a finite-dimensional projection $P \in [E]$ such that*

(i) the family of projections $\{P(t) : t \in \mathbb{R}\}$, where $P(t) := U(t, A)PU^{-1}(t, A)$, is relatively compact in $[E]$, and

(ii) there are positive numbers N and ν such that

$$\|U(t, A)QU^{-1}(\tau, A)\| \leq Ne^{-\nu(t-\tau)}$$

for all $t \geq \tau$, where $Q := I - P$.

Then the linear non-autonomous dynamical system generated by equation (13.77) is an α-contraction.

Proof. Since the family of projections $P(t) = U(t, A)PU^{-1}(t, A)$ is relatively compact in $[E]$, the family $\mathbb{H} = \overline{\{P(t) : t \in \mathbb{R}\}}$ is uniformly completely continuous, where the bar denotes closure in $[E]$. Indeed, let A be a bounded subset of E, $\{x_n\} \subseteq \{QA : Q \in \mathbb{H}\}$, and $\varepsilon_n \downarrow 0$. Then there are $t_n \in \mathbb{R}$ and $v_n \in A$ such that $|x_n - P(t_n)v_n| \leq \varepsilon_n$. Since the sequence $\{P(t_n)\}$ is relatively compact, we can assume that it converges. Let $L := \lim_{n \to +\infty} P(t_n)$. Then L is completely continuous, which implies that the sequence $\{x_n'\} = \{Lv_n\}$ is relatively compact. Note that

$$|x_n - x_n'| \leq |x_n - P(t_n)v_n| + |P(t_n)v_n - Lv_n| \leq \varepsilon_n + \|P(t_n) - L\||v_n|,$$

which implies that $|x_n - x_n'| \to 0$ as $n \to +\infty$. Hence, $\{x_n\}$ is relatively compact.

Assume that $B \in H(A)$ and $\{t_n\} \subset \mathbb{R}$ are such that

$$B = \lim_{n \to +\infty} A_{t_n}, P(B) = \lim_{n \to +\infty} P(A_{t_n}),$$

where $P(A_{t_n}) = U(t_n, A)PU^{-1}(t_n, A)$. The assertions proved above imply that the family $\{P(B) : B \in H(A)\}$ is uniformly completely continuous. Note that $Q(B) = \lim\limits_{n \to +\infty} Q(A_{t_n})$, where $Q(B) := I - P(B)$ and $Q(A_{t_n}) = I - P(A_{t_n})$. Moreover, condition (ii) of Theorem 13.30 implies that

$$\|U(t, A_{t_n})QU^{-1}(\tau, A_{t_n})\| \leq Ne^{-\nu(t-\tau)} \tag{13.83}$$

for all $t \geq \tau$. Passing to the limit in (13.83) as $n \to +\infty$ and taking into account Lemma 13.13, we obtain that

$$\|U(t, B)QU^{-1}(\tau, B)\| \leq Ne^{-\nu(t-\tau)}$$

for all $t \geq \tau$ and $B \in H(A)$. We complete the proof of the theorem by observing that $U(t, B)Q(B) + U(t, B)P(B) = U(t, B)$ and applying Theorem 13.21. $\qquad\square$

Theorem 13.31 *Suppose that the following conditions are satisfied:*

(1) The operator-function $A \in C(\mathbb{R}, [E])$ is recurrent with respect to $t \in \mathbb{R}$.

(2) The linear non-autonomous dynamical system $\langle (X, \mathbb{S}_+, \pi), (Y, \mathbb{S}, \sigma), h \rangle$ generated by equation (13.77) is conditionally α-condensing.

(3) the trivial solution of equation (13.77) is uniformly stable in the positive direction, i.e. there exist a positive number M such that

$$\|U(t, A)U^{-1}(\tau, A)\| \leq M \tag{13.84}$$

for all $t \geq \tau$.

Then there are two vectorial positively invariant sub-fibers (X^0, h, Y) and (X^s, h, Y) of (X, h, Y) such that:

a. $X_y = X_y^0 + X_y^s$ and $X_y^0 \cap X_y^s = \theta_y$ for all $y \in Y$, where $\theta_y = (0, y) \in X = E \times Y$ and 0 is the zero in the Banach space E.

b. The vectorial sub-fiber (X^0, h, Y) is finite dimensional, invariant (i.e. $\pi^t X^0 = X^0$ for all $t \in \mathbb{S}_+$) and every trajectory of a dynamical system $(X, \mathbb{S}_+, \pi)$ belonging to X^0 is recurrent.

c. There exist two positive numbers N and ν such that $|\varphi(t, u, B)| \leq Ne^{-\nu t}|u|$ for all $(u, B) \in X^s$ and $t \in \mathbb{S}_+$, where $\varphi(t, u, B) := U(t, B)u$.

Proof. Assume that $B \in H(A)$ and $\{t_n\} \subset \mathbb{R}$ are such that $B = \lim\limits_{n \to +\infty} A_{t_n}$, then condition (3) of Theorem 13.31 implies that

$$\|U(t, A_{t_n})U^{-1}(\tau, A_{t_n})\| \leq N \tag{13.85}$$

for all $t \geq \tau$. Passing to the limit in (13.85) as $n \to +\infty$ and taking into account Lemma 13.13, we obtain that

$$\|U(t, B)U^{-1}(\tau, B)\| \leq N$$

for all $t \geq \tau$ and $B \in H(A)$ and, consequently,

$$\|U(t, B)\| \leq N$$

for all $t \geq 0$ and $B \in H(A)$. Now to finish the proof of Theorem 13.31 it is sufficiently to refer on Theorem 13.26. $\qquad\square$

Linear partial differential equations. Let Λ be some complete metric space of linear closed operators acting into Banach space E (for example $\Lambda = \{A_0 + C | C \in [E]\}$, where A_0 is a closed operator that acts on E). We assume that the following conditions are fulfilled for equation (13.77) and its $H-$ class (13.78):

a. for any $v \in E$ and $B \in H(A)$ equation (13.78) has exactly one mild solution $\varphi(t, v, B)$ (i.e. $\varphi(\cdot, v, B)$ is continuous, defined on $\mathbb{R}_+$ and satisfies of equation

$$\varphi(t, v, B) = U(t, B)v + \int_0^t U(t - \tau, B)\varphi(\tau, v, B)d\tau. \tag{13.86}$$

and the condition $\varphi(0, v, B) = v$;

b. the mapping $\varphi : (t, v, B) \to \varphi(t, v, B)$ is continuous in the topology of $\mathbb{R}_+ \times E \times C(\mathbb{R}; \Lambda)$;

c. for every $t \in \mathbb{R}_+$ the mapping $U(t, \cdot) : H(A) \to [E]$ is continuous, where $U(t, \cdot)$ is the Cauchy's operator of equation (13.78), i.e. $U(t, B)v := \varphi(t, v, B)$ $(t \in \mathbb{R}_+, v \in E$ and $B \in H(A)$).

Under the above assumptions the equation (13.77) generates a linear non-autonomous dynamical system $\langle (X, \mathbb{R}_+, \pi), (Y, \mathbb{R}, \sigma), h \rangle$, where $X := E \times Y, \pi = (\varphi, \sigma)$ and $h := pr_2 : X \to Y$. Applying the results from the sections 13.6 and 13.7 to this system, we will obtain the analogous assertions for different classes of partial differential equations.

We will consider examples of partial differential equations which satisfy the above conditions a.-c.

Example 13.4 *Consider the differential equation*

$$u' = (A_1 + A_2(t))u, \tag{13.87}$$

where A_1 is a sectorial operator that does not depend on $t \in \mathbb{R}$, and $A_2 \in C(\mathbb{R}, [E])$. The results of [188], [238] imply that equation (13.87) satisfies conditions a.-c.

Example 13.5 *Let H be a Hilbert space with a scalar product $\langle \cdot, \cdot \rangle = |\cdot|^2$, $\mathcal{D}(\mathbb{R}_+, H)$ be the set of all infinite differentiable, bounded functions on $\mathbb{R}_+$ with values into H.*

Denote by $(C(\mathbb{R}, [H]), \mathbb{R}, \sigma)$ a dynamical system of shifts on $C(\mathbb{R}, [H])$. Consider the equation

$$\int_{\mathbb{R}_+} [\langle u(t), \varphi'(t)\rangle + \langle A(t)u(t), \varphi(t)\rangle]dt = 0, \tag{13.88}$$

along with the family of equations

$$\int_{\mathbb{R}_+} [\langle u(t), \varphi'(t)\rangle + \langle B(t)u(t), \varphi(t)\rangle]dt = 0, \tag{13.89}$$

where $B \in H(A) := \overline{\{A_\tau | \tau \in \mathbb{R}\}}$, $A_\tau(t) := (t + \tau)$ and the bar denotes closure in $C(\mathbb{R}, [H])$.

The function $u \in C(\mathbb{R}_+, H)$ is called a solution of equation (13.88), if (13.88) takes place for all $\varphi \in \mathcal{D}(\mathbb{R}_+, H)$.

Let $(H(A), \mathbb{R}, \sigma)$ be a dynamical system of shifts on $H(A)$, $\varphi(t, v, B)$ be a solution of (13.89) with condition $\varphi(0, v, B) = v$, $\tilde{X} := H \times H(A)$, X be a set of all the points $\langle u, B \rangle \in \tilde{X}$ such that through point $u \in H$ passes a solution $\varphi(t, u, A)$ of equation (13.88) defined on $\mathbb{R}_+$. According to Lemma 2.21 in [92] the set X is closed in $\tilde{X}$. In virtue of Lemma 2.22 in [92] the triple $(X, \mathbb{R}_+, \pi)$ is a dynamical system on X (where $\pi := (\varphi, \sigma)$) and $\langle (X, \mathbb{R}_+, \pi), (Y, \mathbb{R}, \sigma), h \rangle$ is a linear non-autonomous dynamical system, where $h := pr_2 : X \to Y := H(A)$.

Applying the results from [5] it is possible to show that for every t the mapping $B \mapsto U(t, B)$ (where $U(t, B)v := \varphi(t, v, B)$) from $H(A)$ into $[H]$ is continuous and, consequently, for this system Theorem 13.22 is applicable. Thus the following assertion takes place.

Theorem 13.32   *Let $A \in C(\mathbb{R}, [H])$ be compact, then the following assertion hold:*

(1) *The trivial solution of equation (13.77) is uniformly exponentially stable, i.e. there exist positive numbers N and ν such that $\|U(t, B)\| \leq Ne^{-\nu t}$ for all $t \geq 0$ and $B \in H(A)$.*

(2) $\limsup\limits_{t \to +\infty} \{\|U(t, B)\| : B \in H(A)\} = 0.$

Proof.   This statement follows directly from Theorem 13.22.   □

Linear functional-differential equations.   Consider the equation

$$u' = A(t)u_t , \tag{13.90}$$

where $A \in C(\mathbb{R}, \mathfrak{A})$. We put $H(A) := \overline{\{A_\tau : \tau \in \mathbb{R}\}}$, $A_\tau(t) = A(t + \tau)$, where the bar denotes closure in the topology of uniform convergence on every compact of $\mathbb{R}$.

Along with equation (13.90) we also consider the family of equations

$$u' = B(t)u_t \ , \tag{13.91}$$

where $B \in H(A)$. Let $\varphi_t(v, B)$ be a solution of equation (13.91) with condition $\varphi_0(v, B) = v$ defined on $\mathbb{R}_+$. We put $Y := H(A)$ and denote by $(Y, \mathbb{R}, \sigma)$ the dynamical system of shifts on $H(A)$. Let $X := C \times Y$ and $\pi := (\varphi, \sigma)$ the dynamical system on X defined by the equality $\pi(\tau, (v, B)) := (\varphi_\tau(v, B), B_\tau)$. The non-autonomous dynamical system $\langle (X, \mathbb{R}_+, \pi), (Y, \mathbb{R}, \sigma), h \rangle$ $(h := pr_2 : X \to Y)$ is linear. The following assertion takes place.

Theorem 13.33   *Let $H(A)$ be compact. Then the following assertions are equivalent:*

(1) For any $B \in H(A)$ the zero solution of equation (13.91) is asymptotically stable, i.e. $\lim\limits_{t \to +\infty} |\varphi_t(v, B)| = 0$ for all $v \in C$ and $B \in H(A)$

(2) The zero solution of equation (13.90) is uniformly exponentially stable, i.e. there are positive numbers N and ν such that $|\varphi_t(v, B)| \leq Ne^{-\nu t}|v|$ for all $t \geq 0, v \in C$ and $B \in H(A)$.

Proof.   Let $\langle (X, \mathbb{R}_+, \pi), (Y, \mathbb{R}, \sigma), h \rangle$ be the linear non-autonomous dynamical system, generated by equation (13.90). According to Lemma 13.9 this system is completely continuous and to finish the proof it is sufficient to refer to Theorem 13.21.   □

Consider the neutral functional differential equation

$$\frac{d}{dt}Dx_t = A(t)x_t \ , \tag{13.92}$$

where $A \in C(\mathbb{R}, \mathfrak{A})$ and $D \in \mathfrak{A}$ is an operator non-atomic at zero [19, p.67]. As well as in the case of equation (13.90), the equation (13.92) generates a linear non-autonomous dynamical system $\langle (X, \mathbb{R}_+, \pi), (Y, \mathbb{R}, \sigma), h \rangle$, where $X = C \times Y$, $Y = H(A)$ and $\pi = (\varphi, \sigma)$. The following statement takes place.

Lemma 13.14   *Let $H(A)$ be compact and the operator D is stable; i.e., the zero solution of the homogeneous difference equation $Dy_t = 0$ is uniformly asymptotically stable. Then the linear non-autonomous dynamical system $\langle (X, \mathbb{R}_+, \pi), (Y, \mathbb{R}, \sigma), h \rangle$, generated by equation (13.92), is conditionally α-condensing.*

Proof.   According to [20, p.119, formula (5.18)] the mapping $\varphi_t(\cdot, B) : C \to C$ can be written as

$$\varphi_t(\cdot, B) = S_t(\cdot) + U_t(\cdot, B)$$

for all $B \in H(A)$, where $U_t(\cdot, B)$ is conditionally completely continuous for $t \geq r$. Also there exist positive constants N, ν such that $\|S_t\| \leq Ne^{-\nu t}(t \geq 0)$. Then the proof is completed by referring to Theorem 13.23. $\qquad\square$

Theorem 13.34 *Let $A \in C(\mathbb{R}, \mathfrak{A})$ be recurrent (i.e. $H(A)$ is compact minimal set in the dynamical system of shifts $(C(\mathbb{R}, \mathfrak{A}), \mathbb{R}, \sigma)$) and D is stable, then the following assertions are equivalent:*

(1) The zero solution of equation (13.90) and all equations

$$\frac{d}{dt}Dx_t = B(t)x_t \ , \tag{13.93}$$

where $B \in H(A)$, is asymptotically stable, i.e. $\lim\limits_{t \to +\infty} |\varphi_t(v, B)| = 0$ for all $v \in C$ and $B \in H(A)$ ($\varphi_t(v, B)$ is the solution of equation (13.93) with condition $\varphi_0(v, B) = v$).

(2) The zero solution of equation (13.92) is uniformly exponentially stable, i.e. there are positive numbers N and ν such that $|\varphi_t(v, B)| \leq Ne^{-\nu t}|v|$ for all $t \geq 0, v \in C$ and $B \in H(A)$.

(3) All solutions of all equations (13.93) are bounded on $\mathbb{R}_+$ and they don't have non-trivial solutions bounded on $\mathbb{R}$.

Proof. Let $\langle (X, \mathbb{R}_+, \pi), (Y, \mathbb{R}, \sigma), h \rangle$ be the linear non-autonomous dynamical system, generated by equation (13.92). According to Lemma 13.14 this system is conditionally α condensing. To finish the proof of Theorem 13.34 it is sufficient to refer to Theorems 13.23 and 13.26. $\qquad\square$

Theorem 13.35 *Let $A \in C(\mathbb{R}, \mathfrak{A})$ be recurrent (i.e. $H(A)$ is compact minimal in the dynamical system of shifts $(C(\mathbb{R}, \mathfrak{A}), \mathbb{R}, \sigma)$), D is stable, and all solutions of all equations (13.93) are bounded on $\mathbb{R}_+$. Let $\langle (X, \mathbb{S}_+, \pi), (Y, \mathbb{S}, \sigma), h \rangle$ be the linear non-autonomous dynamical system generated by equation (13.92), then there are two positively invariant vector sub-fiberings (X^0, h, Y) and (X^s, h, Y) of (X, h, Y) such that:*

a. $X_y = X_y^0 + X_y^s$ *and* $X_y^0 \cap X_y^s = \theta_y$ *for all* $y \in Y$, *where* $\theta_y = (0, y) \in X = E \times Y$ *and 0 is the zero in the Banach space E.*

b. *The vector subfibering (X^0, h, Y) is finite dimensional, invariant (i.e. $\pi^t X^0 = X^0$ for all $t \in \mathbb{S}_+$) and every trajectory of a dynamical system $(X, \mathbb{S}_+, \pi)$ belonging to X^0 is recurrent.*

c. *There exist two positive numbers N and ν such that $|\varphi_t(v, B)| \leq Ne^{-\nu t}|v|$ for all $(v, B) \in X^s$, where $\varphi_t(v, B) := U(t, B)v$ and $U(t, B)$ is the Cauchy operator of equation (13.93).*

Proof. Let $\langle (X, \mathbb{S}_+, \pi), (Y, \mathbb{S}, \sigma), h \rangle$ be the linear non-autonomous dynamical system generated by equation (13.92). In virtue of Lemma 13.14 this non-autonomous dynamical system is conditionally α-condensing. According to Theorem 13.1 there exists a positive number M such that $|\varphi_t(v, B)| \leq M|v|$ for all $t \geq 0, v \in C$ and $B \in H(A)$. To finish the proof of Theorem 13.35 it is sufficient to refer to Theorem 13.26. $\qquad\square$

13.9   Linear periodic systems

This section is devoted to the study of linear periodic dynamical systems, possessing the property of uniform exponential stability. It is proved that if the Cauchy's operator of these systems possesses a certain compactness property, then the asymptotic stability implies the uniform exponential stability. We also give the application to different classes of linear evolution equations, such as ordinary linear differential equations in the space of Banach, retarded and neutral functional differential equations, some classes of evolution partial differential equations.

Let $A(t)$ be a τ-periodic continuous $n \times n$ matrix-function. It is well-known that the following three conditions are equivalent:

1. The trivial solution of equation

$$u' = A(t)u \tag{13.94}$$

   is uniformly exponentially stable.
2. The trivial solution of equation (13.94) is uniformly asymptotically stable.
3. The trivial solution of equation (13.94) is asymptotically stable.

For equations in infinite-dimensional spaces the statements 1.-3. are not equivalent, as shown by the examples in [121, 249].

It is clear that in general for the infinite-dimensional case condition 1. implies 2. and 2. implies 3. In this section we show that if the Cauchy operator of equation (13.94) satisfies some compactness condition, then the condition 3. implies 1. (see Theorem 13.37 below).

Applications to different classes of linear evolution equations (ordinary linear differential equations in a Banach space, retarded and neutral functional-differential equations, some classes of evolutionary partial differential equations) are given.

The exponential dichotomy of asymptotically compact cocycles was studied by R. Sacker and G. Sell [277]. The general case was studied by C. Chicone and Yu. Latushkin [117] (see also their references), Yu. Latushkin and R. Schnaubelt [236], and many other authors.

13.9.1 *Exponential stable linear periodic dynamical systems*

Let X and Y be complete metric spaces, (X, h, Y) be a locally trivial Banach fiber bundle over Y [29], $[E]$ be a Banach space of the all linear continuous operators acting onto Banach space E with the operator norm and $U : \mathbb{S}_+ \times Y \mapsto [E]$ be a mapping with properties: $U(0, y) = I$, $U(t + \tau, y) = U(t, \sigma(\tau, y))U(\tau, y)$ for all $y \in Y$ and $t, \tau \in \mathbb{S}_+$ and the mapping $\varphi(\cdot, u, \cdot) : \mathbb{S}_+ \times Y \to E$ ($\varphi(t, u, y) := U(t, y)u$) is continuous for every $u \in E$.

Definition 13.7 A triplet $\langle [E], U, (Y, \mathbb{S}, \sigma) \rangle$ is called a c_0- cocycle on $(Y, \mathbb{S}, \sigma)$ with fiber $[E]$.

Lemma 13.15 *Let $\langle [E], U, (Y, \mathbb{S}, \sigma) \rangle$ be a c_0- cocycle on $(Y, \mathbb{S}, \sigma)$ with fiber $[E]$ and Y be a compact, then the following assertions hold:*

(1) For every $\ell > 0$ there exists a positive number $M(\ell)$ such that $\|U(t, y)\| \le M(\ell)$ for all $t \in [0, \ell]$ and $y \in Y$;
(2) The mapping $\varphi : \mathbb{S}_+ \times E \times Y \mapsto E$ ($\varphi(t, u, y) = U(t, y)u$) is continuous;
(3) There exist positive numbers N and ν such that $\|U(t, y)\| \le N e^{\nu t}$ for all $t \in \mathbb{S}_+$ and $y \in Y$.

Proof. Let $\ell > 0$ and $u \in E$, then there exists a positive number $M(\ell, u)$ such that $|U(t, y)u| \le M(\ell, u)$ for all $(t, y) \in [0, \ell] \times Y$ because the mapping $(t, y) \to U(t, y)u$ is continuous. According to principle of uniformly boundedness there exists a positive number $M(\ell)$ such that $\|U(t, y)\| \le M(\ell)$ for all $(t, y) \in [0, \ell] \times Y$.

Let now $(t_0, u_0, y_0) \in \mathbb{S}_+ \times E \times Y$ and $t_n \to t_0, u_n \to u_0$ and $y_n \to y_0$, then we have

$$|\varphi(t_n, u_n, y_n) - \varphi(t_0, u_0, y_0)|$$

$$\le |\varphi(t_n, u_n, y_n) - \varphi(t_n, u_0, y_n)| + |\varphi(t_n, u_0, y_n) - \varphi(t_0, u_0, y_0)|$$

$$\le \|U(t_n, y_n)(u_n - u_0)\| + |(U(t_n, y_n) - U(t_0, y_0))u_0|. \tag{13.95}$$

In view of first statement of Lemma 13.15 there exists the positive number M such that

$$\|U(t_n, y_n)\| \le M \tag{13.96}$$

for all $n \in \mathbb{N}$. From inequalities (13.95) and (13.96) follows the continuity of mapping $\varphi : \mathbb{S}_+ \times E \times Y \to E$ ($\varphi(t, u, y) = U(t, y)u$).

Denote by $a := \sup\{\|U(t, y)\| : (t, y) \in [0, 1] \times Y\}$ and let $t \in \mathbb{S}_+, t = n + \tau (n \in \mathbb{N}, \tau \in [0, 1))$, then we obtain

$$\|U(t, y)\| \le \|U(n, y\tau)\| \|U(\tau, y)\| \le a^{n+1} \le N e^{\nu t}$$

for all $t \in \mathbb{S}_+$ and $y \in Y$, where $N := a$ and $\nu := \ln a$. $\qquad\square$

Lemma 13.16 [121, Chapter 9] *Let $m : \mathbb{R}_+ \mapsto \mathbb{R}_+$ be a positive and continuous function. If there exists a positive constant M such that $m(t + s) \leq M m(t)$ for all $s \in [0, 1]$ and $t \in \mathbb{S}_+$, then $\int_0^{+\infty} m(t)dt < +\infty$ implies $m(t) \to 0$ as $t \to +\infty$.*

Theorem 13.36 *Let $\langle [E], U, (Y, \mathbb{S}, \sigma) \rangle$ be the C_0- cocycle on $(Y, \mathbb{S}, \sigma)$ with fiber $[E]$ and $(Y, \mathbb{S}, \sigma)$ be a periodical dynamical system (i.e. there are $y_0 \in Y$ and $\tau \in \mathbb{S}$ ($\tau > 0$) such that $Y = \{y_0 t : 0 \leq t < \tau\}$). Then the following conditions are equivalent:*

(i)

$$\lim_{t \to +\infty} \|U(t, y_0)\| = 0. \tag{13.97}$$

(ii) There exist positive constants N and ν such that for all $t \in \mathbb{S}_+$ and $y \in Y$,

$$\|U(t, y)\| \leq N e^{-\nu t}. \tag{13.98}$$

(iii) There exists $p \geq 1$ such that for all $u \in E$,

$$\int_0^{+\infty} |U(t, y_0)u|^p dt < +\infty. \tag{13.99}$$

Proof. We remark that from equality (13.97) follows the condition

$$\lim_{n \to +\infty} \sup_{0 \leq s \leq \tau} \|U(s + n\tau, y_0)\| = 0. \tag{13.100}$$

In fact, by virtue of Lemma 13.15 there exists a positive constant M such that

$$\|U(s, y)\| \leq M \tag{13.101}$$

for all $s \in [0, \tau]$ and $y \in Y$. Therefore,

$$\|U(s + n\tau, y_0)\| = \|U(s, y_0)U(n\tau, y_0)\| \leq M\|U(n\tau, y_0)\| \tag{13.102}$$

for all $0 \leq s \leq \tau$. Consequently, from (13.97) and (13.102) results the condition (13.100).

We will show that under the condition (13.100) the equality

$$\lim_{t \to +\infty} \sup_{y \in Y} \|U(t, y)\| = 0 \tag{13.103}$$

holds. In fact, let $y \in Y$ then there exists a number $s \in [0, \tau)$ such that $y = y_0 s$ and, consequently, for $t \in \mathbb{S}_+$ ($t = n\tau + \bar{t}, \bar{t} \in [0, \tau)$) we obtain

$$\|U(t,y)\| = \|U(t,y_0s)\| = \|U(n\tau + \bar{t}, y_0s)\|$$

$$= \|U((n-1)\tau + \bar{t} + s, y_0\tau)U(\tau - s, y_0s)\|$$

$$\leq M \max\{ \sup_{0\leq s\leq\tau} \|U((n-1)\tau + s, y_0)\|, \sup_{0\leq s\leq\tau} \|U(n\tau + s, y_0)\|\}. \quad (13.104)$$

From (13.100) and (13.104) results the equality (13.103). For finishing the proof that (i) implies (ii) is sufficient to apply Theorem 13.22.

The fact that (ii) implies (iii) is obvious. Now we prove that (iii) implies (i). Indeed, let $u \in E$ and we consider the function $m(t) = |U(t, y_0)u|^p$ $(t \geq 0)$. We note that

$$m(t + s) = |U(t + s, y_0)u|^p = |U(s, y_0t)U(t, y_0)u|^p$$

$$\leq \|U(s, y_0t)\|^p |U(t, y_0)u|^p \leq M^p m(t)$$

for all $t \in \mathbb{S}_+$ and $s \in [0, 1]$, where $M := \sup_{0\leq s\leq 1, y\in Y} \|U(s, y)\|$. By Lemma 13.16 $m(t) \to 0$ as $t \to +\infty$ and, consequently,

$$\lim_{t\to+\infty} |U(t, y_0)u|^p = 0 \quad (13.105)$$

for all $u \in E$. Let now $y \in Y$, then there exists $s \in [0, \tau)$ such that $y = y_0s$ and for $t \geq \tau - s$ we have

$$U(t, y)u = U(t, y_0s)u = U(t - \tau + s, y_0)U(\tau - s, y_0s)u. \quad (13.106)$$

From equalities (13.105) and (13.106),

$$\lim_{t\to+\infty} |U(t, y)u|^p = 0 \quad (13.107)$$

for all $u \in E$ and $y \in Y$. According to Theorem 13.1 there exists a positive number M such that $\|U(t, y)\| \leq M$ for all $t \in \mathbb{S}_+$ and $y \in Y$. Let $t > 0$ and $u \in E$, then we obtain

$$t|U(t, y_0)u|^p = \int_0^t |U(t, y_0)u|^p ds \leq \quad (13.108)$$

$$\int_0^t |U(t - s, y_0s)|^p |U(s, y_0)u|^p ds$$

$$\leq$$

$$M^p \int_0^t |U(s, y_0)u|^p ds \leq M^p \int_0^{+\infty} |U(s, y_0)u|^p ds = C_u$$

for all $t \geq 0$. By virtue of principle of uniformly boundedness there exists a positive number C such that

$$t\|U(t, y_0)\|^p \leq C$$

for all $t > 0$ and, consequently

$$\|U(t, y_0)\| \le C^{\frac{1}{p}} t^{-\frac{1}{p}} \to 0$$

as $t \to +\infty$. This completes the present proof. $\qquad\square$

Remark 13.8   *a. Theorem 13.36 (the equivalence of assertions (ii) and (iii)) is a variant of the Datko-Pazy theorem (see [121],[133] and [134]) for the cocycle over periodic dynamical systems.*

    b. Periodic, almost periodic and asymptotic almost periodic mild solutions of inhomogeneous periodic Cauchy's problems considered recently by C. J. K. Batty, W.Hutter and F. Räbiger [22] and W. Hutter [191].

Definition 13.8   The operator $U(\tau, y_0)$ is called operator of monodromy for τ-periodic cocycle $U(t, y)$. The number $0 \ne \lambda \in \mathbb{C}$ is called multiplicator of operator of monodromy $U(\tau, y_0)$ if there exists $u_0 \in E$ ($u_0 \ne 0$) such that $U(\tau, y_0)u_0 = \lambda u_0$ (or, what is the same, $U(t + \tau, y_0)u_0 = \lambda U(t, y_0)u_0$ for all $t \in \mathbb{S}_+$).

Remark 13.9   *a. Condition (13.97) and the equality*

$$\lim_{n \to +\infty} \|U(n\tau, y_0)\| = 0. \tag{13.109}$$

are equivalent. We show that (13.109) implies (13.97) as follows. Let now $t = n\tau + s, 0 \le s < \tau$, then $U(t, y_0) = U(s + n\tau, y_0) = U(s, y_0)U(n\tau, y_0)$ and, consequently,

$$\|U(t, y_0)\| \le \max_{0 \le s \le \tau} \|U(s, y_0)\| \|U(n\tau, y_0)\|. \tag{13.110}$$

From conditions (13.109) and (13.110) results (13.97).

    b. Condition (13.98) and the inequality

$$\|U(t, y_0)\| \le N_1 e^{-\nu_1 t} \quad (\forall t \in \mathbb{S}_+) \tag{13.111}$$

are equivalent, where N_1 and ν_1 are some positive constants. Indeed, from (13.111), taking into account (13.106), we obtain (13.98).

    c. Condition (13.109) is satisfied if and only if $\sigma(U(\tau, y_0)) \subset \mathbb{D} = \{z \in \mathbb{C} : |z| < 1\}$, where $\sigma(U(\tau, y_0))$ is a spectrum of operator of monodromy $U(\tau, y_0)$. In fact, from (13.98) results that $r_{U(\tau, y_0)} := \limsup_{n \to +\infty} (\|U(n\tau, y_0))\|)^{1/n} \le e^{-\nu} < 1$, because $U^n(\tau, y_0) = U(n\tau, y_0)$. If $\gamma = r_{U(\tau, y_0)} < 1$, then for all $\varepsilon > 0$ there exists a $n(\varepsilon) \in \mathbb{N}$ such that $(\|U(n\tau, y_0)\|)^{1/n} \le \gamma + \varepsilon$ for all $n \ge n(\varepsilon)$ and, consequently, $\|U(n\tau, y_0)\| \le (\gamma + \varepsilon)^n$ for all $n \ge n(\varepsilon)$. Thus $\|U(n\tau, y_0)\| \to 0$ as $n \to +\infty$.

Theorem 13.37   *Let $\langle [E], U, (Y, \mathbb{S}, \sigma) \rangle$ be a c_0- cocycle on $(Y, \mathbb{S}, \sigma)$ with fiber $[E]$, $(Y, \mathbb{S}, \sigma)$ be a periodic dynamical system and $U(\tau, y_0)$ be asymptotically compact (i.e. if $k_n \to +\infty$ $(k_n \in \mathbb{N})$, the sequences $\{u_n\} \subseteq E$ and $\{U(k_n\tau, y_0)u_n\}$ are bounded;*

then the sequence $\{U(k_n\tau, y_0)u_n\}$ is relatively compact). Then the following conditions are equivalent

1. *Equality (13.97) holds.*
2. *For all $u \in E$,*

$$\lim_{t \to +\infty} |U(t, y_0)u| = 0. \tag{13.112}$$

Proof. It is evidently that 1. implies 2. Now, under the conditions of Theorem 13.37 the mapping $P := U(\tau, y_0) : E \mapsto E$ is asymptotically compact because $P^n = U(n\tau, y_0)$. From condition (13.112) according to uniform boundedness principle it follows that there is a positive constant M such that $\|P^n\| \leq M$ for all $n \in \mathbb{Z}_+$ and, consequently, the set $B = \cup\{P^n x : |x| \leq 1, n \in \mathbb{Z}_+\}$ is bounded and $P(B) \subset B$. Since the mapping P is asymptotically compact in virtue of Lemma 1.3 the set

$$\omega(B) = \bigcap_{n \geq 0} \overline{\bigcup_{m \geq n} P^m(B)}$$

is nonempty, compact, and invariant and $\omega(B)$ attracts B.

Now we will prove that $\lim_{n \to +\infty} \|P^n\| = 0$. If we suppose the contrary, then there are $\varepsilon_0 > 0, \{x_n\}(|x_n| \leq 1)$ and $n_k \to +\infty(\{n_k\} \subset \mathbb{Z}_+)$ such that

$$|P^{n_k}x_k| \geq \varepsilon_0. \tag{13.113}$$

Since P is asymptotically compact without loss of generality we can suppose that the sequence $\{P^{n_k}x_k\}$ is convergent. Let $\bar{x} := \lim_{k \to +\infty} P^{n_k}x_k$, then $\bar{x} \in \omega(B)$ and from (13.113) we have $|\bar{x}| \geq \varepsilon_0 > 0$. According to the invariance of the set $\omega(B)$ there exists a beside sequence $\{w_n\}_{n \in \mathbb{Z}} \subset \omega(B)$ such that: $w_0 = \bar{x}$ and $P(w_n) = w_{n+1}$ for all $n \in \mathbb{Z}$. We note that

$$\inf_{n \in \mathbb{Z}_-} |w_n| = 0. \tag{13.114}$$

Suppose that it is not true, then there is a positive number ℓ such that

$$|w_n| \geq \ell \tag{13.115}$$

for all $n \in \mathbb{Z}_-$. Let $p := \lim_{k \to +\infty} w_{n_k}$ and $\{z_n\} \subseteq \alpha_{w_0}$, where

$$\alpha_{w_0} = \bigcap_{n \leq 0} \overline{\bigcup_{m \leq n} w_m},$$

be a beside sequence such that $z_0 = p$ and $P(z_n) = z_{n+1}$ for all $n \in \mathbb{Z}$. From the inequality (13.115) results that $|z_n| \geq \ell$ for all $n \in \mathbb{Z}$. On the other hand in view of (13.112) $\lim_{n \to +\infty} |w_n| = \lim_{n \to +\infty} |P^n w_0| = 0$. The obtained contradiction proves the equality (13.114).

Let now $n_r \to -\infty$ and $|w_{n_r}| \to 0$, then $w_0 = P^{-n_r} w_{n_r}$ for all $r \in \mathbb{N}$ and, consequently, $|w_0| = 0$ because $|w_0| \le \|P^{-n_r}\| |w_{n_r}| \le M|w_{n_r}|$. On the other hand $|w_0| = |\bar{x}| \ge \varepsilon_0 > 0$. The obtained contradiction finishes the proof of our assertion. The Theorem is proved. $\qquad\square$

Remark 13.10   C.Buşe wrote several papers [36],[37]and [38] on evolution periodic processes that are in the spirit of the current section. In particularly, in [38] it is proved that a trivial solution of equation $u'(t) = A(t)u(t)$ with $p-$periodic coefficients on a separable Hilbert space H is uniformly exponentially stable if the mild solution $u_{\mu x}$ of a well-posed inhomogeneous Cauchy's problem $u'(t) = A(t)u(t) + e^{i\mu t}x(t \ge 0), \mu \in \mathbb{R}, u(0) = 0$ satisfies the following condition $\sup\limits_{\mu \in \mathbb{R}} \sup\limits_{t>0} |u_{\mu x}(t)| < +\infty, \forall x \in H$.

13.9.2   *Some classes of linear uniformly exponentially stable periodic differential equations*

Let Λ be the complete metric space of linear operators that act on Banach space E and $C(\mathbb{R}, \Lambda)$ be the space of all continuous operator-functions $A : \mathbb{R} \to \Lambda$ equipped with open-compact topology and $(C(\mathbb{R}, \Lambda), \mathbb{R}, \sigma)$ be the dynamical system of shifts on $C(\mathbb{R}, \Lambda)$.

Linear ordinary differential equations.   Let $\Lambda = [E]$ and consider the linear differential equation

$$u' = A(t)u , \qquad (13.116)$$

where $A \in C(\mathbb{R}, \Lambda)$. Along with equation (13.116), we shall also consider its $H-$class, that is, the family of equations

$$v' = B(t)v , \qquad (13.117)$$

where $B \in H(A) := \overline{\{A_s : s \in \mathbb{R}\}}, A_s(t) := A(t + s)$ $(t \in \mathbb{R})$ and the bar denotes closure in $C(\mathbb{R}, \Lambda)$. Let $\varphi(t, u, B)$ be the solution of equation (13.117) that satisfies the condition $\varphi(0, v, B) = v$. We put $Y := H(A)$ and denote the dynamical system of shifts on $H(A)$ by $(Y, \mathbb{R}, \sigma)$, then the triple $\langle [E], U, (Y, \mathbb{R}, \sigma) \rangle$ is the linear cocycle on $(Y, \mathbb{R}, \sigma)$, where $U(t, B) := \varphi(t, \cdot, B)$ for all $t \in \mathbb{R}$ and $B \in Y$.

According to Lemma 13.13

(i) The mapping $(t, u, A) \mapsto \varphi(t, u, A)$ of $\mathbb{R} \times E \times C(\mathbb{R}, [E])$ to E is continuous, and

(ii) the mapping $\mathcal{U} : A \to U(\cdot, A)$ of $C(\mathbb{R}, [E])$ to $C(\mathbb{R}, [E])$ is continuous, where $U(\cdot, A)$ is the Cauchy's operator [93] of equation (13.116).

Theorem 13.38   *Let $A \in C(\mathbb{R}, \Lambda)$ be τ-periodic (i.e. $A(t + \tau) = A(t)$ for all $t \in \mathbb{R}$), then the following conditions are equivalent:*

1. *The trivial solution of (13.116) is uniformly exponentially stable, i.e. there exist positive numbers N and ν such that $\|U(t,A)U(\tau,A)^{-1}\| \le Ne^{-\nu(t-\tau)}$ for all $t \ge \tau$.*

2. *There exist positive numbers N and ν such that $\|U(t,B)U(\tau,B)^{-1}\| \le Ne^{-\nu(t-\tau)}$ for all $t \ge \tau$ and $B \in H(A) = \{A_s : s \in [0,\tau)\}$.*

3. $\lim\limits_{t\to+\infty} \|U(t,A)\| = 0.$

4. *There exists $p \ge 1$ such that $\int_0^{+\infty} |U(t,A)u|^p dt < +\infty$ for all $u \in E$.*

Proof. Applying Theorem 13.36 to the cocycle $\langle [E], U, (Y, \mathbb{R}, \sigma)\rangle$, generated by equation (13.116) we obtain the equivalence of conditions 2., 3. and 4. According to Lemma 3 [60] the conditions 1. and 2. are equivalent. The theorem is proved. $\qquad\square$

Theorem 13.39 *Let $A \in C(\mathbb{R}, \Lambda)$ be τ-periodic and $U(\tau, A)$ be asymptotically compact, then the following conditions are equivalent:*

(1) The trivial solution of equation (13.116) is uniformly exponentially stable.

(2) $\lim\limits_{t\to+\infty} |U(t,A)u| = 0$ for every $u \in E$.

Proof. Applying Theorem 13.37 to non-autonomous system $\langle (X, \mathbb{R}_+, \pi), (Y, \mathbb{R}, \sigma), h\rangle$ generated by equation (13.116), we obtain the equivalence of conditions 1. and 2. The theorem is proved. $\qquad\square$

Linear partial differential equations. Let Λ be some complete metric space of linear closed operators acting into a Banach space E (for example $\Lambda = \{A_0 + B | B \in [E]\}$, where A_0 is a closed operator that acts on E). We assume that the following conditions are fulfilled for equation (13.116) and its H-class (13.117):

(a) for any $v \in E$ and $B \in H(A)$ equation (13.117) has exactly one mild solution defined on $\mathbb{R}_+$ and satisfies the condition $\varphi(0, v, B) = v$;

(b) the mapping $\varphi : (t, v, B) \to \varphi(t, v, B)$ is continuous in the topology of $\mathbb{R}_+ \times E \times C(\mathbb{R}; \Lambda)$;

Under the assumptions above, (13.116) generates a linear cocycle $\langle [E], U, (Y, \mathbb{R}, \sigma)\rangle$, where $U(t, B) := \varphi(t, \cdot, B)$.

Applying the results from subsections 13.9.1 and 13.9.2 to this cocycle, we will obtain the analogous assertions for different classes of partial differential equations.

We will consider examples of partial differential equations which satisfy the above conditions a. and b.

Consider the differential equation

$$u' = (A_0 + A(t))u \, , \qquad (13.118)$$

where A_0 is a sectorial operator that does not depend on $t \in \mathbb{R}$, and $A \in C(\mathbb{R}, [E])$. The results of [122] imply that equation (13.118) satisfies conditions a. and b.

Under the assumptions above, (13.118) generates a linear cocycle $\langle [E], U, (Y, \mathbb{R}, \sigma) \rangle$, where $Y := H(A)$ and $U(t, B) := \varphi(t, \cdot, B)$. Applying the results from subsections 13.9.1 and 13.9.2 to this system, we will obtain the following results.

Theorem 13.40 *Let A_0 - be the sectorial operator and $A \in C(\mathbb{R}, \Lambda)$ be τ -periodic, then the following conditions are equivalent:*

(1) *The trivial solution of equation (13.118) is uniformly exponentially stable, i.e. there exist positive numbers N and ν such that $\|U(t, A_0 + A)U(\tau, A_0 + A)^{-1}\| \leq Ne^{-\nu(t-\tau)}$ for all $t \geq \tau$.*

(2) *There exist positive numbers N and ν such that $\|U(t, A_0 + B)U(\tau, A_0 + B)^{-1}\| \leq Ne^{-\nu(t-\tau)}$ for all $t \geq \tau$ and $B \in H(A)$.*

(3) $\lim_{t \to +\infty} \|U(t, A_0 + A)\| = 0.$

(4) *There exists $p \geq 1$ such that $\int_0^{+\infty} |U(t, A_0 + A)u|^p dt < +\infty$ for all $u \in E$.*

(5) $\sigma(U(\tau, A_0 + A)) \subset \mathbb{D}.$

Theorem 13.41 *Let A_0 - be the sectorial operator with compact resolvant and $A \in C(\mathbb{R}, \Lambda)$ be τ - periodic, then the following conditions are equivalent:*

1. *The trivial solution of equation (13.118) is uniformly exponentially stable.*

2. $\lim_{t \to +\infty} |U(t, A_0 + A)u| = 0$ *for every $u \in E$.*

3. $|\lambda| < 1$ *for every multiplicator λ of operator of monodromy $U(\tau, A_0 + A)$.*

Proof. Since the sectorial operator A_0 admits a compact resolvant, then in view of Lemma 7.2.2 [122] the operator $U(\tau, A_0 + A)$ is compact and, consequently (see, for example [328, p.391-396]), every $0 \neq \lambda \in \sigma(U(\tau, A_0 + A))$ is a multiplicator for operator of monodromy $U(\tau, A_0 + A)$. Applying Theorem 13.40 (see also Remark 13.8) to linear cocycle $\langle [E], U, (Y, \mathbb{R}, \sigma) \rangle$ generated by equation (13.118), we obtain the equivalence of conditions 1., 2. and 3. The theorem is proved. $\qquad \square$

Linear functional-differential equations. Let $\mathfrak{A} = \mathfrak{A}(C, \mathbb{R}^n)$ be the Banach space of all linear continuous operators acting from C into $\mathbb{R}^n$, equipped by operator norm. Consider the equation

$$u' = A(t)u_t , \qquad (13.119)$$

where $A \in C(\mathbb{R}, \mathfrak{A})$. We put $H(A) := \overline{\{A_\tau : \tau \in \mathbb{R}\}}, A_\tau(t) := A(t + \tau)$ and the bar denotes the closure in the topology of uniform convergence on compacts of $\mathbb{R}$.

Along with equation (13.119) we also consider the family of equations

$$u' = B(t)u_t , \qquad (13.120)$$

where $B \in H(A)$. Let $\varphi_t(v, B)$ be a solution of equation (13.120) with condition $\varphi_0(v, B) = v$ defined on $\mathbb{R}_+$. We put $Y := H(A)$ and denote by $(Y, \mathbb{R}, \sigma)$ the dynamical system of shifts on $H(A)$. Let $\mathcal{C} := C([-r, 0], \mathbb{R}^n)$, $X := \mathcal{C} \times Y$ and $\pi := (\varphi, \sigma)$ the dynamical system on X, defined by the equality $\pi(\tau, (v, B)) := (\varphi_\tau(v, B), B_\tau)$. The non-autonomous dynamical system $\langle (X, \mathbb{R}_+, \pi), (Y, \mathbb{R}, \sigma), h \rangle$ $(h := pr_2 : X \to Y)$ is linear. The following assertion takes place.

Lemma 13.17 [93] *Let $H(A)$ be compact in $C(\mathbb{R}, \mathfrak{A})$, then the non-autonomous dynamical system $\langle (X, \mathbb{R}_+, \pi), (Y, \mathbb{R}, \sigma), h \rangle$ generated by equation (13.119) is completely continuous, i.e. for every bounded set $A \subset X$ there exists a positive number ℓ such that $\pi^\ell A$ is relatively compact.*

Theorem 13.42 *Let A be τ-periodic. Then the following assertions are equivalent:*

(1) The trivial solution of equation (13.119) is uniformly exponentially stable.
(2) $\lim\limits_{t \to +\infty} |U(t, A)u| = 0$ for every $u \in E$.
(3) $|\lambda| < 1$ for every multiplicator λ of operator of monodromy $U(\tau, A)$.

Proof. Let $\langle (X, \mathbb{R}_+, \pi), (Y, \mathbb{R}, \sigma), h \rangle$ be the linear non-autonomous dynamical system, generated by equation (13.119). According to Lemma 13.17 this system is completely continuous and, consequently, there exists a number $k \in \mathbb{N}$ such that $U^k(\tau, y_0) = U(k\tau, y_0)$ is relatively compact. By virtue of theory of Riesz-Schauder (see for example [328, p.391-395]) every $0 \neq \lambda \in \sigma(U(\tau, A))$ is a multiplicator of operator of monodromy $U(\tau, A)$. To finish the proof it is sufficient to refer to Theorems 13.36, 13.37 and Remark 13.8. $\qquad\square$

Consider the neutral functional differential equation

$$\frac{d}{dt} Du_t = A(t)u_t \, , \tag{13.121}$$

where $A \in C(\mathbb{R}, \mathfrak{A})$ and $D \in \mathfrak{A}$ is non-atomic at zero operator [175, p.67]. As well as in the case of equation (13.119), the equation (13.121) generates a linear non-autonomous dynamical system $\langle (X, \mathbb{R}_+, \pi), (Y, \mathbb{R}, \sigma), h \rangle$, where $X = \mathcal{C} \times Y, Y := H(A)$ and $\pi := (\varphi, \sigma)$. The following statement holds.

Theorem 13.43 *Let $A \in C(\mathbb{R}, \mathfrak{A})$ be τ-periodic and D is stable, then the following assertions are equivalent:*

(1) The trivial solution of equation (13.121) is uniformly exponentially stable;
(2) $\lim\limits_{t \to +\infty} |U(t, A)u| = 0$ for every $u \in E$;
(3) $|\lambda| < 1$ for every multiplier λ of operator of monodromy $U(\tau, A)$.

Proof.   Let $\langle(X,\mathbb{R}_+,\pi),(Y,\mathbb{R},\sigma),h\rangle$ be the linear non-autonomous dynamical system, generated by equation (13.121). According to Lemma 13.10 this system is asymptotically compact. According to results of $[175,\text{ Chapter }12]$ every $0\neq\lambda\in\sigma(U(\tau,y_0))$ is a multiplier of operator of monodromy $U(\tau,y_0)$. To finish the proof of Theorem 13.43 it is sufficient to refer to Theorems 13.36, 13.37 and Remark 13.8. The theorem is proved. $\square$

Remark 13.11   *The equivalence of conditions 1. and 3. in Theorem 13.41 (Theorem 13.42, Theorem 13.43) was proved in* $[188,\text{ p.219}]$ *(resp. in* $[188,\text{ p.233}]$, $[175,\text{ p.365}])$.

Chapter 14

Triangular maps

Chapter 14 is devoted to the study of quasi-linear triangular maps: chaos, almost periodic and recurrent solutions, integral manifolds, chaotic sets etc. This problem is formulated and solved in the framework of non-autonomous dynamical systems with discrete time. We prove that such systems admit an invariant continuous section (an invariant manifold). Then, we obtain the conditions of the existence of a compact global attractor and characterize its structure. We give a criterion for the existence of almost periodic and recurrent solutions of the quasi-linear triangular maps. Finally, we prove that quasi-linear maps with chaotic base admits a chaotic compact invariant set.

This chapter is organized as follows.

In Section 1 we establish the relation between triangular maps and non-autonomous dynamical systems with discrete time.

In Section 2 we study linear non-autonomous dynamical systems with discrete time and prove that they admit a unique continuous invariant section – invariant manifold (Theorem 14.2).

Section 3 is devoted to the study of the existence of invariant sections of quasi-linear non-autonomous dynamical systems with discrete time (Theorems 14.3 and 14.4).

In Section 4 we prove the existence of compact global attractors of quasi-linear dynamical systems (Theorems 14.5 and 14.7) and give the description of the structure of these attractors (Theorems 14.9 and 14.10).

Section 5 is devoted to the study of almost periodic and recurrent solutions of quasi-linear difference equations (Theorem 14.13 and 14.14).

In section 6 we introduce the notion of pseudo-recurrent solution and prove that quasi-linear dynamical systems with pseudo-recurrent base admit a pseudo-recurrent solution (Theorem 14.15).

Section 7 is devoted to the study of chaos in triangular maps and non-autonomous dynamical systems with discrete time (Theorem 14.17).

14.1 Triangular maps and non-autonomous dynamical systems

Let W and Ω be two complete metric spaces and denote by $X := W \times \Omega$ its cartesian product. Recall (see, for example, [214, 222]) that a continuous map $F : X \to X$ is called triangular if there are two continuous maps $f : W \times \Omega \to W$ and $g : \Omega \to \Omega$ such that $F = (f, g)$, i.e. $F(x) = F(u, \omega) = (f(u, \omega), g(\omega))$ for all $x =: (u, \omega) \in X$.

Consider a system of difference equations

$$\begin{cases} u_{n+1} = f(\omega_n, u_n) \\ \omega_{n+1} = g(\omega_n), \end{cases} \tag{14.1}$$

for all $n \in \mathbb{Z}_+$, where $\mathbb{Z}_+$ is the set of all non-negative integer numbers.

Along with system (14.1) we consider the family of equations

$$u_{n+1} = f(g^n \omega, u_n) \ (\omega \in \Omega), \tag{14.2}$$

which is equivalent to system (14.1). Let $\varphi(n, u, \omega)$ be a solution of equation (14.2) passing through the point $u \in W$ for $n = 0$. It is easy to verify that the map $\varphi : \mathbb{Z}_+ \times W \times \Omega \to W$ $((n, u, \omega) \mapsto \varphi(n, u, \omega))$ satisfies the following conditions:

(1) $\varphi(0, u, \omega) = u$ for all $u \in W$;
(2) $\varphi(n + m, u, \omega) = \varphi(n, \varphi(m, u, \omega), \sigma(m, \omega))$ for all $n, m \in \mathbb{Z}_+, u \in W$ and $\omega \in \Omega$, where $\sigma(n, \omega) := g^n \omega$;
(3) the map $\varphi : \mathbb{Z}_+ \times W \times \Omega \to W$ is continuous.

Denote by $(\Omega, \mathbb{Z}_+, \sigma)$ the semi-group dynamical system generated by positive powers of the map $g : \Omega \to \Omega$, i.e. $\sigma(n, \omega) := g^n \omega$ for all $n \in \mathbb{Z}_+$ and $\omega \in \Omega$.

Definition 14.1   Recall [102] that a triplet $\langle W, \varphi, (\Omega, \mathbb{Z}_+, \sigma) \rangle$ (or briefly φ) is called a cocycle (or non-autonomous dynamical system) over the dynamical system $(\Omega, \mathbb{Z}_+, \sigma)$ with fiber W.

Thus, the reasoning above shows that every triangular map generates a cocycle and, obviously, vice versa. Taking into consideration this remark we can study triangular maps in the framework of non-autonomous dynamical systems (cocycles) with discrete time.

Definition 14.2   A map $\gamma : \mathbb{Z} \to \Omega$ (respectively $\alpha : \mathbb{Z} \to X$, where $X := W \times \Omega$) is called an entire trajectory of the dynamical system $(\Omega, \mathbb{Z}_+, \sigma)$ (respectively, of the skew-product dynamical system $(X, \mathbb{Z}_+, \pi)$, where $\pi := (\varphi, \sigma)$ and $\langle W, \varphi, (\Omega, \mathbb{Z}_+, \sigma) \rangle$ is a cocycle over $(\Omega, \mathbb{Z}_+, \sigma)$ with fiber W) passing through the point $\omega \in \Omega$ (respectively, $x := (u, \omega) \in X$), if $\gamma(0) = \omega$ (resp. $\alpha(0) = x$) and $\gamma(n + m) = \sigma(m, \gamma(n))$ (resp. $\alpha(n + m) = \pi(m, \alpha(n))$ for all $n \in \mathbb{Z}$ and $m \in \mathbb{Z}_+$.

Denote by Φ_ω the set of all the entire trajectories of the semi-group dynamical system $(\Omega, \mathbb{Z}_+, \sigma)$ passing through the point $\omega \in \Omega$ at the initial moment $n = 0$ and $\Phi := \bigcup \{\Phi_\omega \mid \omega \in \Omega\}$.

Definition 14.3 A map $\nu : \mathbb{Z} \to W$ is called an entire trajectory of the cocycle $\langle W, \varphi, (\Omega, \mathbb{Z}_+, \sigma) \rangle$ passing through the point $u \in W$ if there are $\omega \in \Omega$ and $\gamma_\omega \in \Phi_\omega$ such that the map $\alpha : \mathbb{Z} \to X$ $(X := W \times \Omega)$ defined by by the equality $\alpha := (\nu, \gamma_\omega)$ is an entire trajectory of the skew-product dynamical system $(X, \mathbb{Z}_+, \pi)$, i.e. $\alpha \in \Phi_x$ and $x := (u, \omega)$.

14.2 Linear non-autonomous dynamical systems

Let Ω be a compact metric space and $(\Omega, \mathbb{Z}_+, \sigma)$ be a semi-group dynamical system on Ω with discrete time.

Definition 14.4 Recall that a subset $A \subseteq \Omega$ is called invariant (positively invariant, negatively invariant) if $\sigma^n A = A$ $(\sigma^n A \subseteq A, A \subseteq \sigma^n A)$ for all $n \in \mathbb{Z}_+$, where $\sigma^n := \sigma(n, \cdot) : \Omega \to \Omega$.

Below we will suppose that the set Ω is invariant, i.e. $\sigma^n \Omega = \Omega$ for all $n \in \mathbb{Z}_+$. Let E be a finite-dimensional Banach space with the norm $|\cdot|$ and W be a complete metric space. Denote by $L(E)$ the space of all linear continuous operators on E and by $C(\Omega, W)$ the space of all the continuous functions $f : \Omega \to W$ endowed with the compact-open topology, i.e. the uniform convergence on compact subsets in Ω. The results of this section will be used in the next sections.

Consider a linear equation

$$u_{n+1} = A(\sigma^n \omega) u_n \quad (\omega \in \Omega, \ \sigma^n \omega := \sigma(n, \omega)) \tag{14.3}$$

and an in-homogeneous equation

$$u_{n+1} = A(\sigma^n \omega) u_n + f(\sigma^n \omega), \tag{14.4}$$

where $A \in C(\Omega, L(E))$ and $f \in C(\Omega, E)$.

Definition 14.5 Recall that a linear bounded operator $P : E \to E$ is called a projection, if $P^2 = P$, where $P^2 := P \circ P$.

Definition 14.6 Let $U(n, \omega)$ be the operator of Cauchy (a solution operator) of linear equation (14.3). Following [117] we will say that equation (14.3) has an exponential dichotomy on Ω, if there exists a continuous projection valued function $P : \Omega \to L(E)$ satisfying:

(1) $P(\sigma^n \omega) U(n, \omega) = U(n, \omega) P(\omega)$;

(2) $U_Q(n, \omega)$ is inversible as an operator from $ImQ(\omega)$ to $ImQ(\sigma^n \omega)$, where $U_Q(n, \omega) := U(n, \omega)Q(\omega)$;

(3) there exist constants $0 < \nu < 1$ and $N > 0$ such that

$$\|U_P(n, \omega)\| \le Nq^n \text{ and } \|U_Q(n, \omega)^{-1}\| \le Nq^n$$

for all $\omega \in \Omega$ and $n \in \mathbb{Z}_+$, where $U_P(n, \omega) := U(n, \omega)P(\omega)$.

Let $\omega \in \Omega$ and $\gamma_\omega \in \Phi_\omega$. Consider a difference equation

$$u_{n+1} = A(\gamma_\omega(n))u_n + f(\gamma_\omega(n)), \tag{14.5}$$

and the corresponding homogeneous linear equation

$$u_{n+1} = A(\gamma_\omega(n))u_n \quad (\omega \in \Omega). \tag{14.6}$$

Let (X, ρ) be a metric space with distance ρ. Denote by $C(\mathbb{Z}, X)$ the space of all the functions $f : \mathbb{Z} \to X$ equipped with a pointwise topology. This topology can be metrized. For example, by the equality

$$d(f_1, f_2) := \sum_1^{+\infty} \frac{1}{2^n} \frac{d_n(f_1, d_2)}{1 + d_n(f_1, d_2)},$$

where $d_n(f_1, d_2) := \max\{\rho(f_1(k), f_2(k)) \mid k \in [-n, n]\}$, there is defined a distance on $C(\mathbb{Z}, X)$ which generates the pointwise topology.

If $x \in X$ and $A, B \subseteq X$, then denote by $\rho(x, A) := \inf\{\rho(x, a) \mid a \in A\}$ and $\beta(A, B) := \sup\{\rho(a, B) \mid a \in A\}$ the semi-distance of Hausdorff.

Theorem 14.1   *Suppose that linear equation (14.3) has an exponential dichotomy on Ω. Then for $f \in C(\Omega, E)$ the following statements hold:*

(1) the set $I_\omega := \{u \in E \mid \exists \gamma_\omega \in \Phi_\omega$ such that equation (14.5) admits a bounded solution ψ_ω defined on $\mathbb{Z}$ with the initial condition $\psi_\omega(0) = u\}$ is nonempty and compact;

(2) $\varphi(n, I_\omega, \omega) = I_{\sigma(n, \omega)}$ for all $n \in \mathbb{Z}_+$ and $\omega \in \Omega$, where $\varphi(n, u, \omega)$ is a solution of equation (14.4) with the condition $\varphi(0, u, \omega) = u$ and $\varphi(n, M, \omega) := \{\varphi(n, u, \omega) \mid u \in M\}$;

(3) the map $\omega \to I_\omega$ is upper-semicontinuous, i.e.

$$\lim_{\omega \to \omega_0} \beta(I_\omega, I_{\omega_0}) = 0$$

for every $\omega_0 \in \Omega$, where β is the semi-distance of Hausdorff;

(4) the set $I := \bigcup\{I_\omega \mid \omega \in \Omega\}$ is compact.

Proof.   Let $\omega \in \Omega$. Since Ω is compact and invariant, the set $\Phi_\omega \ne \emptyset$. We fix $\gamma_\omega \in \Phi_\omega$. Under the conditions of Theorem 14.1 equation (14.6) has an exponential dichotomy on Ω with the same constants N and q that in equation (14.3). Then equation (14.5) admits the unique solution $\nu_\omega : \mathbb{Z} \to E$ with the condition

$$\|\nu_\omega\|_\infty \le N\frac{1+q}{1-q}\|f\|_\infty \le N\frac{1+q}{1-q}\|f\|, \tag{14.7}$$

where $\|f\| := \sup\{|f(\omega)| \mid \omega \in \Omega\}$ and $\|\nu_\omega\|_\infty := \sup\{|\nu_\omega(n)| \mid n \in \mathbb{Z}\}$ (see, for example, [188, 120]). Thus, the set I_ω is not empty. From the continuity of the function $\varphi : \mathbb{Z}_+ \times E \times \Omega \to E$ and inequality (14.7) follows that the set I_ω is closed, bounded and

$$|u| \le N\frac{1+q}{1-q}\|f\|$$

for all $u \in I_\omega$ and $\omega \in \Omega$.

The second statement of the theorem follows from the equality $S_h(\Phi_\omega) = \Phi_{\sigma(h,\omega)}$ ($h \in \mathbb{Z}$), where $S_h\gamma_\omega$ is an h-translation of the trajectory γ_ω, i.e. $S_h\gamma_\omega(n) := \gamma_\omega(n+h)$ for all $n \in \mathbb{Z}$.

We will prove now the third affirmation. Let $\omega_0 \in \Omega$, $\omega_k \to \omega_0$, $u_k \in I_{\omega_k}$ and $u_k \to u$. To prove our statement it is sufficient to show that $u \in I_{\omega_0}$. Since $u_k \in I_{\omega_k}$, there is a trajectory $\gamma_{\omega_k} \in \Phi_{\omega_k}$ such that γ_{ω_k} converges to $\gamma_{\omega_0} \in \Phi_{\omega_0}$ in $C(\mathbb{Z}, \Omega)$ and the equation

$$u_{n+1} = A(\gamma_{\omega_k}(n))u_n + f(\gamma_{\omega_k}(n)) \tag{14.8}$$

has a solution ν_{ω_k} with the initial condition $\nu_{\omega_k}(0) = u_k$ and satisfying inequality (14.7), i.e.

$$|\nu_{\omega_k}(n)| \le N\frac{1+q}{1-q}\|f\|_\infty \le N\frac{1+q}{1-q}\|f\| \tag{14.9}$$

for all $n \in \mathbb{Z}$ and $k \in \mathbb{N}$. It is clear that the sequence $\{\nu_{\omega_k}(n)\}$ converges for every $n \in \mathbb{Z}$. By Tihonoff's theorem the sequences $\{\nu_{\omega_k}\} \subset C(\mathbb{Z}, E)$ is relatively compact. From equality (14.8) and inequality (14.9) follows that every limit point of the sequence $\{\nu_{\omega_k}\}$ is a bounded on $\mathbb{Z}$ solution of the equation

$$u_{n+1} = A(\gamma_{\omega_0}(n))u_n + f(\gamma_{\omega_0}(n)). \tag{14.10}$$

Taking into account that equation (14.10) admits a unique bounded on $\mathbb{Z}$ solution, we obtain the convergence of the sequence $\{\nu_{\omega_k}\}$ in the space $C(\mathbb{Z}, E)$. We put $\nu_0 := \lim_{k \to +\infty} \nu_{\omega_k}$. It is easy to see that $\nu_0(0) = u$ and, consequently, $u \in I_{\omega_0}$.

To prove the fourth affirmation it is sufficient to remark that for every $\omega \in \Omega$ the set I_ω is compact, the map $F : \omega \to I_\omega$ ($F(\omega) := I_\omega$) is upper-semicontinuous and, consequently, the set $I := \bigcup\{I_\omega \mid \omega \in \Omega\} = F(\Omega)$ is compact. The theorem is completely proved. $\qquad\qquad\square$

Lemma 14.1   *Let $\omega \in \Omega$ and $\gamma_\omega^l \in \Phi_\omega$ $(l = 1, 2)$. Under the conditions of Theorem 14.1 we have*

$$|\nu_\omega^1(n) - \nu_\omega^2(n)| \le Nq^n |u^1 - u^2| \tag{14.11}$$

for all $n \in \mathbb{N}$, where ν_ω^l is a bounded on $\mathbb{Z}$ solution of the equation

$$u_{n+1} = A(\gamma_\omega^l(n))u_n + f(\gamma_\omega^l(n)) \tag{14.12}$$

($l = 1, 2$).

Proof.   Denote $\nu(n) := \nu_\omega^1(n) - \nu_\omega^2(n)$, then the sequence ν is a bounded on $\mathbb{Z}_+$ solution of equation (14.6), because $\gamma_\omega^1(n) = \gamma_\omega^2(n) = \sigma(n, \omega)$ for all $n \in \mathbb{Z}_+$. Since equation (14.3) (and, consequently, equation (14.6) too) has an exponential dichotomy, we have $|\nu(n)| \le Nq^n |\nu(0)|$ for all $n \in \mathbb{Z}_+$. Taking into consideration that $\nu(0) = u^1 - u^2$, we obtain the required statement. The lemma is proved. $\quad\square$

Lemma 14.2   *Under the conditions of Theorem 14.1 for every $\omega \in \Omega$ the set I_ω contains a single point u_ω.*

Proof.   Suppose that the statement of the lemma is not true. Then there exists $\omega_0 \in \Omega$ such that I_{ω_0} contains at least two points u_1, u_2 $(u_1 \ne u_2)$. Let ν_i $(i = 1, 2)$ be a bounded on $\mathbb{Z}$ solution of equation (14.10) with the condition $\nu_\omega^i(0) = u_i$ $(i = 1, 2)$. According to inequality (14.11), we have

$$|\nu_\omega^1(-n) - \nu_\omega^2(-n)| \ge Nq^{-n} |u^1 - u^2| \tag{14.13}$$

for all $n \in \mathbb{N}$. On the other hand, since ν_ω^l $(l = 1, 2)$ is a bounded on $\mathbb{Z}$ solution of equation (14.12), we have

$$\sup_{n \in \mathbb{Z}} |\nu_\omega^1(n) - \nu_\omega^2(n)| < +\infty. \tag{14.14}$$

Inequalities (14.13) and (14.14) are contradictory. The obtained contradiction proves our affirmation. The lemma is proved. $\quad\square$

Theorem 14.2   *Under the condition of Theorem 14.1 there exists a unique continuous function $\nu : \Omega \to E$ satisfying the following conditions:*

a. the equality

$$\nu(\sigma(n, \omega)) = \varphi(n, \nu(\omega), \omega)$$

    holds for all $n \in \mathbb{Z}_+$ and $\omega \in \Omega$, where $\varphi(n, u, \omega)$ is the unique solution of equation (14.4) with the initial condition $\varphi(0, u, \omega) = u$.

b.

$$\|\nu\| \le N \frac{1+q}{1-q} \|f\|.$$

Proof. This affirmations follows directly from Theorem 14.1 and Lemmas 14.1 and 14.2. $\qquad\square$

Definition 14.7 A dynamical system $(\Omega, \mathbb{Z}_+, \sigma)$ is said to be invertible, if the map $f := \sigma(1, \cdot) : \Omega \to \Omega$ is invertible and, consequently, on Ω there is defined a group dynamical system $(\Omega, \mathbb{Z}, \sigma)$, where $\pi(-n, \cdot) := (\pi(n, \cdot)^{-1}$.

Remark 14.1 *In case the dynamical system $(\Omega, \mathbb{Z}_+, \sigma)$ is invertible, Theorem 14.2 is well known (see, for example, [122]).*

14.3 Quasi-linear non-autonomous dynamical systems

Let us consider the following quasi-linear equation

$$u_{n+1} = A(\sigma^n \omega)u_n + f(\sigma^n \omega) + F(\sigma^n \omega, u_n), \tag{14.15}$$

where $A \in C(\Omega, L(E))$, $f \in C(\Omega, E)$ and $F \in C(\Omega \times E, E)$.

Theorem 14.3 *Assume that there exist positive numbers $L < L_0 := \frac{1-q}{N(1+q)}$ and $r < r_0 := \varepsilon_0 \frac{N(1+q)}{1-q}(1 - NL_0\frac{1+q}{1-q})^{-1}$ such that*

$$\|F(\omega, x_1) - F(\omega, x_2)\| \le L\|x_1 - x_2\|$$

for all $\omega \in \Omega$ and $x_1, x_2 \in B[Q, r] = \{x \in E \mid \rho(x, Q) \le r\}$, where $Q := \nu(\Omega)$, $\nu \in C(\Omega, E)$ is the unique function from $C(\Omega, E)$ figuring in Theorem 14.2 and $\varepsilon_0 = \max_{\omega \in \Omega} \|F(\omega, \nu(\omega))\|$. Then there exists a function $w \in C(\Omega, B[Q, r])$ such that

$$w(\sigma^n \omega) = \psi(n, w(\omega), \omega)$$

for all $n \in \mathbb{Z}_+$ and $\omega \in \Omega$, where $\psi(\cdot, u, \omega)$ is the unique solution of quasi-linear equation (14.15) with the initial condition $\psi(0, u, \omega) = u$.

Proof. Let $u := v + \nu(\sigma^n \omega)$. Then from equation (14.15) we get

$$v_{n+1} = A(\sigma^n \omega)v_n + F(\sigma^n \omega, v + \nu(\sigma^n \omega)).$$

Let $0 < r < r_0$ and $\alpha \in C(\Omega, B[Q, r])$, where $B[Q, r] := \{u \in E \mid \rho(u, Q) \leq r\}$. Consider the equation

$$v_{n+1} = A(\sigma^n \omega)v_n + \tilde{f}(\sigma^n \omega),$$

where $\tilde{f}(\omega) := F(\omega, \alpha(\omega) + \nu(\omega))$ for all $\omega \in \Omega$. According to Theorem 14.2, there exists a unique function $\nu_\alpha \in C(\Omega, E)$ with properties a. and b. In virtue of Theorem 14.2, we have

$$\|\nu_\alpha\| \leq N \frac{1+q}{1-q} \max_{\omega \in \Omega} \|F(\omega, \alpha(\omega) + \nu(\omega))\|$$

$$\leq N \frac{1+q}{1-q} \max_{\omega \in \Omega} |F(\omega, \alpha(\omega) + \nu(\omega)) - F(\omega, \gamma(\omega))| + N \frac{1+q}{1-q} \max_{\omega \in \Omega} |F(\omega, \gamma(\omega))|$$

$$\leq N \frac{1+q}{1-q} L\|\alpha\| + N \frac{1+q}{1-q}\varepsilon_0 \leq N \frac{1+q}{1-q}Lr + \frac{2N}{\nu}\varepsilon_0$$

$$\leq N \frac{1+q}{1-q}L_0 r_0 + N \frac{1+q}{1-q}\varepsilon_0 = r_0$$

and, consequently, $\mathcal{F}(C(\Omega, B[Q, r_0])) \subseteq C(\Omega, B[Q, r_0])$, where $\mathcal{F} : C(\Omega, B[Q, r])) \to C(\Omega, E)$ is the map defined by the equality $\mathcal{F}(\alpha) := \nu_\alpha$.

Now we will show that the map $\mathcal{F} : C(\Omega, B[Q, r_0]) \to C(\Omega, B[Q, r_0])$ is a Lipschitzian one. In fact, according to Theorem 14.2 we have

$$\|\mathcal{F}(\alpha_1) - \mathcal{F}(\alpha_2)\| \leq N \frac{1+q}{1-q} \max_{\omega \in \Omega} |F(\omega, \alpha_1(\omega) + \nu(\omega)) - F(\omega, \alpha_2(\omega) + \nu(\omega))|$$

$$\leq N \frac{1+q}{1-q}L \max_{\omega \in \Omega} |\alpha_1(\omega) - \alpha_2(\omega)|.$$

We note that $N\frac{1+q}{1-q}L \leq N\frac{1+q}{1-q}L_0 < 1$. Thus, the map $\mathcal{F}$ is a contraction and, consequently, by Banach fixed point theorem, there exists a unique function $\alpha \in C(\Omega, B[Q, r_0])$ such that $\mathcal{F}(\alpha) = \alpha$. To finish the proof of the theorem it is sufficient to put $w := \gamma + \alpha$. $\qquad\square$

We now consider a perturbed quasi-linear equation

$$u_{n+1} = A(\sigma^n \omega)u_n + f(\sigma^n \omega) + \lambda F(\sigma^n \omega, u_n), \tag{14.16}$$

where $\lambda \in [-\lambda_0, \lambda_0]$ $(\lambda_0 > 0)$ is a small parameter.

Theorem 14.4 *Assume that there exist positive numbers r and L such that*

$$|F(\omega, u_1) - F(\omega, u_2)| \leq L|u_1 - u_2|$$

for all $\omega \in \Omega$ and $u_1, u_2 \in B[Q, r]$. Then for λ small enough there exists a unique function $\nu_\lambda \in C(\Omega, B[Q, r])$ such that

$$\nu_\lambda(\sigma^n \omega) = \psi_\lambda(n, \nu_\lambda(\omega), \omega)$$

for all $n \in \mathbb{Z}_+$ and $\omega \in \Omega$, where $\psi_\lambda(\cdot, u, \omega)$ is the unique solution of equation (14.16) *with the initial condition $\psi_\lambda(0, u, \omega) = u$. Moreover,*

$$\lim_{\lambda \to 0} \max_{\omega \in \Omega} |\nu_\lambda(\omega) - \nu(\omega)| = 0, \tag{14.17}$$

where $\nu \in C(\omega, E)$ is defined in Theorem 14.2.

Proof. We can prove the existence of ν_λ by a slight modification of the proof of Theorem 14.3.

To prove (14.17) we note that

$$|F(\omega, \nu_\lambda(\omega))| \le |F(\omega, \nu_\lambda(\omega)) - F(\omega, \nu(\omega))| + |F(\omega, \nu(\omega))| \le Lr + \varepsilon_0.$$

Denote $\Delta_\lambda(\omega) := \nu_\lambda(\omega) - \nu(\omega)$ $(\omega \in \Omega)$. It is easy to see that the sequence $\{\Delta_\lambda(\sigma^n\omega)\}_{n\in\mathbb{Z}}$ is the unique bounded on $\mathbb{Z}$ solution of the equation

$$z_{n+1} = A(\sigma^n\omega)z_n + \lambda F(\sigma^n\omega, \nu_\lambda(\sigma^n(\omega)))$$

and, consequently,

$$|\nu_\lambda(\sigma^n\omega) - \nu(\sigma^n\omega)| = |\Delta(\sigma^n\omega)| N \frac{1+q}{1-q} \tag{14.18}$$

$$\times \sup_{n\in\mathbb{Z}} |\lambda F(\sigma^n\omega, \nu_\lambda(\sigma^n\omega))| \le |\lambda| N \frac{1+q}{1-q} (Lr + \varepsilon_0)$$

for all $\omega \in \Omega$, $n \in \mathbb{Z}$ and $\lambda \in [-\lambda_0, \lambda_0]$. Thus, from inequality (14.18) we obtain

$$|\nu_\lambda(\omega) - \nu(\omega)| \le |\lambda| N \frac{1+q}{1-q} (Lr + \varepsilon_0) \tag{14.19}$$

for all $\omega \in \Omega$ and $\lambda \in [-\lambda_0, \lambda_0]$. Passing to limit in inequality (14.19) as $\lambda \to 0$, we obtain (14.17). $\qquad\square$

14.4 Global attractors of quasi-linear triangular systems

Consider a difference equation

$$u_{n+1} = f(\sigma^n\omega, u_n) \ (\omega \in \Omega). \tag{14.20}$$

Denote by $\varphi(n, u, \omega)$ a unique solution of equation (14.20) with the initial condition $\varphi(0, u, \omega) = u$.

Definition 14.8 Equation (14.20) is said to be dissipative, if there exists a positive number r such that

$$\limsup_{n \to +\infty} |\varphi(n, u, \omega)| \le r$$

for all $u \in E$ and $\omega \in \Omega$.

Consider a quasi-linear equation

$$u_{n+1} = A(\sigma^n \omega)u_k + F(\sigma^n \omega, u_k), \tag{14.21}$$

where $A \in C(\Omega, [E])$ and the function $F \in C(\Omega \times E, E)$ satisfies to "the condition of smallness".

Denote by $U(k, \omega)$ the Cauchy's matrix for the linear equation

$$u_{n+1} = A(\sigma^n \omega)u_k.$$

Theorem 14.5   *Suppose that the following conditions hold:*

(1) there are positive numbers N and $q < 1$ such that

$$\|U(n, \omega)\| \le Mq^n \qquad (n \in \mathbb{Z}_+); \tag{14.22}$$

(2) $|F(\omega, u)| \le C + D|u|$ $(C \ge 0,\ 0 \le D < (1 - q)N^{-1})$ for all $u \in E$ and $\omega \in \Omega$.

Then equation (14.21) is dissipative.

Proof. Let $\varphi(\cdot, u, \omega)$ be the solution of equation (14.21) passing through the point $u \in E$ for $n = 0$. According to the formula of the variation of constants (see, for example, [170] and [188]) we have

$$\varphi(n, u, \omega) = U(k, \omega)u + \sum_{m=0}^{n-1} U(n - m - 1, \omega)F(\sigma^m \omega, \varphi(m, u, \omega)),$$

and, consequently,

$$|\varphi(n, u, \omega)| \le Nq^n |u| + \sum_{m=0}^{n-1} q^{n-m-1}(C + D|\varphi(m, u, \omega)|). \tag{14.23}$$

We set $u(n) := q^{-n}|\varphi(n, u, \omega)|$ and, taking into account (14.23), obtain

$$u(n) \le N|u| + CNq^{-1} \sum_{m=0}^{n-1} q^{-m} + DNq^{-1} \sum_{m=0}^{n-1} u(m). \tag{14.24}$$

Denote the right hand side of inequality (14.24) by $v(n)$. Note, that

$$v(n + 1) - v(n) = q^{-n}\frac{CN}{q} + \frac{DN}{q}u(n) \le \frac{DN}{q}v(n) + \frac{CN}{q}q^{-n},$$

and, hence,

$$v(n + 1) \le \left(1 + \frac{DN}{q}\right)v(n) + \frac{CN}{q}q^{-n}.$$

From this inequality we obtain

$$v(n) \le \left(1 + \frac{DN}{q}\right)^{n-1}v(1) + \frac{CN}{q}\frac{1 - q^{n-1}}{1 - q}.$$

Therefore,

$$|\varphi(n, u, \omega)| \leq (q + DN)^{n-1} qN|u| + \frac{CN}{q-1}(q^{n-1} - 1), \qquad (14.25)$$

because $v(1) = N|u|$. From (14.25) follows that

$$\limsup_{k \to +\infty} |\varphi(n, u, \omega)| \leq \frac{CN}{1-q}$$

for all $u \in E$ and $\omega \in \Omega$. The theorem is proved. $\qquad\square$

Let $\langle E, \varphi, (\Omega, \mathbb{Z}_+, \sigma)\rangle$ be a cocycle over $(\Omega, \mathbb{Z}_+, \sigma)$ with the fiber E.

Definition 14.9 Recall that a family $\{I_\omega \mid \omega \in \Omega\}(I_\omega \subset E)$ of nonempty compact subsets is called a compact global attractor of the cocycle φ, if the following conditions are fulfilled:

(1) the set $\bigcup\{I_\omega \mid \omega \in \Omega\}$ is relatively compact;
(2) the family $I := \{I_\omega \mid \omega \in \Omega\}$ is invariant with respect to the cocycle φ, i.e.
$\varphi(n, I_\omega, \omega) = I_{\sigma^n \omega}$ for all $n \in \mathbb{Z}_+$ and $\omega \in \Omega$;
(3) the equality

$$\lim_{n \to +\infty} \sup_{\omega \in \Omega} \beta(\varphi(n, K, \omega), I) = 0$$

takes place for every $K \in C(E)$, where $C(E)$ is a family of all compacts from E.

Definition 14.10 A cocycle $\langle E, \varphi, (\Omega, \mathbb{T}, \sigma)\rangle$ is said to be dissipative, if there exists a positive number r such that

$$\limsup_{n \to +\infty} |\varphi(n, u, \omega)| \leq r$$

for all $u \in E$ and $\omega \in \Omega$.

Theorem 14.6 *([102]) Every dissipative cocycle $\langle E, \varphi, (\Omega, \mathbb{T}, \sigma)\rangle$ admits a compact global attractor $\{I_\omega \mid \omega \in \Omega\}$ such that:*

(1) the component I_ω $(\omega \in \Omega)$ is connected;
(2) the set $I = \bigcup\{I_\omega \mid \omega \in \Omega\}$ is connected, if the space Ω also is.

Theorem 14.7 *Under the conditions of Theorem 14.5, equation (14.21) (i.e. the cocycle φ generated by equation (14.21)) admits a compact global attractor $\{I_\omega \mid \omega \in \Omega\}$ with the following properties:*

(1) the component I_ω $(\omega \in \Omega)$ is connected;
(2) the set $I = \bigcup\{I_\omega \mid \omega \in \Omega\}$ is connected, if the space Ω also is.

Proof. This statement follows directly from Theorems 14.5 and 14.6. $\square$

Definition 14.11   Recall that a cocycle $\langle E, \varphi, (\Omega, \mathbb{T}, \sigma) \rangle$ is said to be convergent, if it admits a compact global attractor $\{I_\omega \mid \omega \in \Omega\}$ such that every component I_ω $(\omega \in \Omega)$ contains a single point u_ω, i.e. $I_\omega = \{u_\omega\}$ $(\omega \in \Omega)$.

Theorem 14.8   [102, Ch.2] *Let $\langle E, \varphi, (\Omega, \mathbb{T}, \sigma) \rangle$ be a cocycle admitting a compact global attractor $I = \{I_\omega \mid \omega \in \Omega\}$. If*

$$\lim_{n \to +\infty} \max_{|u_1|, |u_2| \le r, \omega \in \Omega} |\varphi(n, u_1, \omega) - \varphi(n, u_2, \omega)| = 0 \qquad (14.26)$$

for every $r > 0$, then the set I_ω contains a single point for every $\omega \in \Omega$.

Theorem 14.9   *Let $A \in C(\Omega, [E])$ and $F \in C(\Omega \times E, E)$ and the following conditions be fulfilled:*

(1) there exist positive numbers N and $q < 1$ such that inequality (14.22) holds;
(2) $|F(\omega, u_1) - F(\omega, u_2)| \le L|u_1 - u_2|$ $(0 \le L < N^{-1}(1 - q))$ for all $\omega \in \Omega$ and $u_1, u_2 \in E$.

     Then equation (14.21) is convergent, i.e. the cocycle generated by this equation is convergent.

Proof. First step. We will prove that under the conditions of Theorem 14.9 equation (14.21) admits a compact global attractor $I = \{I_\omega \mid \omega \in \Omega\}$. In fact,

$$|F(\omega, u)| \le |F(\omega, 0)| + L|u| \le C + L|u|$$

for all $u \in E$, where $C := \max\{|F(\omega, 0)| \mid \omega \in \Omega\}$. According to Theorems 14.5 and 14.6, the equation (14.21) admits a compact global attractor $I = \{I_\omega \mid \omega \in \Omega\}$.

     Second step. Let φ be the cocycle generated by equation (14.21). In virtue of the formula of the variation of constants, we have

$$\varphi(n, u, \omega) = U(n, \omega)u + \sum_{m=0}^{n-1} U(n - m - 1, \omega)F(\sigma^m \omega, \varphi(m, u, \omega)).$$

Consequently,

$$\varphi(n, u_1, \omega) - \varphi(n, u_2, \omega) = U(n, \omega)(u_1 - u_2) +$$

$$\sum_{m=1}^{n-1} U(n - m - 1, A)[F(\sigma^m \omega, \varphi(m, u, \omega)) - F(\sigma^m \omega, \varphi(m, u_2, \omega))].$$

Thus,

$$|\varphi(n, u_1, \omega) - \varphi(n, u_2, \omega)| \le Nq^n(|u_1 - u_2|$$

$$+ Lq^{-1} \sum_{m=0}^{n-1} q^{-m}|\varphi(m, u_1, \omega) - \varphi(m, u_2, \omega)|). \tag{14.27}$$

Let $u(n) := |\varphi(n, u_1, \omega) - \varphi(n, u_2, \omega)|q^{-n}$. From (14.27) follows that

$$u(n) \le N\left(|u_1 - u_2| + Lq^{-1} \sum_{m=0}^{n-1} u(m)\right). \tag{14.28}$$

Denote by $v(n)$ the right hand side of (14.28). The following inequality

$$v(n + 1) - v(n) = LNq^{-1}u(n) \le LNq^{-1}v(n). \tag{14.29}$$

holds. From (14.29) we obtain

$$v(n) \le (1 + LNq^{-1})^{n-1}v(1)$$

and, since $v(1) = N|u_1 - u_2|$, we get

$$u(n) \le (1 + LNq^{-1})N|u_1 - u_2|. \tag{14.30}$$

From (14.30) we have

$$|\varphi(n, u_1, \omega) - \varphi(n, u_2, \omega)| \le (q + LM)^{n-1}qN|u_1 - u_2|$$

for all $u_1, u_2 \in E$ and $\omega \in \Omega$.

To finish the proof of Theorem 14.9 it is sufficient to refer to Theorem 14.8. The theorem is proved. $\qquad\square$

Remark 14.2 *Simple examples show that under the conditions of Theorem 14.7 the compact global attractor $\{I_\omega \mid \omega \in \Omega\}$, generally speaking, is not trivial, i.e. the component set I_ω contains more than one point. This statement can be illustrated by the following example: $u_{n+} = \frac{1}{2}u_n + \frac{2u_n}{1+u_n^2}$.*

Theorem 14.10 *Let $A \in C(\Omega, [E]), F \in C(\Omega \times E, E)$ and the following conditions be fulfilled:*

(1) there are positive numbers N and $q < 1$ such that

$$\|U(n, \omega)\| \le Nq^n \quad (n \ge 0, \omega \in \Omega);$$

(2) $|F(\omega, u)| \le M + \varepsilon|u|$ ($u \in E$, $\omega \in \Omega$) and $0 \le \varepsilon \le \varepsilon_0 < \frac{1-q}{N(1+q)}$;

(3) the restriction F_0 of the function F on $\Omega \times B[0, r_0]$, where $r_0 := NM\frac{1+q}{1-q}(1 - \varepsilon_0 N\frac{1+q}{1-q})^{-1}$, satisfies the condition of Lipschitz with the Lipschitz's constant $Lip(F_0) < \frac{1-q}{N(1+q)}$.

Then the following statements hold:

a. *equation (14.21) admits a compact global attractor $I = \{I_\omega \mid \omega \in \Omega\}$;*

b. *there exists a continuous function $\nu : \Omega \to E$ satisfying the following conditions:*

   b1. the equality

$$\nu(\sigma(n,\omega)) = \varphi(n, \nu(\omega), \omega)$$

   holds for all $n \in \mathbb{Z}_+$ and $\omega \in \Omega$, where $\varphi(n, u, \omega)$ is the unique solution of equation (14.21) with the initial condition $\varphi(0, u, \omega) = u$.

   b2.

$$\nu(\omega) \in I_\omega \quad \text{for all } \omega \in \Omega.$$

Proof. The first statement follows directly from Theorems 14.5 and 14.7.

We will prove now the second affirmation. Define an operator $\mathcal{F}$: $C(\Omega, B[0, r_0]) \to C(\Omega, E)$ in the following way. Let $\alpha \in C(\Omega, B[0, r_0])$, then under the conditions of Theorem 14.10 we have $\tilde{f}(\cdot) := F(\cdot, \alpha(\cdot)) \in C(\Omega, B[0, r_0])$. According to Theorem 14.2, there exists a unique function $\nu_\alpha \in C(\Omega, E)$ with properties a. and b. In virtue of Theorem 14.2, we have

$$\|\nu_\alpha\| \le N \frac{1+q}{1-q} \max_{\omega \in \Omega} |F(\omega, \alpha(\omega))|$$

$$\le N \frac{1+q}{1-q}(M + \varepsilon_0 \|\alpha\|) \le N \frac{1+q}{1-q}(M + \varepsilon_0 r_0) \le r_0$$

and, consequently, $\mathcal{F}(C(\Omega, B[Q, r_0])) \subseteq C(\Omega, B[Q, r_0])$, where $\mathcal{F} : C(\Omega, B[Q, r])) \to C(\Omega, E)$ is the map defined by the equality $\mathcal{F}(\alpha) := \nu_\alpha$.

Now we will show that the map $\mathcal{F} : C(\Omega, B[Q, r_0]) \to C(\Omega, B[Q, r_0])$ is Lipschitzian. In fact, according to Theorem 14.2 we have

$$\|\mathcal{F}(\alpha_1) - \mathcal{F}(\alpha_2)\| \le N \frac{1+q}{1-q} \max_{\omega \in \Omega} |F(\omega, \alpha_1(\omega)) - F(\omega, \alpha_2(\omega))|$$

$$\le N \frac{1+q}{1-q} L \max_{\omega \in \Omega} |\alpha_1(\omega) - \alpha_2(\omega)|.$$

We note that $N \frac{1+q}{1-q} L \le N \frac{1+q}{1-q} L_0 < 1$. Thus, the map $\mathcal{F}$ is a contraction and, consequently, in virtue of the Banach's fixed point theorem there exists a unique function $w \in C(\Omega, B[Q, r_0])$ such that $\mathcal{F}(w) = w$. It is easy to see that w satisfies the first condition. To finish the proof of the theorem it is sufficient to remark that the set $A := \bigcup \{A_\omega \mid \omega \in \Omega\}$, where $A_\omega := \{w(\omega)\} \times \{\omega\}$, is a compact invariant set in the skew-product dynamical system $(X, \mathbb{Z}_+, \pi)$ ($X := E \times \Omega$ and $\pi := (\varphi, \sigma)$) and $J := \bigcup \{J_\omega \mid \omega \in \Omega\}$ is a maximal compact invariant set of this system. Thus, we have $A \subset J$ and, consequently, $w(\omega) \in I_\omega$ for all $\omega \in \Omega$. $\qquad\square$

Remark 14.3 *Under the conditions of Theorem 14.10, dissipativity does not reduce to convergence.*

We will give an example which confirms this statement.

Example 14.1 Let $k \in (\frac{1}{2}, 1)$. Consider a scalar equation

$$u_{n+1} = ku_n + \alpha F(u_n) , \tag{14.31}$$

where

$$F(x) := \begin{cases} \frac{x^2}{2} & \text{for } |x| \leq 10 \\ 50 + 10[1 - \exp(10 - |x|)] & \text{for } |x| > 10 \end{cases}$$

for $k < 1 < k + 5\alpha$. We can take $r_0 := \frac{2(1-k)}{\alpha}$, then it is easy to check that for equation (14.31) all the conditions of Theorem 14.10 are fulfilled for chosen α, k, and F. In addition, its Levinson's center (compact global attractor) contains at least two fixed points (in fact, there are 3 of them) and, consequently, equation (14.31) is not convergent.

14.5 Almost periodic and recurrent solutions

Let $(X, \mathbb{Z}_+, \pi)$ be a dynamical system, $x \in X$, $m \in \mathbb{Z}_+$, $m > 0$, $\varepsilon > 0$.
  Denote $\mathfrak{M}_x = \{\{t_n\} \mid \{xt_n\} \text{ is } convergent\}$.

Theorem 14.11 *([300], [302]) Let $(X, \mathbb{Z}_+, \pi)$ and $(Y, \mathbb{Z}_+, \sigma)$ be two dynamical systems. Assume that $h : X \to Y$ is a homomorphism of $(X, \mathbb{Z}_+, \pi)$ onto $(Y, \mathbb{Z}_+, \sigma)$. If a point $x \in X$ is stationary (m-periodic, almost periodic, recurrent), then the point $y := h(x)$ is also stationary (m-periodic, almost periodic, recurrent) and $\mathfrak{M}_x \subset \mathfrak{M}_y$.*

Definition 14.12 A solution $\varphi(n, u, \omega)$ of equation (14.20) is said to be stationary (m-periodic, almost periodic, recurrent), if the point $x := (u, \omega) \in X :=$ $E \times \Omega$ is a stationary (m-periodic, almost periodic, recurrent) point of the skew-product dynamical system $(X, \mathbb{Z}_+, \pi)$, where $\pi := (\varphi, \sigma)$, i.e. $\pi(n, (u, \omega)) :=$ $(\varphi(n, u, \omega), \sigma(n, \omega))$ for all $n \in \mathbb{Z}_+$ and $(u, \omega) \in E \times \Omega$.

Lemma 14.3 *Suppose that $u \in C(\Omega, E)$ satisfies the condition*

$$u(\sigma(n, \omega)) = \varphi(n, u(\omega), \omega) \tag{14.32}$$

for all $n \in \mathbb{Z}_+$ and $\omega \in \Omega$. Then the map $h : \Omega \to X$, defined by

$$h(\omega) := (u(\omega), \omega) \tag{14.33}$$

for all $\omega \in \Omega$, is a homomorphism of $(\Omega, \mathbb{Z}_+, \sigma)$ onto $(X, \mathbb{Z}_+, \pi)$.

Proof. This assertion follows from equalities (14.32) and (14.33) $\qquad\square$

Remark 14.4 *A function $u \in C(\Omega, E)$ with property (14.32) is called a continuous invariant section (or an integral manifold) of non-autonomous difference equation (14.20).*

Theorem 14.12 *If a function $u \in C(\Omega, E)$ satisfies condition (14.32) and a point $\omega \in \Omega$ is stationary (m-periodic, almost periodic, recurrent), then the solution $\varphi(n, u(\omega), \omega)$ of equation (14.20) also is stationary (m-periodic, almost periodic, recurrent).*

Proof. This statement follows from Theorem 14.11 and Lemma 14.3. $\qquad\square$

Using Theorem 14.12 and the results from sections 3 and 4we obtain some criterions of the existence of periodic (almost periodic, recurrent) solutions of equation (14.15). For example, the following statements hold.

Theorem 14.13 *Let $(\Omega, \mathbb{Z}_+, \pi)$ be a dynamical system and Ω consists of m-periodic (almost periodic, recurrent) points. Then under the conditions of Theorem 14.3 the equation (14.15) admits an invariant integral manifold consisting from m-periodic (almost periodic, recurrent) solutions.*

Theorem 14.14 *Let $(\Omega, \mathbb{Z}_+, \pi)$ be a dynamical system and Ω consists of m-periodic (almost periodic, recurrent) points. Under the conditions of Theorem 14.4 for a sufficiently small λ the equation (14.16) admits a unique invariant manifold ν_λ consisting from m-periodic (almost periodic, recurrent) solutions of equation (14.16).*

Example 14.2 Consider the equation

$$u_{n+1} = f(n, u_n) \tag{14.34}$$

where $f \in C(\mathbb{Z}_+ \times E, E)$; here $C(\mathbb{Z}_+ \times E, E)$ is the space of all continuous functions $\mathbb{Z}_+ \times E \to E)$ equipped with a compact-open topology. This topology can be metrized. For example, by the equality

$$d(f_1, f_2) := \sum_{1}^{+\infty} \frac{1}{2^n} \frac{d_n(f_1, d_2)}{1 + d_n(f_1, d_2)},$$

where $d_n(f_1, d_2) := \max\{\rho(f_1(k, u), f_2(k, u)) \mid k \in [0, n], |u| \leq n\}$, there is defined a distance on $C(\mathbb{Z}_+ \times E, E)$ which generates the topology of pointwise convergence with respect to $n \in \mathbb{Z}_+$ uniformly with respect to u on every compact from E.

Along with equation (14.34), we will consider the H-class of equation (14.34)

$$v_{n+1} = g(n, v_n) \quad (g \in H(f)), \tag{14.35}$$

where $H(f) = \overline{\{f_m \mid m \in \mathbb{Z}_+\}}$ and the over bar denotes the closure in $C(\mathbb{Z}_+ \times E, E)$, and $f_m(n, u) = f(n + m, u)$ for all $n \in \mathbb{Z}_+$ and $u \in E$. Denote by $(C(\mathbb{Z}_+ \times E, E), \mathbb{Z}_+, \sigma)$ the dynamical system of translations. Here $\sigma(m, g) := g_m$ for all $m \in \mathbb{Z}_+$ and $g \in C(\mathbb{Z}_+ \times E, E)$.

Let Ω be the hull $H(f)$ of a given function $f \in C(\mathbb{Z}_+ \times E, E)$ and denote the restriction of $(C(\mathbb{Z}_+ \times E, E), \mathbb{R}, \sigma)$ on Ω by $(\Omega, \mathbb{Z}_+, \sigma)$. Let $F : \Omega \times E \to E$ be a continuous map defined by $F(g, u) = g(0, u)$ for $g \in \Omega$ and $u \in E$. Then equation (14.35) can be rewritten in such form:

$$u_{n+1} = F(\sigma^n \omega, u_n),$$

where $\omega := g$ and $\sigma^n \omega := g_n$.

Definition 14.13 The function $f \in C(\mathbb{Z}_+ \times E, E)$ is said to be periodic (almost periodic, recurrent), if $f \in C(\mathbb{Z}_+ \times E, E)$ is a periodic (almost periodic, recurrent) point of the dynamical system of translations $(C(\mathbb{Z}_+ \times E, E), \mathbb{Z}_+, \sigma)$.

If the function $f \in C(\mathbb{Z}_+ \times E, E)$ is periodic (almost periodic, recurrent), then the set $\Omega := H(f)$ is the compact minimal set of the dynamical system $(C(\mathbb{Z}_+ \times E, E), \mathbb{Z}_+, \sigma)$ consisting from periodic (almost periodic, recurrent) points.

14.6 Pseudo recurrent solutions

Definition 14.14 Recall that an autonomous dynamical system $(\Omega, \mathbb{Z}_+, \sigma)$ is said to be pseudo recurrent, if the following conditions are fulfilled:

a) Ω is compact;
b) $(\Omega, \mathbb{Z}_+, \sigma)$ is transitive, i.e. there exists a point $\omega_0 \in \Omega$ such that $\Omega = \overline{\{\sigma^n \omega_0 \mid n \in \mathbb{Z}_+\}}$;
c) every point $\omega \in \Omega$ is stable in the sense of Poisson, i.e.

$$\mathfrak{N}_\omega = \{\{n_k\} \mid \sigma^{n_k} \omega \to \omega \text{ and } n_k \to +\infty\} \neq \emptyset.$$

Lemma 14.4 *Let* $\langle (X, \mathbb{Z}_+, \pi), (\Omega, \mathbb{Z}_+, \sigma), h \rangle$ *be a non-autonomous dynamical system and the following conditions be fulfilled:*

1) *$(\Omega, \mathbb{Z}_+, \sigma)$ is pseudo recurrent;*
2) *$\gamma \in C(\Omega, X)$ is an invariant section of the homomorphism $h : X \to \Omega$.*

Then the autonomous dynamical system $(\gamma(\Omega), \mathbb{Z}_+, \pi)$ is pseudo recurrent too.

Proof. It is evident that the space $\gamma(\Omega)$ is compact, because Ω is compact and $\gamma \in C(\Omega, X)$. We note that on the space $\gamma(\Omega)$ we have the dynamical system $(\gamma(\Omega), \mathbb{Z}_+, \widehat{\pi})$ defined by the homomorphism $\gamma : \Omega \to \gamma(\Omega)$. Namely, $\widehat{\pi}^n \gamma(\omega) :=$

$\gamma(\sigma^n \omega)$ for all $n \in \mathbb{Z}_+$ and $\omega \in \Omega$. Hence, $\widehat{\pi}^n \gamma(\omega) = \pi^n \gamma(\omega)$ for all $n \in \mathbb{Z}_+$ and $\omega \in \Omega$. Now we will show that $\gamma(\Omega) = \overline{\{\pi^n \gamma(\omega_0) \mid n \in \mathbb{Z}_+\}}$. Really, let $x \in \gamma(\Omega)$. Then there exists a unique point $\omega \in \Omega$ such that $x = \gamma(\omega)$. Let $\{n_k\} \subset \mathbb{Z}_+$ be a sequence such that $\sigma^{n_k} \omega_0 \to \omega$. Then $x = \gamma(\omega) = \lim_{k \to +\infty} \gamma(\sigma^{n_k} \omega_0) = \lim_{k \to +\infty} \pi^{n_k} \gamma(\omega)$ and, consequently, $\gamma(\Omega) \subset \overline{\{\pi^n \gamma(\omega_0) \mid n \in \mathbb{Z}_+\}}$. The inverse inclusion is trivial. Thus, $\gamma(\Omega) = \overline{\{\pi^n \gamma(\omega_0) \mid n \in \mathbb{Z}_+\}}$. To finish the proof of the lemma it is sufficient to note that $\mathfrak{N}_\omega \subseteq \mathfrak{N}_{\gamma(\omega)}$ for every point $\omega \in \Omega$ and, consequently, every point $\gamma(\omega)$ is stable in the sense of Poisson. The lemma is proved. $\qquad\square$

Lemma 14.4 implies that under the conditions of Theorem 14.2 (respectively of Th. 14.3, 14.5 or Th. 14.9) equation (14.32) or equation (14.15)) admits a pseudo recurrent integral manifold.

Therefore, we have the following result.

Theorem 14.15 *Assume the driving dynamical system $(\Omega, \mathbb{Z}_+, \sigma)$ is pseudo recurrent, and assume the conditions of Theorem 14.2 (respectively of Th. 14.3, 14.5 or Th. 14.9) are satisfied. Then equation (14.32) (respectively, equation (14.34) or equation (14.15)) admits a pseudo recurrent integral manifold.*

14.7 Chaos in triangular maps

Let (X, ρ) be a metric space and $(X, \mathbb{Z}_+, \pi)$ be a discrete dynamical system generated by positive powers of the map $f : X \to X$, i.e. $\pi(n, x) = f^n x$ for all $x \in X$ and $n \in \mathbb{Z}_+$, where $f^n := f^{n-1} \circ f$.

Definition 14.15 A subset $M \subseteq X$ is called transitive, if there exists a point $x_0 \in X$ such that $H(x_0) := \overline{\{\pi(n, x_0) \mid n \in \mathbb{Z}_+\}} = M$.

Definition 14.16 $\{p, q\} \subseteq X$ is called a Li-Yorke pair, if simultaneously

$$\liminf_{n \to +\infty} \rho(\pi(n, p), \pi(n, q)) = 0 \text{ and } \limsup_{n \to +\infty} \rho(\pi(n, p), \pi(n, q)) > 0.$$

Definition 14.17 A set $M \subseteq X$ is called scrambled, if any pair of distinct points $\{p, q\} \subseteq M$ is a Li-Yorke pair.

Definition 14.18 A dynamical system $(X, \mathbb{Z}_+, \pi)$ is said to be chaotic, if X contains an uncountable subset M satisfying the following conditions:

(1) the set M is transitive;
(2) M is scrambled;
(3) $\overline{P(M)} = M$, where $P(M) := \{x \in M \mid \mathfrak{N}_x \neq \emptyset\}$ (i.e. $x \in P(M)$, if x is stable in the sense of Poisson) and by bar we denote the closure in X.

Theorem 14.16 *Let $(X, \mathbb{Z}_+, \pi)$ and $(\Omega, \mathbb{Z}_+, \sigma)$ be two dynamical systems and $\nu :$ $X \to \Omega$ be a homeomorphism of $(\Omega, \mathbb{Z}_+, \sigma)$ onto $(X, \mathbb{Z}_+, \pi)$. Assume that $(\Omega, \mathbb{Z}_+, \sigma)$ is chaotic. Then the dynamical system $(X, \mathbb{Z}_+, \pi)$ is chaotic too.*

Proof. Let $(\Omega, \mathbb{Z}_+, \sigma)$ be chaotic and $\nu : X \to \Omega$ be a homeomorphism of $(\Omega, \mathbb{Z}_+, \sigma)$ onto $(X, \mathbb{Z}_+, \pi)$. Then there exists a subset $A \subseteq \Omega$ satisfying all the conditions of Definition 14.17. Denote $M := \nu(A)$. Then the set M is transitive and $\overline{P(M)} = M$ (see the proof of Lemma 14.4). Note that the set $M := \nu(A)$ is uncountable, because the homeomorphism $\nu : \Omega \to X$ transforms different points to different points. Finally, to finish the proof of Theorem 14.16 it is sufficient to note that the homeomorphism $\nu : \Omega \to X$ maps a Li-Yorke pair in $(\Omega, \mathbb{Z}_+, \sigma)$ to a Li-Yorke pair in $(X, \mathbb{Z}_+, \pi)$. The theorem is proved. $\square$

Using Theorem 14.16 and the results from sections 3 and 4 we will obtain some criterions of the existence of chaotic sets for triangular maps. For example the following statements are held.

Theorem 14.17 *Let $(\Omega, \mathbb{Z}_+, \pi)$ be a chaotic dynamical system. Then under the conditions of Theorem 14.3 equation (14.15) (respectively triangular map) admits a compact invariant chaotic set.*

Bibliography

[1] Abergel F., Existence and Finite Dimensionality of the Global Attracor for Evolutin Equations and Unbounded Domains. *Journal of Differential Equations*, 83(11):85–108, 1990.

[2] Alekseev V. M., *Symbolic Dynamics*. Naukova Dumka, Kiev, 1986. The 11th Mathematical School.

[3] Alekseev V. M. and Yakobson M. V., *Symbolic Dinamics and Hyperbolic Dynamic Sistems*. In R. Bowen. *Methods of Symbolic Dynamics*, 196–240. M., 1979.

[4] Andreev A. S., On Asymptotical Stability and Unstability of Zero Solution of Nonautonomous Systems. *Prikladnaya Matematika i Mehanika.*, 48(2):1225–232, 1984.

[5] Ararktsyan B. G., Asymptotic Almost Periodic Solutions of certain Linear Evolutionary Equations. *Matematicheskii Sbornik*, 133(175), 1(5):3–10, 1988.

[6] Arnold L., *Random Dynamical Systems*. Springer-Verlag, 1998.

[7] Arnold V. I., *Additional Chapters of Theory of Ordinary Differential Equations*. Nauka, Moscow, 1978.

[8] Artstein Z., Uniform Asymptotic Stability via the Limiting Equations. *Journal of Differential Equations*, 27(2):172–189, 1978.

[9] Atrashenok P. V., Some Questions of Stability of Motion. *Vestnik LGU, Matematika.*, (8):79–106, 1954.

[10] Aulbach B. and Garay B. M., Discretization of Semilinear Equations with an Exponential Dichotomy. *Computers Math. Applic.* **28** (1994), 23–35.

[11] Auslander J. and Seifert P., Prolongation Stability in Dynamical Systems. *Ann. Inst. Fourier*, Grenoble, 14(2):237–268, 1964.

[12] Babin A. V. and Vishik M. I., Attractors of Evolutionary Equations with Partial Derivatives and Estimation of Their Dimensionality. *Uspekhi Matematicheskih Nauk*, 38(4(232)):133-187, 1983.

[13] Babin A. V. and Vishik M. I., Upper and Lower Estimation of Dimensionality of Attractors of Evolutionary Partial Equations. *Sibirskii Matematicheskii*

Zhurnal., 24(5):15–30, 1983.

[14] Babin A. V. and Vishik M. I., Attractors of Parabolic Equations and Navier–Stocks Systems and Estimation of Their Dimensionality. In collection. *Obschaya Teoriya Granichnyh Zadach. Kiev*, 1983. pages 14–25.

[15] Babin A. V. and Vishik M. I., Regular Attractor of Semigroups and Evolutional Equations. *Journal Math. Pures et Appl.*, 62:172–189, 1983.

[16] Babin A. V. and Vishik M. I., Unstable Invariant Sets of Semi-groups of Nonlinear Operators and Their Perturbations. *Uspekhi Matematicheskih Nauk*, 41(4(250)):3–34, 1986.

[17] Babin A. V. and Vishik M. I., On Unstable Sets of Evolutionary Equations in Neighbourhood of Critical Points of Stationary Curve. *Izvestiya AN SSSR, Matematika.*, 51(1):44–78, 1987.

[18] Babin A. V. and Vishik M. I., Semi-Groups Depending on Parameter, their Attractors and Asymptotical Behaviour. In collection *Globalnyi Analiz i Nelineinye Uravneniya*. Iz-vo VGU, 1988.

[19] Babin A. V. and Vishik M. I., Attractors of Parabolic and Hyperbolic Equations, Character of their Compactness and Attractions to Them. *Vestnik MGU, Series I, Matematika, Mehanika*, (3):71–73, 1988.

[20] Babin A. V. and Vishik M. I., *Attractors of Evolutionary Equations*. Nauka, Moscow, 1989; English translation, North–Holland, Amsterdam, 1992.

[21] Barbashin E. A., *Introduction in Theory of Stability*. Nauka, Moscow, 1967.

[22] Batty C. J. K., Hutter W. and Räbiger F., Almost Periodicity of Mild Solutions of Inhomogeneous Periodic Cauchy Problems. *Journal of Differential Equations*, 156(1999), no.2, pp.309–327

[23] Bellman R., *Introduction to Matrix Analysis*. McGraw-Hill, New York-Toronto-London 1960

[24] Bhatia N. P. and Szegö G. P., *Local Semi-Dynamical Systems. Lecture Notes in Mathematics*, v.80. Springer, Berlin–Heidelberg–New York, 1969.

[25] Bhatia N. P. and Szegö G. P., *Stability Theory of Dynamical Systems*. Lecture Notes in Mathematics. Springer, Berlin–Heidelberg–New York, 1970.

[26] Billotti J. E. and LaSalle J. P., Dissipative Periodic Process. *Bull. Amer. Math. Soc.*, 77(6):1082–1088, 1971.

[27] Bogolyubov N. N. and Mitropolsky Yu. A., *Asymptotic Methods in the Theory of Non-Linear Oscilations*. Fizmatgiz, Moscow 1963; English transl., Gordon and Dreach, New York 1962.

[28] Bourbaki N., *Espaces Vectoriels Topologiques*. Hermann, Paris, 1955.

[29] Bourbaki N., *Variétés Différentielles et Analitiques (Fascicule de résultats)*. Herman, Paris, 1971.

[30] Brezis H., *Operateurs Maximaux Monotones et Semigroupes de Contractions dans les Espaces de Hilbert. Math.Studies*, v.5. North Holland, 1973.

[31] Bronstein I. U. and Gerko A. I., On Inclusion of Some Topological Semi-

groups of Transformations in Topological Groups of Transformations. *Izvestiya AN Moldavskoi SSR*, seriya fiz.-tehn. i matem. nauk, 3:18-24, 1970.

[32] Bronsteyn I. U., *Extensions of Minimal Transformation Group.* Noordhoff, 1979.

[33] Bronstein I. U., *Nonautonomous Dynamical Systems.* Stiintsa, Chişinău, 1984.

[34] Bronstein I. U. and Kopanskii A. Ya., *Smooth Invariant Manifolds and Normal Forms.* World Scientific, Rive Edge, NJ, 1994.

[35] Browder F. E., Some New Asymptotic Fixed Point Theorems. *Proc. Acad. Sci.*, 71:2734–2735, 1974.

[36] Buşe C., Asymptotic Stability and Perron Condition for Periodic Evolution Families on the Half Line. Preprint in Evolution Equations and Semigroups (http://math.lsu.edu/tiger/evolve/archiv/98-00013ps)

[37] Buşe C. and Giurgiulescu M., A New Proof for a Barbashin's Theorem in the Periodic Case. Preprint in Evolution Equations and Semigroups (http://math.lsu.edu/tiger/evolve/archiv/99-00099ps)

[38] Buşe C., An Exponential Stability Criterion for q-Periodic Evolution Families on Hilbert Spaces. Preprint in Evolution Equations and Semigroups (http://math.lsu.edu/tiger/evolve/archiv/00-00001ps)

[39] Cameron R. H., Almost Periodic Properties of Bounded Solutions of Linear Differential Equation with Almost Periodic Coefficients. *J. Math. Phys*, 15:73–81, 1936.

[40] Caraballo T., Langa J. A. and Robinson J., Upper Semicontinuity of Attractors for Small Random Perturbations of Dynamical Systems. *Comm. in Partial Differential Equations*, 23:1557-1581, 1998.

[41] Caraballo T. and Langa J. A., On the Upper Semicontinuity of Cocycle Attractors for Nonautonomous and Random Dynamical Systems. *Dynamics of Continuous, Discrete and Impulsive Systems*, 10(2003), no. 4, pp.491-514.

[42] Cartan H., *Elementary Theory of Analytic Functions of one and Several Complex Variables [Russian translation].* Mir, Moscow, 1963.

[43] Carvalho A. N., Cholewa J. W. and Dlotko T., Global Attractors for Problems with Monotone Operators. *Boll. Unione Mat.Ital.* Sez.B Artic.Ric.Mat.(8)2(1999), no.3, pp.693-706.

[44] Cartwrigth M. L. and Littlewood J. E., On Non-Linear Differential Equations of the Second Order, I. The Equation $\ddot{y} - k(1 - y^2)\dot{y} + y = b\mu k \cos(\mu t + \alpha)k$ for Large k. *J. Math. Soc.*, 20:180–189, 1945.

[45] Cartwrigth M. L. and Littlewood J. E., On Non-Linear Differential Equations of the Second Order, II. The Equation $\ddot{y} - k(1 - y^2)\dot{y} + y = b\mu k \cos(\mu t + \alpha)k$ for Large k and its Generalizations. *Acta Math.*, 97:3–4, 1957.

[46] Cartwrigth M. L. and Littlewood J. E., On Non-Linear Differential Equations of the Second Order, III. The General Equation. *Acta Math.*, 98:1–2, 1957.

[47] Cerons S. S. and Lopes O., α-Contractions and Attractors for Dissipative

Semilinear Hiperbolic Equations and Systems. *Annali di Matematica Pura ad Applicata*, CLX(4):193–206, 1991.

[48] Cheban D. N., Poisson Asymptotic Stability of the Solutions of Operational Equations. *Differentsial'nye Uravneniya*, 13(8):978–983, 1977.

[49] Cheban D. N., Comparability of the Points of Dynamical Systems by Character of Recurrence in the Limit. *Matematicheskie Nauki*, Ştiinţa, Chişinău, 1:66-71, 1977.

[50] Cheban D. N., Linear Differential Equations Satisfying the Condition of Favard. In *Proceedings of the 5th All-Union Conference on the Qualitative Theory of Differential Equations*, pp.181–182, Chişinău, 1979.

[51] Cheban D. N., *Theory of Linear Differential Equations (Selected Topics)*. Ştiincta, Chişinău, 1980.

[52] Cheban D. N., Stability of the Levinson Center of Nonautonomous Dissipative Dynamical Systems. *Differentsial'nye Uravneniya*, 20(11):2016–2018, 1984.

[53] Cheban D. N., Nonautonomous Dissipative Dynamical Systems. *Differentsial'nye Uravneniya*, 21(5):913–914, 1985.

[54] Cheban D. N., ℂ-Analytic Dissipative Dynamical Systems. In *Proceedings of the 3rd International Conference of Differential Equations and Applications*, p.141, Roose (Bulgaria), 1985.

[55] Cheban D. N., Dissipative Dynamical Systems. In *Proceedings of the Extended Sessions of Seminar of the I. N. Vekua Institute of Applied Mathematics*, v. 1, no. 3, pp.154–157, Tbilisi, 1985.

[56] Cheban D. N., Lyapunov's Second Method in the Theory of Stability of Dynamical Systems. In *Mat. Issled. No. 80, Differentsial'nye Uravneniya i Dynamicheskie Sistemy*, pp.139–147. Ştiinţa, Chişinău, 1985.

[57] Cheban D. N., Quasiperiodic Solutions of Dissipative Systems with Quasiperiodic Coefficients. *Differentsial'nye Uravneniya*, 22(2):267–278, 1986.

[58] Cheban D. N., Nonautonomous Dissipative Dynamical Systems. *Dokl. Akad. Nauk SSSR*, 286(1986), pp.136-143. Translation in *Soviet Math. Dokl*, 33(1):207–210, 1986.

[59] Cheban D. N., A criterion for Convergence of Nonlinear Systems in Hilbert Space. In *Mat. Issled., No. 88, Differentsial'nye Uravneniya i ih Invarianty*, pp. 136–143. Ştiinţa, Chişinău, 1986.

[60] Cheban D. N., A criterion for Convergence of Nonlinear Systems with Respect to First Approximation. In *Mat. Issled., No. 88, Differentsial'nye Uravneniya i ih Invarianty*, pp. 144–150. Ştiinţa, Chişinău, 1986.

[61] Cheban D. N., Nonautonomous Dynamical Systems. Method of Lyapunov's Functions. In *Proceedings of the 6th All-Union Conference of Qualitative Theory of Differential Equations*, pages 197–198, Irkutsk, 1986.

[62] Cheban D. N., ℂ-Analytic Dissipative Dynamical Systems. *Differentsial'nye Uravneniya*, 22(11):1915–1922, 1986.

[63] Cheban D. N., On Nonautonomous Dissipative Dynamical Systems. *Uspekhi Matematicheskih Nauk*, 41(4(250)):169, 1986.

[64] Cheban D. N., Nonautonomous Dissipative Dynamical Systems. Method of Lyapunov's Functions. *Differentsial'nye Uravneniya*, 23(3):464–474, 1987.

[65] Cheban D. N., Boundedness, Dissipativity and Almost Periodicity of the Solutions of Linear and Quasi-Linear Systems of Differential Equations. In *Mat. Issled. No. 92, Dinamicheskie Sistemy i Kraevye Zadachi*, pages 143–159. Ştiinţa, Chişinău, 1987.

[66] Cheban D. N., Nonautonomous Dissipative Dynamical systems. Method of Lyapunov's Functions. In *Problems in the Qualitative Theory of Differential Equations*, "Nauka", Sibirsk. Otdel., Novosibirsk, 1988, pp. 56–64.

[67] Cheban D. N., Impulse and Difference Dissipative Systems with Periodic Coefficients. In *Issledovaniya po Differentsialnym Uravneniyam i Matematicheskomu Analizu*, pages 127–142. Ştiinţa, Chişinău, 1988.

[68] Cheban D. N., On the Structure of Levinson's Centre of Dissipative Dynamical Systems. *Differentsial'nye Uravneniya*, 24(6):1086, 1988.

[69] Cheban D. N., On the Structure of Levinson's Centre of Dissipative Dynamical Systems. *Differentsial'nye Uravneniya*, 24(9):1564–1576, 1988; Translation in *Differential Equation*, 24 (1989), no.9, pp. 1031–1040.

[70] Cheban D. N., Principle of Averaging on Semi-Axis for Dissipative Systems. In *Dinamicheskie Sistemy i Uravneniya Matematicheskoi Fiziki*, pages 149–161. Ştiinţa, Chişinău, 1988.

[71] Cheban D. N., Nonautonomous Dissipative Dynamical Systems with Hyperbolic Subset of Centre (One-Dimensional Case). *Mat. Zametki*, 45(6):93–98, 1989; translation in Math. Notes 45 (1989), no. 5-6, pp.493–498.

[72] Cheban D. N., On One Problem of J.Hale. In *Proceedings of the 7th All-Union Conference of Qualitative Theory of Differential Equations*, pp. 227, Riga, 1989.

[73] Cheban D. N., Some Problems of Theory of Dissipative Dynamical Systems, I. In *Differentsial'nye Uravneniya i Matematicheskaya Fizika*, pages 106:147–163. Ştiinţa, Chişinău, 1989.

[74] Cheban D. N., On One Problem of J.Hale. *Mat. Zametki*, 46(1):120–121, 1989.

[75] Cheban D. N., Nonautonomous Dynamical Systems with Convergence. *Differentsial'nye Uravneniya*, 25(9):1633–1635, 1989.

[76] Cheban D. N., On the Structure of Levinson's Centre of Dissipative Dynamical System with Condition of Hyperbolicity on Closure of the Set of Recurrent Motions. *Differentsial'nye Uravneniya*, 26(5):913-914, 1990.

[77] Cheban D. N., On the Structure of Levinson's Centre of Dissipative Dynamical System with Condition of Hyperbolicity on Closure of the Set of Recurrent Motions. *Bulletin of Academy of Sciences of Republic of Moldova, Mathematics*, (2):34–43, 1990.

[78] Cheban D. N., Dissipative Functional Differential Equations. *Bulletin of Academy of Sciences of Republic of Moldova. Mathematics*, (2(5)):3–12, 1991.

[79] Cheban D. N., *Nonautonomous Dissipative Dynamical Systems*. Thesis of Dr. of Sc., Institute of Mathematics, Minsk, 1991.

[80] Cheban D. N., Some Problems of Theory of Dissipative Dynamical Systems, I. In *Mat.Issled. No. 124, Funktsionalnaye Metody v Teorii Differentsial'nyh Uravnenii*, pp.106–122. Ştiinţa, Chişinău, 1992.

[81] Cheban D. N., Locally Dissipative Dynamical Systems and Some Their Applications. *Bulletin of Academy of Sciences of Republic of Moldova. Mathematics*, (1):7–14, 1992.

[82] Cheban D. N., Global Attractors of Infinite-dimensional Dynamical Systems, I. *Bulletin of Academy of Sciences of Republic of Moldova, Mathematics*, (2(15)):12–21, 1994.

[83] Cheban D. N. and Fakeeh D. S., *Global Attractors of Disperse Dynamical Systems*. Sigma, Chişinău, 1994.

[84] Cheban D. N., On the Structure of Compact Asymptotic Stable Invariant Set of ℂ-analytic Almost periodic Systems. *Differential Equations*, v.31, No.12, pp.1995-1998, 1995 [Translated from Differentsial'nye Uravneniya, v.31, No.12, pp.2025-2028, 1995].

[85] Cheban D. N., On the Conversibility of Theorem of Lyapunov about a Stability Asymptotic by First Approximation. *Mat, Zametki* 57(1995), no.1, 139–142; Translation in *Mathematical Notes*, v.57, no.1-2, 1995, 100–102.

[86] Cheban D. N., Global Attractors of Infinite-Dimensional Dynamical Systems, II. *Bulletin of Academy of Sciences of Republic of Moldova, Mathematics*, 1(17):28–37, 1995.

[87] Cheban D. N. and Fakeeh D. S., Global Attractors of Infinite-Dimensional Dynamical Systems, III. *Bulletin of Academy of Sciences of Republic of Moldova, Mathematics*, 2-3(18-19):3–13, 1995.

[88] Cheban D. N., Global Attractors of Infinite-Dimensional Nonautonomous Dynamical Systems,I. *Bulletin of Academy of Sciences of Republic of Moldova. Mathematics*. 1997, N3 (25), pp. 42-55

[89] Cheban D. N., Bounded Solutions of Linear Almost Periodic Differential Equations. *Izv. Ross. Akad. Nauk, Ser. Mat.* 62(1998), no.3, 155-174. Translation in Izvestiya: Mathematics, 3(62):581–600, 1998.

[90] Cheban D. N., Global Attractors of Infinite-Dimensional Nonautonomous Dynamical Systems, II. *Bulletin of Academy of Sciences of Republic of Moldova, Mathematics*, (2(27)):111–126, 1998.

[91] Cheban D. N., The Asymptotics of Solutions of Infinite Dimensional Homogeneous Dynamical Systems. *Mat. Zametki* 63 (1998), No.1, 115-126; Translation in *Mathematical Notes*, 1998. v. 63, No.1 , pp.115-126.

[92] Cheban D. N., The Global Attractors of Nonautonomous Dynamical Systems

and Almost Periodic Limit Regimes of some Classes of Evolution Equations. *Anale Fac. de Mat. si Inform.*, Chişinău, v.1, 1999, pp.1-26.

[93] Cheban D. N., Relations Between the Different Type of Stability of the Linear Almost Periodical Systems in the Banach Space. *Electron. J. Diff. Eqns.* v.1999 (1999), No.46, pp.1-9.

[94] Cheban D. N., Global Attractors of Quasi-Homogeneous Nonautonomous Dynamical Systems. *Proceedings of the International Conference on Dynamical Systems and Differential Equations.* May 18-21, 2000, Kennesaw, USA. pp.96-101.

[95] Cheban D. N., Uniform Exponential Stability of Linear Almost Periodic Systems in Banach Space. *Electron. J. Diff. Eqns.*, v.2000 (2000), No.29, pp.1-18

[96] Cheban D. N., Kloeden P. E. and Schmalfuss B., Pullbac Attractors under Discretization. Proceeding EQUADIFF99. Berlin 1999, vol.2 (Edited by B. Fiedler, K. Groger and J. Sprekels). World Scientific 2000, pp.1024-1029.

[97] Cheban D. N., Schmalfuss B. and Kloeden P. E., Pullback Attractors in Dissipative Nonautonomous Differential Equations under Discretization. *Journal of Dynamics and Differential Equations*, v.13, No.1, 2001, pp. 185-213.

[98] Cheban D. N., Global Pullback Attractors of $\mathbb{C}$-Analytic Nonautonomous Dynamical Systems. *Stochastics and Dynamics*, v.1, No.4, 2001, pp.511-536.

[99] Cheban D. N., Global Attractors of Quasi-Homogeneous Nonautonomous Dynamical Systems. *Electron. J. Diff. Eqns.*, v.2002(2002), No.10, pp.1-19.

[100] Cheban D. N., Kloeden P. E. and Schmalfuss B., Relation Between Pullback and Global Attractors of Nonautonomous Dynamical Systems. *Nonlinear Dynamics and Systems Theory*, v.2, No.2(2002), pp.9-28.

[101] Cheban D. N., Upper Semicontinuity of Attractors of Non-autonomous Dynamical Systems for Small Perturbations. *Electron. J. Diff. Eqns.*, v.2002(2002), No.42, pp.1-21.

[102] Cheban D. N., Global Attractors of Nonautonomous Dynamical Systems. *Chişinău, University Press of USM*, 2002 (in Russian).

[103] Cheban D. N., Asymptotic Almost Periodic Solutions of Differential Equations. *Chişinău, University Press of USM*, 2002.(in Russian)

[104] Cheban D. N. and Duan J., Recurrent Motions and Global Attractors of Nonautonomous Lorenz Systems. *Dynamical Systems: An International Journal*, v.19, No.1, 2004, pp.41-59.

[105] Cheban D. N., Duan J. and Gherco A. I., Generalization of the Second Bogolyubov's Theorem for Non-Almost Periodic Systems. *Nonlinear Analysis: Real World Applications, v.4, No.4, 2003, pp.599-613.*

[106] Cheban D. N., Kloeden P. E. and Schmalfuss B., Global Attractors of V-monotone Nonautonomous Dynamical Systems. *Bulletin of Academy of sciences of Republic of Moldova. Mathematics, 2003. No.1(41), pp.47-57.*

[107] Cheban D.N. and Duan J., Almost Periodic Solutions and Global Attractors of Non-Atonomous Navier-Stokes Equations. Journal of Dynamics and Differential Equations, v.16, No.1, 2004, pp.

[108] Cheban D.N. and Mammana C., Invariant manifolds, global attractors and almost periodic solutions of non-autonomous difference equations. Nonlinear Analysis, series A v.56, No.4, 2004, pp.465-484.

[109] Chepyzhev V. V., Unbounded Attractors of Some Parabolic Systems of Differential Equations and Bound of their Dimensionality. *Dolady Akademii Nauk SSSR*, 301(1):46-49, 1988.

[110] Chepyzhov V. V. and Vishik M. I., A Hausdorff Dimension Estimate for Kernel Sections of Non-autonomous Evolutions Equations. *Indian Univ. Math. J*, 42(3):1057–1076, 1993.

[111] Chepyzhov V. V. and Vishik M. I., Attractors of Non-Autonomous Dynamical Systems and their Dimension. *J. Math. Pures Appl.*, **73** (1994), 279–333.

[112] Chepyzhov V. V. and Vishik M. I., *Attractors for Equations of Mathematical Physics*. Amer. Math. Soc., Providence, RI, 2002.

[113] Cheresiz V. M., *V*-Monotone Systems and Almost Periodic Solutions. *Sibirskii Matematicheskii Zhurnal*. 1972, v.13, No.4, pp.921–932.

[114] Cheresiz V. M., Uniformly *V* - Monotone Systems. Almost Periodic Solutions. *Sibirskii Matematicheskii Zhurnal*. 1972, v.13, No.5, pp.11107–1123.

[115] Chernikova O. S., On Dissipativity of Systems of Differential Equations with Impulse Effect. *Ukrainskii Matematicheskii Zhurnal*, 35(5):656–660, 1983.

[116] Chernikova O. S., On Boundedness of Solutions of Differential Equations with Impulse Effect. *Ukrainskii Matematicheskii Zhurnal*, 38(1):124–127, 1986.

[117] Chicone C. and Latushkin Yu., Evolution Semigroups in Dynamicals Systems and Differential Equations. *Amer. Math. Soc.*, Providence, RI, 1999.

[118] Chueshov I. D., Finite-Dimensionality of Attractor in Some Problems of Nonlinear Theory of Envelopes. *Matematicheskii Sbornik*, 133(4):419–428, 1989.

[119] Chueshov I. D., Strong Solutions and Attractor of System of Karman's Equations. *Matematicheskii Sbornik*, 181(1):25–36, 1990.

[120] Chueshov I. D., *Introduction to the Theory of Infinite-Dimensional Dissipative Systems*. Acta Scientific Publishing Hous, Kharkov, 2002.

[121] Clement Ph, Heijmans H. J. A. M., Angenent S., C.V. van Duijn and B. de Pagter, *One-Parameter Semigroups. CWI Monograph 5.* North-Holland, Amsterdam.New York.Oxford.Tokyo, 1987

[122] Coddington E. A. and Levinson N., *Theory of Ordinary Differential Equations.* McGraw Hill, New York, 1955.

[123] Conley C. C. and Miller R. K., Asymptotic Stability without Uniform Stability: Almost Periodic Coefficients. *Journal of Differential Equations*, 1:333–336, 1965.

[124] Cooperman G. D., α-*Condensing Maps and Dissipative Systems. PhD thesis,*

Brown University, 1978.

[125] Corduneanu C., Systems Differentiels Admettant des Solutions Bornées. *C.R. Acad. Sci., Ser.A–B.*, 245:21–24, 1957.

[126] Crauel H. and Flandoli F., Attractors for Random Dynamical Systems. *Prob. Theor. Related Fields*, **100** (1994), 365-393.

[127] Crauel H., Debussche A. and Flandoli F., Random Attractors. *Journal of Dynamics and Differential Equations*, **9** (1997), 307–341.

[128] Dafermos C. M., An Invariance Principle for Compact Processes. *Journal of Differential Equations*, (9):239–252, 1971.

[129] Dafermos C. M., Uniform Processes and Semicontinuous Lyapunov Fonctionals. *Journal of Differential Equations*, (11):401–415, 1972.

[130] Dafermos C. M., Semiflows Associated with Compact and Uniform Processes. *Math. Syst. Theory*, 8(2):142–149, 1974.

[131] Dafermos C. M., Almost Periodic Process and Almost Periodic Solutions of Evolution Equations. In *Dynamical Systems. Proceedings of a University of Florida International Symposium*, pp.43–58, 1977.

[132] Daletskii Yu. L. and Krein M. G., *Stability of Solutions of Differential Equations in Banach Space*. Moscow, "Nauka", 1970. [English transl., Amer. Math. Soc., Providence, RI 1974.]

[133] Datko R., Extending a Theorem of A. M. Lyapunov to Hilbert Space. *J. Math. Anal. Appl.* 32 (1970), pp. 610-616

[134] Datko R., Uniform Asymptotic Stability of Evolutionary Processes in a Banach Space. *SIAM, J. Math. Anal.* 3 (1973), pp. 428-445

[135] Demidovich B. P., On Dissipativity of Certain Nonlinear Systems of Differential Equations, I. *Vestnik MGU*, (6):19–27, 1961.

[136] Demidovich B. P., On Dissipativity of Certain Nonlinear Systems of Differential Equations, II. *Vestnik MGU*, (1):3–8, 1962.

[137] Demidovich B. P., *Lektsii po Matematicheskoi Teorii Ustoichivosti*. Moscow, "Nauka", 1967.

[138] Denjoy A., Sur les Courbes Definiés par les Equations Differentielles a la Surface du Tore. *Journal Math. Pures et Appll (Ser.9)*, (11):333–375, 1932.

[139] Deyseach L. G. and Sell G. R., On the Existence of Almost Periodic Motions. *Michigan Mathematic Journal*, 12(1):87–95, 1965.

[140] Dymnikov V. P. and Filatov A. N., Mathematics of Climate Modeling. *Birkhaüser*, Boston, MA, 1997.

[141] Duan J. and Kloeden P. E., Dissipative Quasigeostrophic Motion under Temporally Almost Periodic Forcing. *J. Math. Anal.Applns.* **236** (1999), 74-85.

[142] Engel K. J. and Nagel R., *One-Parameter Semigroups for Linear Evolution Equations*. Springer-Verlag, Berlin, 1999.

[143] Eskman J. P., Roads to Turbulence in Dissipative Dynamical Systems. *Rev.Mod.Phys*, 53:643–654, 1981.

[144] Fakeeh D. S., Levinson's Center of Disperse Dissipative Dynamical Systems. *Izvestiya AN SSRM*, (3):55–59, 1990.

[145] Fakeeh D. S., On Structure of Levinson's Center of Disperse Dynamical Systems. *Izvestiya AN SSRM*, (1):62–67, 1991.

[146] Fakeeh D. S., Analogue of Levinson–Pliss' Theorem for Differential Inclusions. In collection *Funktsionalnye Metody v Teorii Differentsialnyh Uravnenii, Matematicheskie Issledovaniya*. 124:100–105, Chişinău, 1992.

[147] Fakeeh D. S. and Cheban D. N., Connectedness of Levinson's Center of Compact Dissipative Dynamical System without Uniqueness. *Izvestiya AN RM, Matematika*, (1):15–22, 1993.

[148] Fang S., Global Attractor for General Non-Autonomous Dynamical Systems. *Nonlinear World*, 2 (1995), pp.191-216.

[149] Fang S., Finite Dimensonal Behaviour of Periodic and Asymptotically Periodic Processes. *Nonlinear Analysis TMA*, **28** (11) (1997), 1785–1797.

[150] Favard J., Sur les Equations Differentielles a Coefficients Presque-Periodiques. *Acta Math.* 51 (1927), pp.31-81.

[151] Favard J., *Leçons sur Fonction Presque-Periodiques*. Paris, Gauthier-Villars, 1933.

[152] Fink A. M. and Fredericson P. O., Ultimate Boundedness Does not Imply Almost Periodicity. *Journal of Differential Equations*, (9):280–284, 1971.

[153] Flandoli F. and Schmalfuß B., Random Attractors for the Stochastic 3D Navier–Stokes Equation with Multiplicative White Noise, *Stochastics and Stochastics Reports*, **59** (1996), 21–45.

[154] Flandoli F. and Schmalfuß B., Weak Solutions and Attractors of the 3D Navier Stokes Equation with Nonregular Force, *Journal of Dynamics and Differential Equations*, v. 11, No. 2(1999), pp. 355–398.

[155] Foias C., Sell G. R. and Temam R., Inertial Manifolds of Nonlinear Evolutionary Equations. *Journal of Differential Equations*, 73(4):309–353, 1988.

[156] Furstenberg H., The Structure of Distal Flows. *Amer. J. Math.*, **85** (1963), 477–515.

[157] Fusco G. and Oliva M., Dissipative Systems with Constraints. *Journal of Differential Equations*, 63:362–388, 1986.

[158] Gerko A. I., *Extensions of Topological Semi-Groups of Transformations*. University Press of USM, Chişinău, 2001.

[159] Gerstein V. M. and Krasnoselskii M. A., Structure of Set of Solution of Dissipative Equations. *Doklady Akademii Nauk SSSR*, 183(2):267–269, 1968.

[160] Gerstein V. M., On Dissipativity of One Two-Dimensional System. *Differentsial'nye Uravneniya*, 5(8):1438–1444, 1969.

[161] Gerstein V. M., Note on Dissipative Flows. *Trudy Matematicheskogo Fakultete Voronezhskogo Gosuniversiteta*, (1):26–34, 1970.

[162] Gerstein V. M., On Theory of Dissipative Differential Equations in Banach

Space. *Funktsionalnyi Analiz i Ego Prilozheniya*, 4(3):99–100, 1970.

[163] Gerstein V. M., On Nonperiodical Dissipative Systems. *Trudy Matematicheskogo Fakulteta Voronezhskogo Gosuniversiteta*, (6):12–20, 1972.

[164] Ghidaglia J. M. and Temam R., Attractor for Damped Nonlinear Hiperbolic Equations. *J. Pures et Appl.*, 66:273–319, 1987.

[165] Glavan V. A., About One Problem of Johnson–Sell. Discontinuous Dynamical Systems. *Reports of the Scientific Conference*, p.9, Ivano–Frankovsk, Ukraine, 1990.

[166] Gobbino M. and Sardella M., On the Connectedness of Attractors for Dynamical Systems. *Journal of Differential equations*, 133(1):1-14, 1997.

[167] Gunning C. R. and Rossi H., *Analytic Functions of Several Complex Variables.* Prentice–Hall, Englewood Cliffs, New Jersey, 1965.

[168] Hahn W., *The Present State of Lyapunov's Method. In "Nonlinear Problems" (R.E.Langer, ed.)* University of Wisconsin,1959.

[169] Hahn w., Stability of Motion. *Springer-Verlag*, Berlin, 1967.

[170] Halanay A. and Wexler D., *Teoria Calitativă a Sistemelor cu Impulsuri.* Bucureşti, 1968.

[171] Hale J. K., LaSalle J. P. and Slemrod M., Theory of a General Class of Dissipative Processes. *Journal Math. Appl.*, 39:177–191, 1972.

[172] Hale J. K., α-Contraction and Differential Equations. In *Actes de la Conference Internationalle "Equa–Difs–73"*, Actualites Scientifiques et Industrielles 1361, pp.16–41, Bruxelles et Louvain, 1973. Hermann.

[173] Hale J. K. and Lopes O. F., Fixed Point Theorems and Dissipative Process. *Journal of Differential Equations*, 13:391–402, 1973.

[174] Hale J. K., Some Recent Results on Dissipative Processes. *Lecture Notes in Mathematics,* 1980, v. 799, pp. 152–172.

[175] Hale J. K., *Theory of Functional-Differential Equations.* Mir, Moscow, 1984.

[176] Hale J. K., Asymptotic Behavior and Dynamics in Infinite Dimensions. In *Nonlinear Differ. Equations*, pages 1–42. Boston e. á, 1985.

[177] Hale J. K., Asymptotically Smooth Semigroups and Applications. *Lecture Notes in Math.*, (1248):85–93, 1987.

[178] Hale J. K. and Raugel G., Upper Semicontinuity for a Singulary Perturbed Hyperbolic Equation. *Journal of Differential Equations*, 73(2):197–214, 1988.

[179] Hale J. K., *Asymptotic Behaviour of Dissipative Systems.* Amer. Math. Soc., Providence, RI, 1988.

[180] Hale J. K. and Raugel G., Lower Semicontinuity of Attractors of Gradient Systems and Applications. *Ann. Math. Pura Appl.*, 4(154):281–326, 1989.

[181] Hale J. K., Asymptotic Behavior of Dissipative Systems. *Bulletin of the American Mathematical Society*, (1):175–183, 1990.

[182] Halmosh P., *A Hilbert Space Problem Book.* D. Van Nostrand Company, Inc., Toronto–London, 1967.

[183] Hapaev M. M., *Averaging in Stability Theory.* Nauka, Moscow (1986); English transl. Kluwer Dordrech (1992).

[184] Haraux A., Attractors of Asymptotically Compact Processes and Applications to Nonlinear Partial Differential Equations. *Commun. in Partial Differ. Equat.*, 13(11):1383–1414, 1988.

[185] Haraux A., *Systémes Dynamiques Dissipativs et Applications.* Masson, Paris–Milan–Barselona–Rome, 1991.

[186] Hartman P., *Ordinary Differential Equations.* Birkhauser, Boston–Basel–Stuttgart, 1982.

[187] Hassani N., Systems Dynamiques Nonautonomes Contractants et leur Applications. Theése de magister. Algerie, USTHB, 1983.

[188] Henry D., *Geometric Theory of Semilinear Parabolic Equations.* Lecture Notes in Mathematics, No.840, Springer-Verlag, New York 1981.

[189] Hitoshi I., On the Existence of Almost Periodic Complete Trajectories for Contractive Almost Periodic Processes. *Journal of Differential Equations*, 43(1):66–72, 1982.

[190] Husemoller D., *Fibre Bundles.* Springer–Verlag, Berlin–Heidelberg–New York, 1994.

[191] Hutter W., Hyperbolicity of Almost Periodic Evolution Families. Preprint (http://michelangelo.mathematik.uni-tuebingen.de/tuebingerichte/1999.html)

[192] Iftrode V., A-Dissipativity for Periodic Solutions in Ordinary Differential Equations. *Rev. Roum. Pures et Appl.*, 23(6):513–521, 1983.

[193] Ignatiev A. O., Some Generalizations of Barbashin–Krasovskii's Theorems. *Matematicheskaya Fizika*, 34:19–22, 1983.

[194] Ilyashenko Yu. S., Weakly Contractive Systems and Attractors of Galerkian Approksimations of Navier–Stocks' Equation. *Uspehi Matematicheskih Nauk*, 36(3):243–244, 1981.

[195] Ilyashenko Yu. S., Weakly Contractive Dynamical Systems and Attractors. *Uspehi Matematicheskih Nauk*, 37(1):166, 1982.

[196] Ilyashenko Yu. S. and Chetaev A. I., On Dimensionality of One Class of Dissipative Systems. *Prikladnaya Matematika i Mehanika*, 46(3):374–381, 1982.

[197] Ilyashenko Yu. S., On Dimensionality of Attractors k-Contractive Systems in Infinite-Dimensional Space. *Vestnik MGU, Series I, Matematika, Mehanika*, 3:52–59, 1983.

[198] Ilyashenko Yu. S., The Concept of Minimal Attractor and Maximal Attractor of Partial Differential Equations of the Kuramoto–Sivashinsky Type. *Chaos*, 1(2):168–173, 1991.

[199] Ilyin A. A., Averaging Principle for Dissipative Dynamical System with Rapidly Oscillating Right-Hand Sides. *Mathematicheskii Sbornik*, 187(1996), No.5, 15-58: English Transl. in Sbornik: Mathematics, 187(1996).

[200] Ilyin A. A., Global Averaging of Dissipative Dynamical System. *Rendiconti Academia Nazionale delle Scidetta dli XL. Memorie di Matematica e Applicazioni.* 116(1998), v.XXII, fasc.1, pp.165-191.

[201] Izé A. F. and J.G.Reis J. G., Contributions of Stability of Neutral Functional Differential Equations. *Journal of Differential Equations*, 29:58–65, 1978.

[202] Izé A. F., On a Topological Method for the Analysis of the Asymptotic Behavior of Dynamical Systems and Processes. *Complex Analysis, Functional Analysis and Approximation Theory*, North-Holland 1986, pp. 109-128.

[203] Johnson R. A., Ergodic Theory and Linear Differential Equations Systems. *Journal of Differential Equations*, 28:23–34, 1978.

[204] Johnson R. A., On a Floquet Theory for Almost Periodic Two-Dimensional Linear Systems. *Journal of Differential Equations*, 37:184–205, 1980.

[205] Jones G.S., The existence of Critical Points in Generalized Dynamical Systems. In Springer-Verlag, editor, *Lecture Notes in Math.*, v.60, pp.7–19, 1968. Seminaire on Differential Equations and Dynamical Systems.

[206] Kapitanskii L. V. and Kostin I. N., Attractors of Nonlinear Evolutionary Equations and Their Approximation. *Algebra i Analiz*, 2(1):114–140, 1990.

[207] Kartan A., *Basic Theory of Analitic Functions of One and Several Complex Variables.* Mir, Moscow, 1963.

[208] Kato J. and Nakajima F., On Sacker–Sell's Theorem for a Linear Skew Flow. *Tôhoku Math. Journ.*, 28:79–88, 1976.

[209] Kato J., Martynyuk A. A. and Shestakov A. A., *Stability of Motion of Nonautonomous Systems (Method of Limiting Equations).* Gordon and Breach, Luxembourg, 1996.

[210] Katok A. B., *Dynamical Systems with Hyperbolic Structure.* Kiev, 1972, pages 125–211. 9th Mathematical School.

[211] Keller H. and Schmalfuß B., Random Attractors for Stochastic Sine Gordon Equation via Transformation into Random Equations. Institut für Dynamische System, Universität Bremen, 1999.

[212] Keller H. and Schmalfuß B., Attractors for Stochastic Differential Equations with Nontrivial Noise. *Izvestiya Akad. Nauk. RM.*, 1(26):43–54, 1998.

[213] Kelley J., *General Topology.* D. Van Norstand Company Inc., Toronto–New York–London, 1957.

[214] Kloeden P. E., On Sharkowsky's Cycle Coexistence Ordering, *Bull. Austr. Math. Soc.* 20 (1979), 171-177.

[215] Kolyada S., On Dynamics of Triangular Maps of Square, Ergodic Theory and Dynamical Systems, 12(1992), p.749-768.

[216] Kloeden P. E. and Lorenz J., Stable Attracting Sets in Dynamical Systems and in their One-Step Discretizations. *SIAM J. Numer. Analysis* **23** (1986), 986-995.

[217] Kloeden P. E. and Schmalfuß B., Lyapunov Functions and Attractors

under Variable Time-Step Discretization. *Discrete and Continuous Dynamical Systems*, 2(2):163–172, 1996.

[218] Kloeden P. E. and Schmalfuß B., Cocycle Attractors of Variable Time Step Discretizations of Lorenzian Systems. *J. Difference Eqns. Applns.*, **3** (1997), 125–145.

[219] Kloeden P. E. and Schmalfuss B., Nonautonomous Systems, Cocycle Attractors and Variable Time–Step Discretization. *Numer. Algorithms* **14** (1997), 141–152.

[220] Kloeden P. E. and Stonier D. J., Cocycle Attractors in Nonautonomously Perturbed Differential Equations. *Dynamics of Continuous, Discrete and Impulsive Systems.* 4(1998),211-226.

[221] Kloeden P. E. and Kozyakin V. S., The Inflation of Nonautonomous Systems and their Pullback Attractors. *Transactions of the Russian Academy of Natural Sciences, Series MMMIU.* **4**, No. 1-2, (2000), 144-169.

[222] Kolyada S., On Dynamics of Triangular Maps of Square, Ergodic Theory and Dynamical Systems, 12(1992), p.749-768.

[223] Krasnosel'skii M. A., *The Operator of Translation along Trajectories of Differential Equations*. Translations of Mathematical Monographs, Volume 19. American Math. Soc., Providence, R.I., 1968.

[224] Krasnosel'skii M. A. and Zabreiko P. P., *Geometrical Methodes of Nonlinear Analyses*. Nauka, Moscow, 1975

[225] Kuratovskii K., *Topology*. Mir, Moscow, 1966 2 volumes, v. 1.

[226] Kuratovskii K., *Topology*. Mir, Moscow, 1969 2 volumes, v. 2.

[227] Ladis N. N., Asymptotics of the Solutions of Quasihomogeneous Systems. *Differential Equations*, 9(12):2257-2260, 1973.

[228] Ladyzhenskaya O. A., *The Mathematical Theory of Viscous Incompressible Flows*. Nauka, Moscow, (1970); English transl. Gordon and Breach, New York, 1969.

[229] Ladyzhenskaya O. A., On Dynamical System Generated by Navier–Stocks' Equations. In *Zapiski Nauchnyh Seminarov LOMI*, 27:911–115, 1972.

[230] Ladyzhenskaya O. A., On Finite-Dimensionality of Invariant Sets for Navier–Stocks' System and Other Dissipative Systems. In *Zapiski Nauchnyh Seminarov LOMI*, 112:137–155, 1982.

[231] Ladyzhenskaya O. A., On Attractors of Nonlinear Evolutionary Problems with Dissipation. In *Zapiski Nauchnyh Seminarov LOMI*, 152:72–85, 1986.

[232] Ladyzhenskaya O. A., On the Determination of Minimal Global Attractors for Navier–Stocks' Equations and Other Equations with Partial Derivatives. *Uspekhi Mat. Nauk*, 42(6(258)):25–60, 1987.

[233] Ladyzhenskaya O.A., Some Additions and Concretizations to My Works on Theory of Attractors for Abstract Semi-groups. In *Zapiski Nauchnyh Seminarov LOMI*, 182:102–112, 1990.

[234] Ladyzhenskaya O. A., *Attractors for Semigroups and Evolution Equations.* Lizioni Lincei, Cambridge Univ. Press, Cambridge, New-York, 1991.

[235] LaSalle J. P., Dissipative Systems. In *Proccedings of the Conference on Ordinary Differential Equations*, pp.165–174, Washington, D.C., 1971. New York.

[236] Latushkin Yu. and Schnaubelt R., Evolution Semigroups, Translation Algebras, and Exponential Dichotomy of Cocycles. *Journal of Differential Equations* 1999, v.159, pp.321-369

[237] Levinson N., Transformation Theory of Nonlinear Differential Equations of the Second Order. *Ann. Math.*, 45(4):723–737, 1944.

[238] Levitan B. M. and Zhikov V. V., Almost Periodic Functions and Differential Equations. Cambridge Univ. Press, Cambridge, 1982.

[239] Lihtenberg A. and Liberman N., *Regular and Stochastic Dynamics.* Mir, Moscow, 1984.

[240] Loins J. L., *Quelques Methodes de Résolution des Problèmes aux Limites non Linéaires.* Dunod, Paris, 1969.

[241] Liubich Yu. I., Note on Stability of Complex Dynamical Systems. *Izvestiya VUZov, Matematika*, (10(257)):49–50, 1983.

[242] Lorenz E. N., Deterministic Nonperiodic Flow. *Journal of the Atmospheric Sciences.* 1962, 20, pp.130-141.

[243] Mallet-Paret J., Morse Decomposition for Delay-Differential Equations. *Journal of Differential Equations*, 72(2):270–315, 1988.

[244] Mallet-Paret J. and Sell G. R., Inertial Manifolds for Reaction Diffusion Equations in Higher Space Dimentions. *Journal Amer. Math. Soc.*, (1):805–866, 1998.

[245] Manfredi Bianca, Dissipativity, quasiperiodicity and recurrence. *Riv. Mat. Univ. Parma* (4) 10 (1984), pp.449–458.

[246] Massat P., Stability and Fixed Points of Point-Dissipative Systems. *Journal of Differential Equations*, 40:217–231, 1981.

[247] Massat P., On Obtaininig Ultimate Boundedness for α- Contraction Volterra and Functional-Differential Equations. In *Lecture Notes in Pure and Appl. Math.*, pp.273–280. Dekker, New York, 1983.

[248] Massat P., Attractivity Properties of α-Contractions. *Journal of Differential Equations*, 48:326–333, 1983.

[249] Massera J. L. and Schäffer J. J., Linear Differential Equations and Functional Analysis, I. *Ann. of Math.*, 1958, v.67, No.2, pp. 517-573

[250] Massera J. L. and Shäffer J. J., *Linear Differential Equations and Function Spaces.* Academic Press, New York–London, 1966.

[251] Matrosov V. M., Principle of Comparison with Vector Function of Lyapunov. *Differentsial'nye Uravneniya*, 5(12):2129–2143, 1969.

[252] Miller R. K., Almost Periodic Differential Equations as Dynamical System with Applications to the Existence of Almost Periodic Solutions. *Journal of*

Differential Equations, 1:337–345, 1965.

[253] Millionshchikov V. M., Recurrent and Almost Periodic Limit Trajectories of Nonautonomous Systems of Differential Equations. *Doklady Akademii Nauk SSSR*, 161(1):43–44, 1965.

[254] Millionshchikov V. M., On Recurrent and Almost Periodic Limit Solutions of Nonautonomous Systems. *Differentsial'nye Uravneniya*, 4(9):1555–1559, 1968.

[255] Millionshchikov V. M., Linear Systems of Ordinary Differential Equations. *Differentsial'nye Uravneniya*, 7(3):387–390, 1971.

[256] Mitropolskii Yu. A., Perestiuk N. A. and Chernikova O. S., Convergence of Systems of Differential Equations with Impulse Effect. *Doklady Akademii Nauk SSSR*, A(11):11–15, 1983.

[257] Mitropolskii Yu. A. and Kulik V. L., Bounded Solutions of Nonlinear Systems of Differential Equations. *Ukrainskii Matematicheskii Zhurnal*, 36(6):720–729, 1984.

[258] Nemytskii V. V. and Stepanov V. V., *Qualitative Theory of Differential Equations*. Nauka, Moscow, 1949.

[259] Nemytskii V. V., On Some Methods of Global Qualitative Research of Multi-Dimensional Autonomous Systems. *Trudy Moskovskogo Matematicheskogo Obschestva*, 5:455–482, 1956.

[260] Nemytskii V. V., Some Modern Problems of Qualitative Theory of Ordinary Differential Equations. *Uspekhi Matematicheskih Nauk*, 22(4):3–36, 1965.

[261] Nitecki Z., *Differentiable Dynamics*. The MIT Press, Cambridge–Massachusetts–London, 1971.

[262] Ochs G., Weak Random Attractors. *Technical Report, Institut für Dynamische Systeme, Universität Bremen*, 1999.

[263] Opial Z., Sur une Equation Diffirentielle Presque-Periodique sans Solution Presque-Periodique. *Bull. Acad. Polon. Sci., Ser. Math. Astron. Phys.*, 9:673–676, 1961.

[264] Pavel N., On Dissipative Systems. *Bolletino Un. Mat. Ital.*, 4(5):701–707, 1971.

[265] Pavel N., A Generalization of Ultimately Bounded Systems. *An. Sti. Uni. "Al.I.Cuza", Iaşi. Sect. I.a, Mat.*, 18:81–86, 1972.

[266] Pavel N., On the Boundedness of Solutions of Systems of Differential Equations. *Tohoku Math. Journal*, 24:21–32, 1972.

[267] Perov A.I. and Trubnikov Yu. V., Monotone Differential Equations, I. *Differentsial'nye Uravneniya*, 10(5):804–815, 1974.

[268] Perov A.I. and Trubnikov Yu. V., Monotone Differential Equations, II. *Differentsial'nye Uravneniya*, 12(7):1223–1237, 1976.

[269] Piliugin S. Yu., *Structure of Attracting Sets of Hyperbolic Systems of Differential Equations*. Thesis of Dr. of Sc., Leningrad, 1983.

[270] Pliss V. A. *Nonlocal Problems in the Theory of Oscilations*. Academic Press,

1966.

[271] Pliss V. A., *Integral Sets of Periodic Systems of Differential Equations.* Nauka, Moscow, 1977.

[272] Reissig R., Sansone G. and Conti R., *Qualitative Theorie Nichthlinearer Differentialgleichungen.* Edizioni Cremonese, Roma, 1963.

[273] Rouche N., Habets P. and Laloy M., *Stability Theory by Lyapunov's Direct Method.* Springer–Verlag, New York–Heidelberg–Berlin, 1977.

[274] Rudin U., *Functional Analysis.* Mir, Moscow, 1975.

[275] Sacker R. J. and Sell G. R., Existence of Dichotomies and Invariant Splittings for Linear Differential Systems, I. *Journal of Differential Equations*, 15:429–458, 1974.

[276] Sacker R. J. and Sell G. R., *Lifting Properties in Skew–Product Flows with Applications to Differential Equations.* Memoirs oft he American Math. Soc., v.190, Providence, R.I., 1977.

[277] Sacker R. J. and Sell G. R., Dichotomies for Linear Evoltionary Equations in Banach Spaces. *Journal of Differential Equations* 1994, v.113, pp.17-67

[278] Sadovskii B. N., Limit Compact and Condensing Operators. *Uspechi Matem. Nauk*, 27(1(163)):81–146, 1972.

[279] Samoilenko A. M. and Perestiuk N. A., *Impulsive Differential Equations.* World Scientific Publishing Co., Inc., River Edge, NY, 1995.

[280] Samoilenko A. M., and Trofimuchiuk S. I., Unbounded Functions with Almost Periodic Differences. *Ukrainian Math. J.*, **43** (1991), 1409-1413.

[281] Samoilenko A. M. and Trofimuchiuk S. I., On the Space of Piecewise Continuous Almost Periodic Functions and Almost Periodic Sets on the Line. *Ukrainian Math. J.,***43** (1991), 1613–1619.

[282] Saperstone S. N., *Semidynamical Systems in Infinite Dimensional Spaces.* Springer-Verlag, New York-Heidelberg-Berlin 1981.

[283] Scheutzow M., Comparison of Various Concepts of a Random Atractor: A Case Study. *Technical report, Technische Universität Berlin,* 2000.

[284] Schmalfuß B., Backward Cocycles and Attractors of Stochastic Differential Equations. in *International Seminar on Applied Mathematics–Nonlinear Dynamics: Attractor Approximation and Global Behaviour,* V. Reitmann, T. Riedrich, and N. Koksch (eds.), TU Dresden, 1992, pp 185–192.

[285] Schmalfuß B., Attractors for the Non-Autonomous Dynamical Systems. In K. Gröger B. Fiedler and J. Sprekels, editors, *Proceedings EQUADIFF99,* pages 684–690. World Scientific, 2000.

[286] Schmalfuss B., Attractors for the Non-Autonomous Navier-Stokes Equation (to appear).

[287] Schwartz L., *Analyse Mathématique,* v. 1. Hermann, 1967.

[288] Schwartz L., *Analyse Mathématique,* v. 2. Hermann, 1967.

[289] Seifert G., Almost Periodic Solutions for Almost Periodic System of Ordinary

Differential Equations. *Journal of Differential Equations*, 2:305–319, 1966.

[290] Sell G. R., Non-Autonomous Differential Equations and Topological Dynamics, I. The basic theory. *Trans. Amer. Math. Soc.*, 127:241–262, 1967.

[291] Sell G. R., Non-Autonomous Differential Equations and Topological Dynamics, I. Limiting equations. *Trans. Amer. Math. Soc.*, 127:263–283, 1967.

[292] Sell G. R., *Lectures on Topological Dynamics and Differential Equations*, volume 2 of *Van Nostrand Reinhold math. studies*. Van Nostrand–Reinbold, London, 1971.

[293] Sell G. R., W. Shen and Y. Yi, Topological Dynamics and Differential Equations, *Contemporary Mathematics*, volume 215, 1998, pp.279-297.

[294] Sell G. R. and You Y., Dynamics of Evolutionary Equations. Springer-Verlag, New York, 2002.

[295] Shabat B. V., *Introduction into Complex Analysis*. Moscow, "Nauka", 1969.

[296] Sharkovskii A. N., Attractors of Some Nonlinear Boundary Problems. *Proceedings of 6th All-Union Conference on Qualitative Theory of Differential Equations*, pp.205, Irkutsk, 1986.

[297] Sharkovskii A. N, Maistrenko Yu. L. and Romanenko E. Yu., *Difference Equations and Their Applications*. Naukova Dumka, Kiev, 1986.

[298] Shchennikov V. N., Research of Convergence in Non-Autonomous Differential System with the Help of Vector Function of Lyapunov. *Differentsial'nye Uravneniya*, 19(11):1902–1907, 1983.

[299] Shchennikov V. N., Convergence of Complicated Systems of Differential Equations. *Differentsial'nye Uravneniya*, 19(11):1568–1571, 1984.

[300] Shcherbakov B. A., *Topologic Dynamics and Poisson Stability of Solutions of Differential Equations*. Ştiinţa, Chişinău, 1972.

[301] Shcherbakov B.A. and Cheban D. N., Asymptotically Poisson Stable Motions of Dynamical Systems and Comparability of Their Reccurence in Limit. *Differentsial'nye Uravneniya*, 13(5):898–906, 1977.

[302] Shcherbakov B. A., *Poisson Stability of Motions of Dynamical Systems and Solutions of Differential Equations*. Ştiinţa, Chişinău, 1985.

[303] Shestakov A. A., Generalized Lyapunov's Method for Abstract Semi-Dynamical Processes, I. Semi-dynamical Processes as Semi-Dynamical Systems. Localisation of Limite Set of Autonomous and Asymptoticaly Autonomous Semi-Dynamical Processes. *Differentsial'nye Uravneniya*, 22(9):1475–1490, 1986.

[304] Sibirsky K. S., *Introduction to Topological Dynamics*. Noordhoff, Leyden, 1975.

[305] Sibirskii K. S. and Shube A. S., *Semidynamical Systems*. Stiintsa, Kishinev 1987. (Russian)

[306] Sinai Yu. G. and Shilnikov L. P. (eds.), *Strange Attractors (Collection of Articles)*. Mir, Moscow, 1981.

[307] Skovronski J. and Ziemba S., The Problem of Vibrations of Nonautonomic Systems with Strong Nonlinearity. *Arch. Mech. Stosowanej*, 10(4):517–523, 1958.

[308] Stuart A. M. and Humphries A. R., *Numerical Analysis and Dynamical Systems*. Cambridge University Press, Cambridge 1996.

[309] Talpalaru P., Sur les Systems Dissipatifs. *Ann. Stiint. Univ. Iasi*, Sec. I.a, 13(1):43–47, 1967.

[310] Tchéban D. N., Systemes Dynamiques Dissipatifs. In *II-Rencontre Nationale sur les Equations Differentielles Ordinaires. Resumes des Exposes*, pages 40–41, Alger, 1983.

[311] Temam R., *Navier–Stokes Equation–Theory and Numerical Analysis*. North-Holland, Amsterdam, 1979.

[312] Temam R., Induced Trajectories and Approximate Inertial Manifolds. *Math. Modelling. Numer. Anal.*, 23:541–561, 1989.

[313] Temam R., Inertial Manifolds. *Math. Intelligencer*, 12(4):68–74, 1990.

[314] Temam R., *Infinite–Dimensional Dynamical Systems in Mechanics and Physics*. 2nd edition, Springer Verlag, New York,1997.

[315] Titi E. S., On Approximate-Inertial Manifolds to the Navier–Stokes Equations. *Journal Math. Anal. Appl.*, 149:540–557, 1990.

[316] Trubnikov Yu. V. and Perov A. I., *The Differential Equations with Monotone Nonlinearity*. Nauka i Tehnika. Minsk, 1986 (in Russian).

[317] Vishik M. I. and Chepyzhev V. V., Attractors of Non-Autonomous Dynamical Systems. *Matematicheskie Zametki*, 51(6):141–143, 1992.

[318] Vishik M. I., *Asymptotic Behaviour of Solutions of Evolutionary Equations*. Cambridge Univ. Press, Cambridge, 1992.

[319] Vuillermot Pierre-A., Attracteurs Presque-Priodiques pour une Classe d'Equations Paraboliques non Linaires du Type Raction-Diffusion sur R^N. (French. English summary) [Almost-Periodic Attractors for a Class of Nonlinear Parabolic Reaction-Diffusion Equations in $\mathbf{R}^N$] *C. R. Acad. Sci.* Paris Sr. I Math. 311 (1990), no. 10, pp.583–588.

[320] Vuillermot Pierre-A., Almost-Periodic Attractors for a Class of Non-Autonomous Reaction-Diffusion Equations on R^N.I. Global Stabilization Processes. *J. Differential Equations* 94 (1991), no. 2, 228–253.

[321] Vuillermot Pierre-A., Almost-Periodic Attractors for a Class of Non-Autonomous Reaction-Diffusion Equations on R^N. II. Codimension-One Stable Manifolds. *Differential Integral Equations* 5 (1992), no. 3, 693–720.

[322] Vuillermot Pierre-A., Global Exponential Attractors for a Class of Almost-Periodic Parabolic Equations in R^N. *Proc. Amer. Math. Soc.* 116 (1992), no. 3, 775–782.

[323] Vuillermot Pierre-A., Almost-Periodic Attractors for a Class of Non-Autonomous Reaction-Diffusion Equations on R^N. III. Center Curves and

Liapounov Stability. *Nonlinear Anal.* 22 (1994), no. 5, 533–559.

[324] Xavier M., Finite-Dimentional Attracting Manifolds in Reaction-Diffusion Equations. *Contemp. Math.*, 17:540–557, 1983.

[325] Yoshizawa T., Note on the Boundedness and the Ultimate Boundedness of Solutions $x' = f(t, x)$. *Memoirs of the College of Science. University of Kyoto, Serie A*, 29(3):275–291, 1955.

[326] Yoshizawa T., Lyapunov's Function and Boundedness of Solutions. *Funkcialaj Ekvacioj*, 2:95–142, 1959.

[327] Yoshizawa T., *Stability Theory by Lyapunov's Second Method.* The Mathematical Series of Japan. Tokyo, 1966.

[328] Yosida K., *Functional Analysis.* Springer–Verlag, Berlin, 1995.

[329] Zadorozhnyi V. G., On Nonlocal Solutions of V-Dissipative Differential Equations. *Ukrainskii Matematicheskii Zhurnal*, 35(3):303–308, 1983.

[330] Zhang S., Functional Differential Equations and Topological Dynamics. *Chinese Science Bulletin*, 35(17):1441–1448, 1990.

[331] Zhikov V. V., On Problem of Existence of Almost Periodic Solutions of Differential and Operator Equations. *Nauchnye Trudy VVPI, Matematika*, (8):94–188, 1969.

[332] Zhikov V. V., On Stability and Unstability of Levinson's centre. *Differentsial'nye Uravneniya*, 8(12):2167–2170, 1972.

[333] Zhikov V. V., Monotonicity in the Theory of Almost Periodic Solutions of Non-Linear operator Equations. *Mat. Sbornik* 90 (1973), 214–228; English transl., Math. USSR-Sb. 19 (1974), 209-223.

[334] Zinchenko I. L., On Structure of Asymptotically Stable Invariant Sets of Periodical Complex Analytic Systems of Differential Equations. *Vestnik LGU, Matematika, Mehanika, Astronomiya*, 1(1):23–25, 1980.

[335] Zinchenko I. L., Theorem on Convergence of Complex Analitic Systems. *Differential Equations*, 25(10):1809–1810, 1989.

[336] Zubov V. I., *The Methods of A. M. Lyapunov and Their Application.* Noordhoof, Groningen, 1964.

[337] Zubov V. I., *Stability of Motion.* Vyshaya Shkola, Moscow, 1973.

[338] Zubov V. I., *Theory of Oscillations.* Nauka, Moscow, 1979.

Index